COMETS

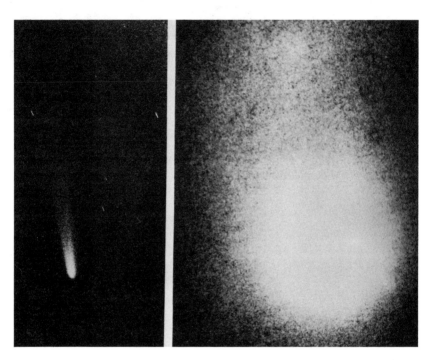

The extent of the H I Lyman-alpha envelope is seen on the right in the ultraviolet image of Comet Kohoutek 1973 XII obtained by an electrographic camera on board a sounding rocket on 8.1 January 1974 (Opal et al., *Science,* 185:702, 1974). For comparison, a visible image taken by a Nikon-F camera with an f/2.8, 180-mm lens on a similar rocket three days earlier (Feldman et al., *Science,* 185:705, 1974), is shown to the same scale on the left.

COMETS

Edited by
LAUREL L. WILKENING

With the assistance of
MILDRED SHAPLEY MATTHEWS

With 48 collaborating authors

THE UNIVERSITY OF ARIZONA PRESS
TUCSON, ARIZONA

SPACE SCIENCE SERIES
Tom Gehrels, Space Sciences Consultant
PLANETS, STARS AND NEBULAE, STUDIED WITH PHOTOPOLARIMETRY, T. Gehrels, ed., 1974, 1133 pp.
JUPITER, T. Gehrels, ed., 1976, 1254 pp.
PLANETARY SATELLITES, J. A. Burns, ed., 1977, 598 pp.
PROTOSTARS AND PLANETS, T. Gehrels, ed., 1978, 756 pp.
ASTEROIDS, T. Gehrels, ed., 1979, 1181 pp.
COMETS, L. L. Wilkening, ed., 1982, 766 pp.
THE SATELLITIES OF JUPITER, D. Morrison, ed., 1982, 972 pp.
VENUS, D. M. Hunten et al., eds., 1983, 1143 pp.

The cover is a photograph of Comet West 1976 VI taken by S. Larson from the Mt. Lemmon Observatory (elevation 2790 m) on 1976 March 9.5 UT with an 80 mm (f/2) Zeiss lens on Kodak Tri-X panchromatic film. The type I and type II tails are evident. Striations in the type II tail result from earlier fragmentation of the nucleus.

Second printing 1983

THE UNIVERSITY OF ARIZONA PRESS

Copyright © 1982
The Arizona Board of Regents
All Rights Reserved

This book was set in 10/12 IBM MTSC Times Roman
Manufactured in the U.S.A.

Library of Congress Cataloging in Publication Data
Main entry under title:

Comets.

  (Space Science Series)
  Includes index.
  1. Comets. I. Wilkening, Laurel L.
II. Matthews, Mildred Shapley. III. Series.
QB 721.C648   523.6   81-21814

ISBN 0-8165-0769-4   AACR2

CONTENTS

COLLABORATING AUTHORS	viii
PREFACE	ix

Part I—OVERVIEW

OVERVIEW OF COMET OBSERVATIONS *S. Wyckoff*	3
COMET DISCOVERIES, STATISTICS, AND OBSERVATIONAL SELECTION *Ľ. Kresák*	56

Part II—NUCLEUS

CHEMICAL COMPOSITION OF COMETARY NUCLEI *A.H. Delsemme*	85
WHAT ARE COMETS MADE OF? A MODEL BASED ON INTERSTELLAR DUST *J.M. Greenberg*	131
THE INFRARED SPECTRAL PROPERTIES OF FROZEN VOLATILES *U. Fink and G.T. Sill*	164
STRUCTURE AND ORIGIN OF COMETARY NUCLEI *B. Donn and J. Rahe*	203
THE ROTATION OF COMET NUCLEI *F.L. Whipple*	227
THE PROBLEM OF SPLIT COMETS IN REVIEW *Z. Sekanina*	251
RADAR DETECTABILITY OF COMETS *P.G. Kamoun, G.H. Pettengill, and I.I. Shapiro*	288
RELATIONSHIPS BETWEEN COMETS, LARGE METEORS, AND METEORITES *G.W. Wetherill and D.O. ReVelle*	297

Part III—DUST

OPTICAL AND INFRARED OBSERVATIONS OF BRIGHT
    COMETS IN THE RANGE 0.5 μm TO 20 μm   323
  E.P. Ney

INTERPRETING THE THERMAL PROPERTIES OF
    COMETARY DUST   341
  H. Campins and M.S. Hanner

DUSTY GAS-DYNAMICS IN REAL COMETS   357
  M.K. Wallis

COMETARY DUST IN THE SOLAR SYSTEM   370
  H. Fechtig

LABORATORY STUDIES OF INTERPLANETARY DUST   383
  P. Fraundorf, D.E. Brownlee, and R.M. Walker

Part IV—COMA

COMET HEAD PHOTOMETRY: PAST, PRESENT,
    AND FUTURE   413
  D.D. Meisel and C.S. Morris

SPECTROPHOTOMETRY OF COMETS AT OPTICAL
    WAVELENGTHS   433
  M.F. A'Hearn

ULTRAVIOLET SPECTROSCOPY OF COMAE   461
  P.D. Feldman

LABORATORY STUDIES OF PHOTOCHEMICAL AND
    SPECTROSCOPIC PHENOMENA RELATED TO COMETS   480
  W.M. Jackson

PHOTOCHEMICAL PROCESSES IN THE INNER COMA   496
  W.F. Huebner, P.T. Giguere, and W.L. Slattery

Part V—ION TAILS AND SOLAR WIND INTERACTIONS

OBSERVATIONS AND DYNAMICS OF PLASMA TAILS   519
  J.C. Brandt

PLASMA FLOW AND MAGNETIC FIELDS IN COMETS   538
  H.U. Schmidt and R. Wegmann

SOLAR WIND INTERACTION WITH COMETS:
    LESSONS FROM VENUS   561
  C.T. Russell, J.G. Luhmann, R.C. Elphic, and M. Neugebauer

THEORIES OF PHYSICAL PROCESSES IN THE COMETARY
    COMAE AND ION TAILS   588
W.-H. Ip and W.I. Axford

Part VI—ORIGIN, EVOLUTION, AND INTERRELATIONS

DYNAMICAL HISTORY OF THE OORT CLOUD   637
P.R. Weissman

EVOLUTION OF LONG- AND SHORT-PERIOD ORBITS   659
E. Everhart

DO COMETS EVOLVE INTO ASTEROIDS? EVIDENCE FROM
    PHYSICAL STUDIES   665
J. Degewij and E.F. Tedesco

COMETS AND ORIGIN OF LIFE   696
C. Ponnamperuma and E. Ochiai

Part VII—APPENDIX

BASIC INFORMATION AND REFERENCES   707
B.G. Marsden and E. Roemer

GLOSSARY, ACKNOWLEDGMENTS, AND INDEX

GLOSSARY   737
LIST OF PARTICIPANTS   753
INDEX   761

COLLABORATING AUTHORS

M.F. A'Hearn, *433*
W.I. Axford, *588*
J.C. Brandt, *519*
D.E. Brownlee, *383*
H. Campins, *341*
J. Degewij, *665*
A.H. Delsemme, *85*
B. Donn, *203*
R.C. Elphic, *561*
E. Everhart, *659*
H. Fechtig, *370*
P.D. Feldman, *461*
U. Fink, *164*
P. Fraundorf, *383*
P.T. Giguere, *496*
J.M. Greenberg, *131*
M.S. Hanner, *341*
W.F. Huebner, *496*
W.-H. Ip, *588*
W.M. Jackson, *480*
P.G. Kamoun, *288*
L. Kresák, *56*
J.G. Luhmann, *561*
B.G. Marsden, *707*

D.D. Meisel, *413*
C.S. Morris, *413*
E.P. Ney, *323*
M. Neugebauer, *561*
E. Ochiai, *696*
G.H. Pettengill, *288*
C. Ponnamperuma, *696*
J. Rahe, *203*
D.O. ReVelle, *297*
E. Roemer, *707*
C.T. Russell, *561*
H.U. Schmidt, *538*
Z. Sekanina, *251*
I.I. Shapiro, *288*
G.T. Sill, *164*
W.L. Slattery, *496*
E.F. Tedesco, *665*
R.M. Walker, *383*
M.K. Wallis, *357*
R. Wegmann, *538*
P.R. Weissman, *637*
G.W. Wetherill, *297*
F.L. Whipple, *227*
S. Wyckoff, *3*

PREFACE

The noted explorer and colonizer of northern Sonora and southern Arizona, Father Eusebio Francisco Kino, had no sooner arrived on the American continent in 1681 than he became involved in a debate with Mexican scholar Sigüenza y Gongora on the significance of comets. Always an inquiring natural scientist, Kino's interest in comets had been stimulated by observations which he made in Spain while awaiting passage to the new world. After arriving in Mexico City, he published a book in 1681, "Astronomical exposition on the comet which in the months of November and December of 1680 and in the months of January and February of 1681 was seen in all the world and observed in the city of Cadiz." In this book Kino took the traditional view of comets as evil omens. Sigüenza y Gongora, who had a much more modern view of comets, countered with another book, "Astronomical and Philosophical Balance in which Don Carlos de Sigüenza . . . examines not only the objection to his *Philosophical Manifest* against Comets raised by the same Reverend Father Eusebio Francisco Kino . . . but also what the same Reverend Father pretended to have demonstrated in his *Exposición Astronomica de el Cometa.*" A splendid account of the debate is given in Bolton's biography[a] of Kino. Historian Bolton, who wrote with a wonderfully dry wit about the Spanish exploration of the West, concluded his chapter on this episode with the remark, "From [the episode] we learn something about the comet and a good deal about astronomers."

---

[a]Bolton, H.E. 1960. *Rim of Christendom, a Biography of Eusebio Francisco Kino, Pacific Coast Pioneer.* (New York: Russell and Russell).

PREFACE

Three hundred years later in Kino's territory in Tucson, Arizona 175 scientists from 13 nations gathered to renew the scientific debate about comets at the International Astronomical Union's Colloquium No. 61, "Comets: Gases, Ices, Grains and Plasma." The meeting provided the basis for this book; from it we learned a good deal more about comets and perhaps less about astronomers.

As demonstrated by Kino and Sigüenza y Gongora, comets have for centuries captured the imagination of mankind. But this is a special time—our generation will send robot spacecraft to visit comets. This possibility has stimulated new studies of comets: observations of faint and distant comets, new models of cometary activity, new data concerning the nature of cometary solids. There have been other stimuli; the long and successful operation of the International Ultraviolet Explorer satellite made possible the crucial observations of H and OH in the extended comae of comets, collection of interplanetary dust in the Earth's stratosphere yielded new clues to the nature of cometary dust, and the Pioneer Venus orbiter gave a detailed picture of the interaction of solar wind with an atmosphere not protected by a magnetic field.

These new theories and data are collected here as an introduction to comets for newcomers as well as a reference volume for specialists.

Many people made valuable contributions to the meeting and the book; since it is unfortunately impossible to thank them individually here, they are listed in the Acknowledgments. I would especially like to thank E. Roemer for advice and guidance and S. Wyckoff for undertaking the enormous job of summarizing the papers presented at the meeting and reviewing the current literature on cometary observations in her introductory chapter. T.S. Smith and M.A. Matthews deserve special thanks for their assistance in completing the book. It is a pleasure to acknowledge essential support from the University of Arizona Press, the National Science Foundation, the National Aeronautics and Space Administration, and the International Astronomical Union.

Laurel L. Wilkening

PART I
Overview

OVERVIEW OF COMET OBSERVATIONS

SUSAN WYCKOFF
Arizona State University

Observations presently indicate that the nuclei of comets are ice-dust conglomerates with masses $\sim 10^{13}$ to 10^{19} g, radii $\sim$ few km, average rotation periods ~ 15 hr and tensile strengths $\sim 10^5$ dyne cm^{-2}. The latter indicates that cometary nuclei are very fragile entities. All observations support the basic concept of a comet nucleus based on Whipple's icy conglomerate model of H_2O ice plus an admixture of other ices and dust. Observational support for the icy conglomerate model comes mainly from ultraviolet observations of H and OH in comets, as well as H_2O^+ observed in the 5000-7000 Å region. Although water has been established as an abundant constituent in cometary nuclei, observations indicate that a volatile other than H_2O may occasionally control the vaporization of cometary ices as a comet approaches the Sun. Radicals other than dissociation products of H_2O are observed in the spectra of comets which gives credence to the hydrate clathrate modification of Whipple's model. The atoms and molecules identified in cometary spectra indicate that the cosmically abundant species H, C, N, and O are well represented as are the abundant metals. After the coma forms, a dust and/or plasma tail may develop. The type II tails are comprised of dust released from the nucleus with the sublimed gases. Observed dust-to-gas ratios are in the range 0.1 to 1 by mass. Solar radiation pressure acting on the 1-µm diameter dust particles exceeds that of solar (and cometary) gravitational attraction. Hence the dust particles entrained in the vaporizing gases from the nucleus are forced from the coma by solar radiation pressure and, due to Keplerian motion, lag the antisolar direction by tens of degrees. The type I or

> *plasma tail results from the interaction of the solar wind with the cometary ionosphere, which arises from ionization of coma gases. The magnetic field carried by the solar wind accelerates the cometary ions from the coma in the antisolar direction forming the plasma tail. The observed distribution of the orbits of the long-period comets supports their hypothesized origin in the Oort Cloud at the periphery of the solar system. Due in part to their small masses, comets have undergone virtually no internal and little external processing during their 5-billion yr history. They therefore must represent the purest samples of the primordial solar nebula.*

Comets provide tantalizing clues to the early history of the solar system. Due to both their small masses and long orbital periods, they have probably undergone less internal and external evolution than other members of the solar system. Opportunities to observe bright comets ($V \lesssim 5$ mag) at close approach to the Earth ($\lesssim 1$ AU) arise only every few years, while several faint ($V > 15$ mag) comets are almost always available for observation. Since virtually no hyperbolic orbits have been observed (Marsden 1974, 1979), comets are probably part of the original solar nebula (Öpik 1932; Oort 1950). The perihelia of their orbits range from $q = 0.005$ to 7 AU. Comets have been observed out to 12 AU from the Sun, and as close as 0.02 AU from the Earth. Their periods of revolution are conventionally termed short for $P < 200$ yr and long for $P > 200$ yr. Comets have also been categorized as "old" or "new" with the implication that a comet evolves during many solar passages (Oort 1950; Oort and Schmidt 1951). This terminology, however, is based on orbital characteristics rather than observed physical properties of comets (cf. Marsden 1974; see Table I in Marsden and Roemer's Appendix to this book).

The icy conglomerate model proposed by Whipple (1950, 1951) provides the basis for our present conception of a comet nucleus, namely, that it is a solid but fragile, ice-dust mixture of diameter 0.1 to 10 km. As this so-called dirty snowball transits the solar system toward perihelion the surface is subjected to a variety of erosive processes, with the most prolonged exposure probably being to cosmic rays, solar wind ions and surface photoelectron currents (cf. Mendis et al. 1981). At heliocentric distances, $r \gtrsim 3$ AU, comets are also occasionally activated for as yet unexplained reasons by perhaps internal mechanisms, cometary ices more volatile than H_2O, or electrostatic processes (cf. Mendis et al. 1981; Houpis and Mendis 1981c). After a comet passes within $r \sim 3$ AU of the Sun, volatiles sublimed from the surface by solar radiation form an expanding atmosphere of gas and dust called the coma. The formation of a dust and/or plasma tail may or may not follow. The type II tail is comprised of dust formerly embedded in the nucleus and released as the surface ices vaporize in the solar radiation field. The type I, plasma or ion tail is formed by the interaction of ions in the coma with the solar wind magnetic field which folds onto the comet ionosphere, and is presumed to accelerate ions from the coma in the antisolar direction to form the plasma tail.

In the following sections observations of the nucleus, dust, coma, and plasma tail of a comet are reviewed. Additional references can be found in reviews by Wurm (1963), Roemer (1963,1966), Antrack et al. (1964), Swings (1965), Brandt (1968), Marsden (1974), Arpigny (1965,1976), Herzberg (1976), Herbig (1976), Keller (1976), Whipple and Huebner (1976), Delsemme (1977), Whipple (1978a), Sekanina (1981) as well as in the chapters and the appendix of this book.

I. NUCLEUS

A. Radius

Possibly the only direct observation of a comet nucleus has been achieved using radar ranging techniques. Kamoun et al. (1981) have used the Arecibo telescope to observe radar echoes from P/Encke during its perihelion passage in 1980. The radar observations indicate a radius in the range 0.4 to 4.0 km after adopting a rotation period of 6.5 hr and an orientation of the spin axis of the nucleus derived by Whipple and Sekanina (1979). Unfortunately radar techniques can be applied to very few comets due to the small cross section of the nucleus and to the Δ^{-4} dependence of the returned signal, where Δ represents the geocentric distance (chapter by Kamoun et al.).

The direct observation of a bare comet nucleus using conventional groundbased optical telescopes is extremely difficult due to the presence of highly volatile ices which vaporize under the influence of solar radiation at heliocentric distances at least as large as 7 AU. Perhaps the most convincing cases of bare nuclei detections are for P/Neujmin 1 and P/Arend-Rigaux. Marsden (1974) notes that both comets are relatively inert as evidenced by the lack of nongravitational effects in their orbits and by brightness variations (at large heliocentric distances) which conform to the inverse square law, $I \sim r^{-2} \Delta^{-2}$. The relative inertness of these two comets is attributed to the lack of surface ices. Spinrad et al. (1979) have claimed a probable spectroscopic detection of the bare nucleus of P/Tempel 2 at a heliocentric distance, $r \sim 3$ AU ($V \gtrsim 19$ mag). However, other observers have not succeeded in the direct spectroscopic detection of other comet nuclei (Degewij 1980; A'Hearn et al. 1981).

If subsequent investigations confirm that the bare nucleus is observed in some comets at large heliocentric distances, then photometric measurements can be used to estimate radii of comets. At large heliocentric distances, in the absence of a coma, the observed magnitude of the nucleus depends on the heliocentric distance r, the geocentric distance Δ, the geometric albedo A, the comet radius R, the phase function $\phi(\alpha)$ (cf. Spinrad et al. 1979)

$$R^2 = r^2 A^{-1} \phi^{-1}(\alpha) 10 \exp\left\{0.4[M_\odot - (m - 5 \log \Delta)]\right\} \quad (1)$$

where $M_\odot$ is the absolute magnitude of the Sun in the same passband as the observed magnitude m of the comet nucleus. This method of determining the radii of cometary nuclei requires the assumption of the albedo A. Several observers have estimated these radii from photometric magnitudes to be in the range 1 to 10 km for a range in assumed albedos (Roemer 1966; Delsemme and Rud 1973; Sekanina 1976; Fäy and Wisniewski 1978). In these cases the comets were observed at large heliocentric distances and the coma assumed to be absent.

B. Mass

No direct determination of the mass of a comet has yet been made, since no perturbations of another body by a comet during close approaches have been detected. Also for split cometary nuclei, no obvious gravitational perturbations have been observed among the fragments (Sekanina 1977,1979b). Rather Sekanina (see his chapter) has identified the dominant force governing the dynamics of split nuclei as a differential nongravitational force.

The mere appearance of the coma indicates that a significant fraction of the nucleus is in the form of volatiles. As discussed below the dominant ice is probably H_2O for most observed comets and the dust-to-gas ratio by mass approximately one. For an assumed mean density of a comet nucleus, $\rho \sim 1.3$ g cm^{-3}, a spherical nucleus with radius 2 km, has a mass,

$$M = 4/3 \, \pi \rho \, R^3 \sim 4 \times 10^{16} \text{ g}. \qquad (2)$$

Donn and Rahe (in their chapter) have estimated cometary masses to be in the range of 10^{13} to 10^{19} g.

Whipple and Sekanina (1979) determined an upper limit for the mass of P/Encke to be 10^{16} g and the sublimation mass-loss rate as 10^{13} g/orbital revolution. Thus an upper limit to the expected lifetime of P/Encke, is 3×10^3 more revolutions or $\sim 10^4$ yr.

For a mass of 10^{16} g, the escape velocity from a comet ($R = 2$ km) is 1 m s^{-1} and the gravitational binding energy $\sim 10^{19}$ erg.

C. Rotation

Part of the rationale for Whipple's (1950,1951) icy conglomerate model of the nucleus was to explain the nongravitational effects long observed in cometary motions. Whipple (1950) suggested that the afternoon side of a rotating icy conglomerate nucleus would experience a larger vaporization rate than the rest of the nucleus thereby giving rise to a net impulse in the opposite direction. This impulse would tend to accelerate or decelerate the comet in its orbit depending on the sense of rotation and orientation of the spin axis relative to the comet's orbital plane. Marsden et al. (1973) showed that if the nongravitational forces were due to rotation, then the directions of rotation of cometary nuclei were random (they found equal numbers of accelerating and decelerating comets). Whipple's (1950) model can explain nongravitational forces very well.

The effects of nonuniform sublimation predicted by Whipple's (1950) model might be expected to modulate the morphology of the cometary coma with the rotation period of the (unobserved) nucleus at heliocentric distances $\lesssim 3$ AU. Indeed changes in the fan-shaped comae and near-nuclear jets have been associated with rotation of a nucleus with localized icy spots that give rise to peak vaporization rates with each rotation (Sekanina 1979a,1981).

Another morphological phenomenon probably associated with the rotation of comet nuclei is the repeated ejection of expanding halos due presumably to icy spots on the nucleus which are periodically exposed to the Sun. The expanding halos were particularly well documented for several bright 19th century comets for which Whipple (see his chapter) has derived periods. The technique involves the assumption of a velocity of expansion for the halos and the existence of only one active surface region. Rotation periods for ~ 50 comets have been determined by the expanding halo method (see Whipple's chapter). The orientations of the rotation axes have been determined for ~ 7 comets (Sekanina 1979a).

Spiral structure has been observed in short exposures of Comet Bennett 1970 II, which is shown in Fig. 1 (Larson and Minton 1972; Larson 1981). Because the structure observed is not wavelength dependent, Larson (1981)

Fig. 1. High-resolution, short-exposure photograph of the coma of Comet Bennett 1970 II, taken on 28 March 1970 by S.M. Larson. Note pinwheel structure probably caused by discrete jets of ejected dust from a rotating nucleus.

concludes that it represents the dust component of the coma. The multiple jet model which he developed to explain the spiral structure leads to a rotation period of 34 hr (and a dust expansion velocity ~ 0.6 km s^{-1}), in fair agreement with the 28 hr period derived by Whipple using the halo method. The pronounced asymmetry of the coma of Comet Bennett 1970 II seen in Fig. 1 is in the sunward direction with a phase lag from the subsolar point of a few degrees.

Whipple (see his chapter) has found that the median period of rotation of comet nuclei is 15 hr, and that the directions of rotation and the orientations of the spin axes are random. The shortest period of rotation is 4 hr, which is near the critical period expected for disruption of the comet. For two split comets for which the periods of rotation have been determined, P is $\lesssim 4$ hr. One would expect a spinup of the rotation rate as a function of time, but no such effect has yet been observed.

A cautionary note might be made in regard to interpreting expanding halos of comet comae. If the halos are correctly identified with the dust component in the coma, observations demonstrate for some comets that the dust tails are coupled to the plasma tails in the coma (Belton 1965; Belton and Brandt 1966; Finson and Probstein 1968a,b). This interaction between the dust and plasma tails would alter the velocity flow of the dust. Also it is conceivable that the halo structure may be modulated by unknown mechanisms other than, or in addition to rotation (cf. Houpis and Mendis 1981c).

For a nonspherical nucleus with off-center asymmetric vaporization, one would expect precession of the spin axis. Indeed Whipple and Sekanina (1979) have very convincingly demonstrated that the rotation axis of P/Encke precesses, thus explaining the large secular changes in the nongravitational motions observed for many years for this comet (Marsden et al. 1973). The observations of P/Encke used by Whipple and Sekanina (1979) covered 59 perihelion passages during the interval 1786 to 1977. Their analysis also yielded a period of rotation of 6.5 hr and showed that the comet nucleus must be nonspherical and the ices unevenly distributed. Both short- and long-period comets have been observed to have expanding halos, implying that the surface layers of the ices are distributed unevenly, regardless of the age of the comet.

D. Structure

For many years the two most popular models for a comet nucleus were the sandbank model of a swarm of ice-coated dust grains (Lyttleton 1953,1972) and the icy conglomerate mixture of ices (predominantly H_2O) with embedded dust (Whipple 1950). Observational evidence clearly favors Whipple's (1950) solid nucleus model, with the probable modification that the nucleus is a water (or other ice) clathrate which contains other trapped compounds (Delsemme and Swings 1952). The clathrate model is required to explain the appearance of a variety of molecular and atomic spectral features at similar heliocentric distances. However, there is presently considerable question whether H_2O is always the dominant species in the nuclear ices (cf. A'Hearn and Cowan 1980; Houpis and Mendis 1981b).

Support for the icy conglomerate model is found in many observations of comets. The nongravitational forces affecting cometary motions can be explained nicely by asymmetric vaporization from a rotating ice-covered solid body (Marsden 1974; Yeomans 1977). In addition the observed regularity of nongravitational effects attests to a solid nucleus (Marsden 1974). Furthermore the rotation period of P/Encke determined from observations over a 140 yr interval (Whipple 1980) is consistently the same within 10%. It is difficult to understand how a loose swarm of particles would present such consistent periodic behavior. Evidence that a comet nucleus is solid is further given by the survival of some sungrazing comets with perihelion distances $q \sim 0.005$ AU, and by the repeated apparitions of periodic comets. The recent radar detection of the nucleus of P/Encke also attests to the solid-body nature of the central condensation (see chapter by Kamoun et al.). Hence there seems to be little doubt that comet nuclei are basically monolithic dust and ice conglomerates (cf. Axford and Keller 1978).

The observational evidence also indicates that the nucleus of a comet is homogeneous, although the surface is probably nonuniform. The gas production rates observed for different species in different comets (Feldman's chapter; A'Hearn and Millis 1980; Weaver et al. 1981b) are indicative of undifferentiated nuclei (though not necessarily the same chemical mixtures in all comets). One important inference from the lack of evidence for differentiation of a comet nucleus is the absence of significant internal heating, for example, by the presence of radioactive ^{26}Al (half-life $\sim 10^6$ yr) which might have been present in the primordial protosolar nebula if contaminated by ejecta from a supernova (Irvine et al. 1980). Split nuclei provide the most sensitive tests for homogeneity, particularly if observed both before and after breakup. There were no obvious inhomogeneities observed in Comet West 1976 VI (cf. Ney's chapter; Opal and Carruthers 1977).

As discussed above, the surfaces of comets are quite nonuniform even though the interiors evidently are not. Studies of many comets make clear that periodic comets have small regions where vaporization rates are greater than in other areas of the nucleus. Also relatively new comets have long been known to exhibit more activity than short-period comets at large heliocentric distances ($r \gtrsim 5$ AU), presumably due to the presence of a larger proportion of highly volatile surface ices (although observational selection effects cannot be ruled out [Whipple and Huebner 1976]). One explanation (Brin and Mendis 1979; Brin 1980) for the difference in the observed pre- and post-perihelion magnitudes of comets (Whipple 1978b) involves the gradual formation of a crust or insulating mantle on the surface of a nucleus as the volatile ices sublime in the solar radiation field when the comet is at very large heliocentric distances.

Other erosive processes can affect a comet surface during its long exposure in the outer regions of the solar system. Galactic cosmic rays (energies

~ 1 MeV) have been suggested as one mechanism (cf. Moore and Donn 1981; chapter by Donn and Rahe). Also erosion by solar wind ions and electrons as well as ions trapped in the extensive Jovian magnetotail could have a significant effect on the surfaces of cometary nuclei at large heliocentric distances. In regard to sputtering of cometary ices by solar wind protons, energy is deposited by incident protons into the ices via an electronic rather than a nuclear process (Brown et al. 1980a) which acts to greatly increase the erosion coefficients. For 1 keV protons incident on H_2O ice, the eroison coefficient is ~ 0.3 sputtered H_2O molecules per incident ion (Brown et al. 1980b). At $r \sim 5$ AU, a solar wind flux $\sim 10^7$ particles cm^{-2} s^{-1} impinging on a comet of pure H_2O ice with 10-km radius would produce 10^{19} sputtered H_2O molecules s^{-1} or $\sim 10^4$ g yr^{-1}. The sputtered molecules would have energies ~ 1 eV corresponding to velocities ~ 3 km s^{-1}, well above escape velocity from the comet. Compared with vaporization by solar radiation which at 5 AU produces $\sim 7 \times 10^{22}$ H_2O molecules s^{-1} from a comet of radius 10 km (Houpis and Mendis 1981c), sputtering is a relatively inefficient mass-loss mechanism for quiet solar wind conditions. It should be noted, however, that the sputtering yield at 5 AU can increase by $\sim 10^3$ in high-speed solar wind streams (Houpis and Mendis 1981c), making the sputtering and vaporization H_2O production rates comparable. Hence depending on the specific solar wind conditions and the particular cometary orbit, the erosive and sublimation processes may compete as mass loss and surface erosion mechanisms ($r \gtrsim 5$ AU). Significant electrostatic effects may also alter the surfaces of comet nuclei (Mendis et al. 1981). It is obviously important to study by both photometric and spectroscopic methods, the surface characteristics of a variety of comets at large heliocentric distances, especially comparing pre- and postperihelion properties so that long-term surface erosion processes and associated time scales may be better understood.

An additional way to investigate the structure of a cometary nucleus is to recover and study samples. Presumably the interplanetary material originates from comets and asteroids. Comets are observed to shed mass in the form of both dust and gas. They are also known to fragment and to disintegrate. Some of these cometary debris are in Earth-crossing orbits, in which case they can be observed as meteor showers or fireballs (see chapter by Wetherill and ReVelle). An example is the Taurid stream associated with P/Encke. Unfortunately no recovered meteorite has ever been associated with a known cometary meteor stream. It should be noted that the identity of a recovered meteorite with a comet nucleus will always be somewhat moot, since asteroid interlopers may have orbits like short-period comets (see chapter by Wetherill and ReVelle). However, by limiting observations of fireballs to known cometary meteor showers the sampling problems can be minimized. Information on the physical properties of cometary debris can be extracted from the trajectories and integrated magnitudes of fireball tracks observed in photographic surveys such as the Prairie Network. The luminous efficiencies of meteoroidal material impacting the Earth's atmosphere are unknown, thereby preventing reliable photometric mass estimates. However, compressive strengths can be determined from the dynamics of the fireball trajectories. For the most abundant form of cometary debris in the mass range 10^2 to 10^6 g, compressive strengths $\sim 10^5$ dyne cm^{-2} have been inferred from the observations (chapter by Wetherill and ReVelle).

Another approach to obtaining samples of cometary material has been to gather interplanetary dust from collectors aboard a U-2 airplane cruising at altitudes ~ 20 km above the Earth's surface. A newly discovered class of fragile interplanetary particles, described as fluffy aggregates, has solar abundances indicating an extraterrestrial origin (Brownlee 1978). These particles have very low densities and are candidates for recovered cometary samples (see chapter by Fraundorf et al.).

Various estimates of the cohesive strength of the comet nucleus indicate that it must be a very fragile entity. The low mass (inferred from the small radius and basic icy nature) produces a central pressure $\lesssim 10^5$ dyne cm^{-2}, not enough to compact crystalline snow (Donn 1963; chapter by Donn and Rahe). An estimate for the tensile strength of a comet nucleus can be obtained from the sungrazing comets, some of which have survived disruption at perihelion distances as small as 0.005 AU (chapter by Sekanina). For comets with mean densities ~ 1 g cm^{-3}, the tensile strength against tidal disruption is in the range $F \sim GM\rho R^2 r^{-3} \sim 10^4$-$10^5$ dyne cm^{-2} (Whipple 1963; Sekanina's chapter) for radii in the range of $R = 1$ to 10 km.

Similar lower limits on the tensile strengths of cometary nuclei can be set by the radii and periods of rotation. Even though the sample of comets studied by Whipple (see his chapter) is incomplete, if it is assumed that the periods of rotation are representative of the larger population of comets, the shortest period of rotation is $P \sim 4$ hr for Comet Honda 1955 V. For an assumed radius $R \sim 1$ km, $\rho \sim 1$ g cm^{-3} the disruptive force due to rotation is (chapter by Sekanina), $F \sim 2\pi^2 \rho R^2 P^{-2} \gtrsim 10^3$ dyne cm^{-2}, which is consistent with the above estimate and with previous results (Whipple 1963; Sekanina's chapter). These estimates of tensile strengths for cometary nuclei are in the range of compressive strengths estimated for fireballs which have been identified with comets, namely, $F \sim 10^5$ dyne cm^{-2} (chapter by Wetherill and ReVelle). Thus the observations indicate that the comet nuclei are quite fragile.

E. Split Nuclei

Over twenty comets have been observed to have multiple nuclei consisting of from 2 to 5 fragments (see Sekanina's chapter). Although the actual disruption process of the nucleus has never been observed directly, correlations have been established in five cases between either a brightness increase or the appearance of dust streamers in the tail and the estimated time of fragmentation. The best evidence for an association between cometary activity and the time of splitting was for Comet West 1976 VI. Both a dust burst and a 2-mag increase in brightness were observed at the estimated time that the nucleus split into four fragments (see Figs. 2 and 3). At least twelve

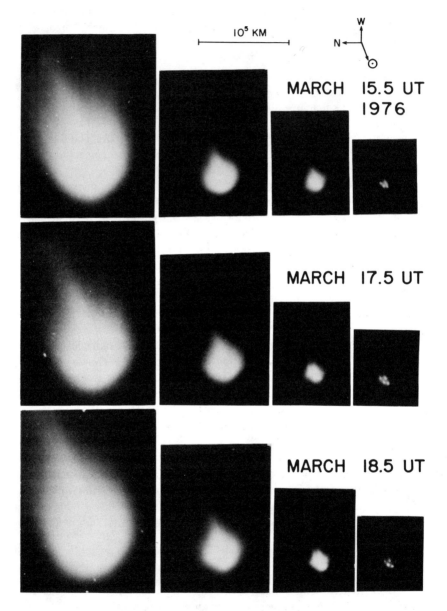

Fig. 2. High-resolution, short-exposure photograph taken with the 154-cm telescope at Mt. Lemmon Observatory by S.M. Larson of the split nucleus of Comet West 1976 VI. The four components of the nucleus moved apart over three nights, and each varied in brightness and displayed an individual tail.

OVERVIEW OF OBSERVATIONS

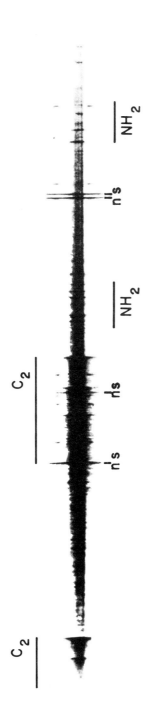

Fig. 3. Spectrogram (42 Å mm^{-1}, 60 arcsec mm^{-1}) of Comet West 1976 VI taken with the 2.5-m Isaac Newton Telescope by S. S. Wyckoff and P.A. Wehinger. Note separate continua for split nucleus fragments and the extension of NH$_2$ emission perpendicular to dispersion, indicating a common gaseous envelope for the fragments.

comets are known to have fragmented at heliocentric distances remote from perihelion or Jupiter (Whipple and Huebner 1976). One may have split at a heliocentric distance of 9 AU. On the other hand, the nuclei of some sungrazing comets have been known to split near perihelion presumably due to solar tidal forces. P/Brooks 2 apparently fragmented due to Jovian tidal forces during a near encounter in 1886 (Sekanina's chapter). In view of the small tensile strength of a cometary nucleus and the above evidence, it seems that more than one process can cause fragmentation of a comet nucleus. Breakup mechanisms which have been suggested include gravitational tidal effects, rotation (the rates of which should increase with time), and internal exothermic reactions. Collisions with other solar system objects are highly improbable. The fragmented nuclei are in most cases probably not gravitationally bound. They are observed to separate at velocities of a few m s^{-1} (also indicative that an explosive process is probably not involved in splitting the nuclei), which is larger than escape velocity of a monolithic (2-km radius) comet, namely $\sim$ 1 m s^{-1}. Marsden and Sekanina (1974) attribute the relative motions of the split nuclei to a differential nongravitational acceleration, presumably due to asymmetric outgassing by the irregular, rotating fragments (Wallis 1980).

The split nuclei generally appear enveloped in a common coma. There is no indication for large spatial variations in the abundance ratios of various molecular (C_2 and NH_2) species in the coma of Comet West 1976 VI, as is evident in Fig. 3 where spectra of the individual nuclear fragments are resolved; nor is there evidence for time variations in abundances (A'Hearn et al. 1980).

Independent evidence for fragmenting nuclei is found in the orbital elements of several sungrazing comets which have appeared during the past century (cf. Marsden 1967). The orbital elements for 8 bright comets which appeared from 1843 to 1970 are nearly identical. This fact coupled with the small perihelion distances of $q \sim 0.005$ for all 8 comets probably indicates a common origin and a splitting by tidal interaction(s) with the Sun. In fact the sungrazers Great September Comet (CRULS) 1882 II and Comet Ikeya-Seki 1965 VIII did split near the time of perihelion passage.

Comets have occasionally been observed to disintegrate; while one comet, Great Southern Comet (THOME) 1887 I, may even have dissipated upon collision with the solar photosphere.

F. Outbursts

Brightness flares of 1 to 2 mag are frequently reported in both parabolic and periodic comets (cf. Roemer 1962). The most famous example is P/Schwassmann-Wachmann 1 (P = 16.5 yr) which has been observed to brighten as much as 8 mag. At least 100 flares have been observed for this comet over the past fifty years (Whipple 1980). Such outbursts are particularly interesting because the orbit of P/Schwassmann-Wachmann 1 is nearly circular (e = 0.13) with the heliocentric distance ranging between 5 and 6 AU.

The most energetic outburst observed for a comet was in P/Tuttle-Giacobini-Kresák (P = 5.6 yr) which brightened by 9 mag twice in 1973 (Whipple and Huebner 1976).

Flaring activity is not confined to periodic comets. Both comets Morehouse 1908 III and Humason 1962 VIII showed brightness increases as they approached the Sun presumably for the first time. Comet Humason 1962 VIII showed unusual physical activity at a heliocentric distance as large as 5 AU when the comet was described as diffuse with a diameter of 1 arcmin and an asymmetry extending $\sim$ 5 arcmin (Roemer and Lloyd 1966). Comet Humason 1962 VIII also was described by Roemer (1961a, 1962) at that time as having a crab-like appearance (see photographs in Roemer [1962] and Brandt and Mendis [1979]). According to Roemer (personal communication) Comet Humason remained active during its entire apparition, even undergoing a flare of 6 mag at a heliocentric distance $\sim$ 6 AU in 1964.

Suggestions have been made that outbursts in comet nuclei may be caused by chemical heating, the presence of ices more volatile than H_2O or by amorphous H_2O (Donn and Urey 1957; Whipple 1980). Smoluchowski (1981) has elaborated on the suggestion that the crystallization of amorphous ice is associated with a rapid exothermic reaction accompanied by an abrupt density change as the temperature of the ice rises $>$ 153 K. The effect might explain outburst of comets provided a crystalline or crustal surface had previously been blown off. Smoluchowski's (1981) model predicts the crystallization of surface layers ($\sim$ few cm to a few m deep) of amorphous ice to occur at heliocentric distances in the range 3 to 6 AU, depending on the orientation of the spin axis of the nucleus with respect to the Sun. Comets in outburst have shown enhanced strengths of CO^+ bands (cf. Greenstein 1962) in emission. If a sudden phase change in H_2O ice is the outburst source in cometary nuclei, H_2O^+ emission might be observed at or near the time of outburst.

G. Composition

Whipple's (1950) original model contained a mixture of ices of the abundant species, H, C, N and O with H_2O being the most abundant constituent in the nucleus. He also posited that dust was embedded in the icy conglomerate. The inference of the relative abundances of species in the nuclear ices from the observed spectra is presently an unsolved problem. Molecules released directly from the subliming ices have a variety of (short) lifetimes against photodissociation in the solar radiation field. Hence only dissociation and ionization products of the sublimed parents are observed in the optical spectra. Moreover, any polyatomic molecules which might be released directly as a result of the vaporization of the cometary ices (e.g. H_2O, CO_2) have their resonance transitions outside the ultraviolet and optical windows ($\sim$ 0.12 to 1 μm). Until direct samples of cometary ices or direct observations of parent molecules are obtained, any knowledge of the chemical composition of the nuclear ices must remain extremely model-dependent at best (see chapters by Delsemme and by Huebner et al.; Festou 1981a).

It now seems clear that in many comets, though not in all, water ice plays a dominant role in the vaporization processes of the nuclear material. The strong OH, Lyman-α and [OI] (6300, 6364 Å) lines observed in the spectra of comets constitute convincing evidence for a large abundance of water vapor in the coma (Biermann and Trefftz 1964; Code et al. 1972; Code and Savage 1972; Carruthers et al. 1974; Keller 1976; Festou and Feldman 1981; Weaver et al. 1981a,b; Spinrad 1981). Moreover, the tentative detection of a H_2O 1.35-cm feature (Jackson et al. 1976; see, however, Crovisier et al. 1981; Hollis et al. 1981) and the definitive identification of the H_2O^+ emission spectrum in more than one comet (Herzberg and Lew 1974; Wehinger et al. 1974; Wehinger and Wyckoff 1974; Miller 1980) corroborate the evidence that H_2O ice is abundant in comets. Also the measurement of H and OH emission spectra in Comet Bennett 1970 II with an abundance ratio 2:1, indicates an origin in water ice (Code et al. 1972; Keller and Thomas 1975; chapter by Feldman), while the identification of a high-velocity component of neutral H expanding from the coma is consistent with an origin from the photodissociation of H_2O (Keller and Thomas 1975).

It should be remarked, however, that the identification of H_2O ice as controlling the vaporization process in comets has by no means been proven to be a universal characteristic. There seems to be some observational evidence that CO_2 (or CO) ice may play a significant role in controlling the gas production rates in some comets (A'Hearn and Cowan 1980). Indeed large variations in H_2O^+/CO^+ emission band strength ratios from one comet to another (at comparable heliocentric distances) might possibly indicate abundance differences in H_2O and CO_2 (or CO) (Wyckoff and Wehinger 1976a; Miller 1980).

As the nuclear ices vaporize in a comet, embedded dust is released and becomes entrained in the expanding gases. The spectra of most comets show the reflected solar spectrum, presumably from the expanding dust halo, while the type II tails of comets have long indicated that copious amounts of dust are present to varying degrees in many comet nuclei. Accurate measurements of the dust-to-gas ratios in comets are not yet available (Sekanina and Miller 1973; Donn 1977; A'Hearn et al. 1979). Various estimates have given dust-to-gas ratios by mass in the range 0.5 for gas-rich to 1.0 for dusty comets. A'Hearn et al. (1979) have stressed the lack of and the need for accurate and homogeneous measurements. Measurements of the total gas production rates and dust production rates are important for determining the mass fractions of ices and solids in comet nuclei.

II. DUST

A. Observations

Dust has long been known to be an important constituent in comets, primarily due to the extensive type II tails developed particularly by bright comets near perihelion. Evidence for dust in the coma and the first estimate that the particle sizes were 1 μm was determined from the reddened spectrum of reflected sunlight observed by Liller (1960) in Comet Arend-Roland 1957 III. An extended dust coma of Comet Bowell 1980b has been observed at a heliocentric distance as large as 7.2 AU by Cochran and McCall (1980) who noted a coma diameter of 12 arcsec, but detected no gaseous emission spectrum after subtraction of the solar scattered light spectrum. Hence any explanation of the release mechanisms of dust from comet nuclei must explain observations at $r > 3$ AU where H_2O ice from the nucleus is not expected to vaporize in the solar radiation field. Houpis and Mendis (1981c) have suggested that electrostatic blow-off or sublimation of CO_2 may effect the release of dust from Comet Bowell 1980b at large heliocentric distances. For small heliocentric distances ($r \lesssim 3$ AU) the dust becomes entrained by drag forces of the expanding, vaporized coma gases.

The dust tails which develop in most (but not all) comets are observed to lag the comet-Sun axis by angles $\lesssim 60$ deg (Osterbrock 1958; Belton 1965; Brandt and Belton 1966). Bessel (1836) first explained the development of type II tails by solar radiation pressure on the dust particles expanding from the nucleus. The orbital lag of the particles is a result of Keplerian motion under the gravitational influence of the Sun (cf. Finson and Probstein 1968a).

Ney (1974) has pointed out the large differences in dust activity among various comets, both new and old. Observed dust-to-gas ratios range from 0.1 to 1 by mass, but are evidently not indicators of the prominence of a type II tail. As Ney (see his chapter) points out, Comet Kobayashi-Berger-Milon 1975 IX had a dust-to-gas ratio ~ 1, yet did not develop a dust tail. Why some comets develop strong, dominant type II tails, others have both plasma and dust tails, and some have predominantly plasma tails is presently unexplained. The grain size distribution and perihelion distances are undoubtedly important factors in the development of a dust tail.

The presence of dust in cometary comae has been inferred from spectrophotometric observations ranging from visible to infrared wavelengths (Liller 1960; Becklin and Westphal 1966; Maas et al. 1970; Ney's chapter). The main spectral characteristics of the dust for both periodic and parabolic comets may be summarized in various wavelength intervals (cf. Hanner 1980; chapter by Campins and Hanner): $\lambda < 1.6$ μm: scattered solar spectrum; $3 \mu m < \lambda < 5 \mu m$: color temperature greater than a blackbody temperature; $8 \mu m < \lambda < 13 \mu m$: broad emission feature; $\lambda \sim 18 \mu m$: weaker emission feature.

Optical and infrared observations of comets provide information on the composition and size distribution of the dust, the dust albedo and its dependence on the scattering phase angle (Sun-comet-Earth), grain temperatures, dust production rates and the particle velocities in the coma and tail. No spectral signatures due to frost (e.g., the 2-μm absorption features) have yet been detected in cometary dust or nuclei (A'Hearn et al. 1981).

B. Inferred Physical Properties

Dominant 1 μm-sized particles in comets are inferred from the scattered sunlight observed at visible wavelengths (Liller 1960; Ney's chapter), polarization measurements (Blackwell and Willstrop 1957; Clark 1971), and broadband infrared photometry (see Ney's chapter). Observations of the scattering phase function of the dust in two comets are consistent with slightly absorbing dielectric particles (Ney and Merrill 1976). The 10-μm and 18-μm emission features observed in some (not all) comets probably indicate the presence of silicate particles (Maas et al. 1970). In general, dusty comets appear to have strong silicate features. The silicate feature in Comet Bradfield 1974 III was observed to disappear as the comet's brightness dropped abruptly at $r \sim 0.6$ AU, indicating a sudden decrease in grain production (Ney's chapter). The 10-μm feature was observed both before and after the fragmentation of Comet West 1976 VI for which the dust production rate per fragment remained unchanged (see Ney's chapter). Spatially resolved polarization measurements of Comet West 1976 VI (Isobe et al. 1978) indicate variations in polarization from a few to 20% over the several nuclear fragments. These observations could indicate particle size variations due to differential radiation pressure or due to internal inhomogeneities in the comet nucleus (Hanner 1980).

Other polarimetric observations indicate variations in the particle size distribution along the dust tails (Krishna Swamy 1978). In this regard imaging data of dust tails exhibiting striae have been explained as fragments of larger dust particles ejected from the nucleus (Sekanina and Farrell 1980; Hill and Mendis 1980).

Evidence for dust particles with sizes $\gtrsim 30\,\mu$m was first suggested by Spinrad and Miner (1968) to explain the presence of Na I atoms at distances $\sim 10^5$ km in the tail of Comet Ikeya-Seki 1965 VIII. Additional evidence for large dust grains has been found from thermal emission observed in Comet Kobayashi-Berger-Milon 1975 IX (Ney's chapter), which developed no type II tail, nor exhibited a 10-μm emission feature. Also the 10-μm silicate feature was not observed in the antitail of Comet Kohoutek 1973 XII, which may be indicative of particles > 30-μm. Gibson and Hobbs (1981) have suggested that the likely source of microwave emission detected in comets is an icy-grain halo having characteristic grain sizes of 1 cm.

The albedo of the dust measured by comparing visible scattered light with infrared thermal emission from the same volume is consistent from comet to comet, with observed values from 0.1 to 0.4.

C. Kinematics

Studies of the observed surface brightness distributions of type II comet tails reveal the particle size distributions, dust production rates and the velocities of ejection of the particles as a function of size and time (Finson and Probstein 1968a,b; Sekanina and Miller 1973). Dust is continuously released into the coma as the nuclear ices vaporize. Both dust and plasma tails develop at $r < 3$ AU, although tail-like asymmetries have often been observed at larger heliocentric distances (cf. Roemer 1962; Roemer and Lloyd 1966). The dust particles released into the coma are accelerated by drag forces of the expanding gas to terminal velocities ~ 1 km s^{-1} (Finson and Probstein 1968b). They are entrained by the coma gases until they decouple at radial distances from the nucleus ~ 10 to 10^2 km (see Houpis and Mendis 1981c). The initial velocities of the dust particles and directions of ejection into the tail depend to a certain extent on the presence of a plasma tail. Evidently, in the inner coma, the trajectories of the ions and dust tend to couple, giving rise to a smaller initial lag angle for the dust tail. Beyond 10^4 km the neutral particles are influenced only by solar radiation pressure and solar gravitation where the forces are given by

$$F_{\rm rad} = \epsilon c^{-1} \pi s^2 F_\odot (4\pi r^2)^{-1}$$
$$F_{\rm grav} = GM_\odot r^{-2} (4/3) \rho \pi s^3 \tag{3}$$

where ϵ is scattering efficiency of the dust particles, s the mean radius of the dust particles, and ρ the mean density of the dust particles. The ratio of the two forces gives

$$F_r/F_g \sim (\rho s)^{-1} \tag{4}$$

from which we see that the radiation pressure from the Sun is more significant the smaller the dust particle. The above ratio has been estimated from observed tail curvatures of comets to be in the range

$$F_r/F_g \sim 0.1 \text{ to } 1.0 \ . \tag{5}$$

Evidence for the release of relatively large particles is found in observations of antitails for several comets. These sunward spikes can be explained by relatively large particles ($> 30 \mu$m) released from nuclear ices with velocities of only $\sim$ few m s^{-1} which hover near the nucleus. When the Earth crosses the comet's orbital plane, we observe these zero-velocity particles edge-on distributed in the comet orbit plane and, due to projection effects on the sunward side of the coma, hence the appearance of the spike (Öpik 1958; Gary and O'Dell 1974). Infrared photometry appears to corroborate this explanation for the antitails of comets.

D. Dust Production

Entrainment of the smaller dust grains by the expanding coma determines the ejection velocities of these particles that are ~ 1 km s^{-1}, well in excess of escape velocity from the nucleus (few m s^{-1}). Mass lost in the form of dust is observed to vary continuously with heliocentric distance with the production rates proportional to r^{-2}, as determined from the observed total energy emitted by the grains in the wavelength range 0.5 to 20 μm for $r \lesssim 1$ AU. The infrared luminosity is a measure of the dust production rate, because there is no dependence on particle size (for small particles) (see chapter by Ney). Mass-loss rates can also be determined from photometric measurements in the visible, but the size distribution of particles and the scattering phase function must be known.

Dust mass-loss rates (at 1 AU) determined for 4 long-period comets are 4×10^6 g s^{-1} and for P/Encke, 2×10^4 g s^{-1} (Ney's chapter). The dust-plus-gas production rates (by mass) for the same long-period comets, $Q(\text{dust}+\text{H}_2\text{O}) \sim 1.4 \times 10^7$ g s^{-1} and for P/Encke $\sim 2.2 \times 10^5$ g s^{-1}. Thus the mass-loss rate from the short-period comet is less than for the long-period comets, a difference which could reflect a difference in the sizes of the nuclei.

III. COMA

As a comet approaches the Sun the evaporation rate of the nuclear ice increases, and an expanding atmosphere of gas and dust develops around the nucleus. This atmosphere is loosely referred to as the coma and has been subjected to thorough studies over the years, yet is still poorly understood. The observed coma is comprised of dust (possibly ice-coated grains), molecules, neutral radicals, atomic species, and molecular ions which are released from the inner coma with velocities $\sim 0.5 - 1$ km s^{-1}, well above the escape velocity from the nucleus as discussed above.

Hence as the coma develops, the comet rapidly begins to lose mass. Solar radiation is the primary factor controlling the development of the coma, and the vaporization rate depends on the latent heats of sublimation of the dominant nuclear ices and on the physical structure of the nucleus (shadowing by grains, rotation rate, etc.). Although the observational statistics are somewhat biased by selection effects due to outburst activity in distant comets and the dearth of observations of faint comets, the coma is usually well developed at $r \lesssim 3$ AU, as evidenced by a diffuse image, as well as by the emission spectra of CN and C_2 observed in the optical wavelength region. However, it should be noted that the average seeing resolution (~ 2 arcsec) of a groundbased optical telescope corresponds to 4×10^3 km for a geocentric distance ~ 3 AU (only relatively well-developed comae would be resolved at $\Delta \gtrsim 3$ AU). This point has an important bearing on inferences with regard to distances at which volatile activation begins, and hence the composition of the surface ices of the nucleus. Also spectroscopic observations of faint comets at $r \sim 3$ AU have in the past been essentially at the detection limits of optical telescopes (V mag > 18). Observations of faint comets such as those by Spinrad et al. (1979), Cochran and McCall (1980), Larson (1980), A'Hearn and Millis (1980) are needed to clarify precisely how and when the comae of comets develop in the solar radiation field. Complications such as the removal of a dust mantle (Brin and Mendis 1979; Brin 1980), phase transitions in water ice (Smoluchowski 1981), presence of an icy-grain halo, and dominant volatiles other than water ice (A'Hearn and Cowan 1980) will have to be considered when studying the development of the coma at large heliocentric distances.

The appearance of an obvious, extended coma at $r \lesssim 3$ AU has long been considered evidence that water ice is an important constituent in the nuclei of comets, since the solar radiation temperature at $r \sim 3$ AU is approximately that required to vaporize ice. The clathrate model of the nucleus was suggested to explain the presence in the coma of radicals other than H_2O derivatives (Delsemme and Swings 1952).

The extent and the luminosity of the coma are determined by the gas and dust production rates which change continuously with time and the comet's heliocentric distance. The production rates of gas and dust depend predominantly on the solar radiation field, the size of the nucleus and the vaporization properties of the constituent ices. One purpose of studying the coma is to infer the chemical composition and physical properties of the nucleus which requires both detailed observations of the coma as well as complex modeling of the chemical, radiative, and collisional interactions known to occur in the inner coma. At $r \sim 1$ AU, observed gas production rates for bright comets are $Q \sim 10^{30}$ molecules s^{-1} and gas expansion velocities, $v \sim 1$ km s^{-1}. The density $n(\ell)$ at a distance ℓ from an assumed spherical nucleus of radius R is (cf. Oppenheimer 1975)

$$n(\ell) = QR^2 \, (4\pi R^2 \, v \, \ell^2)^{-1} \qquad (6)$$

and the mean free path L of a molecule in the coma is therefore

$$L = [n(\ell)\sigma]^{-1} = 4\pi v \ell^2 \, (\sigma Q)^{-1} \qquad (7)$$

where σ is the collisional cross section. For a typical neutral molecule, $\sigma \sim 10^{-15}$ cm^2 so that $\ell = L$ at 10^4 km. Hence the inner coma ($\ell \lesssim 10^4$ km) is generally considered the collision zone of a cometary atmosphere. Lifetimes against photodissociation cover a wide range, but are generally longer than a few hr, so that collisions in the inner coma may be significant in altering the abundances of evaporating molecules from the nucleus. Photoionization is known to produce molecular ions in the inner coma, and large rate coefficients measured for ion-neutral reactions indicate that abundances can be significantly changed as the coma gases expand into the interplanetary medium (Oppenheimer 1975; Giguere and Huebner 1978). Observational evidence indicates that ion chemistry can be important (Feldman 1978), but the degree to which the overall relative abundances are altered is probably minimal as evidenced by spectral similarities from one comet to another (chapter by Feldman; A'Hearn and Millis 1980). The chemical activity expected in the coma of a comet is directly related to the gas production and ionization rates, and hence the heliocentric distance as well as the effective geometric cross section of the nucleus.

A. Morphology

The coma of an average comet generally appears spherically symmetric on direct, blue-sensitive photographs. However, its photo center probably does not coincide with the center of mass of the underlying nucleus (cf. Roemer 1961b) due to the unidirectional evaporation process. Indeed, evidence that the coma is brighter on the sunward side of the nucleus is found in the surface brightness distribution observed for Comet Bennett 1970 II where a phase lag with respect to the subsolar point in the symmetry axis of the isophotes was attributed to rotation (Larson 1981). Whipple (see his chapter) has discussed similar asymmetries in the form of fan-shaped comae or near-nucleus jets which are presumably partly caused by a spotty distribution of the nuclear ices. The photograph (Fig. 4) of P/Encke illustrates the long-known pronounced fan-shaped asymmetry of the coma, oriented toward the Sun.

On the other hand, an extended Lyman-α coma ($\sim 10^6$ km) appeared elongated in the antisolar direction in Comet West 1976 VI possibly attributed to either the fragmented nucleus (see Fig. 2) or to optical depth effects (Opal and Carruthers 1977). The Lyman-α image of Comet Kohoutek 1973 XII showed an asymmetry in the antisunward direction which was attributed to solar radiation pressure (Opal et al. 1974).

The only accurate monochromatic surface brightness distributions of coma gases which have been mapped have been of H and OH in the ultraviolet (cf. Keller 1976; Festou 1981b). Such observations provide information on the velocity field (particularly at large distances from the collision region where the energetics of the photodestruction processes could significantly alter the systematic and random gas motions in the expanding coma), as well as the lifetimes of cometary radicals in the solar radiation field.

B. Integrated Magnitudes

Although integrated magnitudes have been measured of the coma + nuclear components of many comets over a large range in heliocentric distances (see chapter by Meisel and Morris), the broadband observations are difficult to interpret in terms of physical processes occurring in the coma. Nevertheless A'Hearn and Millis (1980) have shown that the broadband total magnitudes correlate with their narrowband photometry, which can be related to individual species in the coma. The total magnitude is expressed as (cf. Marsden and Roemer in the Appendix)

$$m_1 = H_1 + 2.5 \, n \log r + 5 \log \Delta \tag{8}$$

where m_1 is the observed magnitude in some passband, H_1 is the absolute magnitude reduced to standard distance (usually $r = \Delta = 1$ AU), Δ is the geocentric distance of the comet, and r the heliocentric distance. n is determined empirically and is presumably a function of the evaporation rate, the albedo

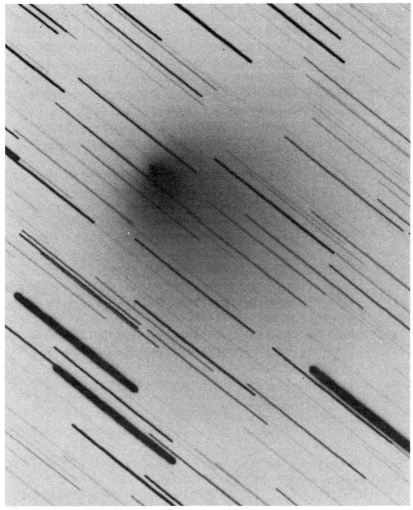

Fig. 4. Photograph of P/Encke obtained in 1980 with the Kitt Peak 4-m telescope by H. Spinrad and J. Stauffer. Note the asymmetrical surface brightness of the coma. The fan-shaped coma is roughly aligned with the Sun-comet axis.

of the dust or icy grains, the fluorescence rate in the relevant passband, the abundances of emitting species and their lifetimes. The quantities H_1 and n are called the photometric parameters. Values of n in the range 2 to 8 have been observed for various comets (cf. Whipple 1978a,b). There is some evidence for a secular decrease in n for P/Encke. Whipple (1978b) has shown that the average value of n is different before perihelion compared with postperihelion observations of the same comet. It would be interesting to attempt to model the parameters n and H_1 with physical processes occurring in the coma. A'Hearn and Millis (1980) utilize narrowband filters to isolate

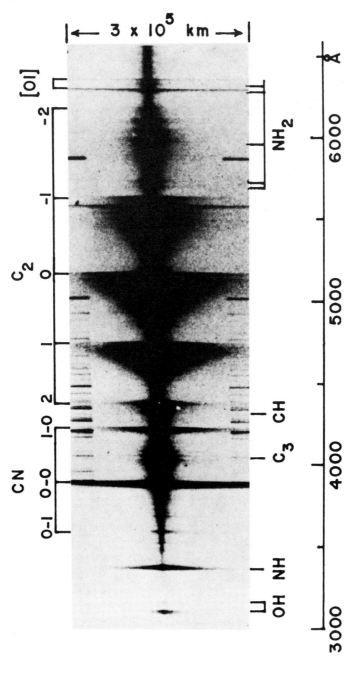

Fig. 5. Spectrogram of Comet Kobayashi-Berger-Milon 1975 IX obtained with Wise Observatory 1-m telescope by S. Wyckoff and P.A. Wehinger on 7 August 1975 ($r = 0.8$ AU, $\Delta = 0.6$ AU). The molecular spectrum from the neutral gas was much stronger than ions. H_2O^+ features are weakly present (6200 Å); CO^+ bands (4200 Å) are absent.

different species in their photometric observations of comets. To facilitate comparisons with models of the secular vaporization behavior of comets, the total magnitudes must be measured to a specific surface brightness level.

C. Observed Species

Most of the observations of the comae of comets over the past century have been devoted to optical spectroscopy (3000 to 8000 Å) and in particular to the identification of various atomic and molecular emission features which appear at heliocentric distances $r \lesssim 3$ AU. Because of detector limitations, the bulk of past spectroscopic observations were virtually only of bright comets at small heliocentric distances (cf. Swings and Haser 1956; Arpigny et al. 1982). In the last decade improved detectors in the optical spectral region, satellite observations in the ultraviolet, and radio studies have significantly increased the number of identified species (by 25%) and revealed additional transitions of molecules or atoms previously detected. A summary of the atomic and molecular species identified in the spectra of comets (projected distances from the nucleus $< 10^4$ km) is given in Table I, together with the observed wavelengths or frequencies of characteristic features, and the associated transitions. The references in Table I cite discoveries, identifications or significant studies of the species indicated. Colons after entries in the table signify uncertainty. The cosmically abundant elements, H, C, N, O, and S are well represented in this table as are the most abundant metals. It should be noted, however, that the latter have been observed in emission only in sungrazing comets where radiation temperatures in excess of 10^3 K can vaporize the refractory components of the dust (Preston 1967). As can be seen essentially all of the transitions listed arise from the ground states of the atoms or molecules, although there are some exceptions (see the chapters by A'Hearn and by Feldman for discussions).

TABLE I

Observed Species in Comet Spectra

Atom or Molecule	Characteristic Band or Line	Transition	Reference
H I	1216 Å	$^2P^0 \to {}^2S$	Code et al. 1972
	6563 Å	$^2D \to {}^2P^0$	Lanzerotti et al. 1974
			Huppler et al. 1975
O I	1304 Å	$^3S^0 \to {}^3P$	Feldman et al. 1974
[O I]	2972 Å:	$^1S \to {}^3P$:	Feldman et al. 1980
	6300 Å	$^1D \to {}^3P$	Swings and Greenstein 1958
			Festou and Feldman 1981

TABLE I (Continued)

Observed Species in Comet Spectra

Atom or Molecule	Characteristic Band or Line	Transition	Reference
C I	1561 Å	$^3D^0 \rightarrow {}^3P$	Feldman et al. 1974
	1657 Å	$^3P^0 \rightarrow {}^3P$	Feldman et al. 1974
[C I]	1931 Å	$^1D \rightarrow {}^1P^0$	Feldman and Brune 1976
C II	1335 Å	$^2D \rightarrow {}^2P^0$	Feldman and Brune 1976
S I	1814 Å	$^3P \rightarrow {}^3S^0$	Smith et al. 1980
Na I	3303 Å	$^2P^0 \rightarrow {}^2S$	Preston 1967
	5890 Å	$^2P^0 \rightarrow {}^2S$	Preston 1967
			Oppenheimer 1980
K I	7665 Å	$^2P^0 \rightarrow {}^2S$	Preston 1967
	4044 Å	$^2P^0 \rightarrow {}^2S$	Preston 1967
Ca II	3934 Å	$^2P^0 \rightarrow {}^2S$	Preston 1967
Cr I	3579 Å	$^7P^0 \rightarrow {}^7S$	Preston 1967
Mn I	4031 Å	$^6P^0 \rightarrow {}^6S$	Preston 1967
Fe I	3441 Å	$^5P^0 \rightarrow {}^5D$	Preston 1967
	3570 Å	$^3G^0 \rightarrow {}^5F$	Preston 1967
	3581 Å	$^5G^0 \rightarrow {}^5F$	Preston 1967
	3720 Å	$^5F^0 \rightarrow {}^5D$	Preston 1967
	3749 Å	$^5F^0 \rightarrow {}^5F$	Preston 1967
	3813 Å	$^3P^0 \rightarrow {}^5F$	Preston 1967
	3816 Å	$^3D^0 \rightarrow {}^3F$	Preston 1967
	3820 Å	$^5D^0 \rightarrow {}^5F$	Preston 1967
	3860 Å	$^5D^0 \rightarrow {}^5D$	Preston 1967
	4046 Å	$^3F^0 \rightarrow {}^3F$	Preston 1967
Ni I	3381 Å	$^1P^0 \rightarrow {}^1D$	Preston 1967
	3446 Å	$^3D^0 \rightarrow {}^3D$	Preston 1967
	3458 Å	$^3F^0 \rightarrow {}^3D$	Preston 1967
	3462 Å	$^5F^0 \rightarrow {}^3D$	Preston 1967
	3525 Å	$^3P^0 \rightarrow {}^3D$	Preston 1967
	3566 Å	$^1D^0 \rightarrow {}^1D$	Preston 1967
	3619 Å	$^1F^0 \rightarrow {}^1D$	Preston 1967

TABLE I (Continued)

Observed Species in Comet Spectra

Atom or Molecule	Characteristic Band or Line	Transition	Reference
Cu I	3248 Å	$^2P^0 \to {}^2S$	Preston 1967
C_2	2313 Å	$D^1\Sigma_u^+ \to X^1\Sigma_g^+$	A'Hearn and Feldman 1980
	5165 Å	$d^3\Pi_g \to a^3\Pi_u$	Huggins 1882
	7715 Å	$A^1\Pi_u \to X^1\Sigma_g^+$:	Danks and Dennefeld 1981
$^{12}C^{13}C$	4745 Å	$d^3\Pi_g \to a^3\Pi_u$	Swings 1943
			Stawikowski and Greenstein 1964
C_3	4040 Å	$\tilde{A}^1\Pi_u \to \tilde{X}^1\Sigma_g^+$	Huggins 1882
			Douglas 1951
			Gausset et al. 1965
CH	3889 Å	$B^2\Sigma^- \to X^2\Pi_r$	Dufay 1940
	4315 Å	$A^2\Delta \to X^2\Pi_r$	Nicolet 1938
	3.3 GHz	$X^2\Pi_{1/2}$ (F=1→1)	Black et al. 1974
CH_3CN	110.7 GHz	$\tilde{X}^1A_1$ (J=6→5)	Ulich and Conklin 1975
CN	3883 Å	$B^2\Sigma^+ \to X^2\Sigma^+$	Huggins 1882
	7873 Å	$A^2\Pi_i \to X^2\Sigma^+$	Swings and Page 1948
			O'Dell 1971
CO	1510 Å	$A^1\Pi \to X^1\Sigma^+$	Feldman and Brune 1976
CS	2576 Å	$A^1\Pi \to X^1\Sigma^+$	Smith et al. 1980
HCN	88.6 GHz	$\tilde{X}^1\Sigma^+$ (J=1→0)	Huebner et al. 1974
NH	3360 Å	$A^3\Pi_i \to X^3\Sigma^-$	Swings et al. 1941
NH_2	5700 Å	$\tilde{A}^2A_1 \to \tilde{X}^2B_1$	Swings et al. 1943
			Greenstein and Arpigny 1962
OH	3090 Å	$A^2\Sigma^+ \to X^2\Pi_i$	Swings et al. 1941
	1.665 GHz	$X^2\Pi_{3/2}$ (F=2→2)	Biraud et al. 1974
	1.667 GHz	$X^2\Pi_{3/2}$ (F=1→1)	Giguere et al. 1980
			Turner 1974
CH^+	4225 Å	$A^1\Pi \to X^1\Sigma^+$	Swings 1941b
			Greenstein 1962

TABLE I (Continued)

Observed Species in Comet Spectra

Atom or Molecule	Characteristic Band or Line	Transition	Reference
CN^+:	2181 Å:	$f^1\Sigma \to a^1\Sigma$:	Smith et al. 1980
	3185 Å:	$c^1\Sigma \to a^1\Sigma$:	
CO^+	2190 Å	$B^2\Sigma^+ \to X^2\Sigma^+$	Feldman and Brune 1976
	3954 Å	$B^2\Sigma^+ \to A^2\Pi_i$	Swings and Page 1950
			Greenstein 1962
	4273 Å	$A^2\Pi_i \to X^2\Sigma^+$	Fowler 1910
			Pluvinel and Baldet 1911
			Swings and Page 1950
			Greenstein 1962
CO_2^+	2890 Å	$B^2\Sigma_u^+ \to X^2\Pi_g$	Feldman and Brune 1976
	3509 Å	$\tilde{A}^2\Pi_u \to \tilde{X}^2\Pi_g$	Swings and Page 1950
	3674 Å		Greenstein 1962
H_2O^+	6198 Å	$\tilde{A}^2A_1 \to \tilde{X}^2B_1$	Herzberg and Lew 1974
			Wehinger et al. 1974
			Benvenuti and Wurm 1974
N_2^+	3914 Å	$B^2\Sigma_u^+ \to X^2\Sigma_g^+$	Fowler 1910
			Swings and Page 1950
OH^+	3565 Å	$A^3\Pi_i \to X^3\Sigma^-$	Swings and Page 1950
Silicate	10 μm	—	Maas et al. 1970
			Day 1974
			Rose 1979

The ultraviolet spectrum of a comet is dominated by Lyman-α and the OH (0-0) 3080 Å emission features (Feldman's chapter). The optical spectra of comets (3000 to 8000 Å) are dominated by the neutral radicals CN and C_2 (cf. Swings and Haser 1956; A'Hearn's chapter; Arpigny et al. 1982). A spectrogram of the 3000 to 6000 Å region of Comet Kobayashi-Berger-Milon 1975 IX is shown in Fig. 5 to illustrate the dominance of the C_2 and CN emission features. NH, C_3, and NH_2 are also frequently prominent (see Fig. 5). Figures 6 and 7 illustrate the spectrum of P/Encke which has been corrected for the scattered solar spectrum by ratioing the comet spectrum with that of a G2V star. This spectrum shows essentially only emission from atoms and molecules in the coma. (A'Hearn has discussed in his chapter spectro-

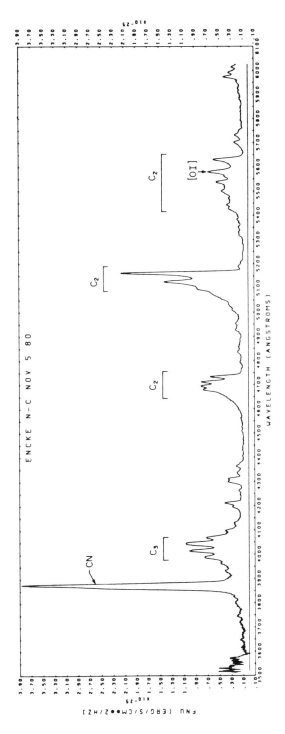

Fig. 6. Image Dissector Scanner spectrum of P/Encke obtained by H. Spinrad using the 3-m Lick Observatory telescope with scattered solar spectrum from dust removed. Note strengths of CN, C_2 and C_3.

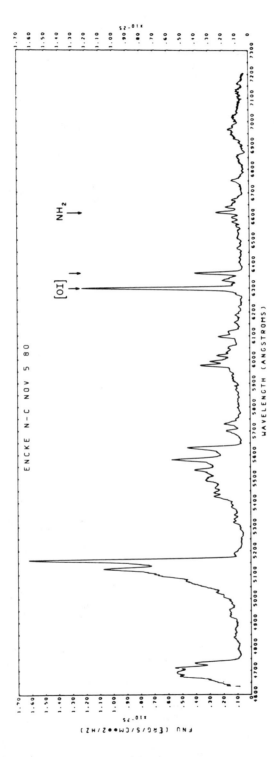

Fig. 7. Image Dissector Scanner spectrum of P/Encke obtained by H. Spinrad using the 3-m Lick Observatory telescope with scattered light from dust removed. The C_2 Swan bands are strong. Also present are the $O(^1D)$ lines and NH_2 $\tilde{A}^2A_1 \to \tilde{X}^2B_1$ bands.

scopic details of various atomic and molecular features observed in coma spectra.) A further example of an optical spectrum of a comet is shown in Fig. 8 for P/Stephan-Oterma 1980g. This figure shows a direct image of the comet (top); an untrailed image-tube spectrogram of the comet (middle), and density traces (bottom) of the nucleus (A → A), and 2.5×10^4 km from the nucleus (B → B). Note the strength of the CN 3883 Å band in all spectra (Figs. 5 through 8). The resolution of these spectra is $\sim$ 5-8 Å. To illustrate a high-resolution spectrum, the CN 0-0 band observed in the spectrum of Comet Bennett 1970 II is shown in Fig. 9 (Aikman et al. 1974). The spectrogram has a spectral resolution $\sim$ 0.2 Å and a spatial resolution (normal to the direction of dispersion) of 2000 km at the comet ($\Delta \sim$ 1 AU). The rotational structure of the CN band is well resolved, the resolution being more than adequate to separate the isotopic features of ^{13}CN from ^{12}CN. However, an overlapping 1-1 ^{12}CN band makes the identification of isotopic features uncertain according to Aikman et al. (1974). A measurement of ^{12}C/^{13}C $\sim 70 \pm 15$ was obtained from vibrational shifts in the C_2 Swan system by Stawikowski and Greenstein (1964) who concluded that the ratio was essentially terrestrial, as did Danks et al. (1975).

Some comets develop very prominent tail spectra due to molecular ions. As can be seen in Fig. 10, the spectrum of H_2O^+ is asymmetric with respect to the coma spectrum (the center of which coincides with the central condensation of the comet). The spectrograph slit was aligned along the comet tail axis. Rarely does the emission spectrum of molecular ions extend beyond $\sim 10^3$ km on the sunward side of the coma (Wyckoff and Wehinger 1976a). This means that once formed, the ions must be accelerated tailward on a very rapid time scale (compared to neutral radical expansion time scales). One measurement of H_2O^+ expansion velocities in the coma was $\sim$ 20 to 40 km s^{-1} (Huppler et al. 1975).

D. Excitation Mechanisms

Although collisions occur in the inner comae of comets, the kinetic temperatures are evidently too low to populate even the lowest levels of the observed radicals and ions. Exceptions may be found in the spectra of CN and CH (Malaise 1970; Arpigny 1965,1976). Most species observed in the ultraviolet and optical spectra of comets are excited by resonance fluorescence in the solar radiation field (Swings 1941a; Arpigny 1965; Wyckoff and Wehinger 1976a,b; Feldman's chapter). Notable exceptions are the forbidden O(^{1}D) 6300 and 6364 Å lines which are excited as a result of photodissociation of H_2O and CO_2 (the dominant parent not yet determined) (Delsemme and Combi 1979; Festou and Feldman 1981); and the metastable C(^{1}D) 1931 Å line thought to arise from dissociative recombination of CO$^+$ (Feldman 1978). The CO $a^3\Pi_r - X^1\Sigma^+$ band system tentatively identified in Comet West 1976 VI (Smith et al. 1980) at 2100 Å is a forbidden intercombination transition, so that excitation of the $a^3\Pi$ state must be photochemically produced provided the CO identification is correct. Also the CII 1335 Å doublet may be

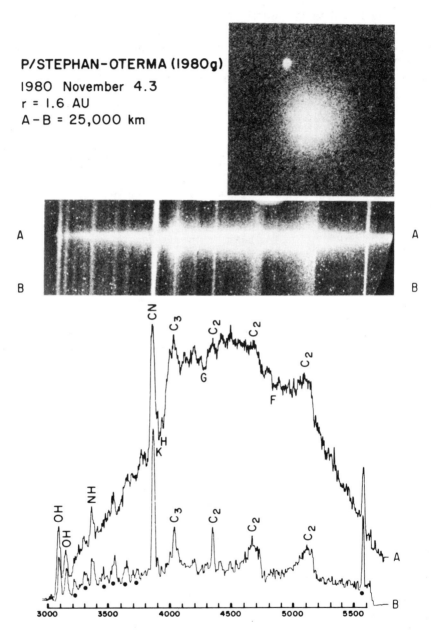

Fig. 8. Image and spectrogram of P/Stephan-Oterma 1980g obtained by S.M. Larson. Below are given density traces of photographic spectrum. Trace A → A shows strong continuum contribution. Trace B → B corresponds to a distance 2.5×10^4 km from the nucleus.

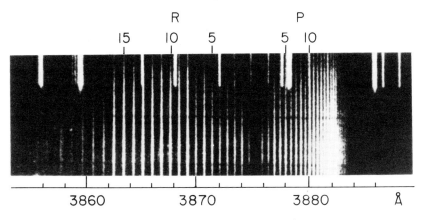

Fig. 9. High-resolution spectrogram of the CN (0-0) 3883 Å band in Comet Bennett 1970 II taken with 1.8-m telescope of the Dominion Astrophysical Observatory by J.B. Tatum (Aikman et al. 1974). Selected rotational lines in the P and R branches are labeled by the quantum numbers of the ground $X^2\Sigma^+$ state. This is one of the strongest bands observed in optical spectra of comets (see e.g., Fig. 6).

excited by some mechanism other than resonance fluorescence (see Feldman's chapter).

E. Gas Production Rates

After the gases sublime from the nucleus of a comet under the influence of solar radiation, they are subsequently subjected to a variety of solar radiation and wind interactions which dissociate and ionize the parent molecules in a cascade process. The abundances of the species observed in the coma therefore depend on the initial abundances of released molecules plus the lifetimes against the various decomposition processes which occur in the coma. For a dissociation energy of 5 eV, H_2O vaporized from the nucleus photodecomposes in $\sim 10^5$ s at 1 AU from the Sun, while at an expansion rate of 1 km s^{-1} an H_2O molecule is not likely to survive its transit of the coma ($\sim 10^5$ km). Hence the species observed in the spectra of comets are a mixture of byproducts of decomposed parents, radicals possibly released from the clathrate hydrate ices, and possibly parents released directly. One aim of studying spectra of the coma is to reconstruct this complex process and infer the nuclear ice abundances. Obviously any inference based on observations of the coma will be highly model dependent.

Only recently have routine quantitative measurements of emission line and band strengths become possible (O'Dell 1971; Code et al. 1972; Keller and Lillie 1974; Feldman's chapter). Both multiaperture narrowband photometry (cf. A'Hearn et al. 1977; A'Hearn and Millis 1980; Newburn et al. 1981) and digital spectrophotometry (Cochran and McCall 1980; Spinrad et al. 1979; Feldman's chapter) have been used successfully to determine band

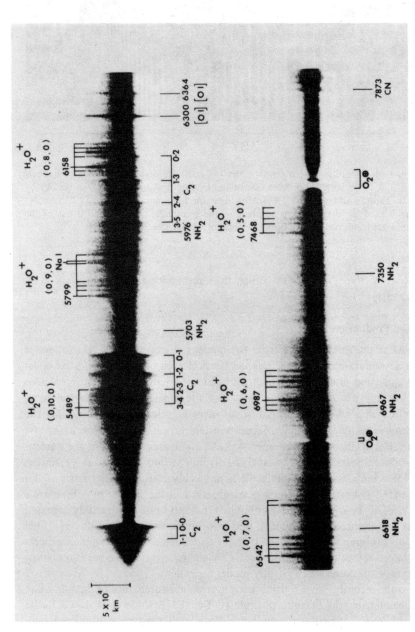

Fig. 10. Spectrogram (150 Å mm^{-1}, 150 arcsec mm^{-1}) of Comet Kohoutek 1973 XII taken with 1-m Wise Observatory Telescope by P.A. Wehinger and S. Wyckoff on 11 January 1974 ($r = 0.5$ AU). Note the strength and asymmetrical distribution of the H$_2$O$^+$ spectrum. Slit of spectrograph is centered on the coma, and aligned with the tail axis (extending upward) of the comet.

and line emission rates in comets. Previous band emission rates determined from photographic spectra were crude estimates at best. Spectroscopic observations, in particular, of emission rates from comets remain limited by aperture effects (cf. Delsemme 1973). If the projected scalelength, $\beta = v\tau$ of a given species is greater than the projected entrance aperture of the instrument used, all of the emission from the comet cannot be observed. For a geocentric distance ~ 1 AU, 10^5 km corresponds to a projected angular diameter of 3 arcmin, which of course varies continuously as the comet-Earth distance changes.

An observed species' emission rate (total from all spectral lines or bands) L is related to the total number of emitters of the given species N in the aperture by

$$L = gN \qquad (9)$$

where L is usually expressed in photon s^{-1} and g is the excitation rate g-factor for an optically thin coma (generally the case). If all the emission from a given species is measured, then the production rate Q of that species is given by

$$Q = L(g\tau)^{-1} \text{ molecules s}^{-1} \qquad (10)$$

where τ is the lifetime of the species. For excitation by resonance fluorescence the g-factor or photon scattering coefficient depends on the solar flux incident at the comet $\pi F_\odot(\lambda)$, the band (or line) oscillator strength of the particular transition f, the albedo for a single scattering $\tilde{\omega}$, and various parameters (λ = wavelength of transition, e = electron charge, m = mass of species, c = velocity of light) expressed in photons s^{-1} molecule^{-1} (cf. Chamberlain 1978)

$$g = \pi F_\odot(\lambda) \frac{\pi e^2}{mc^2} \lambda^2 f \tilde{\omega} \qquad (11)$$

where $\pi e^2/mc^2 = 8.83 \times 10^{-21}$ cm^2 Å^{-1}. For a single resonant scattering and no collisional deactivation (generally true for distances from the nucleus $\gtrsim 10^4$ km), the albedo is simply the branching ratio for cascades to ground state,

$$\tilde{\omega} = \frac{A_{v'v''}}{\Sigma_{v''} A_{v'v''}}, \qquad (12)$$

where the $A_{v'v''}$ values represent the relevant Einstein emission transition probabilities. For excitation by resonance fluorescence, accurate measurements (with high spectral resolution) of the solar flux (preferably of the integrated disk) are required and must be folded with the comet's heliocentric radial velocity for accurate calculation of the g-factor. This is particularly true for the CN 3883 Å, and the OH 3090 Å and 18-cm bands, as well as for the OI 1304 Å line (Mies 1974; Turner 1974; Huebner and Giguere 1980; Tatum and Gillespie 1977; Jackson 1980; Weaver et al. 1981a). The sensitivity of the g-factor to the solar flux is due to the Swings (1941a) effect and in some cases to the fact that the ultraviolet flux from the Sun ($\lambda < 2000$ Å) is largely in the form of emission lines. The Swings effect arises from sharp variations as a function of wavelength of the solar flux received by the comet, due to Fraunhofer absorption lines in the solar spectrum. Thus the intensity of the fluorescing solar radiation received by the comet which excites the atoms and molecules can be a very sensitive function of the radial velocity of the comet with respect to the Sun. The Greenstein (1962) effect arises from non-constant outflow velocities of gases in the coma which give rise to a second-order Swings effect. The Greenstein effect may significantly affect the g-factors of molecular ions which obviously undergo significant accelerations in the coma (e.g., see Fig. 10). Oppenheimer and Downey (1980) have discussed the effects of variations in the solar flux with the 11-yr solar cycle. There may also be variations in the solar flux with rotation (Tousey 1971), and with solar flare activity, which will affect the g-factors.

In order to determine production rates from the observed band strengths, the lifetimes of the observed emitters must be determined either empirically or by calculation. Very few lifetimes have been derived observationally from monochromatic surface brightness profiles (cf. Festou 1981b). To determine the lifetimes from observed scalelengths $\beta = v \tau$, v must be known or assumed. Expansion velocities of coma gases have been determined for P/Halley 1910 II (Bobrovnikoff 1931) where v = 0.6 km s^{-1} at $r = 1$ AU was inferred from the morphological changes as a function of time. Malaise (1970) utilized the Greenstein effect to determine the expansion rate, and Arpigny (1976) has discussed the method. However, additional observations are needed because the velocity field of an expanding coma is probably not uniform or isotropic once a significant amount of decomposition has taken place especially beyond the collision region, since energy excesses involved in molecular dissociation and ionization are transferred as kinetic energy to the fragments. In this regard at least two velocity components of H I (Lyman-α) have been observed at 8 and 20 km s^{-1} (Keller and Lillie 1974). These two components are believed to arise from the photodissociation of OH and H_2O, respectively (Keller 1976). Since vaporization from the nucleus is known to be non-uniform (Figs. 1 and 4), the velocity fields of expanding comet comae are not well known (cf. Ip 1981). This is particularly true for atomic and molecular ions which are strongly coupled to the interplanetary magnetic field.

On the other hand, if τ is calculated, the decomposition processes must be assumed and various interaction cross sections known. Usually solar wind interactions are ignored (which is justified of course only for those observed species that photodecompose). The lifetimes for most radicals and ions in cometary comae can therefore be computed from the photodestruction rates (ionization plus dissociation) usually expressed in s^{-1} molecule and given by

$$J = \int_{\lambda_c}^{0} \sigma(\lambda)\, \pi F_{\odot}(\lambda)\, d\lambda \qquad (13)$$

where $\sigma(\lambda)$ is the photodestruction cross section, λ_c is the dissociation (or ionization) threshold (usually < 2000 Å for most observed radicals), and $\pi F_{\odot}(\lambda)$ is the incident solar flux. Since the dissociation energies are generally smaller than ionization potentials for the observed radicals, decomposition of a molecule in the solar radiation field is usually determined by photodissociation for most species. A notable exception, however, is CO which has a relatively high dissociation potential (11 eV), comparable with its ionization potential (14 eV). The lifetime in s of a given species is given by

$$\tau = J^{-1} \,. \qquad (14)$$

Note that the lifetime is a function of the heliocentric distance of the comet, so that the solar flux (usually given for $r = 1$ AU) must be scaled accordingly $(F_{\odot}(\lambda) \sim r^{-2})$ in the above equation. Photodecomposition rates for quiet solar conditions for a variety of species have been given by Huebner and Carpenter (1979).

A suitable model is required in order to derive lifetimes from surface brightness distributions. Traditionally a two-step decay point source vaporization model has been utilized by various investigators, the Haser model (Haser 1957,1966; A'Hearn and Cowan 1975; Festou 1978,1981a). However, this model clearly does not apply to ions that do not expand isotropically or with constant velocities, nor does the Haser model apply to species produced via photochemical processes (cf. Feldman 1978; Ip and Mendis 1976; Huebner and Giguere 1980; Mitchell et al. 1981). Combi and Delsemme (1980c) have recently developed a model which takes into account a more realistic velocity field as well as radiation pressure with application to the CN 3883 Å band. Festou (1981a) discussed the attributes and problems of the Haser model, concluding that it applies only when the velocities of the parent and the dissociation product are equal. It should be remarked that it is not yet certain that a point-source vaporization model is correct. The icy-grain halo model (Delsemme and Miller 1971) may fit the observations better (Hobbs et al. 1975,1977; A'Hearn and Cowan 1980; Newburn et al. 1981). For comparison of observations with models Huebner and Giguere (1980) have emphasized the need for observers to express their results in terms of column densities, which can be obtained by column integrations using a suitable model.

Gas production rates based on homogeneous sets of observations have been determined for ~ 15 comets (A'Hearn and Millis 1980; chapter by Feldman). The ultraviolet observations obtained by Feldman and his collaborators (cf. Weaver et al. 1981a,b) give the production rate of water, generally Q (H_2O) $\sim 10^{28}$ to 10^{29} molecules s^{-1}, as inferred from the OH band emission rates ($\sim 90\%$ of the vaporized H_2O molecules dissociate into OH). The narrowband photometry of C_2, C_3 and CN in the optical region gives production rates for these radicals which are a few percent of those measured for water (A'Hearn and Millis 1980). (All measurements are reduced to $r = 1$ AU.) Significant variations in production rates among comets are observed, but the range is rather narrow (factor ~ 10) indicating similarities in sizes and surface structures of observed cometary nuclei (Feldman's chapter; A'Hearn et al. 1977; A'Hearn and Millis 1980; Weaver et al. 1981a,b). The observed range in H_2O production rates corresponds to typical mass-loss rates at 1 AU from 10^5 to 10^6 g s^{-1}, which agrees within an order of magnitude with the mass-loss rates observed for dust (Ney's chapter).

Both the gas (2×10^5 g s^{-1}) and dust (2×10^4 g s^{-1}) production rates measured for P/Encke (see chapters by Feldman and by Ney) give a total production rate, $Q(H_2O + \text{dust}) \sim 2.2 \times 10^5$ g s^{-1} and a dust-to-gas ratio ~ 0.1. The total mass lost during one orbital revolution ($P = 3.3$ yr) of P/Encke is $\sim 10^{13}$ g.

F. Chemical Abundances

Observed gas production rates or column densities give a measure of the abundances of the observed species in the coma. Since chemical, radiative, and collisional processes are computed to alter the gas mixture in regions near the nucleus on relatively short time scales one might expect the abundances to differ considerably from one comet to another. However, abundances derived for the comae of many comets are remarkably homogeneous.

Emission rates have not been reported for all of the species listed in Table I. In fact the relative abundances of only CN, C_2, C_3, OH, H, O, C have been reliably measured in several comets (cf. Keller 1976; Opal and Carruthers 1977; A'Hearn et al. 1977; A'Hearn and Millis 1980; chapter by Feldman).

Feldman has emphasized in his chapter that the similarities between the ultraviolet spectra of 7 comets implies a common, homogenous composition. Except for the absolute and relative strengths of molecular ion band emission the same is generally true for spectra in the visible region. The production rates of CN, C_2, C_3, OH are not correlated with dust production rates, which vary appreciably from comet to comet, nor with the periods of orbital revolution. Hence the observations indicate that comet nuclei are remarkably homogeneous and that their outer surface layers are undifferentiated (A'Hearn and Millis 1980; Feldman's chapter; Delsemme's chapter; Weaver et al. 1981b).

Observations of Comet West 1976 VI indicate that a species more volatile than water ice, possibly CO_2, may have controlled the evaporation from the nucleus at $r < 1$ AU (Feldman 1978; A'Hearn and Cowan 1980; Houpis and Mendis 1981b). Indeed the optical spectrum of Comet West 1976 VI displayed one of the strongest CO^+ spectra yet observed in a comet, as shown in

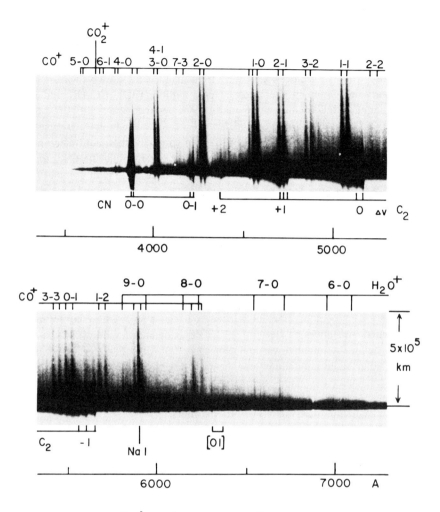

Fig. 11. Spectrogram (120 Å mm^{-1}, 150 arcsec mm^{-1}) of Comet West 1976 VI obtained with Wise Observatory 1-m telescope by E. Leibowitz on 10 March 1976 (r = 0.5 AU, Δ = 0.9 AU). Slit of spectrograph is centered on the coma with the tail axis extending upward. Note exceptional strength of CO$^+$ emission bands, and in particular the overlap with the H$_2$O$^+$ 9-0 and 8-0 bands.

Fig. 11. The CO$^+$ spectrum was so strong that for the first time bands arising from the v'' = 3 levels ($\sim$ 0.8 eV above ground state) were observed. Note the extension of the CO$^+$ spectrum into the 6000 Å region in this figure. A'Hearn and Cowan (1980) have suggested, on the basis of abrupt changes in the $Q(C_2)/Q(CN)$ ratio at $r \sim$ 2 AU that the icy-grain halo model could better explain their observations of Comet West 1976 VI than a simple clathrate hydrate. They suggest that the icy grains released after evaporation from the nucleus would produce H_2O while vaporization of the nucleus would produce CO_2.

Only three molecules have been tentatively detected which might have sublimed directly from the nuclear ices, namely HCN, CH_3CN and H_2O (Huebner et al. 1974; Ulich and Conklin 1975; Jackson et al. 1976; Crovisier et al. 1981). It should be emphasized that all of these single detections remain unconfirmed (see Crovisier et al. 1981), in spite of subsequent attempts to observe them (cf. Hollis et al. 1981). Another parent molecule CS_2, though not observed directly, is confidently inferred to be present in cometary ices by Jackson (see his chapter) due to (1) the very small observed spatial concentration of CS in comet comae (Feldman et al. 1980), and (2) the extremely short lifetime ($\sim 10^2$ s) expected for CS_2 in the solar radiation field (Jackson's chapter). Probably the best opportunity to detect H_2O directly, since the detection of radio 1.35-cm features is dubious (Crovisier et al. 1981; Hollis et al. 1981), will be to observe the 1240 Å $\tilde{C}^1B_1 - \tilde{X}^1A_1$ line in absorption (Smith et al. 1981), frost bands in the infrared $\sim 2\,\mu$m, or the 3_{13}-3_{20} transition at 183 GHz (Hollis et al. 1981).

Only a few comets have been observed over a sufficiently large range of heliocentric distances to determine the dependence of the production rates on r. The most straightforward relation expected for pure water ice subliming in the solar radiation field would be $Q(H_2O) \sim r^{-2}$. Weaver et al. (1981a) found $Q(H_2O) \sim r^{-3.7}$ for Comet Bradfield 1979 X, which is at variance with simplistic theory, and must be explained by any model of the nucleus, if verified by future observations of other comets.

IV. PLASMA TAILS

Comets have long been observed to have two types of tails extending from the coma. The type I (ion or plasma) tails are aligned more nearly with the radius vector of the cometary orbit than are the type II (dust) tails, the lag angles measuring ~ 5 deg and $\lesssim 60$ deg, respectively. Both ultraviolet and visible spectroscopy of plasma tails show that they are comprised entirely of ions (other than a few neutrals, such as CN radicals and Na I atoms which due to radiation pressure, extend asymmetrically in an antisunward direction). Prior to 1951 ion tails of comets were known not to be explicable by effects of either solar radiation or gravitation. From motions of features in comet plasma tails, Biermann (1951,1957) inferred the existence of the solar wind, and Alfvén (1957) suggested that a magnetic field dragged by the solar wind wraps field lines around the comet ionosphere and is pulled into the ion tail by interaction with the comet ions. This scenario is the basis of all plasma tail models today (cf. Brandt and Mendis 1979; chapter by Ip and Axford). At $r \sim 1$ AU the quiet solar wind plasma density is ~ 5 cm^{-3} and the flow velocity ~ 400 km s^{-1}.

A. Morphology

Nearly all photographs of comets have been on blue sensitive emulsions. Hence it is likely that they show the distribution and variations of the CO^+ ion in comet plasma tails, since the CO^+ spectrum dominates the relevant spectral band pass (3500 to 5000 Å). The large variety of forms and inhomogeneities in plasma tails is evident in the *Atlas of Cometary Forms* (Rahe et al. 1969). Features described as wavy motions, kinks, knots and helices distort the ion tails and vary both in brightness and spatially (usually a systematic outward flow) on time scales of hours (cf. Biermann 1962; Jockers et al. 1972; Brandt and Mendis 1979; Brandt's chapter; Brandt et al. 1980).

The axis of the plasma tail has been measured (cf. Belton and Brandt 1966) to lag the comet's orbital radius vector by an aberration angle of ~ 5 deg. Whether or not the plasma tail lies in the comet's orbital plane has not been observationally established.

Probably the most prominent morphological phenomenon observed to be associated with plasma tails is the ray structure. Narrow ($< 10^3$ km), straight tail rays extend ($\sim 10^5$ to 10^7 km) from the coma in an antisunward direction. The rays make angles $\lesssim 60°$ with the tails axis and are observed to turn toward the tail axis, coalescing with it on time scales of 20 hr (Wurm 1968; Wurm and Mammano 1964; chapter by Brandt).

Another feature characteristic of the plasma tails is a disconnection event, a phenomenon which was noted even by visual observers early in this century (cf. Barnard 1920). Seventy-two such events have now been documented (Niedner 1981). Disconnection events appear to be a cyclical process involving the abrupt disconnection of the plasma tail in the coma, followed by the reforming of the tail, as tail rays converge toward the tail axis (chapter by Brandt). Ordinarily the detached plasma tail accelerates in an antisolar direction, but for Comet Kohoutek 1973 XII a deceleration was observed (Niedner and Brandt 1980). One explanation for plasma tail disconnection events is that a high-velocity stream of the solar wind passes the comet disrupting a cross-tail current, which causes the plasma tail to disconnect (Ip and Mendis 1976). On the other hand, Niedner and Brandt (1978; Brandt's chapter) have noted that disconnection events are associated with magnetic sector boundaries in the interplanetary field and suggest that cyclical polarity reversals in the solar wind may explain the severed plasma tails. Ip and Axford have also discussed disconnection events in their chapter. No correlation between solar activity and the morphology of plasma tails has yet been found (Hyder et al. 1974; Miller 1979), though it is clear that interaction of the solar wind with the cometary ions must give rise to plasma tails in comets.

Several bright comets have been distinguished by spectacular displays of plasma tail activity, for example comets Burnham 1960 II, Morehouse 1908 III, and Humason 1962 VIII (cf. Brandt and Mendis 1979). Miller (1980) has reported that the H_2O^+/CO^+ intensity ratio in Comet Burnham 1960 II was large. Comet Humason 1962 VIII was active during its entire apparition from $r > 6$ AU to perihelion at $r \sim 2$ AU (Greenstein 1962; van Biesbroeck 1962; Roemer 1962; Roemer and Lloyd 1966). Dossin (1966) observed CO^+ in the spectrum of Comet Humason 1962 VIII at $r \sim 5$ AU; Greenstein's (1962) spectrogram (~ 3000 to 5100 Å) of the comet at $r \sim 2.8$ AU (preperihelion) showed the spectrum to be dominated by the emission spectrum of CO^+, essentially as strong as that observed in Comet West 1976 VI (Fig. 11). Houpis and Mendis (1981b) have suggested that the dominant volatile in the nucleus of Comet Humason 1962 VIII was either CO_2 or CO. Their models can account for both the irregular structure of Comet Humason 1962 VIII (see photographs in Roemer [1962] and Brandt and Mendis [1979]) and ionization by solar wind interactions. It is interesting for observers to note that for CO-dominated comets, Houpis and Mendis (1981b) predict that at large heliocentric distances the ions and neutrals in the coma would couple and give rise to a neutral molecule tail. It should be remarked that no observational evidence directly supports the presumption that Comet Humason 1962 VIII or Comet Morehouse 1908 III had a relatively high CO/H_2O abundance ratio. Models of cometary ionospheres generally espouse the concept that mass loading of the interplanetary magnetic field by the coma plasma creates a standing bow shock toward the Sun ($\sim 10^5$ to 10^6 km) (cf. Biermann et al. 1967). A contact surface at $\sim 10^2$ to 10^4 km from the nucleus has also been suggested to exist where the solar wind pressure is balanced by the dynamical pressure of the expanding coma (cf. Ip and Mendis 1976; Houpis and Mendis 1980; chapter by Ip and Axford). The mere existence and morphology of the plasma head of a comet attests to some such stand-off distance between the coma and the solar wind, but its exact location relative to the nucleus has not yet been observationally determined. Another loading point has been suggested by Houpis and Mendis (1981a) where the ions and neutrals decouple. Monochromatic imaging with spatial resolution $\sim 10^3$ km can be achieved for comets with geocentric distances $\Delta \lesssim 1$ AU. Such observations would be particularly important for determining the location of the ionopause (Houpis and Mendis 1981a).

B. Velocities

Proper motions of features in plasma tails of comets have been measured from direct photographs to be in the range 20 to 250 km s^{-1} (cf. Biermann and Lüst 1963; Jockers et al. 1972; Brandt and Mendis 1979). Accelerations of cometary ions in the antisolar direction are measured to be 10 to 10^3 cm s^{-2} (Eddington 1910; Miller 1969; Chernikov 1975; Ip 1980). The question of whether the observed motions represent hydromagnetic waves propagating down the tails (Alfvén 1957; Ness and Donn 1966) or mass motions of cometary ions has not yet been settled (Brandt and Mendis 1979). The radial velocity measurement of H_2O^+ ions in the tail (projected distance

2×10^5 km) of Comet Kohoutek 1973 XII indicated tailward mass motions ~ 20 to 40 km s^{-1} (Huppler et al. 1975). Such high-resolution measurements at various locations in the plasma tail would settle the question of wave versus mass motions.

C. Observed Species

The plasma tail is dominated by the emission spectra of CO$^+$ and H$_2$O$^+$ (cf. Swings 1965; Wehinger et al. 1974; Wyckoff and and Wehinger 1976a). Although CO$^+$ emission dominates in the 3800 to 4800 Å region (Fig. 11) and H$_2$O$^+$ at 5600 to 7000 Å (Fig. 10), the two molecular ion spectra have been observed to overlap in the 6000 Å region in Comet West whenever the CO$^+$ spectrum is particularly strong. An objective prism spectrum of Comet West ($\sim$ 3000 to 6500 Å) in Fig. 12 also illustrates the great strength of both CO$^+$ and H$_2$O$^+$ emission. Comparisons of Figs. 11 and 12 indicate that the contri-

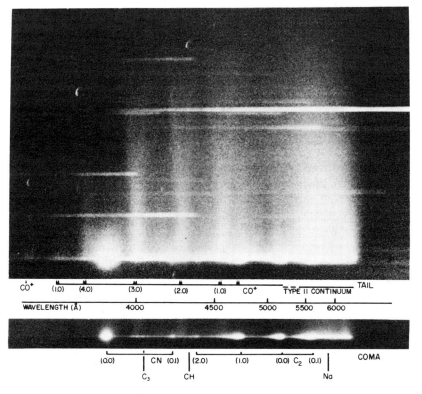

Fig. 12. Objective prism spectrum of Comet West 1976 VI obtained by S.M. Larson at Mt. Lemmon Observatory 1976 March 14.5 UT. Tail axis of the comet extends upward, displaying strong bands of CO$^+$ (4000 to 6000 Å), and H$_2$O$^+$ (5800 to 6500 Å). The latter is contaminated with dust and neutrals spectrum close to the coma; compare with Figs. 11 and 13.

bution of the dust tail to the objective prism spectrum may have been comparable to H_2O^+ emission at projected distances from the nucleus $> 10^4$ km in Comet West 1976 VI. A photograph of the comet is shown in Fig. 13.

Miller (1980) has estimated the H_2O^+/CO^+ intensity ratios for the spectra of 13 comets. He finds that the ratio is highly variable from comet to comet and that the ratio is not correlated with age, presence of dust tails, or perihelion distances; nor does Miller (1980) find evidence for spatial variations in the H_2O^+/CO^+ ratio along a plasma tail. It is noteworthy that he estimates the emission strengths of H_2O^+ and CO^+ to be about equal in P/Encke at $r \sim 0.5$ AU. Measurement of the H_2O^+/CO^+ ratio in this well-observed comet would be a valuable diagnostic for models of cometary comae (Shimizu 1975; Huebner and Giguere 1980; Giguere and Huebner 1978; Mitchell et al. 1981; chapter by Ip and Axford).

Several other ions, both diatomic and triatomic have been identified in the plasma tail spectra of comets and are included in Table I. The spectra of these ions are not as extensive (in wavelength) nor as strong as are those of CO^+ and H_2O^+, and hence have not been well studied.

A'Hearn and Feldman (1980) have detected CO_2^+ but not CO^+ in the ultraviolet spectrum of Comet Bradfield 1979 X, which they explain by a difference in the scalelengths of the two ions.

D. Ionization Mechanisms

Although a dominant ionization mechanism has never been established for molecular ions observed in spectra of comets, it is likely that for quiet solar conditions and heliocentric distances $\lesssim 2$ AU, photoionization of molecules in the coma prevails (Wyckoff and Wehinger 1976a,b). The molecules from which the observed ions are derived have ionization potentials ~ 12 eV, so that the photoionization rates are determined by the solar flux shortward of ~ 1000 Å. Calculated photoionization rates are generally $\sim 10^{-6}$ s^{-1} (Wyckoff and Wehinger 1976a,b), and lifetimes of the ions $\gtrsim 10^6$ s.

Because of uncertainties in the physical conditions resulting from the interaction of the solar wind with the cometary coma, it is very difficult to evaluate the importance of other ionization mechanisms. If solar wind electrons are thermalized by the comet ionosphere bow shock, and the electron energies increased to 1 keV, then collisional ionization with cometary molecules would be important (Axford 1964; Beard 1966). Both the thermalizing and penetration efficiencies of solar wind electrons in a comet ionosphere are unknown and would probably depend on variable solar wind conditions (Houpis and Mendis 1981b).

In the collision region of the coma (distances from the nucleus $< 10^4$ km) many possible sources of ionization have been suggested, among them photoionization, charge exchange with solar wind protons, electron impact by solar wind electrons or by photoelectrons (average energies $\lesssim 30$ eV) and photodissociative ionization. The latter two mechanisms may be significant sources of CO^+ ions in the inner coma (Cravens and Green

Fig. 13. Photograph of Comet West 1976 VI taken by S.M. Larson one day prior to the objective prism spectrogram in Fig. 12. Note the prominent dust and plasma tails.

1978; Huebner and Giguere 1980). Ion-neutral reactions are probably significant (Oppenheimer 1975), but not as a source of ions, nor even perhaps for altering relative abundances of neutrals (chapter by Feldman; Weaver et al. 1981b).

Because rapid variations in the intensity of tail rays are observed in some comets, Wurm (1961) suggested that an internal ionization mechanism might be important in comets. During high-velocity stream encounters of the solar wind with the comet, magnetic field strengths $\sim 100\,\gamma$ might arise, which could give rise to a cross-tail current flowing through the coma. Electrons with energies ~ 1 to 10 keV might be generated which could ionize the coma on time scales $\sim 10^3$ s (Ip and Mendis 1976; Houpis and Mendis 1980,1981a,b).

While the ionization process in comets is not well understood, there is little question that ions in comets are excited by resonance fluorescence (Arpigny 1965; Wyckoff and Wehinger 1976a,b). If the ions are accelerated along the tail, there should be a very pronounced Greenstein (1962) effect in the rotational line intensity distributions of, particularly, the CO^+ bands. Appropriate modeling plus spatially resolved measurements could provide velocities and accelerations of ions along the plasma tail.

E. Ion Emission Rates

The only measurements of ion emission band strengths have been obtained from photographic spectra. Although the relative intensities may be fairly accurate, the absolute measurements should be regarded as highly uncertain (Wyckoff and Wehinger 1976a,b; Delsemme and Combi 1979; Combi and Delsemme 1980a,b). Measurements of production rates for H_2O^+ in Comet Bennett 1970 II (Delsemme and Combi 1979) and for CO^+ in Comet West 1976 VI (Combi and Delsemme 1980d) differ by several orders of magnitude from production rates derivable from data for comets Kohoutek 1973 XII and Bradfield 1974 II (Wyckoff and Wehinger 1976a,b).

Column densities of several ions at projected distances $\sim 10^4$ km from the nucleus of comets Kohoutek 1973 XII and Bradfield 1974 III were estimated by Wyckoff and Wehinger (1976a,b) and the ratio $Q(CO^+)/Q(H_2O^+) \sim 30$ was measured for Comet Kohoutek, which may imply a relatively large CO_2 (or CO) abundance (Huebner and Giguere 1980).

Since the tail is spatially resolved, a particularly useful monochromatic study of the surface brightness distribution of the ion tail of a comet can be obtained using long-slit spectroscopy. Because the dust-tail spectrum must be removed from the measurements (and corrections applied for vignetting), photographic spectra are again poorly suited for such measurements. Surface brightness profiles of ion emission have been derived for two comets from photographic spectra, Comet Bennett 1970 II (H_2O^+ and CO^+) (Combi and Delsemme 1980d; Delsemme and Combi 1979), and Comet West 1976 VI (Wyckoff 1981). A bright comet with a prominent plasma tail which approaches the Earth with $\Delta < 1$ AU might reveal the fine structure in the comet ionosphere and the location of the ionopause, if that comet is carefully mapped or imaged with suitable interference filters.

V. SUMMARY

The icy conglomerate model of the nucleus seems firmly established by the observation of nongravitational effects, production rates of OH, H, and identification of H_2O^+. Dust is an abundant component of the nucleus. Both the dust and CO_2 (or CO) content appear to vary from comet to comet, while the water production rates differ by only an order of magnitude from one comet to another. Preliminary models of the coma indicate that the elemental abundances correspond roughly to an interstellar mix (Huntress et al. 1981). Rotation rates and spin axis orientations have been inferred from observations of a number of comets, while the spin axis precession rate has been determined for P/Encke. The first direct measurement of the diameter of a comet has been made using radar echo techniques. Although the general mechanisms for formation of the dust and plasma tails seem well understood, many important details remain unexplained. Quantitative measurements of emission rates from plasma tails, in particular, are badly needed for comparison with extant models.

Quantitative observations of comets are needed to determine production rates, column densities, and surface brightnesses. Either spectroscopic or narrowband imaging techniques are required to segregate the dust, neutral molecules, and ions. More observations are needed of comets at heliocentric distances beyond 3 AU to clarify the role of volatiles other than water ice in comet nuclei, to study the surface structure of the nucleus, and outburst phenomena. Time resolved imaging of both the plasma tail and the coma are needed to investigate mechanisms giving rise to the observed morphologies and associated changes.

Acknowledgments: Comments on an early draft of this chapter by H. Campins, P.D. Feldman, H.L.F. Houpis, D.M. Hunten, S.M. Larson, B.G. Marsden and E. Roemer are gratefully acknowledged.

REFERENCES

A'Hearn, M.F., and Cowan, J.J. 1975. Molecular production rates in Comet Kohoutek. *Astrophys. J.* 80:852-860.

A'Hearn, M.F. and Cowan, J.J. 1980. Vaporization in comets: The icy grain halo of Comet West. *Moon and Planets* 22:41-52.

A'Hearn, M.F.; Dwek, E.; and Tokunaga, A.T. 1981. Where is the ice in comets? *Astrophys. J.* 248:L147-L151.

A'Hearn, M.F., and Feldman, P.D. 1980. Carbon in Comet Bradfield 1979l. *Astrophys. J.* 242:L187-L190.

A'Hearn, M.F.; Hanisch, R.J.; and Thurber, C.H. 1980. Spectrophotometry of Comet West. *Astron. J.* 85:74-80.

A'Hearn, M.F., and Millis, R.L. 1980. Abundance correlations among comets. *Astron. J.* 85:1528-1537.

A'Hearn, M.F.; Millis, R.L.; and Birch, P.V. 1979. Gas and dust in some recent periodic comets. *Astron. J.* 84:570-579.

A'Hearn, M.F.; Thurber, C.H.; and Millis, R.L. 1977. Evaporation of ices from Comet West. *Astron. J.* 82:518-524.
Aikman, G.C.L.; Balfour, W.J.; and Tatum, J.B. 1974. The cyanogen bands of Comet Bennett 1970 II. *Icarus* 21:303-316.
Alfvén, H. 1957. On the theory of comet tails. *Tellus* 9:92-96.
Antrack, D.; Biermann, L.; and Lüst, Rh. 1964. Some statistical properties of comets with plasma tails. *Ann. Rev. Astron. Astrophys.* 2:327-340.
Arpigny, C. 1965. Spectra of comets and their interpretation. *Ann. Rev. Astron. Astrophys.* 3:351-376.
Arpigny, C. 1976. Interpretation of comet spectra: A review. In *The Study of Comets*, eds. B. Donn, M. Mumma, W. Jackson, M. A'Hearn, and R. Harrington, (Washington: NASA SP-393), pp. 797-838.
Arpigny, C.; Dossin, F.; Donn, B.; Rahe, J.; and Wyckoff, S. 1982. *Atlas of Comet Spectra* (NASA SP). In preparation.
Axford, W.I. 1964. The interaction of the solar wind with comets. *Planet. Space Sci.* 12:719-720.
Axford, W.I., and Keller, H.U. 1978. Are comets dirty snowballs or dust swarms? *Nature* 273:427-428.
Barnard, E.E. 1920. On Comet 1919b and on the rejection of a comet's tail. *Astrophys. J.* 51:102-108.
Beard, D.B. 1966. A theory of type I comet tails. *Planet. Space Sci.* 14:303-311.
Becklin, E.E., and Westphal, J.A. 1966. Infrared observations of Comet 1965f. *Astrophys. J.* 145:445-453.
Belton, M.J.S. 1965. Some characteristics of type II comet tails and the problem of the distant comets. *Astron. J.* 70:451-465.
Belton, M.J.S., and Brandt, J.C. 1966. Interplanetary gas XII. A catalogue of comet tail orientations. *Astrophys. J. Suppl.* 13:125-331.
Benvenuti, P., and Wurm, K. 1974. Unidentified bands in Comet Kohoutek. *Astron. Astrophys.* 31:121-122.
Bessel, F.W. 1836. Bemerkungen uber Moglich Unzulanglich keit der die Anziehungen Allein Berucksichtigenden Theorie der Kometen. *Astr. Nach.* 13:345-350.
Biermann, L. 1951. Kometenschweife und solare Korpuskular Strahlung. *Zs. f. Astrophys.* 29:274-286.
Biermann, L. 1957. Solar corpuscular radiation and the interplanetary gas. *Observatory* 77:109-110.
Biermann, L. 1962. The plasma tails of comets and interplanetary plasma. *Space Sci. Rev.* 1:553.
Biermann, L.; Brosowski, B.; and Schmidt, H.U. 1967. The interaction of the solar wind with a comet. *Solar Phys.* 1:254-284.
Biermann, L., and Lüst, Rh. 1963. Comets: Structure and dynamics of tails. In *The Moon, Meteorites and Comets*, eds. B.M. Middlehurst and G.P. Kuiper, (Chicago: Univ. of Chicago Press), pp. 618-638.
Biermann, L., and Trefftz, E. 1964. Uber die Mechanismen der Ionisation und der Anregung in Kometen Atmospheren. *Zs. f. Astrophys.* 59:1-28.
Biraud, F.; Bourgois, G.; Crovisier, J.; Fillit, R.; Gerard, E.; and Kazes, I. 1974. Discovery of 18-cm OH emission in Comet Kohoutek. *Astron. Astrophys.* 34:163-166.
Black, J.H.; Chaisson, E.J.; Ball, J.A.; Penfield, H.; Lilley, A.E. 1974. A radio frequency emission from CH in Comet Kohoutek (1973f). *Astrophys. J.* 191:L45-L47.
Blackwell, D.E., and Willstrop, R.V. 1957. A study of the monochromatic polarization of Comet Arend-Roland (1956h). *Mon. Not. Roy. Astron. Soc.* 117:590-599.
Bobrovnikov, N. 1931. On the spectrum of Halley's Comet. *Astrophys. J.* 66:145-169.
Brandt, J.C. 1968. The physics of comet tails. *Ann. Rev. Astron. Astrophys.* 6:267-286.
Brandt, J.C.; Hawley, J.D.; and Niedner, M.B. 1980. A very rapid turning of the plasma-tail axis of Comet Bradfield 1979ℓ on 1980 February 6. *Astrophys. J.* 241:L51-L54.
Brandt, J.C., and Mendis, D.A. 1979. The interaction of the solar wind with comets. In *Solar System Plasma Physics*, Vol II, eds. C.F. Kennel, L.J. Lanzerotti, and E.N. Parker (Amsterdam: North Holland Publ. Co.), pp. 253-292.
Brin, G.D. 1980. Three models of dust layers on cometary nuclei. *Astrophys. J.* 237:265-279.

Brin, G.D., and Mendis, D.A. 1979. Dust release and mantle development in comets. *Astrophys. J.* 229:402-408.
Brown, W.L.; Augustyniak, W.M.; Lanzerotti, L.J.; Johnson, R.E.; and Evatt, R. 1980a. Linear and nonlinear processes in the erosion of H_2O ice by fast light ions. *Phys. Rev. Letters* 45:1632-1635.
Brown, W.L.; Augustyniak, W.M.; Brody, E.; Cooper, B.; Lanzerotti, L.J.; Ramirez, A.; Evatt, R.; and Johnson, R.E. 1980b. Energy dependence of the erosion of H_2O ice films by H and He ions. *Nuclear Instr. and Methods* 170:321-325.
Brownlee, D.E. 1978. Microparticle studies by sampling techniques. In *Cosmic Dust,* ed. J.A.M. McDonnell (New York: J. Wiley), pp. 295-336.
Campins, H.; Rieke, G.H.; and Lebofsky, M. 1981. Infrared observations of faint comets. In *Modern Observational Techniques for Comets,* ed. J.C. Brandt (Pasadena: Jet Propulsion Lab.), pp. 83-89.
Carruthers, G.R.; Opal, C.B.; Page, T.L.; Meier, R.R.; and Prinz, D.K. 1974. Lyman-α imagery of Comet Kohoutek. *Icarus* 23:526-537.
Chamberlain, J.W. 1978. *Theory of Planetary Atmospheres* (New York: Academic Press).
Chernikov, A.A. 1975. On a possible mechanism of ion acceleration in type I comet tails. *Sov. A.J.* 18:505-508 (English trans.).
Clarke, D. 1971. Polarization measurements of the head of Comet Bennett (1969i). *Astron. Astrophys.* 14:90-94.
Cochran, A.L., and McCall, M.L. 1980. Spectrophotometric observations of Comet Bowell (1980b). *Publ. Astron. Soc. Pacific* 92:854-857.
Code, A.D.; Houck, T.E.; and Lillie, C.F. 1972. Ultraviolet observations of comets. In *The Scientific Results From Orbiting Astronomical Observatory (OAO-2),* ed. A.D. Code, (Washington: NASA SP-310).
Code, A.D., and Savage, B.D. 1972. Orbiting astronomical observatory: Review of scientific results. *Science* 77:213-221.
Combi, M.R., and Delsemme, A.H. 1980a. Neutral cometary atmospheres I. An average random walk model for photodissociation in comets. *Astrophys. J.* 237:633-640.
Combi, M.R., and Delsemme, A.H. 1980b. Neutral cometary atmospheres II. The production of CN in comets. *Astrophys. J.* 237:641-645.
Combi, M.R., and Delsemme, A.H. 1980c. Neutral cometary atmospheres. III. Acceleration of cometary CN by solar radiation pressure. *Astrophys. J.* 241:830-837.
Combi, M.R. and Delsemme, A.H. 1980d. Brightness profiles of CO^+ in the ionosphere of Comet West (1976 VI). *Astrophys. J.* 238:381-384.
Cravens, T.E., and Green, A.E.S. 1978. Airglow from the inner comas of comets. *Icarus* 33:612-623.
Crovisier, J.; Despois, D.; Gerard, E.; Irvine, W.M.; Kazes, I.; Robinson, S.E.; and Schloerb, F.P. 1981. A search for the 1.35-cm. line of H_2O in comets Kohler (1977 XIV) and Meier (1978 XXI). *Astron. Astrophys.* 97:195-198.
Danks, A.C., and Dennefeld, M. 1981. Near infrared spectroscopy of Comet Bradfield 1979l. *Astron. J.* 86:314-317.
Danks, A.C.; Lambert, D.L.; and Arpigny, C. 1975. The $^{12}C/^{13}C$ ratio in Comet Kohoutek (1973f). In *Comet Kohoutek,* ed. G.A. Gary (Washington: NASA SP-355), pp. 137-143.
Day, K.L. 1974. A possible identification of the 10-micron silicate feature. *Astrophys. J.* 192:L15-L17.
Degewij, J. 1980. Spectroscopy of faint asteroids, satellites, and comets. *Astron. J.* 85:1403-1412.
Delsemme, A.H. 1973. The brightness law of comets. *Astrophys. Letters* 14:163-167.
Delsemme, A.H. 1977. The pristine nature of comets. In *Comets, Asteroids, Meteorites,* ed. A.H. Delsemme, (Toledo, Ohio: Univ. Toledo), pp. 7-14.
Delsemme, A.H., and Combi, M.R. 1979. $O(^1D)$ and H_2O^+ in Comet Bennett 1970 II. *Astrophys. J.* 228:330-337.
Delsemme, A.H., and Miller, D.C. 1971. Physical-chemical phenomena in comets. III. The continuum of Comet Burnham (1960 II). *Planet. Space Sci.* 19:1229-1257.
Delsemme, A.H., and Rud, D.A. 1973. Albedos and cross-sections for the nuclei of comets 1969 IX, 1970 II and 1971 I. *Astron. Astrophys.* 28:1-6.
Delsemme, A.H., and Swings, P. 1952. Hydrates de gaz dans les noyaux cométaires et les grains interstellaires. *Ann. d' Astrophys.* 15:1-6.
Donn, B. 1963. The origin and structure of icy cometary nuclei. *Icarus* 2:396-402.

Donn, B. 1977. A comparison of the composition of new and evolved comets. In *Comets, Asteroids, Meteorites*, ed. A.H. Delsemme (Toledo, Ohio: Univ. Toledo), pp. 15-23.
Donn, B., and Urey, H. 1957. Chemical heating of comet nuclei. *Astrophys. J.* 123:339-342.
Dossin, F.V. 1966. Emission spectrum of a comet at large heliocentric distance. *Astron. J.* 71:853-854.
Douglas, A.E. 1951. Laboratory studies of the λ 4050 group of cometary spectra. *Astrophys. J.* 114:466-468.
Dufay, J. 1940. CH bands in comet spectra. *Astrophys. J.* 91:91-102.
Eddington, A.S. 1910. The envelopes of Comet Morehouse (1908c). *Mon. Not. Roy. Astron. Soc.* 70:442-458.
Fäy, T.D., and Wisniewski, W. 1978. The light curve of the nucleus of Comet d'Arrest. *Icarus* 34:1-9.
Feldman, P.D. 1978. A model of carbon production in a cometary coma. *Astron. Astrophys.* 70:547-553.
Feldman, P.D., and Brune, W.H. 1976. Carbon production in Comet West (1975n). *Astrophys. J.* 209:L45-L48.
Feldman, P.D.; Takacs, P.Z.; Fastie, W.G.; and Donn, B. 1974. Rocket UV spectrophotometry of Comet Kohoutek (1973f). *Science* 185:705-707.
Feldman, P.D.; Weaver, H.A.; Festou, M.C.; A'Hearn, M.F.; Jackson, W.M.; Donn, B.; Rahe, J.; Smith, A.M.; and Benvenuti, P. 1980. IUE observations of the UV spectrum of Comet Bradfield. *Nature* 286:132-135.
Festou, M. 1978. Hydrogen Atoms and Hydroxyl Radicals in Comets. Thesis, Université Paris VI.
Festou, M.C. 1981a. The density distribution of neutral compounds in cometary atmospheres. *Astron. Astrophys.* 95:69-79.
Festou, M.C. 1981b. Monochromatic observations of Comet Kobayashi-Berger-Milon (1975 IX). *Astron. Astrophys.* 96:52-61.
Festou, M.C., and Feldman, P.D. 1981. The forbidden oxygen lines in comets. *Astron. Astrophys.* In press.
Finson, M.L., and Probstein, R.F. 1968a. A theory of dust comets. I. Model and equations. *Astrophys. J.* 154:327-352.
Finson, M.L., and Probstein, R.F. 1968b. A theory of dust comets. II. Results for Comet Arend-Roland. *Astrophys. J.* 154:353-380.
Fowler, A. 1910. Investigations relating to the spectra of comets. *Mon. Not. Roy. Astron. Soc.* 70:484-496.
Gary, G.A., and O'Dell, C.R. 1974. Interpretation of the anti-tail of Comet Kohoutek as a particle flow phenomenon. *Icarus* 23:519-525.
Gausset, L.; Herzberg, G.; Lagerquist, A.; and Rosen, B. 1965. Absorption spectrum of C_3. *Astrophys. J.* 142:45-76.
Gibson, D.M., and Hobbs, R.W. 1981. On the microwave emission from comets. *Astrophys. J.* 248:863-866.
Giguere, P.T., and Huebner, W.F. 1978. A model of comet comae. I. Gas-phase chemistry in one dimension. *Astrophys. J.* 223:638-654.
Giguere, P.T.; Huebner, W.F.; and Bania, T.M. 1980. Radio observations of Comet Meier (1978f) in 18-cm OH lines. *Astron. J.* 85:1276-1280.
Greenstein, J.L. 1962. The spectrum of Comet Humason (1961e). *Astrophys J.* 136:688-690.
Greenstein, J.L., and Arpigny, C. 1962. The visual region of the spectrum of Comet Mrkos (1957d) at high resolution. *Astrophys. J.* 135:892-905.
Hanner, M. 1980. Physical characteristics of cometary dust from optical studies. In *Solid Particles in the Solar System*, eds. I. Halliday, and B.A. McIntosh (Dordrecht: D. Reidel), pp. 223-236.
Haser, L. 1957. Distribution d'intensite dans la tete d'une comète. *Bull. Acad. Roy. Belgique*, Classe des Sciences 43:740-750.
Haser, L. 1966. Calcul de distribution d'intensite relative dans une tete comètaire. *Mém. Soc. Roy. Liège*, Ser. 5, 12:233-241.
Herbig, G. 1976. Review of cometary spectra. In *The Study of Comets*, eds. B. Donn, M. Mumma, W. Jackson, M. A'Hearn, and R. Harrington (Washington: NASA SP-393), pp. 136-158.

Herzberg, G. 1976. Cometary spectra and related topics. *Mém. Soc. Roy. Sci. Liège*, Ser. 6, 9:115-132.
Herzberg, G., and Lew, L. 1974. Tentative identification of the H_2O^+ ion in Comet Kohoutek. *Astron Astrophys.* 31:123-124.
Hill, J.R., and Mendis, D.A. 1980. On the origin of striae in cometary dust tails. *Astrophys. J.* 242:395-401.
Hobbs, R.W.; Brandt, J.C.; and Maran, S.P. 1977. Microwave continuum radiation from Comet West 1975n. *Astrophys. J.* 218:573-578.
Hobbs, R.W.; Maran, S.P.; Brandt, J.C.; Webster, W.J. Jr.; Krishna Swamy, K.S. 1975. Continuum radiation from Comet Kohoutek 1973f: Emission from the icy-grain halo? *Astrophys. J.* 201:749-755.
Hollis, J.M.; Brandt, J.C.; Hobbs, R.W.; Maran, S.; and Feldman, P.D. 1981. Radio observations of Comet Bradfield (1979*l*). *Astrophys. J.* 244:355-357.
Houpis, H.L.F., and Mendis, D.A. 1980. Physiochemical and dynamical processes in cometary ionospheres. I. The basic flow profile. *Astrophys. J.* 239:1107-1118.
Houpis, H.L.F., and Mendis, D.A. 1981a. On the development and global oscillations of cometary ionospheres. *Astrophys. J.* 243:1088-1102.
Houpis, H.L.F., and Mendis, D.A. 1981b. The nature of the solar wind interaction with CO_2/CO dominated comets. *Moon and Planets*. In press.
Houpis, H.L.F., and Mendis, D.A. 1981c. Dust emission from comets at large heliocentric distanses. 1. The case of Comet Bowell. *Moon and Planets*. In press.
Huebner, W.F.; Snyder, L.E.; and Buhl, D. 1974. HCN radio emission from Comet Kohoutek (1973f). *Icarus* 23:580-584.
Huebner, W.F., and Carpenter, C.W. 1979. Solar Photo Rate Coefficients. *Los Alamos Report* LA-8085-MS.
Huebner, W.F., and Giguere, P.T. 1980. A model of comet comae. II. Effects of solar photodissociative ionization. *Astrophys. J.* 238:753-762.
Huebner, W.F., and Giguere, P.T. 1981. Observational data needs useful for modeling the coma. In *Modern Observational Techniques for Comets*, ed. J.C. Brandt (Pasadena: Jet Propulsion Lab.), pp. 14-18.
Huggins, W. 1882. Preliminary notes on the photographic spectrum of Comet 1881b. *Proc. Roy. Soc.* 33:1.
Huntress, W.T.; Prasad, S.S.; and Mitchell, G.F. 1981. Chemical models of cometary comae. *Icarus* (special Comet issue). In press.
Huppler, D.; Reynolds, R.J.; Roesler, F.L.; Scherb, F.; and Trauger, J. 1975. Observations of Comet Kohoutek (1973f) with a ground-based Fabry-Perot spectrometer. *Astrophys. J.* 202:276-282.
Hyder, C.L.; Brandt, J.C.; and Roosen, R.G. 1974. Tail structures far from the head of Comet Kohoutek. I. *Icarus* 23:601-610.
Ip, W-H. 1980. On the acceleration of cometary plasma. *Astron Astrophys.* 81:260-262.
Ip, W-H. 1981. Expanding haloes in cometary comae. *Nature* 289:269-271.
Ip, W-H., and Mendis, D.A. 1976. The structure of cometary ionospheres. I. H_2O dominated comets. *Icarus* 28:389-400.
Irvine, W.M.; Leschine, S.B.; and Schloerb, F.P. 1980. Thermal histories, chemical composition and relationship of comets to the origin of life. *Nature* 283:748-749.
Isobe, S.; Saito, K.; Tomita, K.; and Maehara, H. 1978. Polarization of the head of Comet 1976 VI West. *Publ. Astron. Soc. Japan* 30:687-690.
Jackson, W.M. 1980. The lifetime of the OH radical in comets at 1 A.U. *Icarus* 41:147-152.
Jackson, W.M.; Clark, T.; and Donn, B. 1976. Radio detection of H_2O in Comet Bradfield (1974b). In *The Study of Comets*, eds. B. Donn, M. Mumma, W. Jackson, M. A'Hearn, and R. Harrington (Washington: NASA SP-393), pp. 272-280.
Jockers, K.; Lüst, R.; and Nowak, Th. 1972. The kinematical behavior of the plasma tail of Comet Tago-Sato-Kosaka 1969 IX. *Astron. Astrophys.* 21:199-207.
Kamoun, P.; Campbell, P.B.; Ostro, S.J.; Pettengill, G.H.; and Shapiro, I.I. 1981. Radar observations of the nucleus of the Comet P/Encke. *Science*. In press.
Keller, H.U. 1976. The interpretation of ultraviolet observations of comets. *Space Sci. Rev.* 18:641-684.
Keller, H.U., and Lillie, C.F. 1974. The scale length of OH and the production rates of H and OH in Comet Bennett (1970 II). *Astron. Astrophys.* 34:187-196.

Keller, H.U., and Thomas, G.E. 1975. A cometary hydrogen model: Comparison with OGO-5 measurements of Comet Bennett (1970 II). *Astron. Astrophys.* 39:7-19.
Krishna Swamy, K.S. 1978. On the observed polarization of Comet Ikeya-Seki (1965 VIII). *Astrophys. Space Sci.* 57:491-497.
Lanzerotti, L.J.; Robbins, M.F.; and Tolk, N.H. 1974. High resolution scans of Comet Kohoutek in the vicinity of 5015, 5890, 6563 Å. *Icarus* 23:618-622.
Larson, S.M. 1980. CO^+ in Comet Schwassmann-Wachmann 1 near minimum brightness. *Astrophys. J.* 238:L47-L48.
Larson, S.M. 1981. A rotation model for the coma structure in Comet Bennett (1970 II). *Icarus.* In press.
Larson, S.M., and Minton, R.B. 1972. Photographic observations of Comet Bennett. In *Comets: Scientific Data and Missions*, eds. G.P. Kuiper and E. Roemer (Tucson: Lunar and Planetary Lab.), pp. 183-208.
Liller, W.C. 1960. The nature of grains in the tails of comets 1956h and 1957d. *Astrophys. J.* 132:867-882.
Lyttleton, R.A. 1953. *The Comets and their Origin*. (Cambridge: University Press).
Lyttleton, R.A. 1972. Does a continuous solid nucleus exist in comets? *Astrophys. Space Sci.* 15:175-184.
Maas, R.W.; Ney, E.P.; and Woolf, N.J. 1970. The 10 micron emission peak of Comet Bennett 1969i. *Astrophys. J.* 160:L101-L104.
Malaise, D.J. 1970. Collisional effects in cometary atmospheres. I. Model atmospheres and synthetic spectra. *Astron. Astrophys.* 5:209-227.
Marsden, B.G. 1967. The sungrazing comet group. *Astron. J.* 72:1170-1183.
Marsden, B.G. 1974. Comets. *Ann. Rev. Astron. Astrophys.* 12:1-12.
Marsden, B.G. 1979. *Catalogue of Cometary Orbits* (3rd ed.), IAU Central Bureau for Astr. Telegrams.
Marsden, B.G., and Sekanina, Z. 1974. Comets and nongravitational forces. VI. Periodic Comet Encke 1786-1971. *Astron. J.* 79:413-419.
Marsden, B.G.; Sekanina, Z.; and Yeomans, D.K. 1973. Comets and nongravitational forces. V. *Astron. J.* 78:211-225.
Mendis, D.A.; Hill, J.R.; Houpis, H.L.; Whipple, E.C. 1981. On the electrostatic charging of the cometary nucleus. *Astrophys. J.* In press.
Mies, F.H. 1974. Ultraviolet fluorescent pumping of OH 18-cm radiation in comets. *Astrophys. J.* 191:L145-L148.
Miller, F.D. 1969. Comet Arend-Roland (1957 III) on 5 May 1957. I. Development and kinematics of the type I tail. *Astron. J.* 74:248-255.
Miller, F.D. 1979. Comet Tago-Sato-Kosaka 1969 IX: Tail structure 25 December 1969 to 12 January 1970. *Icarus* 37:443-456.
Miller, F.D. 1980. H_2O^+ in the tails of 13 comets. *Astron. J.* 85:468-473.
Mitchell, G.F.; Prasad, S.S.; and Huntress, W.T. 1981. Chemical model calculations of C_2, C_3, CH, CN, OH and NH_2 abundances in cometary comae. *Astrophys. J.* 244:1087-1093.
Moore, M., and Donn, B. 1981. Energetic proton irradiation of ice mixtures. In *Comet Halley Workshop*, ed. J. Brandt (Greenbelt: NASA). In press.
Ness, N.F., and Donn, B. 1966. Concerning a new theory of type I comet tails. *Mém Soc. Roy. Liège*, Ser. 5, 12:343-362.
Newburn, R.L.; Bell, J.F.; and McCord, T.B. 1981. Interference filter photometry of periodic Comet Ashbrook-Jackson. *Astron. J.* 86:469-475.
Ney, E.P. 1974. Multiband photometry of comets Kohoutek, Bennett, Bradfield and Encke. *Icarus* 23:551-560.
Ney, E.P., and Merrill, K. 1976. Comet West and the scattering function of cometary dust. *Science* 194:1051-1053.
Nicolet, M. 1938. Les bandes de CH et la presence de l'hydrogene dan les comètes. *Zs. Astrophys.* 15:154-159.
Niedner, M.B. 1981. Interplanetary gas. XXVII. A catalogue of disconnection events in cometary tails. *Astrophys. J. Suppl.* 46:141-157.
Niedner, M.B., and Brandt, J.B. 1978. Interplanetary gas. XXIII. Plasma tail disconnection events in comets: Evidence for magnetic field line reconnection at interplanetary sector boundaries? *Astrophys. J.* 223:655-670.

Niedner, M.B., and Brandt, J.C. 1980. Structures far from the head of Comet Kohoutek. *Icarus* 42:257-270.
O'Dell, C.R. 1971. Spectrophotometry of Comet 1969g. *Astrophys. J.* 164:511-520.
Oort, J.H. 1950. The structure of the cloud of comets surrounding the solar system and a hypothesis concerning its origin. *Bull. Astron. Inst. Netherlands* XI, 408:91-110.
Oort, J.H., and Schmidt, M. 1951. Differences between new and old comets. *Bull. Astron. Inst. Netherlands* 11:259-270.
Opal, C.B.; Carruthers, G.R.; Prinz, D.K.; and Meier, R.R. 1974. Comet Kohoutek: Ultraviolet images and spectrograms. *Science* 185:702-705.
Opal, C.B., and Carruthers, G.R. 1977. Lyman-alpha observations of Comet West (1975n). *Icarus* 31:503-509.
Öpik, E. 1932. Note on stellar perturbations of nearly parabolic orbits. *Proc. Amer. Acad. Arts Sci.* 67:169-183.
Öpik, E. 1958. The spike of Comet Arend-Roland 1956h. *Irish Astron. J.* 5:37-50.
Oppenheimer, M. 1975. Gas phase chemistry in comets. *Astrophys. J.* 196:251-259.
Oppenheimer, M. 1980. Sodium D-line emission in Comet West (1975n) and the sodium source in comets. *Astrophys. J.* 240:923-928.
Oppenheimer, M., and Downey, C.J. 1980. The effect of solar-cycle untraviolet flux variations on cometary gas. *Astrophys. J.* 241:L123-L127.
Osterbrock, D.E. 1958. A study of two comet tails. *Astrophys. J.* 128:95-105.
Pluvinal, A.B., and Baldet, F. 1911. Spectrum of Comet Morehouse (1908c). *Astrophys. J.* 34:89-104.
Preston, G. 1967. The spectrum of Comet Ikeya Seki (1965f). *Astrophys. J.* 147:718-742.
Rahe, J.; Donn, B.; and Wurm, K. 1969. *Atlas of Cometary Forms*. (Washington: NASA SP-198).
Roemer, E. 1961a. Comet Humason (1961e). *I.A.U. Circ.* No. 1777.
Roemer, E. 1961b. Astrometric observations and orbits of comets. *Astron. J.* 66:368-371.
Roemer, E. 1962. Activity in comets at large heliocentric distance. *Publ. Astron. Soc. Pacific* 74:351-365.
Roemer, E. 1963. Comets: Discovery, orbits, astrometric observations. In *The Moon, Meteorites and Comets,* eds. B.M. Middlehurst, and G.P. Kuiper (Chicago: Univ. Chicago Press), pp. 527-549.
Roemer, E. 1966. The dimensions of cometary nuclei. *Mém. Soc. Roy. Sci. Liège* 12:23-28.
Roemer, E., and Lloyd, R.E. 1966. Observations of comets, minor planets, and satellites. *Astron. J.* 71:443-457.
Rose, L.A. 1979. Laboratory simulation of infrared astrophysical features. *Astrophys. Space Sci.* 65:47-67.
Sekanina, Z. 1976. A continuing controversy: Has the cometary nucleus been resolved? In *The Study of Comets,* eds. B. Donn, M. Mumma, W. Jackson, M. A'Hearn, and R. Harrington, (Washington: NASA SP-393), pp. 537-587.
Sekanina, Z. 1977. Relative motions of fragments of the split comets. I. A new approach. *Icarus* 30:574-594.
Sekanina, Z. 1979a. Fan-shaped coma, orientation of rotation axis and surface structure of a cometary nucleus. I. Test of a model on four comets. *Icarus* 37:420-442.
Sekanina, Z. 1979b. The split comets: Gravitational interaction between the fragments. In *Dynamics of the Solar System,* ed. R.L. Duncombe (Dordrecht: D. Reidel), pp. 311-314.
Sekanina, Z. 1981. Rotation and precession of cometary nuclei. *Ann. Rev. Earth Planet. Sci.* 9:113-145.
Sekanina, Z., and Farrell, J.A. 1980. Evidence for fragmentation of strongly nonspherical dust particles in the tail of Comet West (1976 VI). In *Solid Particles in the Solar System,* eds. I. Halliday and B.A. McIntosh (Dordrecht: D. Reidel), pp. 267-270.
Sekanina, Z., and Miller, F.D. 1973. Comet Bennett (1970 II). *Science* 179:565-567.
Shimizu, M. 1975. Ion chemistry in the cometary atmosphere. *Astrophys. Space Sci.* 36:353-361.
Smith, A.M.; Stecher, T.P.; and Casswell, L. 1980. Production of carbon, sulfur and CS in Comet West. *Astrophys. J.* 242:402-410.

Smith, P.L.; Black, J.H.; and Oppenheimer, M. 1981. Ultraviolet absorption studies of H_2O and other species in Comet Halley with space telescope. *Icarus.* In press.
Smoluchowski, R. 1981. Amorphous ice and the behavior of cometary nuclei. *Astrophys. J.* 244:L31-L34.
Spinrad, H. 1981. Spectroscopy of Comet P/Encke in 1980-81. *JPL Report.*
Spinrad, H., and Miner, E.D. 1968. Sodium velocity fields in Comet 1965f. *Astrophys. J.* 153:355-366.
Spinrad, H.; Stauffer, J.; and Newburn, R.L. 1979. Optical spectrophotometry of Comet Tempel 2 far from the sun. *Publ. Astron. Soc. Pacific.* 91:707-711.
Stawikowski, A., and Greenstein, J.L. 1964. The isotope ratio C^{12}/C^{13} in a comet. *Astrophys. J.* 140:1280-1291.
Swings, P. 1941a. Complex structure of cometary bands tentatively ascribed to the contour of the solar spectrum. *Lick Obs. Bull.* XIX, 408:131-136.
Swings, P. 1941b. Considerations regarding cometary and interstellar molecules. *Astrophys. J.* 95:270-280.
Swings, P. 1943. Reports on the progress in astronomy: Cometary spectra. *Mon. Not. Roy. Astr. Soc.* 103:86-111.
Swings, P. 1965. Cometary spectroscopy. *Quart. J. Roy. Astron. Soc.* 6:28-69.
Swings, P.; Elvey, C.T.; and Babcock, H.W. 1941. The spectrum of Comet Cunningham 1940c. *Astrophys. J.* 94:320-343.
Swings, P., and Greenstein, J. 1958. Presence des raies interdites de l'oxygene dans les spectres comètaires. *C.R. Acad. de Paris* 246:511-513.
Swings, P., and Haser, L. 1956. *Atlas of Representative Cometary Spectra.* (Liège: Inst. d'Astrophysique).
Swings, P.; McKellar, A.; and Minkowski, R. 1943. Cometary emission spectra in the visual region. *Astrophys. J.* 98:142-152.
Swings, P., and Page, T.L. 1948. The spectrum of Comet 1947n. *Astrophys. J.* 108:526-536.
Swings, P., and Page, T.L. 1950. The spectrum of Comet Bester (1947k). *Astrophys. J.* 111:530-534.
Tatum, J.B., and Gillespie, M.I. 1977. The cyanogen abundance of comets. *Astrophys. J.* 218:569-572.
Tousey, R. 1971. Survey of new solar results. In *New Techniques in Space Astronomy,* ed. F. Labuhn and R. Lust. (Dordrecht: D. Reidel), pp. 233-250.
Turner, B.E. 1974. Detection of OH at 18 cm wavelength in Comet Kohoutek (1973f). *Astrophys. J.* 189:L137-139.
Ulich, B.L., and Conklin, E.J. 1975. Discovery of CH_3CN in comets. *Nature* 248:121-122.
Van Biesbroeck, G. 1962. Tail activity of Comet Humason (1961e). *Astrophys. J.* 136:1155-1156.
Wallis, M.K. 1980. Hydrodynamic forces in splitting comets. *Mon. Not. Roy. Astron. Soc.* 190:467-478.
Weaver, H.A.; Feldman, P.D.; Festou, M.C.; and A'Hearn, M.F. 1981a. Water production models for Comet Bradfield (1979X). *Astrophys. J.* In press.
Weaver, H.A.; Feldman, P.D.; Festou, M.C.; A'Hearn, M.F.; and Keller, H.U. 1981b. IUE observations of faint comets. *Icarus.* In press.
Wehinger, P.A., and Wyckoff, S. 1974. H_2O^+ in spectra of Comet Bradfield (1974b). *Astrophys. J.* 192:L41-L45.
Wehinger, P.A.; Wyckoff, S.; Herbig, G.; Herzberg, G.; and Lew, H. 1974. Identification of H_2O^+ in the tail of Comet Kohoutek (1973f). *Astrophys. J.* 190:L43-L47.
Whipple, F.L. 1950. A comet model. I. The acceleration of Comet Encke. *Astrophys. J.* 111:375-394.
Whipple, F.L. 1951. A comet model. II. Physical relations for comets and meteors. *Astrophys. J.* 113:464-474.
Whipple, F.L. 1963. On the structure of the cometary nucleus. In *Moon, Meteorites and Comets,* eds. B.M. Middlehurst and G.P. Kuiper (Chicago: Univ. Chicago Press), pp. 639-664.
Whipple, F.L. 1978a. Comets. In *Cosmic Dust,* ed. J.A.M. McDonnell (New York: J. Wiley and Sons), pp. 1-73.

Whipple, F.L. 1978b. Cometary brightness variation and nucleus structure. *Moon and Planets* 18:343-359.
Whipple, F.L. 1980. Rotation and outbursts of Comet P/Schwassmann-Wachmann 1. *Astron. J.* 85:305-313.
Whipple, F.L., and Huebner, W.F. 1976. Physical process in comets. *Ann. Rev. Astron. Astrophys.* 14:143-172.
Whipple, F.L., and Sekanina, Z. 1979. Comet Encke: Precession of the spin axis, nongravitational motion, and sublimation. *Astron. J.* 84:1894-1909.
Wurm, K. 1961. Structure and development of the gas tails of comets. *Astron. J.* 66:362-367.
Wurm, K. 1963. The physics of comets. In *The Moon, Meteorites, and Comets,* eds. B.M. Middlehurst and G.P. Kuiper (Chicago: Univ. Chicago Press), pp. 573-617.
Wurm, K. 1968. Structure and kinematics of cometary type I tails. *Icarus* 8:287-300.
Wurm, K., and Mammano, A. 1964. The axes of comets. *Icarus* 3:1-7.
Wyckoff, S. 1981. Ground-based cometary spectroscopy. In *Modern Observational Techniques for Comets,* ed. J.C. Brandt (Pasadena: Jet Propulsion Lab.), pp. 129-137.
Wyckoff, S., and Wehinger, P.A. 1976a. Molecular ions in comet tails. *Astrophys. J.* 204:604-615.
Wyckoff, S., and Wehinger, P.A. 1976b. On the ionization and excitation of H_2O^+ in Comet Kohoutek (1973f). *Astrophys. J.* 204:616-625.
Yeomans, D. 1977. Comet Halley and nongravitational forces. In *Comets, Asteroids, Meteorites,* ed. A.H. Delsemme (Toledo, Ohio: Univ. Toledo), pp. 61-64.

COMET DISCOVERIES, STATISTICS, AND OBSERVATIONAL SELECTION

Lubor Kresák
Astronomical Institute of the Slovak Academy of Sciences

The ensemble of known comets, which apparently constitutes less than one millionth of all such objects revolving in the solar system, is strongly biased by observational selection. The factors partaking in this selection can be divided into three categories: (1) absolute brightness and perihelion distance: (2) revolution period, perihelion date, and development of observing techniques: (3) plane and orientation of the orbit, relative position of the Earth at the time of the comet's perihelion passage, and geographic distribution of potential discoverers. Unless the interplay of all these effects is properly taken into account, it is impossible to draw correct conclusions about the real dynamical and evolutionary properties of the system of comets. This chapter is an overview of the distribution of cometary orbits in their individual parameters, to distinguish between real and apparent deviations from randomness and to specify those features that are relevant to the origin, evolution, and constitution of the system of comets. The problems referred to include the degree of completeness of comet discoveries; the total number of comets and their size distribution; the large-scale structure of the comet cloud in relation to the galactic structure, to the environment of the solar system, and to the system of major planets; the small-scale structure in relation to a common origin of comet groups and pairs; and the effects of physical aging of comets. In the subsystem of short-period comets, structural features impressed by the orbit, and position in orbit, of Jupiter are discussed. The results on the observational biases are used to put forward some suggestions on how the efficiency of comet searchers could be optimized.

According to tentative estimates (Oort 1963; Öpik 1973; Weissman's chapter in this book) the solar system is populated by $\sim 10^{10}$ to 10^{12} comets, a number comparable with the number of stars in a large galaxy. Only a negligible fraction of all these objects has actually been observed. There are records of ~ 1300 different comets, and a few hundred cases where the nature of the object cannot be specified. For < 700 comets orbital data of varied accuracy are available; only for about one half of these is the accuracy better than $\pm 0°.01$ in angular elements, ± 0.0001 AU in perihelion distance, and ± 0.001 AU^{-1} in reciprocal semimajor axis. Our current knowledge of the physical processes in comets is essentially based on detailed observations of a few dozen bright objects.

The sample of known comets is by no means representative of their whole population. This situation is caused primarily by the vast dimensions of the comet cloud. Aphelia of the new comets, in Oort's sense, concentrate around heliocentric distances of 40,000 to 50,000 AU (Marsden et al. 1978), i.e., 1000 times farther from the Sun than the outermost known planet. Yet comets are only observable at heliocentric distances of a few AU. Only ~ 30 of them have so far been observed beyond the distance of Jupiter, and five at the distance of Saturn.

The history of three millennia of comet observations is about the average survival time of a comet revolving in a short-period orbit (Kresák 1981b), which would experience hundreds of perihelion passages during that period. At the same time it covers less than 1/1000 of a revolution of a comet coming from, and returning to, the outskirts of the Oort Cloud. Moreover, the orbits of an indefinite but great majority of comets are such that they remain undetectable even when they are passing perihelion, at large distances from the Sun. This is why we know neither the largest members of the comet cloud and their upper size limit, nor the approximate total number and mass of comets. The answer is closely tied with the problem of the origin of comets (for a critical review see Delsemme 1977). A total number of comets up to 10^{15} is not out of question (Whipple 1975; Marsden 1977), and this may be but a small fraction of the original population. On the other hand, theories of a more recent origin of the observed comets would be compatible with much smaller values. There is actually no direct evidence that the present flux of comets through the inner solar system is representative for longer time spans.

When a comet happens to pass sufficiently near the Sun and become bright enough for observation, there are still different biases deciding whether or not it will be detected. In fact, the observed distribution in each of the six orbital elements is subject to a complex of selection effects varying with the development of observing techniques. There is a substantial difference in this respect between the long-period comets, that would become detectable only once during the time span covered by astronomical observations, if at all, and the short-period comets, that would make a number of apparitions spaced by several years and taking place under different observing geometry.

The occurrence rate of comets accessible to exhaustive visual searches is illustrated in Fig. 1. It is interesting to note to what extent our knowledge of comets has been affected by the present density of the comet cloud. If the mean distance between neighboring comets were 10 times its actual value, there would be no cometary astronomy. We would scarcely know of more than one comet apparition, and even this enigmatic object would probably be thought a visitor from interstellar space. On the other hand, if the mean distance were 10 times smaller (which appears quite possible for the earlier evolutionary phases of the solar system), observable comets would be overabundant. Several bright comets with long tails would be seen every twilight; dozens of comets would be visible by unaided eye on every clear moonless night and every exposure with a wide-field camera would record a number of traces of fainter ones. Thus the comet apparitions are both frequent enough to permit statistical studies, and rare enough to substantiate efforts in detecting and following up all new objects with the interest they deserve as main clues to the early history of the solar system.

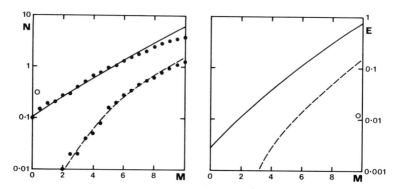

Fig. 1. The cumulative number of comets N passing the perihelion per year and becoming brighter than total apparent magnitude M; and the expectation E that, at a given moment, there will be an observable comet brighter than M anywhere in the sky. Full line, all comets; dashed line, short-period comets, both corrected for the incompleteness of discoveries; dots, observed mean values 1880-1980; open circles, comets with tail length exceeding $10°$. Note that the scales of N and E are logarithmic.

I. THE HISTORY OF COMET DISCOVERIES

Owing to the spectacular appearance of bright comets, their observing history is nearly 10 times longer than that of planetary satellites, and 20 times longer than that of asteroids. The authenticity of the earliest observations dating back to the 23rd century B.C. (Pingré 1783; Baldet 1949) is open to doubt; the first reliable record seems to be that from 1095 B.C.

(Ho Peng Yoke 1962). After 400 B.C. the records become more frequent, most of them from the Far East and from the Mediterranean area. The Chinese chronicles alone maintain for two millennia a mean rate of more than 20 different comet apparitions per century. According to Fig. 1 this is equivalent to a complete record of all comets reaching 2nd apparent magnitude, in spite of the geographic limitation of the source. The ancient verbal descriptions of individual comets are generally inadequate for determining their orbits even approximately, but a few can be exploited for this purpose, as recently demonstrated by Hasegawa (1979).

Marsden's Catalogue (1979) lists five parabolic orbits of comets observed prior to 1000 A.D., 18 between 1000 and 1500, and 31 between 1500 and 1700, in addition to a complete sequence of all apparitions of periodic Comet Halley since 87 B.C. The history after 1700 B.C. is illustrated by Fig. 2. Note

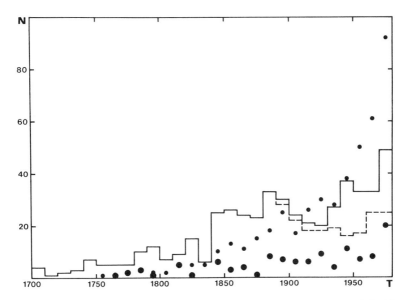

Fig. 2. The numbers of comets N for which the orbits have been determined, in 10-yr intervals of the perihelion date T. Histogram, long-period comets; dashed line, division between visual discoveries (below) and photographic discoveries (above); small dots, all apparitions of short-period comets; large dots, their discovery apparitions.

the major disproportion between the steep increase in the number of apparitions of short-period comets, which doubled within the last 25 years, and the much slower and irregular development in the long-period comets, which required as much as 150 years to double. The source of this difference is the higher efficiency of photographic techniques in detecting short-period comets. Unlike the long-period comets, these concentrate in a strip along the ecliptic and tend to become brightest in the opposition region which is most frequently covered by photographic observations for other purposes. Moreover, the photographic technique, with its lower brightness threshold, is much better suited for ephemeris-aided rediscoveries. Among all comet apparitions for which orbits are available, photographic discoveries constitute 16% for long-period comets, but 52% of first observations and 69% of later rediscoveries for short-period comets. The corresponding figures after 1950 are 42%, 94% and 97%, respectively, the 3% remainder in the latter group being due to the visual rediscoveries of lost objects.

Another difference between these two classes of objects is borne out by Figs. 3 and 4 showing the apparent brightness of individual objects at the time of discovery or rediscovery. It should be noted that the magnitude scales for the visual and photographic observations are not quite consistent, as the latter generally only refers to the image of the central condensation recorded on the plate. The correction depends both on the appearance of the comet and on the telescope used, and may often exceed -2 or -3 mag.

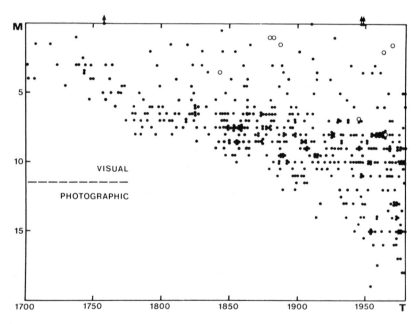

Fig. 3. The apparent magnitudes of long-period comets at the time of their discoveries M plotted against the perihelion date T. The dashed line represents a rough division between two different magnitude scales: visual, referring to total magnitudes, and photographic, mostly referring to the central condensation. Open circles, members of the Kreutz group. These comets are also distinguished by open circles or omitted in Figs. 7-9, because they are products of tidal splitting of a single sungrazing parent object.

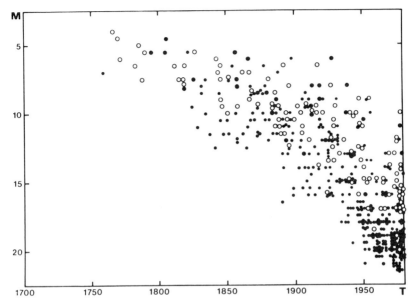

Fig. 4. The same as in Fig. 3 for short-period comets, $P < 200$ yr. Open circles, first observations; larger dots, independent rediscoveries at later returns; small dots, recoveries according to predictions.

The distribution of the data points in Figs. 3 and 4 reveals marked irregularities due to the instrumental and human factors involved in the historical development of the discipline. While the first telescopic observation of a comet was that of 1618 I by Kepler, and the first telescopic discovery that of 1680 by Kirch, telescopic searches for new comets did not begin until around 1760. This can be clearly recognized in Fig. 3 as a displacement of the discovery magnitudes from around $M = 4$ to $M = 7$. Between 1760 and 1840 the majority of comets were discovered by very few observers — the French comet searchers Messier, Méchain, Gambart, and in particular, Pons. While Pons was the single discoverer or independent codiscoverer of 36 of the 48 comets observed between 1800 and 1830 (26 of them bear his name), the number of people engaged in this type of work increased greatly after 1840, in the decade of the discovery of Neptune and the start of the series of discoveries of minor planets. This time it was improved sky surveillance rather than change in the limiting brightness that brought the rate of visual comet discoveries very close to its present value. Fainter comets became detectable after 1880, in the period of activity of the American observers Swift, Barnard, and Brooks, producing a peak in the discovery rate that remained unrivaled for more than 50 yr. Some kind of saturation was reached, in that discoveries at $M = 5$ to 7 became rare after 1900. However, naked-eye objects are still being discovered at small solar elongations even now, some of them from high-flying aircraft where they stand out against a fainter twilight glare. At extremely small elongations it sometimes happens that the end of the tail is discovered first. Most of the recent visual discoveries are to the credit of Japanese amateur observers. A good illustration of their diligence is the fact that Comet Suzuki-Saigusa-Mori 1975 X was independently discovered by five comet searchers from five different locations within half an hour. More details on the visual search techniques and instrumentation can be found elsewhere (e.g., Proctor and Crommelin 1937; Kresák 1966; Bortle 1981).

The first scientific photograph of a comet was obtained on the exposures of the 1882 total solar eclipse in Egypt; the object was never reobserved. (There apparently was a photograph of Comet Donati 1858 VI by some unknown portrait-photographer, but its quality and history are in doubt.) The first true photographic discovery of a comet was that of P/Barnard 3 in 1892, but this technique remained rather inefficient for unpredicted apparitions until the advent of the large Schmidt telescopes. With 32 discoveries on record, the Palomar Big Schmidt has been the most successful instrument in this respect. Large instruments with smaller usable fields have proved very suitable for ephemeris-aided searches for periodic comets, as demonstrated by the notable score of 77 such recoveries by Roemer and her collaborators in 1953-1976. For more information about photographic searches see Roemer (1963).

Considerable biases in comet statistics are produced, not only by the varying number and equipment of potential discoverers, but also by their distribution in geographic latitude. It is obvious that the predominance of observations from one hemisphere exerts influence on the distribution of orbital planes and perihelion directions of the known comets. There were periods during which discoveries from the southern hemisphere were completely lacking, and others during which they were rare. The present improvement of this situation is mainly due to Bradfield in the area of visual discoveries, and to the observers at the UK Schmidt Telescope Unit in Australia and the ESO Observatory in Chile in the area of photographic discoveries.

Assuming that progress in observing and computing techniques will continue at its present rate, a rough prediction of the comet statistics in the year 2000 can be attempted. Unless an extraordinary object appears, the upper limit of perihelion distances of known comets should increase to 8 AU. The number of long-period comets with known orbits should increase by 20 to 25%, and the number of known short-period comets by 35 to 40%. Since the statistical significance of a sample is proportional to the square root of its population, there is little hope for a significant enhancement of the present-day statistics in the near future.

DISCOVERIES, STATISTICS, OBSERVATIONAL SELECTION   63

II. THE SOURCES OF SELECTION EFFECTS

By their influence on the determination of different parameters of the comet system, the variety of selection effects can be divided into three broad categories.

1. *Absolute Brightness and Perihelion Distance.* These two parameters determine the peak apparent magnitude reduced to a uniform distance from the Earth, which is the dominant factor for the detectability of a comet.

2. *Revolution Period and Perihelion Date.* Since the observability of comets is limited to the innermost part of the comet cloud, our knowledge of the vast outer regions rests on the determination of the revolution periods (or semimajor axes or aphelion distances) of comets with sufficiently small perihelion distances. The determination of the size and mass of the cloud is just a matter of extrapolation in space and time. The statistical sample usable for this purpose is rather limited due to the high requirements on the duration of accurate observations, even for an approximate location of the aphelia of comets moving in nearly parabolic orbits.

3. *Plane and Orientation of the Orbit.* Statistics of these properties, as expressed by the three conventional angular elements (or by the axes of symmetry, directions of perihelia and poles) are relevant to the shape and fine structure of the cloud. Here the observing biases are smallest but formally most complicated, being affected by the orbital motion of the Earth and by the geographic distribution of the observers.

The effects of observational selection on the statistics of the orbits and absolute magnitudes of comets have been investigated by a number of authors: Holetschek (1890), Bourgeois and Cox (1932, 1933), Vsekhsvyatskij (1958), Everhart (1967a,b), Pittich (1969a,b), Kresák (1974a, 1975), Revina (1977), Kresák and Pittich (1978) etc. Individual results relevant to different aspects of the problem will be referred to in Secs. III-V.

III. ABSOLUTE BRIGHTNESS AND PERIHELION DISTANCE

The main consequence of the changing brightness threshold with the improvement of instrumentation (Figs. 3 and 4) is the progressive increase of the heliocentric distance to which comets can be observed. This is demonstrated by a strong time dependence of the distribution in perihelion distance q, illustrated in Fig. 5. Some saturation for $q < 1.5$ has already been reached a century ago. The present counts of active comets seem to be at least 60% complete at $q < 1$, 25% at $q = 1.5$, and 10% at $q = 2$ (Kresák and Pittich 1978). The missing objects of small q mostly escape observation due

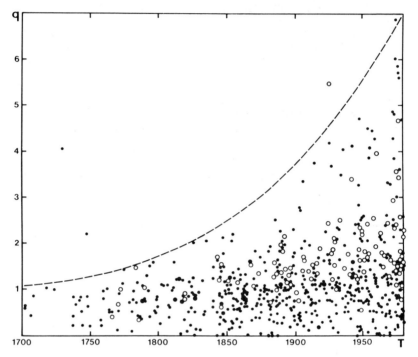

Fig. 5. Perihelion distances q of comets plotted against the perihelion date T. Small dots, long-period comets ($P > 200$ yr or indeterminate); large dots, short-period comets of Halley type (20 yr $< P < 200$ yr); open circles, short-period comets of the Jupiter family ($P < 20$ yr).

to an unfavorable configuration with respect to the Sun and Earth around the perihelion passage.

The slope of dN/dq exhibits a sudden change around $q = 2.5$, with the observed distribution remaining nearly uniform up to $q = 6$. This is probably caused by the steep drop of the vaporization rates, especially for water snow (Whipple 1978), and by the disappearance of the molecular emissions for which the heliocentric luminosity index is much higher than for the continuum (Donn 1977). Furthermore, essentially all discoveries at $r > 2.5$ are photographic, and photographic surveillance of the sky is rather incomplete. This can also be demonstrated by the decreasing proportion of preperihelion to postperihelion discoveries with increasing q.

The upper limit of the q/T area occupied by known comets can be approximated by the empirical formula

$$q = 1 + 1.5 \times 10^{-8} (T - 1600)^{10/3} \tag{1}$$

as indicated by the dashed curve in Fig. 5. The three objects situated clearly outside are two of the three brightest long-period comets observed since 1600, namely Comet Sarabat 1729 and Comet Chéseaux 1747, and the brightest short-period comet P/Schwassmann-Wachmann 1 (according to the list of Vsekhsvyatskij [1958]). If the trend of instrumentation improvement continues consistently with the above formula, comets with perihelia at the distance of Saturn should become detectable about 40 years from now, and those with perihelia at the distance of Uranus 150 years hence. Obviously, this does not apply to exceptional objects like 2060 Chiron, the nature of which is still obscure (Kowal 1979).

Figure 5 also bears out a considerable difference between the q-distributions of long-period and short-period comets. This is apparently due to the combination of three overlapping effects:

1. A step-wise capture into short-period orbits prefers larger perihelion distances;

2. Short-period comets of small perihelion distance are subject to more rapid aging and disintegration;

3. Repetition of perihelion passages makes them more easily detectable at favorable configurations even when q is large, and shifts their detection limit to larger heliocentric distances (see Sec. VII).

The bias-free distribution of comets in perihelion distance is still an open problem. If the analysis is restricted to comets accessible to visual comet seekers, as done by Everhart (1967a,b), the results become rather indeterminate for $q > 2$. For fainter comets of larger q the degree of completeness of the discoveries is difficult to estimate. Neither the data on the total sky coverage and magnitude limit of all contributing photographic programs nor the portion of plates searched through for moving diffuse objects can be assessed, nor does the apparent photographic magnitude simply refer to the visual scale of total magnitudes. Two competing models are $N(q) = $ const. and $N(q) \propto q^{1/2}$. The former is supported by dynamical arguments and model computations (Weissman 1977), but observational verification is lacking because of a depletion by disintegration, which depends on q within the observable range of perihelion distances. The latter model is in better agreement with observational evidence (Kresák and Pittich 1978) and consistent with a uniform number density in the central part of the comet cloud.

While the absolute magnitude is a reasonable measure of relative comet sizes at low q values, it becomes rather problematical if q is large. Such comets never approach the reference configuration (1 AU from the Sun, 1 AU from the Earth) and each error in the photometric exponent builds up with increasing q. This exponent is subject to serious biases as shown, e.g., by its strong correlation with T (Kresák 1974a).

The distribution of long-period comets in standard absolute magnitude H_0 (assuming that the luminosity varies with the 4th power of the heliocentric distance) is definitely unlike that of the asteroids or meteoroids. Both Vsekhsvyatskij (1958) and Everhart (1967b) find an abrupt change of the distribution function $F = d(\log N)/dH_0$ at $H_0 = 6$, but they disagree in the slopes of the two branches: $F = 0.40$ for $H_0 < 6$ and $F = 0.10$ for $H_0 > 6$ according to Vsekhsvyatskij, but $F = 0.60$ for $H_0 < 6$ and $F = 0.25$ for $H_0 > 6$ according to Everhart. There is strong evidence for a cutoff near $H_0 = 12$, with practically no long-period comets being fainter than this (Kresák 1978).

Everhart's figures imply a very low completeness of our comet records, with four out of five comets observable in small telescopes being missed (Everhart 1967a,b). This point, very important for the problems of observational selection, deserves comment. Everhart's treatment was undoubtedly the most sophisticated one applied so far. With the use of modern computing techniques he combined the time variations of a number of relevant parameters into one, the time integral of the excess magnitude over the detection limit. By comparison of the integral from the beginning of observability to the moment of discovery with that from the beginning to the end of observability, he converted the latter quantity into the probability of discovery.

However, there are some doubts whether all the simplifying assumptions involved were sound (Kresák 1975). As a principal conjecture let us consider an alternative approach based on the rate of double and multiple independent discoveries of the same comets. Its applicability is limited by the fact that independent discoveries are only possible within the short time span between the first observation and its announcement by the IAU Central Bureau for Astronomical Telegrams (or other similar communication centers operated in the past). The list of 256 comets used by Everhart includes 44 objects to which he ascribes discovery probabilities in the range $0.00 < p < 0.10$. Even without any time limitation there is an expectation of only two cases of double discovery, and the chance for a single triple discovery would be as low as 1:8. In fact, there were nine double discoveries and six triple discoveries. The discrepancy becomes still greater if one considers the deadline set by the discovery announcement. For $p = 0.05$ Everhart's calibration yields 29 magnitude-days, whereas all but one of the 21 independent discoveries were made within four days of the first discovery. It may also be noted that Everhart's results would imply one comet discovery per ~ 20 to 30 hr of systematic search with a suitable telescope. Experience shows that 10 to 20 times more hours are needed. Thus it appears that the discovery probabilities as determined by Everhart (1967a,b) and adopted by other authors require a substantial revision upward, and the slope of the magnitude distribution and the intrinsic numbers of comets, based on these probabilities (see e.g., Weissman's chapter in this book), require a corresponding revision downward.

IV. REVOLUTION PERIOD AND PERIHELION DATE

The proportion of comets to which specific ranges of revolution periods can be assigned exhibits considerable variations with time. There is a clear-cut dependence on the extent to which different search techniques are applied. For example, the two most lavish current sources of comet discoveries yield an entirely different ratio of the number of short-period and long-period comets — 2:3 for Palomar Mountain (photographic work with Schmidt cameras) but 1:10 for Japan (systematic visual searches with small telescopes). Furthermore, very long revolution periods can only be recognized when the observing arc is long and astrometric observations are both numerous and sufficiently accurate. The first comet which can be identified as new in Oort's sense is from 1849 (Marsden et al. 1978). Marsden's Catalogue (1979) includes 167 long-period comets observed before this date, but for 87% of them only parabolic orbits are available and for others the accuracy in $1/a$ is problematic. This time-dependent selection masks the real fluctuations in the occurrence rate of different dynamical types of comets, which appear quite significant (Kresák 1977, 1981a). Even restriction to orbits above a given accuracy level does not present a coherent picture. As shown by Marsden and Sekanina (1973) and by Marsden et al. (1978), the proportion of new comets depends both on the perihelion distance and on the limit of orbit accuracy adopted. The most pronounced bias is due to the fact that available orbit determinations cover less than 1/20,000 of the revolution period of a typical new comet. All extrapolations are necessarily based on a tacit assumption that the present occurrence rate of short-period comets is representative for time spans of the order of 10^6 yr, which may but need not be true.

The result of such an extrapolation is presented in Fig. 6. The observed distribution, tentatively corrected for the error dispersion at large values of $\log a$, is taken from Kresák (1981a), based on the work of Marsden et al. (1978). The data for short-period comets are normalized to a uniform proportional representation with the accurate long-period orbits. To obtain some approximation of a bias-free distribution, each entry of $P > 180$ yr (log $a > 1.5$) is multiplied by the factor $P/180$. The result refers to the comets which are revolving at the moment in orbits of small perihelion distance. Although this restriction tends to reduce strongly the proportional representation of comets with very long periods, their immense prevalence is visible at first glance. To appreciate fully the role of the cometary reservoir in the Oort Cloud it must be remembered that every passage through the region of major planets would remove a new comet, with a high probability, from the peak of the distribution. The displacement corresponding to the mean negative change in the energy $\Delta(1/a) = 0.0007$ AU^{-1} is indicated by an arrow (Fig. 6).

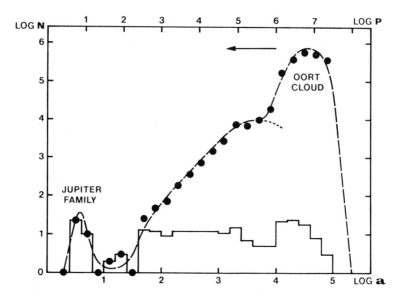

Fig. 6. The number of comets N as a function of their semimajor axis a (lower scale) or revolution period P (upper scale). The histogram shows the observed distribution of high-quality orbits corrected for the error dispersion. The dots and the curve show the values corrected for the probability of observation as a function of the revolution period. Note that all scales are logarithmic.

V. ORBITAL PLANE AND ORIENTATION

Coventional elements defining the plane and orientation of the orbit are the inclination to the ecliptic i, the longitude of the ascending node Ω, and the argument of perihelion ω. Figure 7 shows the observed distributions in these elements based on the statistics of 547 long-period comets ($P > 200$ yr or indeterminate, the Kreutz group omitted) observed to pass perihelion before the end of 1980; namely 16 short-period comets of Halley type ($20 \text{ yr} < P < 200 \text{ yr}$), and 101 short-period comets of the Jupiter family ($P < 20$ yr). For the sake of reference, simplified model distributions are indicated by pairs of lines corresponding to the predicted number of objects in the box plus or minus its square root.

For orbital inclinations, a random state would imply random positions of the poles of the orbital planes, i.e., a distribution function $N(i) \propto \sin i$. The main deviations from this model can be explained by the perturbational transport of comets into short-period orbits. The process of multi-stage capture (Kazimirchak-Polonskaya 1972; Everhart 1973) is responsible for the sharp peak around $i = 8°$ for the Jupiter family. A higher probability of encounter with major planets if i is close to $0°$ or to $180°$, and a higher efficiency of the interaction in the former case, is borne out by the distribution of the comets of Halley type. Some traces of this process can also

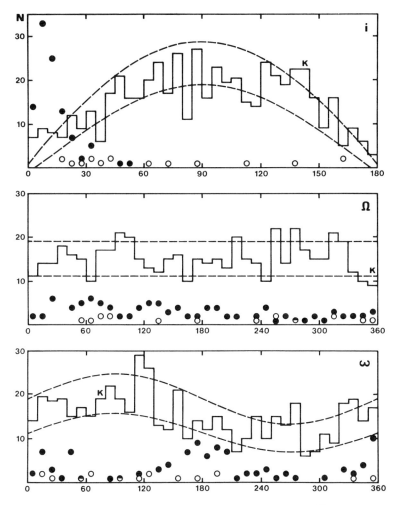

Fig. 7. The distributions of comets in inclination i, nodal longitude Ω, and argument of perihelion ω. Histograms, long-period comets ($P > 200$ yr or indeterminate). The eight members of the Kreutz group are omitted but their mean positions are indicated by the letter K. Open circles, short-period comets of Halley type (20 yr $< P < 200$ yr); solid circles, short-period comets of Jupiter's family ($P < 20$ yr).

be recognized among the long-period comets. A significant excess at $i < 10°$ betrays the presence of some short-period comets which are erroneously classified as long-period because the positional observations were insufficient to reveal deviations from a parabolic motion. Marsden (1979, p. 64) lists five such cases as probable (1585, 1678, 1743 I, 1949 III, 1963 IX) and four as possible (568, 1231, 1702, 1833); a recent addition to the probables is 1979h.

Another deviation from the model is a flatter maximum, with a slight deficiency around $i = 90°$. It is questionable whether this can be regarded as a result of longer average revolution periods of comets most successfully avoiding planetary encounters. A lower detection probability of high-inclination comets with perihelia situated deep in the southern hemisphere can also be involved. Perhaps the most interesting anomaly is the overabundance of comets with $i \sim 140°$ as compared to those with $i \sim 40°$. Yabushita (1972) points out in this connection that direct comets are more rapidly expelled by perturbations than retrograde ones. However, this defect should be counterbalanced by more comets being captured into orbits of shorter revolution period; the net outcome would depend on their poorly known rate of aging. A reasonable explanation is that retrograde comets unfavorably situated with respect to the Earth are more easily detected, because they remain at small solar elongations for shorter time intervals (Kresák 1975). The Kreutz group also falls near $i = 140°$, and Öpik (1966) suggests that it may be genetically associated with a diffuse system of other small-q objects entering our statistics.

The nodal longitudes of long-period comets do not reveal significant deviations from a random distribution. Short-period comets of Jupiter's family exhibit a 63% preference for the half-circle $0° < \Omega < 180°$. This arrangement can be attributed to the fact that the orbital planes of these comets are concentrated towards that of Jupiter rather than towards the ecliptic. The nodal longitude of Jupiter being $100°$, the observed distribution is in qualitative agreement with the expectation.

The distribution of the arguments of perihelia of long-period comets can be roughly approximated by a sinusoid with an amplitude of 2:1, $N(\omega) \propto (3 + \sin \omega)$. This apparently results from a better coverage of the northern hemisphere by comet searches. Since the perihelion latitude B is defined by $\sin B = \sin \omega \sin i$, comets of $\omega \sim 90°$ are more readily observed from the northern hemisphere than those of $\omega \sim 270°$, unless their perihelion distance is very small. In fact, the amplitude becomes still higher when only earlier comet apparitions are taken into account, from the time when southern observations were missing entirely. It is also higher for comets of large perihelion distance. For example, the preference for the interval $0° < \omega < 180°$ is 65% among the comets discovered prior to 1700, and 73% among the comets of $q > 3$. For the whole sample it is 58%, as compared with 61% for the plotted sinusoid and 50% for a uniform distribution. A marked preference for $0° < \omega < 180°$ is also apparent among the comets of Halley type, all 16 discovered from the northern hemisphere.

An entirely different pattern, with two sharp maxima at $\omega = 0°$ and $\omega = 180°$, is displayed by the short-period comets of the Jupiter family. Fifty-three perihelia are situated within ± 20° of these directions, but only six within ± 20° of the perpendicular directions $\omega = 90°$ and $\omega = 270°$. The effect is associated with the capture history of these comets; decisive perturbations at close encounters with Jupiter tend to maintain the line of apsides close to the line of nodes.

Most papers dealing with the cosmogonical implications of the space distribution of comets use as basic quantities the directions of their perihelia. A complete updated version of their two-dimensional distribution in an equal-area projection is plotted in Fig. 8. A series of plots for different ranges of perihelion distance can be found elsewhere (Kresák 1975); these are instructive in demonstrating that the observed distribution also undergoes variations with time, as a consequence of the increasing range of q shown in Fig. 5.

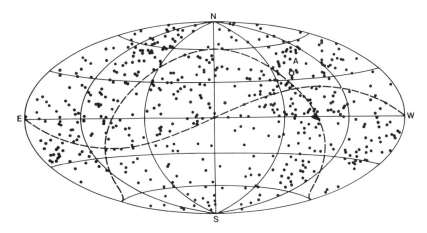

Fig. 8. Directions of the perihelia of long-period comets plotted in an ecliptical equal-area projection. North up, point of vernal equinox in the middle, ecliptical longitude increasing to the left. The two dashed curves show the positions of the equator (sinusoid-like) and of the galactic plane (horseshoe-like). Open circle, eight perihelia of the Kreutz group; A, apex of the motion of the solar system with respect to its stellar environment (basic apex).

The interpretation of the perihelion distribution, with suggested relations to the galactic plane, star streaming, or accretion axes, has been the subject of a great number of papers, but little consensus has been reached so far. For a review of this problem and a list of references see Witkowski (1972). More recent papers are by Tomanov (1973, 1976, 1977), Kresák (1975), Hasegawa (1976), Shteins and Revina (1977), Van Flandern (1977a,b), and Yabushita et al. (1979). A discussion of these highly important problems is outside the scope of this chapter; however some comments on Fig. 8 seem appropriate.

While the expected preference of perihelia for the northern hemisphere is evident, this is not the only nonrandom feature of the distribution. There are both "zones of avoidance," as called in the earliest paper on this issue by Hoek (1866), and regions of local concentration. These, however, do not seem to follow any great circle, which would be the case for a system flattened towards a particular plane. There is indeed some deficiency near the galactic poles, but also in some directions within the galactic plane. The solar motion apex is indeed not far from the most outstanding concentration, but the drop in the density of perihelia in different directions from it is by no means uniform. Although the application of the ellipsoidal analysis represented a methodological progress in making the results more objective (Oppenheim 1924), the actual chaotic pattern does not seem well suited to this type of evaluation. It is problematic whether the two-source model (Yabushita et al. 1979) can cope with this situation. Counts and statistical tests for limited areas, as applied by Öpik (1971) or Hasegawa (1976), appear more promising. The main difficulty is that if selection effects are to be eliminated by a subdivision according to q and a, the samples inevitably become either too large in the extent of the sky areas, or too sparsely populated for statistical tests. It was this type of approach which led Öpik (1971) to suggest that the nonuniformities are due to the common origin of a number of comet groups.

VI. THE REALITY OF COMET GROUPS AND PAIRS

In his extensive analysis Öpik (1971) finds that there is a negligibly low probability, 10^{-39}, of comets not being associated in genetically related groups, and that such groupings of two to seven members constitute at least 60% of the whole known comet population. On the other hand, revision by Whipple (1977), pointing out some incorrect assumptions made by Öpik, suggests that there are very few, if any, real groupings of this type. The only definite exception is the Kreutz group of sungrazing comets. We know more than 20 examples of observed splitting of cometary nuclei, which makes a common origin of comet groups and pairs acceptable in principle (see Sekanina's chapter in this book). However, the existence of numerous observed groups is hardly believable. As already noted by Whipple (1977), an immense number of fragments of a giant parent body would be necessary to allow some of them to appear within the time span of our observations, which is a small fraction of their widely dispersed revolution periods.

A method devised by Southworth and Hawkins (1963) for assessing the meteroid stream membership makes it possible to express the degree of similarity of a given pair of orbits by a single parameter. This method has already been applied to a number of statistical samples, including the asteroid families (Lindblad and Southworth 1971) and some comet groups (Kramer et al. 1979). Recently it was applied by the author to all the long-period comets. A detailed discussion of the results appears elsewhere (Kresák 1981c); their essence is reproduced in Fig. 9.

Three sets of computer-generated models of random orbit distribution were obtained under intentionally conservative assumptions: uniform distribution in Ω; random distribution in ω (no geographic selection effect); random distribution in $\cos i$ (no preference for lower inclinations); and an empirical polynomial distribution function in q. The sets of most similar pairs of model orbits, with the D parameter < 0.30, were compared with those found in a complete search of the real comet sample. A slight overabundance of similar orbits was found, but this was almost entirely due to the earlier comet apparitions. When only apparitions after 1800 are taken into account, the agreement with a random expectation is very good. Also, the pairs are characteristic by a prevalence of poorly determined orbits, smaller perihelion distances, and lower inclinations. There is no tendency towards chain associations of more than two objects.

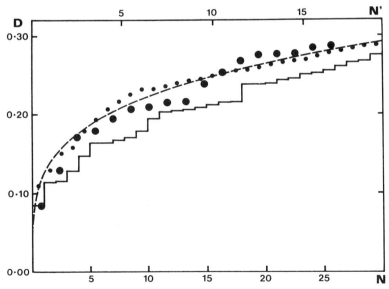

Fig. 9. The distribution of the most similar comet pairs according to the D-criterion. Histogram and the lower scale of N, observed distribution for all long-period comets; solid circles and the upper scale of N', the same for the long-period comets observed after 1800. The Kreutz group is omitted. Dots, a simulated random distribution; dashed curve, the adopted model. D is zero for a pair of identical orbits, and increases with increasing separation in the phase-space of orbital elements ω, Ω, i, q, and e. N and N' are the cumulative numbers of pairs up to the given value of D.

The conclusion is that the slight excess of similar orbits (~ 20% for $D < 0.30$) is due to the following reasons:

1. Nonrandom distribution in ω, produced by the nonrandom geographic distribution of comet discoveries making the phase-space denser for northern perihelion latitudes;

2. Some statistical coupling between individual orbital elements produced by other selection effects;

3. Contamination by some unresolved short-period orbits which makes the phase-space denser near the ecliptic;

4. Very few possible identities of some of the comet pairs for which very rough parabolic orbits are available.

Taking all these effects together, there is no reason to assume any genetic relation for comet pairs or small compact groups, except for the Kreutz group of sungrazing comets.

VII. THE JUPITER FAMILY OF COMETS

The structure of the subsystem of short-period comets is mainly controlled by Jupiter. Some of its features, those manifesting themselves in the distribution of the orbital elements (Figs. 5 and 7) have already been discussed in the preceding sections. The circumsolar ring occupied by the Jupiter family of comets is even more flattened than the asteroid belt (Kresák 1979). Since for $P < 20$ yr the median absolute perihelion latitude is as low as $3°.7$, the distribution of perihelion directions need not be treated in two dimensions. Instead, the perihelion distance q is plotted as the other coordinate in Fig. 10. This figure shows the distribution of perihelion longitudes referred to three coordinate systems rotating with respect to one another. The data points refer to the 97 comets of $P < 20$ yr listed by Marsden (1979), four recent additions to this family (P/Kowal 2, P/Russell 1, P/Russell 2, P/Wild 3), and three probable members for which the periods are indeterminate (1949 III, 1963 IX, 1979h) — 104 objects in all. For each of them the rotating coordinate frames are adjusted to the moment of perihelion passage of the discovery apparition, and distinction is made between visual and photographic discoveries.

The distribution is fairly uniform in the fixed system (panel A in Fig. 10). An 80% excess against the background appears between the dashed lines marked C and J. C is the region of optimum observing conditions in the northern hemisphere, because the ecliptic has the highest northern declination at this point. Moreover, for the optimum perihelion passages near opposition the nights are longest. J denotes the perihelion longitude of

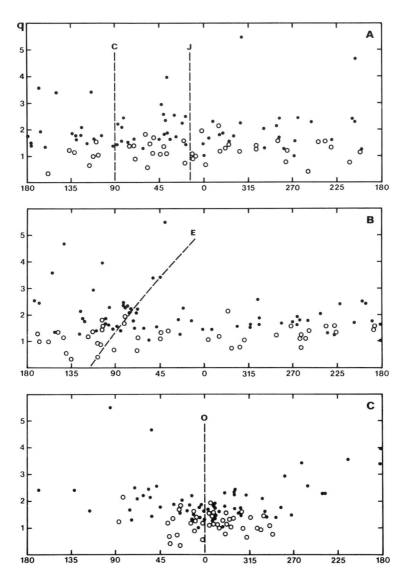

Fig. 10. The distribution of perihelia of short-period comets of the Jupiter family ($P < 20$ yr). Ordinate, perihelion distance q; abscissa, angle on the ecliptic in three reference frames. Panel A, fixed ($0°$ = vernal equinox); Panel B, revolving with Jupiter ($0°$ = longitude of Jupiter at the time of the first observed perihelion passage of the comet); Panel C, revolving with the Earth ($0°$ = longitude of the Earth at the time of the first observed perihelion passage of the comet). Open circles, visual discoveries; dots, photographic discoveries. The dashed lines indicate the expected regions of enhancement as explained in the text.

Jupiter. Owing to the alignment of the lines of apsides produced by a higher capture efficiency near the aphelion of the planet, some real concentration of perihelia towards line J should be present. Thus all the irregularities can be explained by the overlap of perturbations and observational selection. This explanation is worth noting because Rabe (1974) attributed a cosmogonical significance to this asymmetry. Interfacing the prevalence of the perihelia of short-period comets east of the perihelion of Jupiter with the greater population of the Trojan cloud around the preceding Lagrangian point (see Degewij and Van Houten 1979), he suggested a common origin for these two kinds of objects. Such an explanation is apparently unnecessary.

In the system B (panel B in Fig. 10), rotating with the motion of Jupiter, a sharp concentration appears $60°$ to $110°$ ahead of the planet. This is especially pronounced for larger perihelion distances and, hence, for photographic discoveries. The concentration is due to the fact that a significant proportion of short-period comets is found during the perihelion passage immediately after a close encounter with Jupiter. In the case of a perturbational deceleration the perihelion distance decreases, and the discovery probability increases. Moreover, a change in the radiation regime tends to make the comet's absolute magnitude brighter for the next one or two perihelion passages than for the later returns (Kresák 1973). Yet another more subtle mechanism contributes to this effect, the transition into the inner half of the horseshoe-shaped libration curve of the comets librating around the 2:1 resonance with Jupiter (Kresák 1974b) that takes place when the aphelion is not far ahead of the planet. The slanted line E indicates the predicted region of maximum concentration. This is computed under the assumption that the comet has passed the preceding aphelion at the mean heliocentric distance and mean anomaly of Jupiter. A significant concentration of the data-points towards this line is evident.

The largest nonuniformity, this time more outstanding for smaller perihelion distances and visual discoveries, is displayed in the coordinate system referred to the moving Earth (panel C in Fig. 10). A clear-cut concentration of the perihelia at the discovery apparitions towards the opposition point (line O) is due to the geometrical effect already noted by Holetschek (1890); the comets are most easily observable when passing perihelion near opposition and thus coming closest to the Earth on the nighttime side if $q > 1$. Most of the losses of short-period comets have been due to their discovery at such an apparition, followed by a number of returns in less favorable configurations during which the predictions, if any, became useless (Kresák 1981b). It can be concluded that considering the capture process by Jupiter and the effects of observational selection, no unexplained irregularities in the distribution of perihelia of short-period comets are left.

VIII. IMPLICATIONS FOR COMET SEARCHES

Recognition and evaluation of the various selection effects is also important for solving the inverse task: to specify the conditions under which comet searches become most effective. The optimization obviously depends both on the technique applied (visual/photographic) and on the dynamical type of the comets (long-period/short-period).

As shown in Fig. 2, a slight majority of long-period comets is still being discovered visually. Only $\sim 30\%$ of currently discovered long-period comets remain below the detection threshold of small comet seekers (total magnitude $M < 10$) during the whole apparition. According to Fig. 1, there are annually about six comets reaching $M = 10$, three reaching $M = 8$, and one reaching $M = 5$. The long-period comets constitute 80% at $M = 10$, 90% at $M = 5$, and as much as 97% at $M = 2$, with a 50% probability the brightest comet visible at a given moment is brighter than $M = 9$, and the brightest short-period comet brighter than $M = 12$.

It is well known among comet searchers that small wide-field low-power telescopes, preferably binoculars, are best suited for comet searches. With one new comet passing the limit of $M = 10$ every two months, it is advisable to sweep the whole sky once a month, during dark-moon periods. Experience shows that this takes 10-15 hr of net observing time with a 3°-diameter field of view. Since the time spent increases with the inverse square of the field diameter, a complete sky coverage is hardly possible with a field diameter $\leqslant 1°$. This determines the upper limit of the magnifying power, optimum near 20 and preferably below 50, which in turn sets the upper limit of the telescope aperture with an optimum at 10 to 20 cm, yielding a limiting magnitude of $M = 10$ to 12 for diffuse cometary images, and maximum at ~ 30 cm, yielding a limiting magnitude of $M \simeq 13$. Since the number of accessible comets increases approximately with the telescope aperture and the time consumption with its square, the effectiveness of the searches would decrease to one half with each doubling of the aperture. On the other hand, each reduction of the aperture would increase the proportion of those comets which have already been detected. With decreasing magnification, it would also make the distinction between the comets and other objects more difficult. Thus the choice of instrumentation by the most successful comet hunters, beginning with Pons, was indeed the best possible. It resulted in the distribution of the discovery magnitudes shown in Fig. 3, with the present maximum between $M = 8$ and 10 barely one magnitude fainter than a century ago. An improvement would require organized effort, with different areas of the sky surveyed simultaneously by a number of observers.

If larger apertures and magnifications are used, it is advisable to restrict the searches to those regions of the sky where the discovery probability is highest. The same holds for choosing the priorities, say, after a full-moon period or after a spell of bad weather. It has been known for a long time among comet searchers that such regions do exist: the western evening sky and the eastern morning sky. The latter is better, with the discovery probability higher by a factor of two, according to Everhart (1967a).

This asymmetry is explained by Fig. 11 showing the loci of constant difference between the apparent and absolute magnitude, relative to the Earth's orbit and the Sun-Earth line. Brightness variations with the inverse 4th power of the heliocentric distance and with the inverse square of the geocentric distance are assumed, with due corrections for scattered sunlight and atmospheric extinction. The fifth curve from the outside corresponds to $M = H_0$ and each step outward corresponds to a change of $M - H_0$ by +1.5 mag. Interpreting the curves as magnitude thresholds, each step outward is reached by increasing the telescope aperture by a factor of two. Since the H_0-distribution of the long-period comets currently peaks at $H_0 = 7$, the third contour from the outside is characteristic for visual discoveries at $M = 10$, and the outermost contour for $M = 13$. This is why most of the visual discoveries occur at $r < 2$, $\Delta < 1.5$ AU. A three-dimensional pattern of the envelopes can be obtained by a simple rotation of the contours around the Sun-Earth line.

The position angles of the lobes set the optimum conditions for comet discoveries, $M < 10$, at the elongations 30° to 35° from the Sun. Here the depth of the envelope is 2.6 times larger than in the direction of the opposition. The actual ratio of the discovery probabilities is still greater than this, as the region of the lobes can be entered by comets that have increased their brightness at small solar elongations. This state is reflected by the actual distribution of the discovery positions (see Fig. 3 of Everhart 1967a).

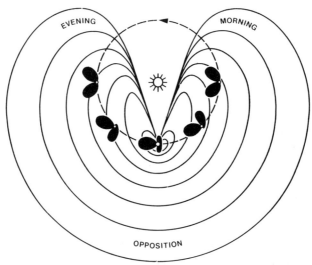

Fig. 11. The geometrical selection affecting comet discoveries. Dashed circle, heliocentric motion of the Earth, with its position marked by the open circle down from the Sun; curves, loci of constant difference between the apparent and absolute magnitude, with the effects of background illumination and extinction subtracted; dark patterns, direction distributions of the geocentric velocity vectors. For detailed explanation see the text.

The evening/morning asymmetry is a result of the orbital motion of the Earth, analogous to the diurnal variation of meteor rates. Assuming that the frequency of perihelion distances changes as $q^{1/2}$ at $q < 1$, the direction distributions of the geocentric motions of comets can be constructed as in Fig. 11 for five selected positions. The polar diagrams centered at each of these positions indicate the proportions of different position angles of geocentric motions within the plane of the ecliptic. In the elongation 45° East (evening sky) prograde comets ($i = 0°$) would generally move away from the Sun and retrograde comets ($i = 180°$) away from the Earth. In the elongation 45° West (morning sky) prograde comets would move towards the Sun and retrograde comets towards the Earth. It is especially the retrograde comets approaching the Earth which make comet discoveries more probable in the morning sky. While the addition of the orbits more perpendicular to the ecliptic makes the difference less pronounced, the general trend remains. Meteor astronomers will notice a striking resemblance of the mushroom-like pattern at small geocentric distances to the distribution of sporadic meteor radiants, with the apex, helion, and antihelion sources. The least promising region is at moderate distances west of the opposition, where many comets would arrive through the opposition area, best covered by photographic observations.

For photographic discoveries of long-period comets the conditions are different. At their mean discovery magnitude $M \sim 15$ the lobes in Fig. 11 are quite inconspicuous. While the depth to the surface of $M = 6$ (for $H_0 = 7$) is more than three times larger at the solar elongation of 30° than at the opposition, the surface of $M = 15$ (for $H_0 = 7$) is much more spherical, with a maximum excess of some 30% at the elongation of 45°. The effect of extinction is already included, but the brighter sky background would make searches closer to the Sun less effective by reducing the usable exposure time. Again, some preference for the hemisphere east of the opposition is produced by the motion of the Earth. In general, however, the discovery probability is rather irrelevant to the position of the area photographed, depending primarily upon the field size and limiting magnitude. Every large-scale photographic program using wide-field instruments, in particular Schmidt or Maksutov cameras, provides an opportunity for detecting comets. Careful and early inspection of such plates, as has been done by Chernykh, Gehrels, Kohoutek, Kowal, Lovas, Russell, and Wild, should be greatly encouraged, since this is the only source of information on comets of large perihelion distance.

Figure 10 shows the special effects operating at the discoveries of short-period comets. The three frames of reference are moving with respect to one another with periods of 1 and 12 yr, producing periodic changes in the discovery conditions. The fact that a considerable proportion of short-period comets have experienced close encounters with Jupiter shortly before their discovery makes it advisable to search those regions where these comets should appear, especially when they are favorably placed with respect to the Earth. The time lapse between the encounter and the next perihelion passage will be normally 3 to 4 yr, during which Jupiter will proceed 90° to 120° in longitude. Since the perturbations are most effective when the planet is at aphelion, the chances for a discovery will to some extent depend on its position in the orbit. They will be highest 2 to 3 yr before Jupiter's perihelion passage, i.e., when Jupiter enters the opposition in or around July. For the northern observers, northern declination of the ecliptic area to be searched will be advantageous, corresponding to Jupiter's oppositions around September and perihelion passages of the comets around December. For the southern observers, these periods would be shifted by half a year.

Keeping all these effects in mind, the optimum strategy can be outlined as follows: (1) give preference to the periods beginning with the time of Jupiter's opposition and ending three months later (e.g., May through July in 1982); (2) search first in the area around the Earth's apex and proceed westwards, ending in the opposition area three months later. For northern observers, the three months following Jupiter's opposition falling in July to September are most favorable, conditions which will next occur around 1985. For southern observers the long-term variations of the discovery probability are smaller, with some preference for the three months following Jupiter's oppositions in April to June, as around 1982. Surveys organized according to these guidelines can increase the reward of the searches for short-period comets, in particular those comets newly captured into the inner planetary system, and provide better insight into their dynamical history.

Acknowledgments. The author is indebted to the reviewers of this chapter, T. Gehrels and B. Marsden, for their most helpful commets.

REFERENCES

Baldet, F. 1949. *Liste générale des Comètes de l'origine a 1948* (Paris: Gauthier-Villars).
Bortle, J.E. 1981. Comets and how to hunt them. *Sky Tel.* 61:123-125.
Bourgeois, P., and Cox, J.F. 1932. Calcul de l'ellipsoide de dispersion des périhélies des comètes connues. *Bull. Astron.* 8:271-304.
Bourgeois, P., and Cox, J.F. 1933. Recherches sur la probabilité de la découverte des comètes et la répartition de leurs orbites. *Bull. Astron.* 9:349-446.
Degewij, J., and Van Houten, C.J. 1979. Distant asteroids and outer Jovian satellites. In *Asteroids*, ed. T. Gehrels (Tucson: Univ. of Arizona Press), pp. 417-435.
Delsemme, A.H. 1977. The origin of comets. In *Comets, Asteroids, Meteorites,* ed. A.H. Delsemme (Toledo, Ohio: Univ. of Toledo), pp. 453-467.
Donn, B. 1977. A comparison of the composition of new and evolved coments. In *Comets, Asteroids, Meteorites,* ed. A.H. Delsemme (Toledo, Ohio: Univ. of Toledo), pp. 15-23.
Everhart, E. 1967a. Comet discoveries and observational selection. *Astron. J.* 72:716-726.
Everhart, E. 1967b. Intrinsic distributions of cometary perihelia and magnitudes. *Astron. J.* 72:1002-1011.
Everhart, E. 1973. Examination of several ideas of comet origins. *Astron. J.* 78:329-337.
Hasegawa, I. 1976. Distribution of the aphelia of long-period comets. *Publ. Astron. Soc. Japan* 28:259-276.

Hasegawa, I. 1979. Orbits of ancient and medieval comets. *Publ. Astron. Soc. Japan* 31:257-270.
Ho Peng Yoke 1962. Ancient and medieval observations of comets and novae in Chinese sources. *Vistas Astron.* 5:127-225.
Hoek, M. 1866. Additions to the investigations on cometary systems. *Mon. Not. Roy. Astron. Soc.* 26:204-207.
Holetschek, J. 1890. Über den scheinbaren Zusammenhang der heliozentrischen Perihellänge mit der Perihelzeit der Kometen. *Astron. Nachr.* 126:75-80.
Kazimirchak-Polonskaya, E.I. 1972. The major planets as powerful transformers of cometary orbits. In *The Motion, Evolution of Orbits and Origin of Comets*, eds. G.A. Chebotarev, E.I. Kazimirchak-Polonskaya and B.G. Marsden (Dordrecht: D. Reidel Publ. Co.), pp. 373-397.
Kowal, C.T. 1979. Chiron. In *Asteroids*, ed. T. Gehrels (Tucson: Univ. of Arizona Press), pp. 436-439.
Kramer, E.N.; Musij, V.I., and Shestaka, I.S. 1979. On one dynamic characteristic of comet orbits. *Astron. Vest.* 13:42-49. (In Russian)
Kresák, Ĺ. 1966. Wie entdeckt man Kometen: Eine Anleitung. *Orion* 11:161-166.
Kresák, Ĺ. 1973. Short-period comets at large heliocentric distances. *Bull. Astron. Inst. Czech.* 24:264-283.
Kresák, Ĺ. 1974a. The aging and the brightness decrease of comets. *Bull. Astron. Inst. Czech.* 25:87-112.
Kresák, Ĺ. 1974b. The effects of resonance with Jupiter in the system of short-period comets. In *Asteroids, Comets, Meteoric Matter*, eds. C. Cristescu, W.J. Klepczynski and B. Milet (Bucharest: Editura Academiei RSR), pp. 193-203.
Kresák, Ĺ. 1975. The bias of the distribution of cometary orbits by observational selection. *Bull. Astron. Inst. Czech.* 26:92-111.
Kresák, Ĺ. 1977. On the differences between the new and old comets. *Bull. Astron. Inst. Czech.* 28:346-355.
Kresák, Ĺ. 1978. The comet and asteroid population of the Earth's environment. *Bull. Astron. Inst. Czech.* 29:114-125.
Kresák, Ĺ. 1979. Three-dimensional distributions of minor planets and comets. In *Dynamics of the Solar System*, ed. R.L. Duncombe (Dordrecht, D. Reidel Publ. Co.), pp. 239-244.
Kresák, Ĺ. 1981a. A strange anomaly in the occurrence rate of old comets. *Bull. Astron. Inst. Czech.* 32. In press.
Kresák, Ĺ. 1981b. The lifetimes and disappearance of periodic comets. *Bull. Astron. Inst. Czech.* 32. In press.
Kresák, Ĺ. 1981c. On the reality and genetic association of comet groups and pairs. *Bull. Astron. Inst. Czech.* In press.
Kresák, Ĺ., and Pittich, E.M. 1978. The intrinsic number density of active long-period comets in the inner solar system. *Bull. Astron. Inst. Czech.* 29:299-309.
Lindblad, B.A., and Southworth, R.B. 1971. A study of asteroid families and streams by computer techniques. In *Physical Studies of Minor Planets*, ed. T. Gehrels (Washington: NASA SP-267), pp. 337-352.
Marsden, B.G. 1977. Orbital data on the existence of Oort's cloud of comets. In *Comets, Asteroids, Meteorites*, ed. A.H. Delsemme (Toledo, Ohio: Univ. of Toledo), pp. 79-86.
Marsden, B.G. 1979. *Catalogue of Cometary Orbits* (Cambridge, Mass.: IAU Central Bureau for Astron. Telegrams).
Marsden, B.G., and Sekanina, Z. 1973. On the distribution of "original" orbits of comets of large perihelion distance. *Astron. J.* 78:1118-1124.
Marsden, B.G.; Sekanina, Z.; and Everhart, E. 1978. New osculating orbits for 110 comets and analysis of original orbits for 200 comets. *Astron. J.* 83:64-71.
Oort, J.H. 1963. Empirical data on the origin of comets. In *The Moon, Meteorites and Comets*, eds. B.M. Middlehurst and G.P. Kuiper (Chicago: Univ. of Chicago Press), pp. 665-673.
Öpik, E.J. 1966. Sun-grazing comets and tidal disruption. *Irish Astron. J.* 7:141-161.
Öpik, E.J. 1971. Comet families and transneptunian planets. *Irish Astron. J.* 10:35-91.

Öpik, E.J. 1973. Comets and the formation of planets. *Astrophys. Space Sci.* 21:307-398.
Oppenheim, S. 1924. Zur Statistik der Kometen und Planeten in Zusammenhang mit der Verteilung der Sterne. In *Probleme der Astronomie,* ed. H. Kienle (Berlin: J. Springer Verlag), pp. 131-143.
Pingré, A.G. 1783. *Comètographie ou Traité historique et théorique des comètes* (Paris: Imprimerie Royale), Tome 1.
Pittich, E.M. 1969a. The selection effects in the discoveries of new comets. *Bull. Astron. Inst. Czech.* 20:85-95.
Pittich, E.M. 1969b. Sudden changes in the brightness of comets before their discovery. *Bull. Astron. Inst. Czech.* 20:251-293.
Proctor, M., and Crommelin, A.C.D. 1937. *Comets* (London: Technical Press Ltd.)
Rabe, E. 1974. Perihelion longitudes and Jacobi constants of Jupiter group comets as indicators of their possible origin from the two equilateral Trojan clouds. In *Asteroids, Comets, Meteoric Matter,* eds. C. Cristescu, W.J. Klepczynski and B. Milet (Bucharest: Editura Academiei RSR), pp. 165-169.
Revina, I.A. 1977. Influence of the selection of discovery in the distribution of comets. *Astronomiya* 12:63-75. (In Russian)
Roemer, E. 1963. Comets: Discovery, orbits, astrometric observations. In *The Moon, Meteorites and Comets,* eds. B.M. Middlehurst and G.P. Kuiper (Chicago: Univ. of Chicago Press), pp. 527-549.
Shteins, K.A., and Revina, I.A. 1977. Selection of discoveries in the concentration of perihelia of cometary orbits. *Astron. Vestn.* 11:67-68. (In Russian)
Southworth, R.B., and Hawkins, G.S. 1963. Statistics of meteor streams. *Smithson. Contr. Astrophys.* 7:261-285.
Tomanov, V.P. 1973. New statistical regularities in the system of long-period comets. *Astron. Vest. 7:83-87. (In Russian)*
Tomanov, V.P. 1976. Checking the hypothesis of interstellar origin of comets. *Astron. Vestn.* 10:44-49. (In Russian)
Tomanov, V.P. 1977. On the asymmetry in the perihelia distribution of cometary orbits. *Astron. Zh.* 54:1346-1348. (In Russian)
Van Flandern, T.C. 1977a. The asteroidal planet as the origin of comets. In *Dynamics of Planets and Satellites and Theories of their Motion,* ed. V.G. Szebehely (Dordrecht: D. Reidel Publ. Co.), pp. 89-99.
Van Flandern, T.C. 1977b. A former major planet of the solar system. In *Comets, Asteroids, Meteorities,* ed. A.H. Delsemme (Toledo, Ohio: Univ. of Toledo), pp. 475-481.
Vsekhsvyatskij, S.K. 1958. *Physical Characteristics of Comets* (Moskva: Gos. izdatelstvo fiz.-mat. literatury). (In Russian)
Weissman, P.R. 1977. Initial energy and perihelion distributions of Oort-cloud comets. In *Comets, Asteroids, Meteorites,* ed. A.H. Delsemme (Toledo, Ohio: Univ. of Toledo), pp. 87-91.
Whipple, F.L. 1975. Do comets play a role in galactic chemistry and γ-ray bursts? *Astron. J.* 80:525-531.
Whipple, F.L. 1977. The reality of comet groups and pairs. *Icarus* 30:736-746.
Whipple, F.L. 1978. Comets. In *Cosmic Dust,* ed. J.A.M. McDonnell (Chichester: J. Wiley), pp. 1-73.
Witkowski, J.M. 1972. On the problem of the origin of comets. In *The Motion, Evolution of Orbits and Origin of Comets,* eds. G.A. Chebotarev, E.I. Kazimirchak-Polonskaya and B.G. Marsden (Dordrecht: D. Reidel Publ. Co.), pp. 419-425.
Yabushita, S. 1972. The dependence on inclination of the planetary perturbations of the orbits of long-period comets. *Astron. Astrophys.* 20:205-214.
Yabushita, S.; Hasegawa, I., and Kobayashi, K. 1979. The distributions of inclination and perihelion latitude of long-period comets and their dynamical implications. *Publ. Astron. Soc. Japan* 31:801-813.

PART II
Nucleus

CHEMICAL COMPOSITION OF COMETARY NUCLEI

A. H. DELSEMME
The University of Toledo

Cometary stuff appears to be made up of an icy conglomerate, with a volatile and refractory fraction. Circumstantial arguments will be shown to be in favor of the undifferentiated character of this icy conglomerate, at least before it starts vaporizing in the solar heat. The first step towards a quantitative understanding of this conglomerate is the assessment of its refractory-to-volatile mass ratio; this is also the dust-to-gas mass ratio of the dusty gas that sublimes from the nucleus, if the latter is undifferentiated throughout, and only when a steady state is reached. It will be shown that this happens, although not often, since an outgassed layer of dust may set in and be blown away repeatedly. The sublimation of the nucleus and the velocity field of the expanding gas are first described theoretically and then confronted with observations. Later, the nature of the cometary dust and the atomic abundances of H, C, N, O, S in the volatile fraction are discussed, and the two fractions are put together to derive elemental abundances. Although there are still many uncertainties, one fact emerges, namely that there is a drastic depletion of hydrogen in the cometary nucleus (probably by three orders of magnitude) with respect to the cosmic abundances whereas oxygen, and possibly nitrogen and sulfur, seem to have essentially cosmic abundances. The apparent depletion of carbon by a factor of more than three is discussed. The missing carbon could be hidden in the dust fraction in the form of heavy organic molecules, but it also may have remained in the primeval solar nebula or in interstellar space.

The physical study of comets has traditionally been more concentrated on a qualitative understanding of the transient phenomena (coma, dust tail, and plasma tail) than on a more quantitative understanding of the underlying permanent features (structure and chemistry of the nucleus). However, the last decades have brought a harvest of quantitative data that can be used as clues for a more fundamental approach to the chemical nature of the nucleus, yielding new insights to its origin and history.

Observations make it clear that the cometary stuff is a mixture of two constituents with very different properties: a volatile fraction, apparently a mixture of molecules from H, C, N, O, S atoms, and a refractory fraction containing mainly fine dust grains, although a small amount of larger grains is sometimes detectable (Sekanina 1974), and the existence of boulders cannot be excluded from the present evidence.

In the chapter by Donn and Rahe the physical structure of the nucleus is reviewed. They confirm the existence of a large consensus that the two constituents mentioned above are present in the nucleus as a single aggregate of ices and meteoritic matter whose cohesive strength is probably very weak but by no means nil. This basic icy conglomerate model was introduced by Whipple (1950,1951). Donn and Rahe also mention briefly the evidence for radial uniformity. Since this evidence is at the core of our argument, it will be discussed in Sec. I in detail. The first step towards a quantitative understanding of the icy conglomerate is the assessment of its refractory-to-volatile mass ratio; in a steady state, this is also the dust-to-gas mass ratio of the dusty gas that sublimes from the nucleus. Since gas drags the dust away, the key to the dust-to-gas ratio is the dust-to-gas velocity ratio established by fluid dynamics (see Sec. V). However, in order to understand the boundary conditions of this fluid dynamics problem, the sublimation of the nucleus must be understood (Sec. II) and the predicted expansion velocities in the coma must be checked with observations (Sec. III). Finally, the nature of cometary dust must be discussed, in order to understand its physical properties: density, albedo, complex index of refraction (Sec. IV). At that time, we shall be ready to discuss the dust-to-gas ratio in Sec. V. Section VI will then deal with the atomic abundances of H, C, N, O, S in the volatile fraction and Sec. VII will conclude with what can be said about atomic and molecular abundances in comets.

I. UNDIFFERENTIATED CHARACTER OF PRISTINE NUCLEI

A. Observational Evidence

No direct information has ever been collected on the internal structure of a cometary nucleus. However, several circumstantial arguments all seem in favor of undifferentiated nuclei (Delsemme 1981b). The basic idea is that, if cometary nuclei were radially differentiated, the decay of their outer layers with aging or the fragmentation of their nucleus by splitting would sooner or later produce observable changes. Nothing of this sort has ever been detected. The following arguments favor the undifferentiated nucleus:

CHEMICAL COMPOSITION OF NUCLEI 87

1. The dust-to-gas ratio of comets does not seem to be influenced by aging. There is a large variation from comet to comet. Some comets appear dusty, some appear almost purely gaseous. Measuring this dust-to-gas ratio is difficult (see the Finson-Probstein method, Sec. V.B) but it is possible to make assessments by estimating in the spectra the intensity ratio of the light reflected by dust (spectral continuum) to that of the molecular emissions. Comparing 85 comets, Donn (1977) concludes that there is no readily apparent difference in the distribution of this ratio between less and more evolved comets.
2. The differences in the molecular emissions detected in cometary spectra do not seem to have anything to do with aging. For instance, A'Hearn and Millis (1980) show that the CN/C_2 ratio follows a dependence law on distance remarkably similar from comet to comet, whatever their age, and Feldman (1981) finds that long-period comets and old short-period comets have a remarkably similar spectrum in the vacuum ultraviolet.
3. Long-period comets or short-period comets fragment at the same rate, $\sim 3\%$ per passage (Pittich 1971; Kresák 1981; Stefanik 1966).
4. The nongravitational acceleration (due to the jet effect of the vaporizing gases) is inversely proportional to the fragments' lifetimes after splitting (Sekanina 1977,1981), i.e., apparently to the fragments' masses.

Taking these observations together, we can infer that cometary material that originated at different depths in the pristine nucleus, shows the same properties, namely: (1) the same dust-to-gas distribution pattern; (2) the same spectral composition for the volatile fraction; (3) the same structural strength against fragmentation; and (4) the same vaporization pattern after fragmentation.

Of course, this implies neither the absence of differences among comets, not the absence of large surface differentiation effects at different times in their aging process. All that is claimed is that the patterns in the observed differences do not change with aging, and minor heterogeneities may or may not be observable. For instance, minor surface differentiation effects could have been induced by galactic cosmic rays in the outermost cm of the nuclear surface of new comets (Shul'man 1972; Donn 1976; Whipple 1977). This phenomenon may be partly responsible for the statistical differences in the brightness dependence on distance of "new" versus old comets (Oort and Schmidt 1951) since it can be assumed that older comets have already lost their outer layers.

The decay of the icy conglomerate in the solar heat corresponds to a loss of a layer of one to several m per passage (Whipple 1950). It is likely that a deep fractionation of the outer layers takes place and that an outgassed mantle of dust is periodically blown away by outbursts (Mendis and Brin 1977,1978; Brin and Mendis 1979; Brin 1980). This may explain the extended range in the dust-to-gas ratio mentioned by Donn, but would not explain why the pattern of its distribution among comets does not change with aging, unless the inside of the nucleus is undifferentiated throughout.

Other phenomena may be linked to cometary decay. For instance, crust forming is likely by agglomeration of outgassed dust and contact welding of solids in vacuum. Later, an irregular fragmentation of the crust—to be associated with cometary orbits suggest pristine material with elements in solar proportions (Millman 1977), whereas there is no reason to associate any origin of these large and fragile meteoroids associated with cometary orbits that give rise to spectacular fireballs (Wetherill 1974). All meteorites associated with cometary orbits suggest pristine material with elements in solar proportions (Millman 1977), whereas there is no reason to associate any other differentiated meteorite with comets. Even if no observational argument seems to contradict the undifferentiated nature of pristine comets, it could be argued that the four basic patterns described above could be simulated by a yet not understood surface process during cometary decay. Nevertheless this simply is not true. If truly differentiated layers were reached during either decay or splitting of the nucleus, not only the dust-to-gas distribution pattern and the nature of volatile material would vary with depth, but the brittleness of the nucleus would vary and the fragments would show volatility differences in the core. The arguments are qualitatively different and no known surface process could simulate all of them.

The conclusion drawn from this section is that the undifferentiated nature of new comets seems a reasonable hypothesis which will be adopted from now on.

B. Recent Speculations

In spite of the previous observational evidence, the undifferentiated character of pristine cometary nuclei has been recently questioned (Irvine et al. 1981; Wallis 1980) in connection with a conjecture on the origin of life (Hoyle and Wickramasinghe 1979). The only connection between these speculations and observation lies in the anomalous abundance of ^{26}Mg in some chondrules of the Allende meteorite (Lee and Papanastassiou 1974; Papanastassiou et al. 1977). It is true that this isotopic anomaly has suggested the possible presence of ^{26}Al in early solar system condensates in sufficient abundance to melt the interiors of minor bodies as small as comets. Since the ^{26}Mg anomaly correlates with the abundance of ordinary aluminum in Ca-rich chondrules of Allende, the decay of ^{26}Al must have happened *after* its imprisonment in the chondrule. The fast decay of ^{26}Al ($\tau = 0.7 \times 10^6$ yr) implies therefore that the chondrules condensed very soon after the ^{26}Al-producing supernova event. However, this does not imply the presence of Al-bearing chondrules in comets. After all, the CI chondrites—which seem to be the most pristine meteorites and the closest to cometary material (Delsemme 1975a)—do not contain chondrules. In the present stage of our knowledge, it is therefore wiser to ignore these speculations and believe in the observational evidence developed earlier.

II. THE VAPORIZATION OF THE NUCLEUS

Even if we accept the undifferentiated character of the pristine nucleus, a fractional vaporization will differentiate its outer layers as soon as the nucleus is heated during its first approach to the Sun. It is therefore essential to discuss and understand first the vaporization theory of the nucleus.

A. Vaporization Theory

From now on, we will use vaporization as the generic term for the change of state, describing as well a sublimation (from solid to gas) as an evaporation (slow) or ebullition (fast) from liquid to gas. In the same way, we will use ice or snow as the generic terms for the compact or fluffy solid state of any compound at least as volatile as water (like CH_4 snows, or CO_2 ice).

When a comet approaches the Sun, its heliocentric distance varies first very slowly and we can assume that its nucleus will reach a surface temperature that represents a slowly shifting steady state. If we assume first a slowly rotating nucleus and if we neglect conduction, the temperature distribution on the sunlit face can be expressed by stating that the fraction of the solar flux being absorbed will be balanced by the sum of two terms, namely the latent heat used to transform the vaporizing ice into gas, and the energy reradiated back to space in the infrared, because of the surface temperature of the nucleus. For a surface area at an angle θ with the solar flux, this steady state can be written

$$F_0(1 - A_0)r^{-2} \cos\theta = \sigma(1 - A_1)T^4 + Z(T)L(T) \tag{1}$$

where F_0 is the solar flux at 1 AU, A_0 and A_1 are the nucleus albedos respectively in the visible and in the infrared between 10 and 20 μm, $(1 - A_1)$ is the emissivity for reradiation, r is the heliocentric distance, σ is Stefan's law constant, T the surface temperature, Z the production rate of gas in molecules cm^{-2}s^{-1} (Z varies very fast with T), and L is the latent heat per molecule for the vaporization of the volatile snows. L is also often (slowly) variable with temperature, but it can usually be found in thermodynamic tables. Z is more difficult to find in tables, because it expresses the vaporization rate of a snow by molecular effusion into the vacuum of space. However, Delsemme and Swings (1952) have shown that it can be deduced from the known vapor pressure p of the subliming snow, at the steady-state temperature T, by using the kinetic theory of gases. Therefore

$$Z = p(2\pi m k T)^{-1/2} \tag{2}$$

where m is the molecular mass of the gas and k is Boltzmann's constant. This approach was used later by Weigert (1959), Watson et al. (1961), Huebner (1965), Delsemme (1966), and Delsemme and Miller (1971). The latter reference gives a lengthy discussion of this vaporization theory. It notes for instance that the latent heat L found in most of the standard thermodynamical tables must be corrected for vacuum, by subtracting the gas expansion work RT from the value of L measured at equilibrium between solid and gas.

Solving Eqs. (1) and (2) by successive approximations gives Z and T for a snow of a specified nature. It shows that the radiative term $\sigma(1-A)T^4$ is negligible in Eq. (1) for most snows at heliocentric distances equal to or smaller than 1 AU.

Table I gives the production rate Z_0 and the temperature T_0 of the subsolar point of a perfectly absorbing nucleus at 1 AU from the Sun; T_1 is the effective temperature of a rotating nucleus ($Z_1 = Z_0/4$ since the total area of the sphere is 4 times its cross section). Taken from Delsemme and Miller (1971), Table I has been completed for a few more molecules for this review. It is important to note that Table I is given for a perfectly absorbing nucleus ($A_0 = 0$); the production rates could easily fall off by a factor of 2 or 3, and the temperatures could be lower by 5 to 10°, for very light objects with A reaching 0.5 or 0.7. It is also conceivable that production rates and temperatures could be shifted by a same amount in the opposite direction, if the

TABLE I

Production Rates for Different Snows of the Cometary Nucleus

Snows Controlling the Vaporizations	Z_0 in 10^{18} mol cm^{-2} s^{-1} [a]	T_0 [b] K	T_1 [c] K	r_0 [d] AU
Nitrogen	14.3	40	35	77.6
Carbon monoxide	13.0	44	39	62.5
Methane	10.6	55	50	38.0
Formaldehyde	5.0	90	82	14.1
Ammonia	3.7	112	99	9.7
Carbon dioxide	3.5	121	107	8.3
Hydrogen cyanide	2.3	160	140	4.8
Gases from ammonia water (aqueous ammonia)	2.7	213	193	2.6
Gases from methane clathrate (same results for all clathrates)	1.9	214	194	2.5
Water	1.7	215	195	2.5

[a] Z_0 is the vaporization rate at subsolar point of a perfectly absorbing nucleus at 1 AU from the Sun.
[b] T_0 is the steady-state temperature of Z_0 (subsolar point, nonrotating nucleus).
[c] T_1 is the effective mean temperature of a rotating nucleus ($Z_1 = 1/4\ Z_0$).
[d] r_0 is the heliocentric distance beyond which the vaporization rate becomes negligible ($\leqslant 2.5\%$ of the solar flux is used for vaporization, $\geqslant 97.5\%$ for reradiation).

nucleus were surrounded by an optically thick haze that absorbs solar light. No complete model of the radiative transfer has been developed for such a case (see, however, Hellmich and Keller 1980), whereas recent models include the effect of diurnal heating and cooling, rotation period and pole orientation, and thermal properties of subsurface layers (Weissman and Kieffer 1981).

B. Confrontation with Empirical Production Rates

The vaporization theory has been successful in predicting the order of magnitude of the total production rate of a few comets, mainly in cases where water controls the vaporization of the nucleus. For instance, the estimates given by Delsemme (1966) are consistent with the production rates of H and OH that were measured later, on the bright comets of the 1970s. However, only orders of magnitude can be obtained, since the total surface area of the nucleus is not known. By photographic photometry (Roemer 1966) only the product $A_0 R^2$ can be deduced, R being the effective radius of the nucleus, i.e., the radius of a sphere of the same cross-section as the (irregular?) nucleus.

Delsemme and Rud (1973) have deduced $A_0 R^2$ from photometry and $(1 - A_0) R^2$ from the vaporization theory applied to the observed production rates of OH and/or H from three comets. Assuming that water is the most volatile major constituent and therefore controls the vaporization rate and temperature, they found consistent results for two comets (see Table II). They interpret the inconsistent results found for the third comet (P/Encke) by the fact that it is one of the oldest short-period comets (whereas the previous two are less evolved comets). The results could be made consistent with empirical data only if $R(\text{ice}) \cong 1/4\, R(\text{nucleus})$, i.e., if the exposed water ice covers only a small fraction of the nucleus of P/Encke. Typically, 90 to 95% of its ices should be hidden by an insulating mantle of dust and the reradiative term would accordingly be larger (effective T larger and more variable with distance). It remains unclear whether the very high albedo found for the two other comets is significant (it is not too high for a moderately clean snow) or whether it should be interpreted as coming from an

TABLE II

Albedo and Nuclear Radius for Two Comets

Comet	Bond Albedo A	Nuclear Radius R
Tago Sato Kosaka	0.63 ± 0.13	2.20 ± 0.27 km
Bennett	0.66 ± 0.13	3.76 ± 0.46 km

extended bright haze with a cross-section larger than the nucleus, in which case the real albedo of the nucleus should be lower and its radius accordingly higher.

C. Dependence of Production Rates on Heliocentric Distance

The results of Eq. (1) integrated over the total area of a spherical nucleus, as a function of the heliocentric distance, have been expressed in Fig. 1 for a sublimation controlled by different snows. For the sake of clarity, let us

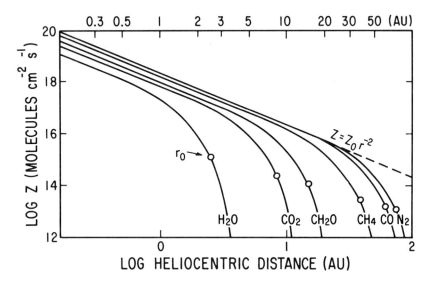

Fig. 1. Vaporization rate, in molecules cm^{-2} s^{-1}, for various snows as a function of the heliocentric distance and assuming a steady-state rotating cometary nucleus, with $A_0 = 2A_1$ and where r_0 is the distance defined in Table I. Variation of the albedos' ratio shifts the curves slightly (see Fig. 2). (From Delsemme 1966 or Delsemme and Miller 1971 completed for large distances.)

specify that a sublimation is controlled by a given snow, when this snow is the most volatile component available, and when its abundance is large enough to stabilize the vaporization temperature (by the effect of its latent heat) in Eq. (1). The curves in Fig. 1 remain almost the same and at the same position whatever the albedo, provided that $A_0 = A_1$ (same albedo in the visible and in the infrared). A difference between A_0 and A_1 keeps the general shape of the curve but shifts its position in the r direction. The largest possible difference between A_0 and A_1 is almost certainly less than a factor of 7 (for instance, $0.1 \leq A_0 \leq 0.7$ and $0.1 \leq A_1 \leq 0.7$). Therefore the curves cannot be shifted horizontally to the left or to the right, by more than ± 0.2 in log r (Fig. 2). This dependence law can be directly compared with the total brightness law of moderately dust-free comets for the following reason; their

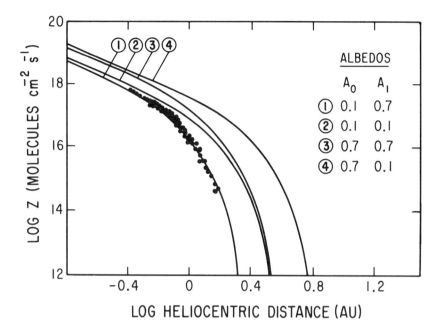

Fig. 2. Lightcurve of P/Encke, fitted to the vaporization curve predicted for a rotating nucleus vaporizing water, assuming the total brightness is in proportion to the production rate of water molecules. The best fit is for $A_0 = 0.7$ and $A_1 = 0.1$, shifting the average curve of Fig. 1 sunwards to $r_0 = 1.5$ AU. None of the other molecules represented in Fig. 1 could be shifted enough without using completely unrealistic values of the albedos. Typically, the albedos for CO_2 that would fit the lightcurve of P/Encke would be $A_0 \geqslant 0.99$ and $A_1 \leqslant 0.1$.

production rate of gas is exactly proportional to their brightness, because the total number of fluorescence cycles per molecule is independent of the heliocentric distance (Levin 1943, 1948). The reason is that the solar flux varies as an inverse-square law, whereas the lifetime of a molecule against photo decay varies as the square of the distance. Of course, if dust were dragged away exactly in proportion to gas, the law could also be checked for dusty comets; however, the presence of dust seems often to be connected with unsteady phenomena; outbursts make the steady state more difficult to observe over a long period.

Periodic comet Encke is one of the best known comets whose spectra show an extremely weak continuum, confirming the practical absence of a dust contribution to its lightcurve. Besides, it has one of the steadiest lightcurves; thanks to its having the shortest period known, it has been observed repeatedly at different passages. Using Beyer's (1950a,b, 1955, 1962) observations, the lightcurves of the different passages of P/Encke are completely identical in shape. They become indistinguishable (Delsemme 1975b) if they are slightly shifted upwards by an average of 0.25 mag per passage. Whether this dimming comes from the decay of P/Encke, or is to be attributed to the downgrading of the sky transparency in the surroundings of Hamburg, Germany, where Beyer observes, is irrelevant for our argument. (Those who are interested in the dimming may look into the diaphragm correction due to the sky brightness, for extended objects; see Delsemme 1973.) The previous results have been rather well confirmed and extended to the 1970 passage by Ferrin and Naranjo (1980) although their model remains simplistic. The high quality and homogeneity of Beyer's work is readily apparent in Fig. 2, where the lightcurves of the three passages have been superimposed. One can even be more convinced by looking at his original work covering his lifetime observations of more than 100 comets, published in 50-odd papers starting in 1921 mainly in Astronomischen Nachrichten. One then sees that outbursts are almost the rule ($>$ 75% of the cases), and that the quiet behavior of P/Encke is really one of a few exceptions (probably due to its old age). Whatever the reason, it is clear that the shape of the lightcurve fits well with the dependence on distance predicted for water ice by the vaporization theory. However, the large sunwards shift of the curve suggests either an uncomfortable difference between the visual and infrared albedo, or rather (see Sec. II.B) that the ground of P/Encke is no longer uniformly covered by water ice, explaining why the radiative term in Eq. (1) corresponds to a mean effective temperature somewhat larger than that of subliming ice.

Using the same technique, Naranjo (1978) has shown that the lightcurves of eleven different comets fit rather well with the vaporization of water. Three of his lightcurves do not cover a large enough range to be conclusive, but the other eight are rather convincing.

D. Dependence on Distance of the Nongravitational Force

Since the nongravitational acceleration stems from a jet effect coming from the asymmetry of the vaporization, its dependence on heliocentric distance should be the same as that of the vaporization rate, at least inasmuch as the anisotropy factor—also called directional factor (Levin 1948,1972)—remains more or less the same for all distances. Parchman (1972) shows that it grows slightly at large distances because the vaporizations concentrate near the subsolar point.

The empirical dependence on distance of the nongravitational force had been established by Marsden (1969), for P/Schwassmann-Wachmann 2, by trying nine different functions. The smallest mean residuals between ephemeris and observation were given by functions 6 and 7. Delsemme (1972) plotted these functions (Fig. 3) and showed that the dependence law for the sublimation of water ice (from Fig. 1) was just in between the two functions 6 and 7. He concluded that between 2.14 AU and 6.51 AU the production rate of gases of P/Schwassmann-Wachmann 2 was indeed totally controlled by the sublimation of water ice. In order to handle the law analytically for orbit computing, Sekanina proposed then to approximate the family of curves defined by Eq. (1) and represented by Fig. 1, by the empirical formula

$$Z = Z_0 \alpha \left(\frac{r}{r_0}\right)^{-m} \left[1 + \left(\frac{r}{r_0}\right)^n\right]^{-k} \qquad (3)$$

with α chosen so that $Z = Z_0$ for $r = 1$ AU. J. Delsemme (1972) computed that for water and for $A_0 = A_1$, $r_0 = 2.8$ AU, $m = 2.15$, $n = 5.09$ and $k = 4.61$, Eq. (3) approximating the solution of Eq. (1) for all relevant distances, within ± 5%. Using these data, Marsden et al. (1973) verified that the vaporization of water ice gave the least residuals for all short-period comets for which the orbits were sufficienty well determined. They also discussed the possible range of all parameters, and made a careful analysis of the uncertainties.

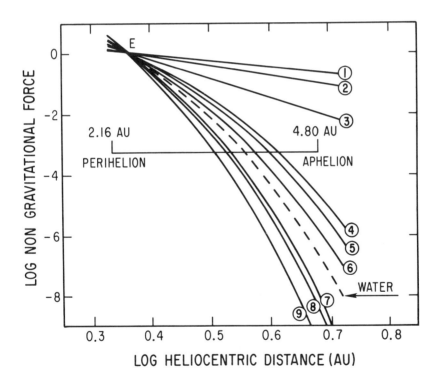

Fig. 3. Dependence on heliocentric distance of the nongravitational force acting on P/Schwassmann-Wachmann 2. Curves 1 to 9 are Marsden's empirical functions. Curves 6 and 7 gave the smallest residuals, from an orbit defined by 163 positions of the comet. The dashed curve represents the mean position for the vaporization curve of water ($A_0 = A_1$). The aphelion and the perihelion of the comet show the effective range of the forces. The nongravitational forces were normalized at E, position of the mean effective force. (From Delsemme 1972.)

E. Conclusions

The vaporization theory, confirmed by brightness laws and by nongravitational force laws, strongly suggests that the sublimation of water ice controls the surface temperature and the vaporization rate of many comets, probably including all short-period comets. It is tempting to believe that it should also be true for all comets that have been through repeated perihelion passages, in particular with small perihelion distances, because more volatile gases would be entirely lost faster than water. For instance, Delsemme (1966) shows that methane would be usually lost in less than one single perihelion passage. The probable exception would obviously be new comets, which by definition come for the first time through the inner planetary system. Delsemme (1975b) explains the infamous lightcurve of new Comet Kohoutek 1973 XII on its leg towards perihelion, by the presence of an amount of carbon dioxide large enough to control the surface temperature and the sublimation rate of its nucleus.

Using the analytic approximation made by Marsden et al. (1973), it is possible to verify that the r_0 of Eq. (3) is a scaling distance at which the solar energy spent in reradiation back to space is ~ 40 times that spent in vaporization. For distances larger than r_0, vaporization subsides and disappears quickly, and therefore the nucleus should become starlike even in large telescopes, whereas for distances much smaller than r_0, vaporization takes over and varies asymptotically with an inverse-square law of the distance. Depending on the nature of the ices, r_0 scales with the inverse square of the effective mean steady-state temperature T_1 of the (rotating) nucleus leading to the last column in Table I. To establish this column, I used an infrared albedo $A_1 = 2A_0$, since most known albedos are larger in the infrared than in the visible. This yielded $r_0 = 2.5$ AU instead of 2.8 AU for water, giving a measure of the uncertainty attached to the albedos. This last column expresses in modern terms what Delsemme and Swings (1952) meant when they introduced the idea that a solid hydrate of gas (clathrate or ionic hydrate)—i.e., in all cases, the binding energy of the lattice of the water ice crystal—controls the vaporizations. It is well known that, apart from very rare exceptions, extensive comae do not appear at large heliocentric distances, but rather between 2 and 3 AU at most. This seems to rule out excesses of volatile gases beyond that amount that could be either absorbed in the water snows (Delsemme and Miller 1970) or imprisoned in solid hydrates (Delsemme and Swings 1952).

F. Interpretation

Since in the Oort Cloud, the outer shell from which new comets come is closer (Marsden et al. 1978) than Oort had anticipated (40 to 60,000 AU instead of beyond 100,000 AU), the perturbation rate of the nearest stars is smaller than expected. For this reason, new comets are being introduced into the inner solar system by smaller perihelia changes. Therefore new comets which come for the first time within the observable range have had their previous perihelion not outside of the planetary system but only slightly farther away than Jupiter's orbit; they probably have had several perihelion passages through the outer giant planets. Their random walk in orbital energy leaves room for exceptions, but most "new" comets will have already lost their most volatile fraction during their previous perihelion passages beyond Jupiter, as is clearly established by the last column in Table I. The exceptions, owing to the random nature of the perturbations, explain the unusually bright comets observed at large distances, like the Great Comet of 1729 whose absolute mag H_{10} was -3, with a perihelion at 4.05 AU, or like the anomalous comets Morehouse 1980 III and Humason 1962 VIII, with a prevailing CO^+ spectrum. As already mentioned, Comet Kohoutek 1973 XII although less distinctive, was also probably one of these comets not controlled by the sublimation of water ice.

The existence of a fraction more volatile than water in some of the "new" comets (in variable amounts due to differences in their last perihelion distance beyond Jupiter) may be the major cause of the statistical differences in their photometric properties found by Oort and Schmidt (1951). The shallow surface irradiation by cosmic rays may play a minor role, as mentioned earlier, but it would rather be in the sense of a volatility decrease, and this is the opposite of what has been observed.

Finally, when comets become observable, the fact that they may have already lost a small, very volatile fraction (like that CO beyond the 15% limit of the CO hydrate) is not necessarily in contradiction with their keeping an essentially undifferentiated character. The large vapor pressure differences between CO and H_2O (typically, a factor of 10^{14} at the same temperature) coupled with the probable porosity and brittleness throughout the nucleus (see the chapter by Donn and Rahe) makes it likely that most of the unbound CO will escape even from deep in the core, in one single perihelion passage. The remaining of some minor pockets of more volatile materials certainly is not inconsistent with the picture developed here; when exposed, these pockets could explain jets and some outbursts.

III. EXPANSION VELOCITIES IN THE COMA

A deep understanding of the gas velocity field in the coma is needed to establish the dust-to-gas ratio (Sec. V) as well as the individual production rates of atoms and molecules (Sec. VI).

A. Theoretical Velocities

The vaporization theory predicts that the subliming gases leave the nuclear snows with a mean radial efflux velocity v_0 which must be between a maximum and a minimum value (Delsemme and Miller 1971):

$$\frac{1}{2}\bar{v} \leq v_0 \leq \frac{2}{3}\bar{v} \qquad (4)$$

where $\bar{v}$ is the mean Maxwellian velocity for the temperature T of the snows $(8\,kT/\pi m)^{1/2}$. The upper limit $2/3\,\bar{v}$ corresponds to the effusive flow of gas into vacuum from a small hole in a pressurized vessel (in this case, the flux polar diagram follows Lambert's law); the lower limit $1/2\,\bar{v}$ corresponds to a solid face, where the escaping molecules must overcome the surface force field (the flux polar diagram becomes an oblate spheroid with minor axis perpendicular to the surface). The latter is true for a subliming solid whose surface is a perfect plane, with no pores down to the molecular level. With the existence of pores the actual v_0 must lie between $1/2$ and $2/3\,\bar{v}$.

However, while expanding in the vacuum, the molecules accelerate radially because of the pressure gradient. Random kinetic energy is therefore transformed into a radial stream. Molecules soon reach sonic velocity (Mach number $M = 1$). For water

$$v_s = \left(\frac{\gamma k T}{m}\right)^{1/2} = 0.71\,\bar{v}. \tag{5}$$

Here γ is the ratio of the specific heats c_p and c_v, k is the Boltzmann constant and m the molecular mass. Gases continue expanding adiabatically going through a shock wave very close to the nucleus, and they eventually reach a radial diffusion velocity corresponding almost to the enthalpy content of the gas. Delsemme and Miller (1971) show, however, that for water vapor, $c_p = 8.67$ cal/mole, with contributions of 8.00 from translational and rotational states and + 0.67 from vibrational states. From the lifetimes of vibrational states, they deduce that their energy will not be transferred by collisions. They conclude that the gas terminal velocity for water is

$$v_\infty = 1.77\,\bar{v}. \tag{6}$$

This yields 860 m s^{-1} at 200 K. Using a fluid dynamics approach, Probstein (1968) finds $v_\infty = 1.54\sqrt{c_p T}$. Translated into the same units, it yields $v_\infty = 1.77\,\bar{v}$, agreeing with Delsemme and Miller, but Probstein assumes $\gamma = 1.40$ and accordingly finds $v_\infty = 740$ m s^{-1} at 200 K. It is clear that $\gamma = 1.40$ is the value for diatomic molecules; for water $\gamma = 1.33$ and so the correct velocity is 860 m s^{-1} at 200 K. This terminal velocity is adiabatic. In practice, it will be changed slightly by sources and sinks of heat, as discussed below.

Before the coma reaches a collisionless effusion in the vacuum beyond 10^4 km, four major phenomena take place:
1. The photoexcitation of molecules into excited states by solar photons;
2. The photodissociation by solar photons of parent molecules, present in the vaporizing gas (but often undetected in our spectra);
3. The photoionization (mainly by vacuum-ultraviolet solar photons) of molecules or of previously dissociated molecular fragments;
4. Fast ion-molecular reactions reshuffling the fragments toward molecular states of lower total internal energies (Cherednichenko 1973; Aikin 1974).

Although ion-molecular reactions in comets are quite accepted now, the fact that spectra of faint comets (with a much smaller collision zone) are not essentially different from those of very bright comets implies that these reactions do not play a major role in building up the radicals and ions observed in cometary spectra. (Mitchell et al., 1981 throw an interesting light on the difficulty of forming new neutral species by any reactions in the coma.)

The coma is being heated by the processes mentioned above, but it cools off easily by reradiating back to space in the rotational wavelengths of molecules. What is then the true terminal velocity v of the gas in the collisional zone? Since water is likely to be the dominant molecule, Shimizu (1976) notes that it is a good infrared radiator because it has a large dipole moment (1.84 Debye). In particular he shows that the spontaneous emission by vibrations (near 6.3 μm) takes a much longer time and can be neglected when compared to rotations (near 50 μm). He then shows that by only changing numerical coefficients, the cooling rate given by Bates (1951) for a linear molecule remains a good approximation for water:

$$R_c \propto T^2 N. \tag{7}$$

Here R_c is the radiative cooling rate by rotational transitions and N is the gas density. The time constant for cooling $\tau = (N \sigma v)^{-1}$ is on the order of 10^2 s. OH also contributes to the cooling at nearly the same rate as H_2O. Since the heating rate R_h comes basically from the solar flux, even if we include a coupling with dust grains and a possible contribution from the solar wind, it remains

$$R_h \propto N r^{-2}. \tag{8}$$

At steady state $R_c = R_h$, therefore

$$T = T_0 r^{-1} \quad \text{or} \quad v = V_0 r^{-0.5}. \tag{9}$$

The coma's final temperature varies in proportion to the reciprocal of the heliocentric distance and the velocity varies therefore with its square root.

B. Observational Data

Bobrovnikoff (1954) has tabulated velocities of expansion for 57 cases of halos observed in cometary comae; Whipple (1980), who has recently used these data, has added an observation of P/Schwassmann-Wachmann 1 by Beyer (1961). Figure 4 represents the results of Whipple's plot. The straight dashed line represents

$$v = 0.535 \, r^{-0.6} \tag{10}$$

in km s^{-1} and it fits the observations well between 0.6 and 7.0 AU. It could be argued that the light of visible halos comes mainly from the finest available dust dragged away by gases because fine dust is dragged away faster than coarse dust. Typically, fine dust would reach $\sim$ 75% of the gas velocity (see Sec. IV). However, Bobrovnikoff (1931) has compared halo velocities, measured in visible light, with some halo velocities observed in CN light on objective-prism spectra of P/Halley, and they were basically the same. It is not surprising that the blue orthochromatic plates used at the beginning of this century recorded more light from CN than from the continuum. Because of the observed dependence on heliocentric distance of the halo velocities, this would imply that CN has been thermalized in the collision zone, and acts as a tracer of the gas velocity, and not a tracer of the dust.

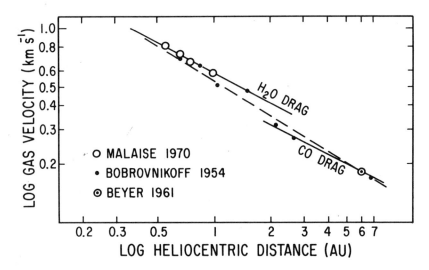

Fig. 4. Observed expansion velocities in cometary comae, as a function of heliocentric distance. Bobrovnikoff's data are averages, per heliocentric distance groupings, of 57 cases of observed expansion velocities in cometary halos. Beyer's is for halos of P/Schwassmann-Wachmann 1. Malaise's data come from the Greenstein effect on CN bands. Solid lines are theoretical laws (slope -0.5). Dotted line is Whipple's (1980) empirical law (slope -0.6).

The hypothesis that CN has been thermalized is confirmed by another empirical source of information which is in principle much more reliable and accurate: a straight measure of the expansion velocity of CN by the differential Swings effect. Swings (1943) explained the anomalous jigsaw appearance of the rotational lines of CN (0-0) band from 3850 to 3890 Å, by the presence of the Fraunhofer lines in the Doppler-shifted solar spectrum exciting the fluorescence of CN. This is the Swings effect. The differential Swings effect, often called Greenstein effect, can be detected sunwards and tailwards and is due to the radial expansion of the coma that modifies the Doppler shift of the solar Fraunhofer lines. There are ~ 270 Fraunhofer lines in the solar spectrum between 3850 and 3890 Å, of which ~ 2/3 come from the rotational levels of the CN (0-0) band itself, but this time in absorption; the rest is mainly absorption lines of iron, including at least six very strong lines. When the excitation of one of the rotational lines of cometary CN is shifted by only a few 10^{-3} Å, by the Greenstein effect, across the profile of any of the solar absorption lines, there are dramatic effects changing the jigsaw appearance of the cometary band. Besides, the many coincidences to be explained by the *same* shift do not leave any ambiguity. For this reason, the Greenstein effect is probably the most reliable and the most accurate technique for establishing the expansion velocity. Unfortunately it involves a large amount of work, and it has been used only once on four spectra coming from three different comets (Malaise 1970). The results have also been plotted on Fig. 4. Their best fit is represented by $v = 580\,r^{-0.5}$ m s^{-1}. The residues in Malaise's analysis probably come merely from his neglect of the CN(1-1) transitions and of the radiation pressure on CN.

Since these data, which are better than those on halos, reproduce the theoretical dependence on distance ($r^{-0.5}$), it is worth discussing the significance of Bobrovnikoff's and Beyer's halo velocities for comets beyond 2 AU. As we have seen earlier, comets showing large expanding halos at distances beyond 3 AU, are unlikely to vaporize water, in particular beyond 6 AU, like P/Schwassmann-Wachmann 1. Since its spectrum shows CO$^+$ only, it rather suggests CO as a parent molecule. The last column of Table I in Sec. II.A completely confirms this choice (with CO_2 and H_2CO as alternate possibilities). Assuming the presence of CO, the same kinetic temperature would diminish the velocity by a factor of $(28/18)^{1/2} = 1.25$, which would bring the last four points exactly on the slope of $r^{-0.5}$. This suggests that the law $T = T_0\,r^{-1}$ is really accurate and that we have been fooled by a change in the molecular weight of the gas vaporizing at larger distances. Whatever the correct interpretation, since the best data show an accurate law in $r^{-0.5}$ in the range between 0.5 and 1.5 AU, it is concluded that the expansion velocity law of CN is understood theoretically and confirmed observationally within ± 5%. It is given (in km s^{-1}) by

$$v_{CN} = (0.58 \pm 0.03)r^{-0.5} . \tag{11}$$

This fact does not seem to have been noticed before the present review. It is worth mentioning here that the Boltzmann temperatures used by Malaise (1970) to explain the pressure effects he detected in the inner coma are completely consistent with the previous velocity and with its dependence on distance. Of course, these temperatures are much less accurate than the previous velocities, because the fitting procedure for the synthetic spectra uses several parameters.

One further remark is needed. Combi and Delsemme (1980a,b) have shown that CN is probably produced by the photodissociation of HCN. This implies that CN is ejected with a random (nonradial) velocity of 1.02 km s^{-1} (independent of the solar distance). But the CN velocities established from the Greenstein effect were measured at one third of the usual definition of the radius of the collision zone (taking into account the proper scaling factors for solar distance and total production rates from the absolute magnitudes of the cometary heads). However, Malaise's results establish clearly that the CN radicals were thermalized; their radial velocity was not only somewhat lower, but clearly dependent on the solar distance. The explanation is that they had been thermalized *before* leaving the collision zone. The scale length for HCN dissociation is such that most CN radicals observed by Malaise undergo from 2 to 20 collisions in their random walk before emerging from the collision zone; not only is their temperature indistinguishable from that of the major molecule, but their radial expansion is the same.

The radial velocity of the water molecule (in km s^{-1}) is therefore

$$v_{H_2O} = 0.58\, r^{-0.5} \,. \tag{12}$$

It is pointed out here that an important verification could be obtained by studying the Greenstein effect on the OH band at 3050 Å for a few comets. Good solar spectrum data are now known for these wavelengths (Kohl et al. 1978). From the line width of the radio line of OH, Crovisier (1981) mentions that the OH expansion velocity indeed diminishes with increasing heliocentric distance.

C. Steady-state and Nonthermal Velocities

The terminal speed of dust particles is reached within 20 radii of the nucleus. From then on the motion of dust is uncoupled from that of the gas. The terminal speed of gas given by Eq. (12) is reached for water (assumed to be the major radiator) in a characteristic radiative time of $\sim$ 100 s (Shimizu 1976). The steady-state temperature comes from the presence of temperature sources and sinks; this is neither the initial nor the adiabatic temperature of the gas subliming from the snows, although it is not too different from the adiabatic in the range of 0.7 to 1.0 AU, where most observations are made (see velocities, Fig. 5). Finally, photodissociations go on outside of the collision zone, depending on the different characteristic times for each individual molecule or radical. At those large distances, the velocities of the molecular fragments result from the energy balance sheet of the photodissociation. They are no longer thermalized but completely uncoupled from the rest of the expanding gas, since there are no more collisions. The best example described in the literature is that of the two velocity components of the Lyman-α halo, moving with mean speeds of 20 and 8 km s^{-1} and attributed to the two-step photodissociations of H_2O, first into OH + H and then OH into O + H (Keller and Lillie 1974; Meier et al. 1976). Apart from a still contested radio identification (Jackson et al. 1976), the water molecule has never been directly detected and could not possibly be observed in comets with optical telescopes, although all of its possible molecular fragments are seen (H_2O^+, OH, O and H). For this reason, the identification of the two velocity components of H as both stemming from water, has come as a welcome further confirmation of its presence in comets.

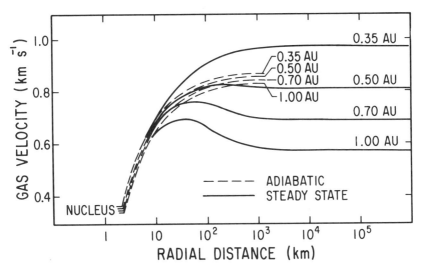

Fig. 5. Expansion velocities of H_2O coma at four heliocentric distances. Dashed lines: adiabatic expansion, as assumed by Finson and Probstein. Solid lines: steady state for water (heat is absorbed in its visual and ultraviolet bands and reradiated in the far infrared). The steady-state asymptotes agree with observations (Eq. 12). Nuclear temperatures are 204, 200, 195 and 188 K, respectively, from Eq. 1 with $A_0 = A_1 = 0.6$ Radius of nucleus is 2.6 km; characteristic time for steady state is 100 s. Coupling with dust disappears beyond 30 km, explaining the success of the 200 K adiabat used by Finson and Probstein.

IV. NATURE OF THE COMETARY DUST

Our knowledge about the nature of dust in comets is fragmentary and circumstantial. However, four indirect lines of evidence are available:

a. The reflection spectrum of the onset of the dust tail is a continuum modified by the two reflection bands of silicates near 10 and 17.5 μm; this signature implies that silicates are at least a major component of the dust grains, although some impurities (probably carbon or organic compounds) seem to diminish their dielectric properties and give them a rather low albedo (Ney 1974a,b).

b. About 200 grains of interplanetary dust have been collected by the U-2 aircraft of NASA (Brownlee et al. 1977). Some 10% had been melted, probably during entry in the atmosphere, and are less interesting for our purpose. The others with only a few exceptions can be presumed to be of cometary origin. These grains are typically from 2-3 to 20-30 μm in size; they are irregular and show a very complex high porosity that could easily go up, in some cases, to 50% of their total volume. They seem basically constituted by a black aggregate of much finer grains, whose size distribution peaks near 0.1 μm. These fine grains have properties closely similar (although not completely identical) to the fine-grained material compacted in CI chondrites and in the matrix of the CM chondrites, at the exclusion of chondrules. In particular, the grains' major components are silicates whose metal abundances are closely chondritic. However, they contain $\sim$ 5% carbon by mass, whereas the median CM chondrite contains 2.5% carbon and the median CI chondrite, 3.2% carbon (Mason 1971). They contain also at least as much sulfur as CI chondrites (whereas sulfur and carbon are much depleted in ordinary chondrites). Since cometary dust contains at least as much as, and possibly slightly more volatile material than the most primitive CI chondrites, it may represent a material even more pristine than the most pristine meteorites.

c. In sun-grazing Comet Ikeya-Seki 1968 I, the assumed vaporization of cometary dust produced an emission-line spectrum due to neutral atoms of metals, namely Ti, V, Cr, Mn, Fe, Co, Ni, Cu. Arpigny (1978,1979) deduced metallic abundances from this spectrum. He found that, compared with solar abundances, the observed gaseous phase is slightly depleted in Al, Si, Ca, Ti, V, and Cr, whereas it contains the right amount of Fe, Co, Ni, and a slight overabundance of Cu. Since the former list of metals form high-temperature condensates (corundum, perovskite, melilite and spinel), he concludes that the temperature was not high enough to vaporize all the dust grains, and that the observed fractionation is consistent with purely solar abundances in the original dust.

d. Millman (1977) studied the chemical composition of meteoroids observed as coming from cometary orbits, and reviewed the evidence from other data. He concludes that the seven elements commonly found in the spectra of cometary meteoroids (Fe, Si, Mg, Ca, Ni, Na, Cr) are consistent only with chondritic abundances; there is also qualitative evidence on the presence of significant quantities of some of the light volatiles.

Henceforth, we will assume that cometary dust is close to CI chondrites; if a difference is assumed in their physical properties, it could be extrapolated along the sequence CO $\rightarrow$ CM $\rightarrow$ CI $\rightarrow$ comet. In particular, Mason (1963) gives 3.6 to 3.3 g cm^{-3} for the density of CV and CO, 2.9 to 2.6 for CM, and 2.3 to 2.2 for CI carbonaceous chondrites. Taking the observed porosity of Brownlee's particles into account, a bulk density of 2.2 to 1.8 is suggested, in agreement with the previous extrapolation. We will adopt 2.0 ± 0.2 in models.

V. THE DUST-TO-GAS MASS RATIO

A. The Fluid Dynamics of Dust Drag

The subliming gases drag away the dust particles imbedded in the icy conglomerate. Starting from zero velocity, dust is accelerated radially outward from the nucleus, as a result of the free molecular drag interaction between dust and expanding gas. Probstein's (1968) model for this interaction describes the dust behavior as a continuous fluid, by fields and by conservation equations that are put in dimensionless form. The reduced radial velocity is expressed by

$$v' = \frac{v}{v_1} \tag{13}$$

where v is the actual radial velocity, and $v_1 = (c_p T)^{1/2}$, with c_p the specific heat at constant pressure and T the initial temperature of the gas. The reduced size of the grain is expressed through a similarity parameter β

$$\beta = \rho_d d(\rho_i R)^{-1} \tag{14}$$

where ρ_d is the dust density, d the dust grain diameter, ρ_i the gas initial density near the nucleus and R the nuclear radius. Probstein also established that β can be expressed in the form

$$\beta = \frac{16}{3} \pi \rho_d dR v_1 / \dot{m} \tag{15}$$

where $v_1 = (c_p T)^{1/2}$ represents the gas velocity units defined in Eq. (13). Finally, the dust-to-gas mass ratio M is

$$M = \dot{m}_d / \dot{m} \tag{16}$$

where $\dot{m}_d$ is the mass flow rate of dust and $\dot{m}$, that of gas. Probstein's results are given in Fig. 6 where the reduced velocity v' of the dust grain is expressed as a function of the reduced size of the grain. The family of curves corresponds to the different values of the dust-to-gas ratio M.

In order to compute the momentum transferred to the dust grains, the variation of the gas velocity with distance is taken into account. Eventually, a terminal velocity is reached asymptotically (Sec. III). Probstein argues rather convincingly that the interaction between gas and dust takes place essentially within 20 radii of the nucleus; the characteristic time to reach this adiabatic velocity is shorter than that needed to reach radiative steady state in the gas. As shown in Fig. 5, the two velocity gradients (adiabatic and radiative steady state) remain rather close during the major dust-to-gas interaction for Probstein's meaningful distances; therefore, Probstein's arguments to adopt the adiabatic velocity gradient near 0.7 AU can be readily accepted.

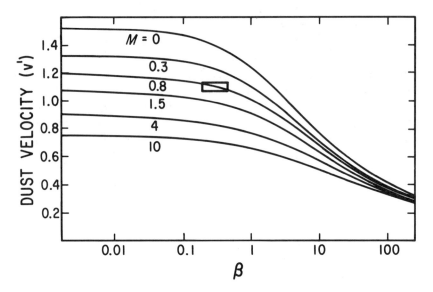

Fig. 6. Reduced terminal dust velocity v', in $\sqrt{C_p T}$ units, expressed as a function of β. β is a similarity parameter measuring the ability of the dust particle to adjust to the gas velocity. Since β is proportional to the size of the particle, each curve for $M =$ constant represents the velocity distribution of the dust particles as a function of their size. The rectangular box represents a typical accuracy in the measure of M. See discussion in the text. (From Probstein 1968.)

B. Isophote Analysis of the Dust Tail

Finson and Probstein (1968a,b) have developed Probstein's (1968) fluid-dynamics approach to explain quantitatively the isophotes observed in the dusty tail of comets, and have applied their theory to four photographs of Comet Arend-Roland 1957 III taken by Ceplecha (1958) on April 27, April 29, May 1, and May 2, 1957. Their extensive treatment can be sketched only briefly here.

A dust particle follows a trajectory defined by its initial velocity combined with an acceleration vector which comes from two terms: the Sun's attraction minus its radiation pressure. The ratio of these two accelerations is usually written $1 - \mu$; it varies with the diameter d and the density ρ of the particle, according to the relation

$$1 - \mu = C(\rho\, d)^{-1} . \qquad (17)$$

The dust distribution in the tail can therefore be easily computed. Let us first consider the idealized case in which the dust leaves the nuclear region with no initial velocity. Then the isophotes of the dust tail (see Fig. 7) represent a two-dimensional resolution of two parameters, completely separated without ambiguity:

1. The particle-size distribution varies along each of the synchrones (traced by all particles emitted the same day by the nucleus);
2. The production rate of dust (of a given size) varies along one of the syndynes (traced by all particles of a given size) as a function of its emission time.

This is a beautifully simple problem of kinematics yielding a unique and accurate solution. In practice however, a third parameter must be introduced: the initial velocity of the grains dragged away by the gas.

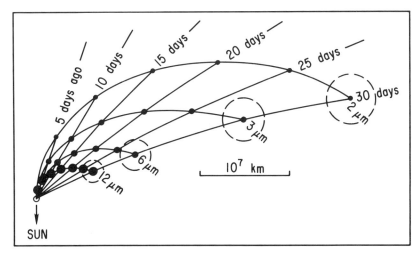

Fig. 7. Dust distribution in cometary tail. Grain trajectories for different sizes are indicated by black dots of different diameters. Emission dates from the nucleus are indicated in days. Dashed circles represent the fuzziness of the trailing edge of the tail, coming from the isotropic initial velocity v_i of the grains.

The interaction between gas and dust takes place essentially within 20 nucleus radii (20 to 100 km). This zone is much smaller than the resolving power of the best photographs; therefore the terminal velocity of the dust grains, resulting from their interaction with the gas, can be identified with the initial velocity of the grains emitted by a pinpoint source representing the nuclear region on the photographs. Finson and Probstein give reasons why the initial velocity vectors may be assumed to be isotropic; however, it is clear from the jets often identified in the coma, that this is an approximation. If this approximation is accepted (and indeed, most of the jets are visual features that are rarely more than 10% brighter than their surroundings), then the initial isotropic velocity of the grains substitutes a sphere with radius growing with time, to each of the points described in Fig. 7.

The deconvolution of the three parameters in the tail isophotes remains straightforward, because the third parameter produces a large widening, mainly of the tail's trailing edge (Fig. 7), that clearly separates the initial velocity from the other two parameters. This initial velocity will specify the dust-to-gas mass ratio M as a function of β (Fig. 6).

There are even cases where the geometry takes care by itself of the separation of the third parameter. If the grain's initial velocity is neglected, a dust tail is essentially a flat 2-dimensional feature in the plane of the cometary orbit. When the Earth is close to this plane during the observations, the tail's apparent widening is essentially due to the grains' initial velocities. This was the case on 27 April 1957 when the tail width showed immediately that the dust grains' velocity v_i was on the order of 0.3 km s^{-1}. A more laborious fitting to the isophotes of the four photographs shows that the best results are consistently reached for a grain velocity beginning at 0.15 km s^{-1} 10 days before perihelion, increasing regularly to 0.33 km s^{-1} at perihelion and decreasing irregularly later to values between 0.3 and 0.2 km s^{-1}. These values are given for grains of an intermediate size ($\sim 7\,\mu$m diameter for a density of 2 g cm^{-3}) defined by a radiative-pressure acceleration equal to 8% of the Sun's attraction (see Eq. 17).

Different approaches can be used to compute the dust-to-gas mass ratio. Finson and Probstein prefer to define a grain velocity which is an *a priori* function of size, as established by fluid dynamics, i.e., following the slopes of the family of curves set for β in Fig. 6. The slopes of their results for different grains are therefore preestablished. Hence, only one point must be computed for a grain of a given size. Its velocity as measured from the isophotes' fitting, is plotted vertically on Fig. 6. Its β (which is its ability to adjust to the local gas velocity) is plotted horizontally. The two coordinates will define a point, that is a value of M on the family of curves. However, only $\beta\dot{m}$ is immediately available, without ambiguity, from Eq. (15). Since $\dot{m}$, the mass production rate of gas and $\dot{m}_d$, that of dust, are linked by their ratio M, either one or the other must be established first to compute β. Finson and Probstein choose $\dot{m}_d$ which is directly available through the tail isophotes, if the dust albedo is known (since the isophotes also yield the size distribution of dust).

C. Discussion of Results

Three comets have been studied so far by the Finson-Probstein method: they are listed in Table III.

Finson's and Probstein's (1968b) study of Comet Arend-Roland 1957 III is by far the most complete. It gives not only the functions (particle-size distribution, relative dust particle emission rate, time dependence of the velocity) that were most successful in duplicating the observed isophotes at four different dates (from the four Ceplecha photographs mentioned above), but they show and discuss all the alternate forms of the functions which were tried and were less successful, allowing the reader to develop a feeling of the possible inaccuracies involved in the method.

One of the major uncertainties remaining in 1968 was the geometric albedo of the dust particles. Another one was the density ρ of the dust particles, but this uncertainty can be avoided in most of the computations by using the product ρd (density times diameter) as the independent variable that defines in particular, the ratio of the Sun's attraction to its radiation pressure (Eq. 17).

TABLE III

Dust Tails Studied by the Finson-Probstein Method

	Arend-Roland 1957 III[a]	Seki-Lines 1962 III[b]	Bennett 1970 II[c]
Nature of orbit	new	new	long-period
Absolute magnitude H_{10}[d]	5.1	6.6	4.5
Perihelion distance	0.32 AU	0.03 AU	0.54 AU
Heliocentric distance at epoch of observations	0.64–0.74 AU	0.64 AU	0.60–0.54 AU
Epoch of emission of observable dust in tail (– before, + after perihelion)	from –10 d to +18 d	from –2 d to +12 d	from –45 d to –2 d
Mean distance of dust emission	0.34	0.16	0.56
Root-mean-square diameter of average dust grain (assuming $\rho = 2$ g cm^{-3})[g]	2.8 μm	13.2 μm	1.0 μm
Diameter of most probable grain in distribution (assuming $\rho = 2$ g cm^{-3})[g]	1.5 μm	10.0 μm	0.7 μm
Dust-to-gas mass ratio	0.8 ± 0.2[e]	meaningless[f]	0.6 ± 0.4[e]

[a]Finson and Probstein (1968b).
[b]Jambor (1973).
[c]Sekanina and Miller (1973).
[d]Useful for scaling the three comets if no other data are available.
[e]Corrected by using homogeneous modern data (see text).
[f]Intense vaporization of dust at 0.03 AU (three solar diameters).
[g]Remarkably the mean grain diameter varies almost as an inverse-square law of the mean heliocentric distance of dust emission.

As far as the albedo is concerned, Finson and Probstein make three estimates for three values of the geometric albedo $A\phi(\alpha) = 0.01$, 0.08 and 0.50. (They define A as the Bond albedo and $\phi(\alpha)$ as the phase function for phase angle α.) Modern values for the albedo of cometary dust make it easy to diminish their uncertainty range. From $(1-A)$ which is reradiated back to space in infrared, O'Dell (1971) gives for three different comets a weighted average of $A = 0.30 \pm 0.15$, much larger than early expectations. Ney (1974a,b) also finds for four comets, that most observations show a geometric albedo $A\phi(\alpha) = 0.18$, with variations from 0.10 to 0.40 at least partially due to the large variation of the phase function $\phi(\alpha)$ with the angle α. Ney and Merrill (1976) have derived the scattering function of Comet West 1976 VI from 34 to 150° phase angle by the same infrared method. They confirm that the Bond albedo is very large, 0.40 ± 0.10, but reflectivity is strongly anisotropic. We conclude that, for a geometric albedo in the general range of a phase angle of 90°, the value that should be adopted is

$$A\phi(\alpha) = 0.20 \pm 0.05 . \tag{18}$$

Since our target is the dust-to-gas mass ratio M, the uncertainty of this ratio is defined by the two uncertainties on β and on v', that define a rectangular box of sides $\Delta\beta$ and $\Delta v'$ to be superimposed on the families of curves defining M in Fig. 6.

From Eq. 15, the $\Delta\beta/\beta$ associated to a given dust diameter d and a given standard velocity v_1 is given by

$$\frac{\Delta\beta}{\beta} = \frac{\Delta\rho}{\rho} + \frac{\Delta R}{R} + \frac{\Delta\dot{m}}{m} . \tag{19}$$

From Sec. IV the conclusion suggests using $\Delta\rho/\rho = 20\%$. Table II shows $\Delta R/R = 12\%$, but the unknown systematic errors are not included. However, the radius of Comet Bennett 1970 II nucleus has been established by two conceptually different methods. It was found to be 3.76 km by Delsemme and Rud (1973) and 2.6 km by Sekanina and Miller (1973) yielding $2\Delta R/\bar{R} = 36\%$ or $\Delta R/\bar{R} = 18\%$ for the average of the two methods. Since $\dot{m}$ (gas) is deduced from $\dot{m}$ (dust), a 25% error on the albedo $A\phi(\alpha)$ yields the same relative error of both $\dot{m}$ (dust) and $\dot{m}$ (gas); altogether, 20% + 18% + 25% = 63% for the inaccuracy on β. Fortunately, β is not a critical parameter for most of the flow regimes, in particular, for $\beta \cong 0.1$, $\Delta M/M \cong 0.1\, \Delta\beta/\beta$. By and large, for $\beta \leq 1$, the major error will not come from β but from the dust velocity.

The emission velocity v of the dust is well determined by the cross-sectional shape of the tail, and the accuracy of the determination is limited only by the quality of isophote fitting. The accuracy for one grain size related to a single syndyne curve could certainly be rather low, but the procedure that links all velocities to a functional shape predicted by gas dynamics improves the fitting considerably. It has also been verified that the tail width can be changed to a somewhat lesser extent by an asymmetrical time variation of v, but the time-averaged v remains constant to a good approximation of the order of 3%. The velocity units v_1 used in Fig. 6 are known with high accuracy if the conclusions of Sec. II are accepted, i.e., if v_1 is defined by $\sqrt{c_p T}$ for water and if T is the sublimation temperature defined by the vaporization theory. However, since water is not a diatomic molecule, its $\gamma = 1.33$ (see Sec. III) and molecular mass $m = 18$, instead of $\gamma = 1.40$ and $m = 25$ used by Finson and Probstein. Taking this correction as well as that of the albedo into consideration in Finson and Probstein's data, at perihelion we read on Fig. 6, $M = 0.9 \pm 0.2$, whereas after perihelion it goes to an asymptotic value of $M = 0.8 \pm 0.2$. Using the same corrections for Finson and Probstein's (1968b) Table IV p. 375, I deduce from different isochrones in the tail pictures the time variation of the dust-to-gas ratio M, before, during, and after passage to perihelion (Fig. 8).

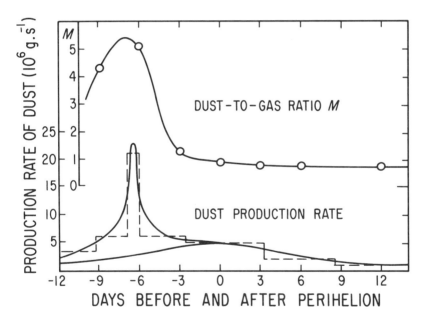

Fig. 8. The variation of the dust-to-gas mass ratio M during the perihelion passage of Comet Arend-Roland. During the anomalous value of M before perihelion, the dust production rate (lower curves in the figure) was also unusually high, particularly during the outburst of day − 6. The dotted lines represent the step function used for convenience by Finson and Probstein (although they recognize the smoothness of the rate variation). To better assess the dissymetry, a curve symmetrical to that after perihelion has been drawn before perihelion. (Data reduced from Finson and Probstein 1968b.)

This huge variation of M may come as a surprise to the reader. However, we also know that there was a big outburst of dust and gas, whose peak lasted less than a day, six days before perihelion. Figure 8 (adapted from Finson-Probstein 1968b) shows this large outburst of dust that culminated on day -6, but probably lasted ~ 10 days before perihelion, if we believe, as suggested by Fig. 8, that it was the major cause of the emission rate dissymmetry before and after perihelion. The outburst is well documented; it also produced a wide isophote distortion conspicuous at first glance in the picture of 27 April.

Finally, the size-distribution of the dust grains also varied; during the culmination of the outburst, the relative number of large grains between $20\,\mu m$ and $100\,\mu m$ was multiplied by a factor of ~ 3. This is visible in Fig. 9 where the size distribution of dust grains is given for Comet Arend-Roland 1957 III both on the average and during the outburst, from Finson-Probstein's data.

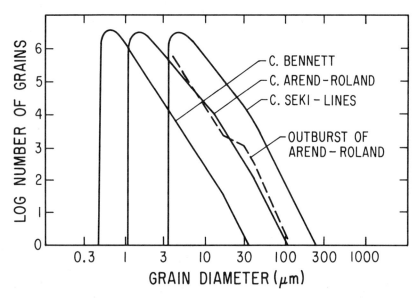

Fig. 9. Size distribution of the dust grains in three comets: Bennett 1970 II, Arend-Roland 1957 III, and Seki-Lines 1962 III. The grains' density is assumed to be 2.0. The three curves have been normalized by bringing their maximum at the same level. The units for the log number of grains are arbitrary. The change in size distribution of the grains during the outburst of Arend-Roland is well documented. The excess of grains of 20 to 100 μm diameter reached at least a factor of 3, suggesting the blow off of a previously outgassed mantle of larger grains.

These three variations (of M, of $\dot{m}_d$, and of the size distribution) are clearly correlated and help our understanding. As a matter of fact, the larger grains of dust were not able to leave the nuclear regions at larger heliocentric distances because the vaporization rate was not large enough to drag them away. During the comet's approach an outgassed mantle of large grains of dust developed, changing the dust-to-gas ratio at the surface of the nucleus, down to a depth of 10 or 20 cm, until the larger dust heating due to the subsiding vaporizations penetrated the dust layer and produced a catastrophic outburst in the gas production rate and blew away the transient mantle of dust. This scenario has been quantitatively explored by models (Mendis and Brin 1977,1978; Brin and Mendis 1979; Brin 1980) that describe very convincingly the observed phenomena. In Comet Arend-Roland 1957 III it therefore seems reasonable to believe that the differentiated mantle has been blown away and that the ratio asymptotically reached after perihelion represents pristine material that has been exposed to the nucleus surface just after the outburst.

Of the two other comets studied by the Finson-Probstein method only Comet Bennett's dust-to-gas mass ratio has been published. As far as Comet Seki-Lines 1963 III is concerned, the published data (Jambor 1973) are too scanty to reconstruct M; besides, it would probably be meaningless because Jambor established that most of the dust was vaporized at a high temperature near perihelion for this sungrazing comet (the perihelion was at only six solar radii). As far as Comet Bennett 1970 II is concerned, a dust-to-gas mass ratio of "about 0.5" was computed by Sekanina and Miller (1973), assuming 0.1 for the geometric albedo of dust. Unfortunately, their paper does not give as many details of the fitting procedure as Finson and Probstein's paper; for instance, it is difficult to assess to what extent their parametric function of the grains' velocity $f(r, 1 - \mu)$ could be changed without affecting the results. They used three photographs but show only one single isophote analysis. Taking into account $\gamma = 1.33$ for water and $A\phi(\alpha) = 0.20$, I find that near perihelion his best value for M is 0.4, but my feeling is that the error bar is large and that the result could be anywhere between 0.2 and 1.0; this is the reason why 0.6 ± 0.4 has been used in Table III.

D. Conclusions about M

The Finson-Probstein method was not developed to reach the dust-to-gas ratio M, but only to explain the dust-tail isophotes and secondarily to deduce approximate production rates of gas and dust. However, the geometrical separation of the particles by sizes, as observed in the tail, allows the separation of the three functions: particle-size distribution, initial velocity of the particles (function of their size), and production rate of particles (function of their epoch of emission). The complete analysis is difficult, and M remains uncertain when established from those large particles for which $\beta > 1$, because then the dust velocity varies much more with β. The method can yield neither direct indication on the presence nor on the absence of particles much smaller than the wavelength of light. The efficiency factor for radiation pressure Q is assumed to be unity in Finson-Probstein's analysis; this is a reasonable approximation for silicates in the range of 0.2 to $2\,\mu\mathrm{m}$ (Hanner 1980) where the bulk of the grains are. However, since the unobserved grains load the gas, they are somewhat taken into account in the ratio of the velocities and therefore in M, even if we cannot represent them in the size distribution of the grains. The method is at its best for the sizes between 0.5 and 2 μm, because C is nearly constant, β usually remains <1, and the large uncertainties on β and the albedo A do not affect the final result in a significant way.

Since the two values accepted here are 0.8 ± 0.2 (Comet Arend-Roland 1957 III) and 0.6 ± 0.4 (Comet Bennett 1970 II), the weighted average of these values suggests adopting M = 0.76 ± 0.3 and using M = 0.8 as a round figure.

VI. ATOMIC ABUNDANCES IN THE VOLATILE FRACTION

A. Measuring Production Rates

The cometary head is not a permanent atmosphere, but an exosphere that is steadily escaping and lost to space, as well as continuously being renewed by the sublimation of the nucleus. The molecular fragments, neutral or ionized, which have been observed in the tail and in the coma of numerous comets (Table IV) make it clear that this volatile fraction is essentially composed of molecules built up with the light elements: H, C, N, O, S. However, most of the parent molecules are unobserved; the observed atoms, radicals, or ions are visible mostly by accident because they happen to have large cross sections to absorb solar photons; they scatter these photons back at the same wavelength (resonance) or at a different wavelength (fluorescence). In order to transform the monochromatic flux of light scattered, for instance, by a radical like CN, into a production rate of the CN radicals by the nucleus, we must know two different parameters:

TABLE IV

Observed Species in Cometary Spectra

Organic[a]	Inorganic[b]	Metals[c]	Ions[d]	Dust[e]
C	H	Na	C^+	Silicates
C_2	NH	K	CO^+	
C_3	NH_2	Ca	CO_2^+	
CH	O	V	CH^+	
CN	OH	Mn	H_2O^+	
CO	H_2O	Fe	OH^+	
CS	S	CO	Ca^+	
HCH		Ni	N_2^+	
CH_3CN		Cu	CN^+	

[a,b] Seen in cometary heads as visual or ultraviolet emission lines or bands except HCN, CH_3CN, and H_2O, as pure rotation lines in radio frequencies (poor S/N ratio for H_2O).
[c] Seen as visual lines in sungrazing comets only (except Na seen in many comets).
[d] Seen in visual or ultraviolet emission lines or bands in onset of plasma tail.
[e] Seen in infrared reflection bands in onset of dust tail.

1. The number of photons scattered per second per radical (this is usually called the emission rate factor g);
2. The (exponential) lifetime τ of the molecule against all decay processes.

The product $g\,r$ establishes the total number of fluorscence cycles during the average lifetime of the radical. The production rate, in number of CN radicals released per second by the nucleus, is therefore given by dividing the total monochromatic flux in the light of CN, by $g\,r$, the average number of fluorescence cycles scattering photons during the lifetime of the radical (before its decay by dissociation or ionization).

The only known exceptions are atoms that do not fluoresce, but are being produced into an excited state, for instance during the dissociation process of their (often unknown) parent molecule. Eventually they release one single photon per atom, and cascade down to their ground state. In this particular case, the production rate is much easier to compute: instead of $g\,r$ photons per atom or radical, we have just one. The only known example so far is the oxygen produced into the ^1D state, which emits the forbidden red doublet $^1D - {}^3P$.

In the general case, the emission rate g depends on the oscillator strength of the transition involved, and on the flux of solar light reaching the radical; both parameters are moderately well known for some radicals. However, the effective lifetime τ is the result of several competitive processes of photoionization and of photodissociation that are often poorly known, not only because of the uncertainties of the cross sections involved, but also because of the poor data from the extreme ultraviolet of the Sun. Since the basic paper of Potter and Del Duca (1964), actual molecular lifetimes can and should certainly be established more often thanks to the exponential scale lengths deduced from the brightness profiles of the cometary head in the light of a given molecule. As expansion velocities are moderately well known in some important cases (see Sec. III), the scale length can then be translated into an effective lifetime against all decay processes.

The complete fluorescence mechanism is well understood, but it is sometimes intricate because it implies the consideration of all the possible absorptions of the solar spectrum (duly Doppler-shifted by the radial velocity of the comet) by the ground state of the relevant atom or molecule, for all those wavelengths which feed energy levels that might eventually cascade to the ground state through the observed transition.

B. Observational Results for Molecules

No production rates have ever been measured from a single comet, on a single date, to produce a complete or even a partial balance sheet for most of the observed molecules, radicals, and ions. The major exception is of course water, which seems to be sufficiently dominant to explain the bulk of H and OH, the ion H_2O^+, and a sizeable fraction of the resonance line and of the forbidden red line of neutral oxygen (Festou 1978; Delsemme 1980).

All that can be said about the other observed radicals, like C_2, C_3 or CN is that they seem to be in the general range of 1 or 2% of the production rate of water. Typically, C_2 = 1.8, C_3 = 2.6, and CN = 0.6 to 2.0 for OH = 100 (A'Hearn and Cowan 1975; A'Hearn and Millis 1980; see also O'Dell 1976 for Comet Kohoutek 1973 XII).

Even if we had a complete balance sheet of all the observed production rates, the observations might still represent a very distorted and biased sample of the actual parent molecules vaporizing from the nucleus. Besides water, the only parent identification that seems reasonably certain is HCN. It was observed only in Comet Kohoutek 1973 XII at mm wavelengths (Huebner et al. 1974; Buhl et al. 1976); however, Combi and Delsemme (1980b) have shown that not only the production rates, but also the kinematics of the CN radical are consistent with the photodissociation of HCN as a parent molecule.

Other parent molecules like CH_3CN are less certain, because the microwave detection is marginal (Ulich and Conklin 1974) although it was also detected by Huebner et al. (1976); its production rate is dubious. Finally, the brightness of the forbidden red doublet of oxygen at 6300 Å and 6364 Å seems to imply a large contribution from the dissociation of CO_2 (Delsemme 1980) but this is a circumstantial deduction, supported by other circumstantial deductions such as the presence of CO_2 controlling the vaporizations in Comet Kohoutek (Delsemme 1975b).

The status of all other possible parent molecules is even in worse shape. I will give only one example; formaldehyde H_2CO has never been observed in cometary spectra, but its numerous absorption bands in the visual guarantee that it would be photodissociated into H_2 + CO in a few hundred seconds after leaving the nucleus. Hence it could not be detected in the coma; admittedly H_2 could be detected in the vacuum ultraviolet, but not necessarily if it is only at the level of the production rates of C_2, C_3, and CN; some H_2 should be produced anyway by the photodissociation of water. It is therefore impossible to decide whether formaldehyde is a parent molecule. Since the same type of reasoning is applicable for a score of other molecules, we conclude that it is too early to try writing a complete balance sheet for molecules.

C. Observed Production Rates of Elements

A less ambitious task may be possible, that is, to assess the different atomic ratios, in order to make a quantitative analysis of the elements present in the volatile fraction, namely H, C, N, O, S.

The rationale comes from the fact that all molecules are usually photodissociated into their atomic constituents, much sooner than the atoms themselves become ionized. In the solar radiation, there is not much energy left in the extreme ultraviolet which is needed to ionize atoms. The ionization potentials of H, C, N, O are all beyond 11.2 eV (S is 10.4 eV). For this reason, at 1 AU, their photoionization lifetimes are, by and large, between 10^6 and 10^7 s, whereas the dissociation potentials of most radicals and molecules are in the 3 to 6 eV range, corresponding to lifetimes in the 10^3 to 10^4 s range.

The most notable exception is CO, one of the strongest molecular bonds, with 11 eV, needing more than 10^6 s at 1 AU. For this reason, it must be considered with the atoms, as well as the next strongest bonds, namely N_2 with 9.8 eV whereas CN with 7.8 eV, will already be broken in not much more than 10^5 s. Charge exchange with the solar wind protons may diminish lifetimes against ionization by factors of 2 or 3 but this does not change the basic separation between atoms and molecules. These facts are consistent with observed brightness profiles in cometary heads. There is an extended zone between 10^5 and 10^6 km where the resonance lines of the atoms contain already all molecular contributions, with the exception of only a partial contribution from CO.

Fortunately the resonance lines of H, C, N, O, S as well as the bands of CO have become accessible in the 1970s by observations in the vacuum ultraviolet through rockets and orbiting telescopes. Only the resonance line of N has not yet been observed, apparently for technical reasons (it is near Lyman α and presumed to be weak).

The most homogeneous set of observations is probably that of Comet West 1976 VI by Feldman and Brune (1976), complemented by Opal et al. (1977a) and by Keller and Meier (1980), whereas the best H/OH ratio can probably be deduced from the long sequence of observations of Comet Bennett 1970 II by the OAO-2 from April 13 to May 13, 1970 (Keller and Lillie 1974; Bertaux et al. 1973). The unsteady behavior of Comet Kohoutek 1973 XII, in particular after perihelion, has enlarged the error bars of its observations. Let us consider the 4 ratios H/O, C/O, N/O, and S/O separately.

H/O. The two production rates of H and OH varied steadily in Comet Bennett 1970 II as $r^{-2.3}$ (r = heliocentric distance) during one whole month; their ratio (in the coma) was therefore constant and equal to H/OH = 1.8. However, Delsemme (1981a) has pointed out the fact that the ratio in the coma is not the ratio of the simultaneous production rates at the nucleus. The average lifetime of H was ~ 13 d in the observed coma, whereas that of OH was only 2 d. Therefore, the H atoms were, on the average, produced by the nucleus 11 d earlier than the OH molecules, at a time when the production rates were growing rapidly. Since the dependence law on distance is known from the observations, the proper correction is easy to make by shifting the production rates of OH and H by 2 and by 13 d, respectively. When reduced to simultaneous production rates, the H/OH ratio is 1.5 ± 0.3 (See Fig. 10). The H/OH ratio for Comet Kohoutek is more difficult to establish. Because there were repeated outbursts after perihelion (see O'Dell 1976) the apparent rates were out of phase due to the different delays of H and OH, yielding wide fluctuations in their ratios. The correction for the OH and H delays diminishes the extreme values and yields 1.8 ± 0.8 after perihelion (Delsemme 1977; Blamont and Festou 1974). For Comet West 1976 VI, since it was closer to the Sun, the shifts are 2 and 0.3 d for H and OH, respectively; using a square law of the distance, the ratio of the simultaneous production rates deduced from Feldman and Brune (1976) is H/OH = 2.2 ± 0.4. The error bars

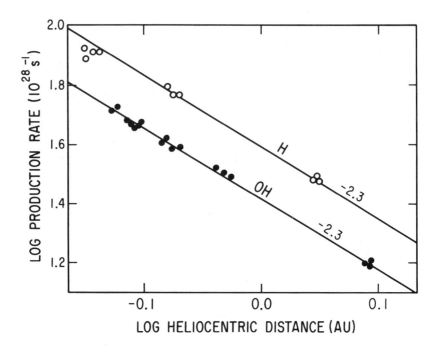

Fig. 10. Production rates of H and OH of Comet Bennett 1970 II, as a function of heliocentric distance. Data is from Keller and Lillie (1974) but shifted by 13 d for H and 2 d for OH, epochs at which the average atom (or molecule respectively) left the vicinity of the nucleus. The H/OH ratio is then 1.5. Compare the slopes with those of Fig. 1; only water ice can explain the − 2.3 slopes in the vicinity of 1 AU.

used are mainly for relative comparison between the three comets; they do not try to assess systematic errors so the results must be taken with a grain of salt. In order to establish the total H/O ratio, we must still consider whether there is another sizeable source of oxygen besides OH. Huppler et al. (1975) give H/O (^{1}D) = 2.5 ± 1.5 for Comet Kohoutek, but it comes from the unusual observation of the H α line in comets, and a major source of uncertainty is the lack of a reliable solar Lyman β profile for that particular day. Delsemme and Combi (1979) give H/O (^{1}D) = 2.5 ± 0.8 for Comet Bennett 1970 II; in their interpretation, 2/3 of the oxygen red line would come from the photodissociation of CO_2 and only 1/3 from that of H_2O. Since Delsemme and Combi used a small dispersion, it could be argued that up to 20% of their O (^{1}D) brightness could be attributed to a blend with NH_2. Besides, it is clear that only a fraction of the total oxygen is produced in the ^{1}D state. However, the best data probably come from Feldman and Brune (1976) revised by Feldman (1978) for Comet West. From their data I deduce O(I)/OH = 1.15 and CO/OH = 0.22. These two ratios are not inconsistent because, for those distances where OI is measured, we have seen that CO is not yet completely dissociated. If the data are taken at face value, the total H/O ratio for Comet West is

$$H/(CO + OH) = 1.8 \qquad (20)$$

If the 15% excess observed in the OI rate comes only from dissociated CO (which is quite a reasonable assumption) then there is no substantial oxygen source that is missing from our final ratio.

If comets were only producing water, their H/OH ratio should be 2 (the H present in OH is also counted as H after OH dissociates, when it reaches the Lyman α coma). The three values of 1.5, 1.8, and 2.2 suggest that the bulk of H and OH indeed comes from water. The total H/O ratio of 1.8, derived from Comet West, makes it clear that there is no room for any sizeable amount of hydrogen coming from another source, like methane or hydrocarbons in general, whereas there is apparently another source of oxygen, probably in the form of CO or CO_2, at the nominal level of 22% of OH (but the error bars are large). A sizeable fraction of CO_2 would help considerably in explaining the anomalously large brightness of the forbidden red line of oxygen (Delsemme 1980) as well as the infamous photometric behavior of Comet Kohoutek 1973 XII (Delsemme 1975b).

C/O. Feldman and Brune (1976) and Opal and Carruthers (1977b) have observed the 1657 Å resonance line of carbon in Comet Kohoutek 1973 XII on 5 and 8 January 1974, respectively. Opal and Carruthers (1977b) have discussed and revised these observations. They find, for the apparent ratio in the coma, C/O = 0.08 on 5 January (r = 0.34 AU) but C/O = 0.58 on 8 January (r = 0.43 AU). Here again the lifetime of O against ionization at 0.34 AU was 5 d whereas that of C was only 2 d. Comet Kohoutek's outbursts at perihelion and 11 d later, correspond to a production rate diminishing by a factor of 2 in the 3 days before 5 January, but growing by a factor of 2 in the 3 days before 8 January (see O'Dell 1976). The wide fluctuations in the apparent C/O coma ratio, diminish considerably when the time lag difference is taken into account. The two ratios, computed for the emission rates of the nucleus, become 0.16 and 0.28, an average of

$$C/O = 0.22 \pm 0.06 \tag{21}$$

but this result neglects the fraction of CO not yet dissociated and is therefore a lower limit.

Let us now consider Feldman's (1978) revision for Comet West 1976 VI. In 10^{28} atom s^{-1}, he gives O = 110 (Feldman and Brune 1976) but C = 19 and CO = 26 (from erratum in Feldman's revision); this yields a total

$$C/O = 0.21 \pm 0.03 . \tag{22}$$

The existence of $\sim$ 1/3 of the carbon atoms in the ^{1}D state must clearly come from their particular production process. Feldman and Brune (1976) show that it is consistent with dissociative recombination of CO$^+$ into C(^{1}D) + O(^{1}D). However, the fact that the brightness profile of O(^{1}D) is completely symmetrical sunwards and tailwards in comets, excludes any sizeable plasma interaction, and therefore seems to exclude any sizeable amount of CO^+ for its parent (Delsemme and Combi 1979). Discussing carbon in Comet Bradfield, A'Hearn and Feldman (1980) recognize the inability of the model of Feldman (1978) which gave a good fit for Comet West, to produce enough atomic carbon by two orders of magnitude to account for the observations of Comet Bradfield. They suggest that an additional source of carbon other than CO is needed. Feldman's (1978) model did not include CO_2 or CO_2^+, nor CS which A'Hearn and Feldman (1980) report in Comet Bradfield. Smith et al. (1980) observing Comet West by rocket on 10 March 1976, also detected CS, a new source of carbon indeed, but only at a level of 1 or 2% of the total carbon content, in other words at the same level as CN and C_2 or C_3. We conclude that the total C/O ratio in the volatile fraction of comets still is rather uncertain, in particular because the mechanism producing $C(^1D)$ is not clearly understood. However, it cannot be very different from C/O = 0.2 ± 0.1.

N/O. Since the resonance line of N has not been observed, the production rate of N must be assessed from the N-bearing radicals and ions, namely CN, NH, NH_2, and N_2^+. The microwave observations of HCN only confirm the order of magnitude of the CN production rate. The transition probability of the NH_2 bands is not known with accuracy. The known *f*-values are $(2.7 \pm 0.3)10^{-2}$ for CN, $(3.5 \pm 0.2)10^{-2}$ for N_2^+, and $(8.0 \pm 1.1)10^{-3}$ for NH (Bennett and Dalby 1962,1959,1960, respectively). Considering the relative intensities of the CN, N_2^+, and NH bands in different representative spectra (Swings and Haser 1956) in connection with their excitation by the solar spectrum, the N_2^+ cannot be neglected with respect to the NH bands; a lifetime of 4×10^5 s for NH (Combi 1978) implies that NH must be produced at a rate about half of that of CN, whereas the slow ionization rate of N_2 suggests that N_2 = 2 CN. Normalizing by using CN/OH = 0.016 from A'Hearn and Millis (1980), I adopt N/O = 0.1, admittedly a very poor assessment.

S/O. The discovery of S and CS in the spectrum of Comet West (Smith et al. 1976,1980) has been confirmed by A'Hearn and Feldman (1980) for the same comet, as well as for Comet Bradfield. Smith et al. (1980) give S/C (^3P) = 0.045 and CS/C (^3P) = 0.008. This can be transformed, from Feldman's (1978) data, as total S/O = 0.003, still a preliminary value that must be clarified in the future.

VII. THE ELEMENTAL ABUNDANCES IN COMETS
A. A Heuristic Model

As established by the previous discussion, it is too early to make a significant chemical model for any of the individual comets mentioned. However, since no spectacular difference has emerged from comet to comet, we can accept the working hypothesis that comets are all made from the same (rather undifferentiated) material, and use the conclusions of the previous sections to build a heuristic model representing the average of the recent bright comets.

TABLE V

Mean Abundance of the Light
Elements in the Volatile Fraction of
Comets (Normalized to O = 1)[a]

Element	Abundance
H	1.5
C	0.2
N	0.1
O	1.0
S	0.003

[a]The H/O ratio is probably known with a ± 30% accuracy; the other ratios may still have error bars larger than a factor of 2.

For this purpose, the volatile fraction is represented by the mean model given in Table V; cometary dust is represented by the mean abundances of the CI chondrites, found in Mason (1971); finally, 0.8 is adopted for the mean dust-to-gas ratio by mass.

The results of this exercise are given in Table VI. Only the 10 most abundant elements need to be represented for our purpose; Ni and Cr are put together with a mean atomic mass of 55. The two columns under "Cometary Dust" give Mason's (1971) abundance data for CI chondrites, in number of atoms and in atomic mass units, normalized to Si = 1 or 28 amu, respectively. The next two columns under "Cometary Gas" use the atomic ratios listed in Table V, normalized to give a dust-to-gas mass ratio of 0.8. The two columns under "Comet Totals" are the sums of the columns labeled "Dust" and "Gas", in atom numbers and in amu respectively; since no metals are assumed to be present in the gas, the "comet totals" remain normalized to silicon = 1 (or 28 amu). Finally, the two last columns express the same results in atom number % and in mass %, respectively. By mass, oxygen is the most abundant element, ~ 47% of the dust and 75% of the gas, whereas hydrogen is 1% of the dust and 7% or the gas.

In Table VII, the previous results are compared with carbonaceous chondrites (Mason 1971) and with cosmic abundances borrowed from Cameron (1981) who incorporates recent solar data for the light elements (Ross and Aller 1976).

B. Discussion

The most significant feature of Table VII is the cometary depletion of hydrogen, in respect to cosmic abundances. Even if the cometary abundance of hydrogen is not known better than within a factor of 2, its depletion by some 3 orders of magnitude is probably one of the best established results of

TABLE VI
Elemental Abundances in Recent Bright Comets
(an average heuristic model)

Name of Element	Atomic Mass	Cometary Dust[a]		Cometary Gas[b]		Comet Totals Normalized to Si		Comet Totals in %	
		Number	Mass	Number	Mass	Number	Mass	Number	Mass
H	1	2.00	2.0	22.30	22.3	24.30	24.3	43.9	4.2
C	12	0.70	8.4	3.03	35.7	3.73	44.1	6.7	7.7
N	14	0.05	0.7	1.46	20.9	1.51	21.6	2.7	3.8
O	16	7.50	120.0	14.80	237.8	22.30	357.8	40.2	62.5
S	32	0.50	16.0	0.05	1.6	0.55	17.6	1.0	3.1
Mg	24.3	1.06	25.8	—	—	1.06	25.8	1.9	4.5
Si	28	1.00	28.0	—	—	1.00	28.0	1.8	4.9
Fe	55.9	0.90	50.3	—	—	0.90	50.3	1.6	8.8
Ni + Cr	< 55 >	0.06	3.3	—	—	0.06	3.3	0.1	0.6
Totals	—	13.77	254.5	41.64	318.3	55.41	572.8	100	100

[a] Assuming a C I chondrite elemental composition; C I data are from Mason (1971) normalized to silicon.
[b] Gas is normalized to dust, assuming dust-to-gas mass ratio of 0.8. The atomic ratios are from Table IV.

TABLE VII

Elemental Abundances in Atom Numbers
(normalized to Si = 1)

Element	Cosmic (Cameron 1981)	Comets (this work)	Chondrites (Mason 1971)[a]		
			CI	CM	CV, CO
H	26600.00	24.30	2.00	1.00	0.10
C	11.70	3.73	0.70	0.40	0.08
N	2.31	1.51	0.05	0.04	0.01
O	18.40	22.30	7.50	5.30	4.10
S	0.50	0.55	0.50	0.23	0.12
Si	1.00	1.00	1.00	1.00	1.00

[a]Data from Mason (1971) are rounded averages; H from his p. 22, C from p. 83, N from p. 96, S from p. 138.

this work. Not only is the old dream of putting liquid or solid hydrogen in the cometary nuclei definitely ruled out, but the H/O ratio is probably lower than 2 and possibly not much higher than 1; such a redox ratio seems to imply that carbon is mainly in CO or CO_2 and that hydrocarbons or other hydrogenated compounds exist in traces only. If comets were condensed out of the solar nebula, this would imply that during cooling off, the CO existing at high temperature was quenched by the slow kinetics of its reduction into CH_4. This view, that I proposed for comets (Delsemme 1976), has been developed for solar nebula models by Lewis and Prinn (1980).

The model proposed in Table VII has another significant feature; it predicts that oxygen is very close to cosmic abundances in comets. Nominally, oxygen would remain lower than its nominal cosmic abundance for $M < 0.6$, and higher for $M > 0.6$. An oxygen abundance higher than its cosmic value would really mean that metals have been depleted with respect to the more volatile fraction, a hypothesis that seems at variance with all previous evidence, whereas a lower abundance would only mean that some oxygen-bearing molecules are missing (for instance, this is a real possibility for CO because of its volatility). It is useful to remember that the model predicts that some 2/3 of the oxygen is in the gas and 1/3 in the dust, and that $M = 0.6$ is within the error bar of the value (0.76 ± 0.30) deduced in Sec. V. The model abundances of N and S differ from their cosmic value, but the difference is not large enough to be significant.

The most puzzling feature of the model probably lies in its carbon depletion by a factor of almost 4. As mentioned previously, the C/O ratio, hence the carbon abundance, is not known with accuracy. However, since a missing factor of 4 seems very high to be attributed to errors only, alternate explanations must also be explored. These explanations range in two categories since one may assume that, either the carbon is really missing in comets, or that it is hidden in the dust where it cannot be detected (Delsemme and Rud 1977).

Let us consider the first category. If 75% of the cosmic carbon is actually missing in comets, it could be interpreted as an indication of the condensation temperature of the solar nebula when comets were formed. If the cosmic abundance of oxygen is accepted at face value, then the missing carbon could not be in CO, but only in CH_4; this implies that the condensation temperature was higher than that of CH_4, specifying a low temperature limit for the condensation of comets (typically 40 K for the usual pressures used in solar nebula models in the 10 to 100 AU range).

In the same category, an alternate possibility is that the recent bright comets used to establish the model had already lost some very volatile carbon-bearing compounds. Comet Kohoutek seems to have lost much CO_2 on its first leg to perihelion, and the observations used here concern its second leg after perihelion. Even comets that are nominally "new" like Comet Arend-Roland (from the binding energy of its original orbit) are likely to have had previous perihelion passages only slightly beyond Jupiter (Sec. II.F) where they would have lost their most volatile molecules (see Table II). Here again, the oxygen abundance seems to imply that no oxygen-bearing molecule has been lost in large quantities.

The second category of explanations would imply that the 3/4 of cosmic carbon is really hidden in the dust; our assumption that dust is chemically identical to CI chondrites can indeed be wrong. The extrapolation proposed in Table VI from carbonaceous chondrites to comets, even if acceptable, is not without ambiguity. The missing organic fraction could be too volatile to be found in museum specimens, but not volatile enough to be detected in gaseous phase in space. Typically, it could be in hydrocarbons of higher molecular weight, or even in nonvolatile polymers like polyformaldehyde, or even polyacetylenes or graphite (Delsemme 1953); it could also be in carbynes, these long polymers of acetylene that are now considered to be a new allotropic form of carbon (Anders, personal communication). Whether this organic fraction does exist (and if it does, what is its chemical nature) is a question directly related to the origin of comets. After all, Brownlee's particles contain already 5% carbon by mass, but the cometary dust should contain 30% carbon for comets to reach cosmic abundances.

C. The Origin of Comets

There is a consensus that the origin of the Oort Cloud of comets is likely to be linked more or less with the origin of the solar system, although there are competing scenarios that will not be described here in any detail. The basic difference boils down to the heliocentric distance of comets' formation.

If they accreted directly from frosty interstellar grains, at distances from 10^3 to 10^5 AU, they may still represent a sample of (more or less compacted) pristine interstellar grains that may never have been processed by heat. Standing in contrast, if interstellar grains followed suit during the gravitational collapse leading to the solar nebula, when the turbulence subsided after formation of the central body, the grains may have sedimented into rings in the midplane of the nebula; gravitational collapse in the rings may have then formed planetesimals of the 1- to 10-km size. Beyond an approximate limit of 5 AU the grains would keep their icy mantles, more or less processed by solar heat. Comets would form beyond this limit and would display a continuous range of volatilities, depending on their temperature during accretion. Grains could also become the condensation centers that collect ices, if the solar nebula reached saturation pressures for volatile materials. In this alternate view, the inhibition of the CO reduction into CH_4 mentioned earlier would produce the high depletion of hydrogen observed in comets. Clues diagnostic of one of these scenarios may be found when better data are obtained, in particular when the apparent depletion of carbon is better understood. It is, however, already tempting to speculate by using a direct comparison with the chemistry of interstellar grains (Delsemme 1981c). Greenberg in his chapter in this book discusses his experiments and shows where the missing carbon is, if indeed comets are basically unprocessed interstellar grains.

D. Conclusions

Thirty years have passed since Whipple's (1950,1951) icy conglomerate model. The major constituent of the volatile fraction is now confirmed to be water, presumed to be present in the form of ice or snow, mixed up with dust in the nucleus, and often controlling the rate of the vaporization of this icy conglomerate. However, the most surprising feature of the elemental abundances in comets is the large depletion of hydrogen, which seems to suggest that there is barely enough H to make water. Therefore the next volatile constituents are probably CO and CO_2 rather than the traditional CH_4. Most parent molecules, although still unidentified, seem to be in the general range of 1 or 2% the abundance of water, whereas CO and CO_2 could be in the 10 to 30% range. CO or CO_2 may indeed be abundant enough to sometimes control the vaporization of the icy conglomerate in unusual comets (Morehouse, Humason, Schwassmann-Wachmann 1, as well as other historic comets); CO_2 seems to have controlled the infamous lightcurve behavior of Comet Kohoutek on its first leg to its perihelion passage.

A final word of caution is needed. Since the results used in this review are scanty, unknown systematic errors may be large. Our conclusions should be critically reviewed as soon as better data are available.

Acknowledgments: C. Chapman made many useful remarks as a referee and convinced the reviewer to develop the beginning of this chapter much more than he originally intended. Comments on the manuscript by P. Feldman, W. Huebner, J. Rahe and Z. Sekanina were very useful. J. Wood corrected a misconception. M. Combi critically proofread the final manuscript. Grants from the National Science Foundation and the Planetary Atmosphere Program of the National Aeronautics and Space Administration are gratefully acknowledged.

REFERENCES

A'Hearn, M.F., and Cowan, J.J. 1975. Molecular production rates in Comet Kohoutek. *Astron. J.* 80:852-860.
A'Hearn, M.F., and Feldman, P.D. 1980. Carbon in Comet Bradfield. *Astrophys. J.* 242:L187-L190.
A'Hearn, M.F., and Millis, R.L. 1980. Abundance correlations among comets. *Astron. J.* 85:1528-1537. (See erratum 86:802.)
Aikin, A.C. 1974. Cometary coma ions. *Astrophys. J.* 193:263-264.
Arpigny, C. 1978. On the nature of comets. In *Proc. Welch Conf. on Cosmochemistry*, ed. W.O. Milligan (Houston: Welch Foundation), pp. 9-57.
Arpigny, C. 1979. Relative abundances of the heavy elements in Comet Ikeya Seki. In *Elements and Isotopes in the Universe*, 22d Colloq. Astrophys. Liège, (Univ. Liège), pp. 189-197.
Bates, D.R. 1951. The temperature of the upper atmosphere. *Proc. Roy. Phys. Soc. London* Sec. B 64:805-821.
Bennett, R.G., and Dalby, F.W. 1959. Experimental determination of the oscillator stength of the first negative band of N_2^+. *J. Chem. Phys.* 31:434-441.
Bennett, R.G., and Dalby, F.W. 1960. Experimental oscillator strength of CH and NH. *J. Chem. Phys.* 32:1716-1719.
Bennett, R.G., and Dalby, F.W. 1962. Experimental oscillator strength of the violet system of CN. *J. Chem. Phys.* 36:399-405.
Bertaux, J.L.; Blamont, J.E.; and Festou, M. 1973. Interpretation of hydrogen Lyman-alpha observations of comets Bennett and Encke. *Astron. Astrophys.* 25:415-430.
Beyer, M. 1950a. Physische Beobachtungen von Kometen VII. *Astron. Nachr.* 278:217-249.
Beyer, M. 1950b. Physische Beobachtungen von Kometen VIII. *Astron. Nachr.* 279:49-61.
Beyer, M. 1955. Physische Beobachtungen von Kometen IX. *Astron. Nachr.* 282:145-167.
Beyer, M. 1961. Physische Beobachtungen von Kometen XI. *Astron. Nachr.* 286:211-262.
Beyer, M. 1962. Physische Beobachtungen von Kometen XII. *Astron. Nachr.* 286:219-240.
Blamont, J.E., and Festou, M. 1974. Observation of Comet Kohoutek (1973f) in the resonance light the OH radical. *Icarus* 23:538-544.
Bobrovnikoff, N.T. 1931. Halley's comet in its apparition of 1909-1911. *Publ. Lick Observ.* 17:309-482.
Bobrovnikoff, N.T. 1954. Physical properties of comets (in Perkins Observatory Report). *Astron. J.* 59:357-358.
Brin, G.D. 1980. Three models of dust layers on cometary nuclei. *Astrophys. J.* 237:265-279.
Brin, G.D. 1980. Three models of dust layers on cometary nuclei. *Astrophys. J. Astrophys. J.* 229:402-408.
Brownlee, D.E.; Rajan, R.S.; and Tomandl, D.A. 1977. A chemical and textural comparison between carbonaceous chondrites and interplanetry dust. In *Comets, Asteroids, Meteorites*, ed. A.H. Delsemme (Toledo, Ohio: Univ. of Toledo), pp. 137-142.
Buhl, D.; Huebner, W.F.; and Snyder, L.E. 1976. Detection of molecular microwave transitions in the 3 mm wavelength range in Comet Kohoutek. In *The Study of Comets*, eds. B. Donn, M. Mumma, W. Jackson, M. A'Hearn, and R. Harrington (Washington: NASA-SP 393), pp. 253-271.

Cameron, A.G.W. 1981. Elementary and nuclidic abundances in the solar system. In *Essays in Nuclear Astrophysics,* eds. C.A. Barnes, D.D. Clayton, and D.N. Schramm (London: Cambridge Univ. Press). In press. (Also *Center for Astrophys. Preprint Ser.* No. 1357, 1980.)

Ceplecha, A. 1958. Photographs of Comet Arend-Roland. *Publ. Czech. Astron. Inst.* 34:13-22.

Cherednichenko, V.I. 1973. Some possible reactions yielding ions N_2^+ and CO^+ by ion-molecular exchange in cometary atmospheres. *Problems of Cosmic Physics* 8:89-92.

Combi, M.R. 1978. Convolution of cometary brightness profiles by circular diaphragms. *Astron. J.* 83:1459-1466.

Combi, M.R., and Delsemme, A.H. 1980a. Neutral cometary atmospheres I. An average random walk model for photodissociation in comets. *Astrophys. J.* 237:633-640.

Combi, M.R., and Delsemme, A.H. 1980b. Neutral cometary atmospheres II. The production of CN in comets. *Astrophys. J.* 237:641-645.

Crovisier, J. 1981. L'observation radioastronomique des comètes. *I'Astronomie* 95:231-240.

Delsemme, A.H. 1953. Discussions sur le chapitre III. In *La Physique des Comètes. Mém.* Soc. Roy. Sci. Liège, 4th ser. 13:196-197.

Delsemme, A.H. 1966. Vers un modèle physico-chimique du noyau comètaire. In *Nature et Origine des Comètes.* Mém. Soc. Roy. Sci. Liège, 5th ser. 12:77-110.

Delsemme, A.H. 1972. Vaporization theory and non-gravitational forces in comets. In *Origin of the Solar System,* ed. H. Reeves, (Paris: C.E.R.N.) pp. 305-310. Also in *Asteroids, Comets, Meteoritic Matter,* ed. C. Cristescu, W. Klepczynski, and B. Milet (Romania: Publ. Acad. Soc. Republ. of Romania, 1974), pp. 315-322.

Delsemme, A.H. 1973. The brightness law of comets. *Astrophys. Letters* 14:163-167.

Delsemme, A.H. 1975a. The volatile fraction of the cometary nucleus. *Icarus* 24:95-110.

Delsemme, A.H. 1975b. Physical interpretation of the brightness law of Comet Kohoutek. In *Comet Kohoutek,* ed. G.A. Gary, (Washington: NASA-SP 355), pp. 195-203.

Delsemme, A.H. 1976. Chemical nature of the cometary snows. *Mém. Soc. Roy. Sci. Liège,* 6th ser. 9:135-145.

Delsemme, A.H. 1977. The pristine nature of comets. In *Comets, Asteroids, Meteorites,* ed. A.H. Delsemme (Toledo, Ohio: Univ. of Toledo), pp. 3-13.

Delsemme, A.H. 1980. Photodissociation of CO_2 into $CO + O(^1D)$. In *Les Spectres des Molecules Simples,* 21st Colloq. Astrophys. Liège, (Univ. Liège), pp. 515-524.

Delsemme, A.H. 1981a. Observing chemical abundances in comets. In *Workshop on Modern Observational Techniques for Comets.* (Pasadena: Jet Propulsion Lab.), pp. 107-109.

Delsemme, A.H. 1981b. Are comets connected to the origin of life? In *Comets and the Origin of Life,* ed. C. Ponnamperuma (Dordrecht: D. Reidel), pp. 141-159.

Delsemme, A.H. 1981c. Nature and Origin of organic molecules in comets. In *Origin of Life,* Proc. 3rd ISSOL Meeting, ed. Y. Wolman, (Dordrecht: Reidel), pp. 33-42.

Delsemme, A.H., and Combi, M.R. 1979. $O(^1D)$ and H_2O^+ in Comet Bennett 1970 II. *Astrophys. J.* 228:330-337.

Delsemme, A.H., and Miller, D.C. 1970. Physico-chemical phenomena in comets, II. Gas absorption in the snows of the nucleus. *Planet. Space Sci.* 18:717-730.

Delsemme, A.H., and Miller, D.C. 1971. Physico-chemical phenomena in comets, III. The continuum of Comet Burnham. *Planet. Space Sci.* 19:1229-1258.

Delsemme, A.H., and Rud, D.A. 1973. Albedos and cross-sections of the nuclei of comets 1969 IX, 1970 II, and 1971 I. *Astron. Astrophys.* 28:1-6.

Delsemme, A.H., and Rud, D.A. 1977. Low-temperature condensates in comets. In *Comets, Asteroids, Meteorites,* ed. A.H. Delsemme (Toledo, Ohio: Univ. of Toledo), pp. 529-535.

Delsemme, A.H., and Swings, P. 1952. Hydrates de gaz dans les Noyaux Comètaires et les Grains Interstellaires. *Annales d'Astrophys.* 15:1-6.

Delsemme, J.A. 1972. An analytic approximation of the dependance on distance for the vaporization of comets. Unpublished preprint, Univ. of Toledo.

Donn, B. 1976. The nucleus: Panel discussion. In *The Study of Comets,* (Washington: NASA-SP 393), pp. 611-619.

Donn, B. 1977. A comparison of the composition of new and evolved comets. In *Comets, Asteroids, Meteorites,* ed. A.H. Delsemme, (Toledo, Ohio: Univ. of Toledo), pp. 15-23.
Feldman, P.D. 1978. A Model of carbon production in a cometary coma. *Astron. Astrophys.* 70:547-553.
Feldman, P.D., and Brune, W.H. 1976. Carbon production in Comet West. *Astrophys. J.* 209:L45-L48.
Ferrin, I., and Naranjo, O. 1980. A possible explanation of the lightcurve of Comet Encke. *Mon. Not. Roy. Astron. Soc.* 193:667-681.
Festou, M. 1978. L'Hydrogène Atomique et le Radical Oxhydrile dans les Comètes. Thèse de Doctorat d'Etat, Université Paris.
Finson, M.L., and Probstein, R.F. 1968a. A theory of dust comets. I. Model and equations. *Astrophys. J.* 154:327-352.
Finson, M.L., and Probstein, R.F. 1968b. A theory of dust comets. II. Results for Comet Arend-Roland. *Astrophys. J.* 154:353-380.
Hanner, M.S. 1980. Physical characteristics of cometary dust from optical studies. In *Solid Particles in the Solar System,* eds. I. Halliday, and B.A. McIntosh (Dordrecht: D. Reidel), pp. 223-236.
Hellmich, R., and Keller, H.U. 1980. On the dust production rate of comets. In *Solid Particles in the Solar System,* eds. I. Halliday, and B.A. McIntosh (Dordrecht: D. Reidel), pp. 255-258.
Hoyle, F., and Wickramashinghe, C. 1979. *Diseases from Space,* (New York: Harper and Row).
Huebner, W.F. 1965. Ueber die Gasproduktion der Kometen. *Z. Astrophys.* 63:22-34.
Huebner, W.F.; Snyder, L.E.; and Buhl, D. 1974. HCN radio emission from Comet Kohoutek (1973f). *Icarus* 23:580-584.
Huebner, W.F.; Buhl, D.; and Snyder, L.E. 1976. Microwave line transitions in the 3-mm wavelength range of Comet Kohoutek. *Astron. J.* 81:671-674.
Huppler, D.; Reynolds, R.J.; Roesler, F.L.; Scherb, F.; and Trauger, J. 1975. Observations of Comet Kohoutek (1973f) with a ground-based Fabry-Perot spectrometer. *Astrophys. J.* 202:276-282.
Irvine, W.M.; Leschine, S.B.; and Schloerb, F.P. 1981. Comets and the origin of life. In *Origins of Life,* Proc. 3rd ISSOL Meeting, ed. Y. Wolman (Dordrecht: D. Reidel), p. 2.
Jackson, W.M.; Clark, T.; and Donn, B. 1976. Radio detection of Comet Bradfield. In *The Study of Comets,* eds. B. Donn, M. Mumma, W. Jackson, M. A'Hearn, and R. Harrington (Washington: NASA-SP 393), pp. 272-280.
Jambor, B.J. 1973. The split tail of Comet Seki-Lines. *Astrophys J.* 185:727-734.
Keller, H.U., and Lillie, C.F. 1974. The scale length of OH and the production rates of H and OH in Comet Bennett (1970II). *Astron. Astrophys.* 34:187-196.
Keller, H.U., and Meier, R.R. 1980. On the Lyman α isophotes of Comet West. *Astron. Astrophys.* 81:210-214.
Keller, H.U., and Thomas, G.E. 1975. A cometary hydrogen model: Comparisons with OGO-5 measurements of Comet Bennett (1970II). *Astron. Astrophys.* 39:7-19.
Kohl, J.L.; Parkinson, W.H.; and Kurucz, R.L. 1978. *Center and Limb Solar Spectrum in High Resolution* (Cambridge: Smithsonian Center for Astrophys.), pp. 300-340.
Kresák, L. 1981. Evolutionary aspects of the splits of cometary nuclei. *Bull. Astron. Czech.* 32:19-40.
Lee, T., and Papanastassiou, D.A. 1974. Mg isotopic anomalies in the Allende meteorite and correlation with O and Sr effects. *Geophys. Res. Letters* 1:225-228.
Levin, B.J. 1943. *Astron. Zh.* 20:48 (in Russian). Also quoted in Levin, B.J. 1972 (see below).
Levin, B.J. 1948. Variation in the brightness of comets. *Astron. Zh.* 25:246-252.
Levin, B.J. 1972. Some remarks on the liberation of gases from cometary nuclei. In *The Motions, Evolution of Orbits and the Origin of Comets,* eds. G.A. Chebotarev, E.I. Kazimirchak-Polonskaya, and B. Marsden (Dordrecht: D. Reidel), pp. 260-264.
Lewis, J.S., and Prinn, R.G. 1980. Kinetic inhibition of CO and N_2 reduction in the solar nebula. *Astrophys. J.* 238:357-364.

Malaise, D.J. 1970. Collisional effects in cometary atmospheres. *Astron. Astrophys.* 5:209-227.
Marsden, B.G. 1969. Comets and nongravitational forces. II. *Astron. J.* 74:720-734.
Marsden, B.G.; Sekanina, Z.; and Everhart, E. 1978. New osculating orbits for 110 comets and analysis of original orbits for 200 comets. *Astron. J.* 83:64-71.
Marsden, B.G.; Sekanina, Z.; and Yeomans, D.K. 1973. Comets and nongravitational forces. V. *Astron. J.* 78:211-225.
Mason, B. 1963. Density of carbonaceous chondrites. *Space Sci. Rev.* 1:621-646.
Mason, B. 1971. *Handbook of Elemental Abundances in Meteorites,* ed. B. Mason (New York: Gordon and Breach), pp. 21-28 for H, 81-91 for C, 93-98 for N, and 99-107 for O.
Meier, R.R.; Opal, C.B; Keller, H.U.; Page, T.L.; and Carruthers, G.R. 1976. Hydrogen production rates from Lyman-α images of Comet Kohoutek (1973XII). *Astron. Astrophys.* 52:283-290.
Mendis, D.A., and Brin, G.D. 1977. Monochromatic brightness variations of comets II. Core-mantle model. *Moon* 17:359-372.
Mendis, D.A., and Brin, G.D. 1978. On the monochromatic brightness variations of comets. *Moon and Planets* 18:77-89.
Millman, P.M. 1977. The chemical composition of cometary meteoroids. In *Comets, Asteroids, Meteorites,* ed. A.H. Delsemme (Toledo, Ohio: Univ. of Toledo), pp 127-132.
Mitchell, G.F.; Prasad, S.S.; and Huntress, W.T. 1981. Chemical model calculations of C_2, C_3, CH, CN, OH, and NH_2 abundances in cometary comae. *Astrophys. J.* 244:1087-1093.
Naranjo, O. 1978. Metodo para la determinación de albedo Bond, radio nuclear y relación α de los cometas. Lic. Thesis Univ. of Los Andes, Venezuela.
Ney, E.P. 1974a. Infrared observations of Comet Kohoutek near perihelion. *Astrophys. J.* 189:L141-L144.
Ney, E.P. 1974b. Multiband photometry of comets Kohoutek, Bennett, Bradfield, and Encke. *Icarus* 23:551-560.
Ney, E.P., and Merrill, K.M. 1976. Comet West and the scattering function of cometary dust. *Science* 194:1051-1053.
O'Dell, C.R. 1971. Nature of particulate matter in comets as determined from infrared observations. *Astrophys. J.* 166:675-681.
O'Dell, C.R. 1976. Physical processes in Comet Kohoutek. *Publ. Astron. Soc. Pacific* 88:342-348.
Oort, J., and Schmidt, M. 1951. Differences between new and old comets. *Bull. Astron. Inst. Netherlands* 11:259-270.
Opal, C.B., and Carruthers, G.R. 1977a. Lyman-alpha observations of Comet West. *Icarus* 31:503-509.
Opal, C.B., and Carruthers, G.R. 1977b. Carbon and oxygen production rates for Comet Kohoutek (1973XII). *Astrophys. J.* 211:294-299.
Papanastassiou, D.A.; Lee, T.; and Wasserburg, G.J. 1977. Evidence for $^{26}A\ell$ in the solar system. In *Comets, Asteroids, Meteorites,* ed. A.H. Delsemme (Toledo, Ohio: Univ. of Toledo), pp. 343-350.
Parchmann, J.E. 1972. A study of the non-gravitational forces in Comet Schwassmann-Wachmann 2. Master's Thesis, University of Toledo.
Pittich, E.M. 1971. Space distribution of the splitting and outbursts of comets. *Bull. Astron. Inst. Czech.* 27:143-153.
Potter, A.E., and Del Duca, B. 1964. Lifetime in space of possible parent molecules of cometary radicals. *Icarus* 3:103-108.
Probstein, R.F. 1968. The dusty gas dynamics of comet heads. In *Problems of Hydrodynamics and Continuum Mechanics,* (Philadelphia: Soc. Industr. Appl. Math.), pp. 568-583.
Roemer, E. 1966. The dimensions of cometary nuclei. *Mém. Soc. Roy. Sci. Liège,* 5th ser. 12:23-31.
Ross, J.E., and Aller, L.H. 1976. The chemical composition of the sun. *Science* 191:1223-1229.
Sekanina, Z. 1974. On the nature of the anti-tail of Comet Kohoutek. *Icarus* 23:502-518.

Sekanina, Z. 1977. Relative motions of fragments of the split comets. I. A new approach. *Icarus* 30:574-594.
Sekanina, Z. 1981. Rotation and precession of cometary nuclei. *Ann. Rev. Earth Planet. Sci.* 9:113-145.
Sekanina, Z., and Miller, F.D. 1973. Comet Bennett 1970II. *Science* 179:565-567.
Shimizu, M. 1976. Neutral temperature of cometary atmospheres, Part 2. In *The Study of Comets,* eds. B. Donn, M. Mumma, W. Jackson, M. A'Hearn and R. Harrington (Washington: NASA-SP 393), pp. 763-772.
Shul'man, L.M. 1972. The chemical composition of cometary nuclei. In *The Motion, Evolution of Orbits and Origin of Comets,* eds. G.A. Chebotarev, E.I. Kazimirchak-Polonskaya, and B.G. Marsden (Dordrecht: D. Reidel), pp. 265-270.
Smith, A.M.; Bohlin, R.G.; and Stecher, T.P. 1976. Ultraviolet objective grating imagery of Comet West. *Bull. Amer. Astron. Soc.* 8:343 (abstract).
Smith, A.M.; Stecher, T.P.; and Casswell, L. 1980. Production of carbon, sulfur, and CS in Comet West. *Astrophys. J.* 242:402-410.
Stefanik, R.P. 1966. On thirteen split comets. *Mém. Soc. Roy. Sci. Liège,* 5th ser. 12:29-32.
Swings, P. 1943. Cometary spectra. *Mon. Not. Roy. Astron. Soc.* 103:83-111. (Swings effect first described in *Lick Obs. Bull.* 19:131, 1941.)
Swings, P., and Haser, L. 1956. *Atlas of Representative Cometary Spectra* (Univ. Liège, Institut d'Astrophysique) *passim.*
Ulich, B.L., and Conklin, E.K. 1974. Detection of methyl cyanide in Comet Kohoutek. *Nature* 248:121-122.
Wallis, M.K. 1980. Radiogenic melting of primordial comet interiors. *Nature* 284:431-433.
Watson, K.; Murray, B.; and Brown, H. 1961. The behavior of volatiles on the lunar surface. *J. Geophys. Res.* 66:3033-3045.
Weigert, A. 1959. Halobildung beim Kometen 1925 II. *Astron. Nachr.* 282:117-129.
Weissman, P.R., and Kieffer, H.H. 1981. Thermal modeling of cometary nuclei. *Icarus* (special Comet issue). In press.
Wetherill, G.W. 1974. Solar system sources of meteorites and large meteoroids. *Ann. Rev. Earth Planet. Sci.* 2:303-331.
Whipple, F.L. 1950. A comet model. I. The acceleration of Comet Encke. *Astrophys. J.* 111:375-394.
Whipple, F.L. 1951. A comet model. II. Physical relations for comets and meteors. *Astrophys. J.* 113:464-474.
Whipple, F.L. 1977. The constitution of cometary nuclei. In *Comets, Asteroids, Meteorites,* ed. A.H. Delsemme (Toledo, Ohio: Univ. of Toledo), pp. 25-35.
Whipple, F.L. 1980. Rotation and outbursts of Comet P/Schwassmann-Wachmann 1. *Astron. J.* 85:305-313.

WHAT ARE COMETS MADE OF? A MODEL BASED ON INTERSTELLAR DUST

J.M. GREENBERG
University of Leiden

A model for pristine comet composition is derived on the basis of aggregated interstellar dust. The chemical and physical constituents of the precomet dust are derived by observing the evolution of clouds of dust and molecules into the molecular cloud phase and extrapolating theoretically to the ultimate composition of the dust. The observational characteristics of the dust are interpreted by a combination of theory with the results of laboratory simulation of the photochemical evolution of interstellar dust materials. Laboratory measures of the infrared absorption of mixtures containing H_2O ice provide a key step in predicting that $\sim 27\%$ by volume of comets is in the form of amorphous H_2O ice. The laboratory results also predict that $\sim 21\%$ of comets consists of complex nonvolatile organic molecules of prebiotic type. Predictions based on the aggregated interstellar dust comet model are consistent as of 1981 with many key observational properties of comets.

The comets which occupy the Oort Cloud (Oort 1950; Oort and Schmidt 1951), a region of space extending to ∼50,000 AU from the Sun, probably have been extremely cold ($\leqslant 10K$) for billions of years. Current theories lean strongly toward the hypothesis that comets originated either far out in the presolar nebula or in a neighboring fragment of the interstellar cloud which formed the solar system (Cameron 1973; Biermann and Michel 1978). In either case the comets were born cold, directly out of the interstellar medium, and unless internally heated retain exactly their internal composition at birth ~4.5 × 10^9 yr ago, i.e. a condensate of dust and gas in the precometary interstellar cloud. If there exists a population of comet-size bodies which accreted in the general planetary region, we may assume that they consist of completely recycled interstellar material in which the original interstellar dust completely evaporated before the total mixture recondensed. The chemical composition of this type of body cannot be related directly to that of an interstellar cloud, except perhaps in terms of cosmic abundance (CA). Although some fraction of the Oort Cloud of comets may have originated in this way, such objects will not be considered here.

The concept of comets as an aggregate of interstellar dust is not new (O'Dell 1971). However, it was not until early work on the photochemical processing of interstellar dust (Greenberg et al. 1972) that it became possible to predict the complex chemistry of primitive comets (Greenberg 1973b,1977).

I apply to this concept the most recent data, as of 1981, on the observational properties of molecular clouds of dust and gas, and the results of new laboratory and theoretical studies of the physical and chemical evolution of the dust just before aggregation into a comet. This provides a basis to predict the original chemical composition of the comet. Then, what can happen to a comet after it forms may be used to provide a theory of the nature of new comets or of the interior of old comets. Although an outer skin of a comet may be significantly modified by cosmic rays in 10^9 yr (Moore and Donn 1981) this layer is quite thin and during the first passage of a new comet we would already expect the original comet material to be completely revealed. I shall not consider in this chapter the consequences of possible metamorphosis due to internal heating of the comet by primordial radioactive elements, or the effects on comet condensation during the T-Tauri phase of the Sun (Schwartz 1978).

If the extent of physical and chemical processing of the comet nucleus is small, the manifestation of a comet as it approaches the Sun can be considered as solar effects on a ball of pristine interstellar material. I attempt to indicate to what extent the observed properties of a comet may be consistent or inconsistent with those of an unmodified conglomerate of primordial precometary interstellar dust.

I. DUST AND MOLECULES IN CLOUDS

There is good evidence that molecular clouds are associated with star formation (Blitz 1980; Falgarone 1980). Since comets are assumed to form within the same cloud complex as stars and planets, the observable ingredients of a molecular cloud may indirectly provide information on the primary comet ingredients. The principal atomic ingredients to consider are those which can either condense directly or combine with each other and with hydrogen to form low-temperature solids. We may never be able to observe the precometary cloud continuously from the molecular cloud phase to the final comet aggregation phase, but I shall try to predict this evolutionary sequence as a continuation of available observations.

The clouds of gas and dust in our Milky Way have a wide range of densities and temperatures. The weakest condensations, called diffuse clouds (Spitzer 1978), have nominal densities (in terms of hydrogen) $n_H = 10$ cm^{-3} and temperatures of the order of 100 K. Generally the key molecule defining a molecular cloud is CO (Andrew 1980), the most abundant molecule seen except for the hydrogen molecule in interstellar space. Carbon monoxide is observed in clouds of density as low as $n_H = 50$ to 100 cm^{-3} and in clouds whose density is inferred to be as high as $\geqslant 10^5$ cm^{-3} although in the highest density regions the CO appears strongly depleted (Snell 1979). Within clouds whose density is $n_H \gtrsim 10^2$ cm^{-3} the hydrogen is mostly in molecular form, but since hydrogen is not a condensible molecule at the normal temperatures and densities of interstellar space it cannot contribute substantially to the final aggregated comet material (Greenberg and de Jong 1969).

The most plausible conclusion from the variety of cloud conditions is that they represent various evolutionary stages, the densest ones corresponding to times just before, during, or just after star formation. A schematic representation is shown in Fig. 1. Diffuse clouds become denser by some mechanism, perhaps collisional combination or external pressure (Oort 1954; Field and Saslaw 1965; Taff and Savedoff 1972a,b; Kwan 1979; Scoville and Hersh 1979). Within the dense clouds critical densities may be reached which lead to instabilities, further contraction, and finally to star formation (Woodward 1978; Bash 1979). If the stars formed are large hot stars, the material they were formed of is ejected by shock waves back into the surrounding space (Blitz and Shu 1980) and much of it, heated and in a tenuous low-pressure environment, expands into diffuse clouds.

Although we have identified $\sim$ 40-50 carbon-bearing molecules containing as many as 11 atoms in the interstellar gas (Mann and Williams 1980), the major repository of condensible atoms is in the solid interstellar particles, the so-called interstellar dust (Greenberg 1978). Note that there is very little direct evidence for significant quantities of nitrogen in either the gas or the dust. Although fully identifying the chemical composition of the dust is not now possible, an approach combining direct and indirect observational characteristics with a laboratory-based physical theory can be used to plausibly estimate its essential chemistry.

One key assumption in all the following is that the relative abundances of the major atomic constituents are uniform over extended regions in space. A summary of some average characteristics of the interstellar medium is given in Table I, where only the cosmic abundances of certain critical chemical elements are included. Although, for example, sulfur is important, I shall assume that it is basically carried along with the O, C, N group.

The gas is seen to contain in molecular form $\leqslant 10\%$ of the available O+C+N constituents (hereafter called the organic condensibles) and considerably less in the diffuse clouds. There is evidence that the density of neutral

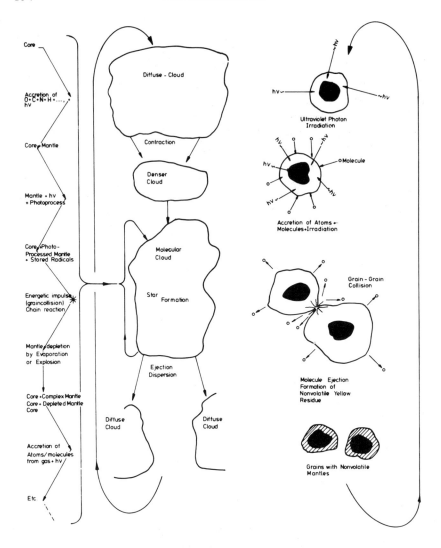

Fig. 1. Schematic diagram of grain and cloud evolution. The sequence on the left corresponds to the molecular cloud phase, the sequence on the right shows how the grains evolve through the molecular cloud and star formation phase and back to the diffuse cloud phase.

carbon itself may be as large as 1/10 that of CO in molecular clouds and "may even be comparable in some regions" (Phillips et al. 1980). On the other hand, the dust contains on the order of 20% of the organic condensibles (undoubtedly in molecular form) in the diffuse clouds and certainly more in the dense clouds, as shown by their observed increase in mean size within such regions (Carrasco et al. 1973) because of accretion from the gas. Although there is some controversy over whether accretion or aggregation

TABLE I
Average Interstellar Medium

Gas

$0.1 \leftarrow \langle n_H \rangle = 1 \text{ cm}^{-3} \rightarrow 10^5$

$\langle n_{O+C+N} \rangle \simeq 10^{-3} \, n_H$

$O : C : N \simeq 6.8 : 3.7 : 1$

$\langle n_{Mg+Si+Fe} \rangle \simeq 10^{-4} \, n_H$

$Mg : Si : Fe \simeq 1 : 1 : 1$

Radiation in Ultraviolet

$\langle n_{\lambda < \lambda_t} \rangle = 3 \times 10^{-3} \text{ cm}^{-3}$

$\lambda_t = 2000 \text{ Å} : h\nu_t = 6 \text{ eV}$

Dust

$\langle n_d \rangle \simeq 10^{-12} \, n_H$

$T_{dust} = 10 \text{ K}$

$\bar{a}_d \simeq 0.12 \, \mu m$

$[a_{core} = 0.05 \, \mu m, \; \bar{a}_{mantle} = 0.12 \, \mu m]$

$a_{bare} \simeq 0.005 \, \mu m$

$n_{bare} \simeq 10^3 \, n_d$

(Mathis and Wallenhorst 1981) is more important in producing the observed changes in grain extinction and polarization in dense clouds, theoretical and observational evidence is in favor of the former (Hong and Greenberg 1978; Breger 1979; McMillan 1978; Wilking et al. 1980). There is observational evidence that all remaining condensible gases in a dense cloud ultimately disappear onto the cold interstellar grains. The Mg, Fe, Si group (hereafter called the stony condensibles) appears entirely bound up in refractory solids so that generally little of these active constituents is seen in the gas phase.

The range of observable molecular densities with respect to hydrogen is really (with CO as the notable exception) quite low when compared with the elemental abundances. A typical example of a moderate size molecule is formaldehyde, H_2CO, with a number density ratio of $[H_2CO]/[H] \leqslant 10^{-8}$ (Sherwood 1980). It is normally difficult to establish for all molecules a definitive value for this sort of ratio, even if their column densities are well known, due to uncertainties in excitation and location along the line of sight. However, we can assume that a mean value $n_{mol}/n_H < 10^{-8}$ for carbon bearing molecules other than CO is a reasonable upper bound for each observed molecule. If we use a representative value of 3 carbon atoms in the ~50 molecules observed (observed range is 0 to 9 carbons), we would estimate that the total observed carbon in all carbon-bearing molecules other than CO is

$$\frac{\frac{[n_C]_{mol}}{[n_H]}}{\frac{[n_C]_{CA}}{[n_H]}} < \frac{50 \times 3 \times 10^{-8}}{3.7 \times 10^{-4}} = 0.4 \times 10^{-2}. \tag{1}$$

This is at least 10 times less than the relative abundance of carbon found in CO, and 20 times less than the carbon in the grain mantles. Similar calculations are possible for individual clouds, and it appears that, although there is a variety of substantially complex molecules in interstellar clouds, they generally contain at most a small fraction of the available atoms of the O, C, N group. Therefore, if we propose that comets are a representative mixture from the primeval or molecular cloud, we are forced to look elsewhere than in the gas for the most important constituents. For example, the H_2O molecule as observed in the gas phase in interstellar space is never as abundant as CO. Towards the BNKL object in Orion, the fraction of O relative to cosmic abundance bound in gas phase H_2O is only $\lesssim 0.014$ (Phillips et al. 1980) but it can be observed as a substantial solid component from its absorption in the dust. Of course, even though the gas exhibits only a small sample of the molecules, it may be that this sample is itself representative only up to a point. In Sec. II, I present a theory for the chemical evolution of the dust which should picture the principal chemical constituents of a premordial comet.

II. THE EVOLUTION OF DUST AND ITS ULTIMATE COMPOSITION

I shall first give a theoretical sketch of interstellar dust, following its chemical and physical changes from birth through growth to destruction. It is necessary to follow through several cycles to show how the grains may look at the stage of maximum size when the properties are most closely related to primordial comet composition. The theoretical sketch will then be related to laboratory analog results obtained in the Leiden Astrophysical Laboratory.

Theoretical Sketch

The left side of Fig. 1 shows a schematic diagram of one of the cyclic elements in the history of a grain. We have broken into the chain by first considering the production of the refractory elements of grains. Cool evolved stars are observed to have an infrared excess, interpreted as emission by heated grains having an absorption band at ~ 10 μm characteristic of the Si-O stretch in materials of silicate type. The examples of absorption spectra of interstellar grains shown in Figs. 2-4 all show a dip at 9.7 μm. Although no known terrestrial or lunar stony material shows this precise absorption position and shape, laboratory production of amorphous silicates (Day 1979) can

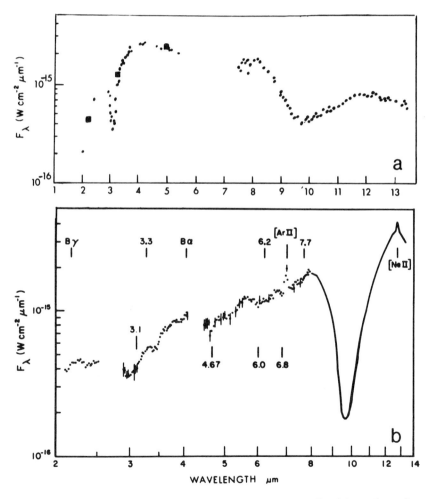

Fig. 2. (a) Infrared absorption spectra of the BN and KL objects. The 3.1 μm absorption is attributed to amorphous H_2O in a complex grain mantle, the 9.7 μm absorption is attributed to an amorphous silicate core of the grain. (b) Absorption spectra in Sgr A-W(N) towards the galactic center (from Willner et al. 1979a).

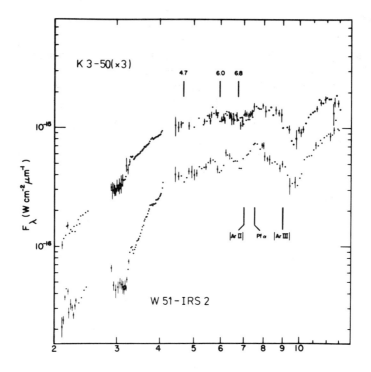

Fig. 3. Absorption spectra in K3-50(X3) and W51-IRS 2 in compact H II region (from Puetter et al. 1979).

simulate the interstellar 9.7 μm absorption quite well. Thus we believe that a significant component of the interstellar dust is in the form of a somewhat undefined silicate. However, the cosmic abundance of the stony condensibles limits the mass fraction of the dust in diffuse clouds in this component to < 20% (Greenberg 1974) and even less in fully grown dust (see below). This is consistent with the fact that in most of the interstellar medium, the observed abundance of stony condensibles in the gas is a very small fraction of the cosmic abundances, thus implying that almost all of these elements are bound up in the silicates (Morton 1974).

Although the stony condensibles cannot account for the observed amount of extinction in the interstellar medium, their obvious existence leads us to use them to provide nucleation cores on which the organic condensibles accrete to form mantles. The first consideration of grains which grew in the interstellar medium, albeit without defined nucleation cores, led to the concept of dirty ice grains; i.e., a mixture of H_2O, CH_4 and NH_3 with trace impurities (van de Hulst 1949). Our knowledge of the interstellar medium has advanced considerably, and we now must view the accreted organics in a more sophisticated way. But the basic idea remains of a grain predominantly consisting of some combination of O, C, and N with H (Greenberg 1974,1978).

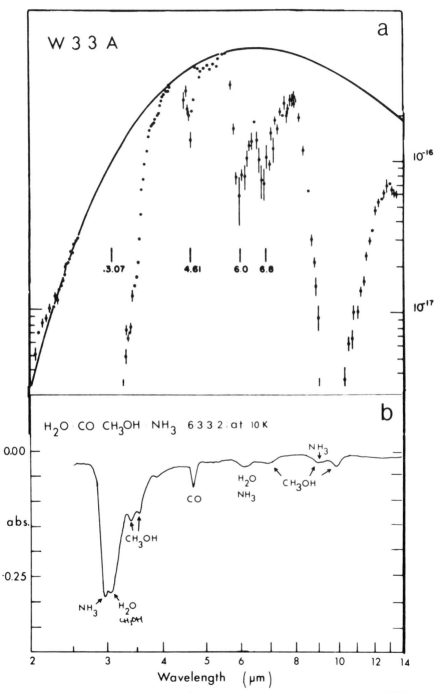

Fig. 4. (a) Absorption spectra in an infrared source in W33A (from Soifer et al. 1979). (b) Comparison spectrum is that of a complex mixture in partial simulation of interstellar grain mantles (no silicates).

Referring again to the left side of Fig. 1, we assume that after the silicate particles are ejected by cool evolved stars they become the cores of core-mantle grains. The subsequent growth by accretion from the interstellar gas is normally accompanied by ultraviolet penetration of the grain. It is readily seen that in diffuse clouds the rate of collision of the ultraviolet photons with the grain is much higher than the rate of collision and accretion of gas atoms and molecules. The ratio of the collision rate of ultraviolet photons ($\lambda < 2000$ Å as defined in Table I) to the collision rate of organic condensibles is

$$\frac{\Phi_{h\nu} A}{\Phi_{mol} A} = n_{h\nu} c / n_{OCN} \bar{v}_{OCN} \approx 10^6 \, n_H^{-1} \qquad (2)$$

where A is grain area, c speed of light, $n_{h\nu}$ and n_{OCN} photon and organic condensible densities; we have assumed velocity of the gas molecules $\bar{v}_{OCN} \simeq 10^5$ cm s^{-1} and ultraviolet from outside the cloud not attenuated. Within the cloud the attenuation of interstellar ultraviolet radiation can be calculated for spherical clouds (Sandell and Mattila 1975; Flannery et al. 1980) and it can be shown that (see Eq. 2) within a rather dense molecular cloud a decrease by a factor of 10^2 (relative to the outside) still provides a flux comparable to the flux of gas molecules. This level is probably a lower limit to the ultraviolet found in active molecular clouds and perhaps even in quiescent dark clouds. In the former there are many other sources of ultraviolet photons such as those produced by winds and shocks (Silk and Norman 1980) as well as direct local sources from nearby or contained young stars. Using this as a lower limit to the effective ultraviolet, we calculate the mean time for such photons to be important in the grain photochemistry by dividing the number of chemical bonds in the grain by the rate of ultraviolet penetration. Thus, an upper limit to the photoprocessing time τ_{pp} for a spherical grain of radius a is obtained by using flux $\phi_{h\nu} = 10^6$ cm^{-2} s^{-1} (compared with a mean interstellar value of $\phi_{h\nu} = 10^8$ cm^{-2} s^{-1}) to obtain

$$\tau_{pp} \approx \frac{\beta \frac{4}{3} \pi a^3}{\Phi_{h\nu} \pi a^2} = \frac{4}{3} \frac{\beta a}{\Phi_{h\nu}} \qquad (3)$$

where β is volume density of bonds (atoms) in the grain. For atomic dimensions we estimate $\beta = 5 \times 10^{22}$ cm^{-3}, and using $a \simeq 10^{-5}$ cm we find 2×10^2 yr $\leqslant \tau_{pp} \leqslant 2 \times 10^4$ yr. Even the longer time is much shorter than the mean lifetime of a molecular cloud, variously estimated between 3×10^7 and 5×10^8 yr and higher (Bash 1979; Blitz and Shu 1980).

It was predicted (Greenberg 1973b, 1976) that as a consequence of the photoprocessing of the grain mantle a sufficient concentration of free radicals could be frozen in the complex mantle matrix so that small triggering events in the grain could initiate chain reactions among them. This would then lead to the production of large complex molecules or even to an explosion of a part of the grain. If this should occur several times during the period of grain residence in a molecular cloud, we would expect a variety of grain mantle molecular types ranging from relatively simple molecules with high radical concentration, to extremely large molecules probably organic in nature. A suggested triggering mechanism arises from grain-grain collisions in the cloud induced by winds, shocks, or turbulence. Based on laboratory evidence we estimate that grain-grain velocity $\gtrsim 30$ m s^{-1} is sufficient to trigger grain explosions by imparting an impulsive increment in temperature $\Delta t \sim 10$ K. At such velocities, an order of magnitude larger than those predicted by Volk et al. (1980) in turbulent clouds, the mean collision time is on the order of 3×10^5 yr. Consequently the cycle shown on the left of Fig. 1 may repeat ~ 100 times while the grains are in the molecular cloud phase, and the ultimate chemical and physical properties of the grains are probably rather homogeneous. At some time during the molecular cloud phase, the grains are in clouds somehow closely associated with formation of a star and solar system. These are the grains we picture as comet progenitors. However, we consider the evolution of grains after ejection from the molecular clouds following the birth of energetic stars because this leads to an understanding, via astronomical observations, of the nature of at least part of the grain mantle material. Before returning to this, a summary of laboratory results on simulated interstellar grain evolution is useful.

Laboratory Results

One of the principal aims of the laboratory at Leiden is to study the chemical and physical evolution of dust, and its interaction with and production of complex interstellar molecules. A detailed description of some of the laboratory work has been given elsewhere (Hagen et al. 1979). A schematic diagram of the laboratory is shown in Fig. 5. A sequence of laboratory steps made to mimic the sequence shown at the left of Fig. 1 is as follows:

1. Deposition on the 10 K cold finger of mixtures of simple molecules such as H_2O, CO, NH_3, and CH_4 generally prepared in cosmic abundance of O:C:N;
2. Irradiation of the frost during deposition by ultraviolet photons from a hydrogen lamp (photon flux $\simeq 10^{15}$ cm^{-2} s^{-1} at 1600 Å);
3. Study of infrared absorption of radicals and molecules produced in the cold sample;
4. Warmup and observations of chemiluminescence and light flashes with vapor pressure spikes showing recombination of frozen radicals and generation of heat from released stored chemical energy;
5. Continued warmup at vacuum to room temperature and study of nonvolatile complex molecular residue, the ultimate product of this sequence.

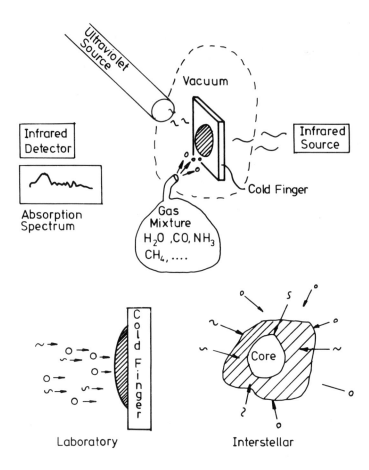

Fig. 5. Schematic diagram of the laboratory analog method for studying interstellar grain evolution. Molecules are deposited on a cold finger in a vacuum while the layer is irradiated by ultraviolet photons (see text). The absorption spectrum in the infrared shows the appearance and disappearance of various molecules and radicals. The cold finger may be an aluminum block (~ 3 cm cube) or a glass, sapphire or LiF window.

Figures 6 through 10 show examples of results of the laboratory analog sequence of dust evolution. The absorption spectra in Fig. 6 are shown first for an unirradiated sample and then compared with the spectrum after irradiation, showing the additional molecules and the appearance of frozen radicals produced. Table II shows results for a variety of samples. As the sample is warmed the spectrum changes dramatically and new features appear which are not readily identified; some are probably produced by complexes of larger molecules. Figure 7 is taken from unpublished data by d'Hendecourt (1980) and shows the light flashes and simultaneous pressure spikes indicating ejection of molecules by the sample as it is slowly allowed to warm above 10 K.

TABLE II

Examples of New Molecular and Radical Species Created in
Various Laboratory Sample Mixtures[a]

Original Mixture	CO CH_4	CO H_2O	CO NH_3	CO H_2O CH_4 NH_3
New Molecules and Radicals	CO_2	CO_2	CO_2	CO_2
	HCO^d	HCO^d	HCO^d	HCO^d
	H_2CO^d	H_2CO^d	H_2CO^d	H_2CO^d
	CH_2CO^d	HOCO	HOCO	HOCO
	C_2H_2	$HCOOH^d$	$HNCO^d$	$HNCO^d$
	C_2H_6	-----[b]	HOCN	$HCOOH^d$
	C_3O_2		$HCONH_2^d$	$HCONH_2^d$
	-----[b]		HNNH	O_3
	CO_3		NH_2	NH_2
	C_3O		NO	C_3O_2
	H_3C-X^c		NCO^d	-----[b]
			$CO(NH_2)_2$	C_3O
			-----[b]	H_3C-X
			N_2O	
			NH	

[a]Mixtures simultaneously deposited at 10 K and irradiated by ultraviolet photons. Note that differences in relative concentration give different molecule production.
[b]Molecules listed below dashed lines are uncertain. In all cases there are many additional species represented by unidentified absorption lines.
[c]H_3C-X means CH_3 containing molecules not completely identified.
[d]Detected in molecular clouds.

We generally observe the analog grain material to explode (a large flash and pressure pulse) at ~27 K. The sequence of photos of a laboratory sample in various stages from cold beginning luminescence, to luminescence, to flash, and finally to residue material is shown in Fig. 8. The infrared absorption spectrum of the photoprocessed residue of a cosmic abundance mixture of $CO:H_2O:NH_3:CH_4$ is shown in Fig. 9 and compared with the spectrum of the original unirradiated mixture. We have obtained many such absorption spectra for the nonvolatile residues from photoprocessing of a variety of mixtures. To compare with the unirradiated mixture in Fig. 9 we have normalized to the total absorption by letting the integrated absorption strengths

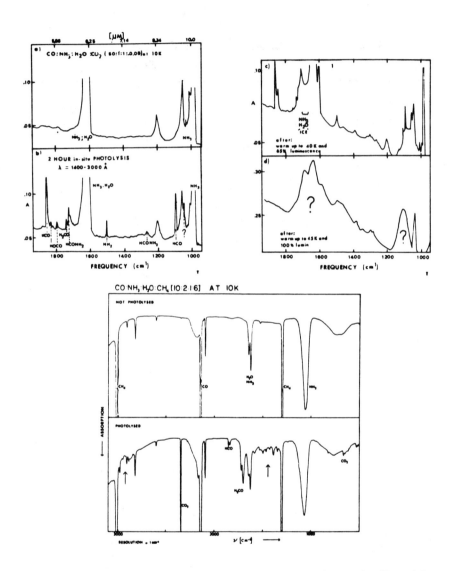

Fig. 6. Infrared absorption spectra for two sample analog grain mantles. Upper left sequence and lower two spectra show production of new molecules and radicals caused by ultraviolet irradiation. Upper right pair indicates appearance of new and unidentified complex molecules during warmup.

of the unirradiated sample and the residue be equal in the range 2000 cm^{-1} $\gtrsim$ $\nu \gtrsim$ 1000 cm^{-1}. This region is chosen to avoid the anomalous effects of H$_2$O which can appear in the 3000 cm^{-1} region. We roughly identify the very broad absorption from 3500 cm^{-1} to 2000 cm^{-1} as due to carboxylic acid groups, and some absorptions between ~1200 cm^{-1} to 1800 cm^{-1} as due to amino groups (Greenberg et al. 1980). Our residues are always water and

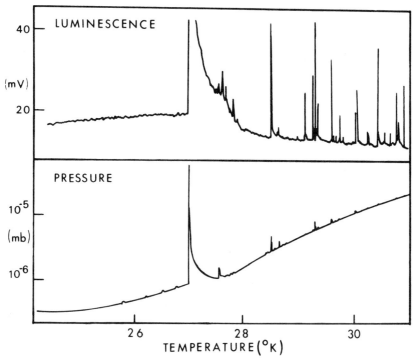

Fig. 7. Light flashes (above) and simultaneous pressure pulses (below) produced as sample irradiated at 10 K is warmed up. The starting composition was $CO:H_2O:NH_3:CH_4 = 10:1:1:4$. The sample was deposited over 10 min and exposed to 2 hr of photolysis (in three cycles). Stored chemical energy is released as light and heat when frozen radicals combine in the solid.

methanol soluble indicating their acid quality. One of our earlier samples gave a single molecular weight of 514 in mass spectrometer analysis, and all our samples do not evaporate at temperatures ≤ 400 K, with at least one pyrolyzing without evaporating at 600 K. A high resolution mass spectrometer analysis of the lowest pressure component in several of the nonvolatile residues after warmup and CO_2 release showed a mass corresponding to that of $C_4H_6N_2$. The intensity ratios of these two components plus trace urea suggest that amino pyroline rings make up a substantial part of the material. One possible representation as part of a polymer is

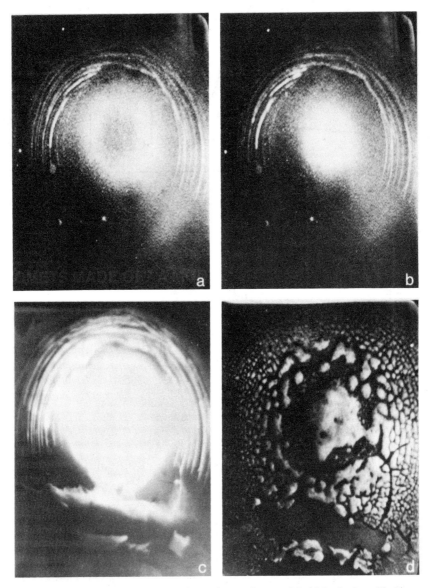

Fig. 8. Sequence of photos of exploding grain mantle material: (a) beginning of luminescence; (b) increasing luminescence; (c) flash; (d) residue (illuminated) after being heated to 180 K in low pressure. Some of this residue evaporated further leaving a final nonvolatile residue which persists in high vacuum at room temperature.

This analysis is the first in a series underway at Leiden. A reasonable solution is that we are dealing with a complex organic molecule with possible prebiotic significance, but this aspect will not be pursued here. However, it is relevant that ~ 2% to 20% of all the original condensate evolves into the complex nonvolatile form in an effective interstellar time scale ~10^7 yr, a

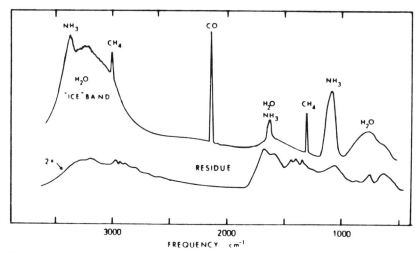

Fig. 9. Comparison of infrared absorption spectra of "yellow stuff" residue with a 10 K mixture containing roughly the same amount of oxygen, carbon, and nitrogen in molecular form. Note the complete absence of an H_2O ice band at 3.07 μm in the residue spectrum.

short time compared with the grain residence time in a molecular cloud. This means that the grains ejected following star formation which appear in the diffuse clouds must contain a substantial amount of nonvolatile residue. An exciting property of the yellow stuff, which gives direct evidence for its ubiquitous presence in space, is its visible absorption spectrum, an excellent likeness in both position and width (F. Baas, as reported in Greenberg 1980) to several unidentified diffuse interstellar bands (Merrill 1934; Herbig 1975) and particularly the famous λ 4430 band (see Fig. 10). It is hypothesized that under interstellar conditions of ultraviolet irradiation a yellow stuff containing the metallic elements Fe, Mg, Mn, Ca will provide a major source of the $\sim$50 observed interstellar bands.

In addition to showing the results of studies of irradiated mixtures it is useful to demonstrate by example the infrared properties of H_2O ice in irradiated and unirradiated mixtures. This will be important in estimating the primeval quantity of H_2O in comets, based on observed dust absorption.

Although H_2O ice was a strongly indicated component of interstellar dust because of the 3.07 μm absorption shown in Fig. 2, it was equally clear that the shape of the crystalline H_2O ice absorption (Bertie et al. 1969) was a poor match to the observation (Gillett et al. 1975). Figure 11a shows a comparison of the Beckler Neugebaur (BN) spectrum with that of crystalline ice and amorphous ice produced by deposition and annealing of pure H_2O vapor on the cold finger in our laboratory (Hagen et al. 1981). A similar amorphous ice band has been produced by Leger et al. (1979). The main point is the broadening of the band from $\sim$150 cm^{-1} to $\sim$300 cm^{-1} in going

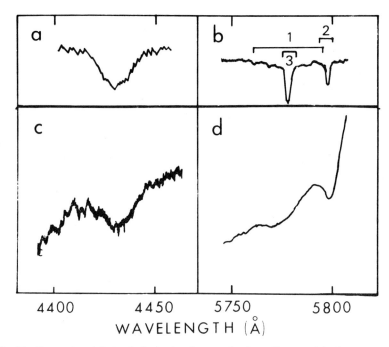

Fig. 10. Comparison of visual absorption features in the yellow stuff (c,d) with several of the strongest astronomically observed "unidentified" diffuse interstellar bands (a,b).

TABLE III

Optical Constants for Various H_2O Mixtures at Peak Absorption ($\sim 3\ \mu m$)

f_{H_2O}[a]	Form	Half Width $\Delta \nu$ (cm^{-1})	m''	m'
1	crystalline	150	0.815	1.37
1	amorphous	300	0.5	1.31
0.75	,,	320	0.29	~ 1.31[b]
0.42	,,	350	0.12	~ 1.31[b]
CA = 0.58[c]	,,	~ 350	~ 0.20	~ 1.31[b]

[a]$f_{H_2O} \equiv$ fraction of H_2O in mixture.
[b]Estimated values of the real part of the index of refraction.
[c]Cosmic abundance fraction. Values of $\Delta \nu$, m'' for this mixture are obtained by interpolation from measured values.

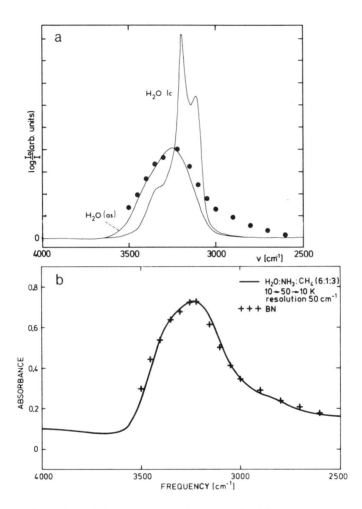

Fig. 11. (a) Comparison of observed shape of the H_2O ice band (BN) with amorphous ice (H_2O) and with pure crystalline ice H_2O. (b) Comparison of observed shape of the H_2O ice band with laboratory absorption by H_2O in a mixture with other molecules in simulation of an interstellar grain mantle.

from the crystalline to the amorphous state. From our unpublished data we deduced the reasonable approximate result that the peak absorption is inversely proportional to the line width; i.e. $m''\Delta \nu \simeq$ constant where m'' is the imaginary part of the complex index of refraction $m = m' - m''i \ (\equiv n - ik)$ and $\Delta \nu$ is the half width (see Table III). The amorphous ice band has the same width as the interstellar ice band but it lacks the long wavelength wing. Although we have readily produced this wing in a variety of irradiated samples containing H_2O (C.E.P.M. van de Bult, unpublished) Fig. 11b shows the result obtained for an unirradiated mixture where the wing is simply produced by H_2O in the presence of molecules which complex with it. We conclude that the observed H_2O ice shape is readily reproduced by mixtures not necessarily exactly as used for Fig. 11b. Note also that the absorption strength per unit H_2O in the mixture is roughly constant, although the reciprocity breaks down at high H_2O dilution so that $m'' \leqslant m''_{H_2O} f_{H_2O}$ where f_{H_2O} is the fraction of H_2O in the mixture.

Precometary Interstellar Dust

By combining laboratory data with observational characteristics of interstellar dust we now can describe dust as it may appear in several key stages of evolution, and ultimately its structure just before coagulation leading to cometesimals.

Diffuse Clouds. All attempts to observe the H_2O ice band absorption in diffuse clouds have failed, implying either that the grains have no mantles or that the mantle material, though it contains substantial oxygen, has a very weak 3.07 μm band. From laboratory studies of H_2O ice in mixtures, we know that the strength of the 3.07 μm (polymeric H_2O) ice band is generally lower than expected because of increased probability that the water molecules remain isolated from each other and that some of them absorb at 2.85 μm, the position of monomeric water absorption. However, this effect is not adequate to explain the fact that even at > 20 mag of visual extinction (Willner et al. 1979a,b; Puetter et al. 1979; Soifer et al. 1979; Whittet 1981) no absorption corresponding to the normal H_2O feature is detected. The evolution of volatile organic molecules into a refractory component during the molecular cloud phase suggests that, following star formation, most of the ejected dust's volatiles sputtered away or evaporated leaving only the refractory organic (OR) mantles on the dust in diffuse clouds. The absorption strength of the OR mixture at 3 μm is not only far less than that of dirty amorphous ice but it is also so much broader (800 cm^{-1} compared with ~ 350 cm^{-1}) that it is difficult to observe. One may roughly estimate from Fig. 9 that the imaginary part of the index of refraction for the OR is $m''_{OR} \simeq 1/6$, that of a cosmic abundance H_2O mixture $m''_{OR} \simeq 0.19/6 = 0.03$.

The absorption for a spherical grain of radius a is given for $a/\lambda \ll 1$ by

$$A_{abs}(\lambda) = 9k \frac{4}{3} \pi a^3 \frac{\epsilon_2(\lambda)}{(\epsilon_1 + 2)^2 + \epsilon_2^2} \quad (4)$$

(Greenberg 1978) where $\epsilon_1 = m'^2 - m''^2$, $\epsilon_2 = 2m'm''$ ($m = m' - im''$), and $k = 2\pi/\lambda$.

The extinction per grain in the visual V is

$$A(V) = Q(V) \pi a^2 \quad (5)$$

where Q is extinction efficiency.

For normal grain sizes $Q(V) = 1.5$ so that the absorption to extinction ratio is

$$\frac{A_{abs}(\lambda)}{A(V)} = 8ka \frac{\epsilon_2(\lambda)}{(\epsilon_1 + 2)^2 + \epsilon_2^2}. \quad (6)$$

The real part of the index of refraction of the OR is probably in the range 1.4 to 1.5 (Handbook of Chemistry and Physics 1966-67) in the visual but its value in the infrared is unknown. I use $m''(3\,\mu) \simeq 1.45$ (the effect of small variations in this is negligible) to obtain

$$\frac{A_{abs}^{OR}}{A(V)} = 0.01 \quad (7)$$

where $a = 0.12\,\mu\text{m}$.

This means that for extinction 20 mag the expected absorption by the OR mantle material is only 0.20. This, combined with the breadth of the absorption should make it difficult but not impossible to detect.

The small structure at $\sim$3.4 μm in the OR spectrum is due to a C-H stretch. Its extra absorptivity above the broad background absorption, peaked at $\sim$3 μm (in Fig. 9), is $\Delta m'' \simeq 0.03/7 = 0.004$. For this value we obtain an extra 3.4 μm absorption strength of

$$\frac{\Delta A_{abs}(3.4)}{A(V)} \simeq 0.001. \quad (8)$$

This value is substantially less than that deduced from observations (Wickramasinghe and Allen 1980) but there are enough uncertainties in both laboratory results (absolute values of m''_{OR} are not well determined) and observational parameters (total extinctions to the various objects are difficult to ascertain; there may be local variations in grain evolution characteristics) to bring the two into agreement. Radiative transfer effects can also contribute to the interpretation of the observations.

A rough estimate of the value of $\Delta A_{abs}(3.4)/A(V)$ based on the data compilation of Fig. 1 in Willner et al. (1979) is $\Delta A_{abs}(3.4)/A(V) \lesssim 0.002$ where the total absorption is obtained from $A(10)/A(V) \lesssim 20$ and $A(10) \simeq 3$.

We conclude that the negative evidence for H_2O in diffuse clouds can readily be ascribed to the existence of complex organic mantles which contain substantial oxygen but have a very different absorption spectrum from H_2O.

Table IV is an attempt to summarize the distribution of the O, C, N and Si, Mg, Fe elements as they would appear in diffuse clouds. It is assumed that no molecules are present in the gas and that the C, N, Si, Mg, Fe abundances are as given in Spitzer (1978). I have used the more recent determination for O by de Boer (1979). The abundances of O, C, Si, Mg, Fe in the core particles and in the bare particles which provide both the 2200 Å hump and the far ultraviolet extinction are estimates based on calculations for not entirely identical but similar grain models (Hong and Greenberg 1980). For the mantle it is assumed that $a \simeq 0.12$ μm is a good estimate for the outer radius of the OR mantle. If the O, C, N atoms were in relative cosmic abundance in the OR, the values would all be ~ 0.22, but we know that carbon should predominate in the complex organics (as is actually seen from the mass spectrometer result). The upper values shown for O and C are based on the assumption that oxygen is as abundant as carbon in the OR and the lower values are for the assumption (probably more accurate) than oxygen is half as abundant as carbon. Note that the total amount of oxygen depleted in the grains is probably $\lesssim 0.20$ and that the carbon depletion in the grains is accordingly ~ 0.69. This means that grains must grow in molecular clouds largely as a result of oxygen, and therefore we may anticipate a substantial amount of H_2O to appear; this is seen below.

Molecular Clouds. Although H_2O is seen in the grain absorption in molecular clouds, it has been difficult to estimate its abundance due to lack of data on appropriate indices of refraction. For an example of a region where the dust may be approaching precometary dust in its observed properties, I have chosen the well-observed BN object. Following the dust grain from its diffuse cloud appearance I assume that its composition in a region of accretion may be represented by a silicate core plus an organic refractory mantle, and a final extra mantle of accreted molecules with H_2O as a constituent. Then, the problem is to determine how much of this outer mantle is H_2O.

We need to know the observed ratio of total extinction to the 3.07 μm absorption. The 3.07 μm optical depth deduced from Fig. 2 is $A(3.07) = 1.46$. The total extinction has been estimated (Bedijn 1977) as consisting of two parts, $\lesssim 30$ mag due to the cool dust in the outer cocoon and ~ 27 mag due to the interior dust heated to ~ 400 K. I assume that only the former dust has any volatile extra mantle constituents, so that

$$\left[\frac{A(3.07)}{A(V)} \right]_{obs} = 0.049 . \qquad (9)$$

Let us start with the cosmic abundance $(CA)^a$ ratio of H_2O on a mantle whose size $a = 0.15$ μm is relative to the standard 0.12 μm, deduced from the higher value of λ_{max} for the polarization (Breger 1977, 1979). Equation (6) can now be rewritten, to a close approximation, as

$$\frac{A(3.07)}{A(V)} = 8k \frac{a_2^3 - a_1^3}{a_2^3} \frac{\epsilon_2}{(\epsilon_1 + 2) + \epsilon_2^2} \qquad (10)$$

---

[a]See note added in proof.

TABLE IV

Elemental Depletion in Diffuse Clouds

	O	C	N	Mg	Si	Fe	Notes
Gas							
Atoms & ions	0.25(0.75)	0.20	0.20	0.03	0.03	0.01	a
Molecules	—	—	—	—	—	—	
Dust							
Core & bare	0.09	<0.27		~1.0	~1.0	~1.0	b
Mantle OR	0.17	0.31	0.22				c
	0.11	0.42					
Volatiles (H$_2$O etc.)	~0	~0	~0				

[a] Depletions in ζ Oph from Morton (1974). For oxygen the depletion we use (in parentheses) is as deduced by de Boer (1979).
[b] The carbon depletion is based on a 0.025 μm radius graphite particle to produce the 2200 Å extinction hump. Should a smaller or differently composed particle be satisfactory, the carbon depletion in the bare particles could be substantially less.
[c] By cosmic abundance the grain model gives equal depletion = 0.22 for O, C, N. If the organic refractory molecule contains equal numbers of oxygen and carbon atoms, the depletions of O and N are the upper values. If, as is probably more accurate, there are twice as many C as O atoms in the OR molecule, the lower pair of depletions is correct.

where $a_2 = 0.15\,\mu m$ is the radius of the outer mantle, containing the H_2O, and $a_1 = 0.12\,\mu m$ is the radius of the inner mantle of OR which contains no H_2O. From Eq. 10 we obtain for m''_{CA} (3.07) = 0.20 (Table III)

$$\left[\frac{A(3.07)}{A(V)}\right]_{CA} = 0.072 \cdot \quad (11)$$

Comparing Eq. (11) with Eq. (9) it appears that ~0.64 of the oxygen by cosmic abundance is in water. Since the volume of the outer mantle is equal to that of the inner mantle, 0.64 × 0.22 = 14% of oxygen is depleted on the grains in the form of H_2O. In other words, ~35% of the total outer mantle is in the form of amorphous H_2O ice. This is not surprising considering that, coming out of the diffuse cloud phase, the available oxygen in the gas exceeds the carbon by perhaps a factor of 7 (see Table IV) rather than the cosmic abundance ratio of ~2.

In Table V, I summarize what may be expected for atomic depletions for the pre- or post-star formation phase. The depletion for CO is based on a mean observed value in molecular clouds. But note that in very dense cores and hot centered clouds the abundance of CO may be substantially less than this, as shown by the values in parentheses (Rowan-Robinson 1979). For the other molecules and perhaps atoms the values given either are estimations or are obtained merely by subtraction of the others from cosmic abundance. It is interesting that the amount of oxygen not accounted for (~55%) is very large and reminiscent of the "depletion mystery" (Greenberg and de Jong 1969).

III. PRECOMET DUST AND GAS

My basic assumption in arriving at Table VI is that essentially all of the condensible atoms and molecules are in the dust. The values derived from CO are probably upper bounds because it is not likely that all the available CO from the gas merely accretes on the grains. Some probably goes to form other molecules. For the nonvolatile component I have simply considered that a small additional amount ~10% has been made in the latest stage. The amount of H_2O is derived by assuming that the 0.64 fraction of CA oxygen exhibited in the BN mantle is continued into the final accretion phase. Using this assumption and the fact that the available oxygen (excluding what has been bound in CO, the cores and bares, and the organic refractory mantle) is ~70% of its cosmic abundance value, the result is 0.64 × 0.70 = 0.45. I have conservatively reduced this to 0.35 to allow for other reactions with oxygen according to the relative abundance of C (or CO) coming in from the gas, ~1/7 of the available oxygen.

IV. COMETS AS AGGREGATED DUST

My basic assumption in the derivation of Table VII from values in Table VI is that the cosmic abundance composition of the molecular cloud of gas

TABLE V

Elemental Depletion in Molecular Clouds*

	O	C	N	Si	Mg	Fe	Notes
Gas							
CO	0.05–0.10	0.11–0.22					a
	(0.01–0.02)	(0.03–0.06)					b
Other molecules	~0.01	~0.01	~0.01				c
	~0.50 [0.45]*		~0.55				d
Dust							
Core & bare	0.09	0.27		~1.0	~1.0	~1.0	
Mantle							
Solid H_2O	0.14 [0.22]*						e
OR	(0.17)	(0.31)	(0.22)				
	0.11	0.42	0.22				
Other mantle solids	0.08 [0.05]*	0.22	0.22				f

[a] Based on "normal" CO abundance in molecular clouds.
[b] Depletion of gas CO in dense cores of molecular clouds. Values in parentheses indicate CO abundances in very dense cores or hot-centered clouds.
[c] Observed (see text and footnote d). This is probably a gross underestimate.
[d] By subtraction. If constituents are not in solids this implies an even larger number of undetected molecules in the gas; N_2 is a prime possibility. Note that the absence of C in the gas is produced by depletion, mostly in refractory organics plus bare particles.
[e] Based on BN absorption. See text.
[f] Assumes extra grain mantle of 0.03 μm as in text.

*See note added in proof.

TABLE VI

Elemental Depletion in Cloud with Precomet Dust

	O	C	N	Si	Mg	Fe	Notes
Gas							
CO	≲ 0.02	≲ 0.04	–	–	–	–	a
Other molecules	–	–	–	–	–	–	
Dust							
Core & bare	0.09	0.27	–	1.0	1.0	1.0	a
Mantle							
OR plus other nonvolatile	0.13	0.45	0.25				b
H_2O	0.35						c
CO	0.05	0.11					d
Other molecules	0.33	0.15	0.75				e
CO_2, N_2, HCN, H_2CO, etc.							

[a] Note, as in Tables IV and V that this may be an overestimate of the carbon depletion in the bare particles.
[b] Estimated extra 10% nonvolatile (not necessarily organic refractory) molecules produced in the molecular cloud.
[c] Estimated ~1/3 of O remaining in gas in late molecular cloud phase goes to making H_2O; i.e. 1/3 × 0.60 (see Table V).
[d] Assumes CO in gas condenses as CO on the grains. It is more likely that some of the remaining gas CO combines with other atoms or molecules on grains.
[e] By subtraction.

and dust is preserved in the final passage to the comet. This constraint must be maintained no matter what model is used for the dust so long as we assume cold coagulation. Astronomical evidence indicates a substantial abundance of amorphous ice in the comet.

The morphological structure of a comet as aggregated dust is a tangle of silicate rods of ~0.1 μm thickness imbedded in a matrix of molecules ranging in volatility from CO to H_2O through various degrees of complexity up to the nonvolatile organics. Scattered throughout the matrix would be inclusions of small (< 0.02 μm) diameter carbon and silicate particles. From Table VII, the nonvolatiles (silicates, carbon, organics) constitute ~46% of the comet by mass and ~29% by volume.

TABLE VII

Suggested Chemical Mass and Volume Distribution of the Principal Condensible Atomic Constituents in a Comet

Component	Mass Fraction	ρ	Volume Fraction
Silicates	0.21	3.5	0.086
Carbon (graphite)	0.06	2.5	0.034
Very complex OR	0.19	~1.3^a	0.21
H_2O	0.19	1	0.27
CO	0.10	1.05	0.13
Other molecules + radicals (CO_2, N_2 HCN, H_2CO...)	0.25	1.3	0.27

[a] Estimated ρ for complex organics in Handbook of Chemistry and Physics (1976-77).

V. SOME COMPARISONS

The current development of a comet theory based on the above model is too preliminary to attempt detailed comparison with comet observations. However, relationships can be established with a few specific observational results and certain broad ideas on comet composition.

1. *H_2O*. The evidence for abundant H_2O in comets is overwhelming. The exact amount cannot be derived from observations, but the predicted volume fraction > 27% appears easily consistent with the data. Of particular interest is the prediction of amorphous ice by the dust aggregation model.

2. *Dust to gas ratio*. This ratio derived from an icy comet model is ~0.7 (Delsemme 1980). From Table VII we find that the sum of the mass fractions of the silicates, carbon and nonvolatile organics is 0.46 while that of the other molecules is 0.54 leading to a reasonably close dust-to-gas ratio 0.85.

3. *Homogeneity.* The evidence for and against the homogeneity of the interior of comets has been collected by Whipple (1981; see also A'Hearn et al. 1981; Donn and Rahe chapter). The impression is strong that comets are the same inside and outside; cold aggregation of interstellar dust and gas (mainly the dust in the final stages) should be homogeneous.
4. *Molecules.* All the molecules and radicals observed in comets are also seen in the interstellar gas or can be detected in the infrared absorption spectrum of dust either in space or in the laboratory.
5. *Silicate.* The observation of 10 μm emission by comet dust (Ney and Merrill 1976) indicates that the same sort of silicates are present in comet dust as in interstellar dust.
6. *Emission of gas by dust.* Since the comet dust as aggregated interstellar dust contains volatile molecules, we expect these to evaporate as the dust travels away from the comet (A'Hearn and Cowan 1980).
7. *Fe emission.* The fact that such heavy elements as Fe are not observed until the comet is in close perihelion passage (Whipple 1978) indicates that the Fe atoms are bound in solid refractory pieces (silicates, oxides) which do not allow the metals to appear until the solids are heated to evaporation. This is consistent with the fact that the Fe, Mg, Si group are almost entirely bound up in the silicates in the interstellar dust and do not exist as isolated molecules. At the other extreme, we may also deduce that the refractories are not totally bound up in large rocks whose rate of evaporation would be much less than that of the very small particle inclusions.
8. *Cosmic abundance.* The relative abundances of the elements in a comet appear consistent with cosmic abundance assumed for interstellar gas and dust. In both interstellar dust and comets hydrogen is greatly underabundant.
9. *Angular scattering by comet dust.* This scattering (Ney and Merrill 1976) is similar to that by models of aggregated interstellar dust (Greenberg and Gustafson 1981) (see Fig. 12), but there is a substantial difference at small scattering angles. There are several ways this may be produced, but detailed calculations are required to justify them. One possibility is that there is an additional component of very fine dust such as that in the small silicate cores.
10. *Comet outbursts.* There have been a number of suggestions for the cause of comet outbursts at large distances from the Sun (Patashnik et al. 1974; Whipple 1978; Smoluchowsky 1981). Among these are trapped radicals, pockets of highly volatile gas, and energy released by phase changes in amorphous ice. I do not know which is the primary source of outbursts but all these possibilities can be predicted on the basis of the aggregated interstellar dust model. It is difficult to imagine small radicals remaining frozen for billions of years but perhaps as part of complex molecules they may have exceedingly long lifetimes.

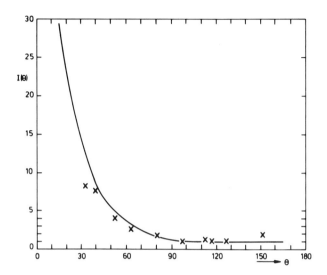

Fig. 12. Comparison of angular scattering by comet dust, as deduced from observations (Ney and Merrill 1976), with theoretical scattering (solid curve) by models of comet debris in the form of "birds nest" particles (Greenberg and Gustafson 1981).

11. *Comet dust morphology.* Closest to direct observation of comet dust morphological structure are the measurements and photographs of interplanetary particles collected in the Earth's upper atmosphere (Brownlee 1978). These particles appear to be made up of much smaller components, but these components are not easy to associate with interstellar dust. Another problem is that silicate absorption in the Brownlee particles resembles that of crystalline silicates (Fraundorf et al. 1981) rather than the amorphous silicates in interstellar dust. Also, rather large clumps of silicates appear in the Brownlee particles; it is difficult to understand how the submicron amorphous silicate particles of the interstellar dust cloud have combined into such large pieces ($\sim$ microns) of crystalline form without a considerable metamorphism of the comets by substantial heating. The amount of heating required appears at first sight inconsistent with the homogeneity of comet nuclei. This may imply that the particles we collect have undergone large changes in structure since leaving the comets.

VI. CONCLUDING REMARKS

For the first time, a detailed model of the chemical composition of interstellar dust has been applied to a prediction of the chemical composition of a comet. All the basic ingredients assumed in the icy comet model appear as a consequence of the composition of aggregated interstellar dust. Some important new predictions are: a lower limit on the volume fraction of $\sim 27\%$ for a comet in the form of amorphous H_2O, the probable presence of substantial CO_2, CO, and perhaps O_2 in the comet; a large fraction of the comet in the form of very complex nonvolatile organic molecules.

Acknowledgments. I wish to thank L. d'Hendecourt, C.E.P.M. van de Bult and W. Hagen for permission to use some of their unpublished data from the Leiden Astrophysical Laboratory. Among the other members of our group I am particularly indebted to L.J. Allamandola and F. Baas for help in frequent discussions. I have also benefited from various discussions with L. Biermann, F. Whipple, A. Delsemme, and B. Donn. The collaboration of J. de Jong on the analysis of the complex molecules is gratefully acknowledged. Some of this work has been supported by grants from the Netherlands Organization for the Advancement of Pure Research and the Organization for Fundamental Research on Matter.

Note added in proof: I have rederived the composition of the dust in molecular clouds using the more consistent assumption that the abundances of the available elements for accretion are those which are not already accreted but remain in the gas rather than the original cosmic abundance. The consequences of this revised treatment are to change the fractional amount: of O in solid H_2O from 0.14 to 0.22; of O in other mantle solids from 0.08 to 0.05; of C in other mantle solids from 0.22 to 0.11; of N in other mantle solids from 0.22 to 0.26. It is anticipated that the ultimate comet composition will not be modified enough to change any of the comparisons in Sec. V significantly.

REFERENCES

A'Hearn, M.F., and Cowan, J.J. 1980. Vaporization in comets: The icy grain halo of Comet West. *Moon and Planets* 23:41-52.

A'Hearn, M.F.; Millis, R.L.; and Birch, T.V. 1981. Comet Bradfield (1979 X): The gaseous comet? *Astron. J.* In press.

Andrew, B.H., ed. 1980. *Interstellar Molecules,* (Dordrecht: D. Reidel Publ. Co.).

Bash, F.N. 1979. Density wave induced star formation: The optical surface brightness of galaxies. *Astrophys. J.* 233:524-538.

Bedijn, P.J. 1977. Studies of Dust Shells around Stars. Ph.D. Dissertation, Leiden University.

Bertie, J.E.; Labbé, H.A.; and Whalley, E. 1969. Absorption of Ice I in the range 4000-30 cm^{-1}. *J. Chem. Phys.* 50:4501-4520.

Biermann, L., and Michel, K.W. 1978. On the origin of cometary nuclei in the presolar nebula. *Moon and Planets* 18:447-464.

Blitz, L. 1980. Large scale mapping of local molecular clouds complexes. In *Giant Molecular Clouds,* eds. P. Solomon and M.G. Edmunds (New York: Pergamon), pp. 1-18.

Blitz, L., and Shu, F.H. 1980. The origin and lifetime of giant molecular cloud complexes. *Astrophys. J.* 238:148-157.

de Boer, K.S. 1979. On the abundance of interstellar oxygen towards Ophiuchi. *Astrophys. J.* 229:132-135.

Breger, M. 1977. Intracluster dust, circumstellar shells, and the wavelength dependence of polarization in Orion. *Astrophys. J.* 215:119-128.

Breger, M. 1979. Interstellar grain size: A look at the deviant four. *Astrophys. J.* 233:97-101.

Brownlee, D.E. 1978. Microparticle studies by sampling techniques. In *Cosmic Dust,* eds. J.A.M. McDonnell (New York: Wiley and Sons), pp. 295-336.

Cameron, A.G.W. 1973. Accumulation processes in the primitive solar nebula. *Icarus* 18:407-450.
Carrasco, L.; Strom, S.E.; and Strom, K.M. 1973. Interstellar dust in the Rho Opiuchi dark clouds. *Astrophys. J.* 182:95-109.
Day, K.L. 1979. Mid-infrared optical properties of vapor-condensed magnesium silicates. *Astrophys. J.* 234:158-161.
Delsemme, A.H. 1980. Workshop on modern observational techniques for comets. *Goddard Space Flight Center* (preprint).
Donn, B. 1981. Comet nucleus: Some characteristics and a hypothesis on origin and structure. In *Comets and the Origin of Life,* ed. C. Ponnamperuma (Dordrecht: D. Reidel Publ. Co.), pp. 21-29.
Falgarone, E. 1980. Contagious B star formation in the Rho Ophiuchi dark cloud. In *Interstellar Molecules,* ed. B.H. Andrew (Dordrecht: D. Reidel Publ. Co.), pp. 183-188.
Field, G.B., and Saslaw, W.C. 1965. A statistical model of the formation of stars and interstellar clouds. *Astrophys. J.* 142:568-583.
Flannery, B.P.; Roberge, W.; and Rybicki, G.B. 1980. The penetration of diffuse ultraviolet radiation into interstellar clouds. *Astrophys. J.* 236:598-608.
Fraundorf, P.; Patel, R.; and Freeman, J.J. 1981. Infrared spectroscopy of interplanetary dust in the laboratory. Submitted to *Icarus* (special Comet issue).
Gillett, F.C.; Jones, T.W.; Merrill, K.M.; and Stein, W.A. 1975. Anisotropy of constituents of interstellar grains. *Astron. Astrophys.* 45:77-81.
Greenberg, J.M. 1973a. The skylab T025 filter photography program for Comet Kohoutek. *Comet Kohoutek Workshop,* Albuquerque, NM.
Greenberg, J.M. 1973b. Chemical and physical properties of interstellar dust. In *Molecules in the Galactic Environment,* eds. M.A. Gordon and L.E. Snyder (New York: Wiley and Sons), pp. 92-124.
Greenberg, J.M. 1974. The interstellar depletion mystery, or where have all those atoms gone? *Astrophys. J.* 189:L81-L85.
Greenberg, J.M. 1976. Radical formation, chemical processing and explosion of interstellar grains. *Astrophys. Space Sci.* 39:9-18.
Greenberg, J.M. 1977. From interstellar dust to comets to dust. In *Comets, Asteroids and Meteorites,* ed. A.H. Delsemme (Toledo, Ohio: Univ. of Toledo), pp. 491-497.
Greenberg, J.M. 1978. Interstellar dust. In *Cosmic Dust,* eds. J.A.M. McDonnell (New York: Wiley and Sons), pp. 187-294.
Greenberg, J.M. 1981. The largest molecules in space (II). *Ned. Tijd. voor Natuurkunde* A 47 (1):24-26.
Greenberg, J.M.; Allamandola, L.J.; Hagen, W.; van de Bult, C.E.P.M.; and Bass, R. 1980. Laboratory and theoretical results on interstellar molecule production by grains in molecular clouds. In *Interstellar Molecules,* ed. B.H. Andrew (Dordrecht: D. Reidel Publ. Co.), pp. 355-363.
Greenberg, J.M., and Gustafson, B.A.S. 1981. A comet fragment model for zodiacal light particles. *Astron. Astrophys.* 93:35-42.
Greenberg, J.M., and de Jong, T. 1969. Can solid hydrogen condense on interstellar grains. *Nature* 224:251-252.
Greenberg, J.M.; Yencha, A.J.; Corbett, J.W.; and Frisch, H.L. 1972. Ultraviolet effects on the chemical composition and optical properties of interstellar grains. *Mém. Soc. Roy. Sci. Liège,* Ser. 6, Vol. III, pp. 425-436.
Hagen, W.; Allamandola, L.J.; and Greenberg, J.M. 1979. Interstellar molecule formation in grain mantles: The laboratory analog experiments, results and implications. *Astrophys. Space Sci.* 65:215-240.
Hagen, W.; Tielens, A.G.G.M.; and Greenberg, J.M. 1981. The infrared spectra of amorphous solid water and ice I_c between 10 and 40 K. *Chem. Phys.* 56:367-379.
Handbook of Chemistry and Physics 1966-1967. (The Chemical Rubber Co., 47th edition).
d'Hendecourt, L. 1980. Laboratory evidence for molecule ejection at low temperatures. *Internal Report Laboratory Astrophysics,* University of Leiden.
Herbig, G.H. 1975. The diffuse interstellar bands IV. The region 4400-6850. *Astrophys. J.* 196:129-160.

Hong, S.S., and Greenberg, J.M. 1978. On the size distribution of interstellar grains. *Astron. Astrophys.* 70:695-699.

Hong, S.S., and Greenberg, J.M. 1980. A unified model of interstellar grains: A connection between alignment efficiency, grain model size, and cosmic abundance. *Astron. Astrophys.* 88:194-202.

van de Hulst, H.C. 1949. The solid particles in interstellar space. *Rech. Astron. Obs., Utrecht* 11:Part 2.

Kwan, J. 1979. The mass spectrum of interstellar clouds. *Astrophys. J.* 229:567-577.

Leger, A.; Klein, J.; de Chevergne, S.; Guinet, C.; Defourneau, D.; and Belin, M. 1979. The 3.1 μm absorption in molecular clouds is probably due to amorphous H_2O ice. *Astron. Astrophys.* 79:256-259.

Mann, A.P.C. and Williams, D.A. 1980. A list of interstellar molecules. *Nature* 283:721-725.

McMillan, R.S. 1978. Predicted color excess ratio versus interstellar grain size. *Astrophys. J.* 255:880-886.

Merrill, P.N. 1934. Unidentified interstellar lines. *Publ. Astron. Soc. Pacific* 46:206-207.

Morton, D.C. 1974. Interstellar abundances towards Zeta Ophiuchi. *Astrophys. J.* 193:L35-L39.

Ney, E.P., and Merrill, K.M. 1976. Comet West and the scattering function of cometary dust. *Science* 194:1051-1053.

O'Dell, C.R. 1971. A new model for cometary nuclei. *Icarus* 19:137-146.

Oort, J.H. 1950. The structure of the cloud of comets surrounding the solar system and a hypothesis concerning its origin. *B.A.N.* 11:91-110.

Oort, J.H. 1954. Outline of a theory on the origin and acceleration of interstellar clouds and O associations. *B.A.N.* 12:177-186.

Oort, J.H., and Schmidt, M. 1951. Differences between new and old comets. *B.A.N.* 11:259-270.

Patashnik, H.; Ruppert, G.; and Schuermann, D.W. 1974. Energy source for comet outbursts. *Nature* 250:313-314.

Phillips, T.G.; Huggins, P.J.; Kuiper, T.B.H.; and Miller, R.E. 1980a. The detection of the 610 micron (392 GHz) line of interstellar atomic carbon. *Astrophys. J.* 238:L103-L106.

Phillips, T.G.; Kwan, J.; and Huggins, P.J. 1980b. Detection of submillimeter line of CO (0.65 mm) and H_2O (0.79 mm). In *Interstellar Molecules,* ed. B.H. Andrew (Dordrecht: D. Reidel Publ. Co.), pp. 21-24.

Puetter, R.C.; Russell, R.W.; Soifer, R.W.; and Willner, S.P. 1979. Spectrophotometry of compact HII regions from 4 to 8 microns. *Astrophys. J.* 225:118-122.

Rowan-Robinson, M. 1979. Clouds of dust and molecules in the galaxy. *Astrophys. J.* 234:111-128.

Sandell, G., and Mattila, K. 1975. Radiation density and lifetimes of molecules in interstellar dust clouds. *Astron. Astrophys.* 42:357-364.

Schwartz, R.D. 1978. A shocked cloudlet model for Herbig-Haro objects. *Astrophys. J.* 223:884-900.

Scoville, N.Z., and Hersh, K. 1979. Collisional growth of giant molecular clouds. *Astrophys. J.* 229:578-582.

Sherwood, W.A. 1980. Cloud-to-cloud variation in H_2CO to H_2 ratios. In *Interstellar Molecules,* ed. B.H. Andrew (Dordrecht: D. Reidel Publ. Co.), pp. 101-102.

Silk, J., and Norman, C. 1980. The interaction of T Tauri stars with molecular clouds. In *Interstellar Molecules,* ed. B.H. Andrew (Dordrecht: D. Reidel Publ. Co.), pp. 165-172.

Smoluchowsky, R. 1981. Amorphous ice and the behavior of cometary nuclei. *Astrophys. J.* 244:L31-L34.

Snell, R.L. 1979. Dissertation (unpublished), The University of Texas at Austin.

Soifer, R.W.; Puetter, R.C.; Russell, R.W.; Willner, S.P.; Harvey, R.M.; and Gillett, F.C. 1979. The 4-8 micron spectrum of the infrared source W33A. *Astrophys. J.* 232:L53-L57.

Spitzer, L., Jr. 1978. *Physical Processes in the Interstellar Medium* (New York: Wiley and Sons).

Taff, L., and Savedoff, M. 1972a. The mass spectrum of interstellar clouds and the assumption of total coalescence. *Mon. Not. Roy. Astron. Soc.* 160:89-97.

Taff, L., and Savedoff, M. 1972b. The mass distribution of objects undergoing collisions with applications to interstellar HI clouds. *Mon. Not. Roy. Astron. Soc.* 164:357-374.

Völk, H.J.; Jones, F.C.; Morfill, G.E.; and Roser, S. 1980. Collisions between grains in a turbulent gas. *Astron. Astrophys.* 85:316-325.

Whipple, F. 1978. Comets. In *Cosmic Dust,* ed. J.A.M. McDonnell (New York: Wiley and Sons), pp. 1-73.

Whipple, F. 1981. The nature of comets. In *Comets and the Origin of Life,* ed. C. Ponnamperuma (Dordrecht: D. Reidel Publ. Co.), pp. 1-20.

Whittet, D.C.B. 1981. The composition of interstellar grains. *O.Jl. Roy. Astron. Soc.* 22:3-21.

Wickramasinghe, D.T , and Allen, D.A. 1980. The 3.4 μm interstellar absorption feature. *Nature* 287:518-519.

Wilking, B.A.; Lebofsky, M.J.; Martin, P.G.; Rieke, G.H.; and Kemp, J.C. 1980. The wavelength dependence of interstellar linear polarization. *Astrophys. J.* 235:905-910.

Willner, S.P.; Russell, R.W.; Puetter, R.C.; Soifer, B.T.; and Harvey, P.M. 1979a. The 4 to 8 micron spectrum of the galactic center. *Astrophys. J.* 229:L65-L68.

Willner, S.P.; Puetter, R.C.; Russell, R.W.; and Soifer, B.T. 1979b. Unidentified infrared spectral features. *Astrophys. Space Sci.* 65:95-101.

Woodward, P.R. 1978. Theoretical models of star formation. *Ann. Rev. Astron. Astrophys.* 16:555-584.

THE INFRARED SPECTRAL PROPERTIES OF FROZEN VOLATILES

UWE FINK and GODFREY T. SILL
University of Arizona

Since Whipple's dirty snowball model of comet nuclei, it has been generally accepted that volatile ices help to explain cometary phenomena. We have been pursuing the infrared spectral properties of many substances that are potential candidates for frozen volatiles in the solar system; indeed some of these frozen materials have been found in the solar system: H_2O, CO_2, and SO_2. We present here a review of laboratory spectra in the range 1 to 20 µm of H_2O, CO_2, SO_2, CH_4, NH_3, H_2S, CO, NH_4HS and $NH_3 \cdot H_2O$. Both reflection spectra of thick frosts and transmission spectra of thin films are shown, and their main characteristics are described. Hydrates, clathrates, and composite spectra are discussed. When it is possible to observe the nuclei of comets at close range, we may be able to identify frozen volatiles by their infrared spectra.

To explain how comets produce dust and gas tails and deviate from Keplerian orbits, Whipple (1950,1951) devised his model of a comet's structure: "a conglomerate made up of ices, such as H_2O, NH_3 and other molecules volatile at normal temperatures, mixed with meteoritic materials." The meteoritic material was considered to be loosely consolidated with ices like H_2O, NH_3, CO_2, CH_4, and perhaps C_2N_2 or HCN. The more volatile the material, the smaller the likelihood that it would remain near the surface of the comet nucleus. A comet that had made many orbits around the Sun,

P/Encke for example, would have an outer shell of nonvolatile meteoritic uncompacted material, and below that a layer of "rotten" ice, then a layer of C_2N_2, followed by NH_3, CO_2, and perhaps CH_4 in the cold interior. A new comet would be less differentiated than an old comet like P/Encke, and could still have its volatiles mixed with the nonvolatile silicates, resulting in a so-called (but not by Whipple) dirty snowball model.

If remote sensing techniques are to be employed to verify the presence of these ices and to identify the various species, laboratory comparison studies are required. Practically all frozen gases or volatiles look white in visible light so that this is not a diagnostic part of the spectrum. Most solids exhibit sharp absorption edges in the ultraviolet, due to electronic transitions with energies of several eV, and differences in the location and shape of these absorptions could be used to distinguish between various ices. The ultraviolet absorption characteristics of a number of materials have been discussed by Hapke et al. (1981). However, the most diagnostic region of the spectrum is probably the infrared. It is in this region that the vibrational and rotational transitions of the molecules absorb radiation at wavelengths characteristic of each substance. By means of infrared reflection spectroscopy, H_2O ice has been identified on the rings of Saturn and the surfaces of the satellites of Jupiter and Saturn. CO_2 ice has been seen on the Martian poles and SO_2 ice on the surface of Io (cf. Sec. VII). If we can penetrate through the outer coma of a comet and observe the surface of the nucleus, infrared spectroscopy should be able to identify the frozen volatiles present.

The spectra of condensables have been investigated for a number of years, but good data directly applicable to solar system studies are not plentiful. There is a fair amount of data in the chemical literature but most of it concentrates on the assignments of the absorptions and the elucidation of the molecular and crystal structure. It contains little quantitative information on intensities or absorption coefficients, little data on reflection spectra of frosts, and almost no spectra shortward of 3 μm. Because quantitative data, such as the index of refraction and absorption coefficient have been required for spacecraft design, spectra of a number of condensables have been published (Mantz et al. 1974; Pipes et al. 1978). Reflection frost studies in the near infrared (1 to 4 μm) specifically addressing the needs of solar system investigations have been performed by Kieffer (1970) for CO_2 and H_2O mixtures and by Kieffer and Smythe (1974) for H_2O, CO_2, CH_4, NH_3, H_2S, and NH_4HS. Some data on SO_2 frost have been published by Fanale et al. (1979) and by Smythe et al. (1979). There has also recently appeared some work of questionable quality by Slobodkin et al. (1978,1981).

The studies of condensable species at the Lunar and Planetary Laboratory had their origin in the need to interpret our solar system observations (cf. Sec. VII). Initial work on H_2O and NH_3 frosts (Kuiper et al. 1970b,c) was followed by a thorough study of the water-frost spectrum as a function of temperature (Fink and Larson 1975). This was extended to reflection spectra of NH_3, CH_4, CO_2, H_2S, and NH_4HS frosts from 1 to 4 μm as a function of temperature (Fink et al. 1982). To obtain more quantitative intensity data and absorption coefficients, we initiated a comprehensive program of transmission spectra of thin ice films of known thickness, in the near infrared (1 to 4 μm), the mid infrared (3 to 20 μm) and the far infrared region. As yet, only a portion of this rather extensive data set has been analyzed and published (Sill et al. 1980,1981; Ferraro et al. 1980).

The choice of volatiles included in this review is governed by what we consider likely candidates for the outer solar system, namely H_2O, CO_2, CO, SO_2, CH_4, NH_3, H_2S, NH_4HS and $NH_3 \cdot H_2O$ (ammonia hydrate). We illustrate their spectra by examples taken from our work, partly because it was more readily available to us, and partly because this allows us to exercise reasonable uniformity in presentation of the data. The spectra shown in the figures have been chosen as representative of the spectrum of a particular compound, after careful inspection of all our data and any published in the literature. Unfortunately we do not have infrared spectra of frozen HCN or CH_3CN, substances that Delsemme and Rud (1977) considered possible candidates for cometary activity.

Two types of spectra are shown: reflection spectra of thick *frosts* of frozen gases in the wavelength region of reflected solar light (~ 1 to 4 μm); and transmission spectra of thin *films* of measured thickness in both the region 1 to 4 μm and the region of the fundamentals (2 to 20 μm). All of the spectra displayed in this chapter were obtained by the technique of Fourier spectroscopy (Fink and Larson 1979). The primary wavelength scale is in cm^{-1} since this unit is standard in molecular spectroscopy, being directly proportional to the energy levels. In many figures a μm scale is also shown at the top for convenience. Uniformity of presentation was attempted in areas such as wavelength coverage, spectral resolution, and intensity and wavelength scales, but not always attained since our data stretched over a number of years, and were taken with different instruments. We have not included tables of wavelengths, assignments, intensities, or absorption coefficients; these are referenced whenever available.

The relative volatilities of the compounds considered important for cometary activity are listed in Table I in order of increasing vapor pressure. Methane clathrate, $NH_3 \cdot H_2O$ and NH_4HS strictly speaking do not have boiling points, so the temperature at which the dissociation pressure reaches 1 atm is listed.

For readers not familiar with infrared spectroscopy a brief explanation is given in Sec. I, with emphasis on the characteristics of solid spectra which are not as widely known. This is followed by a short description of the properties of reflection and transmission spectra (Secs. II and III) and the spectra of individual ices are treated in Sec. IV. Clathrates and composite spectra are discussed in Sec. V and VI, and a brief summary of observations of condensed volatiles is given in Sec. VII.

I. GENERAL INFRARED SPECTRAL CHARACTERISTICS

Most molecules absorb infrared radiation because the energy levels of molecular vibrations and rotations fall in this region of the spectrum. The rotational motions have lower energy and occur roughly in the region of a few to 500 cm^{-1}; the fundamental vibrations extend from a few hundred to a few thousand cm^{-1}. The number of fundamental frequencies that a molecule can have is limited by its internal degrees of freedom and is 3 n-6 (3 n-5 for linear molecules), where n is the number of atoms making up the molecule. The more atoms in a molecule, the more complex the spectrum, with diatomic molecules having the simplest spectrum and possessing only a single fundamental frequency. Because of a molecule's symmetry, some of the vibrations may have the same energy (degenerate) and only some of the vibrations have allowed infrared transitions (infrared active). If a given vibration does not displace the center of negative charge from the center of positive charge, no change in the electric dipole moment occurs and electromagnetic radiation cannot be absorbed by the molecule. Besides the fundamental frequencies, a molecule can also vibrate at the overtones (2 v_1, 3 v_1 etc.) and at combination and difference frequencies ($v_1 + v_3$, $v_1 + 2 v_3 - v_3$ etc.). These bands, however, are generally much weaker than the fundamentals. For more detailed information concerning the spectra and structure of molecules the reader is referred to the series of books by Herzberg (1945, 1952, 1966).

TABLE I

Volatilities of Condensable Gases

Substance	Melting Pt.	Boiling Pt.	Temperature[a]
CO	68.1	81.8	45
CH$_4$	90.6	111.6	59
CH$_4 \cdot$6H$_2$O[b]		(198.2)[d]	108
H$_2$S	187.6	212.7	123
CO$_2$	(215.6)[c]	194.9	125
NH$_3$	195.4	239.5	148
SO$_2$	200.0	263.1	160
NH$_3 \cdot$H$_2$O	193.8	(271.3)[d]	163
NH$_4$HS		(306.4)[d]	(201)
H$_2$O	273.1	373.1	231

[a]Where vapor pressure is 10^{-4} atm.
[b]Methane clathrate.
[c]At 5.2 atm.
[d]Dissociation pressure at 1 atm.

TABLE II
Fundamental Vibrational Frequencies

Molecule	Structure	Observed Frequencies Gaseous State				Observed Frequencies Solid State			
		v_1	v_2	v_3	v_4	v_1	v_2	v_3	v_4
H_2O	Bent	3652	1595	3756		3150	1610	3220	
NH_3	Pyramidal	3336	932	3414	1628	3210	1057	3374	1650
H_2S	Bent	2611	1290	2684		2526	1169	2548	
NH_4^+	Tetrahedral					3033	1697	2925	1418
HS^-	Linear					2570			
CH_4	Tetrahedral	2914	1526	3020	1306		1570	3008	1300
SO_2	Bent	1151	519	1361		1143	522	1323	
CO_2	Linear	1388	667	2349			660	2345	
CO	Linear	2143				2138			

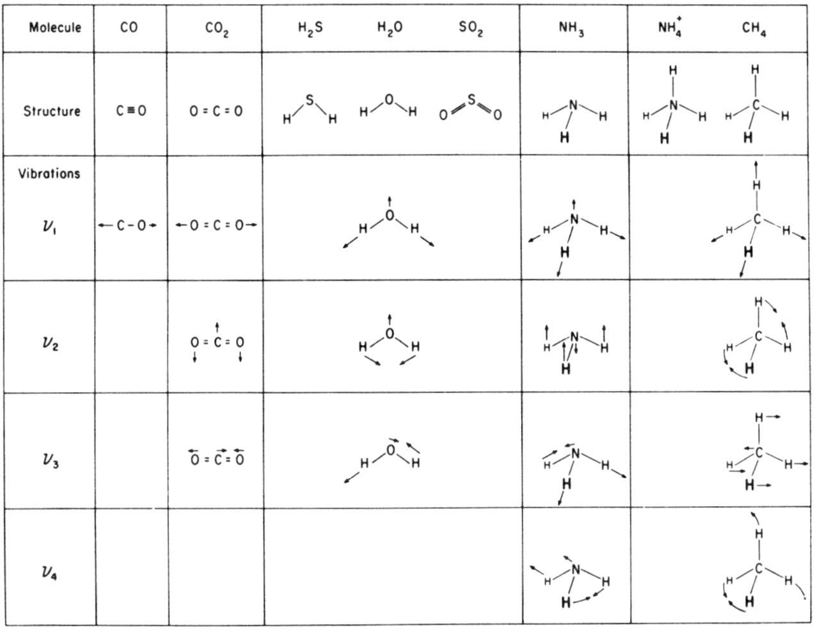

Fig. 1. Molecular structures and fundamental modes of vibration for typical molecules of condensable volatiles. Arrows indicate the direction and relative magnitude of atomic vibrations (after Herzberg 1945,1952).

As an overview, the fundamental vibrational frequencies of the condensed volatiles considered in this chapter are listed in Table II for both the gaseous and the solid state. These molecular species have a variety of geometries: CO and HS$^-$ are simple diatomic (linear) molecules and CO_2 is a triatomic linear molecule; H_2O, SO_2, and H_2S are planar and bent; three are three-dimensional: CH_4 and NH_4^+ tetrahedral and NH_3 pyramidal. These different structures lead to different fundamental models of vibration illustrated in Fig. 1. The diatomic molecules have only one fundamental vibration. The triatomic nonlinear molecules have three; CO_2 has four but the bending vibration v_2 is doubly degenerate which leaves only three; the three-dimensional molecules in Fig. 1 have four modes of which some are degenerate. In general the stretching modes are of higher frequency (or energy) than the bending modes, and the lighter the atoms of the molecules the higher the frequency of vibration (compare H_2O, H_2S and SO_2 in Fig. 1 and Table II).

It might at first be thought that the infrared spectrum of a condensed solid does not bear much resemblance to that of its gaseous counterpart. However, the identity and structure of individual molecules persist within the crystal lattice. This comes about because of the relatively strong *intramolecular* bonding forces that are responsible for forming a molecule, compared to the much weaker *intermolecular* crystal lattice forces, or the even weaker forces in a liquid. Thus the basic mechanisms of absorption and the main features of the gaseous spectrum, rotations at low frequencies and vibrations at higher energies, are preserved in a solid. Only relatively small shifts between the fundamental frequencies of the gas and the solid phase occur, as can be seen in Table II. There are, however, some important spectral changes in going from the gaseous to the solid state which are now discussed.

In the solid state, molecules are often hindered from rotating freely, thus limiting the number of pure rotational transitions observed, and simplifying somewhat the appearance of rotation-vibration bands. This is aided, in our case of condensed volatiles, by the low temperatures involved (~ 100 K) limiting the number of rotational states that are populated. Because of the fixed nature of a molecule in the crystal structure, a translation of a molecule is now a bound energy state so that translational modes can exist in a solid, usually having energy levels lower than the rotational modes.

The centers of the solid state absorption bands are displaced from the band centers in the gaseous state because the chemical bond strengths (or force constants) of the molecule are slightly altered in the electric field of the neighboring atoms. The stronger the intermolecular forces, the larger is the frequency shift. It is especially strong in the cosmochemically important molecule of water vapor. Molecules of liquid or solid water have a very strong coupling between the H^+ of one molecule and the O^- of an adjacent molecule. This dipole attraction is called hydrogen bonding.

Intermolecular forces also have an effect on the widths of the absorption bands. In general, the stronger and the more wide ranging the coupling between molecules, the shorter are the radiative lifetimes for the particular transitions and the broader the absorptions. Because of hydrogen bonding, the infrared spectrum of H_2O ice is dominated by broad absorption bands (several hundred cm^{-1}). For other molecules hydrogen bonding is less pronounced and the spectral features less broad: N-H hydrogen bonding is weaker and the ammonia spectrum shows relatively sharp absorptions (~ 20 cm^{-1}). CH_4 and H_2S have very little hydrogen bonding and their spectral bands are quite sharp (~ 10 cm^{-1}). The weakest interaction occurs for solid CO_2 whose absorption bands are very narrow ($\lesssim 1$ cm^{-1}).

Practically all ice spectra that we have observed show a distinct sharpening of their features with lower temperatures. This is probably due to a combination of causes. The crystal lattice motions decrease at lower temperatures resulting in smaller perturbations of the vibrational energy levels. The vibrational transitions thus have longer lifetimes and produce narrower spectral absorptions. In addition, a number of rotational, librational and translational combination or difference bands are usually included as unresolved components within a vibration band. The number of these states that are populated diminishes with lower temperature so that the main vibrational peak becomes sharper.

Because molecules can be packed in a number of different ways in a crystal lattice, solids possess the complicating property of phase changes with temperature. We have seen this phenomenon in almost every solid that we have studied; it can have marked effects on the appearance of the spectrum. In general, if a volatile is deposited at very low temperatures, a vitreous form results because the molecular motions are too restricted to allow arrangement into a regular crystal lattice. As the solid warms up, conversion into one or possibly several crystal structures takes place. These changes are often irreversible so that on recooling the temperature effects of a particular crystalline structure unhampered by phase changes can be observed. The importance of this annealing process (warming until a phase change occurs and then recooling) cannot be overemphasized, and was not fully recognized by us at the beginning of our investigations.

II. REFLECTION SPECTRA

The main advantage of reflection frost spectra lies in their ability to simulate reasonably closely the spectra of solar system objects observed in reflected sunlight. The scattering process of light reflection enhances weak absorptions over strong ones, which can be an advantage, but by the same token distorts the relative intensities of the features. Frosts are easier to make than thin films and do not require high vacuum, exact metering of deposition rates, or thickness measuring devices. The temperatures for frosts are generally somewhat warmer and more uncertain than those for thin films because the front surface of the frost is heated by the background and illuminating infrared radiation; yet at the same time it is insulated from the cold substrate by the thickness of the layer. On the other hand, frosts can be observed at higher temperatures, because they can suffer a fair amount of evaporation loss, which would completely deplete a thin film. As a consequence frosts allowed us to see high-temperature phase changes that were missed with thin films (e.g. H_2S), yet the opposite was true for low temperatures (e.g. NH_3).

A simple qualitative picture of the absorption process by scattering frost grains is as follows. (The reader is referred to Fig. 2 for a conceptual diagram of the process.) In a transmission spectrum of a thin film, the light at each wavelength must penetrate the same thickness of material. In reflection spectra, light of the strong bands is absorbed within the first few μm of the solid, whereas in the weak bands the light can penetrate more deeply and hence suffer absorption almost equivalent to that occurring in a strong band, simply because of the longer pathlength traversed. Thus reflection spectra show weak absorptions enhanced relative to strong absorptions, and details that are not so obvious in transmission spectra are often visible. The same mechanism enhances the weaker wings of strong features, making the absorptions broader than in transmission. Grain size is, of course, important in the appearance of a reflection spectrum. Light is reflected both from the front and back surface of a frost grain. For strong absorptions, little light is returned from the back

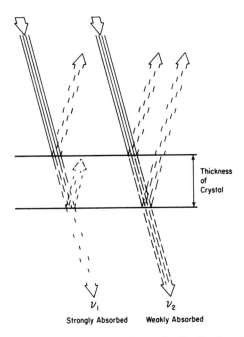

Fig. 2. Schematic illustration of light transmission and reflection for a frost particle. For strong absorptions, light is reflected mainly from the first surface and almost all the light is absorbed. For weak absorptions light is reflected from both the front and back surface and a considerable portion is transmitted to undergo further scattering in other particles. The effects of multiple internal reflections and refraction are omitted in this simplified schematic drawing.

surface of the grain because much of it has been absorbed in the short pathlength in between (see Fig. 2). Most of the light therefore comes from the first outside crystal surface. Except for this small amount of light, bands with large absorption coefficients will display near total absorption. For weak absorptions, an additional amount of light is returned from the back surface of a grain, having traversed a pathlength about twice the size of the grain. On a frost surface with many grains the weakly absorbed photons are reflected from many interfaces, more so in a fine-grained than a coarse-grained frost. Thus, in a fine-grained frost scattering predominates and weak features cannot easily be seen, while in a coarse-grained frost the effective absorption pathlength is much greater relative to scattering and weak absorptions become more obvious. Hapke (1981) treats the diffuse scattering of light from a particulate surface in a more quantitative manner.

A comparison of the reflection spectra of six frosts in the important region of solar reflected light (1 to 4 μm) is shown in Fig. 3. This figure provides a summary of our reflection work (Fink et al. 1982) and an overview of the absorption characteristics of likely frosts in the solar system: H_2O,

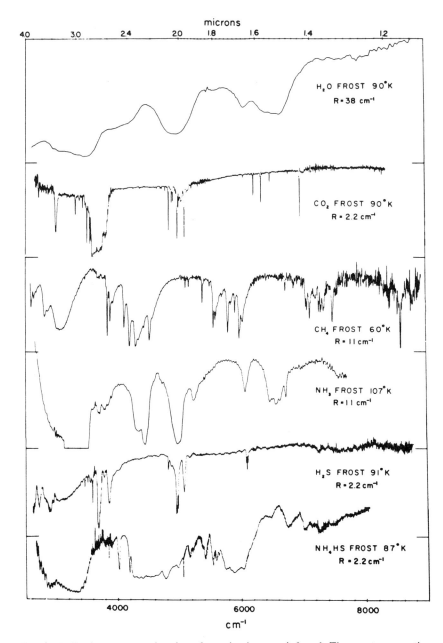

Fig. 3. Reflection spectra of various frosts in the near infrared. The spectra are ratio spectra of the frost to a white, nonabsorbing standard. Intensity scale is linear with zero levels displaced as indicated. Each frost has its own distinctive spectral signature. Most of the absorptions in this region of the spectrum are combination and overtone bands. Solids with hydrogen bonding (H_2O and NH_3) have relatively broad absorption features. Fine structure near 2.7, 1.9, and 1.4 μm is due to incompletely cancelled H_2O vapor in the ambient laboratory atmosphere.

CO_2, NH_3, H_2S, and NH_4HS. The numerous strong absorption features, quite distinct for each molecule, clearly show the diagnostic potential of this region of the spectrum. The spectra displayed are ratios of the frost reflectivity to that of the neutral infrared reflectance standard, sulfur powder. This removes the response of the instrument and the illuminating source. The ordinate scale in all our reflection spectra is linear relative reflectivity normalized to unity at the point of maximum intensity. No absolute intensities are given but from the perfectly white visual appearance of the frosts and the peak values of the interferograms, we estimate that the absolute reflectivities of the frosts in the regions between the absorptions are $\gtrsim 80\%$.

III. TRANSMISSION OF THIN FILMS OF ICES

The study of transmission spectra can provide quantitative optical constants for the condensed volatiles. Although requiring some care in preparation, the process of deposition of thin films allows much greater control over such variables as deposition rate, purity of the deposit, temperature, annealing, and total amount of condensable required and introduced into the dewar. Lower temperatures can be achieved; the actual temperature of the condensed species is measured with greater assurance, but temperatures near the melting point are not possible since the film evaporates before reaching this point.

The Lambert absorption coefficient α is defined by the equation of transmission of light through a parallel homogeneous slab of thickness x, $I = I_0 e^{-\alpha x}$, where I and I_0 respectively are the incident and transmitted intensity. Reflections at the front and back surface of the slab not only weaken the transmitted intensity, but if the film is optically flat, give rise to interference fringes. These channel fringes appear in many of our spectra (e.g. Figs. 9 and 11) and we used this interference phenomena to measure the thickness of the film. Details of our experimental set up can be found in Sill et al. (1980). The films were prepared in a special liquid nitrogen dewar in which vapor from volatiles was released near a cold infrared transparent substrate by a nozzle and condensed as a uniform film. The thickness of the film was monitored during deposition, and measured by observing interference fringes produced with a He-Ne laser. Thicknesses of ~ 100 fringes were readily produced. For thicker deposits the film became optically imperfect washing out the interference fringes. The thickest deposit was a 340-fringe (77-μm thick) CO_2 film.

The optical constants of a solid consist of the real part n and imaginary part k of the complex index of refraction ($\tilde{n} = n - i k$), as a function of wavelength λ. The absorption coefficient can be substituted for the real part of the index since the two are closely related by $\alpha = 4 \pi k/\lambda$. The real and imaginary parts of the index of refraction are also interrelated through the theory of the Kramers-Kronig dispersion relations which allow the determination of either one of these indices if the other is known over a large frequency range. The real part of the index varies rather slowly with wavelength; it is the imaginary part, or absorption coefficient that is dominant in establishing the appearance of a spectrum. With a knowledge of the optical constants, exact transmission or Mie scattering calculations can be carried out. In principle the reflection spectrum of a frost made up of particles of known size, can be synthesized using methods of radiative transfer or the somewhat more empirical approach of Hapke (1981).

All of our transmission spectra in this chapter represent ratios between the spectrum of the substrate material plus ice I divided by the blank spectrum of the substrate material alone I_O. This division produces a simple linear relative transmission I/I_O scale normalized to 1 at the point of maximum transmission (e.g. see Fig. 7). In keeping with the practice of the chemical literature we have also used a logarithmic absorbance scale defined by $-\log_{10} I/I_O$. The minus sign causes inversion of the spectrum making absorptions appear to be in emission (e.g. see Fig. 5). In addition we present plots of absorption coefficient defined above. A summary of absorption coefficients for a variety of ices is shown in Fig. 4. This figure is preliminary and based on only a partial analysis of our data, but shows the major absorption peaks and the magnitude of the absorption coefficient. It is seen that the peak absorptions are rather strong and run typically between 10^4 and 10^5 cm^{-1}.

IV. INDIVIDUAL ICE SPECTRA

H_2O

Doubtlessly, solid H_2O is the most important ice in the solar system. A summary of absorption coefficients and optical constants from 0.7 to 200 μm and a review of the literature has been made by Irvine and Pollack (1968). This work is largely based on the spectra of Ockman (1958) and Bertie and Whalley (1964). Because the absorption coefficients and the spectrum of water ice are reasonably well known, we have not carried out a detailed analysis with our own thin film data, but show a few selected sections of our data in the figures below.

Water ice can exist in a number of different phases, but at ordinary pressures ($\lesssim 2000$ atm) only ice I, to which we will limit our discussion, is stable. (For spectra of the other phases see Bertie et al. 1969, and references therein.) Phase I exhibits three modifications depending on its temperature of formation: if water vapor is condensed below ~ 110 K a vitreous ice I_v, which shows no sharp X-ray diffraction pattern, is produced; as the ice warms up it transforms irreversibly into cubic ice I_c at temperatures between 110 and 150 K, and on further warming the ice changes into its ordinary structure, hexagonal ice I_h at temperatures > 190 K.

A spectrum of a very thin film of water ice in the region of the fundamentals is shown in Fig. 5. The strongest feature is produced by the OH

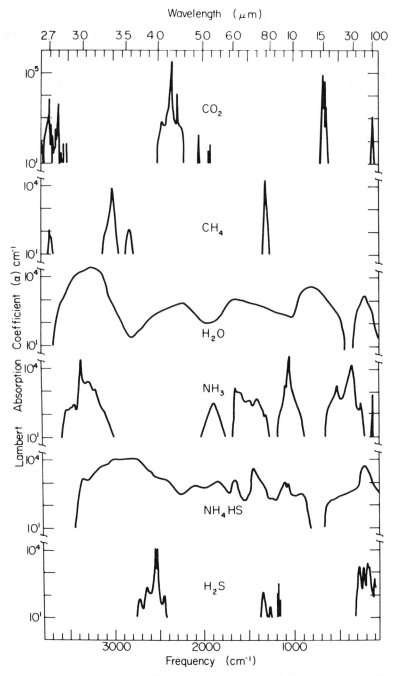

Fig. 4. Lambert absorption coefficients for selected ices. Ordinate scale is logarithmic. The absorption coefficients plotted are preliminary and are meant to convey an idea of their approximate magnitude and the relative band positions. All ices shown are crystalline except NH_4HS which is vitreous.

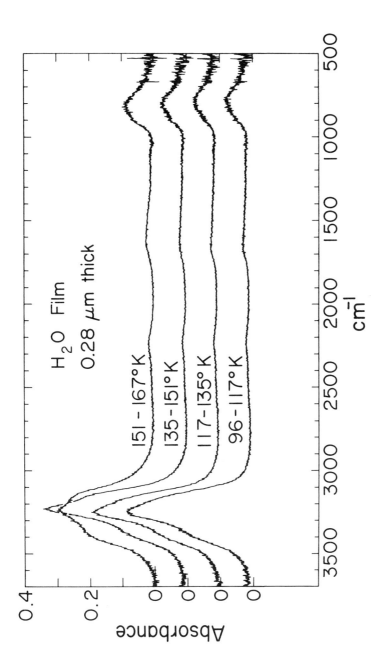

Fig. 5. Warming sequence of thin film H_2O transmission spectra. The ordinate is absorbance ($-\log_{10} I/I_0$). The film was deposited at ~90 K in the vitreous phase. Note frequency shifts, broadening and change in appearance as the film transforms to the crystalline phase near 135-151 K.

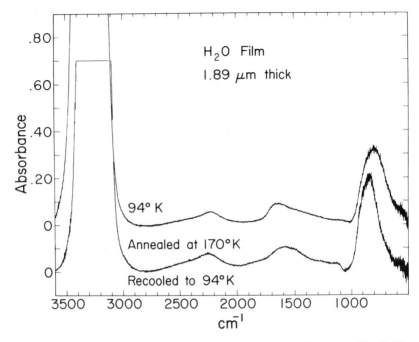

Fig. 6. Transmission spectra of an H_2O ice film somewhat thicker than in Fig. 5. The film was deposited as vitreous at 94 K, then warmed to 170 K, annealed to ice I_c and recooled to 94 K. The bending fundamental at ~ 1600 cm^{-1} shows up on this thicker film. Note the sharper features of the crystalline ice I_c.

symmetric and asymmetric stretching frequencies near 3250 cm^{-1}. The intensity of the bending frequency near 1610 cm^{-1} is considerably weaker. The feature near 800 cm^{-1} is due to librational transitions. Water ice is a classic example of strong hydrogen bonding, and the absorptions are quite broad (~ 300 cm^{-1}); in fact water ice exhibits the broadest absorption feature of any condensable that we have encountered.

The particular film in Fig. 5 was deposited at low temperatures, resulting in vitreous ice, and slowly warmed up. The transition to cubic ice at a temperature of ~ 135 to 150 K is evident in the figure and produces the following changes: the librational frequency shifts ~ 35 cm^{-1} to higher frequencies; the stretching and bending modes shift ~ 40 cm^{-1} to lower frequencies; the region of the stretching frequencies around 3300 cm^{-1} acquires more structure; the absorptions become somewhat sharper. The latter effect is demonstrated more clearly by spectra of a thicker film in Fig. 6, which also shows some of the weaker absorptions. This film was deposited at 94 K, annealed at 170 K and then recooled to 94 K. The differences between the cubic and vitreous forms of ice illustrated by our figures are in good accord with those described by other investigators (Hardin and Harvey 1973, and references

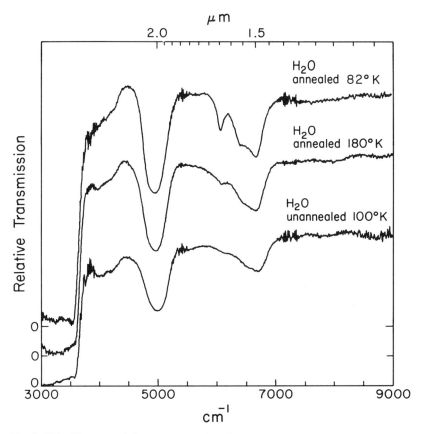

Fig. 7. Thin film transmission spectrum of H_2O in the overtone region. The ordinate is linear intensity. The film is 30 times thicker than that of Fig. 6, so that fundamentals at ~3300 cm^{-1} are saturated. The film was deposited at 80 K, warmed and annealed at 183 K and recooled to 80 K. Note the sharpness of the 6056 cm^{-1} peak and the shape of the whole 1.5 μm absorption which is distinctive of crystalline ice.

therein). Because of evaporation loss of thin films on warming above 200 K, it has been difficult to observe the spectral changes going from the cubic to the hexagonal form, but it appears from our work and that of others (Bertie and Whalley 1964) that the spectra of the two modifications are almost identical.

Transmission spectra of the overtone region from 1 to 3 μm in Fig. 7 also show diagnostic differences between the vitreous and cubic phases. The appearance of a sharp feature at 6056 cm^{-1} is both a function of the ice phase and temperature. For vitreous ice at low temperatures (bottom spectrum) the feature is rather indistinct when compared to cubic ice at the same temperature (top spectrum). On the other hand, the warm cubic phase produces a similarly shallow appearance of the 6056 cm^{-1} feature (middle

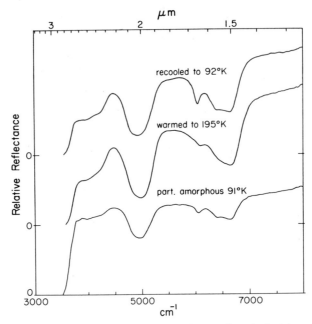

Fig. 8. Reflection spectra of a very fine-grained frost deposited at low temperature (partly amorphous). On warming up, the frost changed to a coarser-grained crystalline structure (middle) increasing the absorption depths. The frost was then recooled (top).

spectrum) as the cold vitreous phase. Temperature effects and phase change are therefore not readily separated. The effect of temperature on the reflection spectrum of crystalline ice is illustrated in Fig. 8. The equivalent width of the 6056 cm^{-1} feature has been calibrated by Fink and Larson (1975) as a function of temperature and applied to solar system spectra. Temperatures of 95 ± 10 K for the ice surface of Europa, 103 ± 10 K for Ganymede, and 80 ± 5 K for Saturn's rings were obtained (cf. Fig. 23 in Sec. VII).

It would certainly be desirable in solar system studies to identify the phase of water ice present on a comet or satellite. Unfortunately the differences in their spectra are subtle and in the overtone region are mixed with changes caused by temperature. For reflection spectra, in the region of the fundamentals, the absorptions become so strong and broad that it is not easily possible to distinguish between the two phases. In order to do so, the object cannot exhibit saturated absorptions, and high-quality spectra with good signal-to-noise characteristics must be available.

NH_3

Gaseous NH_3 is a well-known constituent of Jupiter's atmosphere and because of its relatively high cosmic abundance, it can be presumed in the atmospheres of other major planets and on some of their satellites. Solid NH_3 can be expected wherever the temperature falls below its frost point (see Table I) which should be the case for almost all bodies in the outer solar system, particularly comets. We have completed a rather extensive study of the ammonia spectrum and have obtained absorption coefficients over the range of wavelengths from 50 to 7000 cm^{-1}. Our results together with a detailed description of the method of analysis, thin film preparation, and thickness measurements as well as references to previous investigations are reported in Sill et al. (1980).

The spectrum of solid NH_3 from 2.6 to 20 μm is shown in Fig. 9, for three different thicknesses and different types of substrates. In this region the spectrum is dominated by the NH stretching fundamental at 3374 cm^{-1} and the bending fundamental at 1057 cm^{-1}. Both of these are very strong and sharp, their absorption coefficients reaching peak values of 45,000 and 40,000 cm^{-1}, respectively. The spectrum of NH_3 in the overtone region to $\sim 1.0\ \mu$m is shown in Fig. 10. The regular undulations in this spectrum and those of Fig. 9 are channel interference fringes in the thin NH_3 ice film and should not be construed as NH_3 features. A reflection spectrum of NH_3 can be seen in Fig. 3. An intercomparison of absorption coefficients obtained by three groups of investigators is shown in Fig. 11. For the most part the absorption coefficients are in good accord so that the optical constants for NH_3 can be considered well-determined.

Like water, the spectrum of solid NH_3 is complicated by the existence of several phases. If NH_3 is deposited at very low temperatures ($\sim$ 20 K) a vitreous or amorphous phase results, characterized by rather broad absorptions (half-width $\sim 250\ cm^{-1}$) and peak absorption coefficients of only $\sim 9000\ cm^{-1}$. At intermediate temperatures of 80 to 90 K a metastable phase is observed which transforms to the normal stable cubic phase at temperatures of $\sim$ 100 to 110 K. A detailed discussion of assignments and illustrations of the three phases is given in Ferraro et al. (1980). We believe that the above phases are the only ones for which convincing evidence exists. A variety of spurious phases has been reported in the literature starting with an anomalous NH_3 spectrum by Staats and Morgan (1959). We have shown, unequivocally (Sill et al. 1981), that this spectrum can be assigned to ammonia hydrate (see discussion below). Another series of NH_3 spectra made under poorly controlled conditions and resulting in a considerably different appearance has been published by Slobodkin et al. (1978); we believe their spectra is distorted by contamination and possible hydrate formation.

H_2S

Because of the relatively high cosmic abundance of sulfur, H_2S can be expected as a solid constituent in the solar system if reducing conditions with large amounts of hydrogen prevail. However, H_2S combines rather readily with NH_3 and if an excess of the latter exists it will form NH_4HS (see below).

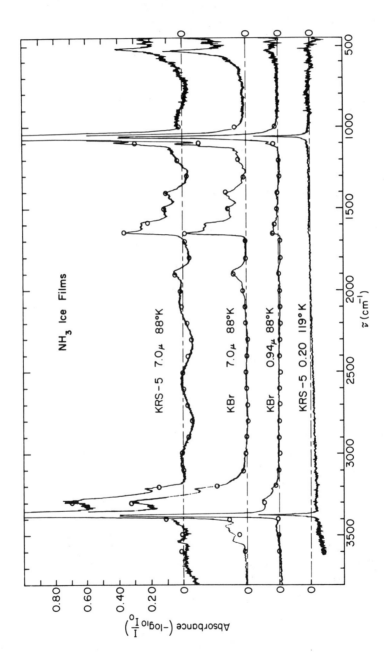

Fig. 9. Transmission spectra of cubic NH$_3$ ice films in the region of the fundamentals for various thicknesses, substrates, and temperatures. Warmer temperature is displayed for the 0.20 μm film to avoid features by the metastable phase. Slow modulations in the spectra are caused by channel fringes in the film. Circles represent calculated points using the absorption coefficients of Sill et al. (1980) and thin film theory.

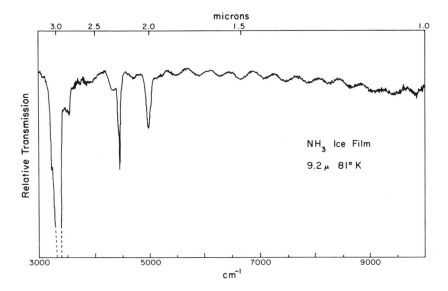

Fig. 10. Transmission spectrum of cubic NH$_3$ in the overtone region. The ordinate is a linear intensity scale. Channel fringes are clearly visible.

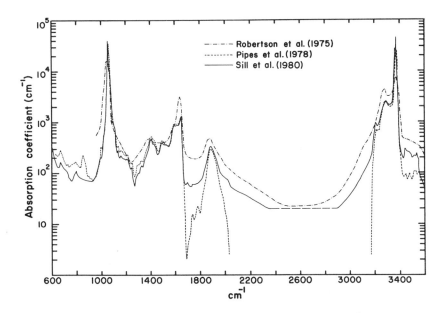

Fig. 11. Comparison of absorption coefficients of cubic NH$_3$ from 600 to 3600 cm^{-1} by various investigators. The figure shows the large range in absorption coefficient covered.

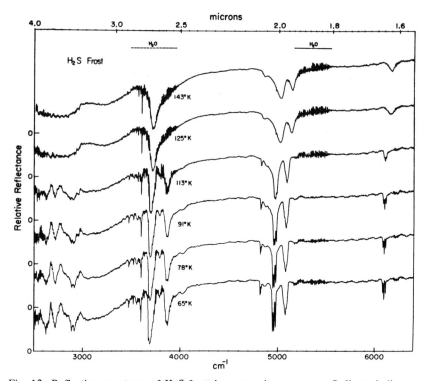

Fig. 12. Reflection spectrum of H_2S frost in a warming sequence. Ordinate is linear intensity scale with zero levels displaced. Note the dramatic change in spectral appearance between 113 and 125 K when the solid changes its crystal structure from tetragonal to cubic. Regions of incomplete H_2O vapor cancellation are marked at the top.

A series or reflection spectra of H_2S from 1 to 4 μm is given in Fig. 12. These spectra display quite a dramatic phase change from a tetragonal crystal structure to a cubic one at a temperature of 125 ± 5 K. Details of the assignments in the two phases are given in Ferraro and Fink (1977). We have obtained thin film spectra of H_2S but have not yet deduced absorption coefficients. An example of a transmission spectrum is given at the bottom of Fig. 13.

NH_4HS and $NH_3 \cdot H_2O$

We have also investigated two compounds of NH_3, NH_4HS and $NH_3 \cdot xH_2O$, which may have important applications in solar system studies. A transmission spectrum of NH_4HS together with spectra of its constituents is shown in Fig. 13. The combined spectrum appears considerably different than either of the two separate compounds. No trace of their fundamental frequencies remains, demonstrating that a new compound has been formed. A similar conclusion holds for reflection spectra in the 1 to 4 μm region which

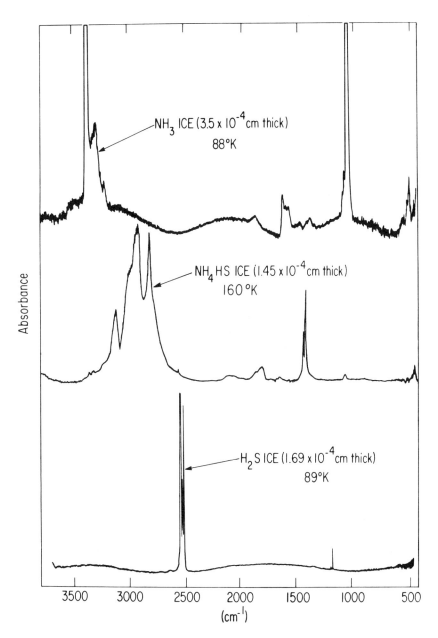

Fig. 13. Transmission spectra of NH_3, H_2S, and NH_4HS ices from 500 to 3800 cm^{-1}. Spectrum of NH_4HS is quite distinct from those of the other two, demonstrating that a new compound has been formed during the codeposition. The NH_4HS spectrum displayed is the crystalline phase.

are illustrated for all three compounds in Fig. 3 above. When NH_4HS is deposited at low temperatures an amorphous phase with broad absorptions results. In order to obtain the crystalline phase in Fig. 13, the ice had to be annealed. An analysis of the infrared and Raman spectrum of NH_4HS and ND_4DS has been published by Bragin et al. (1977) and further details of our own work on NH_3, H_2S and NH_4HS are given in a paper by Ferraro et al. (1980).

The combination of NH_3 with H_2O does not appear to be as strong. While the spectrum of the hydrate is definitely different from either pure molecule, it retains more of the character of its constituents. This is why the spectrum of ammonia hydrate has been confused a number of times with that of pure ammonia (Staats and Morgan 1959; Uyemura and Maeda 1972; Slobodkin et al. 1978). Ammonia is apparently easily contaminated with water vapor, even when the latter is present in only trace amounts. A spectrum of NH_3, H_2O and the hydrate is shown in Fig. 14, taken from Sill et al. (1981). It can be seen that at sufficiently cold temperatures (unannealed film in Fig. 14) the hydrate is not completely formed but a mixture of pure NH_3 ice, H_2O ice and the hydrate exists. Only after annealing is the hydrate completely formed. There are actually three hydrates of ammonia, the hemihydrate $2NH_3 \cdot H_2O$, the monohydrate $NH_3 \cdot H_2O$, and the dihydrate $NH_3 \cdot 2H_2O$. The spectrum in Fig. 14 represents the monohydrate which seems to be the most stable of the group. The identification of the monohydrate in Fig. 14 was made possible by a detailed spectroscopic study of both the hemihydrate and the monohydrate by Bertie and Morrison (1980). We point out that NH_3 is less volatile in the hydrate than in the pure state, so that this would be a likely form in which to find it in a cometary nucleus.

CH_4

Methane is the thermodynamically stable species of carbon in the cooler, hydrogen rich solar nebula and for this reason it has been postulated as an abundant frozen volatile in the nuclei of comets. The large amount of gaseous CH_4 observed in the atmospheres of the major planets provides abundant evidence for this hypothesis. Methane is among a handful of gases with a very low freezing point (see Table I) and is probably not frozen out in the solar system except for the coldest places such as Pluto and comets. Liquid helium needed to be used to produce frosts and thin films of CH_4.

The spectra of two phases of solid CH_4 are shown in Fig. 15, phase I at the top and a quasi-liquid phase at the bottom. At lower temperatures solid CH_4 can also exist in phase II and with slightly elevated pressures in phase III (cf. Medina and Daniels 1979). The quasi-liquid phase was first recognized by us (Fink et al. 1982) and has not been reported in the literature. The change to this phase occurs somewhat below the melting point ($\sim$ 60 to 70 K) and the spectrum of the quasi-liquid phase bears a close resemblance to that of liquid CH_4. The spectral absorptions become very broad, and the peak

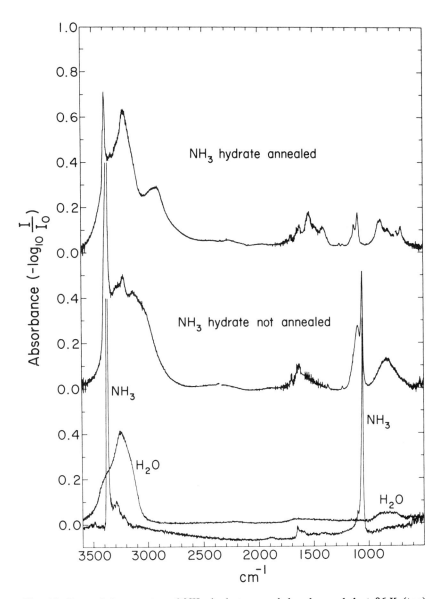

Fig. 14. Transmission spectra of NH$_3$ hydrate annealed and recooled at 96 K (top), unannealed at 105 K (middle), and comparison spectra of NH$_3$ (0.90 μm thickness, 88 K), and H$_2$O (0.30 μm thickness, 90 K) at the bottom. Note the change in appearance between the unannealed and the annealed hydrate, particularly the disappearance of the sharp NH$_3$ peaks. The ratioing process introduces excess noise around 1600 cm^{-1} due to uncancelled water vapor and near the ends of the spectra.

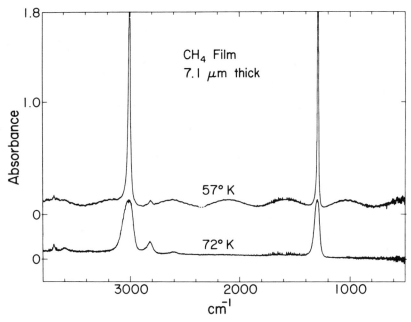

Fig. 15. Transmission spectra of solid CH_4 for the crystalline phase I (top) and the quasi-liquid phase (bottom). Weak absorptions near 3700 cm^{-1} may be due to an impurity.

values decrease considerably (Fig. 15 bottom). The temperature behavior of solid CH_4 is thus different from that of H_2O and NH_3. The latter display a vitreous phase with broad shallower absorptions at low temperatures, changing to sharp crystalline features at higher temperatures. Solid CH_4 has very snarp crystalline features at low temperatures which change to the broad absorptions of the quasi-liquid phase as it is warmed up. A possible explanation for the strange spectral characteristics of this phase is the onset of free rotation of the CH_4 molecules (characteristic of the liquid phase) in place of a locked position characteristic of the crystal structure. The solid and gaseous CH_4 spectra are very similar with only minor frequency shifts of the absorption bands between the two states. Good signal-to-noise ratio and sufficiently high resolution are required to separate the two phases in astronomical observations.

CO_2 and CO

Oxides of carbon at first glance seem less likely components of cometary nuclei. However, CO is thermodynamically more stable than CH_4 for high temperature, low H_2 pressure regimes. Furthermore, CO has been observed as an abundant molecule in certain gas clouds in the galaxy and in Jupiter (Beer 1975). Delsemme and Rud (1977) have proposed that under certain condi-

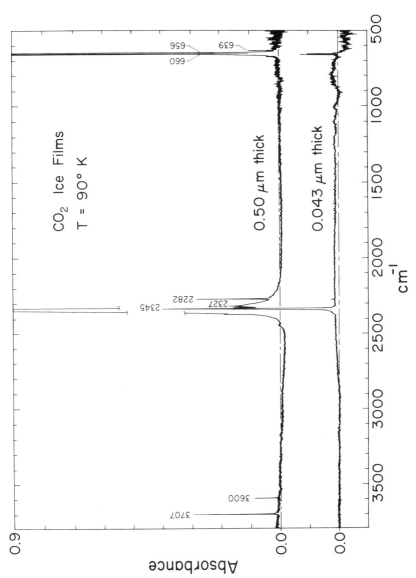

Fig. 16. Transmission spectrum of solid CO_2 for two different thicknesses. Note the extreme sharpness of the absorption features, and the asymmetric shift of the 2345 cm^{-1} peak for the thicker film.

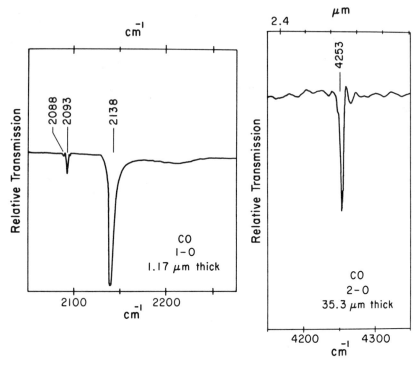

Fig. 17. Transmission spectrum of solid CO in the region of the fundamental (after Pipes et al. 1978), and the 2-0 overtone region from our data. Isotopic bands show up by $^{12}C^{18}O$ at 2088 cm^{-1} and by $^{13}C^{16}O$ at 2039 cm^{-1}. Ordinate is linear intensity scale.

tions CO_2 as well as CO may be incorporated into cometary nuclei as cold condensates.

Transmission spectra of CO_2 are shown in Fig. 16. The spectrum of CO_2 has the sharpest features of any solid we have observed; Fig. 16 displays many of these sharp and intense absorptions. The fundamental v_3 band at 2345 cm^{-1} is the strongest infrared molecular band (both in gaseous and solid spectra) that we are aware of. The band is so intense that CO_2 impurities in other frozen gases can easily be detected at the ppm level (see Figs. 19 and 20). The reflection spectrum of CO_2 from 1 to 4 μm (Fig. 3) also contains many sharp diagnostic features. Early frost studies of mixtures of CO_2 and H_2O by Kieffer (1970) seemed to indicate that the broad H_2O absorptions completely swamp those of CO_2. We believe that this conclusion stems from his low spectral resolution which was inadequate to detect the sharp CO_2 features. In our own spectra of such mixtures taken at moderate resolution (see Fig. 22), the CO_2 features stand out clearly above those of H_2O.

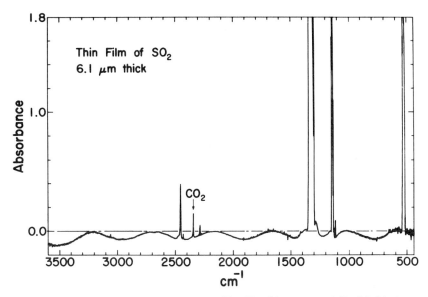

Fig. 18. Thin film transmission spectrum of SO_2. The SO_2 was carefully dried before deposition so that no spurious water absorptions are present. A trace amount of CO_2 impurity is visible.

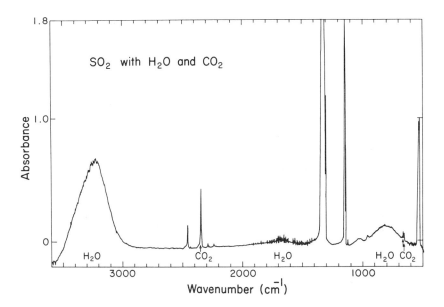

Fig. 19. Transmission spectrum of a thin film of SO_2 that was not purified before deposition. Absorptions by CO_2 and H_2O are marked on the figure and are quite prominent. SO_2 absorptions are unmarked. From the relative absorption depths we can determine the composition of the film as 80% SO_2, 19% H_2O and 1.3% CO_2.

Carbon monoxide, being a simple diatomic molecule, has a much less complex spectrum. Absorptions were found only in the region of the fundamental 1-0 band (2138 cm^{-1}) and the first overtone 2-0 band (4253 cm^{-1}), both of which are displayed in Fig. 17. In order to observe the 2-0 band, a much thicker film had to be employed. Absorptions by various isotopic species are also evident.

SO_2

We studied this solid because it has been observed in spectra of Io (see Sec. VII). Its transmission spectrum in the mid infrared is shown in Fig. 18. The strong fundamental frequencies are quite far in the infrared at 522, 1143, and 1323 cm^{-1}. Evidence for solid SO_2 on Io's surface was provided by the much weaker $v_1 + v_3$ overtone band at 2456 cm^{-1}. Reflection spectra of SO_2 frosts in this region have been published by Fanale et al. (1979) and Smythe et al. (1979). Both of these data sets were badly contaminated with H_2O (as were our own SO_2 reflection spectra which are not shown for this reason). For our transmission data we carefully dried the SO_2 before deposition and removed this impurity from our spectrum. A trace amount of CO_2 can, however, be seen at 2345 cm^{-1}.

V. CLATHRATES

The hydrate compounds discussed above have definite chemical structures but are more weakly bound than their constituent molecules. There also exist even more weakly bound clathrate structures in which guest molecules are more or less physically trapped in roughly spherical cavities of a host lattice. Since H_2O is cosmochemically the most likely and most abundant host material, molecules of spherical symmetry such as CH_4 and the noble gases, may form clathrates during the conditions prevailing in the early solar system cold condensation (Sill and Wilkening 1978). The appropriate proportion of a methane clathrate is one CH_4 molecule per 5.7 H_2O molecules ($CH_4 \cdot 5.7\,H_2O$). Clathrates were originally proposed as constituents of comets (Delsemme and Swings 1952; Miller 1961; Delsemme and Wenger 1970) because they reduce the vapor pressure of the pure guest compound (cf. Table I) so that molecules such as CH_4 are available for release and production of ionization fragments in the inner solar system, rather than being all spent in the outer solar system.

We are not aware of any confirmed spectra of clathrates since they are somewhat difficult to prepare and the existence of the clathrate is difficult to establish. Reflectance spectra of a mixture of H_2O and CH_4 have been published by Smythe (1975) and are dominated by water absorptions not showing any signatures of CH_4. Since the guest molecules are not chemically bound in the lattice, but only trapped, one expects the spectrum to be the composite of the host and the guest with the spectrum of the host dominating due to its greater abundance. The lack of the CH_4 absorptions is therefore puzzling and may be due to the properties of the reflectance spectrum and too low a spectral resolution, since our composite transmission spectra (Sec. VI) show the individual components quite well.

VI. COMPOSITE SPECTRA

We have also formed thin films of mixtures of gases which did not react. This was done both deliberately to see the spectral results of such mixtures and accidentally, because there were impurities present in the gas under study. We show two examples of such mixtures.

The spectrum of SO_2 in Fig. 19 was recorded without first purifying the SO_2 from the gas bottle, so that the signatures of the contaminant H_2O and CO_2 showed up. The three gases were deposited together on the substrate in a random mixture. Very little interaction took place between them because the absorptions of SO_2 and CO_2 retained the same appearance and fell at the same frequencies as the pure substances. The water band at 3 μm was slightly distorted from pure water indicating that some perturbation occured for this molecule, probably due to hydrogen bonding. The final transmission spectrum resulting from such mixtures is simply Lambert's law applied to the sum of their absorption coefficients. Through knowledge of the individual absorption coefficients we can thus determine that this film consists of 80% SO_2, 19% H_2O, and 1.3% CO_2. Because of the added complexity of scattering in reflection spectra, determination and analysis of the composite spectrum will not be quite as straightforward.

We depict in Fig. 20 an unannealed spectrum of a mixture of NH_3 and H_2O with impurities of CO_2 and N_2O which was intended to form ammonia hydrate. Again the spectrum displays the signatures of all the individual components and we determine: 7.0 μm NH_3, 1.8 μm H_2O, and 1.9 μm CO_2 (we did not have absorption coefficients of pure N_2O available). These two examples demonstrate that the analysis of a thin composite film provides a rather sensitive method for determining the identity and composition of an unknown mixture of gases. With suitable instrumentation on board a space probe this method could be employed to sample and analyze the gases emanating from a comet nucleus or existing in a planetary atmosphere.

VII. EXAMPLES OF CONDENSED VOLATILES OBSERVED IN THE SOLAR SYSTEM

It is not the purpose of this chapter to present a comprehensive discussion of condensed volatiles so far observed in the solar system, but we shall mention briefly the major accomplishments in this area.

The compositions of the rings of Saturn had long been a puzzle, until the technique of Fourier spectroscopy provided the first good spectrum of the rings (Kuiper et al. 1970a). Because of the lack of good laboratory data, this spectrum was first erroneously associated with NH_3 ice, as pointed out by

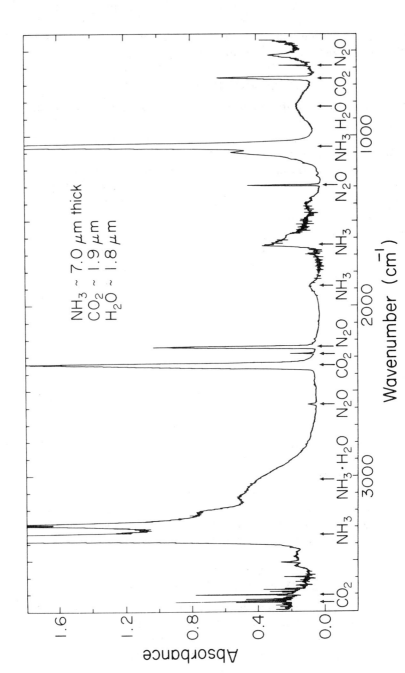

Fig. 20. Composite transmission spectrum of a thin film of codeposited NH_3, H_2O, CO_2, and N_2O. The spectral signatures of each of these constituents are quite distinct and easily recognizable.

IR SPECTRA OF FROZEN VOLATILES 195

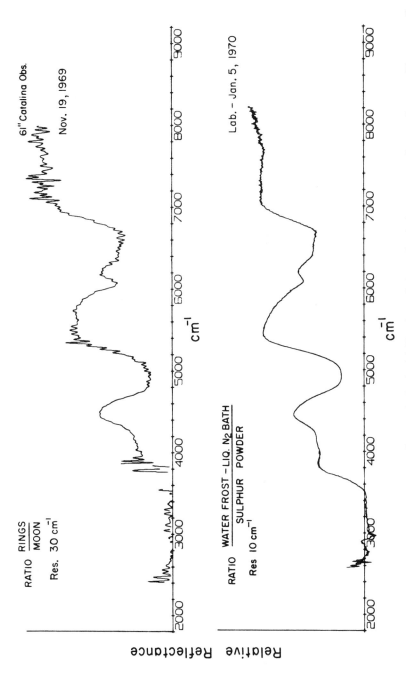

Fig. 21. First high-quality spectrum of the rings of Saturn obtained with Fourier techniques (top). A spectrum of water frost provides an excellent match with the ring spectrum (bottom).

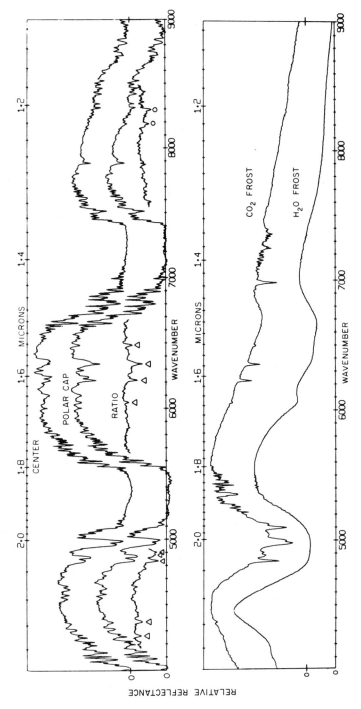

Fig. 22. Spectra of the equatorial region of Mars, the polar caps and their ratio. Prominent CO_2 absorption features are marked with triangles and circles. Below is a spectrum of CO_2 frost contaminated with H_2O and a spectrum of H_2O frost. The sharp CO_2 features, however, stand out clearly (from Larson and Fink 1972).

Pilcher et al. (1970a,b), who showed that water ice might provide a better match. Our own laboratory data (Kuiper et al. 1970b,c) confirmed quite clearly that frozen water vapor was the correct identification of the observed absorptions. This is illustrated in Fig. 21 which shows the laboratory spectrum of water ice, and the ratio spectrum of the rings of Saturn divided by the Moon (the ratio spectrum is utilized to eliminate the instrumental response, the effects of absorption in the Earth's atmosphere, and the changing intensity of sunlight with wavelength). This juxtaposition makes it quite obvious that the major component of the rings is frozen H_2O.

Another interesting example of a surface frost deposit is provided by spectra of the Mars polar cap (Larson and Fink 1972), which unambiguously showed the spectral signatures of dry ice (Fig. 22). Two spectral features of CO_2 had previously been seen by a low resolution infrared spectrometer on board Mariner 7 (Herr and Pimentel 1969), but they were not positively identified as belonging to solid CO_2. During following missions the infrared Fourier spectrometer on Mariner 9 showed spectral absorptions by silicates in a large Martian dust storm (Hanel et al. 1972) and water ice absorptions in the clouds forming above the high plateau region containing the three shield volcanoes (Curran et al. 1973).

During the 1972 apparition of Jupiter, groundbased spectra of the Galilean satellites were acquired simultaneously by Pilcher et al. (1972) and Fink et al. (1973). We show in Fig. 23 the spectra of Europa and Ganymede by the

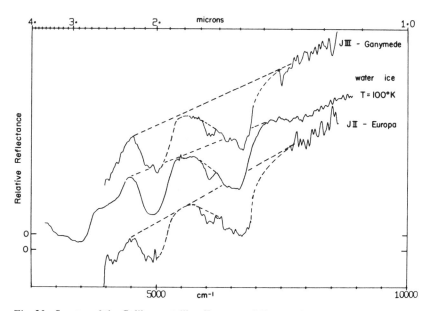

Fig. 23. Spectra of the Galilean satellites Europa and Ganymede compared to laboratory water frost. The temperature of the frost was chosen to match the 6056 cm^{-1} peak on the two satellites.

latter authors, and a laboratory water ice spectrum. The temperature of the laboratory ice was chosen to match the appearance of the 6056 cm^{-1} feature closely, and was used to derive ice temperatures for these two bodies (Fink and Larson 1975). In addition to large water ice absorptions on Europa and Ganymede, Fink et al. (1973) also observed weaker water ice absorptions on Callisto but none on Io. Laboratory spectra of ices mixed with dark silicates show that the ice features are distinctly muted, which may explain the much weaker ice absorptions on Callisto (Sill and Clark 1982).

With further improvements in detector sensitivity, spectra of the much fainter Saturn satellites Tethys, Dione, Rhea, and Iapetus became possible (Fink et al. 1976). All four of these showed deep water ice absorptions. Under certain conditions (discussed below) photometry can also provide evidence for frozen volatiles, and this was accomplished by Morrison et al. (1976) for the same four Saturn satellites. This method has been extended by Cruikshank and others to the fainter satellites of Saturn and Uranus (see Cruikshank 1980, and references therein). The photometry suggests that water ice also exists on the surface of the Saturn satellites Enceladus and Hyperion and the Uranus satellites Umbriel, Titania, and Oberon.

Although it is certainly cold enough in the outer solar system to condense NH_3 and CH_4, either as clouds in the atmospheres of the major planets, or on the satellite surfaces, no convincing direct observation for either has been made. There exists strong indirect evidence, however, through the detection of a CH_4 atmosphere on Pluto (Fink et al. 1980), that CH_4 ice exists on the surface of this body. Absorptions in the 1 to 2.5 μm region have been reported by Cruikshank and Silvaggio (1980) and Soifer et al. (1980). Unfortunately, the solid absorptions of CH_4 are sufficiently similar to those of the gas that fairly good resolution is required to separate the two. Thus, at present, the large gaseous absorptions mask any positive direct evidence for surface CH_4 ice.

An unusual and somewhat unexpected surface frost has recently been detected on Io, which was the one Galilean satellite that did not show water ice bands. An absorption peak on this satellite at 4.1 μm was observed by Pollack et al. (1978) and Cruikshank et al. (1978), but could not be clearly identified and was ascribed to possible evaporite salt deposits. Higher resolution spectra by Fink et al. (1978) showed that there were four features in the 3.5 to 4.0 μm region and that they were too narrow for the proposed salt hypothesis. The identification of this feature came with the Voyager spacecraft, when Pearl et al. (1979) detected gaseous SO_2 absorptions over one of Io's hot spots and the 4.1 μm absorption feature was correlated with SO_2 frost (Fanale et al. 1979; Smythe et al. 1979).

VIII. CONCLUSION

From the above discussion, it is evident that a variety of condensed volatiles can be and have been detected by means of infrared reflectance spectroscopy. To extend these techniques to comets is certainly feasible but not without difficulty. When the comets are bright, they are close to the Sun and the large amounts of dust evolved prevent observation of the nucleus. Thus a spectrum of Comet West 1976 VI, obtained by Fink and Larson showed no indications of any ice absorptions.

When the comet is in the outer solar system it is very faint and difficult to observe. Recent infrared photometry of periodic comets at somewhat greater heliocentric distances shows no indications of ices (Hartmann et al. 1981; chapter by A'Hearn in this book; Veeder and Hanner 1981; chapter by Campins and Hanner). The comets observed appear to be more like C- and S-type asteroids. This is not necessarily evidence against an icy nucleus. It appears that comets develop a more extensive dust coma farther from the Sun than had previously been thought. Moreover, photometry of a diverse class of objects such as comets cannot give unequivocal answers about the presence or absence of ices. To carry this out more definitively, a number of conditions would have to be fulfilled. The absorptions must be strong enough to give a significant deviation in the various filter passbands. There must be one dominant absorber present since photometry would be confused by a mixture of components. The nonicy component cannot have a reflectivity which might simulate the broadband characteristics of ices. To guard against this possibility, it has been customary to compare the photometry of new objects with that of known bodies, such as satellites and asteroids, for which spectroscopy is available. These latter objects may not be appropriate benchmarks for comets whose reflected sunlight comes from an assemblage of particles rather than a solid surface (Veeder and Hanner 1981). There is a good chance, however, that with further advances in detectors and instrumentation, and with space probes to the outer solar system such as a flyby or a rendezvous with a comet, condensed volatiles could be detected and identified.

Acknowledgments: We wish to thank J. Ferraro, who has provided equipment at Argonne National Laboratory, and who has participated in most of these studies. This work was supported by grants from the National Aeronautics and Space Administration.

REFERENCES

Beer, R. 1975. Detection of carbon monoxide in Jupiter. *Astrophys. J.* 200:L167-L169.
Bertie, J.E.; Labbé, H.J.; and Whalley, E. 1969. Absorptivity of ice I in the range 4000-30 cm^{-1}. *J. Chem. Phys.* 50:4501-4520.
Bertie, J.E., and Morrison, M.M. 1980. The infrared spectra of the hydrates of ammonia, $NH_3 \cdot H_2O$ and $2NH_3 \cdot H_2O$ at 95°K. *J. Chem. Phys.* 73:4832-4837.
Bertie, J.E., and Whalley, E. 1964. Infrared spectra of ices Ih and Ic in the range 4000-350 cm^{-1}. *J. Chem. Phys.* 40:1637-1645.
Bragin, J.; Diem, M.; and Guthals, D. 1977. The vibrational spectrum and lattice dynamics of polycrystalline ammonium hydrosulfide. *J. Chem. Phys.* 67:1247-1256.
Cruikshank, D.P. 1980. Near infrared studies of the satellites of Saturn and Uranus. *Icarus* 41:246-258.

Cruikshank, D.P.; Jones, T.J.; and Pilcher, C.B. 1978. Absorption bands in the spectrum of Io. *Astrophys. J.* 225:L89-L92.
Cruikshank, D.P., and Silvaggio, P.M. 1980. The surface and atmosphere of Pluto. *Icarus* 41:96-102.
Curran, R.J.; Conrath, B.J.; Hanel, R.A.; Kunde, V.G.; and Pearl, J.C. 1973. Mars: Mariner 9 spectroscopic evidence for H_2O ice clouds. *Science* 182:381-383.
Delsemme, A.H., and Rud, D. 1977. Low-temperature condensates in comets. In *Comets, Asteroids, Meteorites*, ed. A.H. Delsemme (Toledo, Ohio: Univ. Toledo), pp. 529-535.
Delsemme, A., and Swings, P. 1952. Hydrates de gaz dans les noyaux comètaires et les grains interstellaires. *Ann Astrophys.* 15:1-6.
Delsemme, A., and Wenger, A. 1970. Physico-chemical phenomena in comets. I. Experimental study of snows in a cometary environment. *Planet. Space Sci.* 18:709-715.
Fanale, F.; Brown, H.; Cruikshank, D.P.; and Clark, R. 1979. Significance of absorption features in Io's IR reflectance spectrum. *Nature* 280:761-763.
Ferraro, J.R., and Fink, U. 1977. Near infrared reflectance spectra and analysis of H_2S frost as a function of temperature. *J. Chem. Phys.* 67:409-413.
Ferraro, J.R.; Sill, G.; and Fink, U. 1980. Infrared intensity measurements of cryodeposited thin films of NH_3, NH_4HS, H_2S, and assignments of absorption bands. *Appl. Spectros.* 34:525-533.
Fink, U.; Dekkers, N.H.; and Larson, H.P. 1973. Infrared spectra of the Galilean satellites of Jupiter. *Astrophys. J.* 179:L155-L159.
Fink, U.; Ferraro, J.R.; Sill, G.; and Larson, H.P. 1982. Reflection and transmission spectra of condensed volatiles from 1-4 μm. In preparation.
Fink, U., and Larson, H.P. 1975. Temperature dependence of the water-ice spectrum between 1 and 4 microns: Application to Europa, Ganymede and Saturn's rings. *Icarus* 24:411-420.
Fink, U., and Larson, H.P. 1979. Astronomy: Planetary Atmospheres. *Fourier Transform Infrared Spec.* 2:243-314.
Fink, U.; Larson, H.P.; Gautier III, T.N.; and Treffers, R.R. 1976. Infrared spectra of the satellites of Saturn: Identification of water ice on Iapetus, Rhea, Dione and Tethys. *Astrophys. J.* 207:L63-L67.
Fink, U.; Larson, H.P.; Lebofsky, L.A.; Feierberg, M.; and Smith, H. 1978. The 2-4 micron spectra of Io. *Bull. Amer. Astron. Soc.* 10:595(abstract).
Fink, U.; Smith, B.A.; Benner, D.C.; Johnson, J.R.; Reitsema, H.J.; and Westphal, J.A. 1980. Detection of a CH_4 atmosphere on Pluto. *Icarus* 44:62-71.
Hanel, R.A.; Conrath, B.J.; Hovis, W.A.; Kunde, V.G.; Lowman, P.D.; Pearl, J.C.; Prabhakara, C.; Schlachman, B.; Levin, G.V. 1972. Infrared spectroscopy experiment on the Mariner 9 mission: Preliminary results. *Science* 175:305-308.
Hapke, B. 1981. Bidirectional reflectance spectroscopy. I. Theory. *J. Geophys. Res.* 86:3039-3054.
Hapke, B.; Wagner, J.; Wells, E.; and Partlow, W. 1981. Far-UV, visible and near-IR reflectance spectra of frosts. *Icarus* (special Comet issue). In press.
Hardin, A.H., and Harvey, K.B. 1973. Temperature dependences of ice I—hydrogen bond spectral shifts: The vitreous to cubic ice I phase transformation. *Spectrochim. Acta* 29A:1139-1151.
Hartmann, W.K.; Cruikshank, D.P.; and Degewij, J. 1981. Surface materials on remote comets: Theoretical and observational indications. *Icarus* (special Comet issue). In press.
Herr, K.C., and Pimentel, G.C. 1969. Infrared absorptions near three microns recorded over the polar cap of Mars. *Science* 166:496-499.
Herzberg, G. 1945. *Molecular Spectra and Molecular Structure II. Infrared and Raman Spectra of Polyatomic Molecules.* (Princeton: Van Nostrand).
Herzberg, G. 1952. *Molecular Spectra and Molecular Structure I. Spectra of Diatomic Molecules.* (New York: Van Nostrand).
Herzberg, G. 1966. *Molecular Spectra and Molecular Structure III. Electronic Spectra and Electronic Structure of Polyatomic Molecules.* (New York: Van Nostrand).
Irvine, W.M., and Pollack, J.B. 1968. Infrared optical properties of water and ice spheres. *Icarus* 8:324-360.

Kieffer, H. 1970. Spectral reflectance of CO_2-H_2O frosts. *J. Geophys. Res.* 75:501-509.
Kieffer, H., and Smythe, W.D. 1974. Frost spectra: Comparison with Jupiter's satellites. *Icarus* 21:506-512.
Kuiper, G.P.; Cruikshank, D.P.; and Fink, U. 1970a. The composition of Saturn's rings. *Sky Teles.* 39:14.
Kuiper, G.P.; Cruikshank, D.P.; and Fink, U. 1970b. Letter to the Editor. *Sky Teles.* 39:80.
Kuiper, G.P.; Cruikshank, D.P.; and Fink, U. 1970c. The composition of Saturn's rings. *Bull. Amer. Astron. Soc.* 2:235(abstract).
Larson, H.P., and Fink, U. 1972. Identification of carbon dioxide frost on the Martian polar caps. *Astrophys. J.* 171:L91-L95.
Mantz, A.; Thompson, S.; Arnold, F.; and Sanderson, R. 1974. Optical properties of cryo-deposits on low scatter mirrors. In *Thermophysics and Spacecraft Thermal Control: Progress in Astronautics and Aeronautics.* 35:229-248.
Medina, F., and Daniels, W. 1979. Raman spectrum of phase III of solid CH_4 in the lattice and intramolecular regions. *J. Chem. Phys.* 70:2688-2694.
Miller, S. 1961. The occurrence of gas hydrates in the solar system. *Proc. Nat. Acad. Sci.* 47:1798-1808.
Morrison, D.; Cruikshank, D.P.; Pilcher, C.B.; and Rieke, G.H. 1976. Surface composition of the satellites of Saturn from infrared photometry. *Astrophys. J.* 207:L213-L216.
Ockman, N. 1958. The infra-red and raman spectra of ice. *Advances in Physics* 7:199-220.
Pearl, J.; Hanel, R.; Kunde, V.; Maguire, W.; Fox, K.; Gupta, S.; Ponnamperuma, C.; and Raulin, F. 1979. Identification of gaseous SO_2 and new upper limits for other gases on Io. *Nature* 280:755-758.
Pilcher, C.B.; Chapman, C.R.; Lebofsky, L.A.; and Kieffer, H.H. 1970a. On the identification of ices in Saturn's rings. *Bull. Amer. Astron. Soc.* 2:239(abstract).
Pilcher, C.B.; Chapman, C.R.; Lebofsky, L.A.; and Kieffer, H.H. 1970b. Saturn's rings: Identification of water frost. *Science* 167:1372-1373.
Pilcher, C.B.; Ridgway, S.T.; and McCord, T.B. 1972. Galilean satellites: Identification of water frost. *Science* 178:1087-1089.
Pipes, J.; Roux, J.; and Smith, A. 1978. Infrared transmission of contaminated cryo-cooled optical windows. *Amer. Inst. Aeronaut. Astronaut.* 16:984-990.
Pollack, J.B.; Witteborn, F.C.; Erickson, E.F.; Strecker, D.W.; Baldwin, B.; and Bunch, T.E. 1978. Near-infrared spectra of the Galilean satellites: Observations and compositional implications. *Icarus* 36:271-303.
Robertson, C.W.; Downing, H.D.; Curnutte, B.; and Williams, D. 1975. Optical constants of solid ammonia in the infrared. *J. Opt. Soc. Amer.* 65:432-435.
Sill, G.T., and Clark, R.N. 1982. Surface composition of the Galilean satellites. In *Satellites of Jupiter,* ed. D. Morrison (Tucson: Univ. Arizona Press). In press.
Sill, G.; Fink, U.; and Ferraro, J.R. 1980. Absorption coefficients of solid NH_3 from 50 to 7000 cm^{-1}. *J. Opt. Soc. Amer.* 70:724-739.
Sill, G.; Fink, U.; and Ferraro, J.R. 1981. The infrared spectrum of ammonia hydrate: Explanation for a reported ammonia phase. *J. Chem. Phys.* 74:997-1000.
Sill, G., and Wilkening, L. 1978. Ice clathrate as a possible source of the atmospheres of the terrestrial planets. *Icarus* 33:13-22.
Slobodkin, L.; Buyakov, I.; Cess, R.; and Caldwell, J. 1978. Near infrared reflection spectra of ammonia frost: Interpretation of the upper clouds of Saturn. *J. Quantum Spectrosc. Radiative Transfer* 20:481-490.
Slobodkin, L.; Buyakov, I.; Triput, N.; Caldwell, J.; and Cess, R. 1981. New measurements of the infrared spectrum of solid SO_2: Applications to Io. *J. Quantum Spectrosc. Radiative Transfer* 26:33-38.
Smythe, W.D. 1975. Spectra of hydrate frosts: Their application to the outer solar system. *Icarus* 24:421-427.
Smythe, W.D.; Nelson, R.M.; and Nash, D.B. 1979. Spectral evidence for SO_2 frost or adsorbate on Io's surface. *Nature* 280:766.
Soifer, B.T.; Neugebauer, G.; and Matthews, K. 1980. The 1.5–2.5 μm spectrum of Pluto. *Astron. J.* 85:166-167.

Staats, P., and Morgan, H. 1959. Infrared spectra of solid ammonia. *J. Chem. Phys.* 31:553-554.
Uyemura, M., and Maeda, S. 1972. Infrared intensities of crystalline NH_3 and ND_3. *Bull. Chem. Soc. Japan* 45:2225-2226.
Veeder, G.J., and Hanner, M.S. 1981. Infrared photometry of comets Bowell and P/Stephan-Oterma. *Icarus* (special Comet issue). In press.
Whipple, F.L. 1950. A comet model. I. The acceleration of comet Encke. *Astrophys. J.* 111:375-394.
Whipple, F.L. 1951. A comet model. II. Physical relations for comets and meteors. *Astrophys. J.* 113:465-474.

STRUCTURE AND ORIGIN OF COMETARY NUCLEI

BERTRAM DONN and JURGEN RAHE
NASA Goddard Space Flight Center

There is strong evidence that a comet nucleus consists of a single object whose basic structure is Whipple's icy conglomerate. In this review we consider only such models. Derived radii fall in the range from 0.3 to 16 km. With an adopted density of 1 g cm^{-3}, masses are between 10^{13} and 10^{19} g. Two out of nearly 700 radii appear to be between 50 and 100 km. A number of cometary phenomena indicate that the nucleus is a low-density, fragile object with a large degree of radial uniformity in structure and composition. Details of the ice-dust pattern are more uncertain. A working model is proposed based on theories of accumulation of larger objects from grains. This nucleus is a distorted spherical aggregate of a hierarchy of ice-dust cometesimals. These cometesimals retain some separate identity leading to comet fragmentation when larger components break off. The outer layers of new comets have been modified by cosmic ray irradiation in the Oort Cloud. Current experimental research may account for the observed greater activity of new comets at large heliocentric distances. As a comet ages during successive perihelion passes, an inert dust layer gradually builds up on the surface, changing the characteristics of coma development. This process can ultimately cause comets to become inactive and become the Earth- and Mars-crossing asteroids. Meteoric fireballs are associated with comets and are consistent with the fragile and fragmentizable nature of the nucleus. The evidence for meteorite-comet association is still controversial. Current dynamical studies do not seem to require a cometary source of meteorites. The presence of meteorites in nuclei would require a two-stage accumulation mechanism. The survival of comets in

> the Oort Cloud seems well established although the place of formation
> is uncertain. Various hypotheses for the location have been proposed,
> from the asteroid zone to interstellar clouds. The most likely source
> appears to be the region of the outer planets or interstellar clouds in
> some way associated with the primordial solar nebula.

Comets are best known for the spectacular and nearly always unpredictable appearance of the brighter objects. It is this strange and awesome behavior which has given rise to early myths concerning their role in warning of coming misfortune. In destroying the myth of comets as supernatural objects, modern science has endowed them with an equally engaging and more significant property. The changing appearance of coma and tail tells us of the present environment of interplanetary space with which it interacts. Comets are also indicators of the remote past when they accumulated in a primordial cloud from which the Sun, planets and stars also formed. They have also been proposed as a major stage in the accumulation of the giant planets (Whipple 1964; Öpik 1973).

Because of their small size and remoteness from the Sun, material was stored in comets in relatively unchanged form. The release of this material as they pass through the inner solar system, provides data about their original composition and present environment. Unfortunately, the information we receive from a comet at the present time is akin to transmissions from satellites and space probes whose telemetry data is only partially received, and for which the telemetry code is only partially known. This causes a great deal of uncertainty and confusion in the interpretation of the data.

In this chapter attention is focused on the cometary nucleus. This is generally conceived of as a solid structure which forms the permanent part of a comet as it revolves around the Sun. As it approaches within ~ 5 AU of the Sun, the energy present in the inner solar system causes gases and small grains to be ejected from the nucleus, forming the coma and the tail. Material comprising these features is continuously released and lost to space when the comet is a few AU from the Sun making it part of the interplanetary medium. Consequently, the coma and tail must constantly be replenished during the comet's apparition.

One is faced with devising a model of a permanent object that revolves around the Sun, forms a coma and tail near perihelion and changes little at successive apparitions. Another important characteristic of the nucleus is its ability to survive extremely close solar passage. The sungrazers come within a solar radius of the photosphere where bodies of 30-cm diameter would be destroyed (Russell 1929). At much greater distances all volatiles would be lost and many-body nuclei would not survive at perihelion distances of a few tenths AU typical of many comets. An analysis by O'Dell (1973) concluded that for dynamical reasons the only stable form of a many-body nucleus is to collapse into a single object. Whipple (1963) has discussed models for the structure of the nucleus in more detail and raised several other difficulties with sandbank and many-particle structures. The present analysis reaches the same conclusion as Whipple did earlier, and is now generally accepted, that the best working model is an icy nucleus with embedded solids.

In the following section we first discuss various observational phenomena of comets that are relevant to this review. A brief presentation of basic statistics of comets is followed by a description of cometary phenomena associated with mass loss and disintegration. This leads to conclusions on cometary lifetimes and the preservation of comets over long times. Following this we come to the first details of the nucleus itself; a description of likely internal structures of an icy nucleus is given. Next, the evolutionary changes in structure and composition are considered. We comment briefly on surface structure and composition, including changes with time. The relationships among comets, asteroids, and meteorites are discussed and the review concludes with considerations on cometary origin. The chemical composition is discussed in considerable detail in the chapter by Delsemme and hence is omitted here.

We do not seek a definitive description of the nucleus. Rather, our aim is to present a working model, consistent with present data and which could serve as a sound base for further research. One has, however, always to keep in mind the individual nature of comets and the wide variations among their observed behavior, as well as the incomplete character of cometary data. Our knowledge of the cometary nucleus is fortunately increasing at a significant rate, especially due to ultraviolet observations and observations of comets at large heliocentric distances, many of which are reviewed in other chapters. These recent developments are fortunate for cometary studies, but present problems for the preparation of a review of the nucleus. An accurate description of a comet nucleus will require imaging from a spacecraft with high spatial resolution, remote sensing, and subsurface sounding capabilities.

I. ORBITS AND STATISTICS

Comets have the most variable properties of any objects in the solar system. This is true not only intrinsically, as with changing appearance as a comet approaches the Sun, but also with regard to the range of orbital, luminous, and spectroscopic characteristics.

A distribution of comets among the different categories of orbits taken from Marsden's (1979) catalogue, is given in Table I. This catalogue lists orbital elements for 1027 cometary apparitions of 658 individual comets observed between 87 BC and the end of 1978. The orbits upon which Table I are based are the osculating (instaneous) orbits relative to the Sun. To determine the major axis prior to the comet's approach to the Sun and planets, two corrections are necessary. The change in energy caused by planetary perturbations and by the shift from heliocentric to barycentric orbits must be added. With the further inclusion of nongravitational effects, there is no well-determined original orbit that is definitely hyperbolic (Marsden et al. 1978). Comets are called new when they have extremely large semimajor

TABLE I

Distribution of Orbital Forms for 1027 Apparitions of
658 Individual Comets Observed Between 87 BC and the End of 1978

Orbital Form	Eccentricity	N	%
Elliptical orbits	< 1.0	275	42
Short-periodic comets (P < 200 yr)	< 0.97	113	17
Long-periodic comets (P > 200 yr)	0.97 – 1.0	162	25
Parabolic orbits	1.0	285	43
Hyperbolic orbits	> 1.0	98	15
Strongly hyperbolic orbits	> 1.006	0	0

axes, $1/a < 100 \times 10^{-6}$ AU^{-1} and presumably coming from the Oort Cloud into the planetary system for the first time. Old comets have made some tens of return (Marsden et al. 1978).

The perihelion distance of known comets varies from 0.005 to 6.88 AU (Comet Schuster 1975 II). Long-period comets have an average perihelion distance $\bar{q}$ = 1.08 AU, whereas short-period comets have $\bar{q}$ = 1.61 AU. There are 9 sungrazers with $q \approx 0.01$ AU. Eight of these are clearly related (Table II) and are probably the fragments of a single parent (Marsden 1967).

Alphelion distances Q range from 4 AU to infinity. If there are no comets originally coming from interstellar space (planetary perturbations do eject some from the solar system), the maximum aphelion distance is the outer limit of the Oort Cloud at $\sim 10^5$ AU. Of the 113 short-period comets, 50 (44%) have Q between 5 and 6 AU, corresponding to Jupiter's distance (5.2 AU). The 78 (69%) with $Q < 7$ AU can be classified as belonging to a Jupiter family; the longitude of the nodes of their orbits is strongly concentrated near 0° and 180°, so that near perihelion and aphelion they are also near their nodes, and comets almost always come closer to Jupiter than to any other giant planet. No other planetary family is generally recognized.

Intrinsic brightness variations (bursts) exceeding 1 to 2 mag are frequently observed among all classes of comets. P/Schwassmann-Wachmann 1 (P = 15 yr, q = 5.4 AU) brightens by several mag $\sim$ 2 or 3 times a year; the luminosity of P/Tuttle-Giacobini-Kresák (P = 5.6 yr, q = 1.2 AU) increased by 9 mag twice in 1973.

The size of the nucleus cannot yet be determined by direct measurement. Photographic resolution may be as small as 0″.5 (Dollfus 1961) and visual observations can resolve about 0″.1 (Kuiper 1950). The closest a comet has approached the Earth was 0.015 AU (Comet Lexall 1770 I). In this century the closest approach was 0.04 AU (Comet Pons-Winnecke 1927 VII). With

TABLE II
Orbital Elements of Sungrazing Comets

Comet	T (UT)	q (AU)	e	P (yr)	ω	Ω	i	L	B
1668	Feb. 28.08	0.066604	1.0	—	109°.81	2°.52	144°.38	248°.61	+33°.23
1843 I	Feb. 27.91	0.005527	0.999914	512	82°.64	2°.83	144°.35	281°.86	+35°.31
1880 I	Jan. 28.12	0.005494	1.0	—	86°.25	7°.08	144°.66	281°.68	+35°.25
1882 II	Sep. 17.72	0.007751	0.999907	761	69°.59	346°.96	142°.00	282°.24	+35°.24
1887 I	Jan. 11.63	0.009665	1.0	—	58°.35	325°.50	128°.47	280°.24	+41°.79
1945 VII	Dec. 28.01	0.006305	1.0	—	50°.93	321°.69	137°.02	289°.73	+31°.96
1963 V	Aug. 23.92	0.005161	0.999952	1111	85°.82	6°.77	144°.52	281°.90	+35°.37
1965 VIII	Oct. 21.18	0.007761	0.999918	929	69°.03	346°.25	141°.85	282°.24	+35°.21

modern observing techniques the minimum detectable diameter, visually, would have been 4.5 km and 12 km and photographically, 22.5 and 60 km. However, a serious problem to observing the nucleus occurs for nearby comets because of the surrounding luminous coma which will interfere with the resolution.

At some sufficiently large heliocentric distance the coma becomes apparently nonexistent. The comet then has a sharp, stellar appearance and its luminosity is presumably caused by reflection of sunlight from the nucleus. The luminosity relative to the Sun is readily estimated.

The radius R_c (in km) of the assumed spherical nucleus is given by

$$\log R_c = -0.2 \, m_c - 0.5 \, [\log \alpha + \log P(\Theta)] + \log r + \log \Delta + 2.81 \quad (1)$$

where m_c is the comet magnitude at (r,Δ), r is the heliocentric distance and Δ the geocentric distance in AU, Θ is the phase angle, $P(\Theta)$ is the phase function, and α is the albedo. The radius depends on α and $P(\Theta)$ which must both be assumed. The derivation of the radius is therefore subject to an uncertainty depending on the optical properties of the surface of the nucleus and a possible contribution to the luminosity from a still present coma (Sekanina 1976a). The main uncertainty lies in the value of the albedo. For a clean, rough ice this can be near unity whereas for Whipple's dirty-ice nucleus with a crust, the albedo can be very low; this point is considered later. Roemer (1966) tabulates dimensions for two values, $a = 0.7$ and 0.02. Table III includes her listing and those of Whipple (1978a) and Kresák (1973). Values were recalculated for an assumed albedo of 0.3.

Radii range upward from 0.3 km; they are typically 1 to 2 km for short-period comets and may be up to an order of magnitude larger for long-period comets. Very rare, much larger comets appear to occur with a 50 to 60 km radius and masses of $\sim 10^{21}$ g (e.g., the parent of the sungrazers or the Great Comet of 1729 that could be seen with the naked eye at 4 AU from the Sun).

TABLE III

Nuclear Radii

	N	$<R>$ $a = 0.3$ (km)	R_{min}, R_{max} (km)
Short-period comets			
Roemer	18	1.1	0.3, 2.5
Kresák ($r > 3.2$ AU)	14	2.1	0.7, 6.2
Long-period comets			
Roemer	9	3.5	0.5, 16.5
Whipple	3	4.0	3.1, 5.6

Cometary nuclei are also too small for direct determination of mass, except by some future comet probe. They have produced no detectable effects on the orbital motion of other objects which is the only way to measure masses in the solar system. Upper limits can be obtained in this fashion (Laplace 1805). Roemer (1966) found a radius for Comet Wirtanen 1957 VI before splitting of 16.5 km (with an assumed albedo of 0.3), and of the primary component after splitting of $\sim$ 10 km; with unit density these correspond to masses of $\sim 2 \times 10^{19}$ and 4×10^{18} g, respectively. Masses may be calculated from the radii of Table IV and an assumed density. For an icy nucleus, a density of 1 g cm^{-3} should be reasonable (Donn 1963). The data yield cometary masses between 10^{13} and 10^{21} g. Most nuclei would fall between 10^{14} and 10^{17} g.

II. THE STRUCTURE OF THE NUCLEUS

The concept to which the term nucleus of a comet is applied is often confusing. In this review we mean by cometary nucleus, the permanent structure which orbits the Sun and is the bearer of the cometary mass. Observational descriptions of comets often refer to a stellar nucleus within the diffuse coma. This is best described as the photometric nucleus. Other uses of the term also exist (e.g., Vorontsov-Velyaminov 1946).

The prevailing model prior to 1950 was some form of the sandbank hypothesis, wherein the nucleus was thought of as a diffuse cloud of small particles, traveling together. Lyttleton (1953) has described his version of the sandbank model and an interstellar accretion mechanism for its formation. However, following Whipple's (1950) description and analysis of the nucleus as a single aggregate of ices and meteoritic matter, this icy conglomerate became the generally accepted hypothesis. A detailed analysis of the two models which offers strong support to the icy conglomerate structure and raises serious objections against the sandbank version has been given by Whipple (1964).

The principle arguments favoring the icy conglomerate over the sandbank model are:

1. A relatively large ratio of volatile to nonvolatile material would be required to account for the formation of the coma near the time of each perihelion passage and the repetition of this gas loss for the many revolutions of short-period comets;
2. The occurrence of nongravitational forces is not consistent with a sandbank model;
3. The similarity of pre- and post-perihelion appearances of sungrazing comets, require nuclear aggregates which are at least meters in diameter;
4. The splitting of comet nuclei cannot be reconciled with a sandbank model and is difficult to explain with a collection of large numbers of larger particles;

5. Tidal and Poynting-Robertson forces would disrupt a sandbank nucleus and a many particle model.

These structures are not consistent with the observed splitting of sungrazing comets. The icy conglomerate model on the other hand is reasonably consistent with the observed behavior of comets, and the arguments against the sandbank model seem sufficiently strong that the remainder of this review will be based on the icy-conglomerate model.

Models for Cometary Nuclei

Several structures for an icy nucleus have been proposed since Whipple put forth the ice with embedded dust model, the icy conglomerate, in 1951. O'Dell (1973) showed that a gravitationally-bound, many-particle nucleus in the Oort Cloud would collapse into a single icy-conglomerate structure if it were to survive as a comet. The resultant structure would appear to resemble a Whipple type icy nucleus, except that the volatiles are interstellar material. During successive perihelion passages the volatiles gradually diffuse out and vaporize leaving the grain cores behind. These form a porous matrix as Mendis and Brin (1977,1978), Brin and Mendis (1979), and Weissman and Kieffer (1981) have discussed in detail. O'Dell suggests that the nonvolatile mantle thus formed would be similar to carbonaceous chondrite meteorites. However, the residue left behind would be extremely porous and of low density. Such structures seem better associated with the fragile fireballs observed by the Prairie Network (Ceplecha 1977) than with meteorites.

The model just described resembles that suggested by Sekanina (1972) consisting of a porous matrix of solid material, with ice filling the pores. Initially, the distribution of the two components is uniform throughout the nucleus. Sekanina rejected the ice nucleus with embedded dust for P/Encke because it yields a continuously increasing nongravitational parameter proportional to $\Delta M/M$. Observationally, the nongravitational force has been decreasing with time for P/Encke. To get a more satisfactory fit with various features he proposes a core with an overlying ice mantle. Whipple and Sekanina (1979) derived a rotation period and orientation of rotation axis for P/Encke. They showed that the time variation of the nongravitational force results from precessing of the spin axis rather than a rapid decrease in the nongravitational force. Although the reason for proposing this model no longer applies, the structure has features relevant to possible relationships between comets and asteroids (Sekanina 1971) and is considered in Sec. IV of this chapter.

A modification of the snowball structure has been developed (Donn 1981) based on theories of accumulation of planets (Safronov 1972; Goldreich and Ward 1973; Greenberg et al. 1978); this is a further development of earlier work (Donn 1963). The concept of those investigations, that small solid grains produce gravitational instabilities in a cloud on a planetary scale, was shown by Biermann and Michel (1978) to apply to comet accretion also.

In these unstable zones, gravitational collapse causes small grains to grow into larger aggregates on a short time scale.

Detailed investigations of the accumulation of planetesimals to form planetary objects have been carried out by Safronov (1972) and Greenberg et al. (1978), and reviewed by Wetherill (1980). These analyses show that a size distribution of planetesimals forms as illustrated in Fig. 9 of Greenberg et al. (1978). The mechanism of comet formation can be expected to be similar on a much smaller scale, as the final objects have kilometer dimensions. The size distribution of aggregates leads to a comet nucleus composed of an agglomeration of cometesimals with a size distribution of the form $n(m) \propto m^{-5}$. Figure 1 is an attempt to portray such a nucleus. It is expected that the individual cometesimals retain some degree of their identity. This is exaggerated in the figure.

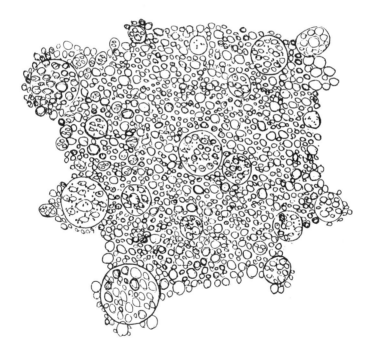

Fig. 1. Proposed model of a comet nucleus. The circular regions schematically represent the larger cometesimals. All cometesimals are aggregates of the basic micron-size ice-dust particles and are expected to have irregular shapes generally similar to that pictured for the nucleus. The identity of the cometesimals is exaggerated in the figure.

The larger cometesimals may be bound to the nucleus only over a fraction of their surface and therefore very weakly attached. Such fragments could readily break off from a vaporizing, rotating nucleus. The larger pieces would become visible as small comets breaking away from the primary nucleus, each fragment having the characteristics of a small comet. The behavior of fragmenting comets according to this model would follow the pattern described by Sekanina (1977). Although obtained in a completely different way from that considered by Whipple (1978b) and described in Sec. III on the outer layer of a nucleus, the present model of nuclear structure appears to lead to a somewhat similar picture, particularly for the outer region.

Internal Structure of an Icy Nucleus

The general considerations on an icy nucleus were combined with views of planet accumulation from small grains to develop a model for the internal characteristics of the nucleus (Donn 1963). Accretion of comets, with their high proportion of volatile species requires low temperatures and low relative velocities. Velocities below ~ 0.05 km s^{-1} would preserve grains of all but the most volatile species, H_2, CO, and CH_4. This would yield aggregates with densities < 0.5 g cm^{-3}, based on the characteristics of wind-blown snow (see e.g., Donn 1963).

Central pressures P_c, surface gravity g, and escape velocity V_e, respectively, are given by

$$P_c = 2/3 \, \pi \, G \, \bar{\rho}^2 \, R^2 \quad (2)$$

$$g = 4/3 \, \pi \, G \, \bar{\rho} \, R \quad (3)$$

$$V_e = (8/3 \, \pi \, G \, \bar{\rho})^{1/2} \, R \quad (4)$$

where G is the gravitational constant, $\bar{\rho}$ is the mean density, and R the radius. From these equations we obtain the data in Table IV, adopting a density of 1 g cm^{-3}. Measurements on the compaction of snow were summarized by Donn (1963) from which we take Fig. 2. These results indicate that no significant compaction occurs for nuclei < 10 km radius. The impact velocity due to gravity also does not affect the structure until the comet grows > 10 km.

TABLE IV

Mechanical and Dynamical Properties of Nuclei

R (km)	P_c (dyne cm^{-2})	g (cm s^{-3})	V_e (km s^{-1})
1	1.4×10^3	0.02	8×10^{-4}
10	1.4×10^5	0.2	8×10^{-3}
100	1.4×10^7	2	8×10^{-2}

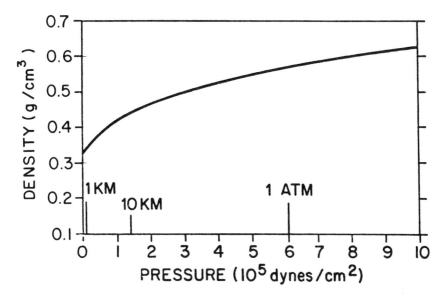

Fig. 2. Compaction of snow. Central pressures for a 1 km and 10 km radius nucleus are marked.

Meteoritic particles are embedded throughout the icy mass as demonstrated by meteor streams and intense showers associated with short-period comets. These meteors have low densities and the latest results have been summarized by Millman (1972,1975); see also the chapter by Wetherill and ReVelle. Verniani (1969,1973) found a populous low-density group, $\bar{\rho} = 0.2$ g cm^{-3}, consisting of 85%, and a high-density group, $\bar{\rho} = 1.4$ g cm^{-3}, consisting of 15%, of the total among sporadic meteors. Fourteen showers had low-density meteors, $\bar{\rho} = 0.2$ g cm^{-3}. The Draconids, associated with periodic Comet Giacobini-Zinner, have extremely low density, 0.01 g cm^{-3}, and are very fragile meteors. For such low-density, fragile aggregates to be embedded in icy masses, porous ices appear necessary as Whipple (1955,1970) suggested.

The occurrence of fragile meteoroids in comets is one of several lines of evidence that indicate cometary nuclei themselves must be low-density, fragile objects. More direct evidence is the tidal disruption of sungrazing comets (Öpik 1966a). Öpik finds the nucleus of the two fragmenting sungrazers (the Great September Comet 1882 II and Comet Ikeya-Seki 1965 VIII) weaker than all materials except meteoritic dust balls. A third indicator is the occurrence of fragmentation among comets at heliocentric distances up to 9 AU (see Sekanina's chapter in this book). He lists 21 split comets or ~ 3% of the total. Thus, theoretical ideas of comet accumulation and observations of cometary phenomena agree in predicting fragile nuclei, with low but finite cohesive strength.

Evidence for Radial Uniformity

Although a noticeable difference between the inner and outer zones of an icy nucleus might be expected, several observational phenomena given below do not suggest this to be the case.

(1) The continuum/emission intensity ratio in the spectra of comets appears to have similar distributions for new and for short-period (old) comets (Donn 1977). These extreme groups, with regard to age, show no difference in the dust/gas ratio or the character of the solid particles.

(2) Further, the emission spectra of new and periodic comets seem to be similar, in the visible as well as in the ultraviolet spectral region. Narrowband filter photometry for 6 comets by A'Hearn and Millis (1980) showed that the CN/C_2 production-rate ratio was remarkably constant ($\pm$ 0.1 in the log) from comet to comet, except for a well-defined variation with heliocentric distance. The relative production rates, specifically of CN, C_2, C_3, and OH, appear to be unrelated to either the emission-to-continuum (gas-to-dust) ratio or the dynamical age of the comet. The ultraviolet spectra (see the chapter by Feldman) of a number of long-period (Seargent 1978 XV, Bradfield 1979 X, West 1976 VI), and periodic (P/Encke, P/Tuttle, P/Stephan-Oterma) comets were found to be remarkably similar, indicating again a homogeneous structure of the nucleus. There are no ultraviolet spectra of new comets.

(3) The 17 fragmenting comets in Sekanina's chapter that have well-determined orbits and were not sungrazers may be classified as, 4 periodic, 5 old, and 8 new or nearly new. The proportion of these values to total comets in each category (Marsden 1979) are 0.04, 0.04, 0.02, respectively. In terms of total appearances rather than individual comets these ratios become 0.01, 0.04, 0.02. In view of the small numbers there does not seem to be any significant difference among the three age categories. Kresák (1981) has given a detailed rediscussion of split comets and reached similar conclusions. If splitting is intrinsic to a comet, the nuclei of new and very old comets behave similarly. Sekanina (1977) also showed that the relative nongravitational effects for the fragments were proportional to the lifetime for all categories. He also suggests (Sekanina 1980) that the behavior of these fragments is similar to that of comets observed to dissipate during their apparition.

III. THE OUTER LAYER OF A NUCLEUS AND ITS EVOLUTION

Our primary concern in this chapter is with initial characteristics of nuclei. During perihelion passage, very pronounced evolutionary effects will occur on the volatile, fragile cometary nuclei. The analysis and interpretation of cometary observations need to consider such changes (see e.g., Shul'man 1972a; Mendis and Brin 1978; Whipple 1978). The initial surface change occurs while a comet is in the Oort Cloud undergoing irradiation by galactic cosmic rays. This process was discussed by Shul'man (1972b), Donn (1976), and Whipple (1977). It was pointed out that significant chemical effects are expected. Shul'man and Donn showed that the chemical composition of the outer layer, to a depth of ∼ 1 m, would be considerably transformed. Production of new species was emphasized by Shul'man. Both Donn and Whipple called attention to changes in the expected behavior of the material, with the first author suggesting the volatile matter may polymerize and become more inert and the latter believing the material would become more reactive. Experimental results of energetic proton irradiation of ice mixtures were obtained by Moore and Donn (see Donn 1981) who found evidence for gas release between ∼ 15 and 40 K and near 150 K. A nonvolatile residue was also produced and compromised a few percent of the original material. Earlier, Patashnik et al. (1974) concluded that energy released by the amorphous to crystalline ice transition could cause enhanced gas release at ∼ 150 K. More recent considerations of the role of amorphous ice have been given by Klinger (1980) and Smoluchowski (1981a,b).

In his initial presentation of the icy model, Whipple (1950) pointed out that the larger, nonvolatile particles will not be carried away. An inert, insulating layer would form and have a large influence on the behavior of the nucleus. This effect was examined in subsequent papers (Whipple 1951,1955). Formation of a crust and its consequence has been examined in several investigations since then (e.g., see Shul'man 1972; Mendis and Brin 1978,1979; Brin and Mendis 1979; Weissman and Kieffer 1981). The last two papers contain the most detailed study. Mendis and Brin assume an initially homogeneous ice-dust nucleus. As the comet approaches perihelion some fraction of the dust is not carried away by subliming ices and remains behind or falls back on the nucleus. Temperatures and vaporization rates were calculated. From the latter, the authors determined monochromatic magnitudes of specific species as a function of heliocentric distance, both before and after perihelion. A significant post-perihelion decrease in luminosity was predicted. In a quantitative investigation of luminosity variation with heliocentric distance, Whipple (1978) found a significant difference between pre- and post-perihelion brightness variation for new and very long-period comets. The exponent of r in the relation $m(r, \Delta = 1) = H_{r,1} + 2.5 n \log r$ increased after perihelion. The reverse was true for shorter period comets except for those with $P < 25$ yr for which the behavior was erratic. As pointed out above Whipple believes new comets have active, irradiated surfaces that are responsible for the excess luminosity of new comets at large distances.

Whipple envisages the outer, surface region of a comet as being irregular in structure with some nonuniformity also in composition. This is the result of formation by the agglomeration of cometesimals. He suggests these cometesimals have cores more cohesive and less volatile than the matrix materials. Whipple traces out an evolutionary process of such a nucleus whereby more coherent, darker clumps, called globs, develop in comets. These globs may give rise to fireballs as those associated with P/Encke or observed by the Prairie Network, some of which are associated with other showers (Wetherill 1974). With the passage of time, surface globs develop into mounds or columns as ices sublime and carry away loose meteoritic material.

Eventually they crumble or perhaps are carried away as objects of considerable size through rotation of the nucleus, after connecting material has sublimed. This picture of a nucleus developed by Whipple from a study of the luminosity variation of comets is strikingly similar to Donn's (1981) model based on a probable mechanism of accumulation. A consequence of these pictures is an expected highly irregular surface structure on meter dimensions or smaller. Plans for a comet rendezvous must take this into account. A mission involving landing on the nucleus must have detailed information on the surface.

Sekanina (1981b) has reviewed studies, mainly by himself and Whipple, of the asymmetrical structures often found in comae. These have been analyzed in terms of nonisotropic ejection from the nucleus. Their results support the picture of a heterogeneous surface consisting mostly of regions of relatively low gas emissions with small, discrete zones of high emissivity producing the jets. According to Sekanina's analysis, the active regions emit for periods of the order of 0.1 day. The bursts of gas and dust tend to reoccur on several successive rotations. The absence of jets in many comets would be indicative of a more uniform surface. There appears to be a variation of surface structure among comets of a given age. Whipple and Sekanina did not discuss the spectral characteristics of the comets, particularly the emission/continuum ratio and its possible connection to surface behavior.

IV. COMETS, ASTEROIDS AND METEORITES

A definitive answer to the relationship between comets and asteroids or comets and meteorites will provide valuable insight into the origin and structure of the nucleus. These relationships are, however, still unresolved problems. Extensive discussions and further references appear in several books and reviews (Gehrels 1971,1979; Delsemme 1977; Wetherill, 1980). The earliest suggestion of such a relationship may have been by Kirkwood following the discovery of asteroids like 132 Althea with eccentric orbits. Following the detection of Earth-crossing Apollo-type asteroids, Öpik (1963) proposed them as possible cometary residues. As few comets are larger than several km, they only seem capable of accounting for the very smallest asteroids. However, the original comets may have been much larger (Sekanina 1971). Earth-crossing Apollo-type or near-Earth Amor-type asteroids are small (Öpik 1963) and only a small fraction is expected to have been discovered. Öpik concluded that these objects cannot have come from the asteroid belt but that a cometary source is not unreasonable. Later work by Wetherill and his associates (see Wetherill 1979) showed that resonance processes could transform asteroid-belt objects to Earth-crossing orbits. It now appears that Apollo-Amor asteroids have a mixed asteroidal-extinct comet source. Kresák (1980) reached a similar conclusion pointing out that it is very unlikely that all types of objects in the solar system have been discovered; he gives examples of recent findings. Chiron (1977 UB) between Saturn and

Uranus and 1978 SB, an asteroid in an orbit closely resembling that of P/Encke are significant examples.

If a comet developed into an asteroid, essentially all volatiles must have been lost and a substantial (> 1 km) residue remain. Such a structure is consistent with the nonvolatile matrix-embedded ice model. However, it is difficult to derive a mechanism to yield such a structure. The nonvolatile matrix must be formed first and have sufficient cohesive strength to stay together. The pores must then be filled with ice. It is this process that presents problems as an outer ice layer will form first and prevent filling the interior. To avoid this and obtain a radially uniform composition would require an implausible temperature distribution and time variation.

An evolutionary history which would convert comets into Apollo-Amor type asteroids appears best explained by the vaporization of Whipple's ice-dust conglomerate (Whipple 1951; Öpik 1963; Levin 1977). An accumultion of nonvolatile and icy grains or perhaps nonvolatile cores with icy coatings forms a dirty snowball. As the ices sublime and larger nonvolatile aggregates are left behind, a crust forms. This gradually encompasses the entire nucleus to a sufficient depth that the deeper-lying volatiles no longer are heated sufficiently to vaporize. The inert residue, appreciably smaller than the original comet, becomes the Apollo-type asteroids. Detailed treatments of dust layer formation by Mendis and Brin (1977,1978), Brin and Mendis (1979) and Weissman and Kieffer (1981) have been referred to above. Öpik (1963), Marsden (1971) and Sekanina (1971) have described a variation of the above procedure. They start with a nucleus consisting of a nonvolatile core and an ice-dust mantle. Complete vaporization of the volatile material leaves an inactive residue.

Öpik (1964,1966a,b,1969) proposed a cometary origin for meteorites because of the dynamical problem of getting objects from the asteroid belt into Earth-crossing orbits. An efficient process seemed necessary in order to account for the short cosmic-ray exposure ages, tens of millions of years for stones and hundreds of millions for irons (Anders 1963). Subsequently, Williams (1973) and Zimmerman and Wetherill (1973) found more efficient mechanisms for perturbing objects from the asteroid belt into Earth-crossing orbits. These schemes can provide the differentiated iron and stony meteorites (Wetherill 1979). Because of short collision lifetimes, the direct source of chondritic meteorites cannot be the asteroid belt. Wetherill and Williams (1979) conclude that the Earth-crossing Apollo-Amor asteroids may be capable of yielding the entire flux of chondritic meteorites. They also conclude that the Apollo-Amor objects themselves may be derived from a mixture of asteroidal and cometary sources. There does not appear to be unambiguous evidence for a cometary source of meteorites although there may be a need to supply chondritic meteorites from comets. The necessity of postulating a cometary source of meteorites to account for the short cosmic ray exposure ages is unclear. The extremely complex dynamical aspects of planetary encounters and perturbations for small objects has not been completely analyzed.

For the purpose of this review we examine the implications for nuclear structure if meteoritic material exists in comet nuclei. It is now well established that fireballs are associated with comets; for example, Taurid shower fireballs can be associated with P/Encke and Prairie Network fireballs (Wetherill 1974) with several other cometary meteor showers. As indicated above such fragile and low-density objects do not appear to present a serious problem. There are, however, serious difficulties with meteorites in comets. Anders (1971,1978) has given several arguments favoring an asteroidal over a cometary source of meteorites. He has emphasized the large size of the parent needed to obtain a sufficiently large regolith to provide the solar-wind induced gas content. He concludes that no stony meteorites, including the carbonaceous chondrites, are derived from comets. Irons and stones (Wood 1967) require large parent bodies to obtain the high temperatures and slow cooling necessary to produce the crystalline structure of these meteorites. Levin (1977) also concludes that the high temperatures required and the complex history needed to account for brecciated meteorites rules out a cometary source. At the low temperatures and pressures found in cometary nuclei, no known physico-chemical processes can yield meteorites. Similarly, as we pointed out about the dynamical processes yielding Earth-crossing fragment, we do not yet understand all the factors entering this problem. However, it is doubtful that the gaps in our knowledge allow for mechanisms producing the complex crystal structure of meteorites at low temperatures (150°C) and pressures. Metallic iron-nickel masses have been produced (Bloch and Muller 1971) by the condensation of vapor produced by the dissociation of iron and nickel carbonyl. Carbon, sulfide, and phosphide phases could be formed by proper vapor composition. These experiments, however, are still far from producing the complex meteoritic minerals.

Because of the inability to produce meteoritic material in comets, Öpik (1966b) proposed a two stage mechanism. Meteorites are broken off fragments of larger bodies, when the pressure and temperature were sufficiently high to cause compaction and melting of the original loose condensate. The fragments become coated with ices and are then ejected to the outskirts of the solar nebula. In this way comet nuclei would have formed. Some comets, according to Öpik's hypothesis, would be first generation products without meteoritic inclusion, those with meteoritic objects are second generation. As the ices subsequently sublimed during perihelion passage, meteorites or small asteroids developed. If the cometary origin of meteorites is accepted, there seems to be no way of avoiding a process of this nature.

V. THE ORIGIN OF COMETS

It is generally assumed that the formation of comets occurred at the time of formation of the solar system (e.g., Whipple 1964; Öpik 1973; Delsemme 1977). It has also been recognized that comets accumulated from a cloud of ice-dust grains at low temperatures and low relative velocities (Levin 1962; Donn 1963; Öpik 1973). What is not generally agreed upon is the region where comets formed. Several authors (e.g., Strömgren 1924; van Woerkom 1948; Sekanina 1976b; Noerdlinger 1977) have shown that the absence of clearly hyperbolic original orbits strongly argues against comets coming from interstellar space; the largest eccentricity observed so far measures $e = 1.006$. It seems likely that observed comets have always been connected with the Sun in some manner. Oort's (1950) proposal for a cloud of comets orbiting the Sun at $\sim 10^4$ to 10^5 AU serves as the basis for investigating the history of comets (e.g. Everhart's chapter and Weissman's chapter). How the Oort Cloud was formed is still a wide open question. Several regions have been proposed for accretion of the nucleus prior to their residence in the Oort Cloud. This subject was reviewed by Delsemme (1977). We discuss below the regions in order of increasing distance from the Sun.

(1) *The asteroid zone.* This region was suggested by Oort (1950) but rejected by subsequent investigators because of problems of ice condensation and subsequent ejection to the Oort Cloud.

(2) *The region of the outer planets, Jupiter to Neptune.* This is a commonly adopted zone (Whipple 1964; Öpik 1973; Safronov 1977). Analysis of the ejection of comets forming between 5 and 30 AU from the Sun has been studied by Öpik (1966,1973). He finds ejection by Jupiter to be inefficient; $\sim 1\%$ of Jupiter-crossing comets are perturbed into the Oort Cloud on a time scale of 10^7 yr. The remainder are ejected into interstellar space or destroyed by close solar passage. Uranus and Neptune are much more efficient because of the small steps by which $1/a$ changes but the time scales become too long and amount to $\sim 10^{11}$ yr. Fernandez (1981) obtains similar conclusion concerning the efficiency but finds time scales of 10^8 yr.

For an average radius of new comets of ~ 3 km and density 1 g cm^{-3} the average mass is 10^{17} g. Weissman (see his chapter) obtains 2×10^{12} comets initially in the Oort Cloud or a mass of 2×10^{29} g = 33 Earth masses. Calculations indicate a combined efficiency of ejection of comets by the giant planets (Fernandez and Ip 1981) and incorporation within the Oort Cloud by stellar perturbation (chapter by Weissman) of about a few tenths percent. These results require a cometary mass in the Jupiter-Neptune region of ~ 0.1 solar masses.

(3) *A comet belt beyond Neptune (30-50 AU).* A residual comet cloud beyond Neptune was suggested by Cameron (1962). Hamid et al. (1968) from

a study of perturbations on P/Halley concluded that there is < 1 Earth mass in this region. Fernandez (1980) analyzed the transformation of comets in the belt beyond Neptune to short-period comets. Close encounters between comets perturb some of them into Neptune-crossing orbits. The Monte Carlo procedure adopts a comet population with the following characteristics: mass distribution $n(m) = Am^{-\alpha}$, with $1.5 < \alpha < 2$; minimum mass of comets = 10^{15} g; maximum mass = $10^{21} - 10^{26}$ g; total mass in the zone amounts to ~ 1 Earth mass. Orbits of Neptune-crossing comets would evolve as described by Everhart (1977). Fernandez concludes that short-period comets can be efficiently produced by a reasonable model for the hypothesized comet belt with the mass at Hamid et al.'s upper limit. Long-period and near-parabolic comets still require the Oort Cloud. There would be two comet sources that are assumed to have much in common because comets from the belt beyond Neptune are fed into the Oort Cloud via Uranus and Neptune. Cameron (1973) has proposed that ejection of comets to large distances would be aided by a mass loss from the solar nebula and Sun during an early T-Tauri stage for the Sun.

(4) *Asteroid belt-interstellar cloud.* O'Dell (1973,1976) has proposed a structure and mechanism of formation of comet nuclei that combines the inner solar system and interstellar clouds. In this hypothesis micron-sized grains are ejected from the solar system by radiation pressure. O'Dell considers an interstellar region with a typical density of one hydrogen atom per cm^3 and relative cosmic abundances. It is assumed that enough grains form a loose, distant cluster to provide a comet of a few km radius when they collapse into a single object during approach to the Sun. In the age of the solar system of $\sim 5 \times 10^9$ yr, each grain has received an ice coating sufficient to provide for the material lost in ~ 100 orbits.

There are several difficulties with this hypothesis. Among them is the difficulty of cluster formation by ejected grains at 10^4 to 10^5 AU and the long time scale for coating grains at interstellar densities. It may be that this process can be modified to provide for cometary meteorites. Instead of grains ejected by radiation pressure, consider fragments of asteroids ejected by planetary perturbations similar to Oort's original proposal. Then let the Sun be formed as part of a star cluster with a number of small clouds very nearly gravitationally bound to it. The ejected meteorites are trapped in these clouds which have relatively high densities. Cameron's (1973) subdisks have densities in the plane of 10^{-11} g cm^{-3} or $\sim 5 \times 10^{12}$ atom cm^{-3}. The accumulation process analyzed by Cameron (1973) and Biermann and Michel (1978) was used by Donn (1977b) to discuss comet formation directly in the Oort Cloud. If meteorite formation and ejection were rapid enough we can conceive of cometary nuclei containing meteorites as well as finer dust occurring in the Oort Cloud. Although this mechanism is speculative and qualitative, several of the steps have been analyzed in some detail. If it is necessary to explain meteorites as fragments of Earth-crossing comets, this process seems to have fewer deficiencies than most other sequence of events leading to cometary meteorites.

(5) *Outlying fragmentary clouds from the presolar nebula.* Cameron (1973) postulated fragments of a few tenths solar mass breaking off from the outer limits of the primordial solar nebula and revolving around it. He suggested that in these smaller clouds numerous small objects could accrete and become the Oort Cloud comets. He applies the mechanism developed for planet accumulation, scaled to the mass and size of the cloud fragment, and concludes that large cometary objects could form in thousands of years. A few subdisks are expected to provide a sufficient number of comets to form the Oort Cloud.

Biermann and Michel (1978) carried out a study of accumulation of comets from ice-dust grains in the outer regions of a primordial solar nebula with a few solar masses. Their analysis is based upon theories of planetary accumulation developed by Safronov (1972) and Goldreich and Ward (1973). In Biermann and Michel's model the cometary aggregates have aphelia $\lesssim 10^4$ AU. The dispersal of approximately half of the nebula not undergoing collapse into the Sun and planets will cause the orbits to enlarge and form the Oort Cloud. The main distinction to Cameron (1973) appears to be the larger nebular mass which permits sufficient density at great distances to form comets in the primordial nebula. In a yet unpublished work, Biermann (1980) concludes from recent studies on cloud collapse that comets could not form in the same cloud as the Sun and planets. He modifies the above process to form comets in a neighboring fragment of the same interstellar cloud as the Sun and planets form. This is generally similar to Cameron's scheme described above and essentially identical with that proposed by Donn (1976). That mechanism is described next.

(6) *In situ formation in the Oort Cloud.* It was pointed out (Donn 1973, 1976) that the tendency of stars to form in clusters provides a means of comet formation in the Oort Cloud. Galactic star clusters contain the order of a hundred stars within a roughly spherical volume of 5 pc diameter (Hogg 1959). The average distance between stars is 0.5 pc. The theory of star formation in clusters is essentially nonexistent, but a considerable empirical-observational base is developing (de Jong and Maeda 1977). Following the procedure of Cameron (1973), we accept the existence of subclouds of small mass that cannot produce stars but are capable of forming much smaller mass objects. In these cloudlets small aggregates form. According to Cameron (1973) and Biermann and Michel (1978) these have masses in the cometary range, i.e., 10^{15} to 10^{20} g. The low-velocity dispersion of cluster stars, < 3 km s^{-1} (Blaauw 1964) and the velocity dispersion of the comet population ensures that many comets will have essentially zero velocity relative to

the Sun. Thus, the Sun will have a comet family moving with it as the cluster disperses.

Greenberg (1977; see also his chapter in this book) and his colleagues (Baas 1980) propose comet formation in dense clouds. The grains consist of refractory cores with icy mantles that have been heavily photolysed as pointed out by Donn and Jackson (1970). These grains may then accrete into cometary size aggregates as in the interstellar hypothesis just described. According to this hypothesis, the icy material exists on the same grains as the nonvolatile component rather than as separate particles. This is probably the case because of the much more favorable path for condensation on an existing grain compared to nucleation of an icy grain from the vapor.

(7) *Periodic formation of solar system comets in interstellar clouds.* The preceding hypotheses make comet formation approximately concurrent with the formation of the solar system. McCrea (1975) has adopted a few million years as the age of comets and proposed a mechanism for satisfying this criterion. As interstellar clouds pass through the shock region in the inner edge of a spiral arm, some are compressed to a degree where dust concentrations are formed and then collapse into comet nuclei. If the Sun is in the vicinity when this happens some are captured and become part of the Sun's family of comets. Important consequences of this hypothesis are:

a. Comet ages are the order of the in-fall time, a few million years;
b. They appear at intervals approximately given by half of the galactic rotation period of 10^8 yr and decay much more rapidly;
c. Sun-comet velocities would have a mean value of several km s^{-1}. The occurrence of parabolic and near parabolic orbits without unambiguous hyperbolic orbits does not seem readily explained. In fact, the observed parabolic orbits are anomalous;
d. There is no way to have an association of meteorites and comets as some authors (Öpik 1966; Wetherill 1977) propose.

In the process of ejecting comets into the Oort Cloud most cometary objects escape from the solar system. If this ejection phenomena occurred for the solar system, it must have been a rather widespread occurrence throughout the galaxy. This would be in addition to the well-established planetary ejection (see e.g., Marsden et al. 1978). There would thus be a considerable number of interstellar comets. The suggestion for comets coming from interstellar space has a long history (van Woerkom 1948). Recent work on this subject is summarized by Noerdlinger (1977).

VI. CONCLUDING REMARKS

A review of the structure and origin of comets must bring together a variety of information from many sources. We have attempted to expose the different phenomena involved in constructing a model for the nucleus and describe those models that seem generally useful. The same plan was used in preparing the section on cometary origin. The material has been presented in a way to permit the investigator working on these problems to take advantage of previous efforts. Although this chapter is not as comprehensive and critical as we initially planned, it is our hope that it will provide cometary scientists with a good starting point for the exciting and often frustrating tasks of studying the structure and origin of comets.

Acknowledgments: We thank D. Brownlee, B. Marsden, Z. Sekanina, P. Weissman and F. Whipple for their critical comments and informative discussions. One of us (J.R.) is a National Academy of Science Senior Resident Research Associate from the Astronomical Institute, University of Bamberg Erlangen.

REFERENCES

A'Hearn, M., and Millis, R.L., 1980. Abundance correlations among comets. *Astron. J.* 85:1528-1537.
Anders, E. 1963. Meteorite ages. In *The Solar System IV: The Moon, Meteorites, and Comets,* eds. B. Middlehurst and G.P. Kuiper (Chicago: Univ. of Chicago Press), pp. 402-495.
Anders, E. 1971. Interrelations of meteorites, asteroids and comets. In *Physical Studies of Minor Planets,* ed. T. Gehrels (Washington: NASA SP-267), pp. 429-446.
Anders, E. 1978. Most stony meteorites come from the asteroid belt. In *Asteroids: An Exploration Assessment,* eds. D. Morrison and W.C. Wells (Washington: NASA, Sci. and Tech. Information Office), pp. 57-75.
Baas, F. 1980. The chemistry of pre-solar dust and related laboratory experiments. In *Cometary Chemistry and Physics,* ed. J.M. Greenberg (Garching: Max Planck Inst. fur Phys. und Astrophys).
Biermann, L. 1980. The smaller bodies of the solar system. In *Planetary Exploration,* Discussion at Royal Society, London.
Biermann, L., and Michel, K.W. 1978. On the origin of cometary nuclei in the presolar nebula. *Moon and Planets* 18:447-464.
Bloch, M.R., and Muller, O. 1971. An alternative model for the formation of iron meteorites. *Earth and Planet. Sci. Letters* 12:134-136.
Brin, G.D., and Mendis, D.A. 1979. Dust release and mantle development in comets. *Astrophys. J.* 229:402-408.
Cameron, A.G.W. 1962. The formation of the sun and the planets. *Icarus* 1:13-69.
Cameron, A.G.W. 1973. Accumulation processes in the primitive solar nebula. *Icarus* 18:407-449.
Ceplecha, Z. 1977. Meteoid populations and orbits. In *Comets, Asteroids, Meteorites,* ed. A.H. Delsemme, (Toledo, Ohio: Univ. of Toledo), pp. 143-152.
de Jonge, T., and Maeda, A. 1977 ed. *Star Formation,* (Dordrecht, Holland: D. Reidel).
Delsemme, A.H. 1977. *Comets, Asteroids, Meteorites,* (Toledo, Ohio: Univ. of Toledo).
Dollfus, A. 1961. Visual and photographic studies of planets at the Pic du Midi. In *Planets and Satellites,* eds. G.P. Kuiper and B. Middlehurst, (Chicago: Univ. of Chicago Press), pp. 154-175.
Donn, B. 1963. The origin and structure of icy cometary nuclei. *Icarus* 2:396-402.
Donn, B. 1976. The nucleus: Panel discussion. In *The Study of Comets* Part 2, eds. B. Donn, M. Mumma, W. Jackson, M. A'Hearn, and R. Harrington, (Washington: NASA SP-373), pp. 611-621.
Donn, B. 1977. A comparison of the composition of new and evolved comets. In *Comets, Asteroids, Meteorites,* ed. A.H. Delsemme, (Toledo, Ohio: Univ. of Toledo), pp. 15-24.

Donn, B. 1981. Comet nucleus: Some characteristics and a hypothesis on origin and structure. In *Comets and the Origin of Life,* ed. C. Ponnamperuma, (Dordrecht, Holland: D. Reidel Pub. Co.), pp. 21-29.
Donn, B., and Jackson, W.M. 1970. Interstellar irradiation and the ultimate properties of icy grains. *Bull. Amer. Astron. Soc.* 2:309-310 (abstract).
Everhart, E. 1977. The evolution of comet orbits as perturbed by Uranus and Neptune. In *Comets, Asteroids, Meteorites,* ed. A.H. Delsemme, (Toledo, Ohio: Univ. of Toledo), pp. 99-104.
Fernandez, J.A. 1980. On the existence of a comet belt beyond Neptune. *Mon. Not. Roy. Astron. Soc.* 192:481-491.
Fernandez, J.A., and Ip, W-H. 1981. Dynamical evolution of a cometary swarm in the outer planetary region. *Icarus* (special Comet issue). In press.
Gehrels, T. Ed. 1971. *Physical Studies of Minor Planets,* (Washington: NASA SP-267).
Gehrels, T. Ed. 1979. *Asteroids,* (Tucson: Univ. of Arizona Press).
Goldreich, P., and Ward, W.R. 1973. The formation of planetesimals. *Astrophys. J.* 183:1051-1061.
Greenberg, J.M. 1977. From dust to comets. In *Comets, Asteroids, Meteorites,* ed. A.H. Delsemme, (Toledo, Ohio: Univ. of Toledo), pp. 491-497.
Greenberg, R.; Wacker, J.F.; Hartmann, W.K.; and Chapman, C.R. 1978. Planetesimals to planets: Numerical simulation of collisional evolution. *Icarus* 35:1-26.
Hamid, S.E.; Marsden, B.G.; and Whipple, F.L. 1968. Influence of a comet belt beyond Neptune on the motions of periodic comets. *Astron. J.* 73:727-729.
Hogg, H.S. 1959. Star clusters. In *Handbuch der Physik* Vol. 53., ed. S. Flugge, (Berlin: Springer Verlag), pp. 129-207.
Klinger, J. 1980. Influence of a phase transition of ice on the heat and mass balance of comets. *Science* 209:271-272.
Kresák, L. 1973. Short period comets at large heliocentric distances. *Bull. Astron. Inst. Czech.* 24:264-283.
Kresák, L. 1980. Dynamics, interrelations and evolution of the systems of asteroids and comets. *Moon and Planets* 22:83-98.
Kresák, L. 1981. Evolutionary aspects of the splits of cometary nuclei. *Bull. Astron. Inst. Czech.* 32:19-40.
Kuiper, G.P. 1950. The diameter of Pluto. *Publ. Astron. Soc. Pacific.* 62:133-137.
Laplace, P.S. 1805. Traite de Mechanique Celeste, Tome IV chez Courcier, Paris: Bronx, New York.
Levin, B.Y. 1962. The structure of icy comet nuclei. *Soviet Astron.* 6:593-595.
Levin, B.Y. 1977. Relationships between meteorites, asteroids and comets. In *Comets, Asteroids, Meteorites,* ed. A.H. Delemme, (Toledo, Ohio: Univ. of Toledo), pp. 307-312.
Lyttleton, R.A. 1953. *The Comets and their Origin.* (Cambridge: Cambridge Univ. Press).
Marsden, B.G. 1967. The sungrazing comet group. *Astron. J.* 72:1170-1183.
Marsden, B.G. 1971. Evolution of comets into asteroids. In *Physical Studies of Minor Planets,* ed. T. Gehrels, (Washington: NASA SP-267), pp. 423-428.
Marsden, B.G.; Sekanina, Z.; and Everhart, E. 1978. New oscillating orbits for 110 comets and analysis of original orbits for 200 comets. *Astron. J.* 83:64-71.
McCrea, W.H. 1975. Solar system as space probe. *Observatory* 95:239-255.
Mendis, D.A., and Brin, G.D. 1977. Monochromatic brightness variations of comets. II. Core-mantle model. *Moon and Planets* 17:359-372.
Mendis, D.A. 1978. On the monochromatic brightness variation of comets. *Moon and Planets* 18:77-89.
Millman, P. 1972. Cometary meteoroids. In *From Plasma to Planet,* ed. A. Elvius, (New York: Wiley-Interscience), pp. 157-168.
Noerdlinger, P. 1977. An examination of an interstellar hypothesis for the source of comets. *Icarus* 30:566-573.
O'Dell, C.R. 1973. A new model for cometary nuclei. *Icarus* 19:137-146.
Oort, J. 1950. The structure of the cloud of comets surrounding the solar system and a hypothesis concerning its origin. *Bull. Astron. Inst. Neth.* XI, pp. 91-110.

Öpik, E. 1963. Survival of cometary nuclei and the asteroids. *Adv. Astron. Astrophys.* 2:219-262.
Öpik, E. 1966a. Sungrazing comets and tidal disruption. *Irish Astron. J.* 1:141-161.
Öpik, E. 1966b. The stray bodies in the solar system II: The cometary origin of meteorites. *Adv. Astron. Astrophys.* 4:302-336.
Öpik, E. 1973. Comets and the formation of planets. *Astrophys. Space Sci.* 21:307-398.
Patashnik, H.; Rupprecht, G.; and Schuerman, D.W. 1974. Energy source for comet outbursts. *Nature* 250:313-314.
Pittich, E.M. 1971. The space distribution of the splitting and outbursts of comets. *Bull. Astron. Inst. Czech.* 22:143-153.
Roemer, E. 1966. The dimensions of cometary nuclei. In *Nature et Origin des Comets, Mém. Sci. Roy. Sci. Liège* Ser. V. 12:23-26.
Russell, H.N. 1929. On meteoric matter near the stars. *Astrophys. J.* 69:49-71.
Safronov, V.S. 1972. *Evolution of the Protoplanetary Cloud and the Formation of the Earth and the Planets.* Trans. from Russian. (Springfield, Virginia: Nat. Tech. Inform. Service).
Safronov, V.S. 1977. Oort's cometary cloud in the light of modern cosmogony. In *Comets, Asteroids, Meteorites,* ed. A.H. Delsemme, (Toledo, Ohio: Univ. of Toledo), pp. 483-484.
Sekanina, Z. 1971. A core-mantle model for cometary nuclei and asteroids of possible cometary origin. In *Physical Studies of Minor Planets,* ed. T. Gehrels, (Washington: NASA SP-267), pp. 423-428.
Sekanina, Z. 1972. A model for the nucleus of Encke's Comet. In *The Motion, Evolution of Orbits and Origin of Comets,* eds. G.A. Chebotarev, E.L. Kazimirchak-Polonskaya, and B.G. Marsden, (Dordrecht, Holland: D. Reidel Publ. Co.), pp. 300-307.
Sekanina, Z. 1976a. A continuing controversy: Has the comet nucleus been resolved? In *The Study of Comets,* eds. B. Donn, M. Mumma, W.M. Jackson, M.F. A'Hearn, and R. Harrington, (Washington: NASA SP-393), pp. 537-587.
Sekanina, Z. 1976b. A probability of encounter with interstellar comets and the likelihood of their existence. *Icarus* 27:127-138.
Sekanina, Z. 1977. Relative motions of fragments of the split comets. I: A new approach. *Icarus* 30:574-594.
Sekanina, Z. 1980. Physical similarities between dissipating comets and short-lived fragments of the split comets. *Bull. Amer. Astron. Soc.* 12:511 (abstract).
Sekanina, Z. 1981. Rotation and precession of cometary nuclei. *Ann. Rev. Earth Planet. Sci.* 9:113-145.
Shul'man, L.M. 1972a. The chemical composition of cometary nuclei. In *The Motion Evolution of Orbits and Origin of Comets.* eds. G.A. Chebotarev, E.I. Kazimirchak-Polonskaya and G.B. Marsden, (Dordrecht, Holland: D. Reidel Publ. Co.), pp. 265-270.
Shul'man, L.M. 1972b. The evolution of cometary nuclei. *ibid.* pp. 271-276.
Smoluchowski, R. 1981a. Amorphous ice and the behavior of cometary nuclei. *Astrophys. J.* 244:L31-L34.
Smoluchowski, R. 1981b. Heat content and evolution of cometary nuclei. *Icarus* (special comet issue). In press.
Strömgren, E. 1924. Uber den Ursprung der Kometen. *Publ. Compenhagen Obs.* No. 19.
van Woerkom, A.F. 1948. On the origin of comets. *Bull. Astron. Inst. Neth.* X, pp. 445-471.
Verniani, F. 1969. Structure and fragmentation of meteroids. *Space Sci. Rev.* 10:230-261.
Verniani, F. 1973. An analysis of the physical parameters of 5759 faint radio meteors. *J. Geophys. Res.* 78:8429–8462.
Vorontsov-Velyaminov, B. 1946. Structure and mass of cometary nuclei. *Astrophys. J.* 104:226-233.
Weissman, P.R., and Kieffer, H.H. 1981. Thermal modeling of cometary nuclei. *Icarus* (special Comet issue). In press.
Wetherill, G.W. 1974. Solar system sources of meteorites and large meteoroids. *Ann. Rev. Earth Planet Sci.* 2:303-331.

Wetherill, G.W. 1979. Steady state population of Apollo-Amor objects. *Icarus* 37:96-112.
Wetherill, G.W. 1980. Formation of the terrestrial planets. *Ann. Rev. Astron. Astrophys.* 18:77-113.
Wetherill, G.W., and Williams, J.G. 1979. Origin of differentiated meteorites. In *The Origin and Distribution of the Elements,* ed. L.M. Ahrens, (New York: Pergamon Press), pp. 19-31.
Whipple, F.L. 1950. A comet model I. The acceleration of Comet Encke. *Astrophys. J.* 111:375-394.
Whipple, F.L. 1951. A comet model II. Physical relations for comets and meteors. *Astrophys. J.* 113:464-474.
Whipple, F.L. 1955. A comet model III. The zodiacal light. *Astrophys J.* 121:750-770.
Whipple, F.L. 1963. On the structure of the cometary nucleus. In *The Solar System IV, The Moon, Meteorites and Comets,* eds. G.P. Kuiper and B. Middlehurst (Chicago: Univ. of Chicago Press), pp. 639-663.
Whipple, F.L. 1964. The history of the solar system. *Proc. Nat. Acad. Sci.* 51:711-718.
Whipple, F.L. 1977. The constitution of cometary nuclei. In *Comets, Asteroids, Meteorites,* ed. A.H. Delsemme, (Toledo, Ohio: Univ. of Toledo), pp. 25-36.
Whipple, F.L. 1978a. Comets. In *Cosmic Dust,* ed. J.A.M. McDonnell, (New York: Wiley-Interscience), pp. 1-73.
Whipple, F.L. 1978b. Cometary brightness variation and nucleus structure. *Moon and Planets,* 18:343-359.
Whipple, F.L., and Sekanina, Z. 1979. Comet Encke precession of the spin axis, nongravitational motion and sublimation. *Astron. J.* 84:1894-1909.
Williams, J.G. 1973. Meteorites from the asteroid belt. *EOS* 54:233 (abstract).
Wood, J.A. 1967. Chondrites: Their metallic minerals, thermal histories and parent planets. *Icarus* 6:1-49.
Zimmerman, P.D., and Wetherill, G.W. 1973. Asteroidal source of meteorites. *Science* 182:51-53.

THE ROTATION OF COMET NUCLEI

FRED L. WHIPPLE
Smithsonian Astrophysical Observatory

The miniscule history of spin-vector research on comet nuclei is reviewed. Major emphasis is placed on actual determinations of rotation period and of spin-axis orientation. The latter is indicated by asymmetrical comae and by the directions of near-nucleus jets. The spin axis is known for seven periodic comets, including P/Encke for which a rapid precession of the spin axis has been established by Whipple and Sekanina. Rotation periods can be measured by photometry (as for asteroids), by near-nucleus jets, and by halo diameters. The halo method, developed by the author, has been by far the most prolific but has been little tested. The spin periods P for 47 comets, almost all new determinations, are presented and compared with those of 41 small asteroids of diameter $\leqslant 40$ km. The median P for the comets is 15.0 hr versus 6.8 hr for these asteroids. The preliminary distribution curve of log P for comets is flatter than for the asteroids and is not gaussian. Slow accumulation at low relative velocities is suggested as the cause, rather than collisional effects as for the small asteroids. The shortest period is 4.1 or 4.6 hr, near the borderline of stability for a low mean density. The period increases statistically with absolute brightness (at the 2.4 σ level). Spinup by collapse from internal heating is not indicated. Spinup by sublimation is slightly indicated but a cometary origin for the Apollo and Amor asteroids remains highly questionable from comparisons of spin periods.

I. HISTORICAL INTRODUCTION

Although the concept of a discrete nucleus of volatile solids in comets goes back at least to Laplace (1813) and Bessel (1836a,b) the idea was generally abandoned for a century because of the association in the 1860s of meteor streams with specific comets. Hence Schmidt's highly successful determinations of rotation periods for the nuclei of comets Donati 1858 VI (Schmidt 1863a,b) and P/Swift-Tuttle 1862 III (Schmidt 1863b) were not recognized as such. For Comet Donati, Schmidt measured the angular velocities of the expanding parabolic envelopes and derived the instants of "Neubildungen" or initiation times of the halos by subtracting the ratios of radius to velocity from the observed dates. These are precisely the zero dates (ZD) that I used in the halo method for P/Schwassmann-Wachmann 1 (Whipple 1977) and for Comet Donati (Whipple 1978a). Apparently Schmidt did not consider the repetition of halo initiation as strictly periodic because in his solution he did not attempt to count sequentially the period numbers of his ZD's, but only used the intervals between them as integral numbers of periods in a few trials. However, his best value of 4.27 hr is quite close to my value of 4.62 hr (Whipple 1978a), which is not definitive. The published observations of this brilliant comet are extensive and were summarized by Bond (1862). Unfortunately Bond did not make long series of measurements each night to confirm Schmidt's evidence for rapid expansions of the envelopes. Probably the near equality of the period and the longitude difference between Harvard Observatory and Western Europe caused Bond to dismiss Schmidt's conclusion. The envelope at the next preceding observation in Western Europe was nearly identical with that seen by Bond; the observations were conveniently made at about the same time of the evening.

In the case of P/Swift-Tuttle, Schmidt (1863c) identified a real periodicity, but did not ascribe it to nuclear rotation. He measured the period by means of three oscilliatory characteristics of the comet: two types of oscillation in the tail, $P = 68$ hr; periodic reappearances (Wiederkehr) of configurations very near the nucleus, $P = 72$ hr; and the position angles of the streaming fans (Strömungfächer) with respect to the tail axis, $P = 65$ hr. This last value is his best determination, in agreement with the extremely sophisticated analysis by Sekanina (1981a) who has identified active areas on the nucleus by the near-nucleus jets and also determined the complete spin vector. It is interesting to note that by the halo method I (Whipple 1978b) obtained precisely half the true period. Sekanina's work indicates that this result was caused by two active areas located on the nucleus nearly 180° apart in longitude.

Horn (1908, and later G. Horn d'Arturo) first specifically identified a spin vector determination for a comet nucleus. For Comet Daniel 1907 IV, Horn assembled 59 photographs and measured the symmetry axis of the nuclear oval with respect to the inner tail axis. This axis oscillated 73 times about the tail axis, with a period close to 16 hr. Horn's involved geometrical solution led him to place the axis of nuclear rotation in a specific direction near the orbital plane. From a very cursory determination of the period by the halo method, I derived a period of 14 hr (Whipple 1978b) but do not consider my result more than tentative. Horn's period for Comet Daniel is included here as statistically significant.

For Comet Brooks 1893 IV, Vorontsov-Velyaminov (1930) concluded that "the movement of the tail and its form can be explained when it is interpreted like a beam continuously emitted by the block which rotates round the radius-vector with the period 3.793 d." This statement clearly indicates the concept of a discrete solid nucleus "that emits the gaseous particles which form the beam." Dubiago (1948) also suggested that nongravitational changes in the orbital periods of comets might arise by the tangential ejection of dust from a discrete nucleus, but did not specify the mechanism.

Larson and Minton (1972) analyzed the spiral structures near the nucleus in a number of photographs of Comet Bennett 1970 II. From the curvature of the spirals and the assumption of 0.6 km s^{-1} outward velocity near 28 Mar. 1970, they derived a period of 1.4 to 1.5 d, direct rotation with respect to the orbital motion. Beyer (1972) published 22 measures of the coma diameter for this comet. The solution for period P is presented in Sec. II as an example. The best solution by the halo method gives P = 1.164 d, in disagreement with the central value of 1.45 d found by Larson and Minton. However, their assumed velocity v of 0.6 km s^{-1} on ~ 28 Mar. is lower than that given by the formula I derived from Bobrovnikov's (1954) solution from 27 comets, namely

$$v = 0.535 \, r^{-0.6} \qquad (1)$$

where v is in km s^{-1} and r is the solar distance of the comet in AU. For Comet Bennett on Mar. 28, r = 0.569 AU, giving v = 0.750 km s^{-1} from Eq. (1). Because the solution for P by Larson and Minton varies as v^{-1}, their corrected period becomes 1.159 d, in excellent agreement with P = 1.164 d by the halo method.

Fay and Wisniewski (1978) applied photoelectric photometry to the nuclear regions of P/d'Arrest on 4, 5, and 6 Aug. 1976, measuring an amplitude of 0.15 ± 0.02 in visual magnitude and a period of 5.17 ± 0.01 hr, as evidenced by a 99.98% peak in their power spectrum. Measures of halo diameters on six Naval Observatory Flagstaff 40-inch reflector plates, made available by E. Roemer, show clearly that the period of P/d'Arrest was short, ~ 0.3 d in 1963-4. Two plates show inner condensation or halos. Periods of 6.7 hr and 7.9 hr fit the few observations extremely well by the halo method. It is interesting to note that Fay and Wisniewski found a weak power peak (98%) at 6.6 hr. Neither the Roemer plates nor 17 measures by Schmidt (1871) in 1870 give the 5.17 hr period by the halo method. Other apparitions of P/d'Arrest may clarify the matter. In the meantime I shall accept P = 5.17 hr as statistically meaningful.

For an excellent summary of the history and present state of our knowledge about the orientation and precession of cometary spin axes, the reader is referred to Sekanina (1981b). The concept of a rotating icy-conglomerate nucleus (Whipple 1950,1951) led finally to new interpretations of the wealth of observational data already published and in photographic plate collections.

In summary, the sublimation of ices on the nucleus of a comet produces a jet force normal to the surface. On the morning face of a rotating icy nucleus there should be a lag in heating so that the mean vector of the sublimation points not directly towards the Sun but produces a force towards the surface on the afternoon side. At least four effects should result.

(1) The outflow of the inner coma is displaced away from the Sun, generally appearing asymmetrical with respect to the comet-Sun line when observed from the Earth. Such asymmetry can be exploited to determine the direction of the spin axis and sense of rotation.

(2) A component of the jet force is directed normal to the radius vector and also along it radially from the Sun. The component perpendicular to the orbit plane affects the inclination of the orbit but is practically impossible to detect in the orbital elements. The component in the orbit plane normal to the radius vector is cumulative and increases or decreases the orbital angular momentum, depending on the sense of the cometary spin with respect to the orbital motion. Hence, it is measured as a secular increase or decrease in the period of revolution. The radial component is generally the largest force, but competes directly against solar gravity. Thus, it can be detected orbitally only for a few comets.

(3) For an oblate nucleus the jet component will not be directed towards the center of gravity except when the subsolar point is at the equator. The result generally is a precession of the spin axis unless it is perpendicular to the orbit plane. Note that most asymmetries in the nucleus will produce a pseudo-oblateness because the spin axis can be expected to shift eventually to coincide with the axis of maximum moment of inertia. In a weak rotating isolated structure as visualized for a cometary nucleus, internal energy dissipation should produce nutational damping until the maximum principal axis is reached.

(4) If the surface of the nucleus is rough, the effect of lag angle in sublimation might cause spinup. Consider A and B in a crater as opposing areas or valleys equal in area and tilt angle. Let A be tilted towards the morning Sun and B turned away from the morning Sun. If the scattered light and sublimed gas from A tends to warm B, sublimation of B might begin at a smaller solar altitude than for A. Furthermore, if the afternoon heating of A by B is less effective than B by A in the morning, B would sublimate more gas and produce a larger forward force component than the negative component of A; spinup could occur. On the other hand, for very rapid rotation and large lag angle, area B might not have time to sublimate as much as A, resulting in a spindown effect or a limit on the shortest period. In any case, the observed existence or absence of spinup effects would be important in comet theory. If

these effects are significant, older comets should rotate statistically faster than new comets. In theory such spinup might produce splitting of comets, but observed splitting appears to favor no age group. Nor is there evidence yet that split comets are rotating rapidly.

The directions of the spin axes and the lag angles in sublimation have now been determined for comets P/Encke, P/Tempel 2, P-Borrelly, P/Schwassmann-Wachmann 3, and P/Swift-Tuttle by Sekanina (1979,1981a) and for P/Schwassmann-Wachmann 1 by Whipple (1977,1980); for Comet Daniel 1907 IV, Horn's results have already been discussed. The few samples show a tendency for the spin axes to be more nearly aligned with the orbital lines of apsides than would be expected by chance, although the effect is not very secure statistically. The lag angles vary from comet to comet and range from 0° to 80°, occasionally being negative for P/Schwassmann-Wachmann 1, which has such a slow rotation period, 5 d.

Our knowledge of the nongravitational motions for more than three dozen comets stems from the superb orbital calculations by Marsden supported by Sekanina and Yeomans (for major results and references, see Marsden et al. [1973], and Sekanina [1981b]). Equal numbers of comets are accelerated and decelerated in their orbits, suggesting a random distribution of spin vectors. A few comets show secular changes and some show reversals of sign.

The phenomenal secular variation in the motion of P/Encke has been quantitatively explained by Whipple and Sekanina (1979) as resulting from the precession of the polar axis caused by sublimation jet forces on a slightly oblate nucleus. The polar axis has precessed more than 100° across the sky since the comet was discovered in 1786. The lag angle appears to remain constant at 45°. Because the calculated oblateness is only $\sim$ 4% and because P/Encke is probably one of the larger short-period comets, one might expect precession to be far more rapid for a highly oblate smaller comet. Sekanina (1981b) indeed suggests that the extremely variable nongravitational motions of comets such as P/Brorsen, P/Comas Solá, P/Finlay, and P/Kopff may arise from rapid toppling of their spin axes by sublimation forces on very irregularly shaped nuclei. The work of interpreting the vast resources of cometary observations is just beginning.

II. THE METHOD OF SOLUTION FOR ROTATION PERIODS

The halo method of determining the rotation periods of comet nuclei depends primarily on the assumption that for many comets the nucleus is spotted, as measured by its activity. The method further presumes that active areas produce conspicuous halos for only a few comets. Thus, for many comets the observer is assumed to have measured the diameter of some outer halo, preferably by a diameter perpendicular to the solar direction, as for parabolic envelopes, seen in Figs. 1 and 2 (Whipple 1981a). Such a measured diameter at heliocentric distance r, corrected for geocentric distance Δ, is divided by twice the velocity given by Eq. (1) to derive a time interval Δt. At $r = \Delta = 1$ AU, the expansion rate in diameter is 127.5 arsec d^{-1}. The zero date ZD of halo initiation is then the observed date minus Δt. The ZD's should be spaced by integer numbers of the period P. The final problem is to find these numbers and P.

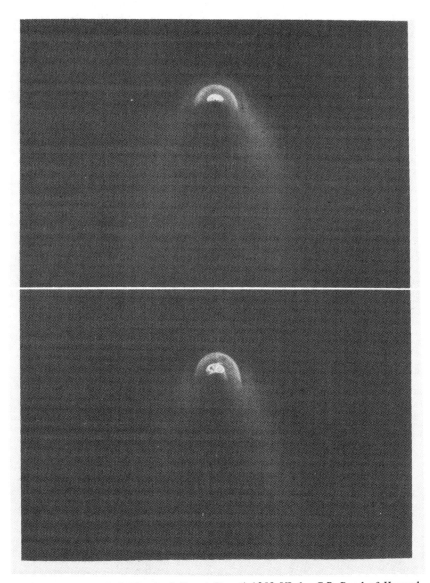

Fig. 1. Drawings of the head of Comet Donati 1858 VI, by G.P. Bond of Harvard Observatory on October 4 (top) and October 5 (bottom) 1858.

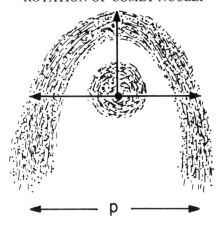

p   Pseudo Latus Rectum

Fig. 2. Idealized coma envelope with diameter p (the pseudo-latus rectum) indicated. The corresponding diameter of the inner coma could be a halo diameter.

The search for P is conducted: (a) by inspection of the differences of the ZD's to suggest possible common integral divisors, and (b) by a calculation based upon a least-squares minimum-residual concept. A series of trial periods P_k are equally spaced between two limits. All the differences between the n ZD's are divided by P_k giving integers $m_{k,i,j}$ in which the residuals $\Delta_{k,i,j}$ satisfy the relation

$$\Delta_{k,i,j} = |(ZD_j - m_{k,i,j}P_k)| \leqslant 0.5 P_k \qquad (2)$$

where $i = 1,2...n-1, j = 2,3...n$ and $j > i$.

For each trial value of P_k, a quantity S_k is given by

$$S_k^2 = \frac{24 \Sigma_i \Sigma_j \Delta_{k,i,j}^2}{n(n-1)-2} \qquad (3)$$

A value of $S_k = 1$ is expected for a random value of P_k. The computer prints out the values of P_k and S_k when $S_k < 0.9$. Least-squares solutions for P are then carried out for minimal values of S_k to give the most probable value of P. Experience shows that for a randomly chosen value of P, a least-squares solution can be found for a nearby value of P in which the ratio $P/\sigma_1 \sim 4$ where σ_1 is the mean error of a single ZD. Thus, values of P/σ_1 below 5 are rarely considered significant.

It is now important to mention some basic weaknesses and uncertainties of the method. There are only one or two real numerical checks on its validity. Other methods of determining the rotation periods for cometary nuclei have been little exploited. It is hoped that this preliminary presentation will spur observers and theoreticians on to greater efforts.

Two major weaknesses are immediately evident: (1) the assumed equivalence of a measured coma diameter with that of a halo, and (2) the lack of discrimination between gas and dust halos in the velocity of expansion. As for (1), the physical significance of a coma diameter, there are a few helpful discriminatory clues, but only for a few comets is the answer definitive. The continued repetition of a derived period among later observations for a given comet is the major source of confidence that the measured coma diameters represent halo diameters produced by active areas. Unsatisfactory, or self-supporting, as such a criterion may be, the alternative is to abandon the solution. When a much more extensive presentation can be made, the reader will be able to judge the validity of the solutions presented.

As for the second major weakness, the use of a single expansion velocity for all halos, a much more extended investigation must eventually be attempted. Some support for the assumption arises from the smoothness of Bobrovnikov's curve giving Eq. (1) and the agreement above for Comet Bennett 1970 II, when Bobrovnikov's velocity is applied to both methods. Other support for the assumption arises chiefly from internal agreement as discussed in the previous paragraph.

A third weakness of the method is sometimes quite evident: the occurrences of multiple active areas. The effect can be devastating to the method for the brightest comets near perihelion. As noted elsewhere, it can lead to a solution that is submultiple of the true period. Experience suggests that this fault is less frequent than might be expected.

Some clues that help assess the validity of a solution for P now follow. The most convincing positive evidence is the actual observation and measurement of two or more simultaneous halos. Next is repetition of a ZD from successive or nearby observations. Another positive indicator is an increase of total brightness of the comet accompanied by a decrease in coma diameter. The observation of a "stellar nucleus" may indicate a ZD. However, this clue cannot be accepted without reservation. Among n observations, where the calculated expansion rate is v'' in arcsec d^{-1}, the number with a stellar nucleus should not exceed $\sim 2n/P\ v''$, assuming that stellar nucleus means a nucleus with diameter $\leqslant 2$ arcsec. When the number of stellar nucleus observations greatly exceeds this number, the comet may have its pole turned too close to the subsolar point, the calculated period may be too long, or the comet may be active over much of its surface; then stellar nucleus observations cannot be accepted as bona fide ZD's.

A calculated period is under suspicion when $\Delta t/P$ is much larger than 3 to 5 times. This means that the 3rd to 5th or larger halo is supposedly being observed. A relatively short period is doubtful when σ_1, the calculated mean error of a diameter measure, is much smaller than the least count, the recording accuracy of the observation. Perfect observations, recorded for example to 0.1 arcmin, would always lie within 3 arcsec of the true value. The mean absolute error would then be 1.5 arcsec and the rms error $6 \times (1/12)^{1/2}$ or

1.7 arcsec. In practice such perfection cannot be achieved so that in this case a solution with $\sigma_1 <$ 3 arcsec is suspect. When the comet is distant, the angular expansion rate becomes small and increases the minimum period that can be derived with confidence from observations of a given angular precision.

Because nearby observations tend to be made at nearly the same time of day, periods near 1.0 d or integral fractions of a day are given suspicious scrutiny. Furthermore, very long periods may tend to combine ZD's from consecutive observations. The result is to reduce the effective number of observed ZD's, and hence to spuriously increase the apparent accuracy of the derived period.

The inherent nature of spottedness on the nucleus of the comet may cause difficulties. In the case of P/Swift-Tuttle mentioned earlier, the halo method led to a derived period equal to one half the true period. Two more examples of such confusion have been found, and others may exist undetected. This type of error leads to an undervaluation of the period. Short periods may be obscured by a large cometary distance or by observations of too great an angular uncertainty. It is difficult to estimate the systematic effects of these weaknesses in the method. I suspect that the greater loss or error lies statistically at the short-period end of the distribution.

The determination of periods by the halo method differs in character from most types of calculations; a period so determined is either almost exactly correct or completely wrong. As an example, Comet Baade 1922 II from about a dozen observations gives solutions for a period of either 81.6 or 10.3 hr; it is not included in this chapter. Because of this characteristic of the halo method, the weights assigned do not have the classical significance. Here the weight listed expresses my subjective opinion as to the probability that the derived period is essentially correct. A weight of 0.2 indicates that in only 1 of 5 such examples should the period be correct. A weight of 1.0 indicates my opinion that the period would be correct in more than 95 of 100 similar cases. This subjective weight is based on internal consistency and the clues mentioned above.

A better than average example of the halo method is used here for illustration, especially because it is more interesting than most. In Table I for Comet Burnham 1960 II, explanation for the successive columns is contained in the table footnotes. The rotation count number n and $O\text{-}C$ residuals (last column) in arcsec from the final solution are given in days by

$$ZD = 4.795 + 0.69965\,n\,. \qquad (4)$$

The perihelion of Comet Burnham occurred on 20 Mar. 1960. The preperihelion observations strongly indicate a period of 0.70 d. A solution with this period using all the observations shows a striking bimodal distribution of residuals with a complete gap between a cluster at +0.25 d and another at −0.18 d. A separation of 0.6 period P was then applied to the n's of the

TABLE I

Comet Burnham 1960 II

Obs.[a]		UT[b] 1960	Coma[c] Dia.	Arcsec[d] d^{-1}	Δt[e] d	ZD[f] d	Sol.[g] n	O-C
vB	1	5.13	20″	82	0.24	4.88	0.0	+ 7
RTL		18.07	1′	79	0.76	17.31	18.0	− 6
RTL		28.08	1′.3	78	1.00	27.08	32.0	− 8
RTL		−	0′.5	−	0.38	27.70	33.0	−14
MB		30.75	4′.6	78	3.54	27.21	32.0	+ 2
MB		−	St	−	−	30.75	37.0	+ 5
MB	2	7.76	4′.2	80	3.16	35.60	44.0	+ 2
RTL		7.09	0′.6	80	0.45	37.64	47.0	− 3
RTL		17.10	0′.7	85	0.50	47.60	61.0	+11
vB	4	10.43	1′	234	0.26	101.17	137.6	+24
vB		12.42	1′	255	0.24	103.18	140.6	+ 3
MB		17.11	∼9′	328	1.65	106.46	145.0	(+71)[i]
MB		18.11	9.0	240	1.54	107.56	147.0	−29
vB		18.41	2′	357	0.34	109.08	149.0	+13
vB		22.41	7′	486	0.87	112.54	154.0	− 1
MB		24.94	15′	588	1.53	114.41	156.6	+29
MB		25.92	18′	610	1.75	115.17	157.6	+67
MB		28.29	18′	606	1.78	118.17	162.0	+19
MB		−	St	−	−	119.95	164.6	− 5
MB		30.87	17′	522	1.95	119.92	164.6	−20
MB	5	1.93	20′	470	2.55	120.38	165.0	+67
MB		1.98	18′.3	467	2.35	120.64	165.6	− 8
MB		2.05	26′	465	2.98	119.69	164.0	(+71)
MB		−	4′.6	−	0.59	122.45	168.0	+53
MB		2.89	16′	426	2.25	121.64	167.0	+ 1
MB		2.94	17′	424	2.40	121.54	167.0	−41
MB		3.97	17′	381	2.67	122.30	168.0	−14
MB		4.93	13′	347	2.25	123.68	170.0	−20
MB		7.92	12′	264	2.73	126.19	173.6	−17
MB		8.90	13′	244	3.20	126.70	174.0	+40
MB		11.91	7′.1:[h]	195	2.19	130.73	180.0	− 1
MB		13.88	6′.0:	171	2.10	132.78	183.0	− 9
MB		16.98	5′.6:	142	2.35	135.63	187.0	− 0
vB		18.17	4′	134	1.79	137.37	189.6	−11
MB		25.96	4′.6:	94	2.94	144.02	199.0	− 1
MB		26.94	3′.9:	90	2.59	145.35	201.0	− 7
vB	6	15.17	20″	51	0.40	166.77	231.6	− 3
RTL		17.22	St	−	−	169.22	235.0	+ 0

NOTES TO TABLE I

[a]Observers: MB, M. Beyer; vB, G. Van Biesbroeck; RTL, E. Roemer, M. Thomas, and R.E. Lloyd.
[b]The date of observation.
[c]The measured coma diameter in arcsec or arcmin, St ± stellar.
[d]Twice the angular expansion rate in arcsec d^{-1} calculated by Eq. (1), from Δ and r.
[e]Δt in days calculated from the ratio of coma diameter to daily expansion rate.
[f]The zero date ZD = Obs. date$-\Delta t$.
[g]The rotation count number n and O-C residuals in arcsec from the final solution given by ZD=4.795 + 0.69965 n in days.
[h]Colon indicates slightly questionable observation.
[i]Parentheses indicate number omitted in the solution.

smaller group, 10 of the 36 usable observations. The resulting solution is satisfactory with $P/\sigma_1 = 8.8$, where σ_1 is the mean error of a single observation. The mean absolute errors for the observations given to accuracies of 0.1 and 1 arcmin are 7 and 23 arcsec, respectively, characteristic of extremely good observations.

The comet appeared to approach perihelion with a single major active area. After perihelion the comet brightened systematically for over a month and a second active area predominated for about a third of the time, persisting at this rate until the end of the measurements. It was displaced about 144° in cometary longitude from the primary active area. The few asymmetries seen in the coma are nearly 90° to the major tail axis and seem more likely to represent a dust tail than a measure of the spin pole of the nucleus. I cannot relate the rotation to the wagging plasma tail observed by Malaise (1963).

The second example, Comet Bennett 1970 II, is a rather mediocre case, but illustrates the method and has interest because of the period determination by Larson and Minton discussed above. The measures in Table II are only restricted by Beyer (1972) to those given to an accuracy of 0.1 arcmin because the expansion rate is small, lying almost entirely within 20 to 45 arcsec d^{-1}. The solution is not definitive, as other diameter measures can surely be found. The maximum-residual calculations were carried out for 22 observations from $P = 0.4$ to 2.0 d with S_k giving three minima that lead to: $P = 0.966$ d, $P/\sigma_1 = 5.5$; $P = 1.165$ d, $P/\sigma_1 = 5.4$; and $P = 1.183$ d, $P/\sigma_1 = 5.0$. The first solution is somewhat suspect because it lies near 1.0 day. The second solution is adopted, 1.165 d or 28.0 hr, for the reason discussed earlier, although the third solution 28.4 hr is almost as good. Were it not for the agreement with the revised value by Larson and Minton, the weight of the determination would be 0.3 rather than 0.5.

TABLE II
Comet Bennett 1970 II[a]

UT[b] 1970	Coma[c]	Δ[d] (AU)	r[e] (AU)	Arcsec[f] d^{-1}	Δt[g] d	ZD[h] d	Δ[i] n arcsec
4 19.10	8.6	1.083	0.859	128.9	4.00	15.10	0 − 0
7 5.99	4.1	2.082	2.496	32.9	7.47	89.51	64 − 5
7 29.94	4.7	2.686	2.420	27.9	10.10	110.84	82 + 6
7 30.93	4.7	2.691	2.435	27.8	10.16	111.78	83 − 1
8 2.91	4.3	2.709	2.478	27.3	9.45	115.46	86 + 4
8 4.91	3.3	2.718	2.502	27.0	7.32	119.59	90 − 10
8 5.93	4.6	2.722	2.514	26.9	10.25	117.68	88 + 1
8 7.93	3.4	2.732	2.541	26.7	7.65	122.28	92 − 0
8 9.92	3.7	2.741	2.567	26.4	8.41	123.52	93 + 2
8 12.95	3.8	2.752	2.608	26.1	8.75	126.20	95 + 11
8 13.91	3.8	2.758	2.621	25.9	8.79	127.12	96 + 4
9 6.88	3.4	2.831	2.933	23.6	8.64	151.24	117 − 4
9 7.88	3.3	2.833	2.945	23.5	8.41	152.47	118 − 3
9 8.92	3.3	2.838	2.959	23.4	8.45	153.47	119 − 6
9 13.89	3.4	2.851	3.022	23.0	8.86	158.03	123 − 9
9 22.84	3.0	2.872	3.134	22.4	8.05	167.79	131 + 1
9 26.81	3.3	2.884	3.184	22.1	8.98	170.08	133 + 0
9 28.83	3.5	2.892	3.208	21.9	9.59	172.24	135 − 3
9 29.92	3.3	2.895	3.220	21.8	9.07	173.85	136 + 6
10 3.90	3.1	2.911	3.271	21.5	8.65	178.25	140 + 0
10 20.78	3.0	3.000	3.475	20.1	8.94	194.84	154 + 6
10 21.80	3.1	3.008	3.486	20.0	9.29	195.55	155 − 3

NOTES TO TABLE II

[a] All observations by M. Beyer (1972).
[b] Date of observation.
[c] Coma diameter in arcmin.
[d] Geocentric distance in AU.
[e] Heliocentric distance in AU.
[f] Twice the angular expansion rate in arcsec d^{-1}.
[g] Δt in days calculated from the ratio of coma diameter to daily expansion rate.
[h] The zero date, ZD = Obs. date $-\Delta t$.
[i] Rotation number n and obs.−comp. ZD in arcsec.

III. THE CALCULATED SPIN PERIODS

Calculations to date have centered about the remarkable series of observations made by Beyer (1969,1972) over 49 years. Beyer summarized and referenced these observations; the uniformity, precision, detail, and beauty of presentation make his observations a source of unique value. G. van Biesbroeck's (see Table I) physical observations of comets have added materially to the solutions, and I have been able to measure a number of his plates. Another major contributor is E. Roemer, both by her published physical observations of comets and by her assistance in providing plates for me to measure, from the Naval Observatory and the Tucson observatories. The results for Comet Timmers 1946 I are based entirely on Flagstaff Observatory plates made by H. Giclas, while my measures of other Flagstaff plates have added materially to the results, particularly for P/Halley. The literature has been augmented by many measures by Schmidt, Bond, and H. Jeffers, M.P. Chofardet, H. Krumpholz and others. Detailed acknowledgments will be made when tables similar to Tables I and II are published in archives such as those of the American Physical Society.

The present effort has been aimed primarily at establishing a baseline among the new and younger comets as a prelude to comparison with the spin periods of the small asteroids and to provide a basis for theories of cometary accumulation processes. The possibility of spinup with comet aging may be evaluated from comparisons with subsequent calculations.

In the framework of the observational sources mentioned above, the study is nearly complete for the Type I (new) comets ($1/a \leq 50 \times 10^{-6}$ AU^{-1}), and somewhat less so for the Type II (nearly new) comets ($50 < 1/a < 2154 \times 10^{-6}$ AU^{-1}), barring large additions from photographic plate collections and observations of recent comets. For the shorter period comets the selection has been unsystematic and by opportunity so that comparison of rotation between new and old comets is somewhat preliminary.

More complete data are presented for the Type I comets in Table III than for the others in Table IV because derivations are more nearly complete.

TABLE III
Type I Comets[a]

	Reid 1921 II	Orkicz 1925 I	Geddes 1932 VI	Cunn. 1941 I	W-B-Ku 1942 IV	Timmers 1946 I
q	1.008	1.109	2.314	0.368	1.445	1.724
i	132	100	125	50	79	73
$1/a \times 10^6$	18 ± 11	40 ± 11	45 ± 3	1 ± 10	16 ± 5	−13 ± 3
H_0	6.4	5.4	3.5	5.8 ± 0.0	6.0	6.1
n				2.0 ± 0.1		
r	1.3 − 1.0	1.1 − 4.6	2.7 − 6.2	1.8 − 0.5	1.8 − 1.5	1.7 − 2.0
Δ	1.5 − 0.6	1.5 − 3.9	2.2 − 6.1	1.4 − 0.7	0.7 − 0.9	1.0 − 2.0
Int	52	343	186	70	52	52
No. Obs.	14	21	21	15	11	19
P/σ_1	6.0	6.5	5.4	6.3	8.2	7.3
σ_1	4 − 11	10 − 2	21 − 5	6 − 28	8 − 11	6 − 3
P	7.75	19.58	93.8	15.02	15.0	12.01
Wt	0.4	0.4	0.3	0.5	0.3	0.8

	Jones 1946 VI	Bester 1947 I	Bester 1948 I	Paj-Mrk 1948 V	Minkow. 1951 I	Peltier 1952 VI
q	1.136	2.408	0.748	2.107	2.572	1.202
i	57	108	141	93	144	46
$1/a \times 10^6$	44 ± 4	−1 ± 5	24 ± 4	34 ± 4	37 ± 4	2 ± 48
H_0	5.0 ± 0.3	3.1 ± 0.4	6.3 ± 0.1	4.4 ± 0.1	7.2 ± 0.6	8.9 ± 0.4
n	3.8 ± 0.3	5.7 ± 0.4	3.0 ± 0.2	4.4 ± 0.1	1.5 ± 0.5	5.7 ± 1.3
r	2.8 − 3.2	2.6 − 2.4	1.0 − 2.1	2.1 − 3.8	3.4 − 3.1	1.2 − 2.2
Δ	2.7 − 2.9	2.0 − 3.0	0.7 − 2.3	1.9 − 4.0	2.5 − 2.9	0.7 − 1.7

ROTATION OF COMET NUCLEI

TABLE III (continued)

Int	40	107	83	328	54	130
No. Obs.	12	17	24	24/111	24/30	22/22
P/σ_1	7.2	5.5	5.3	6.6	5.9	9.1
σ_1	4 – 3	17 – 12	30 – 7	8 – 4	3	7 – 2
P	26.04	63.4	24.8	28.9	33.20	10.08
Wt	0.4	0.3	0.3	0.2	1.0	0.9

	Honda 1955 V	Baade 1955 VI	Burnham 1960 II	Honda 1968 VI
q	0.885	3.870	0.504	1.160
i	108	100	160	143
$1/a \times 10^6$	-727 ± 121	42 ± 3	-135 ± 23	-82 ± 116
H_0	6.8 ± 0.3	3.0 ± 1.2	6.8 ± 0.3	5.8 ± 0.1
n	4.7 ± 0.7	4.3 ± 0.8	7.2 ± 0.7	4.0 ± 0.2
r	0.9 – 2.0	3.9 – 4.3	0.8 – 1.5	1.2 – 1.7
Δ	0.3 – 2.5	4.5 – 3.1	0.2 – 1.8	2.1 – 0.7
Int	104	352	235	98
No. Obs.	39	31	36	39
P/σ_1	5.4	6.2	8.8	4.4
σ_1	16 – 1	3 – 4	49 – 4	3 – 10
P	4.14	32.03	16.8	6.5
Wt	0.7	1.0	0.9	0.5

[a] Units of measure and other information as follows: q, perihelion distance in AU; i, orbit inclination in degrees; $1/a \times 10^6$ AU from Marsden et al. (1978) if given with standard deviation, otherwise from Marsden (1979); $H_0 = \text{mag} -5 \log \Delta - 2.5 n \log r$ if taken from Beyer and identified by σ, or $H_{10}(n=4)$ if taken from Vsekhsvyatskii (1964), no σ given; r, heliocentric distance interval in AU; Δ, geocentric distance interval in AU; Int, interval of observations measured in number of rotations; No. Obs., number of observations used; P/σ_1, period divided by standard deviation of a single zero date; σ_1, standard deviation in arcsec, extreme values; P, period of rotation in hr; Wt, weight as defined in text.

TABLE IV
Period Determination[a]

Comet	Year	q (AU)	i (deg)	$1/a$ or Per.	H_0	n	P (hr)	Wt
				Type II Comets ($50 < 1/a < 2154$) $\times 10^{-6}$ AU^{-1}				
Moreho	1908 III	0.945	140	174 ± 7	4.2	—	15.5	0.8
Finsler	1937 V	0.862	146	124 ± 29	6.1	—	7.1	1.0
DeK-Par	1941 IV	0.790	168	2029 ± 16	5.8	2	18.3	0.8
VanGent	1941 VIII	0.875	95	78 ± 7	6.9 ± 0	3.3	33.6	1.0
Hon-Ber	1948 IV	0.208	23	525 ± 176	7.5 ± 0.1	5.4 ± 0.2	13.4	0.8
Ba-Bo-Ne	1949 IV	2.058	106	735 ± 7	5.6 ± 0.2	5.5 ± 0.2	68.1	0.3
Abell	1954 X	0.970	53	70 ± 25	5.9 ± 0.1	3.6 ± 0.1	19.2	0.6
Burnham	1958 III	1.323	16	256 ± 14	6.8 ± 0.3	7.2 ± 0.7	8.9	0.6
Ta-Sa-Ko	1969 IX	0.473	76	507 ± 4	6.4	3.1	27.0	0.6
				Type III Comets ($2154 < 1/a < 10^{-2}$) $\times 10^{-6}$ AU^{-1}				
Donati	1858 VI	0.578	117	6370 ± 14	3.3	—	4.62	1.0
Coggia	1874 III	0.676	66	3206 ± 300	5.7	—	9.2	0.9
Daniel	1907 IV	0.512	9	2650 ± 7	4.0	—	16	0.8
Whi-Fed	1943 I	1.354	20	7352 ± 6	4.6	—	6.7	0.8
Haring	1953 I	1.665	59	3277 ± 6	0.4 ± 1.7	14 ± 3	30.0	0.5
Ba-Mf-Kr	1955 IV	1.427	50	4355 ± 46	4.8 ± 0.4	7.2 ± 0.6	21.7	0.7
Bennett	1970 II	0.538	90	7334 ± 5	3.4 ± 0.0	4.4 ± 0.1	28.0	0.5

TABLE IV (continued)

Comet	Year	q (AU)	i (deg)	1/a or Per. (yr)	H_0	n	P (hr)	Wt
			Type IV Comets ($10^3 > P > 25$) yr					
Swi-Tut	1862 III	0.963	114	120	4.0	—	66	1.0
Halley	1910 II	0.587	162	76.1	4.6	—	10.3	1.0
Pons-Bro	1954 VII	0.774	74	71	4.7 ± 0.1	4.3 ± 0.2	58	0.3
Olbers	1956 IV	1.179	45	69.5	49 ± 0.3	5 ± 2	16	0.2
Candy	1961 II	1.062	151	930	6.5 ± 0.1	9.0 ± 0.8	11.4	0.6
			Type V Comets ($25 > P$) yr					
Schwas-Wach 1	1952 III	5.52	9	16.1	4?	—	120.0	1.0
Schaumasse	1960 III	1.194	12	8.17	7.4 ± 0.1	6.6 ± 0.4	7.16	0.7
	1932 IX	1.196	12	8.18	7.8 ± 0.3	10.6 ± 0.7		
Faye		1.620	11	7.32	8.9	—	10.52	0.7
	1969 VI	1.616	9	7.41	10.2 ± 0.5	2.6 ± 0.9	10.45	0.9
Sch-Sch	1949 VI	2.234	6	7.27	7.6	—	10.6	0.5
Reinm 2	1947 VII	1.867	7	6.59	10.8 ± 1.6	3.0 ± 2.2	23.9	0.5
d'Arrest		1.17	17	6.2	16	—	5.2	0.5
Pons-Winnecke		1.10	20	6.1	13	—	22.8	0.3
Tempel 2		1.36	12	5.3	12	—	4.8	0.2
Encke		0.339	12	3.3	12	—	6.5	0.7
			Parabolic Comet					
Pa-Ro-We	1946 II	1.018	170		9.9 ± 0.2	2.6 ± 1.1	8.7	0.6

[a] The successive columns are equivalent to the successive lines in Table III with r, Δ, Int, No. Obs., P/σ_1, and σ_1 omitted.

IV. COMPARISONS

Because the Type I ($1/a < 50 \times 10^{-6}$ AU^{-1}) comets are considered to be almost entirely new comets, traversing the planetary system for the first time following their entrance into the Oort Cloud, their periods should provide a type of standard, and represent their rotation state at formation or shortly thereafter. Type II ($50 < 1/a < 2154 < 10^{-6}$ AU^{-1}) comets probably include some of Type I and a number that have made a few passages through the inner solar system. The shorter period comets have made unknown but large numbers of such passages. If spinup by delayed sublimation from rough nuclear surfaces is significant, we should find shorter periods for the older comets of later classes.

To check this possibility and to compare cometary rotation with that of small asteroids, mean periods are presented in Table V. The 41 asteroids are those with diameters $\leqslant 40$ km, from the compilation by Harris and Burns (1979), and the 3-hr period of Apollo by McCrosky (personal communication), all given equal weight. It is immediately evident that the mean periods of the Type I and II comets are almost identical and can be combined for statistical purposes (line 3 of Table V). Then in comparison with all the older comets the mean periods do not differ by as much as one standard deviation. Hence our sample does not prove any effect of spinup with aging, although the difference in mean periods suggests this direction of change. Quite possibly some shorter spin periods are missed among young comets because they probably are less spotted than older ones. Some 9 comets of Types I and II have not yet been amenable to even low-weight solutions for rotation period.

In view of the statistical similarity among the spin distributions of the 47 comets, they can properly be combined (line 5 of Table V) and the means compared with those of the 41 tiny asteroids (line 6). Although Harris and

TABLE V

Weighted Mean Rotation Periods

Types	No[a]	Σ Wt	P[b]	by log P[c]	by $1/P$[d]
I	16	8.8	22.4^h	17.0^h±3.3	13.0^h
II	9	6.5	20.3	17.0±3.5	14.4
I+II	25	15.3	21.5	17.0±2.4	13.6
all>II	22	14.4	24.5	14.8±3.0	10.8
All	47	29.7	23.0	15.9±1.9	12.0
Asteroids	41	41	8.7	6.9±0.6	6.2

[a]Number of determinations.
[b]Mean period.
[c]Antilog of mean log P.
[d]The inverse of mean P^{-1}.

Burns (1979) point out uncertainties in their determinations of photometric rotation periods for asteroids, I think that their data contain less potential for systematic and specific errors than do the data presented for comets. Hence, this comparison and the conclusions should be judged as preliminary. The asteroids appear to be two or more times faster than the comets in the mean. The comparison is better seen in Fig. 3, where the two distributions are normalized and plotted with argument log P weighted for the comets. The dark blocks represent the better determinations with Wt ≥ 0.7. The median period for the comets is 15.0 hr versus 6.8 hr for the small asteroids, consistent with the ratios of the various means in Table V.

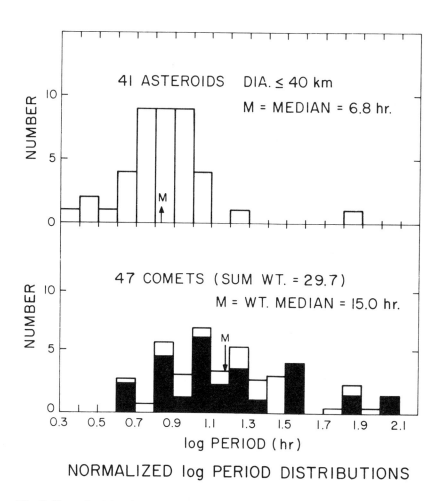

Fig. 3. Normalized log (period) distributions for spin of asteroids (above) and comets (below). The dark blocks represent higher weight determinations.

To search for possible correlations of spin with perihelion distance and inherent brightness, the periods of Tables I and II are divided into four groups, approximately equal in total weight, according to q and H_o as shown in Table VI. The periods from the mean weighted log P are entered in the table along with their summed weights in parentheses, as dependent on q and H_o. This increase in period with q is somewhat biased by the inclusion of P/Schwassmann-Wachmann 1 with a spin period of 120 hr, and the difficulty mentioned earlier in finding short periods for distant comets. A least-squares solutions for correlation of spin period with q does not lead to a significant correlation, a null result that is probably correct.

TABLE VI

Periods from Mean log P Including (Σ Wt)

	$q \leqslant 1.05$ AU	$q > 1.05$ AU	All q
$H_o \leqslant 6.0$	$16.1^h(8.2)$	$28.2^h(6.8)$	$20.8^h(15.0)$
$H_o > 6.0$	12.1(7.0)	12.0(7.7)	12.1(14.7)
All H_o	14.1(15.2)	17.9(14.5)	15.9(29.7)

A weighted least-squares solution for log P versus H_o, omitting q, leads to the result

$$\log P = 1.57 - 0.062 H_o$$
$$\sigma = \pm 0.13 \pm 0.026 \quad (5)$$

where the σ's are indicated, with ± 0.26 that for log P referring to unit weight.

The spin period appears to be inversely related to absolute magnitude, so directly related to absolute brightness at the 2.4σ level. The desired correlation is of course the radius or mass, not brightness, but no satisfactory general relationship is obvious between H_o and the dimensions of cometary nuclei. In one age type we might expect to find dimension and brightness better correlated. This assumption is applied by dividing the comets of each type into two brightness groups and comparing the values of mean $\Delta \log P$ with the values of mean ΔH_o. The Type I comets give $\Delta \log P/\Delta H_o = -0.14$ compared to -0.062 in Eq. (5). Types II, III, and IV all give neutral results while Type V alone gives -0.06. If the brightness varies as a function of radius R, say as R^2, then the period varies as R to the power $-5\Delta \log P/\Delta H_o$ or as $R^{0.7}$ for Type I comets and $R^{0.3}$ for the average of all (Eq. 5) and for Type IV alone. The relation between R and H_o is highly model-dependent while H_o itself depends enormously on the observational technique employed. We may conclude from these data that the spin period of comets probably correlates directly with radius or mass as some small power of radius.

A search for other correlations shows a tendency for the period to decrease with orbital inclination, but not at a statistically significant level. Period versus longitude and latitude of the orbit apsidal line gives no indication of arrangement on the celestial sphere.

Donn (1977) has classified 85 comets according to the general ratio of continuum to emission in their spectra, H for domination by continuum, M where each is prominent, and L for weak or absent continuum. In Donn's compilation periods are calculated for 13 L's and 6 M's, of total weight 13.6. The two sets are very similar and together yield 11.5 ± 1.5 hr as the logarithmic mean. Only 8 H's or high-continuum examples are represented, leading to a corresponding mean period of 32 ± 12 hr. The sample is very small and the long period is produced by three distant and probably large comets: Minkowski 1951 I (33.2 hr); Baade 1922 II (32.0 hr); and P/Schwassmann-Wachmann 1 (120.0 hr). If we ascribe the long periods to large radius and the continuous spectra of these comets to solar distance, any possible correlation of spin period with the dust-gas essentially vanishes. As pointed out by S. Larson (personal communication), this comparison may be specious because of the use of a single velocity expansion law for all comets, whether dusty or gassy.

Delsemme (1979) divides the new comets into a bimodal distribution of absolute magnitudes. Only two of the comets here fall into his faint category, Geddes 1932 VI and Peltier 1952 VI; for the former, Vsekhsvyatskii (1964) lists $H_{10} = 3.5$ in contradiction to Delsemme's $9.1 < H_o < 10$. With only one comet remaining in the faint category ($P = 10.1$ hr) no statement can be made as to whether Delsemme's faint group of new comets may represent small pieces split from larger comets, as he suggests.

IV. COMMENTS AND SUMMARY

The most striking characteristic of the distribution of the derived cometary spin periods (Tables III, IV, V, and Fig. 3) is its difference from that of the 41 small asteroids. The difference is manifest in three respects.

1. The comets systematically appear to rotate more slowly by a factor of ≥ 2, depending on the type of mean used in comparison. Indeed, Degewij (1977) finds that the mean period is 5.2 hr for 53 asteroids of 1-km size, compared to 23.0 hr for the 47 comets.
2. The shortest comet period of 4.14 hr (not completely certain) is considerably longer than the shortest asteroid period, but there are few asteroids of $P < 4$ hr.
3. The distribution function in $\log P$ is much flatter for the comets than for the asteroids, deviating from a Maxwellian distribution. This can be measured, as Harris and Burns (1979) point out, by the ratios of the different means. For a three-dimensional distribution typical of kinetic systems, the theoretical ratio of the mean $\log P$ to the mean P^{-1} is 0.9025. For

the 41 small asteroids this ratio is $6.17/6.94 = 0.889$ whereas for the weighted comet periods the ratio is $12.03/15.87 = 0.758$, far from 0.9025.

If we assume, with Harris and Burns and the common consensus, that the tiny asteroids are collisional fragments from larger bodies and that their distribution of rotation periods is controlled by the collisions, then the comets appear not to be the remnants of collisional breakup. Their distribution of rotational periods may have arisen in their original accretion processes.

For a sphere of mean density ρ (in g cm^{-3}) a particle on the equator will be moving at orbital velocity when $P_{\text{crit}} = 3.30\, \rho^{-1/2}$ hr. As Weidenschilling (1981) shows, bodies of weak structural strength such as badly cracked asteroids and comets having possible low-density nuclei will become oblate Maclaurin spheroids as they are spun up slowly. At about twice P_{crit} they become triaxial Jacobi ellipsoids, and the moment of inertia increases even more. He concludes that the observed 4-hr minimum period for larger ($R > 25$ km) asteroids represents the practical limit at $\rho = 2.5$ g cm^{-3}. The corresponding value for comets is then $P = 5.5$ hr if $\rho = 1.3$ g cm^{-3}. Four of our 47 comets appear to have shorter periods.

We must assume either that some comet nuclei have internal strength or that their densities are >1.3 g cm^{-3}, on the basis of Weidenschilling's (1981) theory. Note that his conclusion fails for three of the Harris and Burns (1979) small asteroids: 321 Florentina ($R = 30$ km, $P = 2.870$ hr); 1566 Icarus ($R = 1.7$ km, $P = 2.27$ hr); and 1978 CA ($R = 1$ km, $P = 3.75$ hr); it also fails for Apollo ($P = 3$ hr). More research on the subject is clearly needed.

Of further interest is that Honda 1955 V is a split comet; its period of 4.14 hr as listed is not definitive. If the period is supported by further analysis of photographic plates, we may have the first evidence that rapid rotation is involved in comet splitting.

Another correlation among the spin periods is an increase in period with absolute brightness, presumably with dimension, at the 2.4σ level, perhaps as the radius to a small power, for instance 0.3. The asteroids show a similar but less striking relationship. Harris and Burns (1979) find by logarithmic means, for example, that 143 asteroids of mean diameter 70.0 km have a mean period of 8.55 hr versus 5.41 for 10 Earth- or Mars-crossers of diameter 3.3 km. On the other hand, Burns and Tedesco (1979) conclude that "There is a change in asteroid characteristics near 175 km: objects with $D < 175$ have $P \sim 11$ hr while larger objects have $P \sim 8$ hr." Table V shows that the corresponding mean period for the 47 comets is 15.9 hr.

Relevant to comet formation and evolution, internal heating by radioactive atoms could significantly reduce the periods from their pristine values. Collapse of the possible snowy structure by internal melting and consolidation could greatly reduce the moment of inertia, and hence the period of rotation. Such heating should be much more effective among large comets, but perhaps they are more compacted because of larger gravity. In any case the data do not support an argument for serious internal heating of large comets versus little heating for small comets. The larger comets statistically show longer spin periods than small comets, which in turn are much larger than those of the large asteroids. It appears that the comets aggregated in a more quiescent region of space than the asteroids, or at least were less disturbed by collisional effects.

Trends in the spin periods of the comets indicate an increase with perihelion distance, and with continuum versus band emission in the spectrum, and a decrease with orbital inclination. None of these trends, however, is yet statistically significant. No trend is found with respect to the orientation of the orbital line of apsides on the celestial sphere.

The question of spinup with age and the development of comets into asteroids is not clearly settled. If we omit P/Schwassmann-Wachmann 1 from the Type V comets, the mean log P goes to $P = 9.5$ hr $\pm$ 1.8 hr versus 5.41 hr for 10 Earth- or Mars-crossing asteroids. It appears tentatively that spinup by sublimation is fairly probable, although size may be the dominant factor for old comets. Even so, they appear to spin more slowly than the Apollo and Amor asteroids. Other general evidence also points tentatively against a cometary origin for these asteroids (Whipple 1981b).

Acknowledgments. I am grateful to Z. Sekanina for consultation and use of his material and to E. Roemer and H.L. Giclas for access to their photographic plates. J.A. Burns and S.M. Larson have contributed valuable suggestions. This research has been supported by Planetary Atmospheres program of the National Aeronautics and Space Administration.

REFERENCES

Bessel, F.W. 1836a. Beobachtungen über die physische Beschaffenheit des Halley'schen Kometen und dadurch veranlasste Bermerkungen. *Astron. Nach.* 13:185-232.
Bessel, F.W. 1836b. Bemerkungen über mögliche un Zulänglichkeit der die Anziehungen allein berücksichtigenden Theorie der Kometen. *Astron. Nach.* 13:345-350.
Beyer, M. 1969. Nachweis und Ergebnisse von Kometen-Beobachtungen aus den Jahren 1921-1968. *Astron. Nach.* 291:257-264.
Beyer, M. 1972. Physische Beobachtungen von Kometen XVII. *Astron. Nach.* 293:241-257.
Bobrovnikoff, N.T. 1954. Physical properties of comets. *Astron. J.* 59:357-358.
Bond, G.P. 1862. Great Comet of 1858. *Ann. Harvard Coll. Obs.* 3:1-372.
Burns, J.A., and Tedesco, E.F. 1979. Asteroid lightcurves: Results for rotation and shapes. In *Asteroids*, ed. T. Gehrels (Tucson: Univ. Arizona Press), pp. 494-527.
Degewij, J. 1977. Lightcurve analyses for 170 small asteroids. *Proc. 8th Lunar Sci. Conf.* 8:145-148.
Delsemme, A.H. 1979. Empirical data on the origin of "new" comets. In *Dynamics of the Solar System,* ed. R.L. Duncombe (Dordrecht: D. Reidel Publ. Co.), pp. 265-269.
Donn, B. 1977. A comparison of the compositions of new and evolved comets. In *Comets, Asteroids and Meteorites,* ed. A.H. Delsemme (Toledo, Ohio: Univ. Toledo), pp. 15-23.

Dubiago, A.D. 1948. On the secular acceleration of motions of the periodic comets. (In Russian). *Astron. J. USSR* 25:361-368.
Fay, T.D., and Wisniewski, W. 1978. The light curve of the nucleus of Comet D'Arrest. *Icarus* 34:1-9.
Harris, A.W., and Burns, J.A. 1979. Asteroid Rotation. *Icarus* 40:115-144.
Horn, G. 1908. Struttura e rotazione della Cometa Daniel (1970d). *Mem. Soc. d Spettroscopisti, Ital.* 37:65-75.
LaPlace, P.S. 1813. *Exposition du système du Monde*, 4th ed. Paris, pp. 130-132.
Larson, S.M., and Minton, R.B. 1972. Photographic observations of Comet Bennett, 1970 II. In *Comets Scientific Data and Missions*, eds. G.P. Kuiper and E. Roemer (Tucson: Lunar and Planetary Lab.), pp. 183-208.
Malaise, D. 1963. Photographic observations of the tail activity of Comet Burnham 1960 II. *Astron. J.* 68:561-565.
Marsden, B.G.; 1979. *Catalogue of Cometary Orbits*, 3rd ed. (Cambridge, Mass: Smithsonian Astrophys. Obs.), pp. 1-88.
Marsden, B.G., Sekanina, Z.; and Everhart, E. 1978. New osculating orbits for 110 comets and analysis of original orbits for 200 comets. *Astr. J.* 83:64-71.
Marsden, B.G.; Sekanina, Z.; and Yeomans, D.K. 1973. Comets and nongravitational forces V. *Astr. J.* 78:211-225.
Schmidt, J.F.J. 1863a. Donati's Cometen. *Astron. Nach.* 59:97-108.
Schmidt, J.F.J. 1863b. Donati's Comet 1858. *Publ. Athens Obs.*, Ser. 1, 1:1-74.
Schmidt, J.F.J. 1863c. Comet II 1862. *Publ. Athens Obs.*, Ser. 1, 1:111-145.
Schmidt, J.F.J.*1871. Beobachtungen auf der Sternwarte zu Athen. *Astron. Nach.* 77:65-70.
Sekanina, Z. 1979. Fan-shaped coma, orientation of rotation axis, and surface structure of a cometary nucleus. *Icarus* 37:420-442.
Sekanina, Z. 1981a. Large-scale nucleus surface topography and outgassing pattern analysis of Comet Swift-Tuttle. *Astron. J.* In press.
Sekanina, Z. 1981b. Rotation and precession of cometary nuclei. *Ann. Rev. Earth Planet. Sci.* 9. In press.
Vorontsev-Velyaminov, B. 1930. Imaginary contradiction of the phenomena observed in the Comet 1893 IV (Brooks) with the mechanical theory of comets. *Russ. Astron. J.* 7:90-99.
Vsekhsvyatskii, S.K. 1964. *Physical Characteristics of Comets* (translation, NASA TTF-80), (Washington: NASA).
Weidenschilling, S.J. 1981. How fast can asteroids spin? *Icarus.* In press.
Whipple, F.L. 1950. A comet model. I. Acceleration of Comet Encke. *Astrophys. J.* 111:374-394.
Whipple, F.L. 1951. A comet model. II. Physical relations for comets and meteors. *Astrophys. J.* 113:464-474.
Whipple, F.L. 1977. Rotation and outbursts of Comet P/Schwassmann-Wachmann 1. *Bull. Am. Astr. Soc.* 9:563 (abstract); *Astron. J.* 85:305-313.
Whipple, F.L. 1978a. Rotation period of Comet Donati. *Nature* 273:134-135.
Whipple, F.L. 1978b. On the nature and origin of comets and their contribution to planets. *Moon and Planets* 19:305-315.
Whipple, F.L. 1980. Rotation and outbursts of Comet P/Schwassmann-Wachmann 1. *Astron. J.* 85:305-313.
Whipple, F.L. 1981a. On observing comets for nuclear rotation. In *Modern Observational Techniques for Comets*. (Pasadena: Jet Propulsion Lab.), pp. 191-201.
Whipple. F.L. 1981b. The nature of comets. *Comets and the Origin of Life*, ed. C. Ponnamperuma (Dordrecht, Holland: D. Reidel), pp. 1-20.
Whipple, F.L., and Sekanina, Z. 1979. Comet Encke: Precession of the spin axis, nongravitational motion, and sublimation. *Astron. J.* 84:1894-1909.

THE PROBLEM OF SPLIT COMETS IN REVIEW

Z. SEKANINA
Jet Propulsion Laboratory

The progress in the investigation of cometary splitting is reviewed from the dynamical and physical standpoints. It is shown that the rate of recession of cometary fragments is determined by the momentum from outgassing, the net differential force thence being of the same nature as the nongravitational perturbations detected in the motions of other comets. The differential radial deceleration γ and the time of splitting are the only parameters of this simple model that successfully represents the positional observations of almost all of the 21 known split comets. The time difference between splitting and final observation, weighted by the varying insolation, is termed the endurance. It measures a fragment's sublimation rate and provides a lower limit to its normalized lifetime. The endurance is found to be highly correlated with γ and either quantity can serve to classify secondary nuclei of split comets into three categories: persistent companions, short-lived companions, and minor fragments. A test of splitting, based on this model, has been devised to check on unconfirmed observations of comet multiplicity. The deceleration γ varies between 10^{-5} and several times 10^{-3} the solar gravity, the endurance between a few and several hundred equivalent days. The points of splitting have a random distribution, the record heliocentric distance being 9 AU before perihelion. Only for the extensively observed split comets need the model incorporate also the velocity of separation, which is never greater than a few meters per second. The interpretation of the separation velocity is complicated by *the gravitational interaction of the fragments for some time after break-up. It is concluded that even in the absence of a net differential non-gravitational force, the existence of gravitationally locked multiple nuclei is at best only remotely possible. In several cases the calculated time of splitting coincides either with a flare-up in the visual and/or infrared brightness or with an outburst detected as an isolated streamer in the dust tail and suggests that the dynamical separation follows the splitting with little time lag. Light variations of fragments are unpredictable, but suggest a pseudo-periodic pattern with a characteristic time constant of 10 to 20 equivalent days (i.e., days normalized to the vaporization rate at 1 AU from the Sun). The short-lived companions show physical similarities with the behavior of a small class of comets that have dissipated literally before the eyes of observers. The identification of the triggering mechanism, the most difficult part of the investigation of split comets, is briefly addressed and several candidates (tidal forces, rotation, dust-mantle dumping, radioactive heating) are discussed. A model that fits both the dynamical data and the physical characteristics suggests that most fragments must be appreciably non-spherical and precessing rapidly.*

A comet model proposed by Whipple (1950) and strong circumstantial evidence from observations accumulating since the time the model was formulated have led to a present consensus that a conglomerate nucleus of ice and dust, averaging a few kilometers in size, is the source of all the mass of a comet and the progenitor of its activity. A low tensile strength of the nucleus and the eruptive character of the activity, also inferred from observational evidence, suggest that comets are prone to self-destruction. Thence, it may be argued that the occasionally observed multiple comets could be products of the splitting of precursor comets. Although this hypothesis is corroborated by the fact that measured projected separations of the components of multiple comets have a strong tendency to increase systematically with time, no comet has ever been observed to break up. Nevertheless, this intuitive notion of cometary multiplicity has affected the current terminology to the point that the expression split comet has become more common than the noncommittal double (or multiple) comet. Similarly, the individual components of a multiple comet are often referred to as fragments, while the presumed original nucleus is a parent. In cases where one of the fragments is dominant in terms of brightness, activity, and/or persistence, it is called a principal (or primary) nucleus, the less conspicuous fragments are labeled companions or secondary nuclei.

Although observed rather rarely, multiplicity is unquestionably one of the most intriguing properties of comets and its understanding should provide vital information for studies of the nature of comets. The solution to the problem has a bearing on the investigations of comet formation, activity, lifetime, and disintegration, on the analysis of the structure and bulk properties of the nucleus, and on the interpretation of comet groups.

A casual inspection does not show any prominent patterns in the properties of the split comets. The number of detected fragments varies generally between two and five. Widely separated components often exhibit their own comas and parallel tails, suggesting that each of them has become an independent comet. Secondary nuclei of compact split comets tend to line up in the tail spine of the primary and some of them show signs of progressive diffuseness and elongation shortly before they fade out of sight. There is a great diversity of the physical behavior of fragments. Their unpredictable lightcurves are characterized by considerable fluctuations, which in some instances make it impossible to distinguish between a principal nucleus and companions.

I. MOTIONS OF COMETARY FRAGMENTS AND THE DIFFERENTIAL NONGRAVITATIONAL FORCE

Positional observations represent a major source of information on the fragments and their formation. The application of standard orbit-determination techniques to some of the best observed split comets has, however, produced disappointing results, as the positions of the individual components extrapolated from their orbits back in time never coincided with one another within about 10,000 km or so (e.g., Jeffers 1922; Marsden and Sekanina 1971). In at least one case (Marsden 1967) it was even possible to link successfully the positions of the parent comet with the positions of either one of its two fragments, in apparent defiance of the dynamical principles. Perhaps out of frustration, it became customary to identify the time of minimum calculated distance between the fragments with the time of splitting, and the relative velocity at that time with the separation velocity vector. Such separation velocities have been shown often to exceed 10 m s^{-1} (Stefanik 1966; Pittich 1972).

A closer examination of the split comets has shown that the position-angle variations of companions as referred to the principal nucleus are by and large restricted to a particular sector. Shortly after breakup, companions tend to move essentially radially away from the Sun relative to the primary, but, as time goes on, their position vectors shift gradually toward the reverse orbital velocity vector, to which their apparent directions of motion appear eventually to converge (Sekanina 1977a). Figure 1 exhibits, as examples, the configurations of the nuclei C and D of Comet West 1976 VI relative to the nucleus A as functions of time. This seemingly peculiar behavior bears a strong resemblance to a characteristic property of particle trajectories in the cometary dust tail. The conservation of momentum law applied to the motions of solid particles subjected to a somewhat reduced gravitational attraction of the Sun due to radiation pressure requires that after their ejection from the nucleus (on the assumption that no impulse is involved) the particles start lagging behind the comet in their radial motions and turning clockwise away from the prolonged radius vector toward the orbit behind the comet in their angular motions.

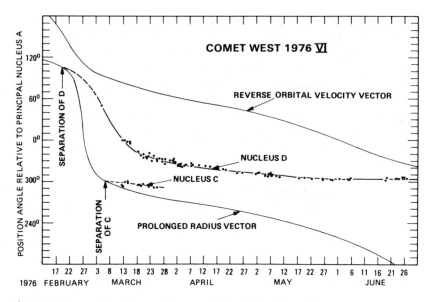

Fig. 1. Position angle of the secondary nuclei C and D relative to the principal nucleus A of Comet 1976 VI versus time. The configurations are restricted to a sector between the directions of the prolonged radius vector and the reverse orbital velocity vector. Nucleus B would run just slightly below nucleus D.

The deceleration hypothesis is strongly supported by information on the orbital periods of fragments, obtained by standard orbit-determination techniques. For example, the orbital period of the primary nucleus of P/Biela between 1846 and 1852 amounted to 2416.05 days, that of the secondary nucleus, 2416.64 days (Marsden and Sekanina 1971). Similarly, Marsden (1967) has found the osculating orbital period of 878 years for the primary component of Comet Ikeya-Seki 1965 VIII, but 1055 years for the secondary component. This must be still an underestimate of the true difference, because the observations of the two separate nuclei after perihelion have been combined with the positions of the single nucleus before perihelion. From the shorter postperihelion arcs alone Sekanina (1966) has obtained less well-defined but dynamically more meaningful values of 830 years for the primary and 1110 years for the secondary. Finally, from an identification of a likely candidate for the presumed parent comet of the sungrazers 1882 II and 1965 VIII, Marsden's (1967) results indicate a probable orbital period of 776.6 years for the first, unquestionably the principal, fragment and 859.7 years for the second comet.

Evidence that companions are measurably decelerated with respect to the principal fragment appears to be overwhelming. The major distinction between the traditional approach and the new hypothesis is that the former assumes that the fragments drift apart because of an impulse acquired *upon* separation, while the latter attributes the effect to continuous action of a radial differential force *after* separation at rest. The strength of the new model lies in the fact that it is more restrictive with only two free parameters to solve for, namely, the time of splitting and the differential deceleration. By contrast, the traditional approach involves four free parameters, the time of splitting and three components of the separation velocity vector.

Comparison with numerous measurements of split comets has shown that the assumption of a net radial force fits generally well both the position angles and the rates of separation (Sekanina 1977a), proving the viability of the two-parameter model. Although the recognition of similarities with the dust-particle dynamics has provided a very convenient formalism, the purely empirical approach could be meaningful only when accompanied by a compatible physical interpretation. The behavior of the split comets suggests that all fragments are rather massive objects that possess a certain degree of internal cohesion and some kind of temporary reservoir of ice and dust to continue to function as comets. Under these circumstances it is impossible to envision a comet fragment, except perhaps shortly before its apparent extinction, as a cluster of loose particles that are measurably affected by radiation pressure. On the other hand, it has now been universally accepted that comets are subjected to nongravitational forces arising from directed outgassing (Whipple 1950). When well established, the radial component of the force is usually one order of magnitude greater than the transverse component (the normal component has never been detected) and typically 10^{-5} the solar attraction, although there are a few comets for which it is as high as 10^{-4} (Marsden 1979). Whipple and Sekanina (1979) have demonstrated that the integrated radial component dominates the other two components even when the vector of effective momentum from outgassing may deviate at times significantly from the radius vector and when the obliquity of the orbit plane to the equator of the rotating comet approaches 90°. It has also been shown theoretically (e.g., Delsemme and Miller 1971) and confirmed observationally (e.g., Keller 1976) that near the Sun the gas production rate often varies nearly as the inverse square of heliocentric distance, so that the associated momentum should follow approximately the same law that is valid for radiation pressure. Generally, this is correct when the part of the absorbed solar energy that is spent on sublimation exceeds the fraction that is spent on thermal reradiation. For water snow the necessary conditions exist at solar distances of up to and somewhat beyond 1 AU and may be fairly well satisfied even beyond 1.5 and 2 AU, depending primarily on the effective surface absorptivity, which is controlled by the degree of contamination by dust. For more volatile snows, the sublimation-dominated regime is in force at still much larger solar distances. The empirically established inverse-square power law is therefore reasonably compatible with the interpretation of the relative motions of fragments in terms of a differential nongravitational force. The physical implications are further discussed in Sec. V.

II. AN IMPROVED MODEL

In spite of the generally high degree of correspondence between observations and the two-parameter model, a close examination has revealed the presence of second-order effects: (a) in one particluar case of a well-observed split comet (Wirtanen 1957 VI), the fit of the companion's measurements has been less than satisfactory; (b) slight systematic trends, amounting for most part to not more than 3 or 4 arcsec, have been detected in the position residuals of companions of several extensively observed comets; and (c) measurements of fragments at the times of the Earth's transit across the orbit planes of some of the split comets have demonstrated the lack of implied coplanarity among orbits of fragments. Of a number of possible causes of the detected departures, two can be offered as most likely: a small impulse at breakup and/or deviations of the deceleration vector from the radial direction and of its magnitude from the adopted power law (Sekanina 1977b). The first option has been pursued, partly because of the lack of information needed to handle sophisticated models of the differential force and partly because of the recognized likelihood that *uneven* fractions of the momentum of the parent nucleus are distributed among the fragments at breakup and show up as velocities of separation that are potentially measurable, though sizably lower than previously estimated.

Accordingly, an elaborate computer program has been written (Sekanina 1978) for an iterative differential-correction procedure to fit positional offsets of a companion in right ascension and declination by a multiparameter model that allows solutions for up to five unknowns: t_s, the time of splitting; γ, the net differential effect in solar attraction between the companion and the primary, the positive sign indicating a deceleration; and V_r, V_t, and V_n, the radial, transverse, and normal components of the companion's velocity of separation from the primary. V_r is positive in the antisolar direction, V_t in the direction perpendicular to V_r in the orbit plane in the sense of the orbital motion, and V_n in the direction of the north orbital pole, from which the comet is seen to orbit the Sun counterclockwise. Provision is made in the computer program whereby V_r, V_t, and V_n can be replaced, if desired, by V_x, V_y, and V_z, the velocity components in the cardinal directions of the ecliptic system. This option can speed up the convergence of the solution in some cases.

Before the results of an application of the multiparameter model to the known split comets are summarized, a note is in order on an important feature of the method. The computer program provides an option to solve for any combination of less than the five unknowns, so that a total of 31 different versions of the program are available. This option turns out to be of vital importance in practice, since the number of unknowns in the differential-correction procedure may usually be increased by not more than one at a time in order to get the solution to converge. The solutions are in the following identified by letter groups in which S stands for the time of splitting, D

for the deceleration of the companion, and R for the radial, T for the transverse, and N for the normal component of its separation velocity. Solution SD is identical with the two-parameter model discussed in Sec. I, solution $SRTN$ with the traditional model.

III. SEPARATION PARAMETERS FOR 21 SPLIT COMETS

An updated list of the separation parameters for 32 secondary nuclei of 21 comets that are believed to have split is presented in Table I. It compiles the data that have already been published (Sekanina 1977a,1978,1979a) as well as recent results communicated here for the first time. Although much of the table is either self-explanatory or understood from the text in Sec. II, some further remarks should be added. Short-period comets, whose orbital periods about the Sun are shorter than 200 yr, are marked by a prefix P/. The designation of fragments corresponds to that prevailing in the literature, the principal component being listed first. Thus, $A \rightarrow B$ means that the results refer to the motion of B relative to A. The number of measurements is strictly that used in the least-squares solution. When such a solution has not been possible, has failed, or has led to results inferior to those obtained by another method (such as trial and error), the number of measurements used by that other method is listed in brackets. If the results are supported by additional evidence, a plus sign follows the number of used measurements. The period of measurement is often shorter than the period of observation, as especially reports on visual observations of companions sometimes state that the objects are too faint for measurement. The range of separation refers to the specified period of measurement. Question marks indicate that no separations have been reported, the results thus depending on position-angle observations only (Sec. VI). The units of the deceleration γ that are adopted are 10^{-5} the solar gravity, equivalent to 5.93×10^{-6} cm s^{-2}, or 2.96×10^{-9} AU day^{-2}, at a heliocentric distance of 1 AU. The column "Solution and Mean Residual" is relevant only to the least-squares solutions. The time of splitting is listed by date and by its difference from the time of perihelion passage T; for short-period comets the T referred to is that of the apparition listed in the first column. The time t_S is accompanied by the corresponding heliocentric distance r_S and the distance from the ecliptic z_S. The separation parameters that are assumed rather than solved for are parenthesized; those assumed to be zero are omitted.

The experimentation with the least-squares procedure has shown that it is preferable to solve for less than the five unknowns, when the precision of the measured positional offsets is inadequate, the period of observation short, or the distribution of observations poor. Difficulties with the convergence have increased noticeably with increasing number of unknowns. The velocity of separation has become virtually indeterminate (and its inclusion in the equations unnecessary) for less extensively observed fragments. To solve for the radial component V_r has been impossible in all but a few cases.

TABLE I
Separation Parameters for 21 Split Comets

Comet/Name	Fragments/Number of measurements	Period of measurement from/to	Separation from/to (arcsec)	Solution/Mean residual	t_s/r_s z_s (U.T.)/(AU)	$t_s - T$/m.e. (days)	γ/m.e. (units)	V_r/m.e. (m s^{-1})	V_t/m.e. (m s^{-1})	V_n/m.e. (m s^{-1})
1846 II P/Biela	A → B 13	1846 Jan. 15 1846 Mar. 19	110 790	SDRTN ±0".22	1840 May 25 3.59 +0.04	−2088 ±56	4.91 ±0.36	−0.26 ±0.03	−0.17 ±0.01	+0.08 ±0.01
1852 III P/Biela	A → B 3	1852 Sept.21 1852 Sept.26	1755 1835	D ±2".95	(1840 May 25) (3.59 +0.04)	(−4505)	10.37 ±0.01	(−0.26)	(−0.17)	(+0.08)
1860 I Liais	A → B 3	1860 Feb. 28 1860 Mar. 10	178 205	ST ±6".31	1859 Sept.18 2.49 +2.17	−152 ±12	(7.0)	.	+5.48 ±0.61	.
1882 II Great September Comet	B → A 8	1882 Oct. 6 1883 Feb. 27	5 49	DR ±1".04	(1882 Sept.17) (0.017 +0.003)	(+0.083)	1.17 ±0.54	.	.	.
	B → C 46	1882 Oct. 5 1883 Mar. 3	5 35	SD ±1".27	1882 Sept.17 0.017 +0.003	+0.083 ±0.023	0.83 ±0.16	.	.	.
	B → D 7	1882 Oct. 13 1883 Feb. 27	14 57	SD ±2".17	1882 Sept.17 0.016 +0.003	+0.080 ±0.053	1.33 ±0.61	.	.	.
1888 I Sawerthal	A → B 3+	1888 Mar. 30 1888 Apr. 16	3 6	SD ±0".14	1888 Mar. 2 0.76 −0.28	−14.9 ±1.4	7.02 ±0.48	.	.	.
1889 IV Davidson	A → B [6]	1889 Aug. 3 1889 Sept. 2	? ~60	.	1889 Jul. 30 1.06 0.00	+11	250	.	.	.
1889 V P/Brooks 2	A → C 112	1889 Aug. 4 1889 Nov. 26	248 358	DRT ±1".57	(1886 Jul. 21) (5.38 +0.09)	(−1168)	6.82 ±0.84	−0.35 ±0.03	−1.26 ±0.06	.
	C → B 20	1889 Aug. 4 1889 Sept. 5	197 272	SD ±1".88	1888 Feb. 10 4.25 −0.22	−598 ±27	22.5 ±1.6	.	.	.
	A → D [1]	1889 Aug. 5 .	333 .	.	(1886 Jul. 21) (5.38 +0.09)	(−1168)	.	.	.	+2.1
	A → E [2]	1889 Aug. 5 1889 Aug. 6	402 437	.	(1886 Jul. 21) (5.38 +0.09)	(−1168)	.	.	.	+4.5
1896 V P/Giacobini	A → B [3]	1896 Sept.26 1896 Sept.28	? ?	.	1896 Apr. 24 2.36 +0.35	−187	.	.	.	.
1899 I Swift	A → B 11	1899 May 12 1899 May 24	12 38	SD ±0".76	1899 Apr. 25 0.48 +0.26	+12.09 ±0.31	41.1 ±1.6	.	.	.
	A → C 4	1899 June 7 1899 June 10	12 17	SD ±0".99	1899 May 28 1.15 +0.53	+44.9 ±1.8	480 ±170	.	.	.

SPLIT COMETS

Comet	Split	Date 1 / Date 2	N1/N2	Type ±	Ref. date	V1	V2	P1	P2	P3
1905 IV Kopff	A→B / 2	1906 Mar. 25 / 1906 Apr. 1	5 / 6	.	1905 Dec. 11 / 3.38 +0.04	+54	260	.	.	.
1914 IV Campbell	A→B / [5]	1914 Sept. 18 / 1914 Oct. 6	19 / 30	.	1914 Aug. 25 / 0.82 −0.59	+20	41	.	.	.
1915 II Mellish	A→B / 17	1915 May 22 / 1915 June 13	32 / 101	SDN ±1″51	1915 Mar. 23 / 2.09 +0.70	−116.6 ±1.8	55.8 ±2.0	.	.	+0.26 ±0.02
	A→C / 38	1915 May 6 / 1915 June 14	18 / 154	SDTN ±1″78	1915 Feb. 26 / 2.38 +1.01	−141.2 ±2.8	57.7 ±2.5	.	+0.23 ±0.06	+0.14 ±0.01
	A→D / 8	1915 June 1 / 1915 June 13	59 / 77	DTN ±1″81	(1915 Feb. 26) / (2.38 +1.01)	(−141.2)	31.3 ±0.5	.	−0.44 ±0.11	+0.34 ±0.03
	A→E / 5	1915 May 24 / 1915 June 7	53 / 120	DTN ±1″38	(1915 Feb. 26) / (2.38 +1.01)	(−141.2)	50.2 ±0.4	.	+0.31 ±0.08	+0.17 ±0.03
1916 I P/Taylor	B→A / 10	1916 Jan. 27 / 1916 Mar. 23	7 / 17	SDTN ±0″59	1915 Dec. 8 / 1.65 −0.27	−53.4 ±8.0	33.9 ±4.8	.	−0.90 ±0.20	−0.47 ±0.09
1943 I Whipple-Fedtke	A→B / 5	1943 Mar. 31 / 1943 Apr. 9	10 / 16	SD ±0″51	1943 Mar. 9 / 1.43 +0.44	+30.8 ±0.8	228 ±16	.	.	.
1947 XII Southern Comet	A→B / 30	1947 Dec. 10 / 1948 Jan. 14	6 / 22	SDN ±0″47	1947 Nov. 30 / 0.15 +0.07	−2.05 ±0.22	4.74 ±0.10	.	.	+1.87 ±0.26
1955 V Honda	A→B / 5	1955 Sept. 21 / 1955 Oct. 19	5 / 7	SD ±0″69	1953 July / 8.2 −3.5	−740 ±260	1.04 ±0.15	.	.	.
1957 VI Wirtanen	A→B / 22+	1957 May 7 / 1959 Sept. 2	7 / 28	SDTRN ±0″56	1954 Sept. 10 / 9.25 −4.97	−1087 ±64	7.10 ±0.75	−0.10 ±0.04	+0.02 ±0.03	−0.24 ±0.01
1965 VIII Ikeya-Seki	A→B / 25	1965 Nov. 5 / 1966 Jan. 14	9 / 58	SD ±1″05	1965 Oct. 21 / 0.008 +0.005	+0.016 ±0.005	0.67 ±0.03	.	.	.
1968 III Wild	A→B / [2]	1968 Nov. 23	4	.	1968 Aug. 3 / 2.92 +1.33	+125	79	.	.	.
1969 IX Tago-Sato-Kosaka	A→B / [2]	1970 Mar. 14	4	.	1970 Feb. 9 / 1.20 +0.20	+50	65	.	.	.
1970 III Kohoutek	A→B / 9	1970 Oct. 31 / 1971 Apr. 1	15 / 34	SD ±1″18	1970 Apr. 29 / 1.79 +0.98	+38.9 ±7.9	28.6 ±3.5	.	.	.
1976 VI West	A→B / 71	1976 Mar. 12 / 1976 Aug. 26	8 / 50	SDTRN ±0″42	1976 Feb. 27 / 0.22 +0.09	+2.47 ±0.23	6.89 ±0.26	+1.44 ±0.20	−0.35 ±0.08	+1.72 ±0.06
	A→C / 12	1976 Mar. 12 / 1976 Mar. 25	2 / 17	SDN ±0″44	1976 Mar. 6 / 0.41 +0.28	+10.29 ±0.34	46.8 ±3.8	.	.	+0.26 ±0.13
	A→D / 89	1976 Mar. 5 / 1976 Aug. 26	3 / 23	SDTN ±0″49	1976 Feb. 19 / 0.30 −0.20	−6.10 ±0.22	4.95 ±0.21	.	−0.87 ±0.08	−0.85 ±0.07

Footnotes to Table I

Column 1: short-period comets with periods about the Sun <200 yr are marked by a prefix P/.

Column 2: the principal component is listed first; $A \to B$, for example, means that the results refer to the motion of B relative to A; the number of measurements is that used in least-squares solution; a bracketed number indicates that another method was used; + indicates that the results are supported by additional evidence.

Column 4: the range of separation refers to the specified period of measurements; ? indicates that no separation distances were reported (see text).

Column 5: relevant only to the least-squares solution. The solutions are identified by letter groups in which S stands for the time of splitting; D for the deceleration of the companion; and R for the radial, T for the transverse; and N for the normal component of its separation velocity.

Column 6: time of splitting t_S listed by date with corresponding heliocentric distance r_S and distance from the ecliptic z_S.

Column 7: time of splitting t_S listed by its difference from the time of perihelion passage T; for short-period comets the T referred to is that of the apparition listed in the first column.

Column 8: net differential effect γ in solar attraction between the companion and primary, the positive sign indicating a deceleration. Adopted units are 10^{-5} the solar gravity, equivalent to 5.93×10^{-6} cm s^{-2}, or 2.96×10^{-9} AU day^{-2}, at a heliocentric distance of 1 AU.

Column 9, 10, 11: the last three columns give V_r, V_t, and V_n, the radial, transverse, and normal components of the companion's velocity of separation from the primary.

Column 6-11: assumed separation parameters (columns 6 through 11) are in parentheses; those assumed to be zero are omitted.

Table I lists only the most preferred solutions among up to a dozen or so made for each of the 32 companions. Selection criteria have included a mean residual as a function of the number of unknowns solved for; a degree of correspondence with observations that could not be used in the least squares (e.g., reports of position angles without separation distances); and a degree of correlation between the individual parameters of the model.

IV. GENERAL CONCLUSIONS

The results of Table I confirm that the differential force has a dominant effect on the observed orientations and recession rates of fragments of the split comets, whether or not a measurable initial velocity of separation is involved. The multiparameter model has removed or reduced the slight systematic trends in the positional residuals of extensively observed split comets and turned a less than satisfactory fit of the companion's motion of Comet Wirtanen 1957 VI into a virtually perfect match. The general quality of fit is illustrated on the companions B, C, and D of Comet West 1976 VI in Fig. 2.

Well-determined decelerations γ range generally between one and ~500 units. They seem to exhibit a bimodal distribution, with one maximum between 5 and 7 units, the other between 30 and 60 units, and the possibility of yet another minor peak between 200 and 300 units. The numbers can be compared with independently determined absolute magnitudes of the non-

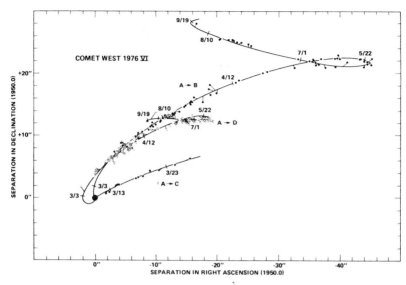

Fig. 2. Separations of the companions of Comet 1976 VI from its principal nucleus (●) in projection onto the sky. Open circles stand for the observations of companion D, dots for the two other nuclei. The curves are the least-squares fits by the multi-parameter model.

gravitational effects in comets. The radial component A_1 of the force is just about 1 unit for Comet Halley (Yeomans 1977) and averages 3 units for short-period comets with well-defined A_1 and 9 units for thirteen nearly parabolic comets with anomalously large deviations from the gravitational law (Marsden et al. 1978). The statistical significance of the peaked γ distribution is unclear.

The nongravitational parameters are known for the principal nuclei of five split comets (Marsden et al. 1973, 1978). The radial components A_1 reduced to the units used in this chapter are listed in Table II. The nongravitational term in Marsden et al.'s equations of motion includes an inverse-square law for the nearly parabolic comets (identical with the law mentioned in Sec. I), but more complicated laws for the short-period comets. The tabulated values for P/Biela and P/Brooks 2 are Marsden et al.'s A_1 at perihelion normalized to 1 AU by an inverse-square law. The absolute radial decelerations of the secondary nuclei are then simply given by $A_1 + \gamma$. Unfortunately, the determinacy of A_1 from the orbital solutions is always rather poor, the uncertainty generally exceeding the formal error; obviously, small relative decelerations γ do not provide much information on the absolute decelerations of such fragments. However, since A_1 does not appear to exceed ~ 10 units, relative decelerations $\gamma \gtrsim 40$ units cannot differ substantially from the absolute decelerations and are therefore more meaningful from the standpoint of a physical interpretation.

TABLE II
Absolute Radial Nongravitational Parameter of the Principal Nuclei of Five Comets

Comet	A_1 (units)
P/Biela (1832-1852)	5.9 ± 0.9
P/Brooks 2 (1889-1904)	5.8 ± 0.6
Swift 1899 I	9.8 ± 1.4
Mellish 1915 II	9.5 ± 0.3
Honda 1955 V	5.1 ± 2.7

For the sungrazing Great September Comet 1882 II and Comet Ikeya-Seki 1965 VIII, which broke up at or very near the perihelion point, the deceleration decreases with heliocentric distance so rapidly that it is virtually impossible to discriminate between its effect and that of a transverse component of the separation velocity. The equivalence of the two, demonstrated by a comparison of the solutions SD and ST (Sekanina 1978), is consistent with the relation between γ and V_t obtained by differentiating the energy integral

$$V_t = 0.149 \, \gamma \left(\frac{1+e}{q}\right)^{1/2} \tag{1}$$

where e is the eccentricity of the orbit, q its perihelion distance in AU, γ is in the adopted units (Sec. III), and V_t in m s^{-1}.

On the other hand, the absolute magnitude of the deceleration for fragments of the split comets of large perihelion distance is very weak. As found from the solutions for more extensively observed split comets, the systematic variation of the separation velocity V with heliocentric distance r (in AU) follows a law (Fig. 3)

$$V = B r^{-b} \tag{2}$$

where $B = 0.70 \pm 0.09$ m s^{-1} and $b = 0.57 \pm 0.10$. Velocity V is thus decreasing much less steeply than γ. As a result, the effect of V is, on the average, at least 100 times as important, relative to the effect of γ, at 5 AU from the Sun as it is at 0.2 AU. This explains why it has been so essential for a successful fit to the nuclei of the distant Comet 1957 VI to solve for the separation velocity, although its magnitude has turned out to be only 26 cm s^{-1}.

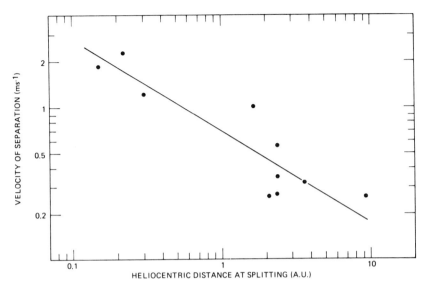

Fig. 3. Velocity of separation versus heliocentric distance for secondary nuclei of the split comets. Only well-determined values are included.

Apparent deviations from the coplanarity of fragments can be used to derive a relation between the point of breakup and the normal component of the separation velocity, because the line of nodes of the orbits of the nuclei is determined by the time t_s. The separation ζ of the companion from the orbit plane of the principal fragment at time t may, with sufficient accuracy, be written as

$$\zeta = V_n p^{-1/2} r_s r \sin(v-v_s) \qquad (3)$$

where r_s and v_s are the heliocentric distance and true anomaly at breakup, r and v at time t, and p the semilatus rectum of the orbit (Sekanina 1978). The off-plane separation follows a sine curve whose amplitude is modulated by the heliocentric distance at observation. One has ζ and r in the same units, when r_s and p are in AU and V_n in units of the Earth's mean orbital velocity or 29.785 km s^{-1}.

Equation (3) has been employed to estimate V_n of companions D and E of the low-inclination periodic Comet Brooks 2 from their isolated measurements; to check on V_n of the various nuclei of Comet 1915 II from the observations made near the Earth's transit across the comet's orbit plane; to perform a similar test for Comet West 1976 VI; and to confirm both r_s and V_n of Comet Wirtanen 1957 VI from the observations made near four of the Earth's transits across its orbit plane in 1957-59 (Sekanina 1978).

The distribution of the times of splitting relative to perihelion and the related distributions of r_s and z_s are, by and large, random. Only three

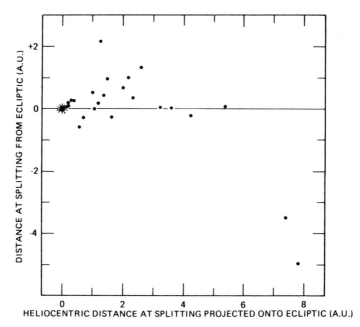

Fig. 4. Distance from the ecliptic versus heliocentric distance projected onto the ecliptic for the points of splitting. The points for the sungrazing comets fall within the Sun's mark on the scale of the figure.

comets, the two sungrazers and P/Brooks 2, broke up due possibly to tidal action of the Sun or Jupiter. Comets seem to split about as often before as after perihelion and ~50% of them break up at heliocentric distances smaller than twice the perihelion distance. Also, about equal numbers of comets split at distances from the ecliptic, as measured from the Sun, that are greater and smaller than is an average deviation given by one half the orbit inclination. The only systematic trend seen in the distribution of the points of splitting is their clear predominance to the north of the ecliptic, an effect most likely due to observational selection. A plot of the points of splitting projected onto the plane normal to the ecliptic and passing through the Sun and each comet's position at breakup is presented in Fig. 4. The breakup of Comet 1957 VI on its first approach to the Sun at a distance comparable with that of Saturn highlights these results, having far-reaching implications for theories of the structure of cometary nuclei.

P/Biela is the only comet whose fragments have been observed at more than one return to the Sun. It has been of considerable interest to learn that while it has not been possible to link the 1852 observations of the companion with its 1846 observations for any combination of t_s, γ, V_r, V_t, and V_n, the solution that fits perfectly the 1846 positions can represent the 1852 positions satisfactorily when the value of γ is approximately doubled.

For most split comets only two fragments have been observed. Although a third fragment has been reported for two more in addition to the five comets listed in Table I (notably 1888 I and 1965 VIII), the positional data for these are insufficient to derive the separation parameters. If any tendency at all can be inferred from a sample of five cases, the comets with multiple nuclei suggest a possible correlation between the times of splitting and the relative decelerations. As illustrated best by comets 1976 VI, 1899 I, and P/Brooks 2, the fragment with the least deceleration appears to have separated first, the one with the largest deceleration last. Although the assumption of nucleus B separating from nucleus D of Comet 1976 VI satisfies the observations practically equally well as the adopted sequence of splitting, P/Brooks 2 has so far been the only comet providing clear-cut evidence for a companion from a previous breakup to become the principal fragment in a subsequent breakup. The initial splitting has been assumed to coincide with the comet's grazing approach to Jupiter on July 21, 1886 after least-squares solutions involving t_s have indicated the likelihood of coincidence of the two events, but the long time between breakup and observations (more than three years) has caused t_s to be rather poorly determined. The involvement of the companion (nucleus C) in the secondary breakup is certain to have slightly affected the fragment's subsequent motion and could have contributed to the less than satisfactory determinacy of the time of the primary breakup. Its adopted timing can be justified by the presumed role of Jupiter's tidal forces (Sec. X). The assumption of coincidence of the two events has clearly had a beneficial effect on the determinacy of the other separation parameters. Similarly, the assumption of a simultaneous separation of fragments C, D, and E of Comet Mellish 1915 II has made it possible to improve the solutions for D and E, the two less well-observed companions.

Finally, the data in Table I suggest that fragments with smaller γ have generally been observed longer. To account for the variation of the sublimation rate with heliocentric distance r, an effect that complicates the comparison of fragments at various distances from the Sun, an endurance E of a fragment has been defined as an interval of time from breakup t_s to its final observation t_f, normalized by the inverse-square law to 1 AU from the Sun (Sekanina 1977a). Expressed in equivalent days, the endurance measures a minimum sublimation lifetime of a companion and is given by

$$E = \int_{t_s}^{t_f} \frac{dt}{r^2} = 1.015\, p^{-1/2}\, A_{sf} \qquad (4)$$

where p is the semilatus rectum of the fragment's orbit in AU and A_{sf} the length of the heliocentric arc of orbit in degrees, swept by the fragment between t_s and t_f. In a few cases where the companion has been seen after perihelion for as long as the primary component, t_f has rather arbitrarily been assumed identical with the time of passage through the subsequent aphelion.

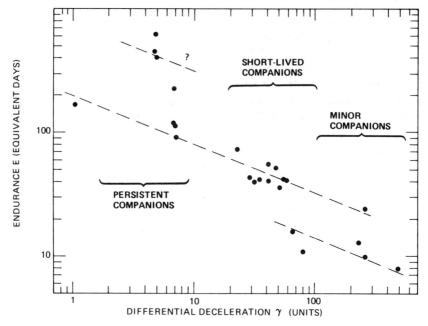

Fig. 5. Endurance E versus differential deceleration γ for the companions of the split comets. This plot is used for the classification of the companions into three categories.

The endurance is plotted versus the deceleration γ in Fig. 5. In spite of large scatter, the inverse correlation between the two independent quantities is striking. There is some indication that the data in Fig. 5 tend to concentrate along three lines, of which the best defined fits a relation

$$E = C\gamma^{c} \qquad (5)$$

where $C = 200 \pm 25$ equivalent days and $c = 0.40 \pm 0.04$, with γ again in the adopted units. The two other groups run nearly parallel, one a factor of ~ 4 above, the other a factor of 2.3 below it.

V. CLASSIFICATION OF COMETARY FRAGMENTS AND THE INTERPRETATION OF THE RELATIVE DECELERATION

The clustering of relative decelerations around particular values in Table I suggests the existence of discrete categories of fragments, while the correlation between the deceleration and the endurance in Fig. 5 indicates that either quantity can be used as a taxonomic parameter.

The first category encompasses *persistent companions*, whose decelerations are always smaller than 10 units and whose endurances are characteristically in excess of 80 equivalent days. The second category consists of *short-lived companions* with decelerations between 20 and 100 units and endurances between 10 and 100 equivalent days. The last category is that of *minor companions*, for which decelerations exceed 100 units, reaching up to ~500 units, and the endurances are $\lesssim$ 30 equivalent days. Decelerations $\gtrsim$ 500 units always turn out to refer to transient phenomena that might on occasions look like nuclei but must differ from them physically. In particular, a bulk of them come from observations with relatively small telescopes and are usually identified with developing jets on high-resolution photographs.

The available sample of the split comets indicates that the short-lived companions must be more numerous than the persistent ones. Considering the difficulties of their detection, minor companions are likely to represent a far more common cometary phenomenon than observations seem to suggest.

Of the 21 split comets in Table I, 19 have sufficiently well-determined orbits to be classified into one of the four categories that Oort and Schmidt (1951) have introduced to express the dynamical age of comets: 5 are new; 3, fairly new; 7, including the two sungrazers, old; and 4, short-period. The fraction of old, periodic comets, though not overwhelming, is decidedly significant. Except for the lack of contribution from the short-period comets to the population of minor companions, no apparent correlation is found between the three categories of companions and the Oort-Schmidt classification.

To examine the relationships among the endurance of a companion, its differential nongravitational deceleration, and the bulk properties of the parent comet, one may write

$$E = \rho \, \Delta R / Z \tag{6}$$

and

$$\gamma = j_{\text{companion}} - j_{\text{principal}} \tag{7}$$

where Z is an average mass vaporization rate per unit surface area, ρ the bulk density, assumed to be constant throughout the parent comet, ΔR an average thickness of the layer of ice evaporated during the time E, and j the net force per unit mass of the fragment exerted by the momentum from nonuniform outgassing on the fragment. Quantities Z and j refer to 1 AU from the Sun. Further,

$$j = \frac{\xi u Z A}{Q \rho} \tag{8}$$

where A and Q are, respectively, the surface area and volume of the fragment, u the initial efflux velocity of the gas, and ξ the momentum transfer coefficient ($\xi < 1$), which accounts for the effects of sublimation anisotropy, projected components of individual molecular impulses into the radial direction, etc. Let all the quantities on the right-hand side of Eq. (8) except for A and Q be the same for both fragments. Since the ratio A/Q varies with time as the surface layer of each fragment evaporates away, one should consider an average value of the deceleration over the time span E:

$$<j> = -\frac{1}{\Delta R} \int_R^{R-\Delta R} j \, dR . \qquad (9)$$

The net effective γ then becomes

$$\gamma = \mu [<(A/Q)_{\text{companion}}> - <(A/Q)_{\text{principal}}>] \qquad (10)$$

where the bracketed expressions are averages over the time E and

$$\mu = \xi u Z/\rho . \qquad (11)$$

As the simplest model one may assume that both the principal nucleus and the companion are spherical, in which case

$$\gamma = \frac{3\mu}{\Delta R} \log_e [(1 - \Delta R/R_{\text{principal}})/(1 - \Delta R/R_{\text{companion}})] \qquad (12)$$

where the subscripted R is the radius of the corresponding fragment at the time of splitting. An order-of-magnitude evaluation of the expressions for γ and E shows that this model has considerable problems especially with the minor companions. Since Z cannot exceed $\sim 10^{-4}$ g cm^{-2} s^{-1} (at 1 AU from the Sun), $\Delta R \lesssim 10^2$ cm from Eq. (6) at a unit density ρ for fragments with an endurance of $\sim 10^6$ equivalent seconds. In order to be subjected to a deceleration of $\sim 10^{-3}$ cm s^{-2} at 1 AU relative to the principal nucleus whose $R_{\text{principal}} \gg \Delta R$, the companion would have to have an initial radius $R_{\text{companion}} \simeq 10 \Delta R \lesssim 10^3$ cm. It is inconceivable that the brightness of a fragment of such dimensions could even remotely or at times be comparable to that of the principal nucleus (of an assumed $R_{\text{principal}} \simeq 10^5$ cm), as its surface area would be more than 10^4 times smaller. In fact, such a fragment would most probably be totally unobservable.

It turns out that a much better model can be offered, based on the assumption of a spherical *parent* nucleus that cracks along a planar section, producing two fragments shaped like spherical segments. Their geometry is fully described by the radius R of the parent nucleus and by the distance X of the plane of fissure from the center of the sphere. The minimum and maximum dimensions of the companion are, respectively, $(R - X)$ and $2(R^2 - X^2)^{1/2}$ at the time of splitting, and $(R - X - 2\Delta R)$ and $2(R + X)^{1/2}(R - X - 2\Delta R)^{1/2}$ at the time of last detection, that is, E equivalent days after splitting. Introducing dimensionless quantities

$$x = X/R \text{ and } \Delta z = \Delta R/R \qquad (13)$$

one can, after some algebra, write Eq. (10) as follows:

$$\gamma = \frac{\mu}{\Delta R} \log_e \left(\frac{1 - \dfrac{\Delta z}{2-x}}{1 - \dfrac{2\Delta z}{2+x}} \right) \left(\frac{1 - \dfrac{2\Delta z}{2+x}}{1 - \dfrac{2\Delta z}{1-x}} \right)^2 . \qquad (14)$$

For $\Delta z \ll 1$ this expression is independent of ΔR and takes a form

$$\gamma = \frac{6\mu}{R} \frac{x(5-x^2)}{(1-x^2)(4-x^2)} . \qquad (15)$$

For $x \gtrsim 0.5$ an excellent approximation is

$$\gamma = \frac{8\mu}{R} \frac{x}{1-x^2} . \qquad (16)$$

With the use of plausible numerical values $R = 1$ km, $\rho = 1$ g cm^{-3}, $u = 0.3$ km s^{-1}, $Z = 5 \times 10^{-5}$ g cm^{-2} s^{-1}, and $\xi = 0.2$, Table III shows the dramatic disparity between the two models for all three categories of companions. The fundamental difference is that the model of nonspherical fragments can achieve the same relative deceleration with much less unevenly distributed mass between the components and also with a surface-area ratio that is more compatible with observations (Sec. IX). The increasing degree of nonsphericity as one proceeds from persistent to minor companions is of course a byproduct of the model. Yet, in the light of the above results, one is inclined to speculate that at least the short-lived and minor companions could be products of parent nuclei that have a tendency to "peel off" rather than break up.

VI. A TEST OF SPLITTING AND COMETS WITH UNCONFIRMED MULTIPLE NUCLEI

The two-parameter model proves most useful in cases of poorly observed split comets. In the relatively narrow range of deceleration γ (compared to the range encountered in dust tails) the loci of constant time of splitting t_s are almost exactly straight lines. Consequently, the position angle of the companion relative to the principal mass measures directly t_s, whereas the separation distance at a given position angle determines the magnitude of the deceleration. A single differential measurement of the companion is thence sufficient to provide both parameters. However, while the position angle is *all* one needs to derive t_s, the separation distance *by itself* supplies no useful information whatsoever.

TABLE III

Segment-shaped and Spherical Models of
Fragments of the Split Comets

Quantity	Relative Deceleration γ (units)		
	7	40	300
Endurance E (equivalent days)	92	46	20
Evaporation loss of surface layer ΔR (km)	0.0040	0.0020	0.0009
Segment-shaped fragments			
Companion's minimum dimension (km)			
Initial	0.25	0.049	0.0067
Final	0.24	0.045	0.0049
Companion's maximum dimension (km)			
Initial	1.32	0.62	0.23
Final	1.30	0.59	0.20
Principal-to-companion surface area ratio			
Initial	4.23	20.6	150
Final	4.35	22.4	204
Principal-to-companion mass ratio			
Initial	22.6	564	$10^{4.47}$
Final	23.9	665	$10^{4.74}$
Spherical fragments			
Companion's diameter (km)			
Initial	0.36	0.073	0.010
Final	0.35	0.069	0.008
Principal-to-companion surface area ratio			
Initial	31.4	748	$10^{4.60}$
Final	32.6	834	$10^{4.77}$
Principal-to-companion mass ratio			
Initial	176	$10^{4.31}$	$10^{6.89}$
Final	186	$10^{4.38}$	$10^{7.15}$

An unconfirmed observation of a secondary nucleus is always suspect. The existing records of comet observation demonstrate that there are at least three major categories of causes for false reports of split comets: (a) coma or tail phenomena whose appearance resembles secondary nuclei, especially when seen with the aid of only a small telescope; (b) optical defects (ghost images) and instrumental and other technical problems (plate flaws, guiding difficulties); and (c) human errors (confusion with noncometary nebulous objects or with blurred images of stars near the horizon).

The two-parameter model is invaluable in that it provides rather a reliable test of splitting (Sekanina 1979a), based in part on the conservation of momentum law requirement already mentioned in Sec. I, in part on the empirical evidence summarized in Sec. IV and on the classification of fragments discussed in Sec. V. If **RV** is the prolonged radius vector and **−V** the reverse orbital velocity vector, the first criterion requires that the position angle of the companion measured from the principal nucleus satisfy a condition

$$P(-\mathbf{V}) < P_{\text{companion}} < P(\mathbf{RV}) \qquad (17)$$

when the Earth is to the north of the comet's orbit plane (from where the comet is seen to orbit the Sun counterclockwise); and

$$P(\mathbf{RV}) < P_{\text{companion}} < P(-\mathbf{V}) \qquad (18)$$

when the Earth is to the south of the plane. This criterion is particularly restrictive for observations made before perihelion, although the range of allowed position angles always spans less than 180°. Only at times of the Earth's transit across the comet's orbit plane is this test either of limited use or none whatsoever, depending on the configuration of the comet relative to the Earth and Sun.

The second criterion, applicable if the separation distance is available, requires that the relative deceleration, derived with the use of the known t_s, not exceed ∼500 units (Sec. V). Finally, with both t_s and γ established, the third criterion tests the formal value of the endurance E, as measured by the time of observation. For a true split comet E should be near the value expected from Fig. 5. Significantly, the application of the test to instances of questionable and/or unconfirmed reports of multiplicity gives, with a few exceptions involving large telescopes, almost invariably negative results (Sekanina 1979a).

VII. GRAVITATIONAL INTERACTION BETWEEN THE FRAGMENTS AND THE PROBLEM OF INTERPRETATION OF THE SEPARATION VELOCITY

One may be tempted to relate the velocity of separation determined from observations to the comet's mass or to use it at least as a measure of an upper limit to the mass. The first attempt to establish the mass of a split comet appears to be due to von Hepperger (1906). Unfortunately, because of the gravitational interaction between the fragments for a short but significant time after separation, there is no straightforward relationship between the two quantities.

The gravitational interaction between two comet fragments in the Sun's field of attraction has been studied numerically as a general three-body problem (Sekanina 1979b). The fragments have been launched at a particular point in the heliocentric orbit of Comet 1976 VI with an initial rate of separation equal to the velocity of escape. The numerical integration of their motions has been carried over the period of 200 days and the following initial conditions have been allowed to vary:

1. Position of the comet in heliocentric orbit at separation;
2. The mass of the principal fragment;
3. The principal-to-companion mass ratio;
4. The bulk density;
5. The initial direction of the companion's motion relative to the principal;
6. The location of the companion's separation on the parent nucleus.

The adopted directions of motion and the separation locations are defined by the six cardinal directions of the coordinate system tied to the orbit plane of the comet and to the Sun-comet direction, the orbital motion of the comet providing the definitions of the forward direction and the leading side of the nucleus.

The variations of the separation distance of the fragments, of their relative velocity, and of the angle subtended by the separation and velocity vectors have been examined as functions of time for 132 different patterns. It has been established that the differential gravitational effect of the Sun becomes noticeable about 0.5 to 1 day after separation at perihelion, but 1 to 2 days after separation at a heliocentric distance twice the perihelion distance, when the comet is 9.5 days before or after perihelion. A great diversity of the companion's dynamical behavior has been found, which illustrates the significance of the circumstances at breakup. The calculated patterns range all the way from escape of the companion along strongly hyperbolic trajectories (relative to the principal fragment, not the Sun) to pursuance of quasi-stable periodic orbits about the principal mass, terminated in exceptional cases by collision of the fragments. Their relative velocity always decreases with time rapidly in the first several hours or so after separation, but starts experiencing complicated variations, once the distance between the nuclei has increased to 10 to 100 km (depending on the mass) and the Sun's perturbations have become more prominent. Figure 6 shows the variations in the perturbed relative velocity of two fragments versus time under six different initial conditions. Table IV summarizes the qualitative character of the relative motion of the fragments after 200 days in orbit. The aspects classified are:

1. The actual energy relative to the binding energy, resulting in either a periodic motion or escape, the latter being divided into vigorous and hesitant depending on the magnitude of the excess velocity over the escape limit;

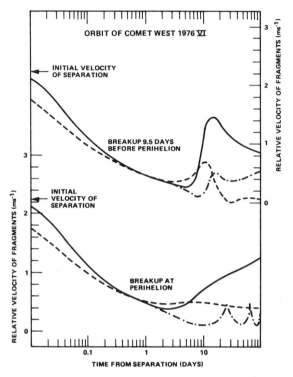

Fig. 6. The relative velocity of two fragments versus time after breakup, derived from theory of a gravitational interaction among the Sun, the principal nucleus, and the companion. The adopted constants and initial conditions: the mass of the primary, 10^{17} g; the primary-to-secondary mass ratio, 100:1; the bulk density, 1 g cm^{-3}; the initial separation velocity equal to velocity of escape; and the direction of separation of the companion, sunward. The solid curves refer to the separation from the leading side of the parent; the dashed curves, from the subsolar location; and the dash-and-dot curves, from the north pole.

2. The inclination of the companion's motion relative to the heliocentric orbit plane of the principal mass (regarded as high when more than 30°);
3. The sense of motion of the companion about the principal fragment.

Only eleven prime patterns are listed, the rest being symmetric or nearly symmetric with respect to one of the eleven.

Since the strongest interaction between the nuclei of a split comet takes place at a close range that is orders of magnitude below the resolution threshold of positional observations, the separation velocity provided by the application of the multiparameter model of Sec. II measures the rate of recession of the fragments at the time of their dynamical separation, which should amount to only a fraction of the rate of recession acquired upon their physi-

TABLE IV

Prime Patterns of the Gravitational Interaction between Two Fragments in the Orbit of Comet West 1976 VI (No Nongravitational Forces Included)

Location of Separation on Parent Nucleus	Initial Direction of Companion Motion	Character of Relative Motion of Companion after 200 days		
		Breakup at True Anomaly of $-90°$ (9.5 days before perihelion)	Breakup at Perihelion	Breakup at True Anomaly of $+90°$ (9.5 days after perihelion)
Subsolar	sunward	periodic orbit zero inclination retrograde motion	vigorous escape zero inclination prograde motion	vigorous escape zero inclination prograde motion
	forward	hesitant escape zero inclination retrograde motion	vigorous escape zero inclination prograde motion	vigorous escape zero inclination prograde motion
	backward	periodic orbit[a] zero inclination retrograde motion	hesitant escape zero inclination prograde motion	vigorous escape zero inclination prograde motion
	northward	periodic orbit high inclination retrograde motion	vigorous escape low inclination prograde motion	vigorous escape low inclination prograde motion
Leading side	sunward	vigorous escape zero inclination prograde motion	vigorous escape zero inclination prograde motion	vigorous escape zero inclination radial motion

Trailing side	forward	vigorous escape, zero inclination, prograde motion	vigorous escape, zero inclination, prograde motion	vigorous escape, zero inclination, radial motion
	northward	vigorous escape, low inclination, prograde motion	vigorous escape, low inclination, prograde motion	vigorous escape, low inclination, radial motion
	sunward	vigorous escape, zero inclination, prograde motion	vigorous escape, zero inclination, prograde motion	periodic orbit, zero inclination, retrograde motion
North pole	sunward	vigorous escape, high inclination, prograde motion	periodic orbit, high inclination, prograde motion	periodic orbit, ~90° inclination, —
	forward	vigorous escape, high inclination, prograde motion	periodic orbit, high inclination, prograde motion	periodic orbit, ~90° inclination, —
	northward	collision	collision	collision

[a]Or collision, depending on bulk density.

cal breakup. Thus, by inference, the initial velocity might be estimated as much as several meters per second for some comets.

The existence of gravitationally bound orbits among the solutions of Table IV does not necessarily mean that the dynamical separation of fragments is not achieved in some cases. It is unlikely that the attraction between the fragments could withstand any differential nongravitational perturbation over a prolonged period of time, even if the initial velocity of separation were somewhat below the parabolic limit. As a result, the existence of gravitationally locked multiple nuclei is at best only very remotely possible.

VIII. CORRELATION BETWEEN SPLITTING AND ERUPTIVE PHENOMENA IN COMETS

If a splitting of the parent nucleus is accompanied by an instantaneous emission of debris observable from the Earth, the time difference between the physical breakup and the dynamical separation, if appreciable, could be tested observationally. In practice it has been possible to pursue this idea with only marginal success.

Whipple (1963) has remarked that whether or not a splitting is accompanied by a brightening of the comet depends on the nature of the disruption mechanism (Sec. X). Considering that at least some aspects of the splitting are now understood, a systematic search for possible correlations with the various signatures of cometary burst activity is clearly justified. Efforts have so far been aimed at brightness flare-ups and at bursts of solid particles evidenced by discrete streamers in the dust tails. Although the progress has been hindered by a lack of relevant observations, some results have recently been obtained (Table V). The duration of the dust bursts is estimated from the breadth of the streamers at not more than ~0.1 day on the average, which is within the uncertainty of observations. The duration of brightness flare-ups is usually much longer and Table V lists the projected times of their onset rather than the times of peak brightness. Also given are estimated amplitudes of the flare-ups.

By far the strongest evidence of correlation is that for the primary breakup of Comet 1976 VI involving the nuclei A and D. Sekanina and Farrell (1978) have shown that the splitting was accompanied both by a flare-up of the total visual brightness (Fig. 7) and by a substantial short-term increase in dust emission, the first of a dozen bursts detected in the course of 2.5 weeks near perihelion (Sekanina 1980a). The event should also have been associated with an increased infrared emission, but observational evidence is in this case less conclusive. The time of a subsequent breakup, which gave birth to nucleus B, falls formally almost exactly midway between two bursts of dust separated from each other by 0.8 day. On the other hand, there is no convincing evidence of any unusual activity connected with the separation of the short-

TABLE V
Correlation of Splitting with Eruptive Events

Comet	Fragments	Companion	Time of Splitting (UT)	Time of Possibly Related Event[a] (UT)	Type of Event
1899 I	A → C	minor	1899 May 28 ± 2	1899 June 1 ± 3	2 mag flare-up
1914 IV	A → B	short-lived	1914 Aug. 25 ± 3	1914 Aug. 18 ± 1	dust burst
1943 I	A → B	minor	1943 Mar. 9.5 ± 0.8	1943 Mar. 5 ± 3	< 0.5 mag brightening
1969 IX	A → B	short-lived	1969 Feb. 9 ± 6	1969 Feb. 6 ± 1 / 1969 Feb. 6.4	1.4 mag flare-up / expanding halo
1976 VI	A → D	persistent	1976 Feb. 19.1 ± 0.2	1976 Feb. 19.4 ± 0.4, 1976 Feb. 19	dust burst / 2 mag flare-up
	A → B	persistent	1976 Feb. 27.7 ± 0.2	1976 Feb. 27.2 ± 0.2 / 1976 Feb. 28.0 ± 0.1	dust bursts
	A → C	short-lived	1976 Mar. 6.5 ± 0.3	1976 Mar. 5.2 ± 0.3 / 1976 Mar. 7.7 ± 0.2	dust bursts

[a] Time of onset of the event, not peak brightness time.

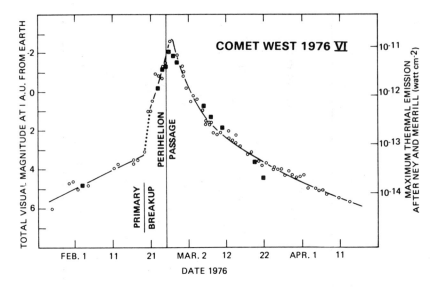

Fig. 7. The lightcurve of Comet 1976 VI. Total magnitudes, estimated with binoculars or the naked eye, are plotted as open circles; the scaled measurements of the maximum thermal emission reported by Ney and Merrill (1976) are plotted as squares. (From Sekanina and Farrell 1978.)

lived nucleus C. The correlation of events is also poor, if not absent, for the minor fragment separating from Comet Whipple-Fedtke-Tevzadze 1943 I. However, this comet had undergone two huge brightness outbursts (Beyer 1948) that peaked 7 and 2 weeks before the splitting. An argument that increased activity is only associated with the formation of persistent companions is invalidated by most entries in Table V and especially by the case of Comet Tago-Sato-Kosaka 1969 IX, where positive evidence is extraordinarily strong in spite of some uncertainty in the time of splitting (for details, see Sekanina [1979a]).

In short, some of the nucleus breakups have undoubtedly been associated with sharply augmented cometary activity with a particularly strong dust signature. The results show no systematic trend in the timings of the events, suggesting that the time of dynamical separation does not differ from the time of physical breakup by more than is the uncertainty in the data. For other splittings only marginal, if any, evidence of such a coincidence of events exists.

IX. PHYSICAL BEHAVIOR OF THE SECONDARY NUCLEI

Physical observations of secondary nuclei of the split comets are very fragmentary and limited to qualitative descriptions of the appearance and to brightness estimates. A reported brightness may refer to the total light, if the components are widely separated and their heads do not overlap. However, the estimates are more often likely to refer to the brightness of the central condensation only. To minimize the effects of heterogeneity of observational data, it is preferable to adopt the principal nucleus as a standard and to refer the lightcurves of companions to that of the main mass. Also for the sake of uniformity, the relative brightness should be plotted versus exposure time, that is, the time interval from splitting weighted by inverse square of heliocentric distance. Such lightcurves are shown for four persistent companions in Fig. 8, for five short-lived and minor companions in Fig. 9.

The lightcurves exhibit profound fluctuations, whose amplitudes indicate intensity variations of up to 20:1 or even more. Although the fluctuations are by no means periodic, a characteristic separation of 10 to 20 equivalent days between two successive peaks is suggested by the four persistent companions and by the nucleus C of Comet 1976 VI. This characteristic spacing appears to be entirely independent of heliocentric distance. Since the exposure time is normalized to 1 AU from the Sun, the fragment brightness actually fluctuates faster when closer to the Sun. For Comet 1976 VI a temporary inverse correlation of lightcurves of the companions B and D was detected by Dawson et al. (1977): when B became brighter, D grew fainter, and vice versa.

Other characteristic physical properties of the secondary nuclei have been perceived. The persistent companions have a tendency of gradually gaining in brightness relative to the principal mass with increasing exposure time before the final fading sets in. The endurance of the other companions may not be long enough to show this effect, should it exist. Most secondary nuclei, particularly the short-lived ones, have, for some time before fading out of sight, repeatedly exhibited a noticeable elongation along the line with the principal nucleus, a progressive expansion of their dimensions, and the lack of central condensation. As shown elsewhere (Sekanina 1980b), this behavior is remarkably similar to that of the so-called dissipating comets, a small class of objects that have disintegrated literally before the eyes of observers.

Although no physical model of cometary fragments can be formulated on the basis of these results alone, it may be significant that the model of strongly nonspherical secondary nuclei suggested in Sec. V can in principle explain most of the listed peculiarities. It is reasonable to assume that the area of the section, along which the companion broke off from the parent, will provide a major, if not dominant, contribution to the fragment's gas production, both because it is a freshly exposed surface of presumably volatile substances formerly protected inside the parent nucleus against solar radiation, and because it may represent a significant fraction of the total surface area of the fragment. Whenever this side is turned to the Sun, the outgassing should be maximum. Because of the apparent dependence of the length of intervals between brightness peaks on heliocentric distance, the light variations cannot be explained by rotation. On the other hand, a strongly nonspherical, subkilometer-sized fragment could be forced by the torque

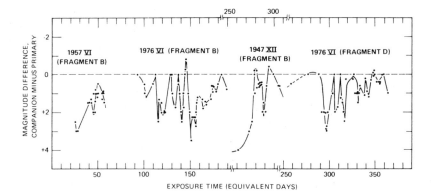

Fig. 8. The lightcurves of four persistent companions. The magnitude difference between the companion and the primary is plotted versus exposure time to account for variations with heliocentric distance.

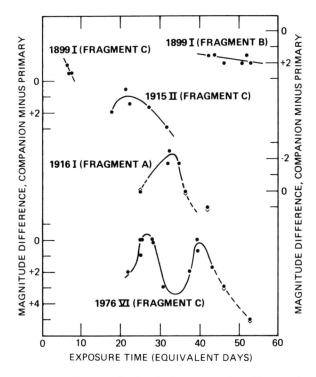

Fig. 9. The lightcurves of five short-lived and minor companions. The magnitude relative to the primary is again plotted versus exposure time of the companion.

from asymmetric outgassing to tumble at a high precession rate. An oblate fragment's spin axis should soon become nearly normal to the surface of the fresh side of the body and the brightness extrema must occur at times when the precession brings the spin axis into alignment with the Sun-comet direction, thus either fully exposing the fresh surface or hiding it for a while. The pattern of light variations has no strict periodicity and no stable amplitude apparently because of the continuously changing factors that determine the precession rate, such as the magnitude and direction of the torque, the shape and dimensions of the body, the orbital geometry, etc.

To find out whether the assumed fast spin-axis motion is in fact dynamically plausible, one may consider a precessing spheroid which rotates about the axis of maximum moment of inertia, a simple model that Whipple and Sekanina (1979) applied successfully to periodic Comet Encke. Its characteristic precession period Π is given by

$$\Pi = \frac{8\pi^2 R_{eq}^2}{5\hat{P}<\ell>j} \tag{19}$$

where R_{eq} is the equatorial radius of the spheroid, P its rotation period, $<\ell>$ an average torque arm, and j the force from asymmetric outgassing per unit mass of the fragment. Obviously, the precession rate generally increases with decreasing heliocentric distance, in accordance with the observed pattern of brightness fluctuations. In the following, Π and j are normalized to 1 AU from the Sun, in which case Π is to be compared directly with the characteristic spacing of brightness peaks in Figs. 8 and 9. The ratio of an average torque arm to the equatorial radius of a randomly oriented spheroid can be defined as

$$<\ell>/R_{eq} \equiv f(\alpha) = \frac{2\alpha}{\pi}(1 - \tfrac{1}{2}\alpha)\int_0^{\pi/2} \frac{\sin 2\psi \, d\psi'}{[1 - 2\alpha(1 - \tfrac{1}{2}\alpha)\sin^2\psi]^{1/2}} \tag{20}$$

where α is the spheroid's oblateness, i.e., a complement to unity of the ratio between the polar and equatorial radii (or diameters), and ψ' and ψ are, respectively, the angular distance from the equator and the angle between the normal to the surface and the equator, of the net vector of the momentum from outgassing. Since

$$\tan \psi' = (1 - \alpha)^2 \tan \psi \tag{21}$$

one can write Eq. (20) as follows:

$$f(\alpha) = \frac{1-\alpha}{\pi(2 - 2\alpha + \alpha^2)^{1/2}} \log_e \left[\frac{1 - \alpha + \alpha^2 + \alpha(2 - 2\alpha + \alpha^2)^{1/2}}{1 - \alpha + \alpha^2 - \alpha(2 - 2\alpha + \alpha^2)^{1/2}}\right]. \tag{22}$$

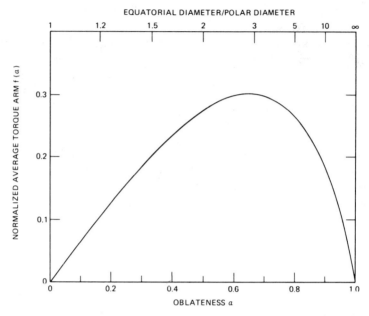

Fig. 10. The normalized average torque arm $f(\alpha)$ versus oblateness α of a spheroid.

Figure 10 shows that this function attains a maximum value of 0.30 at $\alpha =$ 0.65, decreasing steeply for highly oblate spheroids. For an assumed α, R_{eq} (in km), Π (in equivalent days), and $j = A_1 + \gamma$ (in the adopted units) the required rotation period (in days) becomes

$$P = 35.7 \frac{R_{eq}}{(A_1 + \gamma) \Pi f(\alpha)} . \qquad (23)$$

In order to test Eq. (23) on the nonspherical model of short-lived and persistent fragments suggested in Sec. V, it is essential that the error of conversion of the segment-shaped model to the spheroidal model be minimized. This can be accomplished by requiring that the two figures be equivalent in mass and in maximum moment of inertia. Under these assumptions the fragment dimensions listed in Table III give an equatorial radius 0.28 km and an oblateness 0.92 for short-lived companions ($\gamma = 40$ units, $\Pi = 10$ equivalent days), and $R_{eq} = 0.61$ km, $\alpha = 0.81$ for persistent companions ($\gamma = 7$ units, $\Pi = 15$ equivalent days). With an adopted $A_1 = 3$ units the relation Eq. (23) is satisfied for short-lived companions whose $P = 3.5$ hr and for persistent companions whose $P = 13.5$ hr. Although only statistically meaningful, these rotation periods are generally consistent with Whipple's (see his chapter in this book) distribution of cometary spin rates.

The interpretation of the brightness fluctuations as a result of variable outgassing due to a lopsided distribution of volatiles over the surface of a rapidly precessing, highly oblate, subkilometer-sized comet fragment appears to be a feasible hypothesis. Perhaps significantly, the inferred characteristic rotation period of short-lived companions turns out to be comparable to, it not shorter than, a critical period that may suffice to cause a splitting (Sec. X). The temporary inverse correlation of the lightcurves of the companions B and D of Comet 1976 VI can now logically be explained by assuming that the two fragments have had temporarily comparable precession rates and active sides turned to the Sun with a nearly $180°$ shift in phase. The ultimate disappearance of a fragment may either mean its nearly complete disintegration, which could be accelerated by stresses that build up in the body as a result of the precession torque, or it may merely signal the depletion of the reservoir of ice on its most active side. The observed elongation of some of the fragments shortly before their disappearance is likely to be due to a progressively significant loss of relatively large dust particles as the fragments may at this point disintegrate almost spontaneously; if so, the elongation may measure the distribution of accelerations by radiation pressure on the particulate debris relative to the acceleration by outgassing on the remaining body of the fragment. The spectacular Andromedid meteor storms, witnessed as early as three and five revolutions after the two nuclei of the parent Comet Biela had been detected for the last time, illustrate the plausibility of this scenario.

While it seems likely that both fragments of P/Biela have experienced either a complete disintegration or a deactivation accompanied by a partial disintegration, the principal nuclei of two other short-period comets have apparently withstood the splitting relatively unaffected. Statistically, companions have much shorter lifetimes than the primary fragments, a circumstance consistent with the generally high primary-to-secondary mass ratio proposed in Table III and with Whipple's (1977) finding that, with the exception of the sungrazing group and a few pairs, there is no evidence of a physical relationship among the members of other comet groups and that, contrary to previous beliefs, the similarity among orbital elements within such groups is statistically consistent with random expectation.

X. WHY DO COMETS SPLIT?

There are two requisites needed for the understanding of cometary splitting: (a) the knowledge of the tensile strength of a cometary nucleus and the response of cometary material to the development of internal stresses from various processes that the nucleus is subjected to; and (b) the recognition of a mechanism (or mechanisms) that can build in the nucleus stresses sufficiently powerful to disrupt it.

Only five of the 32 secondary nuclei listed in Table I separated from the parent nuclei at times of close approach to either the Sun (1882 II, 1965 VIII) or Jupiter (1889 V) and could result from tidal breakups. Whipple (1963) has shown that a comet of density ρ and radius R that fails to withstand the radial tidal force of a body of mass M_0 at a distance Δ has a tensile strength

$$\sigma < \frac{G M_0 \rho R^2}{\Delta^3} \tag{24}$$

where G is the universal constant of gravitation. For the sungrazing comets 1882 II and 1965 VIII, breaking up at perihelion, the numerical data give a condition

$$\sigma < 840 \, \rho R^2 \tag{25}$$

where R is in km, ρ in g cm^{-3}, and σ in dynes cm^{-2}. For P/Brooks 2's splitting at Jupiter one has in the same units

$$\sigma < 400 \, \rho R^2. \tag{26}$$

Similarly, one can write a condition for the disruptive force resulting from the rotational motion. It reads

$$\sigma < 2\pi^2 \, \rho R^2 / P^2 \tag{27}$$

where P is the rotation period. Comparing it with the condition for the tidal force, one finds an equivalent critical rotation period of 4.3 hr for the two sungrazing comets and 6.2 hr for Comet Brooks 2. Whipple (see his chapter) has shown that some comets, including periodic Comet Encke, spin at comparably high rates. Among the split comets the best candidate for a rotational breakup appears to be 1955 V, for which Whipple gives tentatively $P = 4.1$ hr.

Expressing the ratio of R/P in terms of the equatorial velocity of rotation V_{rot}, one obtains condition (27) in the form

$$\sigma < \tfrac{1}{2} \rho V_{rot}^2 \tag{28}$$

where V_{rot} is to be compared with the *initial* velocity of separation, which, as pointed out in Sec. VII, is estimated at several meters per second. The situation is complicated by a hydrodynamic interaction of the fragments as shown by Wallis (1980), although these forces appear to be somewhat less important than gravity.

With only fragmentary information on the quantities on the right-hand sides of expressions (24) through (28), an upper limit of the tensile strength of cometary material is estimated at anywhere between several times 10^3 and 10^5 dynes cm^{-2}, which in any case is an extremely low value.

Another splitting mechanism has been suggested by Whipple and Stefanik (1966) for new comets of the Oort-Schmidt (1951) classification that, by and large, break up at large heliocentric distance. (Table I indicates that 4 out of the 5 new comets split having never been exposed to insolation rates equivalent to, or higher than, that at 2 AU from the Sun.) The basic idea of this mechanism is the gradual formation, by mass transport due to internal radioactive heating under near-interstellar conditions, of a brittle shell of very volatile materials at the surface of the nucleus. On a first approach to the Sun such a structure may break up failing to withstand heat shock and associated differential expansion caused by the nonuniform distribution of incident solar radiation.

For old and periodic comets, a majority in Table I, an intriguing possibility is to envision a secondary nucleus as a fragment of a jettisoned insulating mantle of debris with a presumably thin layer of dirty ice adhering to its base. Prevailing orientations of cometary spin axes and their precession certainly favor a scenario in which large-sized dust keeps piling up over protected areas of the nucleus surface (Whipple and Sekanina 1979), so that by the time they become exposed to solar radiation the mantle has grown so thick that it can no longer be purged by blowoff mechanisms, described by Whipple (1978), by Brin and Mendis (1979), and by Brin (1980). As it gradually settles down in the growth phase under continuing gentle bombardment by particulate matter in suborbital trajectories, the mantle over shielded areas grows structurally stronger with time. In the meantime, the redistribution of particulate mass over the nucleus surface should start affecting the comet's rotation. Stresses that build up inside a rapidly wobbling nonspherical nucleus may sooner or later overcome the tensile strength along the lines of structural weakness that must exist in a nonhomogeneous dirty-ice mixture and, with some assistance from the spin, the whole mantle including a layer of the underlying ice matrix might eventually detach as essentially one large pancake-shaped fragment from the rest of the nucleus. By assuming that an appreciable part of the pancake does or does not disintegrate into dust particles upon separation, one may account for a splitting with or without an accompanying outburst of dust. Whatever the odds against this scenario may be, the behavior of such fragments is consistent with just about every piece of evidence that has so far been accumulated, including the high decelerations, the short lifetimes (compared to the principal nuclei), the separation velocities, the brightness fluctuations, the random distribution of breakups, the sequence of splitting for comets with more than one companion, and so on.

Acknowledgment. This chapter presents the results of one phase of research carried out at the Jet Propulsion Laboratory, California Institute of Technology, under a contract sponsored by the National Aeronautics and Space Administration, Planetary Atmospheres Program, Office of Space Sciences.

REFERENCES

Beyer, M. 1948. Physische Beobachtungen von Kometen. VI. *Astron. Nachr.* 275:237-250.
Brin, G.D. 1980. Three models of dust layers on cometary nuclei. *Astrophys. J.* 237:265-279.
Brin, G.D., and Mendis, D.A. 1979. Dust release and mantle development in comets. *Astrophys. J.* 229:402-408.
Dawson, D.W.; Knuckles, C.F.; and Murrell, A.S. 1977. An investigation of multiple head structure in Comet West. *Proc. Southwest Reg. Conf.* 2:133-145.
Delsemme, A.H., and Miller, D.C. 1971. Physico-chemical phenomena in comets. III. The continuum of Comet Burnham (1960 II). *Planet. Space Sci.* 19:1229-1257.
Hepperger, J. von 1906. Bestimmung der Masse des Biela'schen Kometen. *Sitzungsber. Math.-Naturwiss. Cl. Kaiserl. Akad. Wien* 115 (Abh. 2a):785-840.
Jeffers, H.M. 1922. An investigation of the orbits of the two components of Taylor's comet, 1916 I. *Lick Obs. Bull.* 10:120-130.
Keller, H.U. 1976. The interpretations of ultraviolet observations of comets. *Space Sci. Rev.* 18:641-684.
Marsden, B.G. 1967. The sungrazing comet group. *Astron. J.* 72:1170-1183.
Marsden, B.G. 1979. *Catalogue of Cometary Orbits.* (3rd ed.) IAU Central Bureau for Astron. Telegrams, pp. 59-60.
Marsden, B.G., and Sekanina, Z. 1971. Comets and nongravitational forces. IV. *Astron. J.* 76:1135-1151.
Marsden, B.G.; Sekanina, Z.; and Everhart, E. 1978. New osculating orbits for 110 comets and analysis of original orbits for 200 comets. *Astron. J.* 83:64-71.
Marsden, B.G.; Sekanina, Z.; and Yeomans, D.K. 1973. Comets and nongravitational forces. V. *Astron. J.* 78:211-225.
Ney, E.P., and Merrill, K.M. 1976. Comet West and the scattering function of cometary dust. *Science* 194:1051-1053.
Oort, J.H., and Schmidt, M. 1951. Differences between new and old comets. *Bull. Astron. Inst. Neth.* 11:259-269.
Pittich, E.M. 1972. Splitting and sudden outbursts of comets as indicators of nongravitational effects. In *The Motion, Evolution of Orbits, and Origin of Comets,* eds. G.A. Chebotarev, E.I. Kazimirchak-Polonskaya, and B.G. Marsden (Dordrecht: D. Reidel Publ. Co.), pp. 283-286.
Sekanina, Z. 1966. Splitting of the primary nucleus of Comet Ikeya-Seki. *Bull. Astron. Inst. Czech.* 17:207-211.
Sekanina, Z. 1977a. Relative motions of fragments of the split comets. I. A new approach. *Icarus* 30:574-594.
Sekanina, Z. 1977b. On the "edge-on" configuration of the nuclei of Comet West (1975n). *Astrophys. Lett.* 18:55-59.
Sekanina, Z. 1978. Relative motions of fragments of the split comets. II. Separation velocities and differential decelerations for extensively observed comets. *Icarus* 33:173-185.
Sekanina, Z. 1979a. Relative motions of fragments of the split comets. III. A test of splitting and comets with suspected multiple nuclei. *Icarus* 38:300-316.
Sekanina, Z. 1979b. The split comets: Gravitational interaction between the fragments. In *Dynamics of the Solar System,* ed. R.L. Duncombe (Dordecht: D. Reidel Publ. Co.), pp. 311-314.
Sekanina, Z. 1980a. Physical characteristics of cometary dust from dynamical studies: A review. In *Solid Particles in the Solar System,* eds. I. Halliday, and B.A. McIntosh (Dordrecht: D. Reidel Publ. Co.), pp. 237-250.
Sekanina, Z. 1980b. Physical similarities between dissipating comets and short-lived fragments of the split comets. *Bull. Amer. Astron. Soc.* 12:511(abstract).
Sekanina, Z., and Farrell, J.A. 1978. Comet West 1976 VI: Discrete bursts of dust, split nucleus, flare-ups, and particle evaporation. *Astron. J.* 83:1675-1680.
Stefanik, R.P. 1966. On thirteen split comets. *Mém. Soc. Roy. Sci. Liège* (Ser. 5) 12:29-32.

Wallis, M.K. 1980. Hydrodynamic forces in splitting comets. *Mon. Not. Roy. Astron. Soc.* 190:467-478.
Whipple, F.L. 1950. A comet model. I. The acceleration of Comet Encke. *Astrophys. J.* 111:375-394.
Whipple, F.L. 1963. On the structure of the cometary nucleus. In *The Solar System IV: The Moon, Meteorites, and Comets,* eds. B.M. Middlehurst, and G.P. Kuiper (Chicago: Univ. of Chicago Press), pp. 639-663.
Whipple, F.L. 1977. The reality of comet groups and pairs. *Icarus* 30:736-746.
Whipple, F.L. 1978. Cometary brightness variation and nucleus structure. *Moon and Planets* 18:343-359.
Whipple, F.L., and Sekanina, Z. 1979. Comet Encke: Precession of the spin axis, nongravitational motion, and sublimation. *Astron. J.* 84:1894-1909.
Whipple, F.L., and Stefanik, R.P. 1966. On the physics and splitting of cometary nuclei. *Mém. Soc. Roy. Sci. Liège* (Ser. 5) 12:33-52.
Yeomans, D.K. 1977. Comet Halley–the orbital motion. *Astron. J.* 82:435-440.

RADAR DETECTABILITY OF COMETS

PAUL G. KAMOUN, GORDON H. PETTENGILL, and IRWIN I. SHAPIRO
Massachusetts Institute of Technology

> *Earth-based radar observations of comets are restricted to the detection of the nucleus since detection of an echo from the dust or from the plasma near a comet seems infeasible. The successful detection of the nucleus can provide information on its size, spin vector, surface-scattering properties, and orbit. Only P/Encke has so far been detected by radar, and that detection was marginal. At most, three other known comets could be detected by radar within the next five years.*

The main motivation for using radar to study comets lies in the fact that, unlike other groundbased astronomical techniques, radar offers the possibility to observe directly the nucleus of a comet and, simultaneously, to obtain information on its surface scattering properties, size, spin rate, and orbit.

The radar detectability of comets is governed by the radar equation:

$$\frac{S}{N} = \left[\frac{P_t G^2 \lambda^{5/2} \tau^{1/2}}{4\pi k T_s}\right] \cdot \left[\frac{\sigma}{32\pi \Delta^4 L}\left(\frac{R^3}{\Omega_p}\right)^{1/2}\right] \cdot \eta \qquad (1)$$

where S/N is the signal-to-noise ratio, R the radius of the (assumed) hard target, σ the ratio of its radar cross section to its geometric cross section, Ω_p the magnitude of its apparent rotation vector projected on the celestial sphere, L the attenuation in the coma, and Δ the geocentric distance of the comet. The effect on the radar detectability of the scattering law obeyed by the comet is characterized by η; it is unity if the spectral density is constant from limb to limb and greater than unity otherwise. The parameters of the radar system are λ the wavelength of the transmitted signals, P_t the transmitted power, G the gain of the antenna relative to an isotropic radiator, T_s the effective system temperature, and τ the integration time. These characteristics differ from one radar system to another; see Jurgens and Bender (1977) for a detailed description. In brief, the two major systems available today for cometary studies are the S-band (12.5 cm) and the B-band (70 cm) radars at the Arecibo Observatory in Puerto Rico and the X-band (3.5 cm) and the S-band (12.5 cm) radars at the Goldstone facility in California. Even for the most sensitive of these systems, the S-band radar configuration at Arecibo, a typical comet nucleus, assumed to be a rotating rigid body, can be detected reliably only if it falls within the declination coverage of that telescope and only when it is at a distance $\lesssim 0.3$ AU, the exact limit depending on the radius, rotation rate, and reflection properties of the nucleus.

In Sec. I, we discuss the radar detectability of the various components of a comet, and in Sec. II we describe primarily the radar detection of P/Encke. A brief conclusion is given in Sec. III.

I. DETECTABILITY OF COMPONENTS OF A COMET

A comet usually consists of two other components, in addition to the nucleus, each of which scatters radio waves differently: the ice and dust particles in the coma and dust tail, and the plasma in the coma and ion tail. We discuss each component in turn.

1. Nucleus

The nucleus is the component of a comet most likely to return a radar echo. As for observations of minor planets (see, e.g., Pettengill et al. 1979), a typical radar experiment to detect echoes from comets consists of the transmission of a continuous-wave, nearly monochromatic signal for the duration of the round-trip echo time delay, then the reception of the echo for about the same amount of time. The detectability of the nucleus is not only a function of the parameters of the radar system, but, as is clear from Eq. (1), is also dependent on the target's surface scattering properties, rotation, and size, and, most importantly, on the inverse fourth power of its geocentric distance. This last factor usually places the principal limit on detectability. Another problem often is the lack of sufficiently accurate ephemerides, even for periodic comets. This problem is especially relevant to radar observations since one needs to know precisely not only the angular position of the object, but also its velocity in order to set the receiver frequency and thus to correct for the Doppler shift introduced by the radial component of the velocity of the target relative to the Earth. Usually inaccuracies of a few seconds of arc for the angular position and of 100 m s^{-1} for the radial velocity are tolerable. For most comets the uncertainty is such that, for S-band radar frequencies, it is necessary to search for the echo in a frequency range of a few kilohertz, corresponding to a radial velocity uncertainty on the order of 100 m s^{-1}. Moreover, the rate of change of the Doppler shift must be known with $< 10^{-3}$ Hz s^{-1} uncertainty in order to be able to combine efficiently the data from several days of observations.

The total broadening of the echo spectrum is related to the size and rotation of the nucleus. Thus, a knowledge of the maximum echo bandwidth and of the spin vector, the latter perhaps obtained from optical observations (Sekanina 1979; the chapter by Whipple in this book), would allow the determination of the nuclear radius. Unfortunately, estimates of rotation rate, spin-axis direction, shape and size have been made for very few comets so that in most cases a radar observation would only help to place some constraint on the values of the physical parameters of the nucleus.

Polarization properties of the radar echoes are useful for determining the surface scattering characteristics of the nucleus. For instance, from echoes received in the sense of circular polarization opposite to that transmitted, one learns mostly about the quasi-specular scattering properties of the target, since a single reflection reverses the sense of circular polarization. Receiving the same sense of circular polarization as that transmitted yields information on the diffuse scattering properties. From Earth, the directions of illumination and observation of the target are nearly identical, so that, for a nearly spherical target, the quasi-specular scattering component of the echo arises mainly from reflections near the subradar point. On the other hand, the diffusely scattered component arises from irregularities widely distributed over the surface.

From the echo power received, the radar (backscattering) cross section is deduced and, when possible, its variation with rotational phase of the target is studied. The ratio of the radar cross section to the geometric cross section is related to the shape and reflectivity of the surface and is generally a function of polarization and radar wavelength. When dual polarization (for instance both left and right circular) receiving systems are available, the geometric albedo can be estimated by combining the power received in the two polarizations.

2. Ice and Dust Particles

The radar detectability of dust depends strongly on the size distribution of the grains. A cloud of particles each of dimension a, small compared to the wavelength λ (λ is usually ~ 10 cm), will present a radar cross section that varies as the fourth power of the ratio a/λ, according to Rayleigh scattering. For a compact cloud of N such spherical particles of complex dielectric constant E_c, the radar cross section is given approximately by

$$4N\pi a^2 \left(\frac{2\pi a}{\lambda}\right)^4 \left|\frac{E_c - 1}{E_c + 2}\right|^2 \quad (2)$$

when multiple scattering is unimportant.

At the shortest of the available radar wavelengths, the number of mm- to cm-sized particles will likely determine the strength of the echo from ice and dust. The number density of such particles probably depends strongly on the particular comet, but so far no useful data have been obtained for this high end of the size distribution. However, it has been inferred that most of the grains in the coma and dust tail have sizes of a few tenths of a micron (Whipple 1978), and consequently it is very unlikely that the dust will return a detectable radar echo.

3. Plasma

For the plasma to be detectable, the electron density must be high enough so that the critical plasma frequency is greater than the radar signal frequency. There is no generally accepted detailed theory for ion production in comets, although photoionization seems to be the primary process involved. The ions appear to be formed in the close vicinity of the nucleus, the source size being quite small. An estimate of the ion density in the coma may be determined approximately by assuming that the coma plasma pressure is comparable to the solar wind ram pressure. This condition at a distance from the Sun of ~ 1 AU requires an ion density of $\sim 10^4$ to 10^5 cm^{-3}, corresponding to a critical plasma frequency lower than a few megahertz. This density is far too low to sustain an echo, the radar signal frequency being restricted to values much higher than this plasma frequency in order to propagate through the Earth's ionosphere. The same considerations should apply to the ion tail where the density should on the average be even smaller. Scattering from irregularities within the ion tail, where knots are known to exist, must be negligible given their small total cross section. Therefore no echo from plasma should be expected in radar observations.

4. Attenuation of Radar Waves in the Coma

The coma of a comet is essentially neutral, the ion concentration being several orders of magnitude lower than the molecular concentration. The predominant gaseous species in the coma at a heliocentric distance of ~ 1 AU is believed to be water with a production rate of $\sim 10^{29}$ to 10^{30} molecules per second for moderately active comets (Whipple and Huebner 1976), or $\sim 10^{27}$ molecules per second for an old comet like P/Encke. The gas concentration is estimated to be $\sim 10^{14}$ molecules per cm^3 near the surface of the nucleus; for typical ambient conditions of ~ 200 K and 1 dyne cm^{-2} in the coma (Delsemme and Miller 1971), the attenuation caused by water vapor at S band is $\lesssim 10^{-6}$ db km^{-1} for such a concentration (Kerr 1951). Therefore, the attenuation of radar waves by the gaseous component of a cometary coma is completely negligible.

II. RADAR OBSERVATIONS OF COMETS

The history of cometary radar astronomy so far includes only three experiments, partly because close approaches of comets to the Earth are rare and partly because sufficient sensitivity in the Arecibo and Goldstone radar systems only became available in the mid-1970s. The first attempt to study a comet by radar was made in January 1974, when Chaisson et al. (1975) attempted to observe Comet Kohoutek 1973 XII, using the Haystack Observatory X-band (3.8 cm) radar. Although the sensitivity of that system was not sufficient to detect the nucleus nor to place a useful upper limit on its radius, Chaisson et al. concluded that the density of mm-sized particles in a coma of diameter 10^4 km was < 1 m^{-3}. The second attempt, to detect P/d'Arrest, was made by Pettengill et al. (unpublished) at the Arecibo Observatory in July 1976. This attempt did not yield positive results, but suggested that the nucleus of d'Arrest was < 1 km in radius.

1. Comet P/Encke

The latest radar experiment resulted in the (marginal) detection of echoes from P/Encke at a geocentric distance of ~ 0.33 AU in November 1980 by Kamoun et al. (1981) using the Arecibo Observatory S-band radar system. P/Encke has the shortest period of any known comet (3.3 yr) and, since it has been studied during 52 apparitions, its orbit is well known (Marsden 1979). Some of its physical parameters such as size, rotation rate, and pole position have also been estimated (Whipple and Sekanina 1979). In Table I we give several estimates of P/Encke's radius and the method used to obtain each. We include the radius estimated from the 1980 radar observations, based on the pole position and rotation rate determined by Whipple and Sekanina (1979); for a spherical target the radius (in km) is given by

$$R = \frac{\lambda B T}{8 \pi \sin\alpha} \times 10^{-3} \tag{3}$$

where λ (in m) is the radar wavelength, B (in Hz) is the total bandwidth of the echo, T (in s) the rotation period of the nucleus, and α (in rad) the angle between the comet's spin axis and the radar line of sight.

From an examination of Table I, it is clear that the different estimates are not mutually consistent, each of the methods used suffering from one or another intrinsic difficulty. It is therefore not now possible to draw reliable conclusions as to which value is the most accurate.

TABLE I

Comet P/Encke: Radius and Albedo of the Nucleus

Radius (km)	Assumed Albedo	Reference: Method Used
0.6 – 3.5	0.7 – 0.02	Roemer (1966): Magnitudes at large heliocentric distances.
1.3	0.2	Delsemme and Rud (1973): Magnitudes at large heliocentric distances and water vaporization rate.
1.7	0.15	Kresák (1973): Magnitudes at large heliocentric distances and comet history (comments on this value by Sekanina, 1976).
< 1.5	--	Whipple and Sekanina (1979): Rate of mass loss from sublimation and rotation of nucleus.
< 0.5	> 0.05	Stauffer and Spinrad (personal communication, 1981): Photometry of the red nuclear continuum.
~ 0.4-4.0	--	Kamoun et al. (1981): Radar observation, and independent (optical) estimate of spin vector.

A radar cross section of 1.1 ± 0.7 km^2 has been estimated for the nucleus of P/Encke, but the rather low signal-to-noise ratio achieved and the availability of only one receiver polarization have prevented a detailed assessment of the scattering properties of the surface and any useful inferences concerning the surface composition of P/Encke's nucleus.

2. Future Opportunities

Since no major improvements in the capabilities of existing radar systems are likely in the next few years, the number of useful opportunities for radar detection of comets is small. Table II presents a summary of the best opportunities for the period 1982-1986, after which, perhaps, radars will yield to spacecraft for closer looks at the central condensations of comets.

TABLE II
Opportunities from 1982 to 1986 for Radar Observations of Comets

Comet	Orbital Elements				Radius (km)	Rotation Period (hr)	Δ^a (AU)	T^a	Suitable Observatories	Estimated Signal-to-Noise Ratio[b]
	P (yr)	e	Q^a (AU)	N^a						
P/Grigg-Skjellerup	5.10	0.67	0.99	13	2^c	$(10)^d$	0.33	May 1982	Arecibo	10
									Goldstone	3
P/Churyumov-Gerasimenko	6.59	0.63	1.31	2	$(2)^d$	$(10)^d$	0.39	Nov. 1982	Arecibo	2
P/Halley	76.09	0.97	0.58	27	4^e	10.3^e	0.62	Nov. 1985	Arecibo	3
							0.42	Apr. 1986	Goldstone	3

[a] Q represents the perihelion distance, N the number of past appearances (Marsden 1979), Δ the geocentric distance at optimum observation, and T the corresponding time.
[b] The signal-to-noise ratio has been calculated for the stated parameter values using a radar cross section equal to one tenth the geometric cross section. Integration of the echo from four days of observation, for a total of 5 hr at Arecibo and 20 hr at Goldstone, has been assumed. The signal-to-noise ratio is expressed in units of the standard deviation of the noise power.
[c] Roemer (1966); median of quoted interval.
[d] Adopted in the absence of an observed value.
[e] Whipple (1981); radius is upper limit of quoted interval.

The best candidates for radar observation among the known periodic comets are P/Grigg-Skjellerup and P/Churyumov-Gerasimenko. The former should yield results slightly better than those obtained for P/Encke in 1980 because of improvements in the Arecibo radar system. The latter is unlikely to yield detectable echoes unless adopted parameters prove too conservative. No other comets whose orbits are well known could be detectable prior to the anticipated spacecraft missions to P/Halley in 1986. P/Halley will lie inside the declination coverage of Arecibo only during its first close approach to the Earth (inbound leg), at which time its minimum geocentric distance will be ~ 0.62 AU. It will fall within Goldstone's coverage at both its close approaches, but will lie close to the southern horizon at its second approach (outbound leg). As may be seen from Table II, the chances for a detection of P/Halley by either facility are slight. There remains, of course, the possibility of a new comet approaching close to the Earth. Indeed, one to two new comets that cross the Earth's orbit are discovered each year. Attempting to observe these, however, would necessitate the preparing of accurate ephemerides and the scheduling of observations with a short lead time. Thus, astrometric determinations must be made of the comet's position before any radar observations could take place. An independent estimate of the spin vector of the nucleus would be required to infer its radius.

III. CONCLUSION

The acquisition and interpretation of radar data from comets is very dependent upon results obtained by observers using visual techniques. There is thus a need for a joint effort by observers using these different techniques. Patrols such as the one for Comet Kohoutek 1973 XII, or the International Halley Watch, set for 1985-1986, could also be of use for the close approach of P/Grigg-Skjellerup and of P/Churyumov-Gerasimenko in 1982.

Acknowledgments: The authors wish to thank F.L. Whipple and Z. Sekanina for many useful discussions, and B.G. Marsden for communication of unpublished results. This research has been supported in part by the National Aeronautics and Space Administration and by the National Science Foundation.

REFERENCES

Chaisson, E.J.; Ingalls, R.P.; Rogers, A.E.E.; and Shapiro, I.I. 1975. Upper limit on the radar cross-section of Comet Kohoutek. *Icarus* 24:188-189.

Delsemme, A.H., and Miller, D.C. 1971. Physico-chemical phenomena in comets-III. The continuum of Comet Burnham (1960 II). *Planet. Space Sci.* 19:1229-1257.

Delsemme, A.H., and Rud, D.A. 1973. Albedos and cross-sections for the nuclei of comets 1969 IX, 1970 II and 1971 I. *Astron. Astrophys.* 28:1-6.

Jurgens, R.F., and Bender, D.F. 1977. Radar detectability of asteroids. *Icarus* 31:483-497.

Kamoun, P.G.; Campbell, D.B.; Ostro, S.J.; Pettengill, G.H.; and Shapiro, I.I. 1981. Comet Encke: Radar detection of nucleus. Submitted to *Science*.

Kerr, D.E. 1951. *Propagation of Short Radio Waves*. First ed. (New York: McGraw-Hill Book Co.), pp. 641-693.

Kresák, L. 1973. Short-period comets at large heliocentric distances. *Bull. Astron. Inst. Czech.* 24:264-283.

Marsden, B.G. 1979. Catalogue of cometary orbits. Central Bureau for Astronomical Telegrams, I.A.U., Smithsonian Astrophys. Obs.

Pettengill, G.H.; Ostro, S.J.; Shapiro, I.I.; and Campbell, D.B. 1979. Radar observations of asteroid 1580 Betulia. *Icarus* 40:350-354.

Roemer, E. 1966. The dimensions of cometary nuclei. *Mém. Soc. Roy. Sci. Liège*, Ser. 5, 12:23-28.

Sekanina, Z. 1976. Has the nucleus been resolved? In *The Study of Comets*, eds. B. Donn, M. Mumma, W. Jackson, M. A'Hearn, and R. Harrington (Washington: NASA SP-323), pp. 555-561.

Sekanina, Z. 1979. Fan-shaped coma, orientation of the rotation axis and surface structure of a cometary nucleus. I. Test of a model on four comets. *Icarus* 37:420-442.

Whipple, F.L. 1978. Comets. In *Cosmic Dust*, ed. J.A.M. McDonnell (New York: John Wiley and Sons), pp. 1-73.

Whipple, F.L. 1981. Notes on the P/Halley nucleus. Scientific and experimental aspects of the Giotto mission. (Paris: European Space Agency SP-163), pp. 101-103.

Whipple, F.L., and Huebner, W.F. 1976. Physical processes in comets. *Ann. Rev. Astron. Astrophys.* 14:143-171.

Whipple, F.L., and Sekanina, Z. 1979. Comet Encke: Precession of the spin axis, non-gravitational motion and sublimation. *Astron. J.* 84:1894-1909.

RELATIONSHIPS BETWEEN COMETS, LARGE METEORS, AND METEORITES

G.W. WETHERILL
Carnegie Institution of Washington

and

D.O. REVELLE
Northern Arizona University

Available data relevant to the use of bright meteors (fireballs) to obtain information regarding the nature of cometary nuclei are reviewed. Fireballs most certainly identifiable with cometary sources are those in retrograde orbits, with aphelia beyond Jupiter, or in meteor streams. The physical properties of fireballs in orbits of this kind are quantitatively characterized by parameters defined by observation of the dynamics of their entry into the atmosphere. This permits comparison of the relative ability of cometary meteoroids, ordinary chondrites, and carbonaceous chondrites to penetrate the atmosphere without total fragmentation. It is found that although many cometary meteoroids have measurable strength, almost all these fireballs are weaker than any recovered meteorites. However, the few exceptions to this rule are adequate to account for the observed number of recovered weak carbonaceous meteorites. These more massive aggregations of relatively strong cometary material are only a small fraction ($\sim 10^{-3}$) of the total cometary meteoroids; the largest portion of material lost from active comets consists of fine particles, recoverable only by stratospheric collection. Fragmentation of inactive cometary nuclei in the form of bodies in Apollo-Amor orbits may be an additional source of cometary fireballs not considered here.

In its journey through space, the Earth collides with debris in its path. Including bodies with masses up to a few tons, it is estimated that $\sim 10^{9\pm1}$ g of this meteoroidal material is swept up by the Earth annually (Hughes 1978). Observation of the association of some meteor streams with the orbits of known periodic comets demonstrates that some debris is derived from cometary nuclei (e.g. Cook 1973). Dynamical theory clearly predicts that a significant portion of the material is also derived from the main asteroid belt, as well as from Earth-approaching Apollo and Amor objects (Wetherill 1974; Wasson and Wetherill 1979).

Observation of the atmospheric flight of meteoroids affords an opportunity to study the physical and chemical properties of the constituent matter in comets, asteroids, and Earth-approaching bodies. This is particularly true for the larger meteoroids (fireballs) $\gtrsim$ 100 g in mass. Atmospheric entry of these bodies produces meteors brighter than −5 mag. These larger and brighter meteors usually penetrate more deeply into the atmosphere than the smaller bodies studied by radar and super-Schmidt cameras.

This deeper penetration permits measurement of mechanical properties under a range of dynamic pressures up to $\sim 10^8$ dyne cm^{-2}. The longer flight path permits obtaining more accurate and useful photometric and deceleration data. The heliocentric orbits of these meteoroids are more stable than those of smaller bodies, hence more diagnostic of origin, because of the relative unimportance of nongravitational forces such as those associated with radiation pressure and the Poynting-Robertson effect. The greater brightness of these fireballs permits spectrometric studies at higher dispersion (Ceplecha 1973). In some cases, survival of atmospheric entry permits connection with meteorites studied in the laboratory.

The most extensive published fireball data is that obtained by the Prairie Network (Ceplecha and McCrosky 1976; McCrosky et al. 1978,1979). Data for the brighter fireballs photographed by the European network can be found in Ceplecha (1977) and references therein. Earlier data were reported by Whipple (1954) and Babadzhanov and Kramer (1965).

Three fireballs have been recovered as ordinary chondritic meteorites: Pribram (Ceplecha 1961), Lost City (McCrosky et al. 1971), and Innisfree (Halliday et al. 1981). Prior to the fall of these objects the prevailing opinion was that almost all of the fireballs were of very weak, low-density material sometimes characterized as dust balls, quite unlike the meteorites in museum collections. The reason for this opinion was the apparent low density of the fireballs calculated from single-body theory and the observation of atmospheric fragmentation at dynamic pressures (10^6-10^7 dynes cm^{-2}) well below those measured on recovered meteorites (Buddhue 1942).

Data on these three recovered fireballs have forced reevaluation of this conclusion. These fireballs also exhibited low apparent density and fragmentation at low dynamic pressures; consequently the possibility exists that a fireball with these characteristics may be recoverable as a common type of meteorite. This reevaluation has led to an evolution in opinion regarding the fraction of the fireballs that are similar to recovered meteorites. McCrosky et al. (1971) recognized the problem and attributed the Lost City result to an unusual flattened shape, but still concluded that such objects were rare and that most fireballs with similar atmospheric trajectories were weak, low-density objects. McCrosky and Ceplecha (1969) suggested that, among other possibilities, the Pribram data could be explained as a consequence of the recovered ordinary chondrite being imbedded in a larger mass of weak material before its entry into the atmosphere.

Recently, the conclusion is more often reached that an appreciable fraction ($\gtrsim 20\%$) of the fireballs is physically indistinguishable from recovered meteorites (Ceplecha and McCrosky 1976; Bronshten 1976). A similar opinion was given earlier by Baldwin and Scheafer (1971). As a result of this work as well as further treatment of the data (Wetherill and ReVelle 1981), the authors believe there is compelling evidence that a large fraction of the fireballs are physically and chemically identical, except for size, to well-known meteorite types, and that in many cases individual fireballs can be identified as being of this kind. At the same time, it has become clear that a similar large fraction of the observed fireballs are unlike known meteorites and the possibility of their surviving atmospheric entry as ponderable bodies is negligible. However, this material is probably represented in micrometeorite collections (Brownlee et al. 1977).

This review considers evidence about the physical nature of the constituents of cometary nuclei obtainable by fireball observations. This evidence is primarily that of the Prairie Network, for which orbital data and atmospheric trajectories have been given for 322 of the 2700 bright meteors photographed (McCrosky et al. 1978,1979). As emphasized by these authors, the data are not a statistically reliable sample of the matter impacting the Earth. In addition to unavoidable observational selection, considerable additional selection was introduced by the choice of data that was reduced. With a few exceptions, fireballs belonging to the more abundant meteor streams other than the Taurids were deliberately rejected. Furthermore, meteors in orbits of cometary origin are statistically more likely to have high atmospheric entry velocities; some objects unmistakably of cometary origin have very brief atmospheric trajectories because of their mechanical properties. In the reduction of the Prairie Network data a strong bias exists against trails of short (< 1 to 2 s) duration. For these reasons the data set is not ideal, but it permits important inferences regarding the nature of cometary material; this discussion may also stimulate further work and thereby lead to more complete discussion in the future.

It is awkward to review the literature on the relationship between fireballs and comets since the subject is not addressed in any detail in recent literature. For this reason, we address this subject in the form of a new synthesis of the data. However it should be recognized that this synthesis is possible only because of extensive data acquisition and reduction by the other workers referred to, and that the relationship of these data to comets and asteroids has been considered by other investigators.

I. IDENTIFICATION OF FIREBALLS OF COMETARY ORIGIN BY THEIR HELIOCENTRIC ORBITS

Generally, it can be argued that almost all fireballs are products of either cometary or asteroidal disintegration. They may be derived immediately from these bodies or through the intermediary of Apollo-Amor objects, themselves of cometary or asteroidal origin. This review will emphasize meteoroids most probably direct products of cometary disintegration. Discussion of the evidence for and against an association between Apollo-Amor objects, their fragmentation debris, recovered meteorites, and comets may be found elsewhere (Öpik 1963; Anders 1964; Levin et al. 1976; Simonenko 1979; Wetherill 1976; Wasson and Wetherill 1979).

The principal criterion for inferring whether a given fireball is of cometary or asteroidal origin must be its heliocentric orbit. If criteria were based on presumed physical and chemical properties of asteroidal and cometary material, this would improperly anticipate the answer to the questions being asked. This point of view is not at variance with the prevailing view that Whipple's (1950) icy conglomerate model of the cometary nucleus is in much better agreement with observation than the alternative "sand-bank" model of Lyttleton (1948). Much more subtle differences in nuclear structure are now discussed; the icy conglomerate model comprises a general class of models involving diverse kinds of conglomerate. The icy conglomerate model does not require that all of the clasts in the conglomerate have the same physical or chemical nature. Evidence for large-scale homogeneity (see chapters by Delsemme, Donn and Rahe, Sekanina, and A'Hearn in this book) of comet nuclei does not preclude heterogeneity on the smaller size scale of fireballs.

If in the future it is found that a cometary origin of certain fireballs requires that they must be derived from extinct Apollo-Amor cometary cores, this might be thought to be inconsistent with evidence for homogeneity of almost all comets, inasmuch as a comet with a core would not be homogeneous. However, long ago Öpik (1963, 1966) pointed out that only a small fraction ($\sim$ 1%) of the periodic comets can contain cores of $\gtrsim$ 1 km in diameter. Otherwise a much larger number of Apollo objects would be expected than is actually observed, even if the increased estimates (Helin and Shoemaker 1979; Wetherill 1976) of the number of these small bodies were used. Therefore, it would be unlikely that a comet with a sizeable core would have been included in the sample of comets upon which the evidence for homogeneity is based.

Use of orbital data to identify fireballs of more immediate cometary origin is not a trivial matter. Observation of P/Encke shows that comets can evolve into orbits well within Jupiter's perihelion (0.9 AU). If conventional opinion identifying all shower meteors with comets is correct, the existence of the Geminid stream and similar more minor streams with aphelia < 2.6 AU represents an even greater intrusion of comets into the domain of the asteroids. On the other hand, in order to impact the Earth an asteroidal meteoroid must be perturbed into an Earth-crossing orbit. Once in such an orbit, the probability is high that it will be perturbed by close approaches to the Earth (and Venus as well) into a Jupiter-crossing orbit (Arnold 1965; Wetherill and Williams 1968). These Jupiter-crossing orbits are unstable; on a time scale of $< 10^5$ yr most of this material will be ejected by Jupiter and Saturn perturbations into hyperbolic solar system escape orbits. Nevertheless, asteroidal debris will transiently reside in these Jupiter-crossing orbits, and have a probability of impacting the Earth during this time.

For this reason identification of fireballs with comets on the basis of their orbits must involve statistical considerations. The probability of an interloping noncometary (asteroidal) object in an orbit considered characteristically cometary must be assessed, and will be different for various cometary orbits.

An extreme case is that of retrograde orbits. Monte Carlo dynamical studies of the evolution of Earth-crossing material of prograde asteroidal origin (e.g. Wetherill 1979) show that the chances of finding asteroidal material in a retrograde orbit are negligible. Data are available for some Prairie Network fireballs in retrograde orbits, in spite of their high entry velocities and consequent short atmospheric trajectories which introduce a strong bias against their inclusion in the reduced data.

Another class of distinctively cometary orbits are those of stream meteors, particularly those associated with an observed periodic comet. The retrograde nature of some stream orbits causes a criterion of stream membership to be redundant for these cases but not for the more common streams in prograde orbits. As mentioned earlier, fireballs belonging to the major streams (Perseids, Leonids, Orionids, Geminids) were rejected in the choice of Prairie Network fireballs reduced (although a few got by). Minor streams could not be well represented. The only major stream not rejected are the Taurids (P/Encke). The Taurid data will be discussed in Sec. III. These data suffer from the disadvantage that the Taurid aphelia at ~4 AU are well within the orbit of Jupiter and of low inclination. Therefore asteroidal material could be found in similar orbits and must be distinguished from true Taurid material.

Considering the difficulties discussed above, the subset of the Prairie Network data most useful for characterizing cometary material are the 66 fireballs with aphelia at or beyond the perihelion of Jupiter (5 AU). To some extent these objects will be contaminated with asteroidal material during their evolution into hyperbolic solar system escape orbits. However the lifetime of asteroidal material transiently in these orbits is ~100 times shorter than that in its precursor orbits. Furthermore, the Earth impact probability per unit time while in these orbits will be about a factor of 10 shorter than while in Earth-crossing orbits with smaller aphelia. In addition, the higher velocity of trans-Jovian asteroidal fireballs compared to that of asteroidal fireballs from the asteroid belt will tend to eliminate them from the data reduction. Thus for each observed trans-Jovian fireball of asteroidal origin there should be ⩾ 1000 times as many observed in orbits with smaller aphelia. Because the set of reduced Prairie Network data includes only 322 fireballs, it is unlikely that > 1 is a stray asteroidal fragment in a trans-Jovian orbit, even in the improbable event that all these fireballs are of asteroidal origin.

Because all Prairie Network fireballs with retrograde orbits also have aphelia beyond 5 AU, the retrograde fireballs constitute a subset of those with aphelia beyond 5 AU. About half the 29 Prairie Network fireballs we identified (as did Ceplecha and McCrosky in some cases) as stream members belong to the group with aphelia beyond 5.0 AU.

II. CHARACTERISTICS OF THE FIREBALLS, ASSUMED TO BE ALMOST ENTIRELY COMETARY, WITH APHELIA GREATER THAN 5.0 AU

McCrosky et al. (1978) have published orbital data for 66 Prairie Network fireballs with aphelia beyond 5.0 AU. Velocity, deceleration, and absolute magnitude (100 km range) as a function of altitude are given in McCrosky et al. (1979). Usable deceleration data are given for 41; two or more values of deceleration are given for 27 objects. These constitute raw data from which inferences can be made about the physical characteristics of the larger cometary meteoroids that impact the Earth. It is difficult to place these inferences in an absolute context, because there is no ground truth in the form of recovered meteorites of this kind with which theoretically based inferences could be tested.

It is possible, however, to discuss whether these cometary fireballs have similar physical characteristics to the recovered ordinary chondrites by comparing cometary fireball data with those of these recovered fireballs, and of other fireballs previously identified to have physical characteristics of recovered meteorites (Wetherill and ReVelle 1981). In this earlier report, four criteria were given for distinguishing meteors as survivable as ordinary chondrites.

(1) Deceleration to velocities < 8 km s^{-1}. Below this velocity ablation will reduce the mass of the meteoroid by < 50%. The meteoroid will also have survived its peak dynamic pressure, even for bodies with atmospheric entry velocities ~ 12 km s^{-1}.

(2) Ratio of photometric mass (determined by integrated luminosity) to dynamic mass (determined by observed deceleration in the atmosphere) not more than twice the ratio of these quantities found for the recovered meteorite Lost City. The dynamic mass will be lower than the actual mass if the

surface area/mass ratio is higher than that of a single sphere of chondritic density as assumed in the calculation of dynamic mass from deceleration data. This can result from various combinations of fragmentation, irregular shape, or low density. The absolute calibration of both the photometric and dynamic mass scales is uncertain, but by normalizing to the known typical ordinary chondrite Lost City, the need for absolute calibration is removed.

(3) Deceleration to final observed velocity at an end height in accordance with single-body meteor theory, assuming the mass to be equal to the observationally determined dynamic mass. This is a property of the recovered chondritic fireballs Lost City and Innisfree, and so can be considered an empirical criterion. Use of dynamic mass in calculating the theoretical end-height automatically corrects for deviations of dynamic mass from actual mass due to fragmentation or irregular shape. In our earlier work, a fireball was judged to pass this criterion if the calculated end-height was within 1.5 km of the observed end-height.

(4) Photometric lightcurve no more irregular than that of the recovered fireballs Lost City, Innisfree and Pribram. This is a qualitative criterion supporting the observation that many fireballs display extensive flaring and terminate their visible flight suddenly while near maximum brightness. This probably indicates that they were so badly disrupted near the end of their visible flight that no recoverable fragments survived. At the same time it takes into account that to some extent these phenomena were also observed for recovered fireballs; to this extent flaring and sudden termination do not preclude survival.

In this chapter, parameters will be defined to quantify the extent to which other fireballs failed to meet the first three criteria.

(1) Criterion ΔZ_8

All but 3 of the 66 fireballs with aphelia > 5 AU ended their visible atmospheric flight at velocities > 8 km s^{-1} and thus failed the first criterion of Wetherill and ReVelle (1981). In a few cases this may have been because they were simply so small and faint that a well-behaved final deceleration was not observable. However in most cases the termination was abrupt and the most likely explanation of the data is that the meteoroid fragmented into many small pieces which then rapidly decelerated without producing observable trails. The fireballs vary greatly in this regard; some almost reached 8 km s^{-1}, whereas others experienced little deceleration and ended their visible flight while still near their entry velocity. These differences can be measured by calculating the difference between the observed end-height and the theoretical height at which a body would slow to 8 km s^{-1}, assuming that it decelerated as the fireball did during the observed portion of its flight.

This is done by calculating the dynamic mass of the body from the deceleration data of McCrosky et al. (1979)

$$M_d(t) = \frac{1}{8} \frac{(C_D S_F \rho_{\text{atm}})^3 V^6}{\rho_M^2 (\dot{V})^3} \tag{1}$$

where $M_d(t)$ is dynamic mass; ρ_{atm} is atmospheric density; $V, \dot{V}$ are velocity and deceleration, respectively; S_F is shape factor (1.209 for a sphere); C_D is drag coefficient (0.92); and ρ_M is density of the meteoroid (taken to be 3.7 g cm^{-3}).

If the fireball is actually a single spherical body with density 3.7 g cm^{-3}, the dynamic mass calculated from the deceleration data should equal the true mass of the body. More frequently, fragmentation, irregular shape, or lower density will increase the effective surface area and the dynamic mass will be less than the true mass. For recovered ordinary chondrites the dynamic mass is ~1/3 to 1/2 the best estimate of the true mass. For more friable objects, the dynamic mass will be even lower.

Dynamic masses were calculated using deceleration values at as many altitudes as given by McCrosky et al. (1979). These were extrapolated to the entry mass by use of the relation

$$M_{d\infty} = M_d(t) e^{\sigma/2 (V_\infty^2 - V^2)} \tag{2}$$

where σ is the average ablation coefficient, taken to be 0.02 s^2 km^{-2} (ReVelle 1979) and V_∞ is observed entry velocity. The average dynamic mass at entry ($M_{d\infty}$) is calculated by averaging the individual values.

The atmospheric density ρ_8 at which the body should have decelerated to 8 km s^{-1} is given by meteor theory (e.g. McIntosh 1970) as

$$\rho_8 = \frac{b \rho_M^{2/3}}{C_D S_F} \cos Z_R \overline{M_{d\infty}}^{1/3} e^{-\sigma V_\infty^2/6} \left[Ei\left(\frac{\sigma V_\infty^2}{6}\right) - Ei\left(\frac{\sigma \cdot 8^2}{6}\right) \right] \tag{3}$$

where b is reciprocal scale height (0.142 km^{-1}); Z_R is entry angle, measured from the vertical; the other variables have been defined. The function $Ei(x)$ is the exponential integral:

$$Ei(x) = \int_{-\infty}^{x} \frac{e^\eta}{\eta} d\eta \tag{4}$$

which is well tabulated (e.g. Federal Works Agency 1940).

It should be noted that when the dynamic mass M_d calculated from Eq. (1) is used in Eq. (3), the dependence of ρ_8 on ρ_M, C_D, and S_F exactly cancels so ρ_8 is independent of the values assumed for these quantities.

The parameter Δz_8 is then obtained by subtracting the altitude corresponding to ρ_8, $z(\rho_8)$ from the observed end-height z_E

$$\Delta Z_8 = Z_E - Z(\rho_8) \ . \tag{5}$$

The values of $Z(\rho_8)$ were obtained from National Oceanic and Atmospheric Administration (NOAA) (1976).

(2) Criterion $M_p/M_{d\infty}$

As discussed above, the dynamic mass is not usually equal to the true mass. For a friable, low-density, or irregular body it will be smaller than the true mass. The photometric mass, determined by the integral of the photometric luminosity, is more likely to be proportional to the true mass at least for bodies of similar physical and chemical nature. Thus similar bodies should have similar ratios of photometric mass to dynamic mass ($M_p/M_{d\infty}$). The absolute value of the photometric mass is uncertain because of uncertainty in the luminous efficiency of the fireball, the fraction of the kinetic energy loss converted into visible radiation. Other studies (ReVelle and Rajan 1979) strongly suggest that the luminous efficiencies used earlier are too low by an order of magnitude, leading to photometric masses about a factor of 10 too high. In order to make the ratios $M_p/M_{d\infty}$ more realistic the photometric masses given by McCrosky et al. have been reduced by a factor of 10. Of course this has no effect on comparison of these trans-Jovian fireballs with one another or with those believed similar to recovered meteorites, because all photometric masses have been reduced by the same factor.

(3) Criterion ΔZ

An expression analogous to Eq. (3) can be written to find the atmospheric density at which the fireball should have been reduced to the observed final velocity V_E

$$\rho_E = \frac{b \rho_M^{2/3}}{C_D S_F} \cos Z_R \, \overline{M_{d\infty}}^{1/3} \, e^{-\sigma V_\infty^2/6} \left[Ei\left(\frac{\sigma V_\infty^2}{6}\right) - Ei\left(\frac{\sigma V_E^2}{6}\right) \right] . \tag{6}$$

This expression differs from Eq. (3) only in the use of V_E in the final term instead of $V = 8$ km s^{-1}. As for Eq. (3) ρ_E is also independent of the values of ρ_M, C_D, and S_F used. Conceptually however, ρ_E and ρ_8 are quite different. The quantity ΔZ,

$$\Delta Z = Z_E - Z(\rho_E) \tag{7}$$

is a measure of the deviation of the observed end-height from the theoretical height at which a body of mass $M_{d\infty}$ would decelerate to the observed final velocity V_E, regardless of how high V_E actually is. Regardless of its density, shape, and degree of fragmentation, a body with initial dynamic mass uniform over entire atmospheric trajectory will have $\Delta Z = 0$. The quantity ΔZ is thus a measure of the constancy of dynamic mass, particularly near the end of the fireball's visible flight where fragmentation is severe. In other words, it is a measure of how well-behaved the fireball was during the observed portion of its flight. In contrast ΔZ_8 is a measure of the fireball's success in safely decelerating to 8 km s^{-1}.

The parameter ΔZ provides information equivalent to that given previously (Wetherill and ReVelle 1981) by the scaled end-height (H_E^*). The relationship between these quantities is

$$\Delta Z = H_E^* - 21.1 \text{ km} . \tag{8}$$

Values of These Parameters for Trans-Jovian Fireballs

The quantity ΔZ_8 is plotted versus M_p/M_d ratio for Prairie Network fireballs with aphelia > 5.0 AU in Fig. 1. The shaded region labeled "meteoritic" is the field occupied by the 27 Prairie Network fireballs previously identified as indistinguishable from recovered meteorites (Wetherill and ReVelle 1981). Two trans-Jovian fireballs lie within this field and a significant number lie near but outside this field. This deviation cannot primarily result from uncertainties in dynamic mass measurements. The appearance of $M_{d\infty}$ in the quantities plotted both on the ordinate and abscissa of Fig. 1 leads to the anticorrelation represented by the dashed line labeled "slope of locus of dynamic mass errors." If points have been misplotted because of error in dynamic mass measurements, they will be displaced along a line of this slope. In all but two cases the points remain outside the meteoritic field even after this displacement.

This anticorrelation permits some use of fireballs for which deceleration data are not given by Ceplecha and McCrosky (1976). These are plotted as open circles in Fig. 1, and are arbitrarily assigned $M_p/M_{d\infty} = 100$. Because this assignment is arbitrary, the points can be freely displaced parallel to the dashed slope.

In contrast to fireballs which cluster near the meteoritic field, many scatter widely and have large values of ΔZ_8 and/or $M_p/M_{d\infty}$. Most of the open-circled points, which have no deceleration data, share this property. Fireballs with high ΔZ_8 or $M_p/M_{d\infty}$ are those most unlike recoverable meteorites. The high $M_p/M_{d\infty}$ values probably result from considerable fragmentation during visible flight, although low density may also contribute to this effect. High values of ΔZ_8 correspond to bodies that ended their flight far higher than the altitude expected if they could penetrate the atmosphere as well as the recoverable meteorites.

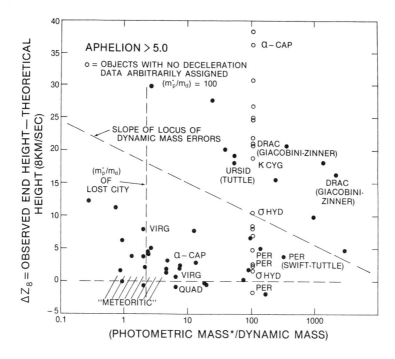

Fig. 1. Data for trans-Jovian fireballs. The parameter ΔZ_8 measures the meteoroid's ability to survive deceleration in the atmosphere. The ratio of photometric mass to dynamic mass $M_p/M_{d\infty}$ primarily measures the degree of fragmentation of the meteroid during visible flight. Objects plotted near the shaded area labeled meteoritic most closely physically resemble recovered chondritic meteorites. Open circles represent fireballs for which deceleration data are not available, arbitrarily assigned $M_p/M_d = 100$.

In Fig. 2 the values of ΔZ are plotted against $M_p/M_{d\infty}$ for trans-Jovian fireballs. Again the points scatter widely. One group clusters in the vicinity of $\Delta Z = 0$ with M_p/M_d between ~1 and 10. For this group the deceleration down to final velocity was well-behaved, i.e. in agreement with meteor theory. The values of $M_p/M_{d\infty}$ are not very different from the value 2.3 found for the recovered meteorite Lost City and the value 3.2 found for Innisfree. Moreover, examination of the individual data points (not identified in Figs. 1 and 2) shows that the same bodies which plot in the cluster near $\Delta Z = 0.1 < M_p/M_{d\infty} < 10$ in Fig. 2 fall in the group near the meteoritic field in Fig. 1. Those with high ΔZ tend to also have high ΔZ_8. Thus the uniformity of deceleration before reaching the final observed altitude and the extent to which the body can safely decelerate to 8 km s^{-1} are related. This is plausible since a friable or otherwise weak meteoroid would be expected to crumble during its flight and be completely disrupted while still at high velocity.

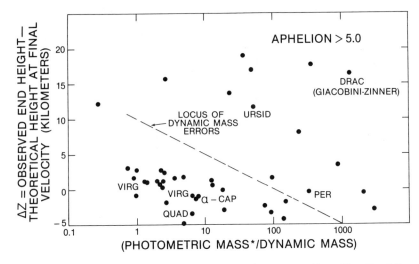

Fig. 2. Data for trans-Jovian fireballs. The parameter ΔZ measures the uniformity of the effective surface area to mass ratio of the meteoroid during its visible flight. Objects with low ΔZ and $M_p/M_{d\infty}$ most closely resemble survivable meteorites.

Figure 3 shows the distribution of peak dynamic pressure P_d at an atmospheric pressure of ρ_{atm}

$$P_d = \frac{1}{2} \rho_{\text{atm}} C_D V^2 \qquad (9)$$

for fireballs that failed to decelerate to 8 km s^{-1}. Assuming that no trail was observed at lower velocity because of disruption of the body, this dynamic pressure represents a measure of the mechanical strength of the fireball. The observed peak pressures range from $\sim 10^4$ to $> 10^7$ dyne cm^{-2}. As expected, the position of a fireball on Fig. 2 and its maximum dynamic pressure are related. The shaded area in Fig. 3 includes fireballs lying above the dashed line labeled "locus of dynamic mass errors" in Fig. 2. These objects have high ΔZ_8, ΔZ, and/or $M_p/M_{d\infty}$, which also indicate mechanical weakness. This group of fireballs also have low dynamic pressures ($< 10^6$ dyne cm^{-2}).

Laboratory measurements of the compressive strength of meteorites range from 6×10^7 to 4×10^9 dyne cm^{-2} (Buddhue 1942). All 8 samples measured were metamorphosed ordinary chondrites (petrographic grades 5 and 6). It seems likely, on the basis of qualitative experience in breaking meteorites for chemical analysis, that CI and possibly CM carbonaceous chondrites are weaker, perhaps with compressive strengths of $\sim 10^7$ dyne cm^{-2}. Based on their mass (~ 10 kg) the single largest fragments of Orgueil (CI) and Murray (CM) should have had end-heights of ~ 20 km. Comparison with Innisfree and Lost City and with meteor theory indicates that these

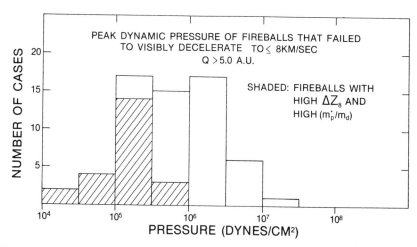

Fig. 3. Peak dynamic pressure of trans-Jovian fireballs.

fragments experienced and survived peak pressures at altitudes of ~ 25 km and velocities of ~ 10 km s^{-1}, corresponding to dynamic pressures of about 1.8×10^7 dyne cm^{-2}. Ordinary chondrites with laboratory compressive strengths $\sim 10^9$ dyne cm^{-2} are observed to break up during atmospheric entry at lower pressures. For example, Lost City was observed to break up at 8×10^6 dyne cm^{-2} and Innisfree at 1×10^6 dyne cm^{-2}. The major difference between the fragmentation history of these recovered ordinary chondrites and that of the objects included in Fig. 3 is not that the latter objects fragmented a low pressures but that no large enough object survived to produce a continuing trail after fragmentation.

We conclude that the mechanical strengths of fireballs with aphelia beyond Jupiter barely overlap those of recovered meteorites. Most of the higher strength members of this population come close in ΔZ_8, ΔZ, $M_p/M_{d\infty}$, and strength to the weakest known meteorites, but seem still weaker. It does not seem appropriate to describe these objects as dustballs; perhaps dirt clods would be more accurate. However it should be remembered that specific descriptions are speculative in the absence of a recovered sample.

It also seems likely that objects which do not decelerate to 8 km s^{-1} are members of a continuous population that grades into survivable meteorites similar to those in collections. If so, the weaker carbonaceous chondrites (CI and CM) seem the best candidates with which to identify these objects. Meteorites of this kind comprise $\sim 2\%$ of recovered meteorites observed to fall. Another study (Wetherill and ReVelle 1981) identified 27 Prairie Network fireballs believed equivalent to survivable stone meteorites such as ordinary chondrites and achondrites, and a similar number of possibly survivable bodies. If these identifications are correct, the strongest trans-Jovian fireballs comprise $\sim 10\%$ of the more clearly meteoritic fireballs. Although fireballs with similar properties and aphelia < 5.0 AU have not been thoroughly examined, a preliminary look leads to an estimate that ~30 objects similar to the strongest trans-Jovian fireballs will be found among these bodies with smaller aphelia. Thus the total number of relatively strong fireballs that failed to decelerate to 8 km s^{-1} is comparable to the number of more obviously survivable meteorites, in contrast to the small abundance (2%) of recovered weak carbonaceous meteorites. Only a small fraction of these fireballs need survive and be recovered to account for the observed frequency of carbonaceous meteorites. If this identification of the strongest trans-Jovian material with CI and CM chondrites is correct, it is likely that one or more of these objects collected arrived from a trans-Jovian orbit and hence is of cometary origin. It should here be noted that in addition to the relatively strong bodies that failed to visibly decelerate to 8 km s^{-1}, three trans-Jovian fireballs (PN39057, PN39972, PN40026A) actually were observed to decelerate to 8 km s^{-1}. PN39057 is one of the 27 objects identified (Wetherill and ReVelle 1981) as a recoverable meteorite, even though its aphelion is well beyond Jupiter at 7.3 AU. In view of all these data, it seems likely that the trans-Jovian fireballs, although usually weak, grade into bodies that should be represented in meteorite collections. The objects falling in the shaded region of Fig. 3 are much weaker; it is hard to believe that any of these objects could be recovered as a meteorite, although their fragmentation debris might be collected from the stratosphere (Brownlee et al. 1977).

Fireballs of varying ability to penetrate the atmosphere were distinguished by Ceplecha and McCrosky (1976) in a classification based on their "PE criterion". The similarities and differences between this criterion and ours is discussed elsewhere (Wetherill and ReVelle 1981). It should however be emphasized that despite conceptual and practical differences between these criteria and with others (Ceplecha 1980), the classification of Prairie Network fireballs made by Ceplecha and McCrosky corresponds closely to that presented here. Almost all the very weak fireballs correspond to their Type IIIB, whereas those that cluster near the meteoritic field are almost all Type II with a small admixture of Type I. Type IIIA, intermediate between II and IIIB, is not well-represented among these trans-Jovian fireballs but those that are found have the appropriate intermediate characteristics. For this reason, the identification of Type II with carbonaceous chondrites made by Ceplecha and McCrosky is compatible with the above discussion, if it is recognized that the recovered carbonaceous meteorites represent unusually survivable endmembers of this group.

As discussed earlier, it is plausible to assume that almost all trans-Jovian fireballs are of cometary origin. A large fraction of the objects belong to the stronger group of fireballs. Because the number of stray asteroidal bodies expected to impact the Earth while in Jupiter-crossing orbits is < 1, it is unlikely that many of these are of asteroidal origin. The principal alternative to cometary bodies would be an undiscovered trans-Jovian asteroid belt that could also be a source of Jupiter-crossing debris. Although it seems better to explain these fireballs in terms of comets rather than appealing to an undiscovered (except possibly for Chiron) class of solar system material, this possibility should not be forgotten.

Assuming cometary origin for almost all trans-Jovian fireballs, to understand the composition of the cometary nucleus the proportion of stronger to weaker trans-Jovian fireballs must be known. Here the difficulty caused by the selected sample is most severe. As discussed above, the Prairie Network fireballs selected for data reduction were strongly biased against trails of short duration; this will obviously discriminate against meteors which have high ΔZ_8 and disintegrate at high altitude. The bias against fireballs identified with meteor streams (except for Taurids) probably also represents a bias against the weaker fireballs, as discussed further in Sec. III.

A rough idea of the extent of this selection effect may be provided by the reduction (McCrosky et al. 1978) during the period JD 39000 to 39500 of a more representative sample including almost all meteors of -8 mag and duration > 2 s. During this period some data are given for fairly faint fireballs (-7 mag) with a duration as short as 1 s. The total number of fireballs reduced during this period is 104; this period represents 12% of the time the network was in operation. A similar reduction effort during the entire operation of the network would have led to a total of ~ 850 reduced orbits instead of the 336 actually published. During the period of more complete reduction 7 of the 14 classifiable observed trans-Jovian fireballs were of the weaker type, corresponding to an expectation of 84 during the entire operation of the network instead of the 23 actually reduced. Thus it seems that observational selection has discriminated against the weaker fireballs and that they are actually at least equal in number to the stronger objects. A total of 2700 meteors were photographed, which leads to 1850 unaccounted for at the higher frequency of reduction. If all these remaining fireballs were weak trans-Jovian objects then the weak objects would outnumber the strong by a factor of ~ 20. This is unlikely, but it could be that the stronger objects represent only 10% of the total number of bright trans-Jovian meteors.

Furthermore, although exact estimates are not possible it is likely that only a very small fraction of the total solid material lost from comets is in the form of these large meteoroids. Most cometary dust is swept out into the dust tail by radiation pressure and easily lost from the solar system. The larger (mm-size range) particles are observed in apparently Sun-directed antitails and impact the Earth as ordinary small meteors. The mass flux of these small meteors probably is greater than that in the fireball mass range (100 to 10^6 g) by an order of magnitude. The stronger cometary fireballs altogether probably comprise $\leqslant 10^{-3}$ of the cometary nucleus. Unless the weaker and more fine-grained cometary solids are of the same chemical and mineralogical composition as the stronger fireballs, studies of cometary surfaces by remote sensing techniques should not be expected to provide information on the composition of these meteoroids.

III. FIREBALLS IN METEOR STREAMS

Because of the well-known association of comets with meteor streams, the most obvious approach to studying the physical characteristics of cometary material would be to examine data for the abundant fireballs belonging to these streams. This is not possible at present because most meteors identified as stream members were rejected in the reduction of Prairie Network data. An adequate discussion of this question will require further reduction of data from the Prairie Network and other networks and possibly acquisition of more data. Nevertheless, some information is available at this time. Trans-Jovian meteors plausibly associated with streams have been marked on Figs. 1 and 2. Many of these are fireballs for which deceleration data are not available, that have been assigned a nominal value of $M_p/M_{d\infty}$ in Fig. 1 (open circles). As was the case for trans-Jovian fireballs in general, these objects scatter widely on this diagram. A few are found in the cluster near the meteoritic field. Those without deceleration data could actually fall anywhere on a line drawn through the open circle points parallel to the dashed line labeled "slope of locus of dynamic mass errors." When this is done several stream objects clearly fall among the weaker trans-Jovian fireballs, and some intermediate cases are also found.

The sole exceptions to the policy of excluding stream fireballs in the reduction of Prairie Network data were the members of the Taurid stream usually attributed to P/Encke. These meteors belong to a more general class of ecliptic streams: meteors of relatively low $< 12°$ inclination with aphelia from 3.5 to 4.5 AU and small perihelia (0.6 AU). Because these orbits have aphelia well inside the orbit of Jupiter the possibility must be considered that asteroidal as well as cometary material will be found in orbits of this kind; this has the effect of blurring the identification of these fireballs with comets.

There are two known mechanisms which appear effective in transporting asteroidal collision debris into Earth-crossing orbits. The first of these involves injection of collision debris into librating orbits in the vicinity of Kirkwood Gaps (Zimmerman and Wetherill 1973; Scholl and Froeschlé 1977). While in these orbits high eccentricities and aphelia of ~ 4.5 AU can be reached. The libration condition precludes close encounters with Jupiter and these orbits are therefore safe from strong perturbations caused by such encounters. However, an inevitable second collision can permit the fragments to escape the librating region while still in orbits of high eccentricity. Successive Jupiter perturbations of the fragments near their aphelia on a 10^6 yr time scale can then cause the perihelion to become Earth-crossing. This is an effective mechanism to provide asteroidal fireballs with perihelia near 1 AU, but only rarely is it possible for the perihelia to diffuse to much smaller values before Earth (or Venus) perturbations cause the aphelia to become Jupiter-crossing and subsequently to be rapidly ejected from the solar system. It is this property of 1 AU initial Earth-crossing orbits that provides good agreement with the observed distribution of times-of-fail and radiants of ordinary chondrites (Wetherill 1968,1969).

The other mechanism involves perturbation into Earth-crossing by the ν_6 secular resonance (Williams 1969; Chapman et al. 1978), probably combined with Mars perturbations (Williams 1973; Wetherill and Williams 1979). Earth-crossing orbits produced by this mechanism tend to have small aphelia, rarely beyond 3.5 AU. (See, for example, the steady-state distribution of orbits produced by this mechanism from asteroids 1204 Renzia and 313 Chaldaea, initially near the inner edge of the belt, in Wetherill 1979.)

Thus it appears that neither of these mechanisms is very effective in placing asteroidal debris into orbits with both large aphelia and small perihelia. Periodic Comet Encke demonstrates that comets can achieve orbits of this kind on a $\sim 10^4$ yr time scale, but the dynamical mechanism for their doing so is unclear. The best candidate remains the "jet effect" resulting from momentum loss associated with the thermal lag in evaporation of a gas from a rotating cometary nucleus (Whipple 1950; Sekanina 1969,1971). However, more recent studies (Sekanina 1981) of cometary rotation and precession suggest that this mechanism may not be as effective in reducing cometary aphelia as was previously believed. It seems likely that these ecliptic-stream orbits are dominantly cometary, but the uncertainties discussed above might introduce some doubt.

The distribution of Prairie Network fireballs in these Encke-like orbits of large aphelia and small perihelia are shown in Fig. 4. The average orbits for well-established meteor streams with orbits of this kind are plotted as star symbols in Fig. 4. All the orbital elements of some fireballs are probably similar enough to those of these streams to identify these fireballs with the streams. The only streams with a significant number of fireballs are the Taurids and the χ-Orionids. The latter, although not a major stream for small meteors (McCrosky and Posen 1961), does appear to be well represented among the fireballs; a total of 9 members appear in data of Ceplecha (1977), Whipple (1954) and Babadzhanov and Kramer (1965), as compiled by Wasson and Wetherill (1979). The boundaries between these two streams are not clear. In the super-Schmidt data of McCrosky and Posen (1961) the boundary seems defined by an absence of data during the period from 25 Nov. to 4 Dec., and thus could be an artifact of the data set used in the statistical identification of streams. In any case, points representing fireballs probably belonging to these streams are plotted as solid circles and enclosed within the dashed area in Fig. 4. These points are surrounded by a number of outliers plotted as open circles.

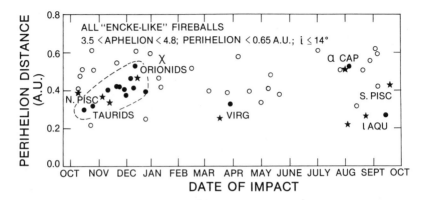

Fig. 4. Perihelion distribution of observed Prairie Network fireballs in orbits similar to that of Comet P/Encke and the Taurid meteor streams (aphelion approaching Jupiter, perihelion near Mercury, low inclination). The points marked with a star identify known meteor streams. The two points for the Taurids and ι Aquarids correspond to the northern and southern branches of these streams. Solid points correspond to fireballs for which all orbital elements support identification with one of these streams.

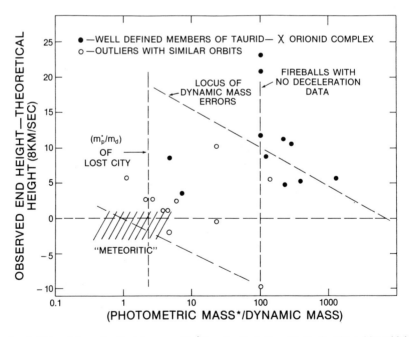

Fig. 5. Plot of Δz_8 (ordinate) versus $M_p/M_{d\infty}$ for fireballs with Encke-like orbits which fell between October and February. The solid points are those lying within the dashed region of Fig. 4 and are plausibly associated with the Taurid and χ-Orionid streams.

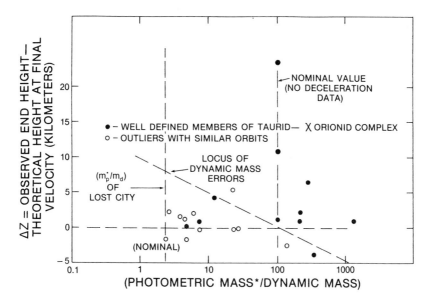

Fig. 6. Plot of ΔZ versus $M_p/M_{d\infty}$ for fireballs with Encke-like orbits which fell between October and February. Solid points lie within the dashed region of Fig. 4.

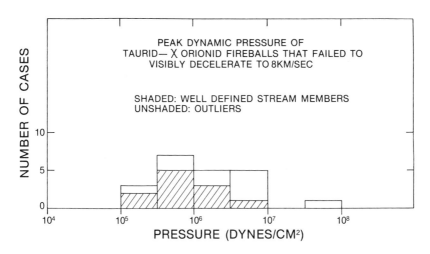

Fig. 7. Peak dynamic pressure for fireballs shown in Figs. 5 and 6. The shaded areas correspond to objects falling within the dashed region of Fig. 4.

The ΔZ_8 criterion was applied to the members of this Taurid and χ-Orionid complex and its outliers (Fig. 5). Although there is some overlap, the more well-defined stream members appear to lie farther from the meteoritic field than the others. In terms of the Ceplecha-McCrosky classification the stream members belong mostly to Type IIIA, whereas the outliers belong to Type II.

The same difference is observed for the ΔZ criterion (Fig. 6) and in the distribution of peak dynamic pressures (Fig. 7). The fireballs most clearly identified with the streams appear weaker than the outliers, yet the stream members are not as weak as the weaker trans-Jovian bodies (Fig. 3). They appear to have been disrupted by dynamic pressures $\sim 10^6$ dyne cm^{-2} (1 atm). When compared to the recovered weak carbonaceous chondrites with strengths of $\sim 10^7$ dyne cm^{-2}, it appears unlikely that these objects in the meteor streams could survive passage through the atmosphere even if they entered at velocities much lower than the actual entry velocities of ~ 27 km s^{-1}. However, in view of the observed spread in ΔZ_8, ΔZ, and peak pressures, it seems possible that stronger objects may occasionally be found in such streams. It also appears incorrect to describe these stream meteors as dustballs, 1 atm (76 cm of mercury) is a heavy load for any object resembling a cobweb.

The outliers appear significantly stronger and therefore more likely to be represented in meteorite collections. The interpretation of this difference is unclear. Meteor streams are dynamically unstable as a result of planetary perturbations. The longitude of the nodes, the quantity that determines the fall date, will be randomized on a time scale of $\sim 10^4$ yr. These low-inclination bodies will undergo small oscillations in e and i as a consequence of secular free oscillations (Kozai 1962). However, the time scale for major changes in a, e, i, and the perihelion and aphelion distances, caused by close encounters to planets will be much longer, 10^6 to 10^7 yr. As a consequence the stream members may be expected to be a better sample of fireball-size material ejected by a comet. Over the longer time scale weak bodies will be eliminated by collisions, particularly for orbits like these which traverse the asteroid belt. Cosmic-ray exposure ages of recovered carbonaceous meteorites are $\sim 10^6$ yr; significantly weaker material should be destroyed in a shorter time. Therefore the stronger fireballs that deviate from the stream may be older cometary stream objects that have survived collisional destruction. On the other hand, the outliers could be stronger because they are from asteroids, in spite of arguments against this possibility. A better study of stream fireballs is required to advance our understanding of this question.

IV. CONCLUSIONS

(1) Available data from the Prairie Network shows that fireballs most probably of cometary origin vary considerably in their physical properties,

especially their ability to penetrate the atmosphere without fragmentation. Although sampling problems are severe, it appears that weaker bodies with compressive strengths $\sim 10^5$ dyne cm^{-2} are the most abundant cometary debris in the 100 to 10^6 g size range. These bodies may be correctly called dustballs only if the term is taken to mean weak dirt clods, not cobwebs.

(2) Perhaps as much as 25% of the cometary fireballs are stronger than these weak bodies. The strongest probably overlap the weakest carbonaceous meteorites in ability to penetrate the atmosphere. The characterization by Ceplecha and McCrosky of these type II fireballs as carbonaceous chondrites may be appropriate, if it is recognized that even though possibly chemically and mineralogically equivalent to these meteorites, most Type II fireballs are weaker than recovered carbonaceous meteorites. The mass flux of these objects appears adequate to supply the observed recovery rate of carbonaceous meteorites, even allowing for their low probability of penetrating the atmosphere and being found on the ground.

(3) These larger and stronger objects probably represent only a small fraction (perhaps 10^{-3}) of the cometary nucleus. Whether they are chemically or mineralogically similar to the more abundant fine cometary dust, depends on the degree to which cometary nuclei are chemically homogeneous and whether or not the processes that create stronger lumps of cometary solids and those that may cause chemical differences are related. For example, aqueous alteration observed in CI chondrites could result in cementing the material into an object of sufficient strength to increase significantly the probability of penetration of the atmosphere and recovery.

REFERENCES

Anders, E. 1964. Origin, age, and composition of meteorites. *Space Sci. Rev.* 3:583-714.
Arnold, J.R. 1965. The origin of meteorites as small bodies. II. The model. III. General considerations. *Astrophys. J.* 141:1536-1556.
Babadzhanov, P.B., and Kramer, E.N. 1965. Orbits of bright photographic meteors. In *Meteor Orbits and Dust,* ed. G.S. Hawkins. Smithsonian Contr. to Astrophys. Vol. 11, (Washington: NASA SP-135), pp. 67-80.
Baldwin, B., and Sheaffer, Y. 1971. Ablation and breakup of large meteoroids during atmospheric entry. *J. Geophys. Res.* 76:4653-4668.
Bronshten, V.A. 1976. The Tunguska meteorite and bolides of the Prairie Network. *Astron. Vestnik* 10:73-80.
Brownlee, D.E.; Rajan, R.S.; and Tomandl, D.A. 1977. A chemical and textural comparison between carbonaceous chondrites and interplanetary dust. In *Comets, Asteroids, Meteorites,* ed. A.H. Delsemme (Toldeo, Ohio: Univ. of Toledo), pp. 137-141.
Buddhue, J.D. 1942. The compressive strength of meteorites. *Contrib. Soc. Res. Meteorites* 3:39-40.
Ceplecha, Z. 1961. Multiple fall of Pribram meteorites photographed. *Bull. Astron. Inst. Czech.* 12:21-47.
Ceplecha, Z. 1973. Evidence from spectra of bright fireballs. In *Evolutionary and Physical Properties of Meteoroids,* eds. C.L. Hemenway, P.M. Millman, and A.F. Cook (Washington: NASA SP-319), pp. 89-102.
Ceplecha, Z. 1977. Fireballs photographed in central Europe. *Bull. Astron. Inst. Czech.* 28:328-340.

Ceplecha, Z. 1980. Observational and theoretical aspects of fireballs. In *Solid Particles in the Solar System,* eds. I. Halliday and B.A. McIntosh (Dordrecht: D. Reidel Publ. Co.), pp. 171-183.
Ceplecha, Z., and McCrosky, R.E. 1976. Fireball and heights: A diagnostic for the structure of meteoric material. *J. Geophys. Res.* 81:6257-6275.
Chapman, C.R.; Williams, J.G.; and Hartmann, W.K. 1978. The asteroids. *Ann. Rev. Astron. Astrophys.* 16:33-75.
Cook, A.F. 1973. A working list of meteor streams. In *Evolutionary and Physical Properties of Meteoroids,* eds. C.L. Hemenway, P.M. Millman, and A.F. Cook (Washington: NASA SP-319), pp. 183-191.
Federal works Agency 1940. Tables of sine, cosine, and exponential integrals (2 Vols). Works Project Administration for the City of New York, conducted under the sponsorship of the National Bureau of Standards. Report of Official Project 765-97-3-10.
Halliday, I.; Griffin, A.A.; and Blackwell, A.T. 1981. The Innisfree meteorite fall: A photographic analysis of fragmentation, dynamics, and luminosity. *Meteoritics* 16:153-170.
Helin, E.F., and Shoemaker, E.M. 1979. The Palomar planet-crossing asteroid survey. *Icarus* 40:321-328.
Hughes, D.W. 1978. Meteors. In *Cosmic Dust,* ed. J.A.M. McDonnell (New York: Wiley and Sons), pp. 123-185.
Kozai, Y. 1962. Secular perturbations of asteroids with high inclination and eccentricity. *Astron. J.* 67:591-598.
Levin, B.J.; Simonenko, A.N.; and Anders, E. 1976. Farmington meteorite: A fragment of an Apollo asteroid? *Icarus* 28:307-324.
Lyttleton, R.A. 1948. On the origin of comets. *Mon. Not. Roy. Astron. Soc.* 108:465-475.
McCrosky, R.E., and Ceplecha, Z. 1969. Photographic network for fireballs. In *Meteorite Research,* ed. P. Millman (Dordrecht: D. Reidel), pp. 600-612.
McCrosky, R.E., and Posen, A. 1961. Orbital elements of photographic meteors. *Smithsonian Contributions to Astrophysics* Vol. 4, no. 2.
McCrosky, R.E.; Posen, A.; Schwartz, G.; and Shao, C.Y. 1971. Lost City meteorite—its recovery and a comparison with other fireballs. *J. Geophys. Res.* 76:4090-4108.
McCrosky, R.E.; Shao, C.-Y.; and Posen, A. 1978. Prairie Network fireball data. I. Summary and orbits. *Meteoritika* 37:44-59.
McCrosky, R.E.; Shao, C.-Y.; and Posen, A. 1979. Prairie Network fireball data. II. Trajectories and light curves. *Meteoritika* 38:106-156.
McIntosh, B.A. 1970. On the end point height of fireballs. *Roy. Astron. Soc. Can. J.* 64:267-281.
National Oceanic and Atmospheric Administration 1976. U.S. standard atmosphere, 1976. *NOAA-S/T 76-1562* (Washington: U.S. Govt. Printing Office).
Öpik, E.J. 1963. Survival of comet nuclei and the asteroids. *Advan. Astron. Astrophys.* 2:219-262.
Öpik, E.J. 1966. The stray bodies in the solar system, Part II. The cometary origin of meteorites. *Advan. Astron. Astrophys.* 4:302-336.
ReVelle, D.O. 1979. A quasi-simple ablation model for large meteorite entry: Theory vs. observations. *J. Atmos. Terr. Phys.* 41:453-473.
ReVelle, D.O., and Rajan, R.S. 1979. On the luminous efficiency of meteoritic fireballs. *J. Geophys. Res.* 84:6255-6262.
Scholl, H., and Froeschlé, C. 1977. The Kirkwood gaps as an asteroidal source of meteorites. In *Comets, Asteroids, Meteorites,* ed. A.H. Delsemme (Toledo, Ohio: Univ. of Toledo), pp. 293-295.
Sekanina, Z. 1969. Total gas concentration in atmosphere of the short-period comets and impulsive forces upon their nuclei. *Astron. J.* 74:944-950.
Sekanina, Z. 1971. A core-mantle model for cometary nuclei and asteroids of possible cometary origin. In *Physical Studies of Minor Planets,* ed. T. Gehrels (Washington: NASA SP-267), pp. 423-428.
Sekanina, Z. 1981. Rotation, precession, and non-gravitational motion of comets. *Ann. Rev. Earth Planet. Sci.* 10. In press.

Simonenko, A.N. 1979. *Meteorites—Fragments of Asteroids*. (Moscow: Nauka).
Wasson, J.T., and Wetherill, G.W. 1979. Dynamical, chemical and isotopic evidence regarding the formation locations of asteroids and meteorites. In *Asteroids*, ed. T. Gehrels (Tucson: Univ. Arizona Press), pp. 926-974.
Wetherill, G.W. 1968. Time of fall and origin of stone meteorites. *Science* 159:79-82.
Wetherill, G.W. 1969. Comments on paper by D.E. Fisher and M.F. Swanson, Frequency distribution of meteorite-earth collisions. *J. Geophys. Res.* 74:4402-4405.
Wetherill, G.W. 1974. Solar system sources of meteorites and large meteoroids. *Ann. Rev. Earth Planet. Sci.* 2:303-331.
Wetherill, G.W. 1976. Where do the meteorites come from? A reevaluation of the earth-crossing Apollo objects as sources of chondritic meteorites. *Geochim. Cosmochim. Acta* 40:1297-1317.
Wetherill, G.W. 1979. Steady-state population of Apollo-Amor objects. *Icarus* 37:96-112.
Wetherill, G.W., and ReVelle, D.O. 1981. Which fireballs are meteorites? A study of the Prairie Network photographic meteor data. Submitted to *Icarus*.
Wetherill, G.W., and Williams, J.G. 1968. Evaluation of the Apollo asteroids as sources of stone meteorites. *J. Geophys. Res.* 73:635-648.
Wetherill, G.W., and Williams, J.G. 1979. Origin of differentiated meteorites. In *Origin and Distribution of the Elements*, ed. L.H. Ahrens (Oxford: Pergamon), pp. 19-31.
Whipple, F.L. 1950. A comet model. I. The acceleration of Comet Encke. *Astrophys. J.* 11:375-394.
Whipple, F.L. 1954. Photographic meteor orbits and their distribution in space. *Astron. J.* 59:300-316.
Williams, J.G. 1969. Secular Perturbations in the Solar System. Ph.D. Dissertation, Univ. California at Los Angeles.
Williams, J.G. 1973. Meteorites from the asteroid belt? *EOS* 54:233 (abstract).
Zimmerman, P.D., and Wetherill, G.W. 1973. Asteroidal source of meteorites. *Science* 182:51-53.

PART III
Dust

OPTICAL AND INFRARED OBSERVATIONS OF BRIGHT COMETS IN THE RANGE 0.5 μm TO 20 μm

E.P. Ney
University of Minnesota

Infrared observations have been obtained on seven bright comets and P/Encke since 1965. Five comets were dusty with pronounced Type II tails; these were Comets Bennett 1970 II, Kohoutek 1973 XII, Bradfield 1974 III, West 1976 VI, and Bradfield 1980t. All these dusty comets showed a silicate signature at 10 and 18 μm indicating the presence of small refractory grains of radius < 5 μm in the comae and tails. The antitail of Comet Kohoutek did not have the silicate feature suggesting the presence of large particles of radius > 30 μm. Comet Kobayashi-Berger-Milon 1975 IX and Comet Ikeya-Seki 1965 VIII had Type I tails. Detailed study of Comet 1975 IX suggests thermal emission from large grains. Measurement of coma colors from visual to infrared wavelengths allows determination of the grain temperature and of the albedo and its dependence on scattering angle. Both Comets 1976 VI and 1980t passed between the Earth and the Sun allowing observations at scattering angles as small as $30°$. Both comets show strong forward scattering as expected for dielectric gains of radius ~ 1 μm. The nature of the size distribution can change abruptly on an individual comet; the visual and infrared brightness of Comet 1974 III decreased in a few days, and the silicate feature disappeared. The dependence of the comet brightness on diaphragm diameter (brightness proportional to diameter) suggests a model for the dust in which the number of grains satisfies the continuity equation. Plots of $(\lambda F_\lambda)_{max}$ show that the total energy radiated by the coma varies as R^{-n} with an exponent n near four when r is < 1 AU. This means than the grain production varies as r^{-2}. The infrared observations allow estimates of the mass ejected in silicate grains and these are consistent with a *quantity between 0.1 and times the mass of water ejected. The number density of particles in the zodiacal cloud is compatible with the fluxes observed for Brownlee particles and supports the hypothesis that the zodiacal cloud is the result of the injection of cometary particles.*

I. HISTORY AND DISCUSSION

Visual observations of comets are as old as astronomy, but the first infrared measurements were made at 1.6, 2.2, 3.4, and 10-μm on Comet Ikeya-Seki 1965 VIII by Becklin and Westphal (1966). They showed that the infrared radiation came from thermal emission by grains, that the color temperature was $\sim$ 30% higher than the gray-body temperature and that the flux detected was proportional to the diaphragm diameter between 20 and 80 arcsec. They concluded that the temperature excess was the same for the particles in the tail and the coma. They suggested a grain material with emissivity decreasing with increasing wavelength and proposed iron as a candidate.

Comet Bennett 1970 II was observed by Maas et al. (1970) who confirmed that thermal emission dominated the infrared and demonstrated further the presence of the silicate emission feature at 10 μm. This feature had been identified in the energy distribution of giant and supergiant circumstellar envelopes by Woolf and Ney (1969). Maas et al. suggested that the comet grains consisted of a mixture of silicates and black material with silicates somewhat in excess; carbonaceous chondrite material was a strong candidate. Studies of R Cr B (Stein et al. 1969) had shown that infrared emitting grains are ejected by carbon stars, and it is now well known (Ney 1972) that oxygen-rich stars copiously produce silicate material. Hackwell (1971) obtained an infrared spectrum of Comet Bennett 1970 II showing that the 10-μm excess resembled the dust bump in μ Cephei.

Comet Kohoutek 1973 XII was observed by the following: Gatley et al. (1974), Rieke and Lee (1974), Ney (1974a,b), Noguchi et al. (1974), Zeilki and Wright (1974), and Merrill (1974). The presence of the silicate signature was verified, but its strength was found to be variable. A medium resolution spectrum ($\lambda/\Delta\lambda$ = 100) obtained by Merrill showed a featureless excess much like that seen in emission in supergiant circumstellar shells and in absorption toward the galactic center. Rieke et al. (1974) pointed out that the silicate feature was weak or missing at heliocentric distances $\gtrsim$ 2 AU. The observations, especially those of Gatley et al., confirmed the diaphragm dependence, that the brightness observed was proportional to the diaphragm diameter. This is consistent with a model in which the dust blowing out from the nucleus satisfies the continuity equation, with the surface brightness proportional to R^{-1} where (R is the radial distance from the nucleus in the coma), and also requires that in correcting observations for the Earth-comet distance Δ, the flux observed in a diaphragm of given angular size is proportional to Δ^{-1}. The total infrared radiation for $r < 1$ AU is well represented by a power law $F \approx r^{-4}$ where r^{-2} comes from the illumination and the remaining r^{-2} from the variation of dust emission with solar heating.

The antitail of Comet Kohoutek was observed near perihelion, (Ney 1974a), and while the silicate signature was strong in the coma and the tail, it was absent in the antitail or sunward spike (Fig. 1). There is direct evidence from the infrared observations that particles of a different nature populate the antitail region. These could be nonsilicates or large particles of the same average composition as the coma. In order for the silicate feature to appear, the grains must be on the order of 1 optical depth or less in a single grain. The opacity of the material varies from 1000 $cm^2 g^{-1}$ to 5000 $cm^2 g^{-1}$ across the band. The feature should be very muted or absent for grains of radius $> 10 \,\mu m$. Hanner (1980) has given a quantitative discussion of the strength of the silicate feature and shows that it should change very little in the 0.5μm to 1μm radius range, but should virtually disappear for 10-μm grains. Laboratory measurements by Rose (1979) using dunite indicated a detectable feature for grains smaller than 37 microns. In the case of Comet Kohoutek, Sekanina (1973) predicted an antitail and showed by the dynamical analyses of Bessel

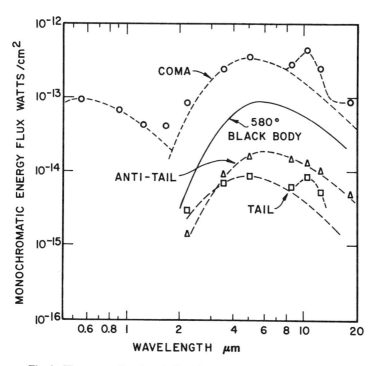

Fig. 1. The coma tail and antitail of Comet Kohoutek January 1-7, 1974.

and Bredichen, and Finson and Probstein (1968) that the antitail should consist of large particles deposited in the orbit plane of the comet. Sekanina's (1974) calculations show that to satisfy the dynamical considerations, the particles in the antitail must have gravitational accelerations ~ 100 times larger than the acceleration due to radiation pressure. For such particles $\rho d \approx 10^{-2}$ and for a density of 1 g cm^{-3} the particle diameters would be 10^{-2} cm or 100 μm; grains of this size would not produce a silicate signature. The large-particles explanation is buttressed by the fact that the antitail energy spectrum showed no temperature excess above the blackbody temperatures; this indicates that the particles are > 10 μm. When the silicate signature is in evidence, small silicate particles are required since large particles, of any composition, will not display the excess.

Comet Kohoutek was observed at a variety of wavelengths with identical geometry to obtain colors from 0.5 μm to 18 μm (Ney 1974b; Rieke and Lee 1974). O'Dell (1971) pointed out that such measurements can lead to a determination of the grain albedo. When the energy spectrum can be fit by a blackbody curve, $(\lambda F_\lambda)_{max}$ is proportional to the total energy under the Planck function, $[(\lambda F_\lambda)_{max} = 1.359 \text{ (total energy)}]$. Since $(\lambda F_\lambda)_{max\ IR}$ for the thermal emission gives the total energy radiated, this is also the total energy absorbed. A fit to the short-wavelength data gives the energy reflected. The albedo γ is determined by the equation

$$S = \frac{(\lambda F_\lambda)_{max\ VIS}}{(\lambda F_\lambda)_{max\ IR}} = \frac{\gamma}{1-\gamma}. \tag{1}$$

The albedo so determined will be an upper limit if the short-wavelength data are seriously contaminated by emission lines or bands. The albedo for the Kohoutek grains was between 0.14 and 0.20 at the scattering angles which were all near 90°, actually 89° to 135°.

Comet Bradfield 1974 III showed a silicate signature at $r = 0.5$ AU, but this disappeared beyond $r = 0.6$ AU and the comet suffered an abrupt decrease in brightness indicating a precipitous drop in grain production.

Comet Kobayashi-Berger-Milon 1975 IX, with a predominant Type I tail (Ney 1975), showed strong thermal emission without a silicate signature, indicating the presence of large or nonsilicate grains. Its temperature excess was only 8% compared with 26% for Kohoutek and 45% for Bennett, suggesting that its grains were large. Figure 2 shows examples of the energy distribution; the data are presented in Table I.

Comet West 1976 VI was the first comet to be observed at forward scattering angles. Because its orbit took it between the Sun and Earth, scattering angles as small as 35° were realized. To make measurements at those forward angles requires observing at small elongations from the Sun at midday. For example, on 26.8 Feb. 1976, the scattering angle was 35° and the elongation only 7°. The forward-scattering enhancement was so great that the coma could be easily seen in the finder telescopes. The ratio of

TABLE I

Comet Kobayashi-Berger-Milon 1975 IX[a]

Date of Obs. 1975	r (AU)	Δ (AU)	Wavelengths (μm) (mag)							
			1.2	1.6	2.2	3.5	4.8	8.5	10.6	12.5
July 27.7	1.02	0.32	–	–	–	–	–	+1.1	0	–0.7
Aug 27.7	0.48	0.99	5.1	5.3	5.6	2.6	+0.5	–2.1	–2.8	–3.1
Aug 31.9	0.44	1.09	5.6	5.7	5.3	2.0	+0.2	–2.3	–3.0	–3.1
Sept 1.9	0.43	1.11	5.3	5.1	5.0	2.1	+0.5	–2.1	–2.8	–3.1
Sept 2.7	0.43	1.12	5.2	5.0	5.2	2.0	+0.3	–2.2	–2.9	–3.2
Sept 3.9	0.43	1.15	5.6	5.7	5.0	1.8	–0.4	–2.6	–3.0	–3.5
Sept 5.3	0.43	1.18	–	–	5.4	2.0	+0.4	–2.1	–2.9	–3.2
Sept 8.8	0.46	1.12	6.5	6.3	6.0	2.0	+0.5	–2.2	–2.9	–3.1

[a]Unpublished data by the author.

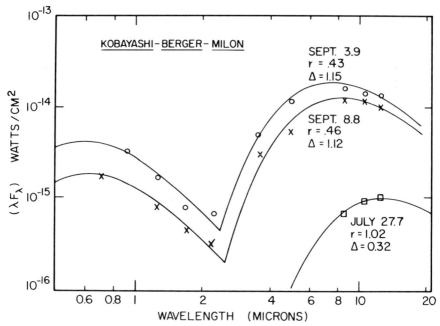

Fig. 2. The energy distribution for Comet Kobayashi-Berger-Milon at several heliocentric distances.

$S = (\lambda F_\lambda)_{\max \text{ VIS}}/(\lambda F_\lambda)_{\max \text{ IR}}$ was 1.1 as opposed to a typical 0.15 for 90° scattering. Observations at a variety of scattering angles gave a phase function strongly peaked in the forward direction and compatible with dirty dielectric grains with optical parameter $2\pi a/\lambda \approx 7$. This would be produced by a grain mixture with dominant size $\sim 1\ \mu$m radius (Ney and Merrill 1976). Giese (1980) has given a recent discussion of the phase function of such particles; see also that of Hanner (1980).

Comet West fragmented near perihelion; the infrared luminosity showed two abrupt increases that together produced a factor of five enhancement of the power radiated by the comet. The 10-μm feature was always in evidence. Note that the fragmentation process produces new surfaces previously shielded in the nuclear interior and that these surfaces produce dust at least as copiously as the original exposed surface. Comet West became four fragments, and the surface exposed should be proportional to $n^{2/3}$ if n is the number of equal mass fragments, which would give an enhancement of $4^{2/3}$ or 6.3. Figure 3 shows the energy spectra as a function of time for Comet West.

Comet Bradfield 1980t was another case of forward-scattering geometry. It fragmented on 15 Jan. 1981, but in this case the forward-scattering observations were obtained between 1 Jan. and 10 Jan. The data are given in Fig. 4 and Table II. Figure 4 shows the variability of the 10-μm excess which was present near perihelion but disappeared as the comet

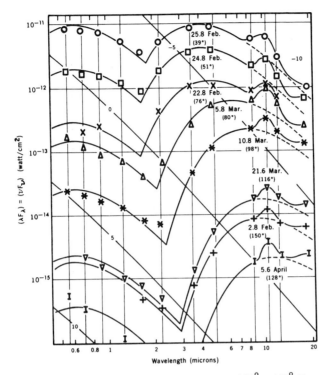

Fig. 3. The behavior of Comet West with scattering angles of 39° to 150°. Dates and scattering angles are indicated.

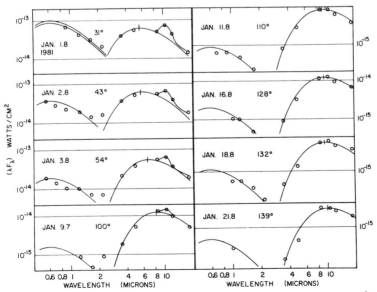

Fig. 4. Energy distribution for Comet Bradfield 1980t at scattering angles of 31° to 139°.

TABLE II
Comet Bradfield 1980t[a]

Date of Obs. 1981	r (AU)	Δ (AU)	Elongation (deg)	Scattering angle (deg)	V (mag)	R (mag)	I (mag)	Wavelength (μm) (mag)								
								1.2	1.6	2.2	3.5	4.8	8.5	10.6	12.5	18
Jan 1.8	0.28	0.73	8.4	31	—	—	2.6	2.2	1.8	1.4	−0.6	−1.9	−3.6	−4.6	−4.6	−4.6
Jan 2.8	0.30	0.74	11.4	41	4.4	4.1	3.8	3.0	2.7	1.7	−0.5	−1.9	−3.7	−4.5	−4.5	−4.9
Jan 3.8	0.31	0.765	14.7	53	5.1	4.9	4.6	3.8	3.3	2.6	0	−1.6	−3.7	−4.4	−4.4	−5.0
Jan 9.7	0.44	0.96	26	102	—	7.8	6.1	6.4	6.2	5.7	+2.7	+0.6	−2.2	−2.9	−2.9	−3.4
Jan 11.9	0.49	1.05	30	105	—	8.2	7.5	7	7	6.3	+3.5	+1.2	−1.7	−2.3	−2.4	−3.2
Jan 16.8	0.63	1.26	30	129	—	—	—	6.2	6.3	6.2	+3.5	+1.1	−2	−2.7	−2.9	−3.7
Jan 17.9	0.63	1.26	30	129	—	8.5	—	7.2	—	6.5	+4	+1.9	−1.5	−2.2	−2.4	−2.7
Jan 18.8	0.655	1.30	30	131	—	—	—	6.8	7.5	6.8	+4.3	+2	−1.2	−1.9	−2.1	−3.2
Jan 21.8	0.72	1.40	30	138	—	—	—	—	—	7.6	+5.4	+3.2	−0.6	−1.2	−1.5	−2.5
Jan 24.8	0.79	1.51	29	144	—	—	—	—	—	—	+6.0	+3.1	+0.4	−0.7	−0.75	—

[a]Unpublished data by the author.

receded from the Sun. The composite scattering function for five comets is shown in Fig. 5. The ratio S is 1.5 at 30° compared with 0.15 at 105°. The back-scattering albedo is slightly enhanced at 150°. It would be highly desirable to see the behavior from 0° to 30° and from 150° to 180°. It is nevertheless significant that the grain albedos for different comets are as similar as they appear to be.

II. MASS LOSS IN PARTICULATES

Many attempts have been made to evaluate the mass loss from observations of scattered light (see Liller 1960; Finson and Probstein 1968; Sekanina and Miller 1973). Mukai (1977) and Krishna Swamy and Donn (1979) have calculated mass-loss rates for Comet Kohoutek using the infrared data. The calculations based on optical scattering depend on the size distribution and the scattering phase function. In the infrared the problem is somewhat simpler because for particles smaller than a few microns the infrared luminosity measures the mass of the grains independent of particle size. This is because $Q_a \approx a\, f(T)$ where a is the grain radius and $f(T)$ is a function of the temperature. However, the exact expression for Q_a depends on the optical constants of the grain material. Krishna Swamy and Donn used the optical constants for several lunar rocks. Mass-loss rates have been determined for the comets in this survey by using the following model.

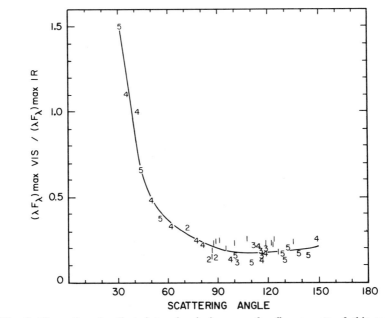

Fig. 5. The ratio of reflected to absorbed energy for five comets of this survey. Identifications are: 1 = 1973 XII, 2 = 1974 III; 3 = 1975 IX; 4 = 1976 VI; 5 = 1980t.

The quantity $(\lambda F_\lambda)_{max}$ is measured in a diaphragm of radius δ and is determined by the mass of grains within this observed geometry. The infrared luminosity is

$$L_{IR} = 4\pi D^2 [1.35 (\lambda F_\lambda)_{max}] = 4\pi N a^2 Q_a \sigma T^4 \tag{2}$$

where D is the Earth-comet distance, N the total number of grains, Q_a the emission efficiency and T the grain temperature. The grain mass is also given by

$$M = N\, 4/3\, \pi \rho a^3 \tag{3}$$

where ρ is the grain bulk density (assumed to be 2 g cm^{-2}). For silicate material we approximate the emission efficiency by

$$Q_a = 0.3a \tag{4}$$

with a measured in microns (Gilman 1974). Solving Eqs. 2, 3, and 4 leads to

$$M = 1.7 \times 10^{35} \frac{(\lambda F_\lambda)_{max} \Delta^2}{T^4} \tag{5}$$

where Δ is the Earth-comet distance in AU, $(\lambda F_\lambda)_{max}$ is in watts cm^{-2} and M is expressed in grams. To determine the mass-loss rate the mass in Eq. (5) is divided by the residence time for the grains in the radius of the coma observed in the diaphragm in which $(\lambda F_\lambda)_{max}$ was measured. The grain ejection velocity is of the order of sonic velocity in the gas (Finson and Probstein 1968) and is taken to be $0.5\, (T_{BB}/300)^{1/2}\text{ km s}^{-1}$ where T_{BB} is the blackbody temperature at the comet-Sun distance.

Equation (5) can be expressed alternatively as

$$\dot M = \text{const } \Delta^2\, \frac{\theta_{BB}}{\delta}\, \frac{1}{x} \tag{6}$$

where x is the opacity of the grains and θ_{BB} is blackbody angular radius of the observed coma. To make Eqs. (6) and (5) agree numerically $x = 1000\text{ cm}^2\text{ g}^{-1}$. $(\theta_{BB}/\delta)^2$ is the average optical depth τ of the coma in the observing diaphragm (for a 26-arcsec diameter diaphragm $10^{-6} < \tau < 10^{-3}$). Although individual grains may be optically thick, they cover a small fraction of the area with a finite diaphragm.

The mass-loss rate calculated for Comet Kohoutek is shown in Fig. 6. A least-squares fit gives $\dot M = 1.25 \times 10^6\, r^{-1.79}$ with correlation coefficient $r^2 = 0.83$. The data are fit almost as well by $\dot M = 10^6\, r^{-2}$. Comet Kohoutek was therefore losing 10^6 g s^{-1} at 1 AU and would have lost 10^8 g s^{-1} at 0.1 AU. The right-hand scale shows the optical depth of the coma in the observing diaphragm of 26 arcsec diameter. Figure 7 shows that the data are fit somewhat better by a Boltzman expression of the form

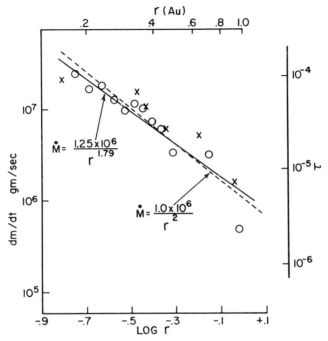

Fig. 6. Mass-loss rate in solids for Comet 1973 XII. The optical depth of the coma in a 26-arcsec beam is shown on the right.

$$\dot{M} = 2.47 \times 10^8 \, e^{-1552/T} = 2.47 \times 10^8 \, e^{-5.56\sqrt{r}} \qquad (7)$$

The correlation coefficient is $r^2 = 0.88$. Interpreted as an activation energy (possibly for releasing water from clathrates), the energy is 0.134 electron volts/molecule or 3.08 Kcal mole^{-1}. The mass-loss rates calculated here are in good agreement with those of Krisha Swamy and Donn (1979). The mass loss in solids can be compared with that inferred for the H_2O of 2×10^{29} molecules s^{-1} at 1 AU (Lillie and Keller 1976; Rahe 1980). The water mass-loss rate inferred from the Lyman α estimates of hydrogen and from the ultraviolet observations of OH in g s^{-1} is therefore

$$\dot{M}_{H_2O} = \frac{9.2 \times 10^6}{r^2} \qquad (8)$$

or 9.2 times the mass-loss rate in solids. The same calculation was done for the other comets; Table III gives the results together with the water mass-loss rates from Rahe (1980).

The ratio of water to solids appears to vary by almost a factor of 10. It is puzzling that Comet Kobayashi-Berger-Milon seems to have about an equal mixture of water and solids, but did not have a dust tail. It may be that the assumptions for calculating mass loss are violated in this comet because of the presence of large grains with emission efficiency approaching unity for radii

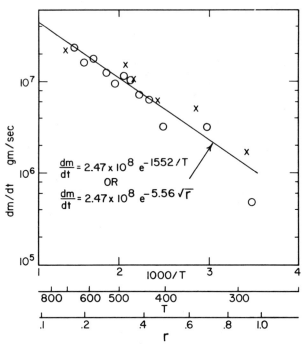

Fig. 7. A Boltzman plot for the mass-loss rate of Comet Kohoutek. This shows dm/dt as a function of $1/T$.

TABLE III
Mass Loss Rates for Water and Solids in Five Comets

Comet	Q_{H_2O} 1 AU molecules s^{-1}	Q_{H_2O} 1 AU g s^{-1}	Solids at 1 AU g s^{-1}	H_2O/Solids
Bennett 1970 II	5×10^{29}	1.5×10^7	3.6×10^6	4.2
P/Encke	7×10^{27}	2.1×10^5	2.2×10^4	9.6
Kohoutek 1973 XII	3×10^{29}	9.2×10^6	1.0×10^6	9.2
K. B. M.[a] 1975 IX	6×10^{28}	6.0×10^5	5.0×10^5	1.2
West 1976 VI	5×10^{29}	1.5×10^7	9.4×10^6	1.6

[a] Kobayashi-Berger-Milon

of $> 10 \mu$m. Although these calculations agree with those of Krisna Swamy and Donn for Comet Kohoutek within 30%, the optical constants of the grains and the grain ejection velocity are uncertain factors in both calculations, adding an uncertainty of at least a factor of two. The relative mass-loss rates for different comets should be somewhat better determined unless the material of the grains varies from comet to comet.

III. THE INFRARED BRIGHTNESS OF COMETS AS A FUNCTION OF HELIOCENTRIC DISTANCE

Figure 8 shows the energy radiated by seven comets. These data are corrected to an Earth-comet distance of 1 AU. Most of the curves are near the canonical $1/r^4$ slope indicating a $1/r^2$ dependence of dust production. Two comets show an abrupt increase in brightness at the time of fragmentation; Comet Bradfield 1974 III, shows an abrupt drop in dust production, and

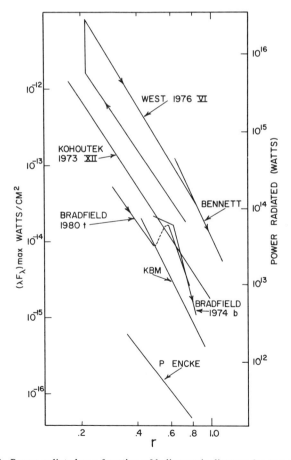

Fig. 8. Power radiated as a function of heliocentric distance for seven comets.

the scant dust production of P/Encke is evidenced by the factor of 100 in power radiated by this comet compared with dust producers like Comet Kobayashi-Berger-Milon 1975 IX, Bradfield 1980t, and Bradfield 1974 III.

IV. THE NATURE OF THE GRAINS

Laboratory measurements by Day (1974) and Rose (1979) have shown that a number of amorphous silicate materials can imitate the cometary 10-μm feature. Some are ruled out, including quartz and montmorillonite, but Type II carbonaceous chondrites are among the many substances allowed. It is still an open question whether the cometary dust is largely silicate with particle size determining the strength of the silicate signature. Silicate strength and temperature excess seem correlated as shown in Fig. 9. Since large particles would mute the signature and come to lower temperatures, this correlation argues for particle size as the primary variant from comet to comet. However, the correlation coefficient in Fig. 8 is only $r^2 = 0.65$.

The fractional silicate excess appears to be twice as large as the fractional temperature excess, but I have not been able to make a satisfactory model to predict this ratio.

V. DUST IN OTHER ASTROPHYSICAL ENVIRONMENTS

Dust has been seen in a variety of astronomical objects, and it is clear that at least two kinds of dust exist. Figure 10 shows examples of astrophysical "look alikes," stars known to be carbon-rich produce grains which do not have a silicate signature but have long-wave excesses. Examples are R Cr B and IRC +10216. Nova shells are seen in which the grains grow from a few molecules to almost a micron in radius without showing any 10-μm excess. They are certainly composed of some other probably carbonaceous material. However, oxygen-rich stars have an ubiquitous silicate feature; some examples of cold blackbody shells of this kind are shown in Fig. 10. Although carbon compounds are surely ejected into the interstellar medium, the great abundance of Mira stars, which produce silicates, probably means that these materials dominate the mass of the interstellar solids. In the carbonaceous chondrites the silicates are dominant suggesting that the presolar nebula was also silicate-rich.

There is a tantalizing connection between the carbonaceous chondrites, the comet grains, the zodiacal cloud, and the extraterrestrial Brownlee particles. The thermal emission from the zodiacal cloud has finally been measured (Price et al. 1980) at 11 and 20 μm. When visual and infrared observations are combined, an albedo similar to the comet grains is indicated. Future rocket experiments should show whether or not a 10-μm excess exists. Even if the particles are predominatly silicates, we would expect the signature to be muted because many of the larger particles (like the antitail of

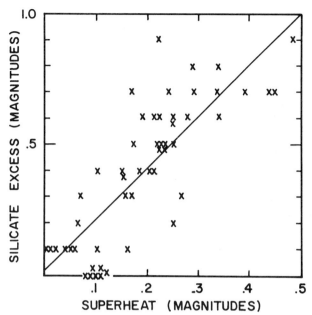

Fig. 9. Strength of the silicate feature as a function of the temperature excess of the thermal emission. The superheat scale is in the fractional temperature excess above the blackbody temperature.

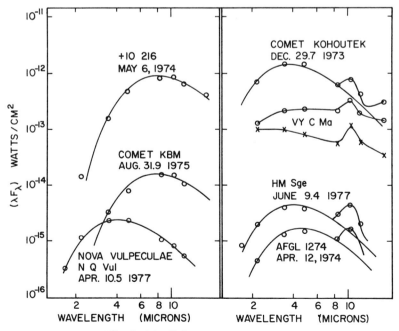

Fig. 10. Astrophysical "look alikes." Comets compared with carbonaceous objects (left panel) and with oxygen-rich shells (right panel).

Kohoutek) will be in transit via the Poynting-Robertson effect. However, there probably are enough grains in the 1- to 10-μm range to give a 10-μm excess if the comets are the source of the zodiacal cloud.

Brownlee's particle fluxes are consistent with the expected mass of zodiacal light particles swept up by the Earth passing through the zodiacal dust (Brownlee 1978). The zodiacal light brightness (Gillett 1966) could be produced by an integral number distribution $n = 5 \times 10^{-13} (125/a)^2$ particles cm^{-3} at 1 AU. For particles of radius 10 μm we expect 3×10^{-16} particles cm^{-3}; Brownlee observes a flux of 10^{-9} particles cm^{-2} s^{-1} of radius $\geqslant 10 \mu$m. At a relative velocity of 30 km s^{-1} and an ambient density of 3×10^{-16} particles cm^{-3}, the Earth should collect 9×10^{-10} particles cm^{-2} s^{-1}, in excellent agreement with Brownlee's flux. Note that this corresponds to 5×10^{-3} particles of > 10-μm radius in a cm^2 column 1 AU long, and only a few grains of all sizes in a cm^2 column between the Earth and the Sun. These tiny densities show the accomplishment of Brownlee in collecting them and the success of the measurement of their thermal emission.

Comets may inject the zodiacal cloud; the total mass in the cloud out to 2 AU is only $\sim 10^{17}$. Comet Kohoutek lost $\sim 2 \times 10^{13}$ g of solids inside 2 AU. It would take 5000 comets like Comet Kohoutek to replenish the cloud; with one such comet every year, 5000 yr would be needed. This is entirely reasonable because the Poynting-Robertson residence time of 10-μm grains from 1 AU is 7000 yr. Certainly within a factor of 10, comets can supply the zodiacal cloud, but any one comet can only be responsible for 10^{-4} of the particles in residence. The zodiacal cloud should be constant and structureless to agree with the observations of Sparrow and Ney (1973) and Leinert et al. (1980) and in contrast to claims for interplanetary spaghetti by Levasseur and Blamont (1973).

Acknowledgment. The author acknowledges research support from the National Aeronautics and Space Administration.

References

Becklin, E.E. and Westphal, J.A. 1966. Infrared observations of comet 1965f. *Astrophys. J.* 145:445-453.

Brownlee, D.E. 1978. Microparticle studies by sampling techniques. In *Cosmic Dust,* ed. J.A.M. McDonnell (New York: John Wiley and Sons), pp. 295-336.

Brownlee D.E.; Pilachowski, L.; Olszewski, E.; and Hodge, P.W. 1980. Analysis of interplanetary dust collections. In *Solid Particles in the Solar System,* eds. I. Halliday and B. McIntosh (Dordrecht: D. Reidel Publ. Co.), pp. 333-342.

Day, K.L. 1974. A possible identification of the 10-micron "silicate" feature. *Astrophys. J.* 192:L15.

Finson, M.L., and Probstein, R.F. 1968. A theory of dust comets: I. Model and equations and II. Results for Comet Arend-Roland. *Astrophys. J.* 154:327.

Gatley, I.; Becklin, E.E.; Neugebauer, G.; and Werner, M.W. 1974. Infrared observations of Comet Kohoutek (1973f). *Icarus* 23:561-565.

Giese, R.H. 1980. Optical investigations of dust in the solar system. In *Solid particles in the solar system,* eds. I. Halliday and B. McIntosh (Dordrecht: D. Reidel Publ. Co.), pp. 1-13.
Gillett, F. 1966. Zodiacal light and interplanetary dust. Ph.D. Dissertation, Univ. of Minnesota.
Gilman, R.C. 1974. Planck mean cross-sections for four grain materials. *Astrophys. J.* Suppl. 268:397-404.
Hackwell, J.A. 1971 10 μ emission spectrum of Comet Bennett. *Observatory* 91:33-34.
Hanner, M. 1980. Physical characteristics of cometary dust from optical studies. In *Solid Particles in the Solar System,* eds. I. Halliday and B. McIntosh (Dordrecht: D. Reidel Publ. Co.), pp. 223-236.
Krishna Swamy, K.S.; and Donn, B. 1979. An analysis of the infrared continuum of Comets II. Comet Kohoutek. *Astron. J.* 84:692-697.
Leinert, C.; Richter, I.; Pitz E.; and Hanner, M. 1980. Four years of zodiacal light observations from the Helios Space Probes: Evidence for a smooth distribution of interplanetary dust. In *Solid Particles in the Solar System,* eds. I. Halliday and B. McIntosh (Dordrecht: D. Reidel Publ. Co.), pp. 15-18.
Levasseur, A.C., and Blamont, J.F. 1973. Satellite observations of intensity variations of the zodiacal light. *Nature* 246:26-28.
Liller, W. 1960. The nature of the grains in the tails of Comets 1956h and 1957d. *Astrophys. J.* 132:867-882.
Lillie, C.F., and Keller, H.U. 1976. Spectrometry of Comet Bennett from OAO-2. In *The Study of Comets,* eds. B. Donn, M. Mumma, W. Jackson, M. A'Hearn and R. Harrington (Washington: NASA SP-393), pp. 322-329.
Maas, R.W.; Ney, E.P.; and Woolf, N.F. 1970. The 10-micron emission peak of Comet Bennett 1969i. *Astrophys. J.* 160:L101-L104.
Merrill, K.M. 1974. 8-13μm spectrophotometry of Comet Kohoutek. *Icarus* 23:566-567.
Mukai, T. 1977. Dust grains in the cometary coma: Interpretation of the infrared continuum. *Astron. Astrophys.* 61:69-74.
Ney, E.P. 1972. Infrared excesses in supergiant stars: Evidence for silicates. *Publ. Astron. Soc. Pacific.* 84:613-618;
Ney, E.P. 1974a. Infrared observations of Comet Kohoutek near perihelion. *Astrophys. J.* 189:L141-L143.
Ney, E.P. 1974b. Multiband photometry of comets Kohoutek, Bennett, Bradfield, and Encke. *Icarus* 23:551-560.
Ney, E.P. 1975. Infrared photometry of Comet Kobayashi Berger Milon. *Bull. Amer. Astron. Soc.* 7:508 (abstract).
Ney, E.P., and Merrill, K.M. 1976. Comet West and the scattering function of cometary dust. *Science* 194:1051-1053.
Noguchi, K.; Sato, S.; Maihara, T.; Okuda, H.; and Uyama, K. 1974. Infrared photometric and polarimetric observations of Comet Kohoutek 1973f. *Icarus* 23:545-550. 23:545-550.
O'Dell, C.R. 1971. Nature of particulate matter in comets as determined from infrared observations. *Astrophys. J.* 166:675-681.
Price, S.D.; Murdock, T.L.; and Marcotte, L.P. 1980. Infrared observations of the zodiacal dust cloud. *Astron. J.* 85:765-771.
Rahe, J. 1980. Ultraviolet spectroscopy of comets. In *Proc. of European I.U.E. Conf.* In press.
Rieke, G.H., and Lee, T.A. 1974. Photometry of Comet Kohoutek (1973f). *Nature* 248:737-740.
Rieke, G.H.; Low, F.J.; Lee, T.A.; and Wisniewski, W.W. 1974. Infrared observations of Comet Kohoutek. In *Comet Kohoutek* (Washington: NASA-SP 355), pp. 175-182.
Rose, L.A. 1979. Laboratory simulation of infrared astrophysical features. *Astrophys. Space Sci.* 65:47-67.
Sekanina, Z. 1973. Comet Kohoutek. *I.A.U. Circ.* 2580.
Sekanina, Z. 1974. On the nature of the anti-tail of Comet Kohoutek (1973f) I. A working model. *Icarus* 23:502-518.
Sekanina, Z., and Miller, F.D. 1973. Comet Bennett 1970II. *Science* 179:565-567.

Sparrow, J., and Ney, E.P. 1973. Temporal constancy of zodiacal light. *Science* 181:438-440.

Stein, W.A.; Gillett, F.C.; Gaustad, J.E.; and Knacke, R.F. 1969. Circumstellar infrared emission from two peculiar objects — R Aquarii and R Coronae Borealis. *Astrophys. J.* 155:L3-L7.

Woolf, N.J., and Ney, E.P. 1969. Circumstellar infrared emission from cool stars. *Astrophys. J.* 155:L181-L184.

Zeilik, M., and Wright, E.L. 1974. Infrared photometry of Comet Kohoutek. *Icarus* 23:577-579.

INTERPRETING THE THERMAL PROPERTIES OF COMETARY DUST

HUMBERTO CAMPINS
University of Arizona

and

MARTHA S. HANNER
Jet Propulsion Laboratory

The characteristics of the thermal emission from cometary dust are discussed and the observations compared with models for absorbing and silicate grains. The observed JHK colors are discussed and suggested observations of future comets are outlined. The observed 4.8 µm/3.5 µm flux ratio can be fit with absorbing grains, with a variation in particle size less than a factor of 2 for all comets observed. The relative number of silicate grains necessary to produce an observable feature at 10 µm is a strong function of their temperature. If they are cold (no absorption at visual wavelengths), a considerable number could be present without producing a detectable feature above the thermal continuum from the hot absorbing grains.

Optical observations of the visual and infrared continuum radiation from comets can be analyzed to derive the composition and size distribution of the solid particles present in cometary comae. The observational data consist of brightness and polarization of the sunlight scattered by the grains as a function of scattering angle and wavelength, and the thermal emission from the grains as a function of wavelength and heliocentric distance (grain temperature). The scattering and polarization data at visual wavelengths have been summarized by Hanner (1980, and references cited therein). Ney and Merrill (1976) have derived the angular scattering function for Comet West 1976 VI. Giese (1980) has shown that this scattering function, as well as the polarization, can be fit with mixtures of irregular dielectric and absorbing grains. The color of the scattered light is generally neutral or slightly red.

A summary of the infrared observations of bright comets is given in the chapter by Ney in this book; a complete review of the physical characteristics of cometary dust from visual and infrared studies has been made by Hanner (1980). Because these two works cover both the observational and theoretical aspects of this particular area of cometary research, we shall concentrate mainly on the latest developments and refer the reader to those articles for a more complete picture.

In this chapter we first consider the infrared scattered light observations, 1 to 2.5 μm. Then we examine the characteristics of the thermal emission in terms of a two-component model of the cometary dust; we conclude with a list of suggested observations. In addition to the observations reported by Ney (his chapter and 1974), several faint (mostly periodic) comets have been observed in the reflected infrared by A'Hearn et al. (1981) and in the reflected and thermal infrared by Campins et al. (1981a,b), and Veeder and Hanner (1981). Special attention will be given to these observations because they represent the first set of infrared observations of periodic comets.

I. SCATTERED LIGHT IN THE RANGE 1 TO 2.5 μm

The absence of gaseous emissions in the near infrared make this an ideal region to observe the light scattered by cometary dust. Broadband photometry at J, H, K (1.25, 1.63, 2.22 μm) has been obtained for a number of bright comets (see Ney 1974 and his chapter; Ney and Merrill 1976; Oishi et al. 1978; Iijima et al. 1975; Noguchi et al. 1974). (Fink and Larson [personal communication] have observed a CN band near 1.1 μm in the spectrum of Comet West 1976 VI. This band may contribute a significant fraction of the J-flux in gassy comets which are near the Sun.) We concentrate here on the periodic comets observed at heliocentric distance $>$ 1 AU, where there is no contamination from thermal emission (A'Hearn et al. 1981; Campins et al. 1981a,b; Veeder and Hanner 1981). The color of the radiation scattered by grains with sizes comparable to the wavelength depends on grain size, composition, and surface roughness or irregularity, as well as on the scattering angle.

First we consider whether *JHK* colors can be diagnostic of the presence or absence of icy grains. Figure 1 compares the observed $J-H$ and $H-K$ colors versus scattering angle with colors computed from Mie theory for icy grains (Hanner 1981). The solid curves correspond to pure water ice with a size distribution of the form $n(a) = 0$; $a < 0.45$ μm; $n(a) \propto (2a - 0.9)a^{-5}$,

THERMAL PROPERTIES OF COMET DUST

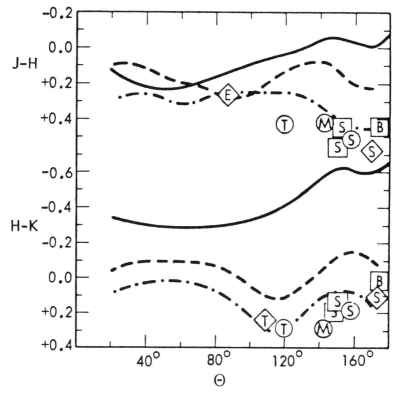

Fig. 1. $J-H$ and $H-K$ colors versus scattering angle for icy grain models (Hanner 1981). The solid curves refer to a size distribution of pure ice grains, (see text); dashed curves represent the 20-μm slightly dirty ice grains, $n'' = 0.0005$; and dash-dot curves, the 20μm slightly dirty ice grains, $n'' = 0.002$. Circles are observations of periodic comets Tuttle, Meier, and Stephan-Oterma from A'Hearn et al. (1981); squares are observations of P/Stephan-Oterma and Comet Bowell from Veeder and Hanner (1981); and diamonds are observations of periodic comets Encke, Tuttle, and Stephan-Oterma from Campins et al. (1981a).

$0.45 \leq a \leq 1.3$ μm; $n(a) \propto a^{-4.2}$, $a > 1.3$ μm, where a = grain radius (see Sec. II). The colors are very blue, as expected for these small grains at infrared wavelengths. The dashed curves represent 20-μm grains with imaginary index of refraction $n'' = 0.0005$ and the dash-dot curves, 20-μm grains with $n'' = 0.002$. Both curves have been smoothed to reduce the effect of resonances in spherical particles; the remaining ripples would probably not be present in the scattering by irregular particles. The effect of this small amount of absorption is to make the colors redder. Clearly, variations with particle size, scattering angle, and impurities in the ice prevent one from defining characteristic JHK colors for the scattering by icy grains. The existence of slightly dirty ice grains cannot be excluded on the basis of observed JHK colors; however, Hanner (1981) concluded that such ice grains would rapidly sublime

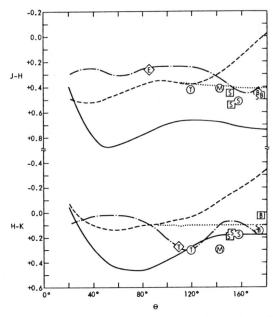

Fig. 2. $J-H$ and $H-K$ colors versus scattering angle for dust grain models (Veeder and Hanner 1981). Solid curves give the size distribution of magnetite grains, $\rho = 1$; dashed curves give the size distribution of silicate grains, $\rho = 1$; dotted curves, the irregular silicate grains; and dash-dot curves, the 20-μm slightly dirty ice grains, $n'' = 0.002$. Observations are of periodic comets Encke, Tuttle, Meier, Stephan-Oterma, and Comet Bowell. Circles: A'Hearn et al. (1981); squares: Veeder and (Hanner 1981); diamonds: Campins et al. (1981a,b).

at heliocentric distances $\lesssim 2$ AU from the Sun. Icy grains might be observable in comets which display a coma at large heliocentric distances, such as Comet Bowell 1980b.

Figure 2 compares the observed colors with theoretical scattering by dust grains (Veeder and Hanner 1981). The dust scattering was computed from Mie theory, using the same size distribution as applied to the icy grains. The solid curves are for magnetite, a typical absorbing material present in meteorites. The red colors are typical of the scattering by absorbing particles. The dashed curves refer to silicate grains. The blue color at large scattering angles is a consequence of the strong backscatter in silicate spheres. This enhanced backscattering is not present in the scattering by irregular dielectric particles, which have a rather flat scattering function at $> 90°$ scattering angles (Hanner et al. 1981). The dotted lines represent the more neutral color one would expect from the scattering by irregularly shaped silicate grains. A small amount of absorbing material in the silicate matrix would also tend to make the colors redder than the colors computed for pure silicate spheres. The dash-dot curves correspond to the scattering by 20-μm grains of slightly dirty ice, from Fig. 1.

The observed *JHK* colors for periodic comets Meier, Tuttle, and Stephan-Oterma at scattering angles ≳ 120° are similar and are compatible with a mixture of absorbing grains and irregular silicate grains of roughly μm size, as discussed below in Secs. II and III, to fit the thermal emission data. The $J-H$ color (0.27) of P/Encke, observed when the comet was at 0.36 AU (Campins et al. 1981b) may indicate a difference in mean particle size or composition or merely the effect of scattering angle. Thermal emission may contribute a few percent to the observed flux from P/Encke at H. Comet Bowell 1980b is a new comet, with a well-defined coma, although it was at a heliocentric distance of ~ 5 AU when these *JHK* observations were made. Sekanina (1981) has concluded that the grains are large, > 100 μm radius.

We come to the conclusion that *JHK* colors of cometary comae are not, by themselves, good indicators of the composition, nor even the size of the dust particles. Only submicron grains can be eliminated as major contributors to the near-infrared scattered light. When combined with the thermal emission, however, the observed *JHK* fluxes can be helpful in determining the albedo of the grains and the dust production rate. One should avoid making conclusions from a comparison of comet and asteroid colors, since the scattering by small particles in a cometray coma is *not* comparable with reflection from a solid surface. Veeder and Hanner have pointed out that comets do not generally occupy the same field as either the S or C asteroid classes in a $J-H$, $H-K$ plot. Furthermore, even among asteroids, objects with similar *JHK* colors can have quite different albedos.

The only spectral scans in the near infrared are by Oishi et al. (1978) for Comet West 1976 VI (λ 2.8 to 3.6 μm) and by A'Hearn et al. (1981) for P/Stephan-Oterma (λ 1.4 to 2.45 μm) and P/Tuttle (λ 2.0 to 2.4 μm). The scan for Comet West, taken when the comet was at a heliocentric distance of 0.53 AU, shows thermal emission equivalent to a blackbody spectrum at 450 K. The scans by A'Hearn et al., taken when the comets were at 1.6 and 1.16 AU respectively, show a smooth scattered light continuum somewhat redder than that of the Sun. The slope of the continuum can be matched with the scattering by magnetite grains using the size distribution given above.

A'Hearn et al. detected no absorption features in their scans at the wavelengths of common ice bands. Hanner (1981) has shown that icy grains would probably sublime too rapidly to be observable even at 1.6 AU. Even if scattering from icy grains were the dominant source of continuum radiation in P/Stephan-Oterma, the lack of a 2-μm absorption feature could have several explanations: scattering by predominantly small grains; broadening due to amorphous ice; masking by absorption in core/mantle grains. However, P/Stephan-Oterma at 1.5 to 1.7 AU showed fairly strong thermal emission, indicating the presence of hot dust grains (Veeder and Hanner 1981; Campins et al. 1981a). Scattering by this dust coma can adequately account for the brightness near 2 μm. A detailed discussion of ice features can be found in the chapter by Fink and Sill in this book. The stronger 3-μm ice band might be detected in comets having a visible coma at large heliocentric distance.

II. THERMAL EMISSION

Thermal emission from cometary dust dominates the observed infrared flux from the coma. The crossover from scattered to thermal emission occurs at $\sim 3\,\mu$m at 1 AU; at 0.75 AU thermal emission may contribute $< 10\%$ to the K bandpass. The total energy emitted by the dust can be well determined by observations in the range 3 to 20 μm, since the wavelength of maximum emission falls within this range for the grain temperatures of interest. Most of the observed comets have shown a grain temperature hotter than a theoretical blackbody at the same heliocentric distance, suggesting small absorbing grains. In this respect, periodic comets Encke, Stephan-Oterma and Tuttle do not differ qualitatively from nonperiodic comets.

A broad emission feature near 10 μm, generally attributed to silicate grains, is often present, superimposed on the thermal continuum. The feature can have variable strength relative to the continuum. Most of the existing photometry has been done with filters $\sim 1\,\mu$m wide, centered near 8.5, 10.3, and 12.5 μm. The single spectrophotometric scan of Comet Kohoutek 1973 XII at 0.31 AU shows a smooth feature similar in width to the interstellar feature (Merrill 1974). A significant recent result is the presence of the 10-μm emission feature in periodic comets Encke, Stephan-Oterma, and Tuttle, observed near perihelion by Campins et al. (1981b). A weak silicate feature was also observed in P/Stephan-Oterma after perihelion (1.6 AU) by Tedesco and Gradie (personal communication). Enhanced emission is also seen in broadband photometry near 18 μm and interpreted as silicate emission (Hanner 1980). Hanner (1980) showed that, even if the silicate grains could be made hot enough to match the thermal emission by introducing a small amount of absorbing material into the silicate matrix, the ratio of the flux at 10 μm to the flux at 3.5 μm would be higher than that observed by more than an order of magnitude.

We are led, then, to a general picture of two dust components: hot absorbing grains and silicate grains. In this section we discuss the conclusions about the size and composition of the absorbing grains which can be drawn from thermal continuum radiation. The silicate component is discussed in Sec. III.

Theoretical modeling of the thermal emission from cometary dust grains has been carried out by Mukai (1977) and Krishna Swamy and Donn (1979) for single-size grains and by Hanner (1980) for a size distribution of grains. We discuss Hanner's models next.

Grain temperatures as a function of particle radius a and heliocentric distance r were computed for minerals likely to be present in cometary dust (and for which the refractive index has been measured as a function of wavelength) by equating the amount of solar energy absorbed by the grain to the thermal energy emitted by the grain

$$\pi a^2 \left(\frac{r_\oplus}{r}\right)^2 \int Q_{\text{abs}}(a,\lambda) S_\lambda \, d\lambda = 4\pi a^2 \int \pi B_\lambda(T) Q_{\text{abs}}(a,\lambda) \, d\lambda \qquad (1)$$

where S_λ = solar flux from Labs and Neckel (1970), $Q_{abs}(a,\lambda)$ = absorption efficiency factor computed by Mie theory from the measured refractive indices for grain radius a and wavelength λ, and $B_\lambda(T)$ = Planck function at grain temperature T.

Magnetite, a mineral found in the micrometeorites collected by Brownlee (1978), will be used here as an example of absorbing grains. Amorphous olivine, an iron-magnesium silicate shown by Hanner (1980) to give a reasonable fit to the observed 10- and 18-μm silicate features, will be adopted for the silicate component, based on refractive indices in the infrared measured by Krätschmer and Huffman (1979). Pure olivine has virtually no absorption at visual wavelengths. However Brownlee frequently found silicate particles embedded in chondritic material or silicate grains with dark material clinging to their surface. A small amount of absorption, corresponding to an imaginary component of the refractive index $n'' = 0.001$, 0.01, or 0.04 will be added here, to approximate the effect on the grain temperature of a small amount of absorbing material mixed in the silicate matrix.

Figure 3 shows the resulting grain temperature versus particle size computed according to Eq. (1) for these two materials at two heliocentric distances (Hanner 1980). Magnetite grains, $a \leqslant 1\,\mu$m, are hotter than a blackbody, since these small grains cannot radiate efficiently in the longer wavelengths. Slightly dirty olivine, $n'' = 0.001$, is $\geqslant 100$ K cooler than a blackbody for $a \leqslant 10\,\mu$m. Addition of $n'' = 0.01$ raises the grain temperature considerably. Note that grains approximating the blackbody temperature at 0.75 AU ($n'' = 0.01$) are considerably hotter at smaller solar distances, and hotter than the magnetite grains, since the peak of the Planck function has shifted away from the middle infrared, where these grains can radiate efficiently.

Dynamical analyses of cometary dust tails have provided the only information on the functional form of the particle size distribution in comets. Accordingly, we will follow Hanner (1980) and adopt the size distribution derived for Comet Bennett 1970 II by Sekanina and Miller (1973), modified for large particles to agree with the analysis of comet antitails (Sekanina and Schuster 1978a,b). This size distribution has the following form:

$$n(a) = 0, \qquad 2a\rho < 0.9 \times 10^{-4}$$

$$n(a) = \frac{0.69\,(2a\rho - 0.9 \times 10^{-4})}{(2a\rho)^5}, \qquad 0.9 \times 10^{-4} \leqslant 2a\rho \leqslant 2.6 \times 10^{-4} \qquad (2)$$

$$n(a) = 0.08656\,(2a\rho)^{-4.2}, \qquad 2a\rho > 2.6 \times 10^{-4}$$

where a = grain radius in cm. In the models discussed in this chapter, ρ will be varied, in order to shift the size distribution to larger or smaller sizes. When applied to Comet Bennett 1970 II, ρ is the grain density in g cm^{-3}. For other comets, however, no dynamical analysis exists and the grain density is not restricted by Eq. (2). In these cases, ρ is simply a dimensionless parameter to describe the shift in mean particle size ($\propto 1/\rho$).

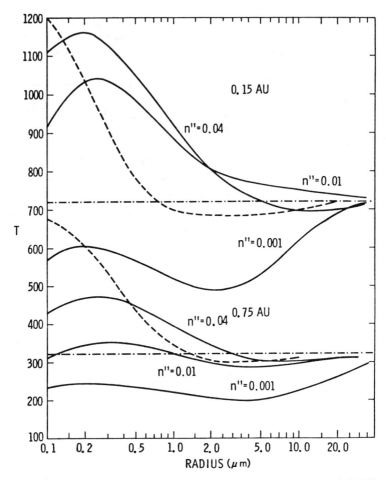

Fig. 3. Computed temperature versus grain radius at heliocentric distance 0.15 AU and 0.75 AU (Hanner 1980). Solid curves represent silicate grains with imaginary index of refraction n'' as indicated and dashed curves represent the magnetite grains. Dash-dot lines show the temperature of a blackbody in equilibrium.

The mean temperature of the grains can be defined from broadband photometry in two ways, either as an effective temperature from a best fit of the data to a blackbody curve (see Ney's chapter) or as a color temperature from the flux ratio (4.8 μm/3.5 μm). When silicate emission contaminates the 8–12 μm and 18–20 μm regions, the two methods are essentially the same, although the effective temperature method may incorporate the 12.5 μm flux if the silicate feature is not too strong.

Figure 4 illustrates that grains of a given temperature cannot radiate efficiently at wavelengths $\gtrsim$ 10 times their radius (Hanner 1980). Therefore, a blackbody fit to observed fluxes at 3 to 6 μm does not necessarily imply a

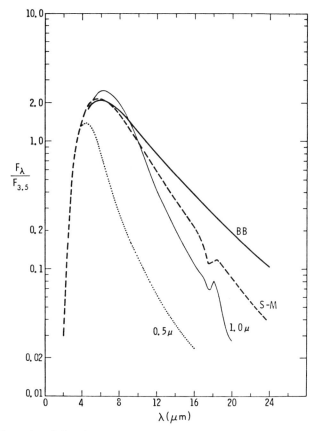

Fig. 4. Thermal emission from magnetite grains at 500 K (normalized to emission at 3.5 μm). The dotted curve represents grain radius $a = 0.5$ μm; solid curve $a = 1.0$ μm; dashed curve gives the size distribution, $\rho = 1$; and the heavy line is the blackbody at 500 K.

smooth blackbody continuum at longer wavelengths. A blackbody continuum out to 18 μm was observed for Comet Bradfield 1974 III at 0.67 AU and in the antitail of Comet Kohoutek 1973 XII (Ney 1974), indicating that the emitting particles were at least a few μm in diameter, consistent with the absence of silicate emission features. In the presence of silicate emission, it is questionable how the underlying continuum should be drawn.

The 4.8/3.5 μm flux ratio has been computed for the grain models by integrating the thermal emission at each wavelength over the Sekanina-Miller size distribution, using the appropriate grain temperature for each grain size and heliocentric distance. The results are plotted in Fig. 5 for magnetite grains with 4 choices of the size parameter ρ, together with available observations of bright comets (Ney 1974, 1981) and the one observation of P/Encke

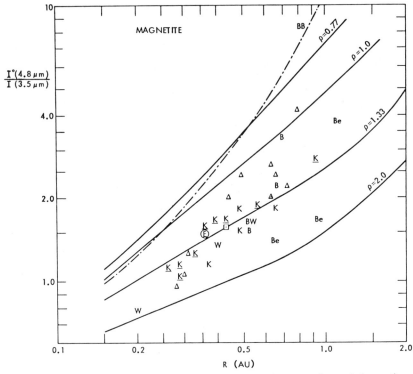

Fig. 5. Flux ratio $I(4.8\,\mu m)/I(3.5\,\mu m)$. The plot shows the comparison of observations with models for size distributions of magnetite grains, varying size parameter ρ in Eq. (2). The dash-dot curve represents the theoretical blackbody (BB). K and $\underline{K}$: Comet Kohoutek 1973 XII pre- and post-perihelion (Ney 1974); B: Bradfield 1974 III (Ney 1974); Be: Bennett 1970 II (Ney 1974); W: West 1976 VI (Ney and Merrill 1976); $\triangle$: Bradfield 1980t (Ney 1981); ▣: Kobayashi-Berger-Milon 1975 IX, average of 5 observations near 0.43 AU (Ney 1981); E: P/Encke (Campins et al. 1981b).

(Campins et al. 1981b). The color ratio for a blackbody in equilibrium is shown as the dash-dot curve. At these wavelengths Ney's individual observations are uncertain by ~ 10% which makes the color ratios uncertain by ~ 14%; the color ratio of P/Encke is uncertain by ~ 5%. Ney's observations of Comet Kohoutek 1973 XII are in close agreement with those of Rieke and Lee (1974).

All of the observations fall between the curves for $\rho = 1.0$ and $\rho = 20$. The extensive observations of Comet Kohoutek (Ney 1974) are matched fairly well by a model with $\rho = 1.33$. The single point for P/Encke falls very close to the same curve. The mean particle size does not change by more than a factor of 2 for abrupt changes in brightness (Comet Bennett 1970 II, at 0.64 versus 0.94 AU), loss of the silicate feature (Comet Bradfield 1974 III at 0.67 AU) or complete absence of a silicate feature (Comet Kobayashi-Berger-Milon 1975 IX).

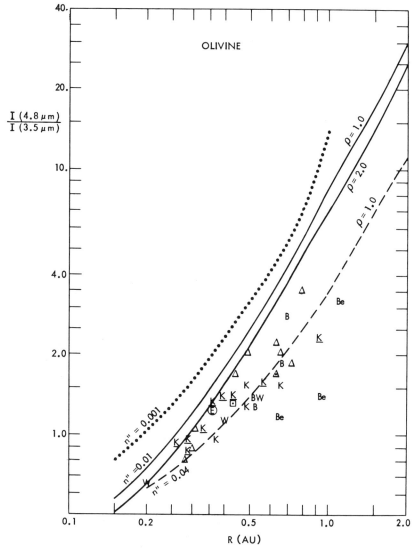

Fig. 6. Flux ratio $I(4.8\,\mu m)/I(3.5\,\mu m)$ for size distribution of silicate grains, compared with the same observations as those in Fig. 5, with imaginary index of refraction n'' and size parameter ρ as indicated. The curve for $n'' = 0.001$ is independent of ρ (see text).

Figure 6 shows a similar plot for a model of amorphous olivine with imaginary index of refraction $n'' = 0.001$ (dotted curve), $n'' = 0.01$ (solid curve), and $n'' = 0.04$ (dashed curve). When $n'' = 0.001$, the small silicate grains are much colder than a blackbody (Fig. 3). The thermal emission at $\lambda < 8\,\mu m$ arises mainly from the larger grains ($\sim 50\%$ from $a \geqslant 100\,\mu m$, if

$n(a) \propto a^{-4.2}$). Therefore, changing the mean size of the smaller grains does not alter the 4.8 μm/3.5 μm brightness ratio. This thermal emission would not, in fact, be observed, if $n(a)$ follows Eq. (2); the scattering of sunlight by the small grains would be stronger than the thermal emission (by a factor of 75 at λ = 3.5 μm at 0.75 AU). For n'' = 0.01 or 0.04, the small grains, being hotter, make a larger contribution and a small effect of cutoff size is present, as illustrated for n'' = 0.01 (ρ = 1 and ρ = 2).

The most significant aspect of Fig. 6 is that the slopes of these curves for olivine differ markedly from the slope of the observations, particularly for Comet Kohoutek 1973 XII, which has the most extensive coverage in heliocentric distance. The only comets for which the slope might be considered comparable to that of the olivine with n'' = 0.04 are the cases in which the silicate feature disappeared. This leads us to conclude that particles with a size distribution peaking in the range depicted in Fig. 5 and composed mostly of a highly absorbing material (magnetite, graphite, or others) are responsible for the thermal continuum of all comets for which data are available. This result is particularly significant because the 7 comets range from "new" (Comet Kohoutek) to very evolved (P/Encke). Mukai (1977) shows a similar plot of the 4.8 μm/3.5 μm flux ratio as well as a plot of 8.5 μm/3.5 μm flux ratio for Comet Kohoutek, compared with models for single-size metal and silicate grains. His plots are particularly instructive because he includes the error bars on the observed flux ratios. Mukai concludes that μm-size absorbing particles such as iron, as opposed to silicate grains, give the best fit to the observations within the estimated uncertainty; this is compatible with our conclusions here.

III. TWO-COMPONENT MODEL

Having established the size distribution of the absorbing grains in Comet Kohoutek 1973 XII from the 4.8 μm/3.5 μm flux ratio, and having concluded that the thermal continuum is produced mainly by these absorbing particles, we can compute the thermal radiation from these grains at wavelengths 8 to 25 μm and estimate the relative number of silicate grains which must be added to produce the observed silicate emission features. The same size distribution is assumed to apply to both components; this is not unreasonable, since the dynamical analysis was based on the observed isophotes in visible light, where the brightness contribution from the two components is roughly comparable (see Table II and explanation below).

Table I gives the ratio of silicate to magnetite grains (by number) obtained by fitting the models to the very complete set of observations by Rieke and Lee (1974) of the 10-μm feature at 0.74, 0.67, and 0.37 AU preperihelion and of the 20-μm feature at 0.37 AU. Observations through filters with a bandwidth of ~ 1 μm centered at 8.8, 10.3, 11.6, and 12.6 μm were used to define the 10-μm band and observations through filters centered at 17, 19, 22.5, and 24.5 μm were used to define the 20-μm band. The ratio is

TABLE I

Relative Silicate to Magnetite Abundance in
Comet Kohoutek 1973 XII[a]

r (AU)	n''	λ (μm)	Silicate: Magnetite
0.74	0.001	10	19
0.74	0.01	10	2.5
0.67	0.001	10	14
0.67	0.01	10	2.5
0.37	0.001	10	2.4
0.37	0.01	10	0.46
0.37	0.001	20	3.5
0.37	0.01	20	0.86

[a]Models are fit to data of Rieke and Lee (1974).

given for olivine models with imaginary index of refraction at visual wavelength $n'' = 0.0001$ and $n'' = 0.01$ in order to illustrate the effect of grain temperature on the derived silicate/magnetite abundance. The abundance ratio for the fit at 10 μm decreases with decreasing heliocentric distance. This may reflect a real variation and/or it may result from a change in mean particle size, affecting both grain temperature and the strength of the 10-μm feature (see Fig. 5b in Hanner 1980) or from a temperature variation with heliocentric distance not accurately predicted by the models with $n'' = $ const.

An interesting result of this analysis is that, for both olivine models, the fit to the 20-μm feature at 0.37 AU required more silicate grains than the fit to the 10-μm band. This result can be interpreted in several ways:

(a) The size distribution for magnetite is not as steep as $a^{-4.2}$, resulting in a higher continuum at 17–25 μm;
(b) The size distribution for the olivine has more large particles;
(c) The olivine grains are colder than the models predict and consequently the maximum of the Planck function is nearer 20 μm;
(d) A silicate mineral in addition to olivine is present, which has its band at 22 μm rather than 18 μm, like that found in meteorite samples (Friedemann et al. 1979).

A combination of (a) and (d) seems to be the most likely explanation in this case, as suggested by the structure of the 20-μm band apparent in Rieke and Lee's observations (d), and by a similar ratio obtained for the 10- and 20-μm bands when they are fit to a blackbody continuum at 17–24 μm rather than

TABLE II
Relative Brightness Contribution at 1 AU:
Silicate to Magnetite Ratio[a]

	λ (μm)	$\theta = 90°$	$\theta = 120°$
V	0.5	2.2	1.1
J	1.25	3.4	2.3
K	2.2	1.6	1.3
L	3.5	0.055	0.03
N	10.6	0.11	0.11

[a]Sekanina-Miller size distribution with $\rho = 1.0$ using equal numbers of silicate and magnetite grains.

to the continuum predicted by the magnetite model (a). Because of these uncertainties it is not possible to use the observed 20-μm/10-μm flux ratio to derive the temperature of the silicate grains. The uncertainty in the temperature of the grains prevents one from deriving their abundance from the thermal spectrum alone.

The presence of both hot absorbing grains and cooler silicate grains means that the albedo derived from the relative strength of the thermal emission and the visual scattering does not refer uniquely to one component. Table II compares the relative contribution of silicate ($n'' = 0.001$) and magnetite grains to the observed flux at visual and infrared wavelengths for a composite model with equal numbers of silicate and magnetite grains at 1 AU. While the silicate grains contribute 50 to 70% of the scattered radiation at $\lambda = 0.5$ μm, they contribute < 10% of the thermal emission. The flux at 3.5 μm consists of thermal emission from the magnetite grains (95%), and scattering from the cooler silicate grains (5%). Further modeling, to incorporate the scattered light continuum in defining the relative abundance of silicate to absorbing grains, is in progress.

IV. SUGGESTED OBSERVATIONS

The availability of more sensitive detectors and telescopes in appropriate sites for infrared work offers a unique opportunity for cometary observations. We stress the importance of near-simultaneous thermal and reflected light measurements. Extensive coverage of the thermal continuum is highly desirable, including L and M, the 8 to 12 μm region, and the 18 to 24 μm region, in order to define the thermal spectrum and the relative contribution of silicate and absorbing grains. Spectral scans across the silicate feature at resolution $\lambda/\Delta\lambda \sim 100$ are desirable for sufficiently bright comets. If only one observation at a thermal bandpass can be obtained because of the faintness of a comet, it is important to make the observation near the expected peak of the thermal curve, to minimize the uncertainties in estimating the integrated thermal flux for temperature and albedo calculations.

A comet bright enough to allow high-spatial-resolution scans through its coma and tail provides a test for possible differences between the particles responsible for scattered and emitted radiation (see Sec. III). Because particles of different optical characteristics and different sizes have different β (the ratio of the force due to radiation pressure to the force due to gravitation) their spatial distribution in the outer coma and in the tail should be different and should become evident in high-resolution scans at reflected and thermal wavelengths.

Observations of the structure of the silicate band in the 18-22 μm region with resolution $\lambda/\Delta\lambda \gtrsim 20$ are highly desirable to discriminate the kind of silicate minerals present in the dust. While amorphous olivine shows an emission band near 18 μm, powdered samples of carbonaceous chondrites show a feature near 22 μm and a weaker one at 16 μm (Zaikowski and Knacke 1975; Friedemann et al. 1979).

The wavelength dependence of the emissivity in the far infrared can be diagnostic of the composition of the cometary dust (Day 1976). Such observations are difficult or impossible to obtain using groundbased telescopes. Observations of this kind can be obtained of P/Halley or any other bright comet using already existing facilities like the Kuiper Airborne Observatory, or the soon to become available Infrared Astronomy Satellite, Spacelab 2 Infrared Telescope, and the German Infrared Laboratory.

Acknowledgments. We wish to thank M. Lebofsky, G. Rieke, and E. Ney for helpful discussions. Part of this research (MH) was performed at the Jet Propulsion Laboratory, California Institute of Technology, under contract with the National Aeronautics and Space Administration and part (HC) was supported by a NASA grant at the University of Arizona.

REFERENCES

A'Hearn, M.F.; Dwek, E.; and Tokunaga, A.T. 1981. Where is the ice in comets? *Astrophys. J.* In press.

Brownlee, D.E. 1978, Microparticle studies by sampling techniques. In *Cosmic Dust,* ed. J.A.M. McDonnell, (New York: John Wiley), pp. 295-336.

Campins, H.; Gradie, J.; Lebofsky, M; and Rieke, G.H. 1981a. Infrared observations of faint comets. In *Modern Observational Techniques for Comets.* (Pasadena: Jet Propulsion Lab.), pp. 83-92.

Campins, H.; Rieke, G.H.; and Lebofsky, M. 1981b. Infrared photometry of periodic comets Encke, Chernyhk, Stephan-Oterma, and Tuttle. Submitted to *Icarus.*

Day, K.L. 1976. Further measurements of amorphous silicates. *Astrophys. J.* 210:614-617.

Friedemann, C.; Gürtler, J.; and Dorschner, J. 1979. The 10- and 20-μm interstellar absorption bands: Comparison with the infrared spectrum of the Nogoya meteorite. *Astrophys. Space Sci.* 60:297-304.

Giese, R.H. 1980. Optical investigation of dust in the solar system. In *Solid Particles in the Solar System,* eds. I. Halliday and B. McIntosh (Dordrecht: D. Reidel), pp. 1-13.

Hanner, M.S. 1980. Physical characteristics of cometary dust from optical studies. In *Solid Particles in the Solar System,* eds. I. Halliday and B. McIntosh (Dordrecht: D. Reidel), pp. 223-236.

Hanner, M.S. 1981. On the detectability of icy grains in the comae of comets. *Icarus.* In press.

Hanner, M.S.; Giese, R.H.; Weiss, K.; and Zerull, R. 1981. On the definition of albedo and application to irregular particles. *Astron. Astrophys.* In press.

Iijima, T.; Matsumoto, T.; Oishi, M.; Okuda, H.; and Ono, T. 1975. Near-infrared observations of Comet Bradfield (1974b). *Publ. Astron. Soc. Japan* 27:507-510.

Krätschmer, W. and Huffman, D.R. 1979. Infrared extinction of heavy ion irradiated and amorphous olivine, with applications to interstellar dust. *Astrophys. Space Sci.* 61:195-203.

Krishna Swamy, K.S., and Donn, B. 1979. An analysis of the infrared continuum of comets. II. Comet Kohoutek. *Astron. J.* 84:692-697.

Labs, D., and Neckel, H. 1970. Transformation of the absolute solar radiation data into the international practical temperature scale of 1968. *Solar Physics* 15:79-87.

Merrill, K.M. 1974. 8–13 μm spectrophotometry of Comet Kohoutek. *Icarus* 23:566-567.

Mukai, T. 1977. Dust grains in the cometary coma: Interpretation of the infrared continuum. *Astron. Astrophys.* 61:69-74.

Ney, E.P. 1974. Multiband photometry of comets Kohoutek, Bennett, Bradfield, and Encke. *Icarus* 23:551-560.

Ney, E.P., and Merrill, K.M. 1976. Comet West and the scattering function of cometary dust. *Science* 194:1051-1053.

Noguchi, K.; Sato, S.; Maihara, T.; Okuda, H.; and Uyama, K. 1974. Infrared photometric and polarimetric observations of Comet Kohoutek 1973f. *Icarus* 23:545-550.

Oishi, M.; Kawara, K.; Kobayashi, Y.; Maihara, T.; Noguchi, K.; Okuda, H.; and Sato, S. 1978. Infrared observations of Comet West (1975n). I. Observational results. *Publ. Astron. Soc. Japan* 30:149-159.

Rieke, G.H., and Lee, T.A. 1974. Photometry of Comet Kohoutek (1973f). *Nature* 248:737-740.

Sekanina, Z. 1981. Comet Bowell (1980b): An active-looking dormant object? Submitted to *Astron. J.*

Sekanina, Z., and Miller, F.D. 1973. Comet Bennett 1970 II. *Science* 179:565-567.

Sekanina, Z., and Schuster, H.E. 1978a. Meteoroids from periodic Comet D'Arrest. *Astron. Astrophys.* 65:29-35.

Sekanina, Z., and Schuster, H.E. 1978b. Dust from periodic Comet Encke: Large grains in short supply. *Astron. Astrophys.* 68:429-435.

Veeder, G.J., and Hanner, M.S. 1981. Infrared photometry of comets Bowell and P/Stephan-Oterma. *Icarus* (special Comet issue). In press.

Zaikowski, A., and Knacke, R.F. 1975. Infrared spectra of carbonaceous chondrites and the composition of interstellar grains. *Astrophys. Space Sci.* 37:3-9.

DUSTY GAS-DYNAMICS IN REAL COMETS

MAX K. WALLIS
University College, Cardiff

The mathematical models of dust grains as dragged out by radially expanding cometary gases are reviewed. The constant speed effusion model leading to grain speeds as convenient functions of size and radial position is criticized as nonphysical. Two-phase gas dynamics treating the dust as a degnerate gas is appropriate for active comets. The set of ordinary differential equations showing the usual singular point at Mach number $M = 1$ have been solved for a single size of grain; subsonic Mach numbers at the nucleus mean that we must analyse and calculate a two-point boundary problem through the X-type singular point. By generalizing the equations to cover realistic cases with isothermal dust grains and a spectrum of grain sizes, it is found that the dust is even more effective in forcing an inner subsonic region in the flow. Terminal speeds for the spectrum of grains for consistency must be found from simultaneous solution for a set of discrete grain sizes. Initial gas speeds for H_2O-ice comets have generally been assumed high by a factor of 2. Though heating by dust and photoprocesses enhances the final speed to observed values, the potentially lower terminal speeds of the grains would mean, for example, a smaller dust hazard zone at P/Halley. Extensions of the model to less active comets including evaporating grains are discussed. Uncertainties on the degree of grain volatility and anisotropy of outgassing limit the quantitative deductions from comet dust tail observations, but new solutions of the dusty gas-dynamic model for realistic isothermal or evaporating grains would enable improvement of current parameters like the gas-dust ratio and the grain size distribution.

I. GAS DRAGGING OF COMETARY DUST GRAINS

It has long been recognized that the forces necessary to overcome cometary gravity and expel dust grains into halos and the tail might be a consequence of outflowing gases (Orlov 1935). The high levels of outgassing now known confirm that this mechanism must operate and is indeed sufficient for active comets within 1 to 2 AU from the Sun. Steady, radially-symmetric outgassing is commonly assumed, so that a grain of radius a and density ρ_d moves radially outwards at velocity v as described by the equation of motion

$$\frac{4}{3}\pi\rho_d a^3 v \frac{dv}{dr} = \frac{1}{2} C_D \pi a^2 (u-v)^2 \rho_o u_o \frac{R^2}{ur^2} - \frac{4}{3}\pi\rho_d a^3 M_{cmt} G. \quad (1)$$

The gas density decreases from its value ρ_O at the nucleus of radius R with inverse square distance r and increasing expansion velocity u. The drag coefficient C_D is $O(1)$ and the final term represents the central gravitational force from the nucleus of mass M_{cmt}.

In early treatments (Whipple 1951; Weigert 1959; Dobrovolskiĭ 1966; Huebner and Weigert 1966), both C_D and u were taken to be constant. Weigert (1959) assumed explicitly that outgassing was so weak that gas particles have long free paths and collide elastically with the dust grains. Then $C_D = 2$ and the equation of motion can be written

$$v\frac{dv}{dr} = \frac{\Lambda}{2a}\left[(u-v)^2 - b^2\right]\frac{R}{r^2} \quad (2)$$

where

$$b = u\,(a/a_{max})^{1/2}, \quad a_{max} = \frac{3}{2}\bar{m}Qu/\rho_d M_{cmt} G$$

$$Q = \rho_o u R^2/\bar{m}, \quad \Lambda = 3\bar{m}Q/\rho_d Ru.$$

As defined here, Q denotes the outflow rate (per second and steradian) of gas with mean molecular weight $\bar{m}$. Evidently there is a maximum radius a_{max} of grains that can escape (Whipple 1951). As a quantitative guide, the following values are considered

$R = 3 \times 10^5$ cm, $M_{cmt} = 1.1 \times 10^{17}$ g (mean density 1 g cm^{-3})

total gas flux $4\pi Q = 3 \times 10^{29}$ s^{-1}, $u = 2.2 \times 10^4$ cm s^{-1}

atomic mass of H$_2$O $\bar{m} = 2.9 \times 10^{-23}$ g, $\rho_d = 3$ g cm^{-3}

whence

$$\Lambda = 1\,\mu\text{m}, \quad a_{max} = 1\,\text{cm}. \quad (3)$$

The ratio of these two lengths is expressible in terms of the escape velocity from the nucleus surface, $\Lambda/a_{max} = (v_{esc}/u)^2$ and is always much less than unity. In physical terms, Λ turns out to be a coupling scale size; grains smaller than Λ approach the gas velocity.

Equation (2) is directly integrable (Dobrovolskiĭ 1966)

$$1 - \frac{R}{r} = \frac{a}{\Lambda b}\left[(V-b)\ln\frac{u-b-v_o}{u-b-v} + (V+b)\ln\frac{u+b-v}{u+b-v_o}\right] \quad (4)$$

for an initial velocity v_o, taken hereafter as zero. Unfortunately this implicit solution for $v(r)$ is not very useful. A piecewise approximation valid over the whole range of sizes a can be expressed in terms of $\eta = (1-R/r)\Lambda/a$ and $\eta_m = (1-R/r)\Lambda/a_{max}$ as

$$v = u \begin{cases} [1+2/(\alpha+\eta)]^{-1} & \eta > \eta_1 \quad (5a) \\ [\beta+(\eta-\eta_m)^{-1/2}]^{-1} & \eta_1 > \eta > \eta_m. \quad (5b) \end{cases}$$

This approximation has the proper behavior at $a \ll \Lambda (\eta \gg \eta_1)$ and $a_{max} > a \gg \Lambda (\eta \ll \eta_1)$. Authors interested in large grains have used Eq. (5b) with $\beta = 0$, but sometimes applied it to micro-sizes (Whipple 1951; Hughes 1979). Delsemme and Miller (1971) derive the behavior $v \propto 1 - (\eta_m/\eta)^{1/2}$ near the a_{max} singularity, but erroneously as seen by neglecting v on the right of Eq. (2). With appropriate numbers α, β, η_1 of order 1, approximation (5a,b) can be chosen continuous at η_1 and a good fit over the full range:

$$\alpha = 2.7, \quad \beta = 0.718, \quad \eta_1 = 2.0. \quad (6)$$

These are fitting parameters only: with the values in Eq. (6), the deviation from the exact solution of Eq. (4) shown as the continuous line in Fig. 1 is under 3% and within the thickness of the line. The results apply as well to the terminal speeds of dust grains v_t, derived by letting $r \to \infty$, $\eta \to \Lambda/a$. Clearly for $a < \Lambda$, the grains approach the gas velocity, while for $a > \Lambda$ they remain much slower. If the grains are evaporating (Huebner and Weigert 1966; Huebner 1970), the solution is a little different; because their radii decrease, the singularity at a_{max} disappears. Other physical processes such as collisional fragmentation or accretion, small initial v_o, rotation of the nucleus, temporal and spatial inhomogeneities in outgassing, etc. may also cause the model Eq. (1) to be inaccurate for the largest grains.

A major uncertainty in such solutions has lain in the choice of u. It was taken as a fraction of the thermal velocity defined by the nucleus temperature, $v_{th} = (8\,kT/\pi m)^{1/2}$. Weigert (1959) assumed that $u = 2\,v_{th}/3$, while Huebner and Weigert (1966) took $u = 0.5\,v_{th}$ for isotropic emission in a half-space, as corresponds to the classical theory of sublimation into vacuum (e.g. Delsemme and Miller 1971). However, in collisionless effusion the mean

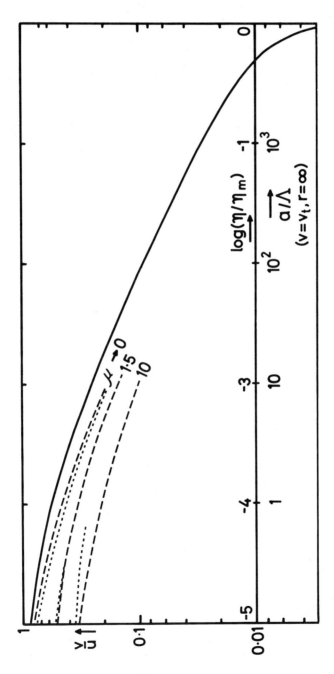

Fig. 1. Grain speeds on the simple effusion model (full line, Eq. 4) as a function of $\eta = (1 - R/r)\Lambda/a$, grain radius a, for $\Lambda = 10^{-4} a_{max}$. Broken lines represent terminal speeds from Probstein's calculations for dust test particles ($\mu \to 0$) and from a single species of grain with $\mu = 1.5$ or 10 times the gas mass flux. Dots represent the empirical fit of Eq. (11).

radial speed increases roughly as $u \sim (1 - R/2r)v_{th}$, evident on geometrical grounds. In any case, the neglect of collisions is inappropriate.

The outgassing rate into vacuum from an icy surface can be written $Z = (1-A) Z_1$, where $Z_1 = 1.7 \times 10^{18}$ molecules/cm^{2}s for H$_2$O-ice at 1 AU, the albedo factor A allowing for nonvolatile contaminants (Delsemme and Miller 1971; Whipple and Huebner 1976). For a molecular collision cross-section σ at 1 AU $\cong$ 10 Å^2, the mean free path is

$$\ell = u/Z\sigma \simeq 13\text{cm}/(1-A) \ . \qquad (7)$$

Even for a largely mineral surface with $1-A \simeq 0.1$, Z has to be reduced by a factor of 10^3 before ℓ is comparable to km-sized comet nuclei. The sublimation curves of Delsemme and Miller (1971) then imply that collisionless effusion from an H$_2$O-ice comet could only be appropriate outside 3 AU, at which distance any active comet is presumed to have more volatile constituents than H$_2$O. The conclusion as first noted by Markovich (1963) is that gas-dynamic methods are needed to determine $u(r)$.

The drag coefficient can also be more precisely evaluated by gas dynamics, using the long mean free path simplification of Eq. (7), $\ell \gg a$. Assuming Maxwellian velocity distributions in the collisional gas and reflection of impacting molecules with a temperature T_{refl}, the drag expressed in terms of a relative Mach number, $\psi = |u-v|/(2kT/m)^{1/2}$, is

$$C_D = \frac{2\sqrt{\pi}}{3\psi}(T_{refl}/T)^{1/2} + \frac{1}{\sqrt{\pi}}(2\psi^{-1} + \psi^{-3})\exp{-\psi^2} + (2+2\psi^{-2}-\psi^{-4})\operatorname{erf}(\psi) \qquad (8)$$

(Probstein 1969; Shul'man 1972a). $C_D = 2$ at $\psi = \infty$ (used above) and increases with decreasing ψ to ~ 5 at ψ of order unity. In practice, the grains are hotter than the gas (Shul'man 1972a), T_{refl} may approach T_{dust} and the first term in Eq. (8) can dominate at small ψ. Consequently, the smallest grains can have velocities closer to u than implied by Eqs. (4) and (5).

II. GAS-DYNAMIC MODELS

For the interesting comets during active phases, when gas molecule free paths from Eq. (7) are 10^{-2} to 10^{-4} times the nucleus radius, hydrodynamic descriptions of the outgassing are appropriate. Shul'man (1969) discussed the simplest adiabatic model in spherical geometry with gas density $\rho \propto r^{-2} u^{-1}$, pressure $p \propto \rho^\gamma$ and the Bernoulli energy integral

$$\frac{1}{2}u^2 + \gamma p/(\gamma-1)\rho = \text{const.} \qquad (9)$$

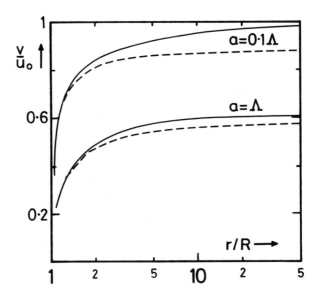

Fig. 2. Acceleration of dust in supersonic flow (from Shul'man 1969). The figure shows the comparison of the effusion model—Eq. (4) with $b = v_0 = 0$—with the adiabatic gas dynamic model in which random gas speeds are converted to radial speeds ($\gamma = 1.4, M_0 = 2$).

Solving numerically for u in terms of the initial Mach number M_0, Shul'man calculated corresponding solutions of Eq. (1) for small dust grains (taking C_D = const. and dropping the gravity terms). His results, reproduced in Fig. 2, show that the accelerating flow gives higher dust speeds. Terminal velocities are reached within 3 to 5 R and are somewhat less than u_0 (e.g., small grains $a = 0.1\Lambda$ have $v_t \cong 0.6\ u_0$). However, the gas dynamics was not fully solved, the initial Mach number being arbitrarily chosen ($M_0 = 2$ as shown). Also, the formulation can apply only for small admixtures of dust (for relative mass flux $\mu = m_d Q_d / \overline{m} Q \ll 1$, i.e., for weakly dusty flows).

The initial Mach number in source flows into vacuum is $O(1)$, indeed equal to unity for the adiabatic case. Probstein (1969) first showed how to choose M_0 in the cometary situation, finding values $M_0 < 1$. The dust grains react back on the flow, invalidating the adiabatic assumption; qualitatively their drag in reducing u near the nucleus reduces a_{max} but increases the dust-gas collisional rate. Rewriting the energy integral Eq. (9) as $u_0 = [1 + 2M_0^{-2}/(\gamma-1)]^{-1}\ v_{th}$ and taking the representative values found by Probstein $M_0 \simeq 0.4$ for $\gamma = 1.4$, one sees that u_0 is smaller by a factor 7 than the ½v_{th} of Huebner and Weigert (1966) or the $4v_{th}/9$ of Shul'man (1969). Effectively the scale Λ is higher and a_{max} is lower by this factor 7, so proper gas-dynamic solutions are quantitatively important.

For small mass fractions of dust ($\mu \ll 1$), $M_0 \to 1$ and the gas-dynamic results are directly comparable with simple effusion. Figure 1 includes some of Probstein's (1969) results (single-size grains, neglecting cometary gravity) for the terminal velocities $v = v_t$ ($\eta = \Lambda/a$; with varying $u(r)$, scaling to a general position is no longer possible). Even on this log-log plot, the $\mu \to 0$ results differ appreciably from collisionless effusion, because the smaller u_0 gives lower accelerations while the higher collision frequency does not compensate for the reduced momentum transfer. If the dust constitutes a significant mass fraction, the $\mu = 1.5$ curve in Fig. 1 being typical, energy is subtracted from the flow on the Probstein model so that all outflow speeds are decreased. Sekanina (1979) uses a semi-empirical fit to Probstein's results, reading in the present units (and with a misprint corrected—Sekanina, personal communication, 1981)

$$v_t/u_\infty = [0.83 + 0.35\mu^{1/2} + 1.37(a/\Lambda)^{1/2}]^{-1}. \tag{10}$$

This is of the same form as Eq. (5b) for $\eta_1 > \eta = \Lambda/a \gg \eta_m$; however, for the constant gas speed model, a function of this form cannot describe the smallest grains properly (it gives $v_t > u_\infty$ for $a \cong 0.01\,\Lambda$ and small μ). In view of nonphysical features of Probstein's model discussed below, more complex fitting functions are hardly worthwhile and over the restricted range $0.01\,\Lambda < a \leqslant a_{\max}, \mu \lesssim 3$, a better fit shown by dots in Fig. 1 is

$$v_t/u_\infty = [0.9 + 0.4\mu^{1/2} + 1.5(a/\Lambda)^{1/2}]^{-1}. \tag{11}$$

One limitation is that Probstein's model treats a single size of dust grain. If the mass distribution peaks significantly, that particular size can be selected to calculate the flow field $u(r)$. The dynamics of smaller and larger grains can subsequently be calculated from Eq. (1), but using Probstein's (1969) results or the approximations such as Eqs. (10) and (11) to describe a range of particle sizes (Sekanina 1979) has never been justified.

A second difficulty lies in choosing the velocity calibration: the value $v_{th} = (8kT_n/\pi m)^{1/2}$ corresponding to the nucleus temperature T_n (which itself is not sensitive to sublimation conditions) is only a guide (Delsemme and Miller 1971). A kinetic equation analysis has recently been used to describe the gas-nucleus boundary (Shul'man 1977). Assuming quasi-Maxwellian solutions, that author confirmed both the sublimation equation for pure gas production and that M_0 is close to unity for small Knudsen numbers ($Kn = \ell/R \ll 1$), but found that particle speeds are somewhat low:

$$M_0 = 1, \quad u_0 = 0.435\,v_{th}, \quad u_\infty^2 \lesssim (\gamma+1)u_0^2/(\gamma-1). \tag{12}$$

The limit on u_∞^2 follows from the energy integral given in Eq. (9), observing that a fraction of the energy is frozen in a finite temperature expansion into vacuum (Shul'man 1972a); the limit also applies if an admixture of dust extracts energy from the flow. In collisionless effusion, energy in the gas molecules' rotational and vibrational modes is frozen in; effectively $\gamma = 5/3$ so $u_\infty < 2u_0$ and a mean value $u = \frac{1}{2}v_{th}$ allowing a small increase with r remains plausible (Sec. 1). For H_2O subliming into a collisional regime, rotational mode energy is convertible (Probstein 1969; Delsemme and Miller 1971) implying values $\gamma = 1.4$ or 1.33; however these authors' values of $u = 1.6 - 1.77\ v_{th}$ come some 50% too high. As a nucleus of subliming H_2O-ice is near 180 K (Delsemme and Miller 1971), the velocities in Eq. (12) are $u_0 = 0.20$ km s^{-1}, $u < 0.54$ km s^{-1}, a factor of at least 2 smaller than commonly deduced for gases in the far cometary coma. This difference is understandable on the basis that heat inputs into the expanding coma are important, indeed probably dominant (Wallis 1974,1975). Photodissociation and ionization processes release energetic particles which thermalize in the gas, while dust grains are kept hot by the solar radiation at ~ 600 K (Shul'man 1972a). Close to the nucleus, heating of the gas by such hot dust may be indirectly significant in dust-grain acceleration, in addition to directly increasing the drag (Eq. 8).

III. DUSTY GAS-DYNAMIC SCHEMA

The two-phase gas dynamic approach introduced by Probstein (1969), Brunner and Michel (1968), and Shul'man (1969) views the dust grains as a much heavier species with negligible thermal motions and pressure, and colliding only with the gas. A range of grain sizes can be important, so one considers a flux distribution with weighting $W(a)$ over discrete classes of particle size, each class having a representative radius and obeying the equation of motion (Eq. 1). The number density of each class for nonevaporating grains is

$$N_a = W(a) R^2/r^2\ v_a(r) \tag{13}$$

and the mass flux related to previously defined parameters by $\Sigma m_a W(a) \equiv m_d Q_d = \mu \bar{m} Q$. Hellmich (1979) has recently used such a discrete grain size approximation. The gas is described by steady state fluid equations with radial symmetry, best written in the conservation form with source terms representing mass gains (P_1), momentum gains including frictional losses (P_2), and energy gains (P_3):

$$\begin{aligned}
P_1 &= r^{-2}\ d(\rho u r^2)/dr \\
P_2 &= r^{-2}\ d(\rho u^2 r^2)/dr + dp/dr \\
P_3 &= r^{-2}\ d\left\{\left[\frac{1}{2}\rho u^2 + \gamma p/(\gamma-1)\right]ur^2\right\}/dr\ .
\end{aligned} \tag{14}$$

If the grains are subliming, or refractory components are recondensing, the mass source P_1 can express this. Gravity is incorporated by including in the sources the terms

$$P_2 = -v_{esc}^2\ \rho R/r^2,\ P_3 = v_{esc}^2\ \rho u R/r^2 \tag{15a}$$

where $v_{esc}^2 = 2M_{cmt}G/R$. However, the gas velocities are relatively very high, $u^2/v_{esc}^2 = a_{max}/\Lambda \ggg 1$, so the terms in Eq. (15a) are negligibly small even for dust exceeding gas production by an order of magnitude.

The momentum loss and energy gain from interaction with the grains are of greater interest and can be written explicitly as

$$P_2 = -\frac{1}{2} \Sigma N_a \pi a^2 C_D \rho (u-v_a)^2$$
$$P_3 = \Sigma N_a 4\pi a^2 C_H \rho (u-v_a) k (T_a - T_{rec})/\bar{m} \qquad (15b)$$

where Σ denotes the sum over the size-classes of grains. The drag coefficient C_D as introduced earlier is a slowly varying function of the relative Mach number as Eq. (8), while the heat transfer coefficient $C_H = \gamma St/(\gamma-1)$ is proportional to the Stanton number St, given by Probstein (1969) in analogous form to C_D. The recovery temperature T_{rec} is considered proportional to the gas temperature, $T = \bar{m}p/k\rho$, and the dust temperature T_0 is taken as independent of grain size and position. The grain density per size-class $N_a \sim v_a^{-1}$ in Eq. (13) is singular both for $r \to R$ and $a \to a_{max}$; through the use of approximation (Eqs. 5a,b) one can average over the largest grain class and over the first step in r to obviate the singularities. Instead of the energy equation in Eq. (14), Probstein used a combined gas-dust energy integral, but this is not valid in the presence of radiatively-heated dust. Shul'man (1972a) and also Hellmich (1979) included the dust heat source and the dependence $C_D(T_0)$, but assumed supersonic initial conditions.

To examine transonic possibilities by the gas-dynamic procedure, write Eq. (14) in terms of the Mach number: for simplicity introduce $\zeta = M^{-2} = \gamma p/\rho u^2$ and put $P_1 = 0$, so that Eq. (14) can be written

$$\frac{d\zeta}{dr} = \left[(\gamma-1+2\zeta)\left(\frac{2\zeta}{r} + \frac{\gamma P_2}{\rho u^2}\right) - (\gamma-1)(\gamma+\zeta)\frac{P_3}{\rho u^3} \right]/(\zeta-1) \qquad (16)$$

$$\frac{\gamma+\zeta}{u}\frac{du}{dr} = \frac{2\zeta}{r} - \frac{d\zeta}{dr} + \frac{\gamma P_2}{\rho u^2}, \quad \rho u r^2 = \text{const.} \qquad (17)$$

The singular Eq. (16) showing the sonic point at $\zeta = 1$ has very similar structure to that discussed by Probstein (1969). The dust drag term in P_2 is negative and dominates near the nucleus, allowing solutions with increasing Mach number (decreasing ζ) only if $M < 1$ ($\zeta > 1$). In the present case, the dust drag effect is enhanced by the energy source P_3. A transition to supersonic flow can occur where the numerator of Eq. (17) reaches zero: at $r = r_c$ where

$$r_c = \frac{1}{2} r_c^2 \left[-\gamma P_2/\rho u^2 + (\gamma-1) P_3/\rho u^3 \right]. \qquad (18)$$

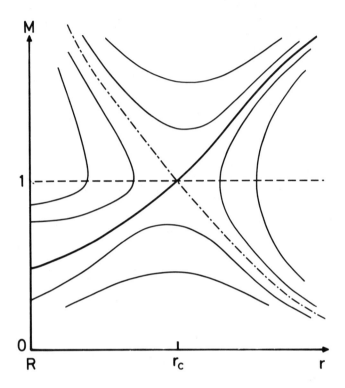

Fig. 3. Possible solutions to the gas-dynamic throat Eq. (16). The appropriate solution satisfying inner and outer boundary conditions is the simple trajectory passing continuously from the subsonic to supersonic regime.

Because of the singularity of $P \sim v_a^{-1} \sim (1-R/r)^{1/2}$ to Eq. (5b), Eq. (18) always has a solution. Notice from Eqs. (13) and (15) that roughly $r^2 P_2 \sim \rho u^2$ and $r^2 P_3 \sim \rho u$. If Eq. (18) holds, it can be shown with the help of auxiliary Eq. (17) that r_c is a saddle point just as in Probstein's case. The structure of solutions to Eqs. (16) and (17) is as shown in Fig. 3 and the one satisfying the boundary conditions passes from subsonic to supersonic conditions along the singular trajectory.

The additional energy source P_3 acts to increase the sonic radius in Eq. (18) and appears potentially large: $P_3/\rho u \sim k(T_a - T_{\text{rec}})/\bar{m}u^2$. In fact, the more effective this term due to larger numbers of grains, the closer the two temperatures become; thus the contribution stays similar to that of P_2 and solutions analogous to Probstein's (1969) should generally be found. Additive inclusion of the gravity terms of Eq. (15a) results in an additional term $Rv_{\text{esc}}^2/4u^2$ on the right of Eq. (18), slightly increasing the transition radius.

IV. DISCUSSION

The above analysis could be extended to cover evaporating cometary grains, which have aroused attention in the past (Huebner and Weigert 1966; Huebner 1970; Delsemme and Miller 1971). The first two papers give the equation of motion modified for reducing mass. Discrete classes of grains can again be introduced, in accordance with the present schema, but N_a in Eq. (13) is replaced by $N(a_0)$. The evaporating gases provide a mass source P_1 in the gas-dynamic Eq. (14) as well as additional energy and momentum sources. The net effect is similar to the grain drag, acting to choke the flow and enhance the subsonic region. However, its radial dependence is different. Even if one is interested in the longer-lived fraction of more refractory grains (Delsemme and Miller 1971), more volatile ice grains would undoubtedly accompany them and probably contribute significantly to the gas flow within a few R. Use of Probstein's (1969) model without the extension to cover this case appears dubious. Brunner and Michel (1968) conducted an experimental simulation with hot evaporating metallic grains in an inert gas flow. Though not giving a detailed analysis, they did point out that the interior flow is still further subsonic and gas-grain coupling is increased.

For less active comets such as P/Encke, having at times gas-free paths (Eq. 7) in excess of $0.1\,R$, gas-dynamical methods become invalid. Hamel and Willis (1966) followed by Shul'man (1970) introduced the kinetic equation to analyze radial expansion of a monatomic gas in vacuum, looking for quasi-Maxwellian solutions but only for large M_0; the main novel feature was the development of temperature anisotropy with a frozen radial temperature $T_r \to T_\infty \neq 0$. There being few collisions, rotational as well as vibrational energies are probably also frozen. Extension of kinetic equation analysis to cover dusty emissions with M_0 of order 1 are feasible but not yet undertaken. One expects results intermediate between the simple effusion and gas-dynamic models, namely higher initial and smaller final gas speeds than from gas dynamics.

Non-isotropic dust and gas emission from the nucleus giving nonradial flow are also likely to be important, for grain-gas coupling depends on their density product giving a large contribution within $3R$ of the nucleus. Even if the patchy model of an aging nucleus (Shul'man 1972b) is not valid and even if there are no localized gas jets and dust sources (Whipple and Huebner 1976), day-night anisotropy is probable. Gas would flow from the dayside to recondense on the nightside making the near-nucleus flow strongly nonradial (Wallis and Macpherson 1981). Generalization of the gas-dynamic analysis to treat nonradial flow is far from easy, for a solution has to be chosen to pass through sonic conditions over a two-dimensional surface: the 2-point boundary problem of Sec. III becomes a two-surface boundary problem with an unfixed outer surface. Relaxation methods to compute continuum solutions are feasible, but a more direct discrete model is perhaps more practicable (Wallis and Macpherson 1981).

In conclusion, the Probstein (1969) approach is valid and powerful but his model was clearly oversimplified. The dynamics of a range of grain sizes need calculation simultaneously with the gas dynamics; the grains need to be treated as near-isothermal, their temperature depending on distance from the Sun (Shul'man 1972a; Ney and Merrill 1976). Initial gas speeds are a small fraction of the speed of sound, commonly taken as representative (Sekanina 1981). The initial enthalpy in the gas should also be taken as lower and gas heating by the grains and photoprocesses should be included in compensation. The possibilities of evaporation from grains and of nonisotropic emission from the nucleus introduce uncertainties, as do optical thickness effects and grain fragmentation-accretion processes close to the nucleus. Observations of comet dust tails do, in principle, allow the deduction of the initial dust speeds, outgassing rate, and dust to gas ratio (Finson and Probstein 1968), but it is premature to claim better than order-of-magnitude accuracy from the over-simplifed model (Sekanina and Miller 1973). In view of the current interest in accurate dust particle distributions and the dust hazard to the Giotto mission, investigations of realistic dusty gas-dynamical models are sorely needed.

REFERENCES

Brunner, W., and Michel, K.W. 1968. Investigation of flow relations in dust-comet comas. *Mitt. Astron. Ges.* 25:220-224.

Delsemme, A.H., and Miller, D.C. 1971. Physico-chemical phenomena in comets III: The continuum of Comet Burnham (1960 II). *Planet. Space Sci.* 19:1229-1257.

Dobrovolskiĭ, O.V. 1966. *Komety* (Moscow: Nauka).

Finson, M.L., and Probstein, R.F. 1968. A theory of dust comets I and II. *Astrophys. J.* 154:327-352, 353-380.

Hamel, B.B., and Willis, D.R. 1966. Kinetic theory of source flow expansion with application to the free jet. *Phys. Fluids* 2:829-841.

Hellmich, R. 1979. Anisotropic multiple scattering in the dust coma and its influence on the sublimation rate of the comet nucleus. Ph.D. Dissertation, Univ. Göttingen.

Huebner, W.F. 1970. Dust from cometary nuclei. *Astron. Astrophys.* 5:286-297.

Huebner, W.F., and Weigert, A. 1966. Ice grains in comet comas. *Zs. f. Astrophys.* 64:185-201.

Hughes, D.W. 1979. The micrometeoroid hazard to a space probe in the vicinity of the nucleus of Halley's Comet. In *The Comet Halley Micrometeoroid Hazard,* (Paris: ESA SP-153), pp. 51-56.

Markovich, M.Z. 1963. *Byull. Komissii po kometam i meteoram.* 8:11.

Ney, E.P., and Merrill, K.M. 1976. Comet West and the scattering function of cometary dust. *Science* 194:1051-1053.

Orlov, S.V. 1935. *Komety*, (Moscow: ONTI).

Probstein, R.F. 1969. The dusty gas dynamics of comet heads. In *Problems of Hydrodynamics and Continuum Mechanics,* ed. M.A. Lavrent'ev (Philadelphia: SIAM), pp. 568-583.

Sekanina, Z. 1979. Expected characteristics of large dust particles in periodic Comet Halley. In *The Comet Halley Micrometeoroid Hazard*, (Paris: ESA SP-153), pp. 25-34.

Sekanina, Z. 1981. Rotation and precession of cometary nuclei. *Ann. Rev. Earth Planet. Sci.* 9:113-145.

Sekanina, Z., and Miller, F.D. 1973. Comet Bennett 1970 II. *Science* 179:565-567.

Shul'man, L.M. 1969. Hydrodynamics of the circum-nuclear region of a comet. *Astromet. Astrofiz.* 4:100-115. (Translation from the Russian, NASA TT F-599, pp. 85-99).

Shul'man, L.M. 1970. Movement of neutral matter in the cometary atmosphere. *Astromet. Astrofiz.* 11:1-15.

Shul'man, L.M. 1972a. Dinamika kometnikh atmsofer-neitralnii gaz. Kiev. Ch. 3.

Shul'man, L.M. 1972b. The evolution of cometary nuclei. In *Proc. IAU Sym. 45, The Motion Evolution of Orbits and the Origin of Comets,* eds. G.A. Chebotarev, E.I. Kazimirchak-Polonskaya, and B.G. Marsden, (Dordrecht, Holland: D. Reidel), pp. 271-276.

Shul'man, L.M. 1977. Theory of the boundary layer of the cometary nucleus. *Astromet. Astrofiz.* 32:24-28.

Wallis, M.K. 1974. Hydrodynamics of the H_2O comet. *Mon. Not. Roy. Astron. Soc.* 166:181-189.

Wallis, M.K. 1975. Expansion velocities of cometary gas *Astrophys. Space Sci.* 30:343-346.

Wallis, M.K., and Macpherson, A.K. 1981. On the outgassing and jet thrust of snowball comets. *Astron. Astrophys.* 98:45-49.

Weigert, A. 1959. Halo production in Comet 1925 II. *Astron. Nachr.* 285:117-128.

Whipple, F.L. 1951. A comet model II: Physical relations for comets and meteors. *Astrophys. J.* 113:464-474.

Whipple, F.L., and Huebner, W.F. 1976. Physical processes in comets. *Ann. Rev. Astron. Astrophys.* 14:143-172.

COMETARY DUST IN THE SOLAR SYSTEM

H. FECHTIG
Max-Planck-Institut für Kernphysik

By two basic methods, studies of microcraters on lunar samples and in situ dust measurements, information has been obtained on cometary dust in the solar system. The diameter-to-depth ratios of lunar microcraters are primarily determined by the densities of the projectiles; ≤25% of all lunar microcraters are produced by low-density projectiles (≤ 1 g cm^{-3}). The results from the HEOS-2 dust experiment showed the existence of so-called swarms, bursts of particles produced shortly before observation. Since these swarms are observed within 10 Earth radii, they probably result from electrostatic fragmentation within the auroral plasma region. The flux of the parent bodies is estimated to be ~30% of the total flux of all bodies traveling through this region, but <50% of the observed dust particles near the Earth are associated with swarms. Differences in events recorded by the two Helios-sensors show the existence of low-density particles. This information indicates that low-density particles probably of cometary origin exist in the solar system but <30% are of this low-density type. This fact, and the low dust production rate of comets, show either that there are other dust sources in the solar system or that we do not understand cometary dust in particular or comets in general. Exploration of the nature of cometary dust particles is an important goal of a cometary mission.

The general theory about the origin of interplanetary dust is a quasi-continuous production of dust mainly by release from comets. Whipple (1967) has estimated that a total production rate of ~10 ton s^{-1} is necessary to maintain the zodiacal dust cloud of the solar system. Of course, there are other possible dust sources in the solar system; the possibility of collisional dust production in the asteroid belt has been discussed, but Dohnanyi (1976) has published calculations which indicate that this source is small compared to the cometary source. The first dust experiments on Pioneer 10/11 seem to justify Dohnanyi's calculations (Humes et al. 1974).

Let us assume that all dust particles in space are of cometary origin; what should we expect to observe in space? There are 3 criteria to discuss: composition, structure and orbital elements.

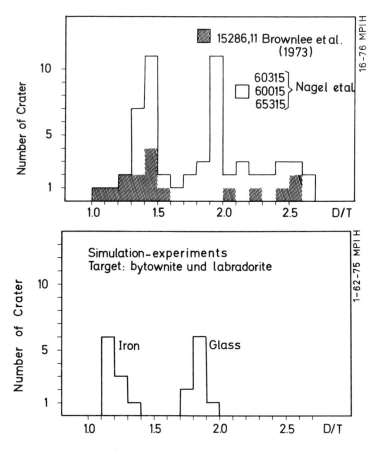

Fig. 1. Diameter/depth D/T measurements of natural and simulated microcraters; (upper) natural craters on lunar samples, (lower) simulated craters in feldspars produced by iron and glass projectiles.

The composition is expected to be twofold, generally carbonaceous chondritic (probably CI type) (Wetherill 1971), but also interstellar, at least for submicron-size individual particles (Biermann and Michel 1978; Greenberg 1979). Yet no one knows which composition to assume for interstellar dust. Depending on its specific origin, deviations from chondritic composition and in particular specific isotopic ratios ($^{12}C/^{13}C$-, $^{6}Li/^{7}Li$-ratios) are expected (Cameron and Fowler 1971; Mitler 1972; Audouze and Truran 1973; Reeves 1974; Starrfield et al. 1978; Vanýsek 1979).

Cometary dust is known to be low in density compared to silicates and chondrites. The photographic meteors (Jacchia 1955; Verniani 1964; Jacchia et al. 1965) and the radio meteors (Verniani 1973) have densities ranging between 10^{-2} and 10 g cm^{-3} with an average of 0.8 g cm^{-3}. The densities of meteor stream particles (Verniani 1967, 1969; Millman 1976) are reported to be very low; they range between 1.06 g cm^{-3} for the Geminids down to 0.01 for the Giacobinids. The mean errors of these measurements are ± 15% (Hughes 1981, personal communication). Since the meteor streams are directly related to comets, there is no doubt that cometary dust is of low density. The particles are believed to be of the so-called fluffy type, aggregates of smaller grains clumped together (Hughes 1978). But meteor observation and interpretation cannot give any information about the grain size of cometary dust particles. Brownlee (1978) has collected dust particles believed to be cometary in the stratosphere which show fluffy or reentrant structures and chondritic compositions. Brownlee (1978) reports that "90% of the chondritic particles are aggregates of 1000 Å sized grains."

Cometary dust can also be identified by comparing orbital elements with those of the producing comet. Dust may interfere with space experiments when the spacecraft is in the orbital plane of the producing comet. Hence, it is important to understand its production. This chapter reviews observations of interplanetary dust in space and estimates the amount of cometary dust.

I. LUNAR MICROCRATERS PRODUCED BY COMETARY DUST PARTICLES

Brownlee et al. (1973) have shown that lunar microcraters show diameter D to depth T ratios between 1.0 and 2.5 with two distinct peaks at 1.4 and 1.9. Smith et al. (1974) suggested that the observed distribution might be due to the different densities of projectiles. These authors have published T/D ratio measurements of lunar microcraters on glassy lunar spherules along with T/D ratios from simulation experiments using aluminum and iron projectiles ranging in speed between 2 and 15 km s^{-1}.

The results of our group (Nagel et al. 1975, 1976a,b; Fechtig et al. 1975) are plotted in Fig. 1, compared to Brownlee's results. The upper histogram shows the distribution for the lunar microcraters on lunar surface samples as a function of D/T. The lower histogram is simulation results using iron and

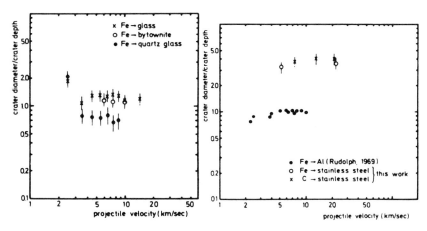

Fig. 2. Experimental results for the velocity dependence of the D/T ratio; (left) for silicate targets, (right) for metal targets.

glass projectiles at various velocities (between 4 and 11 km s^{-1}.) The results show the number of observed craters N plotted as a function of diameter to depth ratios D/T for microcraters on lunar glass. Two peaks are identified as produced by iron meteoroids and meteoroids with the density of glass, indicating production by chondritic silicates. However, the third peak at $D/T =$ 2.5 corresponds to a low-density component of ~ 1 g cm^{-3} as simulated by Vedder and Mandeville (1974). Mandeville (1977) has published similar results. Another paper (Nagel and Fechtig 1980) shows that the D/T ratio is independent of the impact velocity of the projectiles. In Fig. 2 the D/T ratios are plotted as a function of impact velocities; above a threshold velocity of ~ 4 km s^{-1} the ratios D/T are independent of the impact velocity. Yet there is a weak dependence of D/T with the impacting mass when the projectile mass varies by orders of magnitude (Nagel and Fechtig 1980).

A component of low-density dust particles in space is observed that can be considered cometary dust. The percentage of this component with respect to silicate and iron projectiles of densities 3 and 8 g cm^{-3} does not exceed 25%. However, shallow cometary craters may tend to be overlooked more easily. McDonnell and Allison (1981) show that secondary craters are also shallow craters. But even if we assume that all shallow craters on lunar samples are produced by low-density projectiles, certainly $<$ 25% of all projectiles are of low-density type.

II. DUST SWARMS

During the observation period of the Earth satellite HEOS-2 another phenomenon has been explored: the so-called dust swarms (Hoffmann et al. 1975a,b). HEOS-2 has scanned the altitude range between 5,000 and 244,000 km from 6 February 1972 to 2 August 1974.

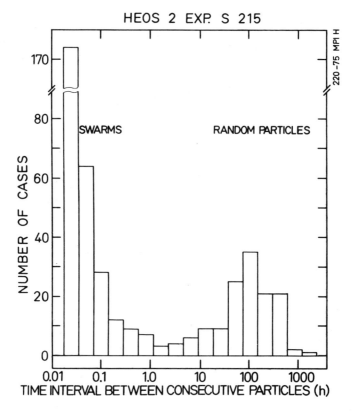

Fig. 3. Temporal distribution of registered dust particles on HEOS-2.

Fig. 3 shows the number of registered impacts of dust particles on the sensor as a function of the time differences between consecutive events. The peak near 100 hr represents the random dust distribution of the zodiacal dust particles. The increase at small time differences was unexpected. In Fig. 4 the swarms are plotted as a function of altitude compared to the random particles. The random particles show a more or less constant distribution above an altitude of ~10 Earth radii. Due to the gravitation of the Earth, an increase of a factor of 3 is recorded below 10 Earth radii for the random particles, but the swarms are exclusively observed within 10 Earth radii. Obviously the experiment has observed dust grains not too distant from each other, meaning that the instrument flew through a dust cloud, a recent production of dust particles from a larger parent body. The cloud must have formed recently, because with time it thins out and finally disappears. Such dust clouds have been observed associated with meteors hitting the Earth's upper atmosphere at grazing incident angles (Rawcliffe et al. 1974; Bigg and Thomson 1969).

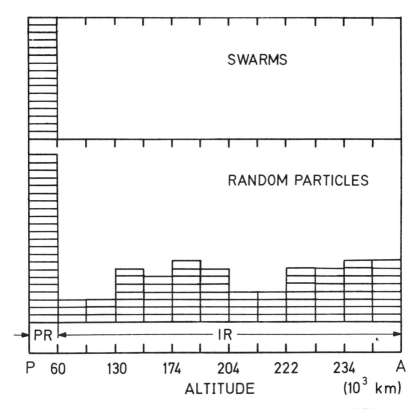

Fig. 4. Rate of swarms and random particles as a function of altitude on HEOS-2.

The HEOS-2 satellite has always operated above any significant atmosphere. In addition the sensor's viewing direction was always directed towards space when the swarms have been recorded. The formation of swarms has been interpreted on the basis of electrostatic fragmentation. As has been considered earlier (Fechtig and Hemenway 1976) such a fragmentation is possible if enough electrostatic charge is collected on the surface of a dust grain so that the tensile strength of the particle is less than the electrostatic repellant forces. In our latest paper on this subject (Fechtig et al. 1979) we have shown that enough negative charges can be collected by solid bodies traveling through the Earth's auroral zones in a high plasma environment. However, only fluffy solids can be disintegrated by this mechanism. It is possible to estimate the amount of mass of the corresponding parent bodies by assuming a spherical distribution of the dust cloud. The results range between 100 and 10^6 g for the 15 parent bodies of the 15 observed swarms. The corresponding fluxes agree well with direct observations of low-density fireballs published by Ceplecha and McCrosky (1976). Because of the fluffy nature of this class of fireballs they are assumed to be cometary.

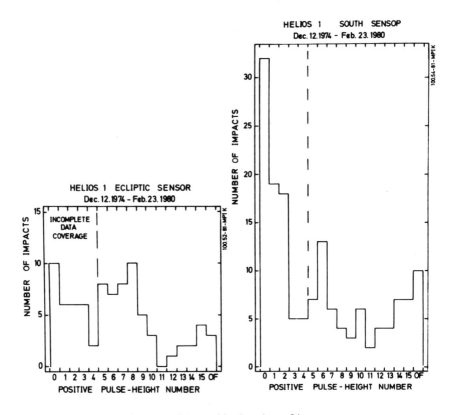

Fig. 5. Pulse height distributions of the positive ion charge IA.

Thus, cometary dust is produced by electrostatic fragmentation from larger cometary material. However, this fragmentation occurs only in areas where a high plasma environment is able to sufficiently charge solid bodies (for example the Earth's auroral zones). It might be possible that the same fragmentation process was responsible for the striae that appeared in the tail of Comet West 1976 VI (Sekanina and Farrell 1980). This particular process of the formation of cometary dust occurs only in specific areas, but the total fraction of cometary dust grains formed in this process near the Earth is $\leqslant$ 50% of all observed particles (see Fig. 3).

III. COMETARY DUST DETECTED BY THE HELIOS DUST EXPERIMENT

Helios, our latest dust experiment in space, scans between 0.3 and 1 AU. The technical details of the experiment have been described by Grün et al. (1980) and Grün (1981). The experiment contains an ecliptic and a tilted sensor (so-called South sensor). The ecliptic sensor is covered by an entrance

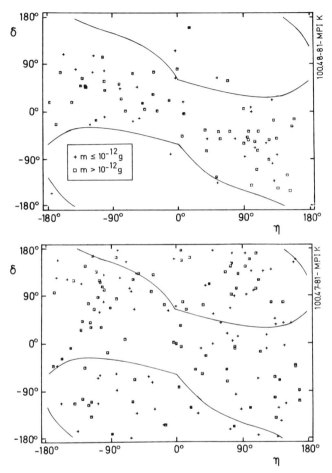

Fig. 6. Sensor azimuth vs. Helios true anomaly for the impacts detected by the Helios sensors for two pulse height number intervals. Contours represent the observational limit of circular orbiting particles.

foil of 0.4 μm thickness, while the South sensor which scans off the ecliptic plane has an open aperture. For the time period December 1974 through January 1980, Helios 1 has registered particles in the size range between 10^{-16} and 10^{-8} g as described in detail elsewhere (Grün et al. 1980; Grün 1981). The ecliptic sensor has registered fewer particles than the South sensor (Fig. 5), contrary to all expectations. This obviously means that a considerable fraction of dust particles is not able to penetrate the foil, because there is no doubt that most particles move in the ecliptic plane. Detailed simulation experiments on the entrance foil show that this foil discriminates between low-density particles and denser particles like silicates or iron particles (Pailer and Grün 1980).

Another interesting property of these low-density particles was found. In Fig. 6 the diagrams show the azimuth (viewing direction) as a function of the true anomaly (location of sensor in the ecliptic plane) for the ecliptic and South sensors. The impacts between the solid lines are interpreted as the so-called apex particles, on quasi-circular orbits around the Sun. The South sensor, however, shows particle impacts that deviate considerably from the apex particle area. As investigated by Grün et al. (1980) and Schmidt and Grün (1980), these particles are moving around the Sun on eccentric orbits ($e > 0.4$). Orbital considerations and calculations confirm the above interpretations and show that low-density particles also hit the ecliptic sensor.

The simulation experiments by Pailer and Grün (1980) have connected the electric charge produced upon impact of a particle on the sensor with the penetration limit through the entrance foil of the ecliptic sensor. The electric charge IA (ion amplitude) is given by

$$IA \sim m v^{2.7} \quad . \tag{1}$$

The penetration limit T reads

$$T \sim m^{0.4} v^{0.88} \rho^{0.33} \tag{2}$$

where m is the mass of projectile, ρ the density of projectile, and v the impact speed, or

$$T^{1/0.4} \sim m v^{2.2} \rho^{0.825} \quad . \tag{3}$$

If we equalize $mv^{2.7} \approx mv^{2.2}$, then IA reads:

$$IA \sim \frac{T^{2.5}}{\rho^{0.825}} \sim \rho^{-0.825} \quad . \tag{4}$$

Therefore the abscissa of Fig. 5 can be replaced by a density scale. Fig. 7 gives the final result for eccentrically orbiting particles as a function of their densities. This means that below a certain density particles are not able to penetrate the foil and that the larger the particles, the smaller their densities must be to be stopped within the foil.

This density scale could only be calibrated within the density range between iron ($\rho = 7.9$ g cm^{-3}) and polyphenylene ($\rho = 1.25$ g cm^{-3}). The whole range down to 10^{-5} g cm^{-3} is an extrapolation based on the derivation given above. There is another problem involved, namely, low-density particles as used for calibration (polyphenylene) are fundamentally different from low-density particles of the Brownlee particle type. While the calibration particles are of compact structure, the Brownlee particles are fluffy in structure. This may mean an unknown deviation from the calibration results,

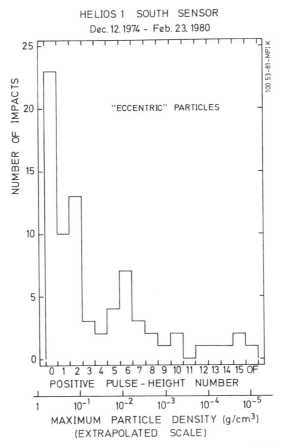

Fig. 7. Pulse height distribution of the eccentric particles observed by the South sensor.

and in this sense the low-density scale must be taken only as relative for the degree of fluffiness of particles. Because the uncertainty is large, it is not possible to quantify the error. As an overall result of the Helios dust experiment we find that these low-density and eccentrically moving particles are of cometary origin, but the total amount of this component is <30% of all registered dust particles of the Helios experiment.

IV. DUST PARTICLES FROM COMET KOHOUTEK

Dust particles when released from a comet either are accelerated by solar radiation pressure and leave the solar system on hyperbolic orbits (<1 μm dust) (Schmidt and Grün 1980) or, in the case of larger particles, are slowly decelerated and give rise to a cometary dust tail (Finson and Probstein 1968a,b). Further influenced by solar radiation pressure, these individual particles slowly spiral towards the Sun (Poynting-Robertson effect).

Hoffmann et al. (1975c) have identified 7 individual dust grains in the orbital plane of Comet Kohoutek 1973 XII by several criteria: enhancement of the particle rate; mass (10^{-13} to 10^{-11} g) and speed ranges (10 to 20 km s^{-1}); and orbital elements relative to Comet Kohoutek and to the sensor.

V. CONCLUSIONS

Returning to our criteria discussed in the introduction, we see dust particles in space undoubtedly of cometary origin. All discussed observations, however, indicate that only a minority of the observed dust particles ($\leqslant 30\%$) are of the low-density type. This result corresponds well with the results of Ceplecha and McCrosky (1976) for the nature of large fireballs.

The general result by Brownlee (1978) is that most fluffy particles are chondritic. This does not necessarily disagree with the expectations of Biermann and Michel (1978) and Greenberg (1979) that small individual dust particles are interstellar, because the compositional results by Brownlee (1978) are always obtained from analyses of many individual grains building up the fluffy-type particle.

Another observation of interest in this context is that the dust production rate of short-period comets is much lower (according to Röser [1976] <300 kg s^{-1}) than Whipple's (1967) estimate of $\sim 10^4$ kg s^{-1}. Delsemme (1976a,b) argued that the zodiacal dust cloud can only be maintained if a sizeable fraction of dust is delivered from long-period and sporadic comets by an unknown orbital mechanism. The difficulty in the latter case is how the dust is kept within the solar system.

The final result that only a minor part of dust is low-density material, i.e., of cometary origin (in agreement with the low-dust production rate of the short-period comets), does not necessarily mean that we should find other sources for the interplanetary dust. Since asteroidal dust source is obviously not important (Humes et al. 1974), it is likely that we still do not know enough about cometary dust; the size range of building blocks for the aggregates may be much wider than currently assumed. In any case, forthcoming missions to Periodic Comet Halley should reveal more information on the nature and dynamics of cometary dust sources.

Acknowledgment: I wish to thank my colleague and friend E. Grün for permission to include the latest results from the dust experiment of the Project Helios in this chapter.

REFERENCES

Audouze, J. and Truran, J.W. 1973. A possible explanation for the origin of lithium, beryllium, and boron. *Astrophys. J.* 182:839-846.

Biermann, L. and Michel, K.W. 1978. On the origin of cometary nuclei in the presolar nebula. *Moon and Planets* 18:447-464.

Bigg, E.K., and Thomson, W.J. 1969. Daytime photographs of a group of meteor trails. *Nature* 222:156-157.
Brownlee, D.E. 1978. Microparticle studies by sampling techniques. In *Cosmic Dust,* ed. J.A.M. McDonnell (Chichester: John Wiley & Sons), pp. 295-336.
Brownlee, D.E.; Hörz, F.; Vedder, J.F.; Gault, D.E.; and Hartung, J.B. 1973. Some physical parameters of micrometeoroids. *Proc. Lunar Sci. Conf.,* 4:3197-3212.
Cameron, A.G.W., and Fowler, W.A. 1971. Lithium and the s-process in red-giant stars. *Astrophys. J.* 164:111-114.
Ceplecha, Z., and McCrosky, R.E. 1976. Fireball end heights: A diagnostic for the structure of meteoritic material. *J. Geophys. Res.* 81:6257-6275.
Delsemme, A.H. 1976a. The production rate of dust by comets. In *Lecture Notes in Physics,* eds. H. Elsässer and H. Fechtig (Berlin: Springer Verlag), Vol. 48, pp. 314-318.
Delsemme, A.H. 1976b. Can comets be the only source of interplanetary dust? In *Lecture Notes in Physics,* eds. H. Elsässer and H. Fechtig (Berlin: Springer Verlag), Vol. 48, pp. 481-484.
Dohnanyi, J.S. 1976. Sources of interplanetary dust: Asteroids. In *Lecture Notes in Physics,* eds. H. Elsässer and H. Fechtig (Berlin: Springer Verlag), Vol. 48, pp. 187-205.
Fechtig, H.; Gentner, W.; Hartung, J.B.; Nagel, K.; Neukum, G.; Schneider, E.; and Storzer, D. 1975. Microcraters on lunar samples. *Proc. Soviet-American Conference on Cosmochemistry,* Moscow. pp. 453-472. (Translation: NASA SP-370, pp. 585-603).
Fechtig, H.; Grün, E.; and Morfill, G. 1979. Micrometeoroids within ten Earth radii. *Planet. Space Phys.* 27:511-531.
Fechtig, H., and Hemenway, C.L. 1976. Near-earth fragmentation of cosmic dust. In *Lecture Notes in Physics,* eds. H. Elsässer and H. Fechtig (Berlin: Springer Verlag), Vol. 48, pp. 290-295.
Finson, M.L., and Probstein, R.F. 1968a. A theory of dust comets. I. Model and equations. *Astrophys. J.* 154:327-352.
Finson, M.L., and Probstein, R.F. 1968b. A theory of dust comets. II. Results for Comet Arend-Roland. *Astrophys. J.* 154:353-380.
Greenberg, J.M. 1979. Pre-stellar interstellar dust. *Moon and Planets* 20:15-48.
Grün, E. 1981. Physikalische und chemische Eigenschaften des interplanetaren Staubes —*Messungen des Mikrometeorintenexperimentes auf Helios.* Habilitationsschrift Universität Heidelberg.
Grün, E.; Pailer, N.; Fechtig, H.; and Kissel, J. 1980. Orbital physical characteristics of micrometeoroids in the inner solar system as observed by Helios I. *Planet. Space Sci.* 28:333-349.
Hoffmann, H.-J.; Fechtig, H.; Grün, E.; and Kissel, J.; 1975a. First results of the micrometeoroid experiment S215 on the HEOS-2 satellite. *Planet. Space Sci.* 23:215-224.
Hoffmann, H.-J.; Fechtig, H.; Grün, E.; and Kissel, J. 1975b. Temporal fluctuations and anisotropy of the micrometeoroid flux in the earth-moon system. *Planet. Space Sci.* 23:985-991.
Hoffmann, H.-J.; Fechtig, H.; Grün, E.; and Kissel, J. 1975c. Particles from Comet Kohoutek detected by the micrometeoroid experiment on HEOS-2. In *The Study of Comets,* eds. B. Donn, M. Mumma, W. Jackson, M. A'Hearn, and R. Harrington (Washington, D.C.: NASA SP-393), pp. 949-961.
Hughes, D.W. 1978. Meteors. In *Cosmic Dust,* ed. J.A.M. McDonnell (Chichester: John Wiley & Sons), pp. 123-185.
Humes, D.H.; Alvarez, J.M.; O'Neal, R.L.; and Kinard, W.H. 1974. The interplanetary and near-Jupiter meteoroid environments. *J. Geophys. Res.* 79:3677-3684.
Jacchia, L.G. 1955. The physical theory of meteors, 8. Fragmentation as a cause of the faint meteor anomaly. *Astrophys. J.* 121:521-527.
Jacchia, L.G.; Verniani, F.; and Briggs, R.E. 1965. An analysis of the atmospheric trajectories of 413 precisely reduced photographic meteors. *Smithson. Contr. Astrophys.* 10:1-139.

Mandeville, J.C. 1977. Impact microcraters on 12054 rock. *Proc. Lunar Sci. Conf.* 8:883-888.
McDonnell, J.A.M., and Allison, R.J. 1981. Lunar microcrater records: Where is the cometary efflux? *Lunar Planet. Sci.* XII:682-684 (abstract).
Millman, P.M. 1976. Meteors and interplanetary dust. In *Lecture Notes in Physics*, eds. H. Elsässer and H. Fechtig (Berlin: Springer Verlag), Vol. 48, pp. 359-372.
Mitler, H.E. 1972. Cosmic-ray production of deuterium, He^3 lithium, beryllium and boron in the galaxy. *Astrophys. Space Sci.* 17:186-218.
Nagel, K., and Fechtig, H. 1980. Diameter to depth dependence of impact craters. *Planet. Space Sci.* 28:567-573.
Nagel, K.; Neukum, G.; Dohnanyi, J.S.; Fechtig, H.; and Gentner, W. 1976a. Density and chemistry of interplanetary dust particles derived from measurements of lunar microcraters. *Proc. Lunar Sci. Conf.* 7:1021-1029.
Nagel, K.; Neukum, G.; Eichhorn, G.; Fechtig, H.; Müller, O.; and Schneider, E. 1975. Dependencies of microcrater formation on impact parameters. *Proc. Lunar Sci. Conf.* 6:3417-3432.
Nagel, K.; Neukum, G.; Fechtig, H.; and Gentner, W. 1976b. Density and composition of interplanetary dust particles. *Earth Planet. Sci. Letters* 30:234-240.
Pailer, N., and Grün, E. 1980. The penetration limit of thin films. *Planet. Space Sci.* 28:321-331.
Rawcliffe, R.D.; Bartky, C.D.; Li F.; Gordon, E.; and Carta, D. 1974. Meteor of August 10, 1974. *Nature* 247:449-450.
Reeves, H. 1974. On the origin of the light elements. *Ann. Rev. Astron. Astrophys.* 12:437-469.
Röser, S. 1976. Can short period comets maintain the zodiacal cloud? In *Lecture Notes in Physics*, eds. H. Elsässer and H. Fechtig (Berlin: Springer Verlag), Vol. 48, pp. 319-322.
Schmidt, K.D., and Grün, E. 1980. Orbital elements of micrometeoroids detected by the Helios I space probe in the inner solar system. In *Solid Particles in the Solar System*, eds. I. Halliday and B.A. McIntosh (Dordrecht: D. Reidel Publ. Co.), pp. 321-324.
Sekanina, Z., and Farrell, J.A. 1980. Evidence for fragmentation of strongly nonspherical dust particles in the tail of Comet West 1976 VI. In *Solid Particles in the Solar System*, eds. I. Halliday and B.A. McIntosh (Dordrecht: D. Reidel Publ. Co.), pp. 267-270.
Smith, D.; Adams, N.C.; and Khan, H.A. 1974. Flux and composition of micrometeoroids in the diameter range 1-10 μm. *Nature* 252:101-106.
Starrfield, S.; Truran, J.W.; Sparks, W.M.; and Arnould, M. 1978. On ^7Li production in nova explosions. *Astrophys. J.* 222:600-603.
Vanýsek, V. 1979. The significance of the determination of Li, B, Be and C isotopic ratios in cometary dust. In *Cometary Missions*, eds. W.I. Axford, H. Fechtig, J. Rahe (Bamberg: Remeis-Sternwarte, Erlangen-Nürnberg: Astronom. Inst. d. Univ.), pp. 159-171.
Vedder, J.F., and Mandeville, J.-C. 1974. Microcraters formed in glass by projectiles of various densities. *J. Geophys. Res.* 79:3247-3256.
Verniani, F. 1964. On the density of meteoroids. II. The density of faint photographic meteors. *Nuovo Cimento* 33:1173-1184.
Verniani, F. 1967. Meteor masses and luminosity. *Smithson. Contrib. Astrophys.* 10:181-195.
Verniani, F. 1969. Structure and fragmentation of meteoroids. *Space Sci. Rev.* 10:230-261.
Verniani, F. 1973. An analysis of the physical parameters of 5759 faint radio meteors. *J. Geophys. Res.* 78:8429-8462.
Wetherill, G.W. 1971. Cometary vs. asteroidal origin of chondritic meteorites. In *Physical Studies of Minor Planets*, ed. T. Gehrels (Washington: NASA SP-267), pp. 447-460.
Whipple, F. 1967. On maintaining the meteoritic complex. In *The Zodiacal Light and the Interplanetary Medium*, ed. J.L. Weinberg (Washington: NASA SP-150), pp. 409-426.

LABORATORY STUDIES OF INTERPLANETARY DUST

P. FRAUNDORF
McDonnell Center for Space Sciences

D.E. BROWNLEE
University of Washington

and

R.M. WALKER
McDonnell Center for Space Sciences

Samples of interplanetary dust have been successfully collected for laboratory study, from the stratosphere in the form of micrometeorites between 2 and 50 µm in size, and from the sea floor in the form of particles between 100 and 3000 µm in size; the latter are in the form of frozen droplets as a result of melting on atmospheric entry. Elemental and isotopic abundances of He, Ne, and Ar in the stratospheric particles suggest a space exposure as small particles in the inner solar system for at least tens of years. Most of the extraterrestrial particles in the stratospheric collections are fine-grained, black aggregates with major and trace element abundances similar to those of chondritic meteorites. High abundances of C, S, Na, Mn, and Zn in the stratospheric particles, and major-element abundances in the chondritic sea sediment particles, suggest a compositional affinity to CI and CM meteorites. On the other hand, two signs commonly attributed to aqueous alteration in such meteorites, hydrated silicates and relative depletion of Na, S, and Ca in the matrix, are evident in only a small fraction of the stratospheric particles. Many of the more typical chondritic particles show a reentrant cluster-of-grapes morphology not typical of meteoritic material. Although pyroxene, olivine, iron sulfide, magnetite, and a noncrystalline carbonaceous material are frequently present, significant differences between individual particles in the oxidation state and structure of the submicron building blocks do exist. While some of the differences are probably due to heating on atmospheric entry, others suggest that the particles contain primitive materials not found in present-day meteorite collections. Because comets are likely to be a major supplier of particles for the solar system dust cloud, many of the collected particles are probably of cometary origin. If so, the particle structures suggest that comets are fine-grained aggregates of nonvolatile building blocks and ice. Some of the submicron building blocks are single crystals, some are noncrystalline or polycrystalline clumps, and others are themselves aggregates of even tinier crystals in a carbonaceous matrix.

Anyone who spends the better part of an hour gazing at the night sky is likely to be impressed by the arrival of visual meteors. In spite of the spectacular display, the amount of material involved is small (typically less than a gram) and the observing area is large ($\sim 10^5$ km^2). Naked eye observation thus establishes two fundamental facts: the Earth is continually bombarded by extraterrestrial material, but the flux of particles is exceedingly low.

The effects of Earth encounter on an interplanetary dust particle depend on the size of the particle, its physical properties, and its entry parameters. Although large objects reach Earth, most particles are destroyed by heating due to friction in the upper atmosphere. However, as pointed out by Öpik (1937) and Whipple (1950), particles that are sufficiently small can survive entry without melting. These particles are called micrometeorites.

For such small particles, the Earth's atmosphere is an excellent collector. Dust particles belonging to the solar system have a minimum Earth-encounter velocity of 11 km s^{-1} and a maximum of 72 km s^{-1}; the average value is probably less than 15 km s^{-1} (Southworth and Sekanina 1973). These velocities exceed the thermal velocities of air molecules, and the initial encounter with the Earth's atmosphere can be approximated as a collision of a moving object with static air molecules. This gives rise to a drag force proportional to ρV^2, and energy dissipation proportional to ρV^3, where V is the velocity of the particle and ρ the density of air. Small particles (< 50 μm in size) totally decelerate from cosmic velocity at altitudes near 100 km. Here the air density is so low that the frictional power generated is small enough to be thermally radiated without the particle being melted. As an example, fewer than half of all 5-μm particles of density 1 g cm^{-3} are likely to be heated above 500°C (Fraundorf 1980), and the duration of heating is only a few seconds.

After deceleration, when the particles descend into the stratosphere where the particle sizes are larger than the mean free path of air molecules, the fall speed is determined by Stoke's law. The settling speed for a 10-μm particle is on the order of 1 cm s^{-1}, and is determined by the particle's size, density, and the viscosity of the ambient air (Kasten 1968). For those particles which survive entry, the particle concentration per unit volume of air mass is proportional to the inverse of the settling velocity. For 10-μm particles the ratio of the cosmic entry velocity to the settling speed is $\sim 10^6$, and thus the concentration of particles in the stratosphere is 10^6 times greater than the concentration found in space.

These facts have been realized for decades, and have been the basis for many searches for extraterrestrial particles. We now know that the concentration of extraterrestrial particles with diameters of $\sim 10\ \mu$m is only 10^{-3} m^{-3} at an altitude of 20 km, meaning that large volumes of air must be sampled and great care be taken to avoid contamination of collectors. Fortunately, in the stratosphere both natural and anthropogenic particles larger than a few microns are so rare that collection of extraterrestrial material is possible. The major contaminant in the range 2 to 15 μm is aluminum oxide spherules generated by solid fuel rockets (Brownlee et al. 1976a). This background contaminant may grow considerably when regular space shuttle flights are established.

The modern era of laboratory studies of extraterrestrial particles started with the stratospheric collection flights of Brownlee and collaborators using balloons and then U2 aircraft. The collectors now in use consist of flat plates 20 to 30 cm^2 in size, coated with viscous silicone oil. These are carried in a wing-mounted pylon built by NASA Ames Research Center, and are opened to the windstream only when the aircraft approaches 20 km altitude. Each hour of flight time at a velocity of 200 m s^{-1} results in the collection of ~ 1 particle larger than 10 μm. Submicron particles are coupled to the air flow around the collector, and generally do not impact. Large particles up to 50 μm in size are found, but are much less common.

The purpose of this chapter is to summarize the current state of knowledge concerning interplanetary dust particles collected in the upper atmosphere, and related particles collected from the sea floor. Initially, the experimental problem was to positively identify extraterrestrial material. This was particularly important in a field which has been characterized by disappointment. For antarctic ice, another potential source of interplanetary dust, work toward positive identification has recently begun (Wagstaff and King 1981). Growing evidence for the extraterrestrial nature of stratospheric and sea sediment particles stimulated other investigations whose primary purpose has been to understand the properties of the particles in the larger context of their relationships to more familiar objects, such as meteorites, comets, and asteroids.

Studies of impact pits on meteorites, lunar rocks, and lunar soil grains have demonstrated conclusively that interplanetary dust has been a feature of the solar system for billions of years (see, e.g., Brownlee and Rajan 1973;

Zinner 1980; Walker 1980). Calculations indicate a short lifetime ($< 10^5$ yr) for dust grains in the inner solar system (e.g., Dohnanyi 1978), and the persistence of the dust cloud implies a continual source of fresh dust. Although other sources certainly exist, comets are widely believed to be the major source of meteoroids ranging at least up to the millimeter sizes that produce visual meteors (Millman 1972; Brownlee 1979a). Active comets are observed to have dust tails, and some meteor showers are known to be associated with specific comets. It is thus likely that many of the interplanetary dust particles (IDP) are cometary. However, the fraction of cometary particles is not really known. For example, Kresák (1980) and others have shown that observed short period comets do not appear to be capable of maintaining the solar system dust cloud in equilibrium.

The study of interplanetary dust particles is interesting in its own right. However, if they are cometary in origin, there is a common presumption that they may also be the best objects in which to find primordial solar system material, i.e., interstellar grains that escaped drastic processing during the formation of the Sun and planets. The current evidence that IDP's contain primordial material is tenuous. Even if the association with comets is correct, comets may not turn out to be the nebular 'Rosetta stones' that we are hoping for, or the experimental problems may be more subtle than we currently envision. Only time and further work will settle these issues.

The evidence that a major subset of stratospheric particles (the chondritic aggregates) and many of the chondritic deep-sea particles are extraterrestrial has been accumulating steadily since they were first observed. In succeeding sections of this chapter, we review the experimental results on the abundance of major and minor elements, the abundances and isotopic compositions of noble gases, the isotopic composition of Mg, and structure and mineralogy leading to this conclusion. Although the presence of solar-flare particle tracks would, by itself, demonstrate beyond any question the extraterrestrial nature of any given particle, the existence of such tracks has not yet been confirmed. Possible reasons for this are discussed briefly. Although all of the chondritic aggregates have common features, detailed electron microscope studies show that individual particles vary. Thus no single, simple explanation is apt to account for the origin of all of the collected particles. They represent a new type of extraterrestrial matter that promises to be challenging to measure and understand.

I. BULK ELEMENTAL AND ISOTOPIC COMPOSITION

The bulk elemental compositions of stratospheric micrometeorites have been measured by electron microprobe techniques for major and minor elements (Brownlee 1978a; Flynn et al. 1978; Brownlee et al, 1980) and by neutron activation for trace elements (Ganapathy and Brownlee 1979). The basic result is that most aggregate particles of 10-μm size have a generally chondritic composition for all elements measured. Within $\sim$ 50% error bars

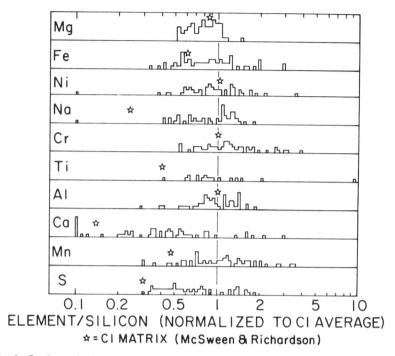

Fig. 1. Semiquantitative element to silicon ratios for 57 stratospheric micrometeorites normalized to the average for CI chondrites. The stars are mean compositions for fine-grained CI matrix material (McSween and Richardson 1977).

there are no systematic deviations from CI composition except for Ca which shows a rather broad range of depletions (Fig. 1). The fact that most of the individual aggregate particles are so close to solar values is a result of the extremely fine-grained nature of the particles. Significant departure from a solar abundance pattern for a given particle is usually attributable to the aggregate containing a single large mineral grain surrounded by fine-grained aggregate material. For example, if 50% of a particle is a large FeS grain then the bulk composition will show enhanced and correlated Fe and S.

The origin of the Ca depletion observed in some of the particles is unknown but may be due to mobilization of Ca in the parent bodies. The Ca correlates with morphology. Particles that are clearly aggregates with high porosity have normal Ca abundances while those that are depleted in Ca are often smooth on a scale of microns. The rare smooth particles have an appearance in the scanning electron microscope (SEM) which is similar to CI or CM meteorite matrix. For some particles, this similarity has been further confirmed by X-ray powder diffraction (Brownlee 1978a), or by infrared spectroscopy (Fraundorf et al. 1981).

Elemental compositions can be measured in a more precise way for sectioned and polished 100-μm to 3-mm chondritic deep-sea meteor ablation spheres (DSS) which have been collected from the sea floor. These particles are homogenized melt droplets of ⩾ 1 mm meteoroids. The dispersion of measured abundance ratios is smaller than that for the stratospheric particles, which probably results from the fact that each deep-sea spherule contains > 3 orders of magnitude more mass than the typical stratospheric micrometeorite. The DSS were melted and devolatilized during entry so that S, Na, and C are highly depleted. Elements less volatile than Mn are not depleted. Of the nonvolatile elements, all are found in chondritic proportions except for Ni which shows a continuous range of depletions up to values ⩾ 0.01. This depletion is believed to be related to the loss of metal or sulfide during atmospheric entry. The other element ratios are well preserved and appear to reveal the elemental abundances in millimeter meteoroids rather precisely. For example, the distribution of Mg/Si ratios for a large number of spheres is strongly peaked at a value coincident with that of CI/CM meteorites and is clearly displaced from that of the ordinary chondrites (Fig. 2). Aluminum is another element that shows a definite peak which again coincides with the CI/CM value and is displaced from that of ordinary chondrites. For the majority of elements there is a correlation with CI/CM meteorites, and for elements fractionated between ordinary and carbonaceous chondrites there is distinction from ordinary chondrites. Even though there is no information on volatile elements, it is clear that the majority of bodies producing millimeter ablation spherules is chemically related to carbonaceous chondrites. This is quite different from normal meteorites where carbonaceous objects are rare (< 5% of falls).

Isotopic measurements of Mg have been made on several of the stratospheric micrometeorites (Esat et al. 1979). The particles analyzed included chondritic aggregates, particles composed of single grains of olivine and pyroxene, and spherules depleted in Fe and enhanced in Ca and Al relative to chondritic proportions. To an uncertainty of 1% all but one particle contained normal solar system ratios for the three Mg isotopes. One sphere had a 1.1% mass fractionation. Although very close to detection limits, three out of four of the chondritic aggregate particles contained a hint of ∼ 0.4% excess of ^{26}Mg.

Measurements of ^{53}Mn, a radioactive isotope produced by cosmic-ray interactions with iron, have also been reported in the deep-sea spherules (Nishizumi et al. 1980). The presence of this isotope clearly and independently establishes the extraterrestrial nature of the deep-sea spherules.

II. NOBLE GAS STUDIES

The measurement of noble gases is one of the more important tools used to study extraterrestrial materials. In mature lunar soils the gas concentrations are large because the individual grains have been exposed to solar wind ions of ∼1 keV/nucleon which results in an implantation of ions to depths of 500 Å. Studies of surface correlated noble gases have in fact been used to

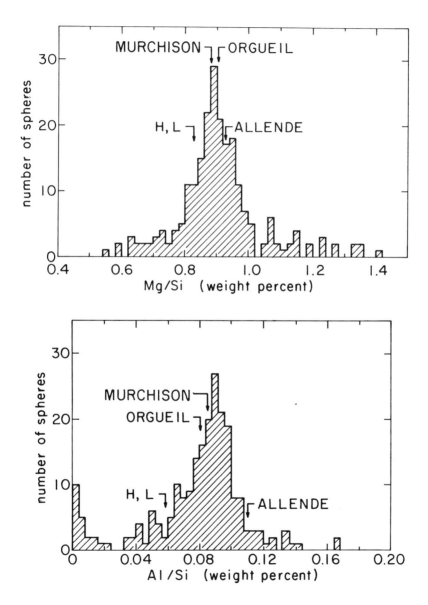

Fig. 2. Element (top, Mg; bottom, Al) to silicon ratios for the unetched interiors of 230 randomly selected stony meteor ablation spheres collected from the ocean floor. The spheres range in size from 200 to 500 μm; the data is compared to the mean compositions of the ordinary H and L chondrites and the composition of 3 carbonaceous chondrites characteristic of the 3 major carbonaceous chondrite groups.

define the long-term elemental abundances and isotopic patterns of solar wind ions (Eberhardt et al. 1970; Podosek et al. 1971). Lunar soil breccias and certain gas-rich meteorites also contain a large solar wind contribution. In the latter case, and in some lunar samples as well, it appears that the solar wind component was added at an early epoch of the solar system when grains that are now found in the interiors of consolidated breccias were exposed to the solar wind.

In general, extraterrestrial samples also contain trapped or indigenous components whose compositions vary considerably from solar values and whose origins reflect a mixture of several processes including, in some cases, the decay of presently extinct isotopes such as ^{244}Pu (Podosek 1978). Remarkably, in some primitive meteorites the trapped noble gas is found primarily in a minor, acid-insoluable phase (e.g., Lewis et al. 1977). The gas concentrations in this minor phase can be very high. High-energy spallation reactions produced by galactic cosmic rays during exposure of meteorites and lunar samples to space as small objects ($\lesssim$ several meters in size) also produce diagnostic noble gas isotopes such as ^{3}He and ^{21}Ne. The concentration of these isotopes can be used to measure the cosmic-ray exposure ages of extraterrestrial materials.

The first noble gas measurements on IDP's were made by Rajan et al. (1977), who determined the ^{4}He concentrations in ten individual chondritic aggregates. Seven of the ten were found to contain measurable amounts of ^{4}He, with estimated concentrations ranging from 0.002 cm^3 (STP)g^{-1} to the very large value of 0.25 cm^3 (STP)g^{-1}. This latter value is the same magnitude as the ^{4}He concentrations found in lunar soils, $\sim$0.1 cm^3 (STP)g^{-1}, due to the bombardment by solar wind. Implanted helium cannot be accumulated indefinitely by a grain surface, and the largest concentration observed may be a reasonable saturation value.

The possibility that the ^{4}He was produced by the α-decay of uranium was rejected on the grounds that chondritic material would be expected to have far too little uranium to produce the observed helium. This was later confirmed by direct measurement by Flynn et al. (1978) who showed that the uranium concentration of a chondritic aggregate was less than 15 parts per billion.

The simplest interpretation of these data is that the helium was implanted into the particles by the solar wind during a recent exposure in space. Since only 10 to 100 yr exposure would suffice to produce the largest concentration seen, and since the expected lifetime of 10-μm IDP's is much longer ($\sim$10^4 yr), this explanation easily accounts for the observed results. This interpretation, if true, has another immediate consequence, that the particles existed as small objects in space and were not produced by fragmentation of larger meteoroids in the atmosphere.

The variability of the ^{4}He concentrations observed by Rajan et al. argues in favor of the recent solar wind origin of the gas. The particles are generally fragile and it is to be expected that the entry, collection, and handling processes would tend to remove the outer layers of the particles, which in the direct solar implantation hypothesis are the only layers that would be enriched in gas. This variable surface removal would result in a considerable variation in gas content from one particle to the next.

Unfortunately the helium observations by themselves do not prove that the 10-μm particles were bombarded by solar wind in the recent past. For one thing, it is conceivable that the individual constituents of the particles were once bombarded in space in the remote past before being assembled into their present form. In this case, they would be analogous to gas-rich meteorites and might have been parts of larger meteoroids prior to entry. Nor is there any *a priori* guarantee that the ^{4}He is of solar origin. We have already remarked that acid-insoluable residues in some meteorites contain large concentrations of trapped gas of planetary, i.e. nonsolar, composition. Finally, the slowing-down process in the Earth's atmosphere itself provides a potential mechanism for implanting helium into the particles.

Stimulated by the ^{4}He results, Hudson et al. (1981) undertook a study of the heavier noble gases (neon, argon, krypton, and xenon) using a noble gas spectrometer of high sensitivity and low background (Hohenberg 1981). A total of 13 chondritic particles weighing $\sim 10^{-8}$ g were combined in a single gas extraction experiment. Krypton was not present at levels above the instrumental blank. Measurable signals were, however, found for Ne, Ar, and Xe.

The results for neon and argon strongly support a solar origin for the noble gases. The ^{20}Ne/^{36}Ar ratio of 9 ± 3 is similar to the ratio of 11 ± 3 found in the fine-grained fraction of a lunar soil and to the ratio of 15 found in the gas-rich meteorite Pesyanoe, but is quite different from the planetary (or terrestrial) value of 0.5 (Fig. 3). The total concentration of ^{20}Ne is intermediate between that of mature lunar soils and gas-rich meteorites. Finally, the measured Ar shows a significant enrichment of ^{36}Ar and ^{38}Ar relative to ^{40}Ar (^{40}Ar/^{36}Ar < 166, ^{36}Ar/^{38}Ar = 4.9 ± 0.9) in comparison with terrestrial, atmospheric Ar(^{40}Ar/^{36}Ar = 295.5, ^{36}Ar/^{38}Ar = 5.35). The light rare gases, therefore, appear to be of solar and not atmospheric (or planetary) composition. The nonatmospheric nature of the gases, as well as their total concentration, effectively eliminate the possibility that the gases were added during atmsopheric entry. The results thus strongly support the extraterrestrial origin of the particles. The solar character of the light rare gases also supports the idea that the gases were added by solar wind bombardment. As in the case of ^{4}He, the exposure time in space necessary to build up the observed concentrations is rather small (15 to 45 yr). While it is tempting to ascribe the solar gases to a recent bombardment in space (with the consequence that the individual particles have not been much changed during entry), this cannot be proven. A primordial bombardment of the individual IDP constituents prior to assemblage remains a possible alternative, and the 10-μm particles may once have been parts of larger meteoroids in space.

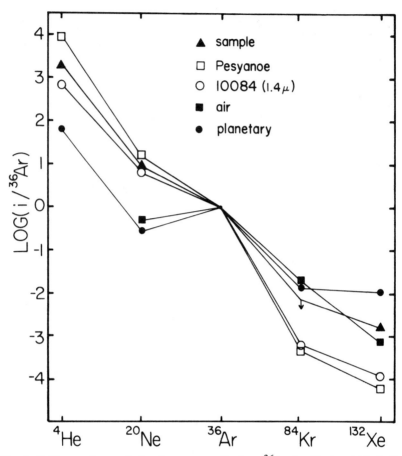

Fig. 3. Noble gas elemental abundances normalized to ^{36}Ar for 13 chondritic IDP's. Planetary gas (carbonaceous chondrites), air, and solar gas (lunar soil 10084 and the gas-rich meteorite Pesyanoe) are shown for comparison. The ^{4}He content was measured by Rajan et al. (1977) on a different set of particles. (Figure from Hudson et al. 1981.)

The observed xenon is a factor of three above blank levels and thus appears to be a small, but measurable, indigenous constituent of the sample. However, the concentration of Xe is far too great relative to Ne and Ar to be attributable to a solar origin. It thus appears to be a trapped, planetary component. While the total amount of gas is small, the inferred concentration ($\sim 10^{-7}$ cm^3 (STP)g^{-1}) is high and is comparable to that found in gas-rich acid residues from primitive meteorites ($\sim 3 \times 10^{-7}$ cm^3(STP)g^{-1}).

The ^{21}Ne/^{20}Ne ratio is solar with no obvious cosmogenic ^{21}Ne. This gives an upper limit on direct cosmic-ray exposure of 130 myr, assuming a production rate of 4×10^{-9} cm^3(STP)g^{-1} myr^{-1} (Hohenberg et al. 1978). Thus the particles were not exposed to cosmic rays at shielding depths of less than several meters, either in meteoroids or in the surface layers of comets, for long periods of time.

Whatever the origin of the noble gases, the presence of species indigenous to the particles show that they were not heated to extreme temperatures during entry. Temperature-release studies on particles implanted with noble gases in the laboratory may help to quantify this conclusion.

III. STRUCTURE AND MINERALOGY

The morphological characteristics of IDP's have been determined largely by scanning and transmission electron microscopy. Mineralogical information has been derived from X-ray and electron diffraction measurements, as well as from elemental abundance determinations on individual grains.

Stratospheric dust particles identified as extraterrestrial are commonly subdivided into four major categories (Brownlee et al. 1976c): chondritic aggregates (~55% by number); nickel-bearing iron-sulfide (FSN) particles (~30%); mafic silicate particles (~10%); and chondritic ablation spheres (~5%). Particles in the size range 2 to 12 μm are much more common than larger ones. In the size range > 10 μm, the FSN particles are almost nonexistent, and for particles > 30 μm in size, mafic silicates are the most abundant (Brownlee 1978b). These observations probably reflect relative densities and melting temperatures for the various particle types, since larger and more dense particles experience more severe heating during deceleration in the atmosphere.

In spite of bulk chemical similarities, especially between the chondritic aggregates, individual particles show a remarkable diversity in their microscopic structures. We first discuss the similarities between typical particles, and then treat some individual particles in more detail.

Gross Morphologies

One of the most striking features of some chondritic aggregates is their reentrant cluster-of-grapes morphology. Brownlee et al. (1976b,c,d, 1980) reported that ~90% of the chondritic particles are aggregates of building blocks between 100 and 1000 nm in size. About 25% of the particles are also highly reentrant, with as much as 50% of the total volume being void space in extreme cases. The range of aggregate morphologies is illustrated in the micrographs of Figs. 4-7.

The more porous aggregates are quite distinct from typical 10-μm meteoritic or lunar particles. The openness and the associated fragility of these aggregates have been variously interpreted to suggest:

1. Minimal thermal alteration of these particles in contrast, e.g., to the spheroidal shape of a melted particle (Brownlee et al. 1976b);
2. Incompatability with a prior history of residence in an asteroidal regolith environment (Brownlee et al. 1977);

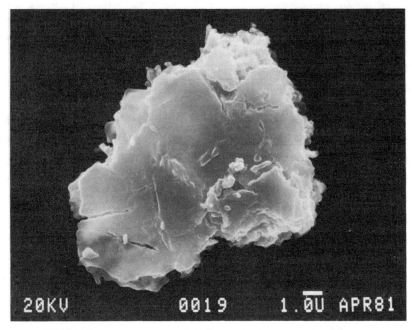

Fig. 4. U2-20A26, a smooth, chronditic IDP, slightly depleted in Ca.

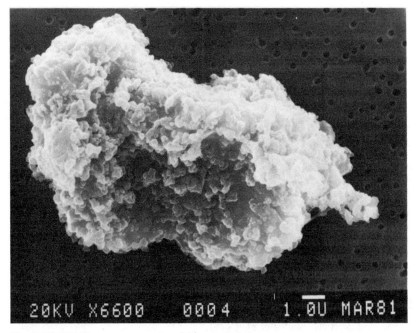

Fig. 5. U2-20A12, a nonporous chondritic aggregate.

INTERPLANETARY DUST IN THE LABORATORY 395

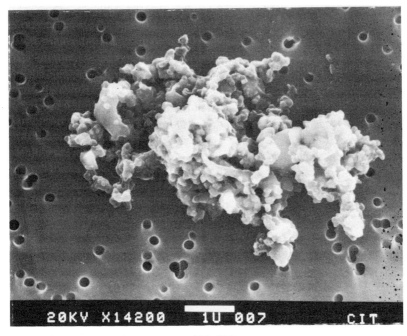

Fig. 6. U2-18A3B, a highly porous chondritic IDP.

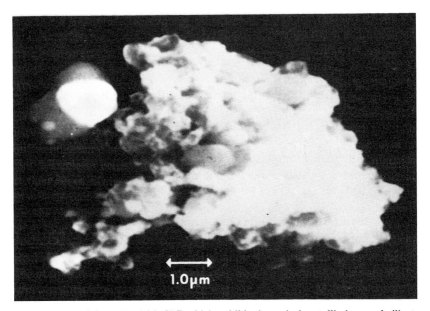

Fig. 7. U2-13-10-3, a chondritic IDP which exhibited atypical metallic iron and silicate droplet crystals in the TEM (Fig. 9).

3. Prior presence of a relatively volatile filler material if the aggregates were at one time stored at depth in a gravitationally bound body (Fraundorf 1981c).

It is presently not known if the range in particle porosities is intrinsic or was caused by alteration in the atmosphere, in space, or on the parent body (Brownlee 1978a). However, annealing studies have shown that particle morphologies are usually insensitive to attempts at thermal alteration (short of melting), and hence many of the observed morphologies are probably similar to those that existed prior to atmospheric entry.

Of the monomineralic IDP's, mafic silicate particles are often subhedral or anhedral crystals, usually with adhering chondritic material. Although the FSN particles are frequently spheres, they have also been found as solid irregular masses, aggregates, well-defined single crystals, and stacks of platelets. The high percentage of spheres probably is due to the fact that 10-μm iron sulfide particles, by virtue of their high densities and low melting temperatures, are much more likely to be melted on atmospheric entry than the other particle types.

Chondritic ablation sphere morphologies imply that they were once molten. Although of roughly chondritic composition, these spheres are often depleted in sulfur with respect to chondritic meteorites (Brownlee et al. 1976b). In composition, texture, and mineralogy, they are similar to fusion crusts of chondritic meteorites (Blanchard and Cunningham 1974) and to ablation debris created in the laboratory from a carbonaceous chondrite (Kyte 1977).

Common Single Crystals

The three major mineral types commonly identified as $\geqslant 0.1$ μm crystals in stratospheric IDP's are olivine, pyroxene, and iron sulfide. These minerals were first identified in the monomineralic particles. They have subsequently been identified as constituents of the chondritic aggregate particles by X-ray powder diffraction, by energy dispersive X-ray (EDX) analysis of individual grains, and by techniques of selected area electron diffraction (Fraundorf 1981b) which allow determination of the complete reciprocal lattice of individual submicron crystals. However, one should be cautioned that chondritic aggregate particles seldom consist mostly of $\geqslant 0.1$-μm single crystals of these minerals.

Individual FSN particles often have compositions consistent with pyrrhotite, ($Fe_{1-x}S$), containing 1 to 5 atomic percent nickel (Brownlee et al. 1976a), although in some of them sulfur is deficient relative to stoichiometric FeS by factors of 50% (Brownlee et al. 1976d). X-ray diffraction work on typical FSN particles indicates a mixture of magnetite (Fe_3O_4) and troilite (FeS). However X-ray diffraction patterns from FSN particles which do not appear to have melted indicates that the major sulfide is pyrrhotite.

X-ray diffraction work also frequently indicated pyrrhotite in the chondritic aggregate particles (Brownlee et al. 1976c). Individual submicron grains of Ni-bearing pyrrhotite and pentlandite (Fe, Ni)$_9$S$_8$ in crushed aggregates were later identified by EDX analysis (Brownlee et al. 1977; Brownlee 1978a); these are both common forms of iron sulfide in CI and CM chondrites (cf. Nagy 1975; Kerridge et al. 1979a; Kerridge et al. 1979b). A distinction between the meteoritic and dust sulfides is that roughly 10% of the latter contain zinc detectable by EDX techniques. EDX analysis in the transmission electron microscope (TEM) by Fraundrof (1981c) also turned up submicron FSN grains in crushed chondritic aggregates. These were evidence for atomic S/Fe ratios ranging between 0 and 1.3, and atomic percent nickel ratios between 1 and 7. The pyrrhotite structure was confirmed by electron diffraction work on several of these grains, and common terrestrial pyrrhotite superstructures were apparently absent in one of them. EDX analysis of a substantial number of micron and submicron grains within a single aggregate showed that most of the grains were polymineralic and composed of pyrrhotite, pentlandite and a high iron phase, either metal or magnetite (Brownlee 1980).

IDP silicates, on the other hand, are often depleted in iron. Brownlee (1978a) reported a uniform distribution of olivine Fe/Mg atomic ratios ranging between 0 and 1 for eleven olivine particles, while eleven pyroxene micrometeorites all exhibited Fe/Mg < 0.2. Although the pyroxene particles are usually enstatite (MgSiO$_3$), occasional calcium-rich pyroxenes have been found (e.g., Esat et al. 1979).

In chondritic aggregate particles, olivine was first detected by an occasional set of weak X-ray diffraction lines (Brownlee et al. 1976c). Submicron grains of olivine (Mg, Fe)$_2$SiO$_4$ and enstatite in crushed aggregates were later identified by EDX analysis (Brownlee et al. 1977). Electron diffraction work by Fraundorf (1981c) on dispersed aggregates confirmed the existence of submicron olivine crystals occurring variously in droplet and subhedral morphologies. EDX data indicated that clear olivine crystals were usually magnesium-rich. Iron-rich silicates in two particles exhibited heavy magnetite decoration, and as discussed below in Sec. V on atmospheric heating, were possibly of secondary origin.

EDX data and electron diffraction from pyroxene crystals in three chondritic aggregates indicated the clinoenstatite crystal structure. It has not yet been determined if the monoclinic structure is primary, and hence, a sign of low metamorphism (cf. Nagy 1975; Buseck and Iijima 1975; Fraundorf 1981a). The remarkable tendency of the crystals to occur in low abundance in many aggregates as long euhedral laths only tens of nm in thickness may bear on this question.

Fine-grained Crystalline Phases

Many chondritic aggregates consist largely of polycrystalline building blocks in which the small crystal size (< 100 nm) prevents the separate study of individual crystals. Selected-area electron diffraction of polycrystalline and poorly crystallized portions of 9 dispersed chondritic IDP's (Fraundorf 1981c) indicated that fine-grained magnetite was a dominant diffracting phase in 6 of the particles. Iron-rich silicate grains in several particles also appeared to be heavily decorated with magnetite. The magnetite was usually most pronounced in IDP's which for other reasons (low-sulfur, large size, ill-defined component boundaries) are likely to have experienced severe heating. Thus, much of the magnetite may be a secondary product resulting from atmospheric entry heating.

The diffraction patterns also indicated that either olivine or pyroxene was usually a dominant diffracting phase in fine-grained portions of examined IDP's. The diffraction patterns did not appear to result from hydrated silicates, like those found in the matrix of the CM meteorite Murchison.

Noncrystalline Phases

Noncrystalline phases, defined as those showing no evidence of Bragg scattering in the TEM, appear to play a significant role in the chondritic aggregate particles. Brownlee et al. (1976c) reported that many of the larger grains are covered with 30 nm coatings, not observed on control samples, of a low atomic weight amorphous material. Fraundorf and Shirck (1979) confirmed this observation with a TEM examination of 6 aggregates. Special care was taken to minimize the chance of artifact coatings resulting from sample collection and preparation. Although the particles differed from one another in texture, submicron crystals in them often appeared to be coated with, or embedded in, a noncrystalline material indigenous to the aggregates. They concluded that noncrystalline material, some of which is probably carbonaceous, is responsible for the characteristic lumpy structure of the more porous aggregates. The noncrystalline material may also be responsible for the fact that the X-ray diffraction patterns for typical aggregates contain only weak lines superposed on a strong background of incoherently scattered X rays.

Fraundorf (1981c) reported that two of eleven particles examined consisted mostly of a second kind of noncrystalline material, one whose major element composition appeared roughly chondritic.

Specific Chondritic Particles

A minor class of the chondritic IDP's contains hydrated silicates identifiable by X-ray diffraction (Brownlee 1978a). These particles do not exhibit the open structures characteristic of more porous aggregates, and are often depleted in calcium (Brownlee et al. 1980). A similar calcium depletion in the matrix of CI meteorites has been attributed to the relocation of calcium by aqueous alteration (McSween and Richardson 1977). One of these particles possesses an infrared absorption spectrum between 4000 and 400 wave

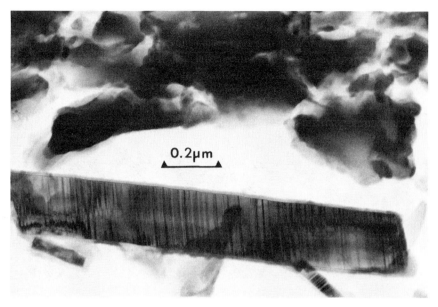

Fig. 8. A transmission electron micrograph of 3 euhedral laths (bottom) in particle U2-13-6-5A, which consisted mostly of irregular clumps of noncrystalline chondritic material like those at the top of the micrograph. The randomly-oriented linear features in the large lath were transient under the electron beam, and may have been partially annealed solar flare tracks (Fraundorf 1981a).

numbers almost identical to that of the CM meteorite Murchison (Fraundorf et al. 1981). Although differences exist, polycrystalline electron diffraction patterns of this particle and Murchison were similar. Thus a fraction of the IDP's apparently consist of primitive carbonaceous chondritic matrix material (cf. Wilkening 1978).

Certainly there is likely to be a class of particles that have been severely heated on atmospheric entry. Larger particles and those with higher densities are more likely to belong to this class, but some particles in all size ranges are probably represented (Fraundorf 1980). As discussed elsewhere, these particles are probably characterized by abundant magnetite and secondary olivine.

On the other hand, considerable diversity seems to exist among chondritic IDP's which do not appear to have severely heated. Of four such particles examined by Fraundorf (1981c), two consisted mostly of a noncrystalline chondritic material with no signs of magnetite decoration. These particles show evidence of atomic S/Fe ratios above 2 and thus the sulfur mineralogy cannot be explained solely by the pyrrhotite and/or pentlandite minerals observed in other particles. Both of these particles also contained occasional enstatite whiskers (Fig. 8), a feature in common with many of the reentrant chondritic particles.

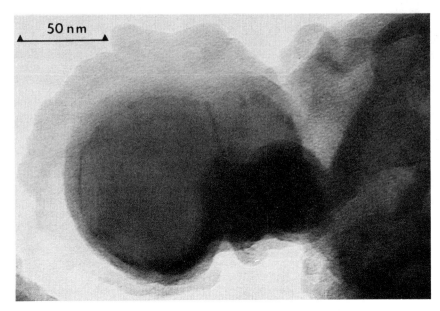

Fig. 9. Cluster of crystalline microdroplets from particle U2-13-10-3 (Fig. 7), which also contained carbide-coated taenite droplets. The largest droplet appears on the basis of EDX and electron diffraction analysis to be a low-iron olivine. Note the flattened contact boundary between droplets (Fraundorf 1981c).

The other two particles are interesting when viewed in the light of astrophysical models for circumstellar dust. Both of these particles were a mixture of submicron grains and a large amount of low-Z material, but the similarities end there. The more chemically reduced of the two particles consisted of low-Ni taenite droplets, clusters of crystalline silicate droplets (Fig. 9), and subhedral pyrrhotite crystals, all typically between 30 and 200 nm in size and embedded in a fluffy low-Z matrix. A coating of iron-carbide platelets 10 to 20 nm in thickness surrounded some of the metallic iron grains. Grains similar to these have been predicted as a universal condensate from circumstellar nebulae with an O/C ratio of < 1 (Lewis and Ney 1979).

In contrast, the other particle contained magnetite-decorated silicate grains, iron-sulfide grains, and one grain in which only tin was detectable by EDX. Almost all of these grains had diameters between 50 and 400 nm, thick low-Z coatings, and disturbed crystal structures. Poorly crystalline silicate grains of this size, decorated with magnetite, provide a potential match to predicted characteristics of silicate cores in core-mantle interstellar dust (cf. Knacke 1978; Greenberg 1977). The wide, nonstoichiometric range of grain chemistries, and low-Z coatings sometimes 100 nm in thickness, argue against significant compositional alterations subsequent to assembly of the aggregate.

Other Stratospheric Particles

The chondritic particles from the stratosphere have commanded by far the most attention, because they have compositions expected in interplanetary dust. However, if there are unexpected components in the flux of 10-μm particles incident on the Earth, these are also likely to be found in the stratosphere. For example, the near-Earth particle measurements of Fechtig et al. (1979) have suggested that some lunar material is likely to be present. Brownlee et al. (1976d) have argued on compositional grounds that any unidentified extraterrestrial component constitutes < 10% of the collected material. On the other hand there are a variety of rarer nonchondritic particles, some of them aggregates, whose origins have not been established (Flynn et al. 1977; Brownlee 1978b; Esat et al. 1979).

Sea-Sediment Particles

The severe atmospheric heating and seawater etching experienced by most sea-sediment particles have severely modified their structures and mineraology. These effects, as well as the magnetic collection techniques, also introduce more potential for sampling bias than is present in the stratospheric collections. For example, fine-grained unmelted material comprising the bulk of the stratospheric collections is likely to be especially susceptible to seawater dissolution.

Because most sea-sediment particles do not directly represent the structures and minerals present in the interplanetary dust cloud, they will not be discussed at length in this section. Some mention, however, should be made of the unmelted relict grains of parent material that are contained in ~10% of the silicate spheres. Usually the relict grains are minerals with high melting points. Forsterite and enstatite are common, although spinel, troilite, pentlandite, diopside, perovskite, and chromite have also been found (Blanchard et al. 1978; Brownlee 1979b). This tendency of primary silicates to have high magnesium compositions appears to be echoed in the stratospheric collections.

IV. THE QUESTION OF SOLAR FLARE TRACKS

Production rates of solar flare tracks in the silicates of 10-μm size and 1 g cm^{-3} density interplanetary dust particles in orbit at 1 AU are probably between ~3 $\times$ 10^5 and 10^7 tracks cm^{-2} yr^{-1} (Fraundorf et al. 1980b). A 10^4 yr lifetime for dust in the inner solar system (e.g., Dohnanyi 1978) would therefore result in at least 30 solar flare tracks per μm^2 of examined silicate. Thus solar flare tracks provide a promising indicator of space exposure for individual particles, if the track has not been erased on atmospheric entry.

Although a simple model of particle heating on atmospheric entry indicates that most 10-μm monomineralic particles are probably heated sufficiently to erase etchable tracks, a sizeable fraction of reentrant aggregates < 10 μm are likely to survive entry with their track records intact (Fraundorf 1980; Fraundorf et al. 1980b). Published searches for these tracks have used transmission electron microscopy (TEM), instead of the usual method of track etching, because of the paucity of micron-sized and larger silicates in collected aggregates. Although accelerator-produced iron tracks are detectable with a conventional TEM in IDP silicates, no unannealed tracks were detected in an examination of > 2 μm² clear silicate from 8 reentrant chondritic particles (Flynn et al. 1977; Fraundorf et al. 1980b). Faint transient structures, with densities between 10^9 and 10^{10} cm^{-2}, were reported in crystals from two particles, but firm identification as partially annealed tracks could not be made. The absence of identifiable tracks in examined silicates could be due to track annealing on atmospheric entry, and may indicate particle densities well above 2 g cm^{-3}, infrared emissivities below 0.3 for these visibly black particles, or fragmentation of the particles on atmospheric entry. Alternatively, the space exposure of these particles as 10-μm objects at 1 AU could be shorter than that given by previous estimates.

V. EFFECTS OF HEATING DURING ATMOSPHERIC ENTRY

It is essential that the nature of thermal alteration of small particles be understood so that features produced by atmospheric entry will not be misinterpreted as primordial properties. This is a more serious problem with dust than with conventional meteorites, because dust particles cannot support large temperature gradients and hence tend to be heated uniformly up to some maximum temperature.

Laboratory heating experiments indicate that typical chondritic IDP's do not undergo major structural changes until they melt, at a temperature around 1200 to 1400°C. Observed morphologies of the chondritic IDP's are therefore probably inherent properties of the particles in space.

The only significant elemental change with heating is the loss of sulfur, which becomes severe above 800°C. Carbon was not measured but probably is also lost, in part, at low temperatures. A normal S/Si ratio, say close to CI values, is probably not compatible with heating above 800°C for even a few seconds. Chondritic aggregates with low S/Si ratios may have intrinsically low sulfur abundances, but they may also represent particles strongly heated during entry.

X-ray diffraction of 4- to 40-μm chondritic particles has shown the following types of patterns to be common: (a) no lines; (b) FeS only; (c) FeS + magnetite; and (d) FeS + magnetite + olivine. These types may represent a heating sequence in which finely divided magnetite and olivine are produced by decomposition of poorly ordered silicate materials. The most pristine, least altered particles probably have weak patterns, with either no lines or only FeS, even though most particles appear to contain primary olivine or pyroxene as a minor phase.

Although magnetite spots are often present in electron diffraction patterns, their strength appears correlated with other signs of severe heating. For example, the strongest magnetite ring patterns reported by Fraundorf (1981c) were found in a particle whose sulfur depletion and ill-defined component boundaries independently suggested thermal alteration. Some of the 5-nm magnetite crystallites responsible for these diffraction rings were found to be oriented in a silicate matrix, thus giving rise to single crystal magnetite patterns.

When individual particles are melted in the laboratory, the major recrystallized phases are olivine and magnetite. These are also the major phases in chondritic spherules found in the stratosphere and on the sea floor.

VI. OPTICAL SPECTROSCOPY

Much of our knowledge of asteroids, comets, and dust clouds is based on measurements of the frequency dependence of optical absorption, emission, or reflection. Laboratory optical data for meteorites and lunar samples of different types are also available. Although the spectra are basically determined by the mineral assemblages present, many parameters affect the spectra, including grain size, degree of compaction, and the presence of complex mixtures that distort or mask individual spectral features. For this reason, spectroscopic comparisons of different objects must be made with caution.

In favorable cases, optical spectroscopy can give valuable information on the mineralogy of the object being studied. One example is the demonstration that the surface of the asteroid Vesta is a differentiated body similar to basaltic achondrites (McCord et al. 1970; Drake 1979).

The first optical spectra of IDP's in the laboratory were reported by Fraundorf et al. (1980a). Using Fourier transform spectroscopy they obtained transmission spectra from the visible into the mid-infrared region. The signal-to-noise ratio left much to be desired, and the most significant observation was a strong spectral feature at 10 μm. A weak absorption, probably due to an iron crystal-field band, was also observed at $\sim$1.02 μm.

The presence of a 10-μm absorption feature is, in itself, hardly surprising; silicate minerals are known to possess strong absorptions at this wavelength. The poor signal-to-noise ratio also left open the possibility that the absorption was due to carbonaceous material in the particles. A better spectrum of the same mount (Fraundorf et al. 1981) clarified the issue by showing a structured absorption feature near 10 μm, more akin to crystalline pyroxene than to olivine or a poorly crystallized silicate. A silicate origin of the feature was further indicated by the observation of a corresponding bending mode absorption near 20 μm.

As previously mentioned, the infrared spectrum of a single low-Ca particle closely matched that of the CM meteorite Murchison. The spectrum contained features commonly attributed to hydrated silicates near 10 and 22 μm, water of hydration near 3 and 6 μm, and carbonates near 7 μm, thus confirming the existance of a class of particles very similar to the matrix material of CI and CM chondrites.

If IDP's originate from comets, their optical spectra should match those of cometary dust tails. A match would not prove cometary origin, but mismatches might cause problems for the cometary origin hypothesis. Neither laboratory measurements nor astrophysical observations are adequate at this time to provide a crucial test.

In an analysis of the available astrophysical data, Hanner (1980) concludes that slightly absorbing disordered olivines can provide a reasonable match to the observed 10- and 18-μm emission features of dust in comets Bennett 1970 II and Kohoutek 1973 XII. Since disordered silicates might be expected to undergo some crystallization during heating on atmospheric entry, particles that have experienced minimal thermal alteration will be most interesting for comparative purposes. Further, detailed spectroscopic measurements and annealing studies may be particularly important in view of the impending arrival of Comet P/Halley in 1986. The nature of the astrophysical observations that should be made may be influenced by the results of such laboratory measurements.

Any connection between collected IDP's and typical interstellar dust is uncertain, but it is conceivable that spectroscopic studies of IDP's may shed some light on the 10- and 18-μm bands seen in other astrophysical settings (e.g., Savage and Mathis 1979). Although normally attributed to amorphous silicates, the origin of the bands is still unsettled (Millar and Duley 1980). Other characteristic interstellar bands (Herbig 1975) and the 2200 Å feature sometimes attributed to graphite (cf. Salpeter 1977) should also be looked for in collected dust particles.

VII. DISCUSSION

Nearly 500 stratospheric micrometeorites of 10-μm size and over 1000 deep-sea cosmic spheres in the 0.1- to 1.0 mm size range have been collected and examined. A small number of these have been extensively analyzed with a variety of techniques. Most of the examined micrometeorites appear to be samples of fine-grained materials with primitive, undifferentiated solar compositions. Significant departures from chondritic composition are attributable in nearly all cases to the inclusion of a large, single mineral grain in the particle. Usually the grain is olivine, pyroxene, or sulfide; rarely is the grain a carbonate, phosphate, or metal. These grains normally have black fine-grained cosmic abundance material adhering to their surfaces, and it is clear that the grains were previously imbedded in a fine-grained material of undifferentiated composition.

At the present state of knowledge it appears likely that more than 85% of interplanetary dust in the 4-μm to 1-mm size range is derived from a parent material similar in elemental composition to CI/CM meteorites. The grain sizes, carbon and sulfur contents, general lack of large metal grains, and presence of Ni-rich sulfide also imply that the dust material is similar to CI or CM meteorites. However, in detail the common types of dust particles differ from CI/CM material A crushed CI meteorite contains abundant Ca and Mg sulfates, and pure magnetite grains ranging in size up to tens of microns. In IDP's, these structures are nonexistent, or at least exceedingly rare. Also, olivine and pyroxene are more abundant in the dust than in CI's. Hydrated silicates are rarely identified in interplanetary dust, even though they are the major silicate phase in CI/CM meteorites.

The structure of common interplanetary aggregate particles is also different from CI/CM material. The meteorites consist mostly of rather compact masses of crumpled sheets and fibers of hydrated layer lattice silicates. In the SEM, 10-μm chunks of this material appear rather smooth on a scale of microns. Typical IDP's, on the other hand, are aggregates of roughly equidimensional building blocks with a range of sizes, often bonded together with a noncrystalline, carbon-rich material. One rare class of interplanetary particles is smooth, contains hydrated silicates, is usually depleted in Ca, and may well be pieces of the CM parent body or bodies. Typical aggregate particles, however, appear to be samples of a type or types of material that have not been found as conventional meteorites. Presumably this is because the material is so friable that it cannot survive atmospheric entry in sizes larger than dust. The maximum dynamic pressures during entry which micrometeorites must survive are several orders of magnitude lower than those experienced by meteorites of conventional size.

If the chondritic aggregate micrometeorites are cometary, then the observations indicate that cometary particles formed by at least two episodes of aggregation. The first step involved assembly of submicron grains of amorphous and crystalline silicate materials, with Ni-bearing iron sulfides and carbonaceous material to form the basic building blocks of cometary solids. These building blocks are nonporous microaggregates typically ranging in size from 0.1 μ to 1 μm. The microaggregates individually often have elemental compositions close to cosmic abundances.

The second aggregation process produced the bunch-of-grapes morphology. This involved aggregation of the basic building blocks, along with lesser numbers of similar-sized or larger grains of monomineralic materials, usually olivine, enstatite, or Ni-bearing iron sulfides. For some reason, the porosity (packing fraction) of these particles varies from extremely porous (bulk density < 1 g cm^{-3}) to compact. This range in porosity could be due to recent parent body processes, such as crust formation, or to the local ice-to-dust ratio at the time of accretion. The void spaces would then previously have been occupied by ice. Perhaps the range in observed porosities is also related to differences in meteoroid density inferred for various meteor showers.

The relationship between these bunch-of-grapes particles and CI/CM meteorites is unknown. Perhaps the meteorites are samples of dust material subjected to aqueous alteration sufficient to form strong rocks capable of surviving atmospheric entry in large chunks. Bunch and Chang (1980) and McSween (1979) have presented evidence for extensive aqueous alteration in both CI and CM meteorites. This raises the possibility that the hydrated silicates in carbonaceous meteorites are secondary, and that the primary silicates in these meteorites, as well as in comets, were anhydrous.

The major solids observed in the dust are poorly-characterized iron-magnesium silicates, Ni-rich sulfides (sometimes with Zn), and organic material. There are rarer grains of enstatite and olivine, and very rare grains of carbonate, phosphate, and metal. In most cases these grains appear to be primary, and accordingly may be products of the solar nebula or presolar galactic environments. The most interesting possibility is that some of the grains are presolar core-mantle interstellar grains whose carbonaceous mantles were modified during the comet formation process. Even though the external morphologies of the bunch-of-grapes particles are very similar, detailed examination shows considerable variation in the basic building blocks from particle to particle. An understanding of all of the particles is thus likely to require more than one explanation.

One final aspect of the aggregate particles is their implication as to the interaction of ice and dust in comets. Cometary dust and ice are often considered to be distinct components. However, the dust particles are often very fine-grained and porous. We would hence expect dust and ice to be so thoroughly mixed that a typical cubic micron of comet is likely to contain both. When this material leaves the comet, it would not be as ice particles and as dust particles, but as fine-grained aggregates of ice and dust. Production of even millimeter-sized pieces of pure ice would probably require a post-accretion separation process, such as melting and recrystallization of the ice component. Since the dust appears similar in size and absorptivity to soot, it probably gives cometary material a rather low albedo, in spite of the presence of ice.

Acknowledgments. We are greatly indebted to G.V. Ferry, N. Farlow, and the Airborne Science Division of the NASA Ames Research Center for building and operating the U2 aerosol sampler that collected nearly all of the stratospheric IDP's used for these studies. We also wish to thank B. Bates, E. Olszewski, R.I. Patel, J. Shirck, W. Siegmund, and M. Wheelock for major contributions to the laboratory work and analysis, and G.J. Flynn for constructive discussions. This work has been largely supported by grants from the National Aeronautics and Space Administration.

REFERENCES

Blanchard, M., and Cunningham, G. 1974. Artificial meteor ablation studies: Olivine. *J. Geophys. Res.* 79:3973-3980.

Blanchard, M.B.; Brownlee, D.E.; Bunch, T.E.; Hodge, P.W.; and Kyte, F.T. 1978. Meteor ablation studies from deep-sea sediments. *NASA TM-78510*. (Springfield, Virginia: National Technical Information Service).

Brownlee, D.E. 1978a. Interplanetary dust: Possible implications for comets and presolar interstellar grains. In *Protostars and Planets,* ed. T. Gehrels, (Tucson, Arizona: Univ. Arizona Press), pp. 134-150.

Brownlee, D.E. 1978b. Microparticle studies by sampling techniques. In *Cosmic Dust,* ed. J.A.M. McDonnell, (New York: Wiley), pp. 295-335.

Brownlee, D.E. 1979a. Interplanetary dust. *Rev. Geophys. Space Phys.* 17:1735-1743.

Brownlee, D.E. 1979b. Extraterrestrial components. In *The Sea 7,* ed. C. Emiliani, (New York: Interscience Publ.), in press.

Brownlee, D.E.; Ferry, G.V.; and Tomandl, D. 1976a. Stratospheric aluminum oxide. *Science* 191:1270-1271.

Brownlee, D.E.; Horz, F.; Tomandl, D.A.; and Hodge, P.W. 1976b. Physical properties of interplanetary grains. In *The Study of Comets,* eds. B. Donn, M. Mumma, W. Jackson, M. A'Hearn, and R. Harrington, (Washington, D.C.: NASA SP-393), pp. 962-981.

Brownlee, D.E.; Pilachowski, L.; Olszewski, E.; and Hodge, P.W. 1980. Analysis of interplanetary dust collections. In *Solid Particles in the Solar System,* eds. I. Halliday and B.A. McIntosh (Dordrecht, Holland: D. Reidel Publ. Co.), pp. 333-342.

Brownlee, D.E., and Rajan, R.S.1973. Micrometeorite craters discovered on chondrule-like objects from Kapoeta meteorite. *Science* 182:1341-1344.

Brownlee, D.E.; Tomandl, D.; Blanchard, M.B.; Ferry, G.V., and Kyte, F.T. 1976c. An atlas of extraterrestrial particles collected with NASA U-2 aircraft: 1974-1976. *NASA TMX-73, 152.* (Springfield, Virginia: National Technical Information Service).

Brownlee, D.E.; Tomandl, D.; and Hodge, P.W. 1976d. Extraterrestrial particles in the stratosphere. In *Interplanetary Dust and the Zodiacal Light,* eds. H. Elsässer and H. Fechtig, (Berlin: Springer-Verlag), pp. 279-284.

Brownlee, D.E.; Tomandl, D.A.; and Olszewski, E. 1977. Interplanetary dust: A new source of extraterrestrial material for laboratory studies. *Proc. Lunar Sci. Conf.* 8:149-160.

Bunch, T.E., and Chang, S. 1980. Carbonaceous chondrites II: Carbonaceous chondrite phyllosilicates and light element geochemistry as indicators of parent body processes and surface conditions. *Geochim. Cosmochim. Acta* 44:1543-1578.

Buseck, P.R., and Iijima, S. 1975. High resolution electron spectroscopy of enstatite. II: Geological application. *Amer. Mineral.* 60:771-784.

Dohnanyi, J.S. 1978. Particle dynamics. In *Cosmic Dust,* ed. J.A.M. McDonnell, (New York: Wiley), pp. 527-605.

Drake, M.J. 1979. Geochemical evolution of the eucrite parent body: Possible nature and evolution of asteroid Vesta? In *Asteroids,* ed. T. Gehrels, (Tucson, Arizona: Univ. Arizona Press), pp. 765-782.

Eberhardt, P.; Geiss, J.; Graf, H.; Grögler, N.; Krähenbühl, U.; Schwaller, H.; Schwarzmüller, J.; and Stettler, A. 1970. Trapped solar wind noble gases, exposure age and K/Ar-age in Apollo 11 lunar fine material. *Proc. Apollo 11 Lunar Sci. Conf.* (Oxford: Pergamon Press), pp. 1037-1070.

Esat, T.M.; Brownlee, D.E.; Papanastassiou, D.A.; and Wasserburg, G.J. 1979. The Mg isotopic composition of interplanetary dust particles. *Science* 206:190-197.

Fechtig, H.; Grün, E.; and Morfill, G. 1979. Micrometeoroids within ten earth radii. *Planet. Space Sci.* 27:511-531.

Flynn, G.J.; Fraundorf, P.; Shirck, J.; and Walker, R.M. 1978. Chemical and structural studies of "Brownlee" particles. *Proc. Lunar Planet. Sci. Conf.* 9:1187-1208.

Fraundorf, P. 1980. The distribution of temperature maxima for micrometeorites decelerated in the earth's atmosphere without melting. *Geophys. Res. Lett.* 10:765-768.

Fraundorf, P. 1981a. Transmission electron microscopy of clinoenstatite "whiskers" in stratosphere-collected interplanetary dust. *Lunar Planet. Sci.* XII:291-293 (abstract).

Fraundorf, P. 1981b. Stereo analysis of single crystal electron diffraction patterns. *Ultramicroscopy* 6. In press.

Fraundorf, P. 1981c. Interplanetary dust in the transmission electron microscope: Diverse materials from the early solar system. *Geochim. Cosmochim. Acta* 45. In press.

Fraundorf, P.; Flynn, G.J.; Shirck, J.; and Walker, R.M. 1980b. Interplanetary dust collected in the earth's stratosphere: The question of solar flare tracks. *Proc. Lunar Planet. Sci. Conf.* 11:1235-1249.

Fraundorf, P.; Patel, R.I.; and Freeman, J.J. 1981. Infrared spectroscopy of interplanetary dust in the laboratory. Submitted to *Icarus* (special Comet issue).

Fraundorf, P.; Patel, R.I.; Shirck, J.; Walker, R.M.; and Freeman, J.J. 1980a. Optical spectroscopy of interplanetary dust collected in the earth's stratosphere. *Nature* 268:866-868.

Fraundorf, P., and Shirck, J. 1979. Microcharacterization of "Brownlee" particles: Features which distinguish interplanetary dust from meteorites? *Proc. Lunar Planet. Sci. Conf.* 10:951-976.

Ganapathy, R., and Brownlee, D.E. 1979. Interplanetary dust: Trace element analysis of individual particles by neutron activation. *Science* 206:1075-1076.

Greenberg, J.M. 1977. Interstellar dust. In *Cosmic Dust,* ed. J.A.M. McDonnell, (New York: Wiley), pp. 187-294.

Hanner, M.S. 1980. Physical characteristics of cometary dust from optical studies. In *Solid Particles in the Solar System,* eds. I. Halliday and B.A. McIntosh, (Dordrecht, Holland: D. Reidel Publ. Co.), pp. 223-236.

Herbig, G.H. 1975. The diffuse interstellar bands IV: The region 4400-6850 Å. *Astrophys. J.* 196:129-160.

Hohenberg, C.M. 1981. High sensitivity pulse-counting mass spectrometer system for noble gas analysis. *Rev. Sci. Instrum.* 51:1075-1082.

Hohenberg, C.M.; Marti, K.; Podosek, F.A.; Reedy, R.C.; and Shirck, J.R. 1978. Comparisons between observed and predicted cosmogenic noble gases in lunar samples. *Proc. Lunar Planet. Sci. Conf.* 9:2311-2344.

Hudson, B.; Flynn, G.J.; Fraundorf, P.; Hohenberg, C.M.; and Shirck, J. 1981. Noble gases in stratospheric dust particles: Confirmation of extraterrestrial origin. *Science* 211:383-386.

Kasten, F. 1968. Falling speed of aerosol particles. *J. Appl. Meteorology* 7:944-947.

Kerridge, J.F.; Macdougall, J.D.; and Carlson, J. 1979. Iron-nickel sulfides in the Murchison meteorite and their relationship to phase Q1. *Earth Planet. Sci. Lett.* 43:1-4.

Kerridge, J.F.; Macdougall, J.D.; and Marti, K. 1979. Clues to the origin of sulfide minerals in CI chondrites. *Earth Planet. Sci. Lett.* 43:359-367.

Knacke, R.F. 1978. Mineralogical similarities between interstellar dust and primitive solar system material. In *Protostars and Planets,* ed. T. Gehrels, (Tucson, Arizona: Univ. Arizona Press), pp. 112-133.

Kresák, L. 1980. Sources of interplanetary dust. In *Solid Particles in the Solar System,* eds. I. Halliday and B.A. McIntosh, (Dordrecht, Holland: D. Reidel Publ. Co.), pp. 211-222.

Kyte, F.T. 1977. On the Origin of Extraterrestrial Stratospheric Particles: Interplanetary Dust or Meteor Ablation Debris? M.S. Thesis, San José University, San José, Calif.

Lewis, J.S., and Ney, E.P. 1979. Iron and the formation of astrophysical dust grains. *Astrophys. J.* 234:154-157.

Lewis, R.S.; Gros, J.; and Anders, E. 1977. Isotopic anomalies of noble gases in meteorites and their origins 2: Separated minerals from Allende. *J. Geophys. Res.* 82:779-792.

McCord, T.B.; Adams, J.B.; and Johnson, T.V. 1970. Asteroid Vesta: Spectral reflectivity and compositional implications. *Science* 168:1442-1447.

McSween, H.Y., Jr. 1979. Are carbonaceous chondrites primitive or processed? A review. *Rev. Geophys. Space Phys.* 17:1059-1078.

McSween, H.Y., Jr., and Richardson, S.M. 1977. The composition of carbonaceous chondrite matrix. *Geochim. Cosmochim. Acta* 41:1145-1162.

Millar, T.J., and Duley, W.W. 1980. Interstellar grains: Constraints on composition from infrared observations. *Mon. Not. Roy. Astron. Soc.* 191:641-649.

Millman, P.M. 1972. Cometary Meteoroids. In *Nobel Symposium No. 21: From Plasma to Planet,* ed. A. Elvius, (New York: Wiley), pp. 157-168.

Nagy, B. 1975. *Carbonaceous Meteorites,* (Amsterdam: Elsevier Scientific).

Nishizumi, K.; Murrell, M.T.; Davis. T.A.; and Arnold, J.R. 1980. Cosmic-ray produced ^{53}Mn in deep sea spheres. *Meteoritics* 15:342.

Öpik, E. 1937. Basis of the physical theory of meteor phenomena. *Tartu Obs. Publ.* 29:51-66.

Podosek, F.A. 1978. Isotopic structures in solar system materials. *Ann. Rev. Astron. Astrophys.* 16:293-334.

Podosek, F.A.; Huneke, J.C.; Burnett, D.S.; and Wasserburg, G.J. 1971. *Earth Planet. Sci. Lett.* 10:199-216.

Rajan, R.S.; Brownlee, D.E.; Tomandl, D.; Hodge, P.W.; Farrar, H.; and Britten, R.A. 1977. Detection of ^{4}He in stratospheric particles gives evidence of extraterrestrial origin. *Nature* 267:133-134.

Salpeter, E.E. 1977. Formation and destruction of dust grains. *Ann. Rev. Astron. Astrophys.* 15:267-193.

Savage, B.D., and Mathis, J.S. 1979. Observed properties of interstellar dust. *Ann. Rev. Astron. Astrophys.* 17:73-111.

Southworth, R.B., and Sekanina, Z. 1973. Physical and dynamical studies of meteors. *NASA CR-2316,* (Springfield, Virginia: National Technical Information Service).

Wagstaff, J., and King, E.A. 1981. Micrometeorites and possible cometary dust from antarctic ice cores. *Lunar Planet. Sci.* XII:1124-1126 (abstract).

Walker, R.M. 1980. Nature of the fossil evidence: Moon and meteorites. In *The Ancient Sun,* eds. R.O. Pepin, J.A. Eddy, and R.B. Merrill (Oxford: Pergamon Press), pp. 11-28.

Whipple, F.L. 1950. The theory of micrometeorites, part I: In an isothermal atmosphere. *Proc. Nat. Acad. Sci.* 36:687-695.

Wilkening, L.L. 1978. Carbonaceous chondritic material in the solar system. *Naturwissenschaften* 65:73-79.

Zinner, E. 1980. On the constancy of solar particle fluxes from track, thermoluminescence and solar wind measurements in lunar rocks. In *The Ancient Sun,* eds. R.O. Pepin, J.A. Eddy, and R.B. Merrill, (Oxford: Pergamon Press), pp. 201-226.

PART IV
Coma

COMET HEAD PHOTOMETRY: PAST, PRESENT, AND FUTURE

David D. Meisel
State University College of New York

and

Charles S. Morris
Prospect Hill Observatory

Until the 1970s, studies of the photometric properties of the comet coma depended heavily on visual or photographic estimates of total brightness. Because of large systematic and random observational errors, most early investigations of comet brightness were empirical or statistical, with little real physical, chemical, or thermodynamical theoretical basis. The introduction of photoelectric techniques has greatly reduced many random errors inherent in earlier methods. Although some systematic effects remain, it is becoming possible to relate the photoelectric work to both the classical methods and to realistic models of comet comae. Recent developments in image processing and spectral analysis of time-series data, greater availability of sophisticated instrumentation for comet studies, and coordinated effort on the part of all observers hold promise for future studies of comet brightness and its relationship to the physical and chemical evolution of the comet coma.

Most early studies of comet brightness centered around establishing the heliocentric variation of brightness, with particular attention to the coma. Lacking a viable physical model, most of these attempts were purely empirical, often simplistic numerical analyses. The simplest of the functional representations, that of an inverse power law in the heliocentric and geocentric distances, is still widely used to represent the main trends of coma brightness. Upon conversion to stellar magnitudes, the most general form becomes

$$H = H_0 + 2.5\, n \log r + 2.5\, K \log \Delta + \phi(D,F,S) \qquad (1)$$

where Δ is geocentric distance, r heliocentric distance, H_0 stellar magnitude at $r = \Delta = 1$ AU and $\phi(D,F,S,)$ the so-called aperture or diaphragm function necessary to make the correct surface brightness to flux conversion and in general is a function of telescope diameter D, focal length F, and projected coma surface brightness S. Assuming Eq. (1), in principle, using techniques for handling correlated variables (r and Δ are highly correlated), one could solve by least squares for H_0, n, K, and parameters needed to specify the ϕ function. Some studies also include time-dependent terms for temporal fluctuations or asymmetries in the lightcurve. Although the assumption has been criticized, most studies assume $K = 2$, the case for a perfect surface brightness to total flux conversion, and then either ignore ϕ or make it linear. This decouples the intrinsic r, Δ dependence that plagues most direct least-squares solutions. In making ϕ linear, the complicated dependence on S is also often ignored for simplicity. In cases in which observations are too few or too poorly distributed to obtain (H_0,n) by conventional least-squares regression techniques, one value of n is selected, sometimes rather arbitrarily, and then the equation is solved for H_0. When dealing, as usual, with observations taken around $r = \Delta = 1$ AU, such an estimate of H_0 probably retains at least statistical significance. The most frequently assumed n value is 4. Since $2.5\, n = 10$ in this case, the H_0 is relabeled as an H_{10} value.

Unless special precautions are taken, photometric observations of the cometary nucleus are complicated by gaseous emission and obscuring dust and ice halos in the central condensation of the coma. For observations in which coma contamination is minimal, a planetary type of magnitude formula is adopted.

$$m = m_0 + 5 \log r + 5 \log \Delta - 2.5 \log \psi(\alpha) \qquad (2)$$

where r is heliocentric distance, Δ geocentric distance, m_0 the nuclear magnitude at $r = \Delta = 1$ AU and $\psi(\alpha)$ phase function with α the angle between Earth and Sun as seen at the comet.

Before 1950, most photometric investigations of comets were based only on visual observation. With improved technology, visual methods gradually gave way to photoelectric or photographic ones. Because of calibration problems, photographic methods are often used only to study the relative variation of coma surface brightness across the image. Because of the relative speed and precision with which photoelectric data can be obtained, the laborious process of deriving photographic magnitudes is often considered unsuitable for routine use.

Steady improvement in the availability and quality of narrowband interference filters that isolate specific molecular, atomic, and ionic features, coupled with photoelectric detectors, have made it possible to study emission behavior in a large number of objects and over a large range of heliocentric distances, with accuracy warranting reduction to a common photometric system.

Because broadband filter colorimetry, photometry, and polarimetry mainly depend on coma dust content, a subject which is covered in the chapter by Ney, in this book, our attention will concentrate on those aspects primarily connected with the gaseous coma. Most aspects of high-resolution spectrophotometry will not be discussed here, as they are also covered in A'Hearn's chapter in this book. The photometric problems upon which we will comment are:

(a) photometric models and properties of the gas coma;
(b) photometry of the nucleus and inner coma;
(c) inferences on compositional differences from photometric behavior;
(d) brightness fluctuations around the heliocentric variation trend line;
(e) secular trends of periodic comets.

I. SUMMARY OF VISUAL MAGNITUDE RESULTS

Although visual magnitude estimates are subject to large systematic effects, they are useful indicators of comet activity, particularly when other observations are lacking. Most photometric reductions of visual total magnitudes use least-squares solutions to derive (H_0,n) estimates.

The largest compilations of H_{10} magnitudes are those by Vsekhsvyatsky (1964), used by Everhart (1967) and more recently by Hughes and Daniels (1980) to derive the intrinsic distributions of comet magnitudes, and by Sekanina (1969) and Wurm (1963) to derive gas production rates of comet atmospheres. Because the derived H_0 is not very sensitive to the n value, these statistical uses of H_{10} values may be justified until a large enough sample of (H_0,n) points is available.

Since comet brightness is usually estimated by comparison with the extrafocal images of stars, it is subject to various systematic and random errors. In his pioneering study of comet brightness, Bobrovnikoff (1941) found an empirical correlation between magnitude and telescope aperture which he used to correct the observations before solving for (H_0,n) sets. This procedure has been severely criticized by Öpik (1963) and Vsekhsvyatsky (1964), but several subsequent investigations (Meisel 1970; Morris 1973; Meisel and Morris 1976) not only confirm the reality of an aperture (or focal-length) correlation, but justify its existence through a Fourier transform analysis. Its dependence on the type of telescope magnification used, the method of comparison, and the degree of coma condensation of each particular comet also can be demonstrated. Kresák (1974) has reviewed these and other problems of visual comet brightness estimates.

In the book *The Study of Comets* we presented a set of 150 comet magnitude solutions for (H_0,n) that could be reduced to a common photometric system (Meisel and Morris 1976). This initial effort to produce (H_0,n) sets reduced for aperture effects has been continued in this chapter by Morris in collaboration with Green. Using observations they have gathered as editors of the *International Comet Quarterly* (ICQ) and supplemented by solutions by Bortle (1977a,b, 1978), they have derived photometric parameters of relatively high accuracy for 41 additional comets observed from 1974 to 1981 (see Table I). In all but one case, the observations were corrected to a standard aperture of 6.78 cm; the exception was P/d'Arrest 1976 XI, for which the standard aperture was 5 cm.

Table I provides values of the comets' absolute magnitude H_0 and power law exponent n. However, for two comets (P/Encke and P/d'Arrest) the standard power-law formula was found inappropriate. P/Encke was analyzed using a formula suggested by Sekanina (1979) which accounts for the significant decrease in n as the comet approaches perihelion. For P/d'Arrest, Bortle (1977a) found that part of the comet's lightcurve was better represented by a linear relationship with time. These special cases are presented in brackets and are not used in computing the average H_0 and n values given below. The computer of each solution and the date (if previously published) are also indicated in the notes section of the table.

We have included 21 cases where the range in r or the paucity of observations prevented a full (H_0,n) solution. For these we have assumed $n = 4$ and the listed H_0 solutions are therefore H_{10} values. The 26 separate "normal" solutions for (H_0,n) have $<H_0> = 7^m\!.1 \pm 2^m\!.5, <n> = 4.0 \pm 2.4$ with $<q> = 0.8 \pm 0.5$ AU and the 21 H_{10} values have $<H_{10}> = 9^m\!.0 \pm 2^m\!.2$ and $<q> = 1.4 \pm 0.5$ AU. Thus Table I contains two distinct groups; the group for which only H_{10} values are available contains over 60% periodic comets while the (H_0,n) group represents only 20% periodic comets.

Finally, Whipple (1978) clearly separated preperihelion from postperihelion observations, and showed an interesting difference among $<n>$ values for solutions grouped according to Oort-Schmidt dynamical age criteria. The effect is particularly strong for the preperihelion sets.

II. SUMMARY OF PHOTOELECTRIC NARROWBAND FILTER RESULTS

Photoelectric methods have obvious advantages over photographic ones in accuracy and speed, but are not without problems. As pointed out by Delsemme (1973), photoelectric measurements made conventionally through

COMET PHOTOMETRY 417

TABLE I
Visual Photometric Parameters of Comets 1974-1981

Comet	q^a	Interval of r^b	H_0^c	n^d	N^e	Notes^f
Bradfield 1974 III	0.50	0.55-0.50-2.70	7.61	2.92	105	Green
P/Encke 1974 V	0.34	0.57-0.34	$[9.5+2.54(r^{1.8}-1)]^g$		6	Morris; 1980 slope assumed
P/Forbes 1974 IX	1.53	1.65-1.78	10.4	4	5	Bortle
P/Honda-Mrkos-Pajdusakova 1974 XVI	0.58	0.74-0.58-0.59	10.62	2.93	31	Green (1980)
P/Boethin 1975 I	1.09	1.09-1.41	10.3	4	10	Morris
P/West-Kohoutek-Ikemura 1975 IV	1.40	1.40-1.41	9.6	4	4	Bortle
Bradfield 1975 V	1.22	1.23-1.22-1.27	6.7	4	2	Morris
Kobayashi-Berger-Milon 1975 IX	0.43	1.41-0.43-1.00	7.34	3.77	155	Green and Morris
Suzuki-Saigusa-Mori 1975 X	0.84	0.84-1.08	9.27	−0.47	33	Green
Bradfield 1975 XI	0.22	1.06-0.22-1.23	8.88	2.91	24	Morris
Mori-Sato-Fujikawa 1975 XII	1.60	1.92-1.60-1.68	5.68	3.25	40	Green (1981)
Sato 1976 I	0.86	0.95-0.86	11.4	4	4	Morris
Bradfield 1976 IV	0.85	0.85-1.16	11.05	5.29	20	Morris
West 1976 VI	0.20	1.58-0.29	5.94	2.42	13	Morris; prior to 4 mag. flare
		0.20-3.16	4.56	3.65	119	Bortle and Morris; after flare
P/Klemola 1976 X	1.77	1.77-1.83	9.7	4	6	Morris

TABLE I (continued)

Visual Photometric Parameters of Comets 1974-1981

Comet	q^a	Interval of r^b	H_0^c	n^d	N^e	Notes[f]
P/d'Arrest 1976 XI	1.16	1.30-1.16-1.21 1.23-1.39 1.40-1.89	$[9.93 - 0.07T]^g$ 8.28 6.5	4 0 5.3	63 23 24	Bortle (1977a); T is the number of days before (−) or after (+) perihelion
P/Faye 1977 IV	1.61	1.78-1.68	8.4	4	4	Bortle
P/Grigg-Skjellerup 1977 VI	0.99	1.05-0.99-1.08	12.5	4	19	Bortle (1977b)
Kohler 1977 XIV	0.99	1.44-0.99-1.75	6.80	5.22	235	Green and Morris
P/Arend-Rigaux 1978 III	1.44	1.44-1.45	10.9	4	5	Morris
P/Chernykh 1978 IV	2.57	2.86-2.63	6.5	4	10	Bortle
Bradfield 1978 VII	0.44	0.99-0.44-0.48	7.19	2.87	26	Morris
P/Wild 2 1978 XI	1.49	1.99-1.50	6.51	5.58	22	Bortle (1978)
Machholz 1978 XIII	1.77	1.82-1.97	6.5	4	22	Morris
P/Ashbrook-Jackson 1978 XIV	2.28	2.31-2.28-2.40	7.1	4	46	Morris
P/Comas-Solá 1978 XVII	1.87	1.97-1.99	8.5	4	2	Morris
Bradfield 1978 XVIII	0.43	0.76	11.1	4	1	Morris
P/Haneda-Campos 1978 XX	1.10	1.19-1.17	12.6	4	8	Morris
Meier 1978 XXI	1.14	3.00-2.14 1.20-4.23	−0.22 2.70	6.64 3.85	26 13	Morris Morris
Bradfield 1979 VII	0.41	0.78-0.41-1.19	11.01	4.47	13	Morris
P/Schwassmann-Wachmann 3 1979 VIII	0.94	0.96-0.94-1.00	11.5	4	5	Morris

TABLE I (continued)

Visual Photometric Parameters of Comets 1974-1981

Comet	q^a	Interval of r^b	H_0^c	n^d	N^e	Notes[f]
Meier 1979 IX	1.43	1.47-1.43-1.69	9.3	4	27	Morris
Bradfield 1979 X	0.55	0.57-1.00	8.4	5.3	19	Morris (1980); graphical solution,
		1.00-1.74	8.0	1.3	35	1 mag flare at break in lightcurve
P/Schwassmann-Wachmann 2 1979k	2.14	2.19-2.15	7.6	4	13	Morris
P/Encke 1980	0.34	1.53-0.43	$[9.79 + 2.54(r^{1.8} - 1)]^g$		40	Green and Morris (1981a)
P/Stephan-Oterma 1980g	1.57	1.93-1.57	3.46	11.92	39	Green and Morris (1981b)
		1.57-1.78	6.34	6.48	24	
P/Tuttle 1980h	1.02	1.43-1.02	7.97	6.01	90	Morris and Green (1981)
		1.02-1.31	8.08	3.88	17	
Cernis-Petraukis 1980k	0.52	1.64-1.70	7.3	4	3	Morris
Meier 1980q	1.52	1.58-1.52	6.2	4	12	Morris
Bradfield 1980t	0.26	0.41-0.28	8.26	3.00	4	Morris
		0.33-0.52	7.31	2.63	15	prior to 1 mag flare
		0.54-0.95	6.74	3.39	14	after flare
Panther 1980u	1.66	1.70-1.66-1.67	5.2	4	5	Morris

[a] Perihelion distance in AU.
[b] Arc of orbit over which calculation is made.
[c] Absolute magnitude.
[d] Power law exponent.
[e] Number of observations.
[f] Computer of the solution and date if previously published.
[g] Special case. See text.

narrow-angle, focal-plane apertures sample only the central portions of the comet coma and therefore are sensitive to the radial variation of brightness across the comet image. Thus, even if a proper photometric calibration and extinction correction is achieved, there is usually no way to satisfactorily extrapolate to the region outside the diaphragm and obtain a value that could be called a total magnitude. It is likewise difficult to compare observations with different apertures and filters. In order to estimate how serious the diaphragm effect is, many observers now provide measurements through more than one diaphragm.

The Working Group on Filter Standards of IAU Commission 15 has addressed itself to standardization of narrowband filters and focal-plane diaphragm sizes with the goal of obtaining future observations which can be reduced to a common photometric system. Narrowband observations made over the last 14 years, representing the closest now possible to a common photometric system, are summarized in Table II in roughly chronological order along with the visual parameters (H_0,n) taken from Table I or Meisel and Morris (1976). When two (H_0,n) values are listed, the first refers to preperihelion, the second to postperihelion. The largest contributors to this compilation are A'Hearn, Millis and their collaborators, who have recently (A'Hearn and Millis 1980) reduced their observations to a common system to obtain estimates of gas production rates at unit distance, and compared them with values of H_{10} with favorable results. We have not yet attempted a similar correlation study with available (H_0,n) sets, because the unit distance gas production rates as computed assume $n = 4$ and thus would require recalculation. Also, there have not been a sufficient number of (H_0,n) sets to make such a study worthwhile, but as Table II illustrates the number of photoelectric narrowband observations with corresponding (H_0,n) sets is rapidly growing. It should therefore be possible in the future to perform such a correlation study, which would provide a connecting link between past and future photometric studies.

There have been several attempts to link the visual and monochromatic observations directly without a detailed physical model. Following the early suggestion by Bobrovnikoff (1951) that the visual brightness of the coma should be dominated by C_2 emission, Wurm (1963) and Sekanina (1969) both have tried to use H_{10} values to obtain estimates of the C_2 density. Even though no attempt was made to remove the part of H_{10} due to dust reflection, these estimates and the total gas densities inferred from them were insufficient to explain the nongravitational force impulses derived from orbit analysis. More recently Bakos (1973) and Svorén and Tremko (1975) have attempted the problem in reverse and tried to derive (H_0,n) parameters from photoelectric data. However, there is more to the problem than the transfer to a common photometric system; there are now indications that Bobrovnikoff's original suggestion of a direct C_2 and (H_0,n) correlation may be incorrect.

TABLE II
Summary of Selected Narrowband Filter Results

Comet	Species Studied	Visual Parameters H_0	n	References
Ikeya-Seki (1968 I)	C_2, CN	3.9	4.9	Vanýsek (1969)
Kohoutek (1973 XII)	C_2, CN, CO^+, Na	5.4[a]	2.5	Kohoutek (1976)
		6.5	2.5	
	C_2, CN, C_3	—	—	Brown (1976)
	C_2, CN, C_3	—	—	A'Hearn and Cowan (1975)
		—	—	Angione et al. (1975)
West (1976 VI)	C_2, CN, C_3	5.9	2.4	A'Hearn et al. (1977)
		4.6	3.6	
P/d'Arrest (1976 XI)	C_2, CN, C_3	*[b]	*	A'Hearn et al. (1979)
P/Encke (1977 XI)	C_2, CN, C_3	*[b]	*	A'Hearn et al. (1979)
P/Grigg-Skjellerup (1977 VI)	C_2, CN, C_3	12.5	(4)[c]	A'Hearn et al. (1979)
P/Ashbrook-Jackson (1978 XIV)	C_2, CN, C_3	7.1	(4)	A'Hearn et al. (1979)
P/Chernykh (1978 IV)	C_2, CN, C_3	6.5	(4)	A'Hearn et al. (1979)
Kohler (1977 XIV)	C_2, CN, C_3	6.8	5.2	A'Hearn and Millis (1980)
P/Wild 2 (1978 XI)	C_2, CN, C_3	6.5	5.6	A'Hearn and Millis (1980)
Meier (1978 XXI)	C_2, CN, C_3	−0.2	6.6	A'Hearn and Millis (1980)
		+2.7	3.9	
P/Haneda-Campos (1978 XX)	C_2, CN, C_3	12.6	(4)	A'Hearn and Millis (1980)
Bradfield (1979 VII)	C_2, CN, C_3	11.0	4.5	A'Hearn and Millis (1980)
Meier (1979 IX)	C_2, CN, C_3	9.3	(4)	A'Hearn and Millis (1980)

TABLE II (continued)
Summary of Selected Narrowband Filter Results

Comet	Species Studied	Visual Parameters H_0	n	References
Bradfield (1979 X)	C_2, CN, C_3, OH, NH	*[b]	*	Millis and A'Hearn (1980) A'Hearn et al. (1980)
P/Encke 1980	C_2, CN, (C_3 + CO^+)	*[b]	*	Neff (1981)
P/Stephan-Oterma 1980g	C_2, CN, (C_3 + CO^+), OH	3.5 6.3	11.9 6.5	Neff (1981); Millis et al. (1981)
P/Tuttle 1980h	C_2, CN, (C_3 + CO^+)	8.0 8.1	6.0 3.9	Neff (1981)

[a]When two sets of measurements are given, the first one is preperihelion, the second postperhelion.
[b]Asterisk indicates complex lightcurve (see Table I).
[c]Parentheses indicate assumed values of n.

III. PHOTOMETRY OF THE INNER COMA AND NUCLEUS

Although large telescopes and highly sensitive detecting devices are required for this work, photometric observations of comets at large heliocentric distances are probably least contaminated by light from the surrounding coma and therefore most likely to reveal light directly from the nucleus. This technique was successfully used by Roemer (1966) to derive photometric diameters for comet nuclei from a long series of her own and her associates' photographic observations. This work was extended by Delsemme and Rud (1973), using additional photographic observations by Roemer. The need to continue this effort, especially with photoelectric equipment, was reiterated by Gehrels (1972). Subsequently, Fäy and Wisniewski (1978) analysed P/d'Arrest in considerable detail. Tedesco and Barker (1981) have analyzed photoelectric observations of P/Tempel 2 obtained by several groups; the results show a minor planet-like phase function. The technique has also been extended to the near-infrared (JHK broadband colors) by A'Hearn et al. (1981) and Degewij et al. (1981). The JHK-filter results so far indicate a dusty rather than icy surface. Spectroscopic separation of the nuclear light from that of the inner coma is facilitated by the use of high-sensitivity, spatially-resolved detectors. Recent examples of the application of such devices are: video cameras (Degewij and Chapman 1981); digicons (Cochran et al. 1981); coudé scanners (Barker 1981); intensified dissector scanners (Cochran and Barker 1981; Stauffer and Spinrad 1981); charge-coupled devices (Johnson et al. 1981); and area scanners (Neff 1981). These cometary applications, if continued, should contribute greatly to our understanding of the photometric behavior of the nucleus and the inner coma. Strauss (1979) has presented a method of separating coma light from the nuclear contribution in drift scans of a comet. The inversion technique is a straightforward application of transform theory, but so far has not found widespread use. In principle, it appears adaptable to a variety of photometric situations.

IV. PHOTOMETRIC MODELS, PHYSICAL THEORIES AND ABUNDANCES

Comparison of several past reviews of cometary physics (see e.g., Bobrovnikoff 1951; Wurm 1963; Whipple and Huebner 1976; Delsemme 1976) shows the steady development and refinement of physical models as bases of photometric reductions. One of the first suggestions for a photometric formula based on physical theory was proposed by Levin (1948) who assumed that total comet brightness was directly related to the production rate of cometary gas. He envisioned this process as gaseous desorption from solids that were in simple thermal equilibrium with the solar radiation field. Under these assumptions, he derived a formula analogous to Eq. (1) but with constants that had a distinct physical interpretation:

$$H = B\sqrt{r} + A + 5 \log \Delta \ldots \qquad (3)$$

where $B = L_0/RT_0$ and $H_0 = A + B$ and T_0 is characteristic temperature at $r = 1$, L_0 heat of desorption, and R universal gas constant. As shown by Bobrovnikoff (1951) and elaborated upon by Meisel (1970) and Meisel and Morris (1976), this equation does not possess any computational advantage since the square root of r can always be expanded as a power series in $\log r$. In the Levin formulation, if the B parameter is a constant then the observationally determined power law exponent must change with heliocentric distance according to the formula $n = 0.43 <\sqrt{r}>B$ where the factor $<\sqrt{r}>$ is the true anomaly average of the square root of r over the range of the observations.

The appeal of a photometric formula with parameters that could be physically interpreted led Schmidt (1951) to reanalyze data for a total of 60 comets to obtain (A,B) sets. These new photometric solutions were used by Oort and Schmidt (1951) to attempt a spectroscopic-photometric-orbital classification of comets. While the orbital distinction between old and new comets is clear enough, Donn (1977) was unable to substantiate any gross spectroscopic difference in emission/continuum ratio. As mentioned earlier, Whipple (1978) has found evidence for a preperihelion/postperihelion photometric distinction that appears to correlate with the Oort-Schmidt criteria. This problem deserves further attention since we (Meisel and Morris 1976), using nearly the same data and corrections, found only a relatively weak photometric distinction between the old and new groups. In any event, the confirmation of the Levin model that the original Oort-Schmidt relationship implied must be regarded as fortuitous. Revisions of the required gas production rates and nuclear size estimates eventually forced major consideration to be given to ice vaporization models (Huebner 1965; Delsemme 1966), in which the solar radiation field is in equilibrium with the energy of the evaporating molecules as well as with the thermal reradiation of the nucleus.

Although vaporization models also predict that the effective n value decreases with geocentric distance, the interpretation of this change in terms of the thermodynamical properties of the nucleus is different from the Levin model and cannot be given in closed analytical form. However, if there is no species decay, the flow velocity is constant, and the comet light is only reradiated solar radiation, a relatively simple asymptotic expression is obtained. Let Q be the production rate, then at any r

$$n = d \log Q/d \log r + 2. \qquad (4)$$

It is not possible from one observation to obtain the instantaneous value of n, so we must use the average derivative $p = <d \log Q/d \log r>$ in our definition of n. Thus $n = p + 2$.

As shown by Huebner (1965), if there were no thermal heating of the nucleus, $p \to 2$ and hence $n \to 4$ as used in the estimates of H_{10}. Representative p values for zero albedo and various values of the heat of vaporization L (assumed constant) can be taken from Huebner's curves. For example, in the range 0.5 AU $< r <$ 1.0 AU, p goes from 1.4 at $L = 2$ kcal mole^{-1} to 3.4 at $L = 12$ kcal mole^{-1}, and in the range 1.0 AU $< r <$ 2.0 AU, p goes from 1.9 to 4.6. Delsemme and Miller (1971) considered the evaporation of specific volatiles and with the exception of H_2O, all species considered show linear slopes ($p \sim 1.3$) in the log Q-log r plot, as long as 0.3 AU $< r <$ 4 AU. Both sets of representative values agree reasonably well with the mean values obtained from visual magnitude estimates.

Although the vaporization curve of H_2O shows a pronounced change of slope at $r = 1.3$ AU (Delsemme and Miller 1971), a clear indication of this rarely appears above the noise level of observing error, dust contamination, and intrinsic comet brightness variations. The one notable exception to this is P/Encke whose behavior clearly does not fit a power law relationship. Instead, as shown by Ferrin and Naranjo (1980), a long series of visual observations by Beyer can be satisfied by an evaporation model with a rapidly rotating nucleus. In Table I, we have acknowledged P/Encke's departure from a power law by using Sekanina's (1979) formula.

A dichotomy in narrowband filter results for the CN/C_2 ratio between those obtained at small r ($r < 1.5$ AU) and those at large r has been noted by A'Hearn and Millis (1980). This seems likely to be a vaporization effect involving mainly the as yet unidentified parent substances of C_2 and CN.

If the coma were in radial, free expansion with no species decay, the apparent surface brightness would be inversely proportional to the projected radial distance from the center. Isophotometric and spectrophotometric studies have clearly shown that monochromatic images of gaseous constituents give a radial dependence that falls off much too steeply for decay to be ignored. As we have pointed out previously (Meisel and Morris 1976), the fact that most visual magnitude estimates statistically satisfy the standard power law formula better than the more general one proposed by Öpik (1963) implies that the average profile for reduction of visual observations is closer to an inverse square apparent distance dependence, not to the inverse apparent distance dependence used in the reduction of broadband photoelectric photometry (Ney, this book). For monochromatic coma images, however, exponential functions rather than a single parameter power law provide a better fit of the data (Delsemme and Moreau 1973). A two-step exponential production-decay model was proposed by Haser (1957) and has gained acceptance in many areas of cometary physics. The critical parameters in the Haser model are the mother and daughter species' lifetimes which, in the case of uniform outward expansion, can be specified in terms of characteristic scale lengths.

To derive production rates, and hence abundances, directly from photometric observations, the Haser model requires an estimate of the two scale lengths. The determination of representative scale lengths and their heliocentric dependence is not trivial even for some conspicuous species (Delsemme and Moreau 1973; A'Hearn and Cowan 1975; Vanýsek 1976; Strauss 1977; Combi 1978; Sivaraman et al. 1979; Bappu et al. 1980; Combi and Delsemme 1980; and A'Hearn and Cowan 1980). These scale-length problems together with uncertainties in the band oscillator strengths represent the greatest sources of error in production rates and therefore in derivation of abundances.

Assuming the canonical $p = 2$ value, A'Hearn and Millis (1980) have calculated unit distance production rates for CN, C_2 and C_3 in 14 comets. This initial study combines the Haser model and the most homogeneous narrowband filter results and provides a firm basis on which future photometric studies can be based. These authors show strong correlations between the production rates of CN, C_2, C_3 and, in objects for which measurements are available, OH. In accord with the spectroscopic results of Donn (1977) and our visual magnitude results, they find no clear abundance difference between old and new comets. However, they do show that the derived CN/C_2 ratio is affected by whether the comets are observed predominantly at $r > 1.5$ AU or at $r < 1.5$ AU. Their comparison of production rates with visual H_{10} values shows the tightest correlation with the CN production and not with the C_2 rate, as would be expected. Since the C_3 and OH productions are also correlated with CN and C_2, one should find H_{10} correlations for these as well. An H_{10} correlation with OH production as determined from 18-cm radio observations has been found by Bockelée-Morvan et al. (1981). The present problem of the choice of appropriate photometric scale lengths should eventually be resolved by simultaneous imagery and spectroscopy.

The application of these two-dimensional devices, such as intensified dissector scanners (IDS) (Cochran and Barker 1981), and charge-coupled devices (CCD) (Johnson et al. 1981) should make it possible to produce consistent photometric calibrations between all the various methods. Modern image restoration techniques applied to the deconvolution of drift scans with slits as discussed by Strauss (1977, 1979) also appear able to produce useful photometric results, but it has only been done for a small number of objects.

The prediction of a solar-cycle variation of the species scale length by Oppenheimer and Downey (1980) introduces additional complications into the Haser model which need to be tested with photoelectric data. It seems unlikely that the present difficulties will lead to rejection of the Haser model or some generalization of it.

V. COMET BRIGHTNESS FLUCTUATIONS

The photometric variability of comets is well known but the underlying physical mechanisms are still poorly understood. Comet variability has been suggested to be of three distinct types: periodic, random, and secular.

Periodic function fitting of visual magnitude residuals was attempted by Bobrovnikoff (1942), but because of a moonlight modulation the results were far from convincing. Because of the random spacing of most cometary observations, it is difficult to prove periodicity by using ordinary Fourier transform or power spectrum techniques. However, several studies of techniques for computing Fourier transforms of unequally-spaced data have developed algorithms to handle data with wide gaps (Meisel 1978, 1979). Although methods of producing power-spectrum estimates from unequally-spaced data continue to be published, only two are known at present to satisfy the necessary orthogonality conditions for minimum distortion estimation. The method developed by Vaníček (1969, 1971) and subsequently tested by Taylor and Hamilton (1972) is useful for producing a single-pass, power-spectrum estimate with improved signal-to-noise ratio. A similar result achieved by a process of successive data filtering has been developed by Ferraz-Mello (1981). Morris and Bortle (1973) found evidence for variations in the visual lightcurve of Comet Ikeya-Seki 1968 I and suggested rotation as the cause, but in view of other possible explanations were not able to prove it conclusively. Ultrashort fluctuations, reminiscent of the 300^s solar oscillation, were reported for Comet Kohoutek 1973 XII by Isserstedt and Schlosser (1975) but such brief variations have apparently not since been searched for in other objects. In a pioneering effort, Fäy and Wisniewski (1978) applied power-spectrum techniques to their nuclear photometry of P/d'Arrest, but since they used an unpublished algorithm, it is not clear whether the orthonormalization criteria for uniqueness were met and which, if any, of their peaks could be spurious. Although there have been a number of studies attempting to connect comet brightness fluctuations with solar activity (see e.g., Meisel 1970; Andrienko 1974; Meisel and Morris 1976; Miller 1976; Dobrovol'skii et al. 1976), the results have been mixed and often unconvincing. The lack of correlation may partly lie with our lack of understanding of solar activity itself. Past studies often dealt with correlation of solar-active zones, when perhaps the coronal holes and their accompanying high-speed particle streams are more important. Indeed, Andrienko (1974) finds in the Bobrovnikoff data a tendency for comets to be brighter when over the heliographic polar regions where the coronal hole phenomenon is a nearly permanent feature. Although the idea of a connection with solar wind is not new, Niedner (1980; see also chapter by Brandt in this book) has considerably strengthened the physical arguments for some types of solar-wind-induced flares. The observational properties of strong cometary outbursts have been well reviewed by Hughes (1975) who concludes that a single mechanism for all types observed is unlikely.

Finally, we mention briefly the controversial subject of secular decreases in comet brightness. In 1974, Kresák presented a critique of published visual magnitudes of periodic comets and gave a careful discussion of the role of instrumental and other systematic effects. He concluded that cometary bursts tend to mask the real secular changes. Svorén (1979) reviewed the problem again using the maximum brightness reduced to unit distance as well as the usual H_{10} value and concluded that these bursts show an average decline of brightness amounting to ~ 0.2 mag per revolution; this value seems somewhat high. In Table I we give a number of H_0 and H_{10} values for periodic comets that have made recent appearances. If these values are compared with those obtained upon discovery, no significant difference is found (see Table III). Thus, unless there is a significant systematic brightening (solar activity?) present in the modern values, we conclude that the average secular decrease is 5 to 10 times smaller than that given by Svorén.

VI. SUMMARY

Comet photometry has reached the point of convergence of new and old methodologies. The present difficulties with scale-length determination do not appear insurmountable so we predict that both modern observational and theoretical techniques will eventually provide a consistent, physically-based photometric model as a basis for studies of brightness fluctuations.

TABLE III
List of Periodic Comets Showing no Significant Secular Decrease

Comet	q (AU)	Period (yr)	No. of Perihelion Passages with no Apparent Fading[a]	Comments
P/Forbes 1974 IX	1.53	6.40	8	No fading since discovery
P/Honda-Mrkos-Pajdusakova 1974 XVI	0.58	5.28	6	No fading since discovery
P/d'Arrest 1976 XI	1.16	6.23	20	1976 H_{10} was 2 mag brighter than discovery H_{10}
P/Faye 1977 IV	1.61	7.4	9	Stable since 10th perihelion passage
P/Grigg-Skjellerup 1977 VI	0.99	5.12	12	Stable since 4th perihelion passage
P/Ashbrook-Jackson 1978 XIV	2.28	7.43	5	No fading since discovery
P/Comas-Solá 1978 XVII	1.87	8.55	7	No fading since discovery
P/Schwassmann-Wachmann 3 1979 VIII	0.94	5.43	10	No fading since discovery
P/Schwassmann-Wachmann 2 1979k	2.14	6.51	9	No fading since discovery
P/Tuttle 1980h	1.01	13.77	9	Very little fading since discovery
P/Encke 1980	0.34	3.30	>24	Wide variation in H_{10} during apparition

[a]Based on a comparison of the calculated H_{10} value (without aperture correction) for the most current apparition and the H_{10} values listed by Vsekhsvyatsky (1964).

Acknowledgments. We thank D.W.E. Green for significant contributions of data and solutions to our compilations of visual-magnitude results; M. A'Hearn for providing several vital preprints of narrowband photoelectric results; J.E. Bortle for supplying four previously unpublished H_{10} values.

REFERENCES

A'Hearn, M.F., and Cowan, J.J. 1975. Molecular production rates in Comet Kohoutek. *Astron. J.* 80:852-860.
A'Hearn, M.F.; Thurber, C.H.; and Millis, R.L. 1977. Evaporation of ices from Comet West. *Astron. J.* 82:518-524.
A'Hearn, M.F.; Millis, R.L.; and Birch, P.V. 1979. Gas and dust in some recent periodic comets. *Astron. J.* 84:570-579.
A'Hearn, M.F., and Cowan, J.J. 1980. Vaporization in comets: The icy grain halo of Comet West. *Moon and Planets* 22:41-52.
A'Hearn, M.F.; Millis, R.L.; and Birch, P.V. 1980. Photometry of Comet Bradfield 1979ℓ *Bull. Amer. Astron. Soc.* 12:730 (abstract).
A'Hearn, M.F., and Millis, R.L. 1980. Abundance correlations among comets. *Astron. J.* 85:1528-1537.
A'Hearn, M.F.; Dwek, K.; and Tokunaga, A.T. 1981. Where is the ice in comets? Submitted to *Icarus*.
Andrienko, D.A. 1974. Comets, solar activity, and the interplanetary. *Problems of Cosmic Physics* 9:104-109.
Angione, R.J.; Roosen, R.G.; and Lanning, H. 1975. Narrowband filter photometry of Comet Kohoutek. *Icarus* 24:116-119.
Bakos, G.A. 1973. Photoelectric observations of Comet Bennett. *J. Roy. Astron. Soc. Canada* 67:183-189.
Bappu, M.K.V.; Parthasarathy, M.; Sivaraman, K.R.; and Babu, G.S.D. 1980. Emission band and continuum photometry of Comet West (1975n)-II. Emission profiles at the neutral coma, lifetimes of molecules and distribution of the molecules and dust within the coma. *Mon. Not. Roy. Astron. Soc.* 192:641-650.
Barker, E.S. 1981. Spatially resolved observations of the inner coma of Bradfield (1979ℓ). Submitted to *Icarus*.
Bobrovnikoff, N.T. 1941. Investigations of the brightness of comets, Part I. *Contrib. Perkins Obs.* No. 15.
Bobrovnikoff, N.T. 1942. Investigations of the brightness of comets, Part II. *Contrib. Perkins Obs.* No. 16.
Bobrovnikoff, N.T. 1951. Comets. In *Astrophysics, A Topical Symposium,* ed. J.A. Hynek (New York: McGraw-Hill), pp. 302-352.
Bockelée-Morvan, D.; Crovisier, J.; Gerand, E.; and Kazés, I. 1981. Observations of the OH radical in comets at 18 cm wavelength. Submitted to *Icarus* (special Comet issue).
Bortle, J.E. 1977a. The 1976 apparition of periodic Comet d'Arrest. *Sky Tel.* 53:152-157.
Bortle, J.E. 1977b. Comet digest. *Sky Tel.* 54:107.
Bortle, J.E. 1978. Comet digest. *Sky Tel.* 56:121.
Brown, L. 1976. Narrowband photometry of Comet Kohoutek. In *The Study of Comets, Part I* eds. B. Donn, M. Mumma, W. Jackson, M. A'Hearn, and R. Harrington (Washington: NASA SP-393), pp. 70-91.
Cochran, W.D.; Cochran, A.L.; and Barker, E.S. 1981. Spectrometric observations of Comet Bradfield (1980t). Submitted to *Icarus*.
Cochran A.L., and Barker, E.S. 1981. Spectrometric observations of Comets Stephan-Oterma and Encke during their 1980 apparitions. Submitted to *Icarus.* (special Comet issue).
Combi, M.R. 1978. Convolution of cometary brightness profiles by circular diaphragms. *Astron. J.* 83:1459-1466.
Combi, M.R., and Delsemme, A.H. 1980. Neutral cometary atmospheres II. The production of CN in comets. *Astrophys. J.* 237:641-645.
Degewij, J., and Chapman, C.R. 1981. Spectrophotometric imagery of P/Schwassmann-Wachmann (2). Submitted to *Icarus*.
Degewij, J.; Hartmann, W.K.; and Cruikshank, D.P. 1981. P/Schwassmann-Wachmann 1 and Chiron: Near infrared photometry. Submitted to *Icarus*.
Delsemme, A.H. 1966. Vers un Modéle physico-chimique de noyan cométaire. *Mém. Soc. Roy. Sci. Liège,* Ser. 5, 28:77-110.

Delsemme, A.H. 1973. The brightness law of comets. *Astrophys. Lett.* 14:163-167.
Delsemme, A.H. 1976. The neutral coma of comets: A review. In *The Study of Comets, Part II,* eds. B. Donn, M. Mumma, W. Jackson, M. A'Hearn, and R. Harrington (Washington: NASA SP-393). pp. 711-732.
Delsemme, A.H., and Miller, D.C. 1971. Physico-chemical phenomena in comets-III. The continuum of Comet Burnham (1960 II). *Planet. Space Sci.* 19:1229-1257.
Delsemme, A.H., and Moreau, J.L. 1973. Brightness profiles in the neutral coma of Comet Bennett (1970 II). *Astrophys. Lett.* 14:181-185.
Delsemme, A.H., and Rud, D.A. 1973. Albedos and cross-sections for the nuclei of Comets 1969 IX, 1970 II, and 1971 I. *Astron. Astrophys.* 28:1-6.
Dobrovol'skii, O.V.; Osherov, R.S.; and Markovich, M.Z. 1976. Statistical investigation of the influence of solar activity on the development of a comet's head. *Komety Meteory* 25:31-43.
Donn, B. 1977. A comparison of the composition of new and evolved comets. In *Comets, Asteroids, Meteorites: Interrelations, Evolution and Origins,* ed. A.H. Delsemme (Toledo, Ohio: Univ. Toledo), pp. 15-23.
Everhart, E. 1967. Intrinsic distributions of cometary perihelia and magnitudes. *Astron. J.* 72:1002-1011.
Fäy, T.D. Jr., and Wisniewski, W. 1978. The light curve of the nucleus of Comet d'Arrest. *Icarus* 34:1-9.
Ferraz-Mello, S. 1981. Estimation of periods from unequally spaced observations. *Astron. J.* 86:619-624.
Ferrin, I., and Naranjo, O. 1980. A possible explanation of the light curve of Comet Encke. *Mon. Not. Roy. Astron. Soc.* 193:667-681.
Gehrels, T. 1972. Shape and orientation of nucleus. In *Comets, Scientific Data and Missions* eds. G.P. Kuiper and E. Romer, (Tucson, Arizona: Lunar and Planetary Lab.), pp. 142-144.
Green, D.W.E. 1980. The 1974/75 apparition of P/Comet Honda-Mrkos-Pajdusakova. *J. Assoc. Lunar Planet. Observers* 28:197-202.
Green, D.W.E. 1981. The apparition of Comet Mori-Sato-Fujikawa 1975 XII. *J. Assoc. Lunar Planet. Observers* 28:217-222.
Green, D.W.E., and Morris, C.S. 1981a. The 1980 apparition of Periodic Comet Encke. *Internat. Comet Quart.* 3:10-12.
Green, D.W.E., and Morris, C.S. 1981b. A report on the 1980-81 apparition of Periodic Comet Stephan-Oterma 1980g. *Internat. Comet Quart.* 3:42-43.
Haser, L. 1957. Distribution d'intensité dans la tête d'une comete. *Bull. Acad. Roy. Belgique* (Cl. Sci.) Ser. 5 18:740-750.
Huebner, W.F. 1965. Über die Entwicklungsraten der Kometen Atmosphären. *Z. Astroph.* 63:22-34.
Hughes, D.W. 1975. Cometary outbursts, a brief survey. *Quar. J. Roy. Astron. Soc.* 16:410-427.
Hughes, D.W., and Daniels, P.A. 1980. The magnitude distribution of comets. *Mon. Not. Roy. Astron. Soc.* 191:511-520.
Isserstedt, J., and Schlosser, W. 1975. Intensity fluctuations in the head of Comet 1973f. *Astron. Astrophys.* 41:9-13.
Johnson, J.R.; Turek, P.; Fink, U.; Larson, S.; Smith, B.A.; and Reitsema, H.J. 1981. Recent results of CCD comet spectroscopy. Submitted to *Icarus.*
Kohoutek, L. 1976. Photoelectric photometry of Comet Kohoutek (1973f). In *The Study of Comets, Part I,* eds. B. Donn, M. Mumma, W. Jackson, M. A'Hearn, and R.S. Harrington (Washington: NASA SP-393), pp. 50-69.
Kresák, L. 1974. The aging and brightness decrease of comets. *Bull. Astron. Inst. Czech.* 25:87-112.
Levin, B.Y. 1948. Variation of comet brightness. *Sov. Astron. J.* 25:246-250.
Meisel, D.D. 1970. An investigation of the brightness of two comets. *Astron. J.* 75:252-257.
Meisel, D.D. 1978. Fourier transforms of data sampled at unequal observational intervals. *Astron. J.* 83:538-545.
Meisel, D.D. 1979. Fourier transforms of data sampled in unequally spaced segments. *Astron. J.* 84:116-126.

Meisel, D.D., and Morris, C.S. 1976. Comet brightness parameters: Definition, determination, and correlations. In *The Study of Comets, Part I* eds. B. Donn, M. Mumma, W. Jackson, M. A'Hearn, and R. Harrington (Washington: NASA SP-393), pp. 410-444.

Miller, F.D. 1976. Solar-cometary relations and the events of June-August 1972. *Space Sci. Rev.* 19:739-759.

Millis, R.L., and A'Hearn, M.F. 1980. Ground-based photometry of comets in the spectral interval 3000 to 5500 Å. In *Modern Techniques for Comets*, (Pasadena: Jet Propulsion Lab.), pp. 57-62.

Millis, R.L.; Thompson, D.T., and A'Hearn, M.F. 1981. Narrowband photometry of P/Stephan-Oterma. Submitted to *Icarus*.

Morris, C.S. 1973. On aperture corrections for comet magnitude estimates. *Publ. Astron. Soc. Pacific* 85:470-473.

Morris, C.S. 1980. Photometric parameters of comets. *Internat. Comet Quart.* 2:24-26.

Morris, C.S., and Bortle, J.E. 1973. The light curve of Comet Ikeya-Seki 1968. I. *Publ. Astron. Soc. Pacific* 85:249-252.

Morris, C.S., and Green, D.W.E. 1981. The 1980-81 apparition of Periodic Comet Tuttle. *Internat. Comet Quart.* 3:44-46.

Neff, J.S. 1981. Filter photometry of comets. Submitted to *Icarus*.

Newburn, R.L.; McCord, T.; and Bell, J. 1981. Interference filter photometry of P/Ashbrook-Jackson. *Astron. J.* 86:469-475.

Niedner, M.B. 1980. Interplanetary gas XXV, a solar wind and interplanetary magnetic field interpretation of cometary light outbursts. *Astrophy. J.* 241:820-829.

Oort, J.H., and Schmidt, M. 1951. Differences between new and old comets. *Bull. Astron. Inst. Netherlands* 11:259-270.

Öpik, E.J. 1963. Photometry, dimensions, and ablation rate of comets. *Irish Astron. J.* 6:93-112.

Oppenheimer, J., and Downey, C.J. 1980. The effect of solar-cycle ultraviolet flux variations on cometary gas. *Astrophys. J.* 241:L123-L126.

Roemer, E. 1966. The dimensions of cometary nuclei. *Mém. Soc. Roy. Sci. Liège. Ser. 5* 12:23-28.

Schmidt, M. 1951. The variation of the total brightness of comets with heliocentric distance. *Bull. Astron. Inst. Netherlands* 11:253-258.

Sekanina, Z. 1969. Total gas concentration in atmospheres of the short-period comets and impulsive forces upon their nuclei. *Astron. J.* 74:944-950.

Sekanina, Z. 1979. Comet Encke ephemeris. In *Handbook of British Astronomical Association for 1980*, ed. G.E. Taylor (London: W1V ONL), p. 93.

Sivaraman, K.R.; Babu, G.S.D.; Bappu, M.K.V.; and Parthasarathy, J. 1979. Emission band and continuum photometry of Comet West (1975n)-I. Heliocentric dependence of the flux in the emission band and the continuum. *Mon. Not. Roy. Astron. Soc.* 189:897-906.

Stauffer, J.R., and Spinrad, H. 1981. The behavior of the nuclear continuum of P/Comets with varying heliocentric distance. Submitted to *Icarus*.

Strauss, F.M. 1977. Strip photometry of diffuse objects. *Astron. Astrophys.* 55:299-302.

Strauss, F.M. 1979. A technique for brightness measurements of cometary nuclei. *Icarus* 39:65-68.

Svorén, J., and Tremko, J. 1975. Integral brightness and photometry in the region of emission band C_2(1-0) and the adjacent continuum of Comet Kohoutek 1973f. *Bull. Astron. Inst. Czech.* 26:342-345.

Svorén, J. 1979. Secular variations in the absolute brightness of short period comets. *Contrib. Astron. Obs. Skalnaté Pleso*, VIII: 105-140.

Taylor, J., and Hamilton, S. 1972. Some tests of the Vaníček method of spectral analysis. *Astrophys. Space Sci.* 17:357-367.

Tedesco, E.F., and Barker, E.S. 1981. Photometric phase function of the Periodic Comet Tempel-2 during its 1978-1979 apparition. Submitted to *Icarus*.

Vaníček, P. 1969. Approximate spectral analysis by least-squares fit. *Astrophys. Space Sci.* 4:387-391.

Vaníček, P. 1971. Further development and properties of the spectral analysis by least-squares. *Astrophys. Space Sci.* 12:10-33.

Vanýsek, V. 1969. Photoelectric measurements of Comet Ikeya-Seki 1969 I. *Bull. Astron. Inst. Czech.* 20:355.

Vanýsek, V. 1976. Photometry of the cometary atmosphere: A review. In *The Study of Comets, Part I*, eds. B. Donn, M. Mumma, W. Jackson, M. A'Hearn, and R. Harrington, (Washington: NASA SP-393), pp. 1-27.

Vsekhsvyatsky, S.K. 1964. *Physical Characteristics of Comets* (English translation and update of 1958 edition NASA TTF-80, OTS 62-11031). Israel Program for Scientific Translation.

Whipple, F.L. 1978. Cometary brightness variation and nucleus structure. *Moon and Planets* 18:343-359.

Whipple, F.L., and Huebner, W.F. 1976. Physical processes in comets. *Ann. Rev. Astron. Astrophys.* 14:143-172.

Wurm, K. 1963. The physics of comets. In *The Moon, Meteorites, and Comets*, eds. B.M. Middlehurst and G.P. Kuiper (Chicago: Univ. of Chicago Press), pp. 573-617.

SPECTROPHOTOMETRY OF COMETS AT OPTICAL WAVELENGTHS

MICHAEL F. A'HEARN
University of Maryland

A review of recent technological advances indicates that we are on the verge of a breakthrough in cometary spectrophotometry. Homogeneous spectrophotometry can now be obtained for many comets, even faint ones. Our understanding of fluorescent emission is improving, but the extent to which other processes such as collisions must be taken into account is still unclear. Quantitative results regarding spatial distributions and abundances are available for several species in enough different comets so that one can make statements about variations, but results of this type are needed for many more species before we can fully understand the nature of comets. The continuous spectrum of cometary grains is slightly redder than that of the Sun, but this is the limit of our knowledge regarding the continuum.

Spectroscopy and spectrophotometry of comets is a wide field which touches on so many aspects of cometary studies that it is not practical to discuss all of them in a single review. This chapter concentrates on spectrophotometric work, measurements of the relative or absolute strength of spectral features by spectrometers or by suitable calibration of conventional spectrographs. I do not consider questions concerning identification of new species, most of which have been answered using exclusively uncalibrated spectrograms. This topic has been reviewed by Delsemme (1980). I also do not consider questions answerable by qualitative study of uncalibrated spectrograms; see Herbig's (1976) review and the study of continuum-to-emission ratios in many comets by Donn (1977). Many high-resolution studies, particularly of relative intensities of rotational lines within molecular bands, were done initially on photographic spectra to show the existence of, e.g., the Swings and Greenstein effects, but have since been done more quantitatively using spectrophotometry or calibrated spectrograms. These data have been fitted with detailed fluorescence calculations including those effects. Much earlier work in this area is reviewed by Arpigny (1976); this chapter concentrates on developments since that review, particularly work at lower spectral resolution.

I first consider different techniques for spectrophotometry. Many new instruments have become available in the last few years and it is important to understand their advantages and limitations. After considering the available techniques, I discuss the physical problems investigated, organizing the material by the chemical species studied. For each species I consider relative line intensities, spatial distribution, abundances, etc. I conclude with discussion of spectrophotometry of the continuum and the relative abundances of different species whose abundances are known.

I. TECHNIQUES FOR SPECTROPHOTOMETRY

The classical method for studying the spectrum of any object is photographic spectroscopy. Most of the history of cometary spectroscopy consists of uncalibrated photographic spectra. Crude spectrophotometry can be carried out on these photographic spectra by tracing them with a microdensitometer, but because intensity calibrations were not put on the plates a microdensitometer tracing yields little more photometric information than visual inspection. Unfortunately, true calibration of a photographic spectrum is extremely difficult. Because the characteristic curve (optical density versus exposure) of a photographic plate is frequently a strong function of wavelength, linearizing the response was not practical until the advent of image intensifiers. Since the phosphor in the image intensifier exposes the entire spectrum using radiation of the same color, the wavelength dependence of the curve no longer matters and it becomes fairly routine to linearize the response of the plate. One can therefore measure spatial profiles of many different emission features by scanning the plate perpendicular to the dispersion and converting the measured optical densities of the plate to relative intensities. One can also determine line profiles and even relative intensities of nearby lines very accurately since the photocathode sensitivity, the atmospheric transmission, and virtually all other relevant factors can be taken to vary slowly with wavelength. Because a photographic plate consists of so many independent detectors, it is an invaluable detector for the types of problems just described; many major results have been based on this sort of data. Despite many new detectors available, there is no other system in routine use to compete with conventional spectroscopy for problems in which both spatial and spectral resolution are required. Spectral resolutions actually used in cometary work range from 50 Å or worse for objective prism system to 0.1 Å or better for Coudé spectra. An excellent example of this type of work is the determination of scale lengths for dissociation of cometary molecules and radicals using long-slit spectrograms as carried out by Combi and Delsemme (1980) and many others. Another example is the study of variation of collisional effects with distance from the nucleus by Malaise (1970). The use of spectrograms to compare intensities of neighboring lines is even more common and is exemplified by the determination of the isotope ratio $^{12}C/^{13}C$. The first such determination was by Stawikowski and Greenstein (1964) who measured widths of lines to minimize the need for accurate intensity calibration.

The major limitation of photographic spectrophotometry is that the calibration of the absolute, or even the relative-absolute, wavelength-dependent sensitivity is almost never carried out. For example, spectroscopists rarely make measurements to determine atmospheric attentuation as a function of wavelength. It is therefore not possible to compare the intensities of emission features at widely separated wavelengths nor to provide a true flux scale. For problems requiring this type of information, one is limited in practice (though not in principle) to photoelectric devices. It seems likely that charge-coupled devices (CCD's) at the focal planes of spectrographs will provide the desired information to supplant photography in problems requiring spatial and spectral resolution and a true flux calibration. These CCD arrays have not yet been applied widely to cometary work, but programs are underway and we expect that by the 1985-86 apparition of P/Halley these arrays may be in common use. The current state of affairs causes most well-calibrated spectrophotometry of comets to have no spatial information.

All the early spectrophotometry of comets with photoelectric detectors involved conventional grating spectrometers scanned slowly in wavelength. Their spectral resolution was generally on the order of 10 to 30 Å. This is sufficient to separate prominent molecular features of most cometary spectra from each other and from the continuum, but not to reliably measure weaker and/or narrower features such as NH_2 bands or sodium-D lines, nor to cleanly resolve the continuum in most cases. Because these instruments are widely available and easy to use, they have been used by many observers in many programs; the bulk of extant data is from such instruments. The kinds of problems typically studied with these instruments involve relative strengths of the stronger molecular emission bands. This information can be used to investigate the vibrational temperature of molecules in comets or to derive column densities of the relevant species. The principal drawback to conventional scanners is their rather low throughput; if one opens the entrance slit enough to include light from a significant fraction of the coma, the spectral resolution degrades seriously. Alternatively, the scanner must sample a small portion of the comet, typically tens of arcsec by a few arcsec for moderate spectral resolution. A scanning spectrophotometer designed specifically to minimize this problem has been described by Mayer and O'Dell (1968) and has been used to observe various comets (e.g., Gebel 1970). They achieved a field of view 8.5 X 3.5 arcmin for 34 Å resolution but at the expense of using only a 25-cm telescope. Scanning can also be done with a Coudé spectrograph to obtain high resolution at the expense of decreasing the field of view. Such an instrument has been described by Tull (1972) and was used by Danks et al. (1974) at resolutions of 0.4 Å and 0.14 Å in their study of the $^{12}C/^{13}C$ ratio in Comet Kohoutek 1973 XII. These scans had a field of view a few by a few tens of arcsec.

Spectrophotometry can be greatly improved by replacing the photomultiplier tube with a one-dimensional multi-element detector, so the many wavelengths in the spectrum can be measured simultaneously. The use of a wide variety of such detectors in spectrophotometry has been reviewed by Ford (1979) along with other systems mentioned below. Two principal advantages have been exploited in cometary work. For example, the large gain in speed due to the many detectors working simultaneously was exploited by Cochran et al. (1980), who used a self-scanned Digicon array detector to obtain spectra of P/Schwassmann-Wachmann 1. Also, the high quantum efficiency at long wavelengths was coupled with the multi-detector advantage by Danks and Dennefeld (1981) who used a Reticon array detector to study the spectral region from 6000 to 10,000 Å. The principal limitations of these systems are that they are only one-dimensional, so all spatial information is lost, and that in practice they view limited portions of typical comets. Although these are one-dimensional detectors, it is common to use two detectors or alternately image two apertures onto one detector for sky subtraction. Using this technique on comets requires care not to subtract one part of the comet from another, since the apertures are frequently spaced insufficiently far apart. On the other hand, the devices are so fast that point-by-point mapping is feasible to a limited extent, which recovers some spatial information normally achievable only through photography.

Another system used quite profitably on comets is the image dissector scanner (IDS) originally developed by Robinson and Wampler (1972), also discussed in the review by Ford (1979). Although this is a scanning instrument since the image dissector scans the spectrum sequentially, it is equivalent to a one-dimensional array detector since all points on the image intensifier phosphor remember the arrival of photoelectrons until scanned by the dissector. The instrument does single-photoelectron counting with quantum efficiency typical of blue-sensitive photocathodes and thus is very fast. It is being applied to cometary problems in a systematic program by Spinrad and Stauffer (personal communication); a similar instrument at Kitt Peak has been used by Cochran and McCall (1980) among others. Both these groups have primarily been observing faint comets. Like the one-dimensional solid-state arrays, this system is in practice used in a one-dimensional mode, but

Newburn, Spinrad, and Stauffer (personal communication) at Lick and Barker and Cochran (personal communication) at McDonald have begun programs involving spatial mapping with the instrument. At least in principle the image dissector can be programmed to scan in wavelength or in the spatial direction along a long spectrograph slit at a few discrete wavelengths, or to sample a mix of spatial and wavelength points. The only inherent limitation is the total number of points scanned within the memory time of the phosphor, but the technique has not been applied in actual observation of comets due to practical difficulties of wavelength/position calibration. The principal limitation of IDS is that at high light levels (e.g., at the peaks of molecular emission bands in comets with large column densities and thus high surface brightnesses) the phosphor's memory time can significantly exceed the time interval between arrival of photons in a particular pixel. Under these conditions the response of the instrument becomes nonlinear and impossible to accurately calibrate.

All the above instruments are types routinely available to numerous investigators. Although many required a great deal of work by dedicated instrumentalists during development, now any experienced observer can use them; they have already been copied at other observatories. CCD array detectors will certainly be similarly available by the mid 1980s.

Several other instruments have been used for cometary spectrophotometry which are still developmental or so difficult to operate that only their developers have used them for cometary work. These instruments are of the Fourier transform and Fabry-Perot types. In an attempt to overcome some throughput limitations of conventional grating spectrometers, A'Hearn et al. (1974) developed a birefringent Fourier spectrometer (FOBOS) which allowed observation of larger portions of a typical comet 1 arcmin on a 102-cm telescope) while retaining spectral resolution up to 5 Å. Because of complexities of operation and data analysis, it has been applied to only a few comets, but has allowed for example the derivation of column densities of species such as NH_2 with only weak emission bands (A'Hearn et al. 1980).

Fabry-Perot spectrometers also have been used to investigate cometary spectra. Perhaps the epitome of these instruments is the PEPSIOS system originally developed by Mack et al. (1963). Like the Fourier spectrometer, the Fabry-Perot has a much higher throughput than the grating spectrometer. It can therefore be used to study a much larger source than can a grating spectrometer of comparable spectral resolution. Also like the Fourier spectrometer, it has been used almost exclusively by or in collaboration with the instrumentation developers. This particular instrument was used by Huppler et al. (1975) to observe line profiles at high resolution in Comet Kohoutek 1973 XII. These data yielded a determination of the hydrogen outflow velocity (from the Doppler profile of Hα), and separation of the cometary component of [O I] from the telluric component by the differential Doppler shift. A Fabry-Perot system has also been used by Barbieri et al. (1974) but in a very unusual manner. They have scanned in wavelength by tilting the etalon, a technique used in other branches of astronomy to optimize the response of very narrowband interference filters but not usually used for scanning. In this case, the instrument was used in isolating continuum regions for dust studies and a negative search for quadrupole lines of H_2.

To summarize the status of these instruments for cometary spectrophotometry, photographic spectrophotometry is probably on the way out. It has contributed significantly, but will soon be entirely supplanted. Even at high spectral resolution (Coudé spectrographs and echelles). The photographic plate will soon be supplanted by photoelectric arrays. Use of conventional grating scanners has probably also peaked but they will continue to be used for cometary problems because they are widely available. The principal effort with these instruments should be to extend the spectral range to the OH bands (3085 Å) at one end and to the red CN bands (1.1 μm) at the other.

Although their high throughput and spectral resolution make them ideal instruments for cometary spectrophotometry, the Fourier transform and Fabry-Perot spectrometer will probably not become widespread due to complexities of operation and data reduction. The future of cometary spectrophotometry probably lies with IDS systems for use on faint comets and with array detectors, particularly two-dimensional CCD arrays, for high-resolution studies and studies requiring information about spatial distributions.

II. EMISSION FEATURES STUDIED SPECTROPHOTOMETRICALLY

In this section I shall organize the material according to the atomic or molecular species studied; first diatomic and triatomic neutral radicals, then atoms, and finally ions.

A spectral scan showing much of the optical region is given in Fig. 1 (A'Hearn 1975) to provide an overview and identify features discussed in subsequent sections. An extension to longer wavelengths is shown in Fig. 2 (Barker, personal communication). The first goal of spectrophotometric study must be to understand the emission mechanism. Although it is known that fluorescence must be dominant in many cometary emission features, we need to explain, for example, the relative intensities of all the emission features for a particular species. In other words, it is important to distinguish between pure fluorescent equilibrium and a combination of fluorescent and collisional equilibrium in the level populations of any particular species. Only through this understanding of emission can we use spectrophotometric results to infer the physical and chemical conditions within the coma and tail of a comet. Once the emission mechanism is understood, one can use spectrophotometry or even filter photometry to investigate relative abundances, spatial distributions, densities, and temperatures within the coma. I therefore first consider for each species the emission mechanism, which can generally be examined only by moderate-to-high resolution spectrophotometry. After that, spatial distribution and relative abundance as deduced from spectro-

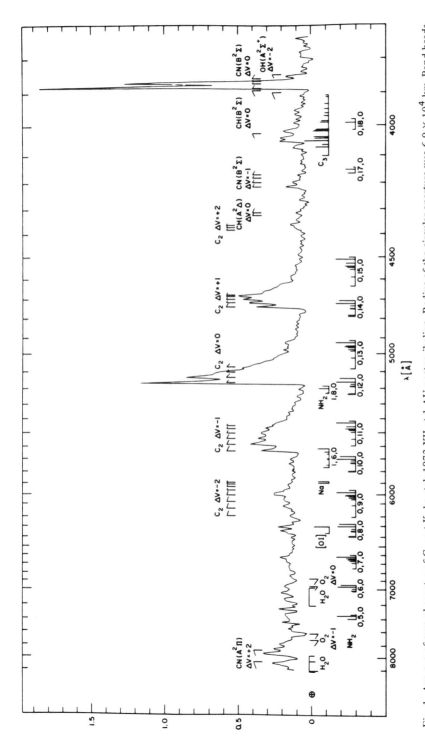

Fig. 1. Average of several spectra of Comet Kohoutek 1973 XII at 1 AU postperihelion. Radius of the circular aperture was 6.0×10^4 km. Band heads of the usually observed diatomic species are shown, as are the strongest lines (from theory and/or higher-resolution studies) of commonly observed triatomics and atoms even though not all are present in this spectrum. Telluric absorption features of O_2 and H_2O are also shown (from A'Hearn 1975).

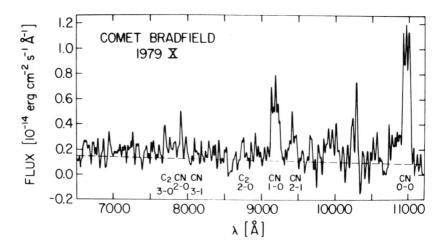

Fig. 2. Near-infrared spectral scan of nuclear region of Comet Bradfield 1979 X at 1 AU postperihelion. Aperture was ~ 1.4 by 2.7×10^3 km. Telluric and solar features have been removed by ratioing with a similar scan of a standard region on the Moon. The bands of the red system of CN and two Phillips bands of C_2 are shown. (Spectrum was provided by E.S. Barker.)

photometry will be considered, supplemented where necessary by results from filter photometric studies. (Some of the results are summarized in Table II at the end of this section.)

C_2

The Swan bands of C_2 (d $^3\Pi_g$ – a $^3\Pi_u$) dominate the visual spectra of comets (cf. Fig. 1) with 5 band sequences measured ($\Delta v = -2$ to $+2$). They were correctly identified > 50 yr ago (see Delsemme 1980, for a review of early history of spectroscopy giving first observations of most species). Because C_2 is homonuclear, nonelectronic (i.e. rotational and vibronic) transitions are strongly forbidden. Thus, absorption of solar photons leads to significant populations in highly excited rotational and vibrational levels that cannot decay back to the ground state, and large rotational and vibrational excitation temperatures are observed which have no physical connection with thermal conditions. A further anomaly of these bands is that the lower state is not the ground state but an excited triplet state (a $^3\Pi_u$) which cannot decay to the ground state (X $^1\Sigma_g^+$) because the intercombination transition is also forbidden. The singlets of C_2 are represented in optical spectra by the Phillips bands (A $^1\Pi_u$ – X $^1\Sigma_g^+$) first suggested by Wyller (1962) and subsequently measured by O'Dell (1971) in Comet Tago-Sato-Kosaka 1969 IX and by Danks and Dennefeld (1981) in Comet Bradfield 1979 X, although the most convincing evidence for singlets is the presence of the Mulliken system

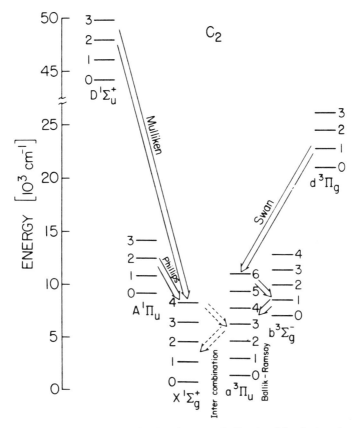

Fig. 3. Energy level diagram for C_2 showing several vibrational levels (numbered by vibrational quantum number) for each electronic state thought important in comets; numerous other states are omitted. The Ballik-Ramsay bands, not yet observed in comets, are important in controlling the vibrational populations because the higher vibrational states of a $^3\Pi_u$ can decay to lower vibrational states of $b^3\Sigma_g^-$. The intercombination transitions have not been observed either in comets or in the laboratory.

in the ultraviolet (Smith et al. 1980; A'Hearn and Feldman 1980). Figure 3 is an energy level diagram showing only levels relevant to the bands observed in comets.

Studies of the vibrational and rotational distribution of C_2 prior to Arpigny's (1976) review showed that the populations were independent of heliocentric distance but unexplained by fluorescence. The vibrational populations were finally explained by Krishna Swamy and O'Dell (1977) who showed that previous fluorescence calculations had not included enough vibrational levels. It turns out that high vibrational levels of the lowest triplet electronic state a $^3\Pi_u$, which cannot decay vibrationally, can decay into lower-lying vibrational levels of the first excited electronic triplet state b $^3\Sigma_g$ via the Ballik-Ramsay bands (see Fig. 3).

Complete calculations to explain the distribution among rotational as well as vibrational levels have never been carried out because the numerous levels that would have to be included make calculation prohibitively expensive in computer time. Numerous investigators, however, have derived rotational excitation temperatures, several of them since Arpigny's (1976) review. A'Hearn (1978), using medium resolution profiles of all 5 band sequences measured in Comet Kohoutek 1973 XII with a Fourier transform spectrometer (FOBOS) and the fluorescence results of Krishna Swamy and O'Dell (1977), has argued for rotational temperature 3000 to 3500 K; Danylewych et al. (1978) via lower resolution scanner spectra have deduced temperatures of 3000 to 4000 K. Most recently, Lambert and Danks (Lambert, personal communication), on the basis of high resolution Reticon spectra of Comet Bradfield 1979 X have suggested rotational temperatures $\sim$ 3500 K. It is generally assumed that the rotational temperature is determined by fluorescence although Woszczyk (1970) has pointed out that some lines cannot be accounted for even qualitatively by fluorescence.

Because the transition moment for the intercombination transition (a $^3\Pi_u$ − X $^1\Sigma_g^-$; see Fig. 3) has not been measured in the laboratory, Krishna Swamy and O'Dell (1981) have calculated equilibrium ratios of various singlet and triplet bands as a function of the transition moment. By assuming that the C_2 had reached equilibrium, A'Hearn and Feldman (1980) were able to use observed intensities of Mulliken and Swan bands in Comet Bradfield 1979 X to deduce that the mean square transition moment, $|R_e^2|$ was 10^{-5} in atomic units (units of $e^2 a_0^2$, where a_0 is Bohr radius and e is electronic charge) and conclude that the ratios of triplets to singlets was $\sim$ 6 (i.e., approximately the ratio of statistical weights), a result in good agreement with that estimated by O'Dell (1971) from the Phillips bands. There is no information available on spatial variations in the triplet/singlet ratios which would imply deviation from equilibrium and might suggest preferential formation of C_2 in either state. Such information would be useful in understanding the production mechanism of this species.

Because the abundance ratio of triplets to singlets is not yet certain, I recommend using optical observations, as in the past, for determining the triplet abundance only. For this purpose, the Swan $\Delta v = 0$ sequence has fluorescence efficiency 2.2×10^{-13} erg s^{-1} per radical. The efficiencies for other sequences can be calculated from the flux ratios given by Krishna Swamy and O'Dell (1977) or by A'Hearn (1978).

Although chemical models of the coma suggest that C_2 is formed primarily by chemical reactions and that its density distribution deviates considerably from a Haser model (Heubner 1981), it is still common practice to use Haser model parameters to empirically describe the distribution. Delsemme and Moreau (1973) first suggested that the parent scale length varied as the first power of heliocentric distance r rather than as r^2, which in itself indicates a deviation from the simple photodissociation process envisioned in the

Haser models. Their scale lengths, however, were underestimated due to vignetting in their spectrograph. More recently Combi (1978) has analyzed data taken by Malaise (1976) on Comet Bennett 1970 II while Bappu et al. (1979) have studied Comet West 1976 VI. Although they derive tolerably consistent scale lengths for C_2 itself, Combi's result for the C_2 parent is $\geqslant 3$ times that of Bappu et al. (when reduced to 1 AU). This difference may be due to different treatment of the effects of aperture size, an effect that works primarily on the smaller scale length. Most recently, Newburn (personal communication) has analyzed IDS spectra of comets 1980g and 1980h. He finds the spatial variations consistent with a parent scale length 1.55×10^4 r. For our own work we have adopted scale lengths 1.7×10^4 r for the parent and 1.2×10^5 r^2 for C_2 itself. Within the errors, these agree with Newburn, Bappu et al., and our estimated corrected values from Delsemme and Moreau.

The only isotope ratio measured in a comet, $^{12}C/^{13}C$, has been measured both photographically and with high-resolution spectral scanning using the 1-0 band of C_2. The results for four comets (Ikeya 1963 I, Tago-Sato-Kosaka 1969 IX, Kohoutek 1973 XII, and Kobayashi-Berger-Milon 1975 IX) have been summarized by Vanýsek and Rahe (1978) who argue that the ratio is systematically higher (> 100) than for Sun and Earth (90). The uncertainties in the result are large, due to the assumption of no chemical fractionation in molecular formation but mostly to the strong contamination of $^{13}C^{12}C$ bands by NH_2 emission (cf. Arpigny 1976); perhaps it is not yet possible to conclude that the isotope ratios differ. More recently, Lambert and Danks (Lambert, personal communication) have carried out spectrophotometry of Comet West 1976 VI and detected lines of the 0-0 band of $^{13}C/^{12}C$. These may yield a completely independent measurement of the isotopic ratio.

CN

This species is also well studied in comets, with two electronic transitions being seen, both the violet system ($B\ ^2\Sigma^+ - X\ ^2\Sigma^+$) and the red system ($A\ ^2\Pi - X\ ^2\Sigma^+$). The spectrophotometry of this radical has been reviewed by Arpigny (1976). This radical shows a very strong Swings effect because the violet bands occur at a wavelength where the solar spectrum is very irregular. Malaise (1970) used the variation of rotational structure of the violet 0-0 band with distance from the nucleus and a detailed calculation of fluorescence to argue that collisions were important in controlling the rotational level populations of this radical. From this he derived total coma densities many orders of magnitude larger than the commonly accepted values. Arpigny (1976) argued that much of the effect seen by Malaise can be understood without collisions by using different values for many atomic parameters, in particular a larger oscillator strength for pure rotational transitions within the ground state. It thus seems that most aspects of the rotational structure of the CN spectrum can be understood by pure fluorescence. Donn and Cody (1978) have suggested that production of CN in excited states by photolysis of HCN should lead to variations in rotational structure of the bands with distance from the nucleus, but because CN relaxes rapidly to fluorescent equilibrium, the effect will be difficult to detect (Arpigny 1976, p. 839). The strong Swings effect does lead to another important effect, that the total intensity of the violet system is a strong function of heliocentric radial velocity. This must be taken into account when converting fluxes into column densities of CN. The effect has been evaluated theoretically by Tatum and Gillespie (1977) who find variation by a factor of two in fluorescence efficiency with a minimum at $\dot{r} = 0$ and maximum at $\dot{r} = +10$ km s^{-1}. Mumma et al. (1978) have obtained results which differ in velocity dependence from those of Tatum and Gillespie by ~ 10%. Since Mumma et al. used a spectrum of the whole solar disk rather than a spectrum of the center of the disk (as well as slightly different atomic parameters), their results should be more reliable (see Fig. 4). Because the time between fluorescence cycles becomes comparable to the radiative lifetime against rotational transitions for $r \lesssim 1$ AU, the fluorescence efficiency also deviates from an r^{-2} dependence by amounts on the order of 10%.

There are also difficulties in understanding the relative band intensities in the red and violet systems. Danks and Arpigny (1973) calculated the relative intensities of various bands in both systems under a wide variety of assumptions regarding poorly known atomic parameters. The only quantitative observational data available for the red system were of the relatively weak 2-0 and 3-1 bands observed by O'Dell in Comet Tago-Sato-Kosaka 1969 IX (cf. Fig. 2). Apparently there were significant discrepancies between laboratory oscillator strengths (all poorly known) and those required to reproduce the observations. Subsequent observations with **FOBOS** of comets Kohoutek 1973 XII and West 1976 VI (A'Hearn 1975; A'Hearn et al. 1980) led to similar conclusions, but again based only on the bands measured by O'Dell. The recent advances in detector technology allow accurate spectrophotometry to longer wavelengths. Reticon spectra of Comet Bradfield 1979 X by Danks and Dennefeld (1981) extended to the 1-0 band of the red system; they concluded that the pure vibrational transitional probability A_{10}^{vib} for ground-state CN is between 0.3 and 1.0 s^{-1}, significantly lower than previously suggested by Danks and Arpigny (1973). Barker (personal communication) has used an InGaAsP photomultiplier to measure all prominent CN bands from the $\Delta v = 0$ sequence of the violet system to the 0-0 band of the red system (Fig. 2). These data should answer the questions originally raised by Danks and Arpigny.

The spatial distribution of CN in the coma has been studied by Combi and Delsemme (1980) using long-slit spectra of comets Bennett 1970 II and West 1976 VI. Correcting errors in their earlier work, they found that the parent scale length for a Haser model was given by $2.19 \times 10^4 \, r^2$ km while the CN scale length was $> 3 \times 10^5$ but varied irregularly due to variations in CN production in times comparable to the time required to establish a

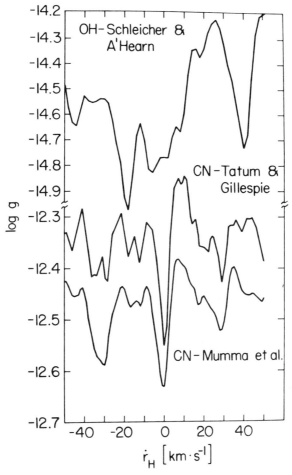

Fig. 4. Variation with heliocentric radial velocity of the logarithm of the fluorescence efficiency for OH (0–0 band) and CN (B–X; $\Delta v = 0$). Note the different ordinate scales for the two species. Because OH exhibits many fewer lines than does CN, the Swings effect is more pronounced. Differences between the two curves for CN are due to different molecular parameters (producing the overall shift) and the difference between the whole-disk solar spectrum used by Mumma et al. (1978) and the disk-center solar spectrum used by Tatum and Gillespie (1977). The shape of the radial velocity variation changes with heliocentric distance by amounts comparable to the difference in shape of the two curves exhibited here (results for OH from Schleicher and A'Hearn 1981).

stationary distribution. They also showed the spatial distribution to be entirely consistent with theoretically expected profiles if CN is produced by photodissociation of HCN. Malaise (1976) has also measured CN profiles using spatial scans with a multichannel polychromator. He argues that time variations make model fitting pointless. Since the details of his observations have not been published it is difficult to assess how much they are affected by the weather; in any case, for many applications, an approximate spatial distribution is still necessary. As working numbers, I have adopted the results of Combi and Delsemme above: $2.19 \times 10^4 \, r^2$ and $3 \times 10^5 \, r^2$ for the parent and daughter respectively.

CH

This radical is represented in cometary spectra by both the $A^2\Delta - X^2\Pi$ and the $B^2\Sigma - X^2\Pi$ systems, both of which occur in the violet. Arpigny (1976) has reviewed the high-resolution work on this species and pointed out the necessity for including collisional effects to completely explain the detailed structure of the bands. These bands should be more sensitive to collisional effects than those of CN discussed above, because CH is more closely confined to the higher density regions near the nucleus than is CN (see below). Arpigny has also used his study of this molecule to show the severe limitations of including collisional effects in a simplifed way, by linearly superimposing Boltzmann and fluorescent populations. From his analysis it is clear that to properly reproduce the spectrum collisional effects in the form of collisional rates for each transition must be included. Doing this, he finds that total densities in the coma must be quite comparable to those usually estimated from other considerations (10^4 to 10^6 cm^{-3} at $R = 10^4$ km from the nucleus) and orders of magnitude less than those found by Malaise from his study of collisional effects on CN.

Lower resolution spectrophotometry of these bands of CH has been carried out for several comets, to determine fluxes relative to those of the stronger bands of CN and C_2. Although CH was very cleanly separated from other species in the calibrated, photographic spectrophotometry of Comet Tago-Sato-Kosaka 1969 IX by O'Dell (1971), he did not discuss this species or give values for the intensity. However, one can estimate from his figures that the $A^2\Delta - X^2\Pi$ transition has an intensity $\sim 5\%$ that of the $\Delta v = 0$ sequence of the C_2 Swan bands in a 40×80 arcsec aperture. Unfortunately, at the lower spectral resolution of most spectrum scanners, this transition of CH becomes blended with the $\Delta v = +2$ sequence of the Swan system while the 0-0 band of the B-X transition is rather close to the strong violet system of CN. Nevertheless, intensities for both the B-X and A-X transitions were measured in comets Kohoutek 1973 XII (A'Hearn 1975) and West 1976 VI (A'Hearn et al. 1980; Sivaraman et al. 1979) using various entrance apertures. Although the numbers are rather uncertain, it seems that the A-X system has intensity ~ 2 or 3% that of the $\Delta v = 0$ Swan sequence. Much higher values quoted, e.g., by Babu (1976), are probably erroneous due to insufficient spectral resolution. A'Hearn et al. (1980) carried out a simplified fluorescence calculation and deduced fluorescence efficiency 9.2×10^{-14} erg s^{-1} per radical for the $\Delta v = 0$ sequence of the A-X system.

The spatial distribution has not been much studied. Malaise (1976) has obtained spatial profiles for CH in Comet Bennett 1970 II but these have not been analyzed in terms of scale lengths. It is clear from his data that the scale lengths for CH must be very short compared to those of CN and C_2, and even shorter than those of C_3. Beyond this qualitative statement little is known. On the other hand, the scale lengths are short enough so that the CH is frequently totally contained within an observing aperture.

OH

This radical, generally thought to be second only to atomic H in production rate in most comets, has been studied in many wavelength regions. The only transitions observed optically from the ground are the 0-0 (3085 Å), and 1-1 (3135 Å) bands of the $A\ ^2\Sigma^+ - X\ ^2\Pi$ transition and although they are inherently the strongest features in the optical spectrum, the strong atmospheric absorption near 3100 Å causes them to appear relatively weak. From space these bands as well as the 1-0 band of the same system, and in the radio region Λ-doublet transitions in the ground state (J = 3/2, v = 0 of $X\ ^2\Pi$), are observed. (See the chapter on ultraviolet observations by Feldman). It has been known since the work of Hunaerts (1953) that fluorescence with a very pronounced Swings effect could qualitatively reproduce the relative line strengths. Other examples of such comparison can be found in Lane et al. (1974) and Depois et al. (1979). Much better calibrated spectra of these bands (as well as of the 1-0 band) have been obtained from space using the International Ultraviolet Explorer (IUE) satellite. These spectra have been fit by Schleicher and A'Hearn (1981) who find a reasonable but far from perfect agreement between pure fluorescence theory and observation. Preliminary inclusion of collisions by the simplified method of Malaise (1970) somewhat improves this agreement. As with CN, much of the emission comes from low-density regions far from the nucleus. Because OH has longer time between fluorescence cycles than does CN, collisions should be more important than for CN but still should be less important than for CH. Furthermore, because very few levels are populated and because OH is much more abundant than most species, optical depth effects can be important for brightest comets.

When studied at lower spectral resolution, with low dispersion spectrophotometry or even filter photometry, OH exhibits a much greater Swings effect than does CN. The total fluorescence efficiency of the 0-0 band varies by more than a factor of 5 over the range of heliocentric velocities in comets (Schleicher and A'Hearn 1981) and can lead to large changes in the relative strength of OH emission. The disparate behaviors of CN and OH are compared in Fig. 4. Because of its much lower fluorescence rate, the OH fluorescence deviates much less from a r^{-2} dependence than does that of CN. The deviations become important only for $r < 0.5$ AU.

The spatial distribution of OH has also been studied from space and the ground using both ultraviolet and radio transitions. There is significant disagreement among determinations of scale length, the radio observations implying systematically larger values than the ultraviolet observations; Jackson (1980) has predicted theoretically that the daughter scale length (i.e. the OH lifetime) should be a function of heliocentric velocity. The most recent determinations based on the ultraviolet bands are those of Festou (1981) from the ground and Weaver et al. (1981) from the IUE satellite. Festou (1981) has discussed his results in terms of a vectorial model physically more correct than the Haser model but impossible to apply to other observed species unless something is known about the excess kinetic energy available at formation. For this reason, I limit my quoted scale lengths to those for an effective Haser model which systematically underestimates the true scale length. A reasonable pair of scale lengths appears to be 4.1×10^4 r^2 for the parent of OH and 1.16×10^5 r^2 for OH (taken from case B of Weaver et al. 1981).

NH

This last neutral diatomic species studied with optical spectrophotometry is infrequently studied, because the only observed transition is in the near ultraviolet (A $^3\Pi_i$ − X $^3\Sigma^-$) at 3360 Å and the band is much weaker than the nearby OH band. This species exhibits a strong Swings effect but detailed calculations of fluorescence efficiency have not been carried out. Because calculations are in progress (Litvak, personal communication) it is pointless to give approximate results for fluorescence efficiency. Malaise (1976) has presented sunward and tailward profiles of this system as measured in Comet Bennett 1970 II with his 6-channel polychromator and Combi (1978) has used these profiles to derive Haser model scale lengths of 2.4×10^3 and 1.8×10^5 km for parent and daughter respectively at $r = 0.644$ AU.

C_3

The $^1\Pi_u - {}^1\Sigma_g^+$ transition of C_3 dominates a rather broad emission feature with strongest peaks near 4050 Å extending roughly from 3900 to 4100 Å. Arpigny (1976) has already reviewed our poor understanding of the fluorescence equilibrium of this molecule which seems to show a low excitation temperature whereas high temperature as for C_2 is expected. At lower spectral resolution, the flux of the band has been measured in many comets (e.g. Gebel 1970; A'Hearn 1975; A'Hearn et al. 1980; Sivaraman et al. 1979) and the intensities are generally on the order of 10 to 20% that of Swan $\Delta v = 0$. Several recent determinations, both theoretical and experimental, of the strength of this transition have been summarized by Cooper and Jones (1979) and a simple average of those data, excluding two very discrepant results, implies a fluorescence efficiency of 1.0×10^{-12} erg s^{-1} at $r = 1$ AU, a value much higher than used, e.g., in our previous work. C_3 is strongly concentrated toward the nuclear region which no doubt contributes to the ease with

TABLE I

Fluorescence Efficiency of NH_2

Band	g^a [10^{-14} erg s^{-1}]
0, 7,0	1.75
0, 8,0	5.34
0, 9,0	3.13
0,10,0	2.99
0,11,0	2.79
0,12,0	5.51

[a]Fluorescent emission per radical at 1 AU.

which this band is seen in comets at large heliocentric distances. We have adopted a scale length for C_3 of 6×10^4 r^2 consistent with the value given by Delsemme (1975); the parent of C_3 is known only to have a much smaller scale length for which we have arbitrarily adopted the value of 1×10^3 r^2.

NH_2

This radical shows numerous bands extending across the entire optical spectrum, all attributed to one electronic transition $^2A_1\Pi_u - ^2B_1\Sigma_g$. Although a complete fluorescence calculation has not been carried out to derive the expected rotational structure of the bands, Arpigny and Woszczyk (1962) showed that fluorescence was dominant. Since then, Gary et al. (1977) have pointed out that a significant number of lines appear unaccountably missing in an echelle spectrogram of Comet West 1976 VI. Whether a complete fluorescence calculation can explain the lines in detail remains to be seen.

Because the emission bands are weak and in many cases blended with stronger features of other species, it is difficult to measure them at low spectral resolution. Consequently there are few measurements of the fluxes of the bands, but several of the bands have been measured in comets Kohoutek 1973 XII (A'Hearn 1975) and West 1976 VI (A'Hearn et al. 1980). Fortunately, the oscillator strengths for NH_2 are known (Halpern et al. 1975) for nearly all the bands observed in comets; A'Hearn et al. used a very simplified calculation, assuming resonance fluorescence separately in each band, to derive fluorescence efficiencies for the 6 bands given in Table I. The lack of observed bands between excited states suggests that the assumption of independent fluorescence is probably valid at the 10% level. i.e. at least at the level of all other uncertainties. As for CH, the spatial distribution of NH_2 is not known quantitatively but the radical appears very strongly concentrated towards the nucleus.

O

This is one of the two atomic species I will consider. Although numerous atomic species are seen in the spectra of comets, most are observed only close to the Sun and have not been much studied. In addition to O and H discussed here, recent work on atomic species has concerned primarily C and S, both of which exhibit spectral lines only in the ultraviolet.

Oxygen is unusual among cometary species because it is represented by forbidden lines, the red doublet at 6300 Å, 6364 Å ($^1D-^3P$) and the green line (less strongly forbidden) at 5577 Å ($^1S-^1D$). Since forbidden lines are not strongly excited by fluorescence, it is generally assumed that O is formed (e.g., by dissociation of a parent molecule) in an excited singlet state (^1D or ^1S) from which it decays to the ground state (^3P). The emission rate in one of these lines then directly gives the production rate of singlet oxygen. There are two major difficulties in studying these lines—contamination by telluric [OI] emission and blending with other cometary emission features, those of NH_2 for 6300 Å, 6364 Å and C_2 for 5577 Å. There are two ways to combat these problems, very high spectral resolution or very long slit spectroscopy which gives spatial information out to the limits of the cometary [OI] emission. Because NH_2 is relatively weak and concentrated to the nuclear region compared to C_2, the red lines are more commonly studied. Festou and Feldman (1981) have pointed out that the ($^1S-^3P$) transition at 2972 Å, which they identified in Comet Bradfield 1979 X, is relatively free from blending with other species and therefore better suited to studies of the production rate of $O(^1S)$. But this line tells us nothing about production of $O(^1D)$ and, because it can be observed only from space, is limited by spacecraft or rocket availability. For these reasons, the optical lines are still important.

Huppler et al. (1975) used PEPSIOS to observe the 6300 Å line at very high spectral resolution in Comet Kohoutek 1973 XII, thereby entirely separating it from both cometary NH_2 and telluric [O I]. They derived halfwidth ~ 3 km s^{-1} for this line suggesting systematic coma expansion ~ 1.5 km s^{-1}. They also found total production rate of $O(^1D)$ near 20% that of H. Delsemme and Combi (1979) have used long-slit spectra of Comet Bennett 1970 II to derive both spatial distributions and production rates. They have suggested from fits of Haser models that $O(^1D)$ must be a granddaughter species with the sum of the two scale lengths being rather short, $\sim 2 \times 10^4$ km at $r = 0.84$ AU. But as pointed out by Festou and Feldman (1981) this line is contaminated by NH_2 near the nucleus and one can therefore question the result. The production rate, determined only by the integrated flux in the line independent of model assumptions, was $\sim 35\%$ that inferred for H_2O. From the observed production rates of $O(^1D)$ relative to H_2O (i.e. H or OH) Delsemme and Combi argued that a large fraction of the $O(^1D)$ must come from CO_2 since only 12% of H_2O dissociates to $H_2 + O(^1D)$). More recent work by Festou (1981) confirms that $\sim 10\%$ of H_2O dissociates to $O(^1D)$ and 1% to $O(^1S)$ (depending somewhat on solar activity). It seems unlikely that NH_2

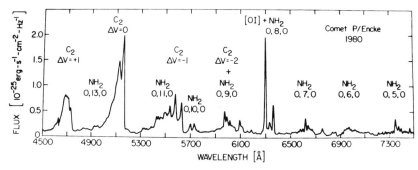

Fig. 5. Portion of an IDS spectrum of nuclear region (~ 4 arcsec) of Comet P/Encke 1980 showing prominent [OI] lines at 6300 Å and 6364 Å. Compare with Fig. 1 for drastic change in relative strength from comet to comet. Note also the difference in field of view. The line at 6300 Å is blended with a strong line of NH_2 but the total [OI] intensity is readily extractable in cases like this (spectrum provided by H. Spinrad).

contamination could lead to an overestimate of [OI] flux by a factor of 3 to 3.5, although it could easily lead to large errors in the deduced scale lengths. So their qualitative conclusion that $O(^1D)$ requires a parent in addition to H_2O seems secure, though the numbers are uncertain as is the identification of the second parent as CO_2. The results would be more certain if the 6364 Å component (intensity $\sim 1/3$ that of 6300 Å) were measured simultaneously. As pointed out by Festou and Feldman, parents of singlet oxygen other than H_2O, such as CO_2, imply different ratios of ^1S to ^1D production and therefore different ratios of 5577 Å to 6300 Å, 6364 Å (or equivalently of 2972 Å to 6300 Å, 6364 Å). Large variations in observed ratios have been reported (see e.g. Swings 1962).

Spinrad and Stauffer (1981) have been obtaining IDS spectra for many comets, primarily faint ones. The separation of cometary from telluric [OI] is clear because of the sky subtraction procedure used, while the separation from NH_2 appears adequate because they generally observed both 6300 Å and 6364 Å as strong compared to NH_2 and in the theoretically expected ratio 3:1 (see Fig. 5). From mapping [OI], they find the emission concentrated toward the nucleus, qualitatively consistent with the short Haser model scale lengths cited above. The most interesting result of their study is that large variations appear in the relative strength of [OI] 6300 Å, 6364 Å relative to the $\Delta v = 0$ sequence of the CN violet system, as measured in a small (4 arcsec) aperture near the nucleus. They find F(CN)/F(OI) ranging from 4 to 75 with typical values 8 for Comet P/Encke 1980 (shown in Fig. 5) and 50 for Comet Bradfield 1979 X. Preliminary examinations suggest the variations are not correlated with heliocentric distance, geometrical effects due to the different scale lengths, gas-to-dust ratio, or dynamical age of the comets. These variations are not yet understood.

H

Since this atom is observed primarily in the ultraviolet and only rarely in the optical, it is discussed primarily in the chapter by Feldman; I will comment only briefly on the one important groundbased observation, that of Hα by Huppler et al. (1975) in Comet Kohoutek 1973 XII using PEPSIOS. From these observations they deduced outflow velocity 7.8 km s^{-1}, in agreement with theoretical expectations and ultraviolet measurements for the component of H produced from OH (cf. Festou et al. 1979) which is a major component only in large apertures such as that of PEPSIOS. They were also able to set an upper limit on the D/H ratio at ~ 0.01.

CO$^+$

Although this species shows several prominent emission bands in the optical identified long ago, there has been relatively little spectrophotometry of these features. Abundances of CO$^+$ and several other ions have been deduced for Comet Kohoutek 1973 XII by Wyckoff and Wehinger (1976), who deduced column density $\sim 4 \times 10^{12}$ cm^{-2} at distance 10^4 km from the nucleus and $r = 0.5$ AU. The intensities of the CO$^+$ bands relative to neutral species are known to vary widely from comet to comet but the source of variation is not understood. The lines have also been observed over a wide range of heliocentric distances. Cochran et al. (1980) have measured CO$^+$ spectrophotometrically in Comet P/Schwassmann-Wachmann 1 during an outburst. This comet was previously thought to exhibit a purely continuous spectrum. Larson (1980) subsequently also detected CO$^+$ in this comet both during and outside of outbrust but his spectra are not calibrated. These measurements have yielded an estimate of 3×10^{30} CO$^+$ ions in the volume sampled by Cochran et al. (cf. Cowan and A'Hearn 1981). This yields average column density $\sim 10^{12}$ cm^{-2} averaged over an 8500 km square. This column density is only a factor of 4 smaller than that observed in Comet Kohoutek 1973 XII by Wyckoff and Wehinger for heliocentric distance 10 times smaller.

H$_2$O$^+$

This ion was first identified by Herzberg and Lew (1974) in spectra of Comet Kohoutek 1973 XII. Wyckoff and Wehinger (1976) have subsequently used calibrated spectra of Comets Kohoutek 1973 XII and Bradfield 1974 III to derive intensities and column densities for H$_2$O$^+$ and several other ions including CO$^+$, N$_2^+$, and CH$^+$. They conclude that H$_2$O$^+$, as well as the two other minor ions, are less abundant by ~ 2 orders of magnitude than CO$^+$ in Comet Kohoutek 1973 XII. It seems likely, however, that the relative abundances of the various ions vary considerably among comets (e.g., Miller 1980); this problem needs to be investigated in more detail. Similarly, quantitative data are needed to describe the spatial variation in the relative column densities of various ions.

Delsemme and Combi (1979) have derived one important quantitative result regarding the spatial distribution of H_2O^+. On the basis of long-slit, calibrated spectrograms of Comet Bennett 1970 II taken only 20 minutes apart, they have shown the rapid motion of density waves of H_2O^+ toward the tail of the comet at speed 17 km s^{-1}. If we combine this information with the Doppler shift of $\sim$ 20 to 40 km s^{-1} measured for a single H_2O^+ line in Comet Kohoutek 1973 XII by Huppler et al. (1975), it appears there is nearly direct evidence that the large accelerations observed photographically in the structures of ion tails are due to mass motions rather than wave phenomena. Similar observations yielding both changes in surface brightness and Doppler shifts for a single comet, rather than two separate comets as in this case, would confirm this key result often assumed but never directly shown.

III. SUMMARY OF RELATIVE ABUNDANCES

In order to discuss relative abundances of various species, it is necessary to combine data from filter photometry for species with strong emission bands with data from spectrophotometry for species with weaker bands. The physically relevant quantities are relative production rates, but in most cases we do not know the lifetimes of the species directly. The measurable quantities are the total abundances in the coma. The amount relative production rates differ between comets is also important, but only for species with prominent emission bands are there enough data to address this question. The relevant parameters used to derive abundances are summarized in Table II.

A'Hearn and Millis (1980) have discussed in detail the relative abundances of CN, C_2, and C_3 in various comets. There is remarkably little variation from comet to comet except for the long-known effects with heliocentric distance. Although A'Hearn and Millis quoted relative production rates, those were derived by assuming lifetimes for the various observed species. If we remove those lifetimes, we get back to estimates for the total molecular content of the coma; these values are given in Table III, column 2. The abundances are quoted relative to OH where we have used our own (i.e. A'Hearn and Millis) photometric data for OH (which agree with the IUE data mentioned above) rather than the radio data which give systematically lower abundances of OH. For heliocentric distances $\lesssim$ 1.5 AU, the variation in all these abundances is small. The other species listed in Table III, CH and NH_2, have been measured for only a few comets so very little is known about variations in their abundance.

We also present in Table III values for the lifetimes of the various species, the Haser-model scale lengths divided by an assumed 1 km s^{-1} expansion velocity. As pointed out by Festou (1981) in his discussion of OH, a Haser model systematically underestimates the true lifetime but the amount can only be determined if the production mechanism is known. These values

TABLE II

Summary of Molecular Parameters[a]

Species	Fluorescence		Haser Model Scales	
	Feature	Efficiency[b] erg s^{-1}	Parent[c] km	Observed Species[c] km
C_2	Swan; $\Delta v = 0$	2.2×10^{-13}	1.7×10^4	1.2×10^5
CN	$B^2\Sigma^+ - X^2\Sigma^+$; $\Delta v = 0$	see Fig. 4	2.2×10^4	3×10^5
CH	$A^2\Delta - X^2\Pi$; $\Delta v = 0$	9.2×10^{-14}	$\lesssim 10^4$	$\lesssim 10^4$
OH	$A^2\Sigma^+ - X^2\Pi$; 0–0 band	see Fig. 4	4.1×10^4	1.16×10^5
NH	$A^3\Pi_i - X^3\Sigma^-$; 0–0 band	unknown	2.4×10^3	1.8×10^5
C_3	$^1\Pi_u - {}^1\Sigma_g^+$	1.0×10^{-12}	$\lesssim 10^3$	$\sim 6 \times 10^4$
NH_2	$^2A_1\Pi_u - {}^2B_1\Sigma_g$	see Table I	$< 10^4$	$< 10^4$

[a]See text under each species for details and references.
[b]Fluorescence efficiency (per molecule) is given for $r = 1$ AU. It varies approximately as r^{-2}, but at least for CN there are ±10% deviations from this dependence.
[c]Scale lengths are based on the standard Haser model, physically incorrect but still useful for field-of-view corrections. All scale lengths are assumed to vary as r^2 except that of the parent of C_2 which is assumed to vary as r^1. Tabulated values are for $r = 1$ AU.

TABLE III

Relative Abundances at $r = 1$ AU[a]

Species	M(x)/M(OH)[b]	$\tau(x)/\tau(OH)$[c]	Q(x)/Q(OH)
C_2 (triplets)	0.4%	1	0.4%
CN	0.4%	3	0.13%
CH	0.005%	< 0.1	$\gtrsim$ 0.05%
NH	?	4	?
C_3	0.007%[d]	0.5	0.015%[d]
NH_2	0.01%	$\lesssim$ 0.1?	$\gtrsim$ 0.1%

[a]See Sec. II for discussion of sources of numbers.
[b]Relative abundances are estimates for total amount of each species integrated over entire coma.
[c]Lifetimes are basically the Haser-model scale lengths and may not refer to true lifetimes.
[d]Previously published data corrected for new fluorescence efficiency is discussed above.

should be taken as lower limits to the lifetimes, and consequently the numbers for relative production rates as upper limits.

We have omitted from the table the two atomic species discussed above. Hydrogen is known primarily from ultraviolet studies (see Feldman's chapter in this book) to have a production rate roughly twice that of OH, as would be expected for dissociation of H_2O. Total oxygen has similary been shown consistent in production rate with dissociation of H_2O (see Feldman's chapter) but for very few comets. The forbidden lines of oxygen observed optically do not yield abundances but rather production rates directly. The production rates of $O(^1D)$ and $O(^1S)$ are apparently highly variable from one comet to the next, quite unlike the production rates of other neutral species. Ions are also highly variable between comets and so are not tabulated here.

IV. STUDIES OF THE CONTINUUM

Continuum spectrophotometry is used to characterize the solid portions of comets, either the solid grains in the coma and tail or the solid nucleus. The nature of the solid particles (or the solid surface) affects the color of the reflected sunlight and, depending on the material, may add discrete but fairly broad absorption features to the optical spectrum. Unfortunately, this area yields many conflicting results. Ney points out in his chapter on broadband photometry from the optical to the infrared that in broadband measurements a comet has colors resembling those of the Sun. On the other hand we know that there are significant emission bands in the optical bandpasses. To draw firm conclusions regarding the color of the solid material in comets we must resort to higher spectral resolution or consider comets which do not show any significant emission features.

There are two comets for which spectrophotometry exists at moderately high resolution which may refer to the bare nucleus of the comet, although even in these optimum cases it is still questionable whether the nucleus was measured. Comet P/Arend-Rigaux 1978 III was nearly stellar in appearance at its last apparition (1977) although a very faint coma was photographed and weak emission features of CN and C_3 were seen on long-slit spectra. In any case the emission features were very weak and spectrophotometry of the nuclear region by D. Allen (cf. Weissman 1978) indicated a nearly neutral color for the nucleus. The reflectivity showed a slope $< 2\%$ per 1000 Å and no indication of the absorption features sometimes seen in asteroidal spectra. (Because the absolute scale of the reflectivity cannot be determined without knowing the size of the reflecting body, only the wavelength dependence of the reflectivity is known and all slopes will be quoted as percentages of the average reflectivity.) Subsequent spectrophotometry of Comet P/Tempel 2 was carried out by Spinrad et al. (1979) with an IDS when the cometary nucleus was thought quiescent. They found the comet redder than the Sun but not by a large amount. The slope of the reflectivity was $\sim 10\%$ per 1000 Å. For both these comets, measured reflectivities were certainly not contaminated by emission features, but in view of other signs of activity (the coma of P/Arend-Rigaux 1978 III, brightness outbursts of P/Tempel 2) it is doubtful that the reflectivities refer to a solid nucleus.

Continuum spectrophotometry of the grains in the coma and tail is much more difficult to assess. Most measurements have been made on comets with prominent emission features, so it is difficult to properly define the continuum level without high spectral resolution. The conflict in these data can be illustrated by reference to two comets. For Comet West 1976 VI A'Hearn et al. (1980) found, using FOBOS at moderate resolution, that the reflectivity increased somewhat toward the red but the slope could not exceed $\sim 20\%$ per 1000 Å. During the same time period, Sivaraman et al. (1979) used a conventional grating scanner and found that the continuum was much redder and varied with phase angle. In the extreme case they found a variation in reflectivity by a factor of 2.5 within only 2000 Å. A similar disagreement exists among the data for Comet Bennett 1970 II, although for that comet the data were taken at slightly different times. Using conventional scanner observations, Babu and Saxena (1972) found the continuum extremely red, showing a factor of 2 variation in reflectivity within 2000 Å. Just a few weeks later, Stokes (1972), using calibrated spectrograms of higher resolution, found the continuum only slightly redder than the Sun, the reflectivity having slope $\sim 5\%$ per 1000 Å. In both cases cited, the higher resolution data give colors slightly redder than the Sun while the lower resolution data give colors much redder than the Sun. It seems likely that the higher resolution data are better and in general the continuum of the cometary coma, at least for dusty comets, has reflectivity increasing only slowly with wavelength over the optical region. Because of the great disparity among observers, it is obviously impossible to make reliable assertions regarding variations of the continuum colors.

Current programs may resolve some of these questions. Because comets at large heliocentric distances typically do not exhibit the Swan bands of C_2, even when they have extensive comae, it is much easier to measure continuum colors for these comets. Unfortunately, such comets are usually faint. Cochran and McCall (1980) have used an IDS to measure such a spectrum for Comet Bowell 1980b at r = 7.2 AU and found it essentially featureless over the spectral range 3600 Å to 5200 Å. The spectrum is somewhat redder than that of the Sun but bluer than that of the S-type asteroid Massinga (Barker and Cochran, personal communication). S-type asteroids typically exhibit slope 10 to 15% per 1000 Å (see, e.g., Fig. 1 in Chapman and Gaffey 1979). Barker and Cochran are extending this program to obtain IDS spectra of all faint comets from McDonald Observatory (Barker et al. 1981). Stauffer and Spinrad (1981) are carrying out similar studies from Lick where their continuum work emphasizes the red region of the spectrum in which the molecular emission features are weaker, although still widespread. Their observations of the nuclear region of P/Encke had the surprising result that the red-continuum brightness of the nuclear region decreased as the comet approached the Sun (clearly not the behavior of a solid body unless there are remarkable geometrical effects due to P/Encke's rotation axis being nearly in its orbital plane) and that any solid nucleus must be remarkably small ($<$ 0.5 km radius) if the albedo is as much as 5%.

We hope these programs will ultimately clarify the nature of the continuum reflectivity in comets, but thus far our conclusions are very limited. We can say that the continuum reflectivity probably increases slowly (0 to 20% per 1000 Å) in the optical region, but cannot yet determine the presence or absence of absorption features or real variations from comet to comet.

Acknowledgment. Much of the author's work has been supported by a grant from the National Aeronautics and Space Administration. The content of this chapter has been influenced by the comments of many people; I especially thank E. Barker and M. Festou for their careful reading of the manuscript.

REFERENCES

A'Hearn, M.F. 1975. Spectrophotometry of Comet Kohoutek 1973f = 1973XII. *Astron. J.* 80:861-875.
A'Hearn, M.F. 1978. Synthetic spectra of C_2 in comets. *Astrophys. J.* 219:768-772.
A'Hearn, M.F.; Ahern, F.J.; and Zipoy, D.M. 1974. A polarization Fourier spectrometer for astronomy. *Applied Optics* 13:1147-1157.

A'Hearn, M.F., and Feldman, P.D. 1980. Carbon in Comet Bradfield 1979*l*. *Astrophys. J.* 242:L187-L190.
A'Hearn, M.F.; Hanisch, R.J.; and Thurber, C.H. 1980. Spectrophotometry of Comet West. *Astron. J.* 85:74-80.
A'Hearn, M.F., and Millis, R.L. 1980. Abundance correlations among comets. *Astron. J.* 85:1528-1537.
Arpigny, C. 1976. Interpretation of comet spectra: A review. In *The Study of Comets*, eds. B. Donn, M. Mumma, W. Jackson, M. A'Hearn, and R. Harrington (Washington: NASA SP-393), pp. 797-838.
Arpigny, C., and Woszczyk, A. 1962. Mechanism d'emission du radical NH_2 dans les comètes. *Bull. Soc. Roy. Sci. Liège* 31:390-395.
Babu, G.S.D. 1976. Spectrophotometry of Comet Kohoutek (1973f) during pre-perihelion period. In *The Study of Comets*, eds. B. Donn, M. Mumma, W. Jackson, M. A'Hearn, and R. Harrington (Washington: NASA SP-393), pp. 220-231.
Babu, G.S.D., and Saxena, P.P. 1972. Spectrophotometry of Comet Bennett. *Bull. Astron. Czech.* 23:346-349.
Bappu, M.K.V.; Parthasarathy, M.; Sivaraman, K.R.; and Babu, G.S.D. 1979. Emission band and continuum photometry of Comet West (1975n) II. Emission profiles of the neutral coma, lifetimes of molecules and distribution of the molecules and dust within the coma. *Mon. Not. Roy. Astron. Soc.* 189:897-906.
Barbieri, C.; Cosmovici, C.B.; Michel, K.W.; Nishimura, T.; and Roche, A.E. 1974. Near infrared observations of the dust coma of Comet Kohoutek (1973f) with a tilting-filter Fabry-Perot photometer. *Icarus* 23:568-576.
Barker, E.S.; Cochran, A.L.; and Rydski, P.M. 1981. Observations of faint comets from McDonald Observatory. In *Modern Observational Techniques for Comets*, ed. J.C. Brandt (Pasadena: Jet Propulsion Lab.), pp. 150-155.
Chapman, C.R., and Gaffey, M.J. 1979. Reflectance spectra for 277 asteroids. In *Asteroids*, ed. T. Gehrels (Tucson: Univ. of Arizona Press), pp. 655-687.
Cochran, A.L.; Barker, E.S.; and Cochran, W.D. 1980. Spectrophotometric observations of P/Schwassmann-Wachmann 1 during outburst. *Astron. J.* 85:474-477.
Cochran, A.L., and McCall, M.L. 1980. Spectrophotometric observations of Comet Bowell (1980b). *Publ. Astron. Soc. Pacific* 92:854-857.
Combi, M.R. 1978. Convolution of cometary brightness profiles by circular diaphragms. *Astron. J.* 83:1459-1466.
Combi, M.R., and Delsemme, A.H. 1980. Neutral cometary atmospheres. II. The production of CN in comets. *Astrophys. J.* 237:641-645.
Cooper, D.M., and Jones, J.J. 1979. An experimental determination of the cross section of the Swings band system of C_3. *J. Quant. Spect. Rad. Trans.* 22:201-208.
Cowan, J.J., and A'Hearn, M.F. 1981. Vaporization in comets: Outbursts from Comet Schwassmann-Wachmann 1. Submitted to *Icarus*.
Danks, A.C., and Dennefeld, M. 1981. Near-infrared spectroscopy of Comet Bradfield 1979*l*. *Astron. J.* 86:314-317.
Danks, A.C.; Lambert, D.L.; and Arpigny, C. 1974. The $^{12}C/^{13}C$ ratio in Comet Kohoutek (1973f). *Astrophys. J.* 194:745-751.
Danks, T., and Arpigny, C. 1973. Relative band intensities in the red and violet systems of CN. *Astron. Astrophys.* 29:347-356.
Danylewych, L.L.; Nicholls, R.W.; Neff, J.S.; and Tatum, J.B. 1978. Absolute spectrophotometry of comets 1933 XII and 1975 IX II. Profiles of the Swan bands. *Icarus* 35:112-120.
Delsemme, A.H. 1980. Pristine nature of comets as revealed by their UV spectra. *Applied Optics* 19:4007-4014.
Delsemme, A.H., and Combi, M.R. 1979. $O(^1D)$ and H_2O^+ in Comet Bennett, 1970II. *Astrophys. J.* 228:330-337.
Delsemme, A.H., and Moreau, J.L. 1973. Brightness profiles in the neutral coma of Comet Bennett (1970II). *Astrophys. Letters* 14:181-185.
Despois D.; Gerard, E.; Crovisier, J.; and Kazes, I. 1979. The OH radical in comets: Observations and interpretations of the 18 cm wavelength radio spectra. Paper presented at IAU General Assembly, Comm. 15, Montreal.

Dessler, K., and Ramsay, D.A. 1959. The electronic absorption spectra of NH_2 and ND_2. *Phil. Trans. Roy. Soc. London A* 251:553-602.

Donn, B. 1977. Comparison of the compositions of new and evolved comets. In *Comets, Asteroids, Meteorites,* ed. A.H. Delsemme (Toledo, Ohio: Univ. Toledo), pp. 15-23.

Donn, B., and Cody, R.J. 1978. On the detection of newly created CN radicals in comets. *Icarus* 34:436-440.

Festou, M.C. 1981. The density distribution of neutral compounds in cometary atmospheres. II. Production rate and lifetime of OH radicals in Comet Kobayashi-Berger-Milon (1975 IX). *Astron. Astrophys.* 96:52-57.

Festou, M.C., and Feldman, P.D. 1981. Forbidden oxygen lines in comets. *Astron. Astrophys.* In press.

Festou, M.C.; Jenkins, E.B.; Keller, H.U.; Barker, E.S.; Bertaux, J.L.; Drake, J.F.; and Upson, W.L., II 1979. Lyman alpha observations of Comet Kobayashi-Berger-Milon (1975 IX) with Copernicus. *Astrophys. J.* 232:318-328.

Ford, W.K., Jr. 1979. Digital imaging techniques. *Ann. Rev. Astron. Astrophys.* 17:189-212.

Gary, G.A.; Fountain, W.F.; and O'Dell, C.R. 1977. Spectrographic observations of Comet West (1975n). *Publ. Astron. Soc. Pacific* 89:97-103.

Gebel, W.L. 1970. Spectrophotometry of comets 1967n, 1968b, and 1968c. *Astrophys. J.* 161:765-777.

Halpern, J.B.; Hancock, G.; Lenzi, M.; and Welge, K.H. 1975. Laser induced fluorescence from $NH_2(^2A_1)$ state: Selected radiative lifetimes and collisional de-excitation rates. *J. Chem. Phys.* 63:4808-4816.

Herbig, G.H. 1976. Review of cometary spectra. In *The Study of Comets,* eds. B. Donn, M. Mumma, W. Jackson, M. A'Hearn, and R. Harrington (Washington: NASA SP-393), pp. 136-154.

Herzberg, G., and Lew, H. 1974. Tentative identification of the H_2O^+ ion in Comet Kohoutek. *Astron. Astrophys.* 31:123-124.

Huebner, W.F. 1981. Chemical kinetics in the coma. In *Comets and the Origin of Life,* ed. C. Ponnamperuma (Dordrecht, Holland: D. Reidel), pp. 91-103.

Hunaerts, J. 1953. Interpretation du spectre d'emission de OH dans les cométes. *Mém. Soc. Roy. Sci. Liège,* 14th Ser. 13:99-136.

Huppler, D.; Reynolds, R.J.; Roesler, F.L.; Scherb, F.; and Trauger, J. 1975. Observations of Comet Kohoutek (1973f) with a ground-based Fabry-Perot spectrometer. *Astrophys. J.* 202:276-282.

Jackson, W.M. 1980. The lifetime of the OH radical in comets at 1 AU. *Icarus* 41:147-152.

Krishna Swamy, K.S., and O'Dell, C.R. 1977. Statistical equilibrium in cometary C_2. I. *Astrophys. J.* 216:158-164.

Krishna Swamy, K.S., and O'Dell, C.R. 1981. Statistical equilibrium in cometary C_2 III. Triplet-singlet, Phillips, Ballik-Ramsay, and Mulliken bands. *Astrophys. J.* In press.

Lane, A.L.; Stockton, A.N.; and Mies, L.H. 1974. Ground-based and near-ultraviolet observation of Comet Kohoutek. In *Comet Kohoutek,* ed. G.A. Gary (Washington: NASA SP-355), pp. 87-94.

Larson, S.M. 1980. CO^+ in Comet Schwassmann-Wachmann 1 near minimum brightness. *Astrophys. J.* 238:L47-L48.

Mack, J.E.; McNutt, D.P.; Roesler, F.L.; and Cahobbol, R. 1963. The PEPSIOS purely interferometric high-resolution scanning spectrometer 1. The pilot model. *Applied Optics* 2:873-885.

Malaise, D.J. 1970. Collisional effects in cometary atmospheres, I. Model atmospheres and synthetic spectra. *Astron. Astrophys.* 5:209-227.

Malaise, D.J. 1976. Untitled contribution to panel discussion. In *The Study of Comets,* eds. B. Donn, M. Mumma, W. Jackson, M. A'Hearn, and R. Harrington (Washington: NASA SP-393), pp. 740-750.

Mayer, P., and O'Dell, C.R. 1968. Emission band ratios in Comet Rudnicki (1966e). *Astrophys. J.* 153:951-962.

Miller, F.D. 1980. H_2O^+ in the tails of 13 comets. *Astron. J.* 85:468-473.

Mumma, M.J.; Cody, R.; and Schleicher, D. 1978. The cyanogen abundance of comets. *Bull. Amer. Astron. Soc.* 10:587 (abstract).
O'Dell, C.R. 1971. Spectrophotometry of Comet 1969g (T-S-K). *Astrophys. J.* 164:511-520.
Robinson, L.B., and Wampler, E.J. 1972. The Lick Observatory image-dissector scanner. *Publ. Astron. Soc. Pacific* 84:161-166.
Schleicher, D.G., and A'Hearn, M.F. 1981. Fluorescence of OH in comets. Submitted to *Astrophys. J.*
Sivaraman, K.R.; Babu, G.S.D.; Bappu, M.K.V.; and Parthasarathy, M. 1979. Emission band and continuum photometry of Comet West (1975n) I. Heliocentric dependence of the flux in the emission bands and the continuum. *Mon. Not. Roy. Astron. Soc.* 189:897-906.
Smith, A.M.; Stecher, T.P.; and Casswell, L. 1980. Production of carbon, sulfur, and CS in Comet West. *Astrophys. J.* 242:402-410.
Spinrad, H., and Stauffer, J. 1981. The production rates of [OI] in recent comets. *Icarus.* In press.
Spinrad, H.; Stauffer, J.; and Newburn, R.L. 1979. Optical spectrophotometry of Comet Tempel 2 far from the Sun. *Publ. Astron. Soc. Pacific* 91:707-711.
Stauffer, J.R., and Spinrad, H. 1981. The behavior of the red nuclear continuum of P/comets with varying heliocentric distance. *Icarus.* In press.
Stokes, G.M. 1972. The scattered light continuum of Comet Bennett 1969i. *Astrophys. J.* 177:829-834.
Swings, P. 1962. Comportement des raies interdites de l'oxygene dans les comètes. *Ann. d'Astrophys.* 25:165-170.
Stawikowski, A., and Greenstein, J.L. 1964. The isotope ratio C^{12}/C^{13} in a comet. *Astrophys. J.* 140:1280-1291.
Tatum, J.B., and Gillespie, M.I. 1977. The cyanogen abundance of comets. *Astrophys. J.* 218:569-572.
Tull, R.G. 1972. The Coudé spectrograph and echelle scanner of the 2.7 m telescope at McDonald Observatory. In *Auxiliary Instrumentation for Large Telescopes,* eds. S. Laustsen and A. Reiz (Geneva: ESO-CERN), pp. 259-274.
Vanýsek, V., and Rahe, J. 1978. The $^{12}C/^{13}C$ isotope ratio in comets, stars and interstellar matter. *Moon and Planets* 18:441-446.
Weaver, H.A.; Feldman, P.D.; Festou, M.C.; and A'Hearn, M.F. 1981. Water production models for Comet Bradfield (1979 X). *Astrophys. J.* In press.
Weissman, P.R. 1978. Observations of a near-extinct cometary nucleus. *Bull. Amer. Astron. Soc.* 10:588 (abstract).
Woszczyk, A. 1970. Le spectre cometaire dans le regions λ3884-3914 et 4180-4752 Å. *Studie Soc. Sci. Torunensis* 4:267-291 (= Nr. 6, pp. 23-47 = *Bull. Astron. Obs. Torun* 46).
Wyckoff, S., and Wehinger, P.A. 1976. Molecular ions in comet tails. *Astrophys. J.* 204:604-615.
Wyller, A.A. 1962. C_2 Phillips bands in emission from Comet Candy (1960n)? *Observatory* 82:73-75.

ULTRAVIOLET SPECTROSCOPY OF COMAE

P.D. FELDMAN
The Johns Hopkins University

Vacuum ultraviolet observations from sounding rockets and satellite observatories of the gaseous comae of several recent comets are reviewed. The earliest of these led to discovery of the hydrogen envelope extending for millions of km from the nucleus. Subsequent observations of H I Lyman α, the OH (0,0) band and the oxygen resonance triplet have provided strong evidence for the water-ice model of the cometary nucleus. Several new species were discovered in the coma, including C, C^+, CO, S and CS. High-resolution spectroscopy and the spatial variation of the observed emissions provide means to elucidate the production and excitation mechanisms of these species. The similarity of the spectra of the half-dozen comets observed to date argues for a common, homogeneous composition (with the exception of dust and CO) of the cometary ice and a minimal effect on the neutral species due to molecular collisions in the inner coma.

Observations of comets in the vacuum ultraviolet have contributed since 1970 to significant progress in understanding cometary comae and the cometary nucleus itself. The first ultraviolet observations, of comets Tago-Sato-Kosaka 1969 IX and Bennett 1970 II, made in 1970 by both Orbiting Astronomical Observatory-2 (OAO-2) and Orbiting Geophysical Observatory-5 (OGO-5), demonstrated the existence of a hydrogen envelope that extended millions of km from the comet's nucleus (Bertaux et al. 1973; Code et al. 1972). Analysis of this H I Lyman-α envelope and the accompanying strong emission from OH at 3085 Å (seen only weakly in groundbased spectra) provided strong confirmation of Whipple's icy conglomerate model proposed two decades earlier (Whipple 1950,1951) on the basis of the noncentral force perturbations of cometary orbits. The observed emissions could be accounted for by photodissociation by sunlight of H_2O evaporated from the surface of the "dirty snowball" nucleus; the derived H_2O production rate, typically on the order of 10^{29} to 10^{30} mol s^{-1}, was exactly the magnitude predicted by Whipple's model (Bertaux et al. 1973; Keller and Lillie 1974). Comet Tago-Sato-Kosaka was also observed in Lyman-α by a rocket experiment (Jenkins and Wingert 1972).

The next opportunity for vacuum ultraviolet observations came with Comet Kohoutek 1973 XII, discovered 10 months before perihelion, whose promise motivated an extensive campaign of coordinated space and groundbased observations. Atomic oxygen and carbon were discovered in the ultraviolet spectra obtained by two rocket experiments (Feldman et al. 1974; Opal and Carruthers 1977a) and direct Lyman-α images of the hydrogen envelope were obtained with rocket (Opal et al. 1974) and Skylab (Carruthers et al. 1974) ultraviolet cameras. The full potential of Comet Kohoutek 1973 XII was realized two years later with Comet West 1976 VI when rocket instrumentation developed for the Kohoutek observations was able to obtain the first comprehensive ultraviolet spectrum of a comet (Feldman and Brune 1976; Smith et al. 1980). The OAO-3 (Copernicus) observatory was used to obtain very high resolution line profiles of the H I Lyman-α emission from Comet West and several other comets (Festou et al. 1979) during this time period.

Since January 1978, the International Ultraviolet Explorer (IUE) satellite observatory has been available for cometary observations and while to date there have been no new comets of the intrinsic brightness of comets Bennett or West, this telescope in Earth orbit has permitted extensive observations of recent comets. Both comets Seargent 1978 XV (Jackson et al. 1979) and Bradfield 1979 X (Feldman et al. 1980) were moderately active comets; the latter was the first comet for which ultraviolet observations were made over a wide range of heliocentric distances, from 0.71 AU to 1.55 AU (Weaver et al. 1981a). Subsequently, several faint comets including P/Encke, P/Tuttle and P/Stephan-Oterma were observed by IUE (Weaver et al. 1981b) and provided a new data base for comparing composition and elucidating physical and chemical processes in the coma dependent on heliocentric distance and gas production rate.

I. THE ULTRAVIOLET SPECTRUM OF A COMET

The earliest results on the ultraviolet observations of comets and their interpretation, with an emphasis on H I Lyman-α emission, were reviewed by Keller (1976). Since that review, the sensitivity of solar-ultraviolet excited

TABLE I

List of Known Species from Rocket
Observations of Comet West

Observed Species	Wavelength (Å)
H I	1216
O I	1304
C I	1561, 1657
C I (^{1}D)	1931
S I	1814
C II	1335
CO	1510
C_2	2313
CS	2576
OH	3085
CO^+	2200
CO_2^+	2890
Upper Limits	
H_2	1608
CO_2	1993
NO	2150

fluorescence in a cometary atmosphere was strikingly demonstrated by the rocket observations of Comet West (Feldman and Brune 1976; Smith et al. 1980) by which several species previously not observed in comets were detected and significant upper limits for several others were obtained. A list of known species is given in Table I. The IUE observations of comets Seargent 1978 XV and Bradfield 1979 X did not contribute to this list though the data from Comet Bradfield on the spatial distribution of CS emission point to CS_2 as the most probable short-lived parent molecule (Jackson et al. 1981). While H_2O is not directly observable in emission, its three dissociation products H, OH, and O are all detectable in the ultraviolet. In contrast, the neutral species seen in the visible spectrum (e.g., CN, NH, C_2, and C_3) are all highly reactive radicals derived from still unknown parent molecules whose abundance in the coma is < 1% that of H_2O. The identification of the $\Delta v = 0$ sequence of Mulliken bands of C_2 near 2313 Å in the spectrum of Comet Bradfield (A'Hearn and Feldman 1980) provides a convenient way to correlate the satellite ultraviolet with visible observations.

Surprisingly, the ultraviolet spectra of all comets observed to date are remarkably similar, despite differences in visual appearance, dust/gas ratio,

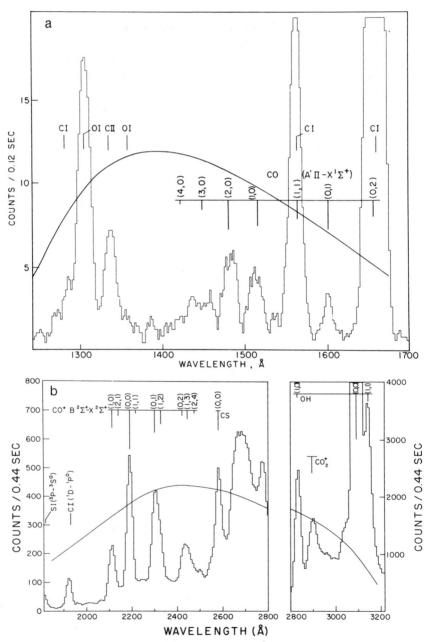

Fig. 1. Ultraviolet spectra of Comet West 1976 VI recorded by sounding rocket instruments on 5 March 1976 (Feldman and Brune 1976). (a) and (b) are short and long wavelength spectra respectively. In (a) a CaF_2 filter was used to attenuate the transmission of HI Lyman α to prevent grating scattered light from masking the weaker emission features.

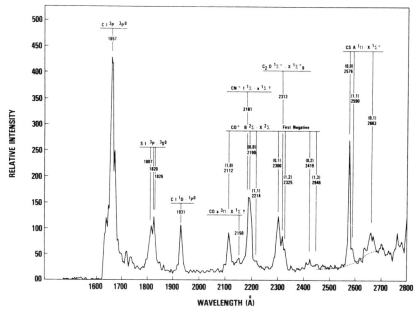

Fig. 2. Objective grating spectrum of Comet West obtained from a rocket launch on 10 March 1976 (Smith et al. 1980).

gas production rate, heliocentric distance and observing geometry. Besides the dust, only the CO^+ abundance appears to vary significantly from one comet to another. The similarity lends further support to the idea of common, homogeneous composition and possibly common origin for most comets. Furthermore, the similarity also suggests that chemical reactions in the inner coma are relatively insignificant in affecting the abundances of observed species largely determined by the photochemistry of the parent species that is evaporated from the cometary nucleus. The data from Comet Bradfield indicate that only the production of metastable $C(^1D)$ atoms (the lower level of the $\lambda 1931$ Å transition) is dependent on inner coma chemistry, probably dissociative recombination of CO^+ ions (Feldman 1978).

Examples of ultraviolet spectra of several comets are shown in Figs. 1 through 5. The rocket spectra of Comet West 1976 VI on 5 March 1976 when the comet was 0.385 AU from the Sun, by Feldman and Brune (1976), are shown in Fig. 1. The short wavelength spectrum has been corrected for atmospheric O_2 absorption, and the relative intensities of the CO fourth positive bands agree much better with expected intensities for resonance fluorescence of sunlight than the published spectrum which included several spectral scans at low altitude that distorted the spectrum due to differential O_2 absorption in the Schumann-Runge bands. The projected slit area at the comet was $1.8 \times 10^5 \times 1.28 \times 10^6$ km^2 for the long wavelength spectrometer and

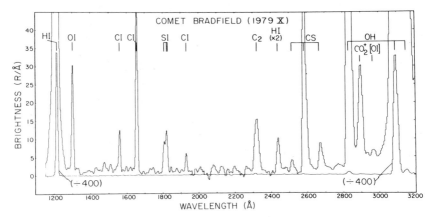

Fig. 3. Composite IUE spectrum of Comet Bradfield 1979 X. The observational parameters for Figs. 3, 4, and 5 are given in Table II.

$3.7 \times 10^5 \times 1.28 \times 10^6$ km² for the short wavelength spectrometer. Spectral resolution was 22 Å and 15 Å, respectively. The spectrum obtained five days later at $r = 0.52$ AU, by Smith et al. (1980), is shown in Fig. 2. The instrument used in this rocket experiment was an objective grating spectrograph with spectral resolution 7 Å permitting definitive identification of the CS(0,0) band at 2576 Å and the S I triplet at 1807, 1820 and 1826 Å. The spatial resolution of these data was ≈ 80,000 km, and indicated a nearly pointlike source for the CS emission.

IUE spectra of comets Bradfield 1979 X, Seargent 1978 XV and P/Encke are shown in Figs. 3, 4 and 5. Heliocentric and geocentric distance and projected area of the 10 × 20 arcsec slit for each observation are given in Table II. The spatial resolution, corresponding to 5 arcsec, is also given. Note the similarity of all three spectra, which differ mainly in relative brightness of O I λ 1304 which depends strongly on heliocentric velocity ($\dot{r}$). These spectra are also quite similar to that of Comet West shown in Figs. 1 and 2, except that the CO⁺ first negative bands in the region 2100-2400 Å are absent. The two features present in this region are the C_2 Mulliken bands mentioned above and H I Lyman-α in second order. The absence of the CO⁺ first negative bands so prominent in the spectrum of Comet West is consistent with the absence of weakness of the CO⁺ comet-tail bands in visible spectra of these comets. The absence of these bands is more puzzling considering the strength of the CO_2 emission at 2890 Å; this might result from the relatively small projected slit of the IUE spectrographs (A'Hearn and Feldman 1980). Note that in the spectra of all these comets, both H I Lyman α at 1216 Å and the OH (0,0) band near 3085 Å are more intense than any other spectral feature, partly due to the large abundance of H and OH in the coma and partly to the large resonance scattering fluorescence efficiencies or g-factors for these transitions.

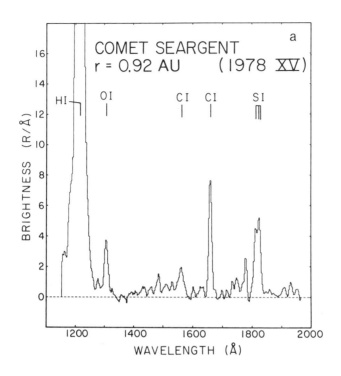
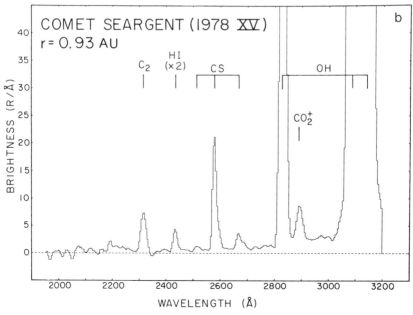

Fig. 4. IUE spectra of Comet Seargent 1978 XV. (a) and (b) are short and long wavelength spectra, respectively.

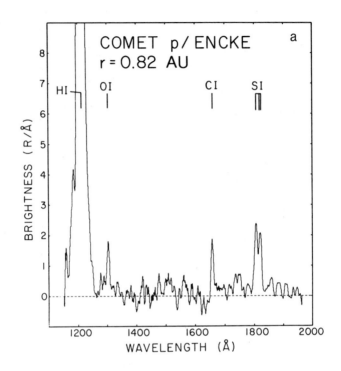
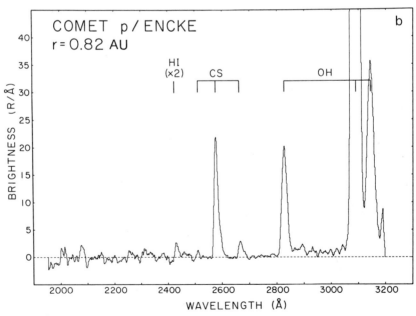

Fig. 5. IUE spectra of P/Encke. (a) and (b) are short and long wavelength spectra, respectively.

TABLE II

Observational Parameters for Figs. 3, 4, and 5

	Comet Bradfield 1979 X	Comet Seargent 1978 XV	Comet P/Encke
Observation Date	10 Jan. 1980	19 Oct. 1978	4 Nov. 1980
Heliocentric Distance (r in AU)	0.71	0.93	0.81
Geocentric Distance (Δ in AU)	0.615	0.76	0.32
Projected Slit Area (km^2)	4500 × 8900	5500 × 11000	2300 × 4600
Spatial Resolution (km)	2200	2800	1200

II. GAS PRODUCTION RATE

In addition to determining which species are present in the coma, a main goal of ultraviolet spectroscopy is to determine the rates at which these species are produced and their variation in response to changes in the solar stimulus. The method of finding the production rate from either surface brightness or total luminosity in a given spectral line is outlined briefly below.

It is generally accepted that almost all visible and ultraviolet emissions of a cometary coma are excited by resonance scattering or fluorescence of solar radiation. For an optically thin emission, the luminosity of the comet in this line or band is proportional to the total number of atoms or molecules in the coma, and the factor of proportionality is known as the g-factor usually expressed as the probability of scattering of a solar photon per unit time per molecule. The g-factor depends on the transition oscillator strength and solar flux. Note that in the ultraviolet below ~1700 Å the solar radiation is mainly line emission, so the Doppler shift due to the comet's heliocentric motion can produce large changes in the g-factor as the comet procedes in its orbit (Feldman et al. 1976). This is particularly important for the O I λ1304 emission, where for $\dot{r} \geq 30$ km s^{-1}, the cometary absorption wavelength is Doppler-shifted outside the solar linewidth and excitation of the oxygen resonance triplet procedes via fluorescent absorption of solar Lyman-β radiation. Figure 6 shows the g-factors for both resonance scattering and fluorescent excitation based on recent measurements of the solar O I λ1302 and Lyman-β fluxes and line shapes; both parameters are subject to variation during the course of a solar cycle (e.g., Mount et al. 1980). The C I λ1657 shows a much

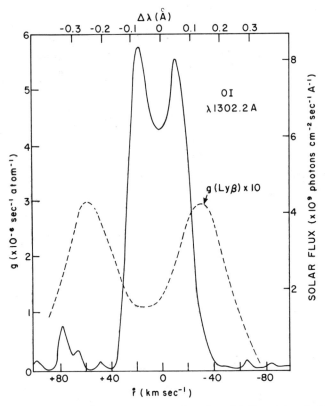

Fig. 6. The g-factor for O I λ1302 shown as a function of heliocentric velocity, with contributions from resonance scattering of solar O I λ1302 and fluorescent excitation by solar HI Lyman β (Feldman et al. 1976).

less drastic effect as the multiplet consists of six separated lines (Feldman et al. 1976). An additional effect, the Greenstein effect, (Greenstein 1958), arises from the differential velocities of atoms on the sunward and antisunward sides of the coma such that the effective g-factor varies across the cometary image. Evidence for such an effect in the O I λ1304 triplet in Comet Bradfield, where $\dot{r}$ shifted the solar line to the steep slope of the solar line profile, has been given by Weaver et al. (1981a).

At wavelengths ≥ 1700 Å where the solar spectrum is continuous with Fraunhofer absorption lines, variation in $\dot{r}$ produces the Swings effect, for example, in the OH(0,0) band at 3085 Å (Mies 1974). This can be seen from a comparison of IUE high dispersion spectra for comets Seargent ($\dot{r} \approx 35$ km s^{-1}) and Bradfield ($\dot{r} \approx 24$ km s^{-1}) (Schleicher and A'Hearn 1981), where the relative intensities of individual lines vary depending on coincidence of the absorption lines with Doppler-shifted absorption features in the solar spectrum. As a consequence, the g-factor for the band taken as a whole (as would be observed at low dispersion) also varies by almost a factor of five with $\dot{r}$, but not in a regular way, and must be accounted for in derivation of gas production rates from observed brightness as described below. There are two other interesting consequences of this effect: the fluorescent pumping of the OH ground-state levels so that the 18 cm transitions can appear either in absorption or emission depending on $\dot{r}$ (Despois et al. 1981), and a variation in the OH lifetime with $\dot{r}$ (Jackson 1980) since the principal photodissociation of OH is through the strongly predissociating v′ ⩾ 2 levels of the $A^1\Pi$ upper state. As yet, the Swings effect on the CS (0,0) band, the S I triplet, the CO^+ first negative bands, and the CO_2^+ bands at 2890 Å remains to be investigated. Also, the C II doublet at 1335 Å in Comet West observed by Feldman and Brune (1976) is not satisfactorily understood since resonance scattering by C^+ ions in the coma or tail does not seem possible due to the large heliocentric Doppler shift at the time of observation.

Thus, if the total flux F_i in a line of the ith species is measured, the production Q_i is given by

$$Q_i = \frac{4\pi\Delta^2 F_i}{g_i \tau_i} \quad (1)$$

where Δ is geocentric distance in cm, g_i is g-factor, and τ_i is lifetime of the species. Note that both g_i and τ_i depend on the solar flux which varies as r^{-2}, but that the product $g_i \tau_i$ is independent of r, and may be conveniently evaluated at 1 AU. Table III gives a list of current values of g-factors and lifetimes used in the interpretation of the observations. As noted above, the g-factors can vary strongly with heliocentric velocity and are also dependent on the solar flux. The largest uncertainty in the application of Eq. (1) is in the species lifetime, which is also a function of the solar cycle (Oppenheimer and Downey 1980).

For most cometary observations, particularly in the ultraviolet where the species lifetimes give scale lengths $\gtrsim 10^6$ km, the field of view is much smaller than the projected size of the coma and instead of the total flux an average surface brightness B_i is measured in Rayleighs

$$B_i = \bar{N}_i g_i \times 10^{-6} \quad (2)$$

where $\bar{N}_i$ is average column density of the species along the line of sight, which must be determined by integrating a suitable model of the density along a line of sight at a projected distance ρ from the nucleus to get $N_i(\rho)$ and then integrating over the instrumental field of view. The simplest model for species density assumes symmetrical radial outflow and exponential decay (Haser 1957, 1966), and introduces another unknown parameter, the species outflow velocity. A description of this model and its extension to the second daughter product has been given by Festou (1981). The outflow velocity has been discussed by Mendis and Ip (1976) for the parent molecules and Festou (1981) and Feldman (1978), among others, for the various atomic fragments. In principle, spatial variation of brightness with ρ is sufficient to give only the

TABLE III

Lifetimes and g-factors at 1 AU (Quiet Sun)

Species	Emission Wavelength (Å)	g-factor $g(s^{-1})$	Lifetime $\tau(s)$
H I	1216	1.4×10^{-3} (a)	2.4×10^6 (a)
O I	1302	$0.3\text{-}6 \times 10^{-6}$ (b)	1.4×10^6 (c)
C I	1657	2.5×10^{-5} (b)	1.7×10^6 (d)
S I	1813	7×10^{-5} (e)	10^6 (e)
C I (^{1}D)	1931	1.2×10^{-4} (f)	3250 (f)(g)
CO	1510	2.2×10^{-7} (f)	1.4×10^6 (d)
H_2	1608	1.6×10^{-7} (b)	9×10^6 (h)
NO	2150	7.7×10^{-6} (j)	3×10^5 (h)
CS	2575	7×10^{-4} (e)	10^5 (e)
OH	3085	$2.5\text{-}10 \times 10^{-4}$ (k)	$0.7\text{-}2.1 \times 10^5$ (l)

(a) Opal and Carruthers (1977b); (b) Feldman et al. (1976); (c) Opal and Carruthers (1977a); (d) Feldman (1978); (e) Jackson et al. (1981); (f) Feldman and Brune (1976); (g) independent of r; (h) Huebner and Carpenter (1979); (j) Cravens (1977); (k) Schleicher and A'Hearn (1981); (l) Jackson (1980).

scale length $v_i \tau_i$ but with dissociation products of H_2O simultaneous observation of H, OH, and O should permit unique determination of both the velocity of the parent and the lifetimes of the daughter products.

The radial outflow model has several limitations. It is only valid for photodestruction products and does not allow for atoms produced by chemical reactions such as carbon by dissociative recombination of CO^+ (Feldman 1978). For hydrogen and oxygen it does not properly account for the spatial distribution resulting from excess velocities of fragment atoms. Festou (1981) has developed a vectorial model of H_2O dissociation and shown that the radial outflow model gives a close approximation to the exact OH brightness profile albeit with an underestimated daughter scale length. An additional difficulty in analyzing resonance scattering of atomic hydrogen and oxygen is that in moderately active comets the column abundances of these species are sufficiently high so that radiation entrapment is significant. However, treatment of the radiative transfer problem is difficult because the exact physical conditions in the gas coma are unknown. Also, the solar extreme ultraviolet fluxes that produce both resonance scattering of the atomic fragments and photodestruction of the parent molecules may vary by a factor of 2 to 4

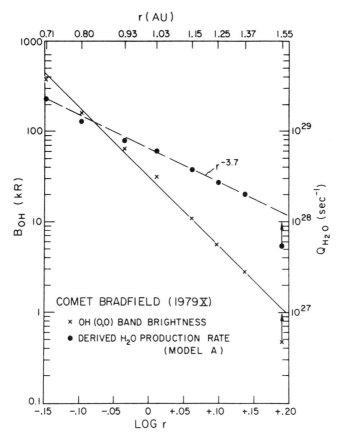

Fig. 7. Variation of the OH (0,0) band brightness and the derived H_2O production rate as a function of heliocentric distance for Comet Bradfield 1979 X (Weaver et al. 1981a).

during the solar cycle (Oppenheimer and Downey 1980). Nevertheless, in our preliminary analysis of HI Lyman-α, OI λ1302, and the OH (0,0) band the observed brightness is completely consistent with a water source (Weaver et al. 1981a).

III. EVOLUTION OF THE COMA

As noted above, the IUE observations of Comet Bradfield made over a wide range of heliocentric distance permit determination of the variation of water vaporization rate from the comet nucleus with time and distance from the Sun. Such data should prove useful for comparison with dirty ice models of the nucleus. The data, the brightness of the OH (0,0) band at 3085 Å averaged over the central 10 $\times$ 15 arcsec of the spectrograph aperture, are shown in Fig. 7 along with the OH production rate ($\approx$ 90% of the H_2O production rate) derived by fitting the data to the predictions of a radial outflow model averaged over the aperture. The model used (A in the figure) assumed H_2O outflow velocity $v_{H_2O} = 1.0$ km s^{-1} and OH lifetime $\tau_{OH} = 5 \times 10^4$ s at 1 AU. The other parameters were the water lifetime $\tau_{H_2O} = 8.2 \times 10^4$ s at 1 AU and the OH outflow velocity $v_{OH} = 1.15$ km s^{-1}. However, a different value of v_{H_2O} can also provide a good fit to the spatial variation of the OH brightness if the OH lifetime is suitably adjusted; a model with $v_{H_2O} = 0.5$ km s^{-1} and $\tau_{OH} = 1 \times 10^5$ s gives a water production rate smaller by a factor of 2 than that shown in Fig. 7. The actual value is probably between these extremes.

The surprising result from Fig. 7 is the dependence of the water production rate on heliocentric distance which varies as $\bar{r}^{-3.7}$. This disagrees with the widely held idea that the controlling influence on vaporization is the total solar energy input which varies only as r^{-2}. Previous OAO-2 results on comets Bennett and Tago-Sato-Kosaka (Keller and Lillie 1974,1978) were in basic agreement with the $\bar{r}^{-2}$ dependence, but the data covered a much narrower range of heliocentric distance. Detailed models by Delsemme (1973) of the evaporation of water from a bare nucleus give a variation of Q_{H_2O} proportional to r^{-n}, with n between 2.4 and 2.9 for a range of visible and infrared albedos. More recently, including effects of the dust coma in such models has led to prediction of an even steeper variation of water vaporization rate with heliocentric distance (P.R. Weissman and H.H. Kieffer, personal communication 1981), in qualitative agreement with the IUE results. Observations of additional comets over a similar range of r are needed to determine whether this behavior is typical of most comets.

IV. COMPARATIVE SPECTROSCOPY

The IUE observatory is capable of detecting and tracking comets whose visual magnitude is as faint as 10. Over its time in space it should permit observations of comets of different types and intrinsic magnitudes from which compositional and evolutionary trends can be studied. Since all observations are made with the same spectrographs uncertainties due to instrumental differences are eliminated. Moreover, by comparing spectra of different comets at the same heliocentric distance (1 AU, for example), the effects of other independent parameters like gas production rate and the consequent size of the collision zone in the inner coma might be detected. In the IUE spectra of Figs. 3 through 5, as well as that of P/Tuttle, the heliocentric velocity dependence of the O I λ1304 g-factor discussed above is clearly demonstrated (Weaver et al. 1981b).

As an example, the water production rates for six comets observed by IUE derived from analysis of the OH (0,0) band brightness using a consistent radial-outflow model at heliocentric distances of ~1 and 1.5 AU, are given in Table IV. Note that the range of production rates near 1 AU is relatively

TABLE IV

Water Production Rates

Comet	Observation Date	Heliocentric Distance r (AU)	Geocentric Distance Δ (AU)	Production Rate Q_{H_2O} (10^{28} s^{-1})
Seargent 1978 XV	16 Oct. 1978	0.87	0.78	33
Bradfield 1979 X	31 Jan. 1980	1.03	0.29	5.1
P/Encke	24 Oct. 1980	1.01	0.29	0.96
P/Tuttle	7 Dec. 1980	1.02	0.50	6.2
Bradfield 1979 X	3 Mar. 1980	1.55	1.45	0.5-1.0
P/Stephan-Oterma	7 Dec. 1980	1.58	0.59	3.0
Meier 1980q	7 Dec. 1980	1.52	1.89	8.5

small, and that the comets whose perihelia lie near 1.5 AU, P/Stephan-Oterma and Meier 1980q, are more active than Comet Bradfield 1979 X at that heliocentric distance. The derived values of Q_{H_2O} appear well correlated with gas production rates derived from groundbased observations of C_2 and CN (Weaver et al. 1981b). For P/Encke, the water production rate is found to be several times higher than the value derived from the OGO-5 observations of H I Lyman-α during the 1970 apparition (Bertaux et al. 1973).

The effect of the 10 × 20 arcsec IUE spectrograph apertures is particularly severe for the CS and S I emissions which appear nearly pointlike at the 5 arcsec instrumental resolution of the spectrographs. Jackson et al. (1981) have demonstrated that this spatial variation is consistent with a common parent for these two species, CS_2, with photochemical lifetime 100 s at 1 AU. A comparison of the observed profile of the CS (0,0) band along the 20 arcsec dimension of the aperture with the expected profile, using a radial outflow model smoothed by instrumental resolution, is shown in Fig. 8. For such emissions, the observed brightness averaged over the slit is strongly dependent on the comet's geocentric distance. Using the model, the CS parent production rate was evaluated for the four comets in Table IV observed near 1 AU and found to be 5 × 10^{-4} of the water production rate, to within a factor of two, for all four comets (Weaver et al. 1981b). Although the statistical sample is too small to draw a conclusion for all comets, CS_2 certainly appears to exist in the same relative amount in the ice of these comets.

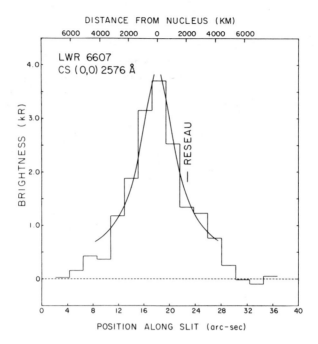

Fig. 8. Variation of the CS (0,0) band brightness in the large aperture of the IUE spectrograph. The smooth lines indicate expected response if the CS was produced by photodissociation of a parent molecule whose lifetime at 1 AU is ∼ 100 s (Jackson et al. 1981).

V. LYMAN-α OBSERVATIONS

An extensive review of the H I Lyman-α observations of comets Bennett 1970 II and Kohoutek 1973 XII and their interpretation was given by Keller (1976). The syndyname model of Keller and Meier (1976) assuming two outflow velocity components (depending on whether the H parent is H_2O or OH) gave an excellent fit to the extensive wide-field Lyman-α images of Comet Kohoutek (Meier et al. 1976). The only other comet for which a direct Lyman-α image was obtained was Comet West (Opal and Carruthers 1977b), but in this case the syndyname model was unable to reproduce the observed isophotes; the most likely source of the discrepancy was conjectured to be irregular gas production related to splitting of the nucleus (Keller and Meier 1980). Despite this problem, the presence of hydrogen in the coma with two different velocity distributions seems well established by the Copernicus observations of the Lyman-α line shape at different positions in the coma of Comet Kobayashi-Berger-Milon 1975 IX (Festou et al. 1979). Unfortunately, the IUE observations are not applicable to this problem.

VI. FUTURE DIRECTIONS

The cometary coma appears quite different in the vacuum ultraviolet from its visual image, both in form and photometric content. Only a handful of comets have been observed from above the Earth's atmosphere and the detailed information on numerous comets available to groundbased observers does not exist for the ultraviolet. However, results to date have clearly demonstrated the importance of ultraviolet observations to understanding cometary phenomena. IUE continues to operate well and will provide an enlarged data base as more comets are observed; hopefully a comet similar to comets Bennett or West will appear during its lifetime. The Space Telescope (ST) to be launched in 1985 will continue to provide ultraviolet spectroscopy and imaging from Earth orbit, although its use in cometary observations will be limited by the small fields of view of the spectrographs and the expected heavy demand for ST observing time. Space Shuttle and Spacelab will provide an additional platform for cometary ultraviolet observations; several instruments now being developed should be available for flight during the 1985-86 apparition of Comet P/Halley. Among these is a 90-cm telescope/spectrograph currently in the definition study phase at Johns Hopkins University which will be sensitive at wavelengths as short as 500 Å, permitting the detection of He I λ584 if helium is present in the coma. Improved instrumentation with higher sensitivity and spectral and spatial resolution should make possible the study of collision processes in the inner coma. The detection of the Lyman-α line of deuterium displaced 0.3 Å from H I Lyman α and currently limited by instrumentally scattered H I Lyman α remains the ultimate challenge to the cometary spectroscopist.

Acknowledgments: The author acknowledges fruitful collaboration on many of the topics discussed above with H.A. Weaver, M.C. Festou and M.F. A'Hearn. The IUE observatory staff have been extremely generous in accommodating their observing schedule to the random apparitions of comets during 1980 and in providing the expertise to track these fast-moving objects. This work was supported by grants from the National Aeronautics and Space Administration.

REFERENCES

A'Hearn, M.F., and Feldman, P.D. 1980. Carbon in Comet Bradfield 1979ℓ. *Astrophys. J.* 242:L187-L190.

Bertaux, J.L.; Blamont, J.E.; and Festou, M. 1973. Interpretation of hydrogen Lyman-alpha observations of Comets Bennett and Encke. *Astron. Astrophys.* 25:415-430.

Carruthers, G.R.; Opal, C.B.; Page, T.L.; Meier, R.R.; and Prinz, D.K. 1974. Lyman-α imagery of Comet Kohoutek. *Icarus* 23:526-537.

Code, A.D.; Houck, T.E.; and Lillie, C.F. 1972. Ultraviolet observations of comets. In *The Scientific Results from Orbiting Astronomical Observatory (OAO-2)*, ed. A.D. Code, (Washington, D.C.: NASA SP-310), pp. 109-114.

Cravens, T.E. 1977. Nitric oxide gamma band emission rate factor. *Planet. Space Sci.* 25:369-372.
Delsemme, A.H. 1973. The brightness law of comets. *Astrophys. Letters* 14:163-167.
Despois, D.; Gerard, E.; Crovisier, J.; and Kazès, I. 1981. The OH radical in comets: Observation and analysis of the hyperfine microwave transitions at 1667 MHz and 1665 MHz. *Astron. Astrophys.* 99:320-340.
Feldman, P.D. 1978. A model of carbon production in a cometary coma. *Astron. Astrophys.* 70:547-553.
Feldman, P.D., and Brune, W.H. 1976. Carbon production in Comet West (1975n). *Astrophys. J.* 209:L145-L148.
Feldman, P.D.; Opal, C.B.; Meier, R.R.; and Nicolas, K.R. 1976. Far ultraviolet excitation processes in comets. In *The Study of Comets*, eds. B. Donn, M. Mumma, W. Jackson, M. A'Hearn, and R. Harrington, (Washington, D.C.: NASA SP-393), pp. 773-795.
Feldman, P.D.; Takacs, P.Z.; Fastie, W.G.; and Donn, B. 1974. Rocket ultraviolet spectrophotometry of Comet Kohoutek (1973f). *Science* 185:705-707.
Feldman, P.D.; Weaver, H.A.; Festou, M.C.; A'Hearn, M.F.; Jackson, W.M.; Donn, B.; Rahe, J.; Smith, A.M.; and Benvenuti, P. 1980. IUE observations of the UV spectrum of Comet Bradfield. *Nature* 286:132-135.
Festou, M. 1981. The density distribution of neutral compounds in cometary atmospheres. I. Models and equations. *Astron. Astrophys.* 95:69-79.
Festou, M.; Jenkins, E.B.; Keller, H.U.; Barker, E.S.; Bertaux, J.L.; Drake, J.F.; and Upson, W.L. 1979. Lyman-alpha observations of Comet Kobayashi-Berger-Milon (1975 IX) with *Copernicus*. *Astrophys. J.* 232:318-328.
Greenstein, J.L. 1958. High-resolution spectra of Comet Mrkos (1957d). *Astrophys. J.* 128:106-113.
Haser, L. 1957. Distribution d'intensite dans la tête d'une comète. *Bull. Acad. Roy. Belgique,* Classe des Sciences 43:740-750.
Haser, L. 1966. Calcui de distribution d'intensité relative dans une tête comètaire. *Cong. Coll. Univ. Liège* 37:233-241.
Huebner, W.F., and Carpenter, C.W. 1979. Solar photo rate coefficients. *Los Alamos Report* LA-8085-MS.
Jackson, W.M. 1980. The lifetime of the OH radical in comets at 1 AU. *Icarus* 41:147-152.
Jackson, W.M.; Halpern, J.; Feldman, P.D.; and Rahe, J. 1981. Production of CS and S in Comet Bradfield (1979 X). Submitted to *Astron. Astrophys.*
Jackson, W.M.; Rahe, J.; Donn, B.; Smith, A.M.; Keller, H.U.; Benvenuti, P.; Delsemme, A.H.; and Owen, T. 1979. The ultraviolet spectrum of Comet Seargent 1978m. *Astron. Astrophys.* 73:L7-L9.
Keller, H.U. 1976. The interpretation of ultraviolet observations of comets. *Space Sci. Rev.* 18:641-684.
Keller, H.U., and Lillie, C.F. 1974. The scale length of OH and the production rates of H and OH in Comet Bennett (1970 II). *Astron. Astrophys.* 34:187-196.
Keller, H.U., and Lillie, C.F. 1978. Hydrogen and hydroxyl production rates of Comet Tago-Sato-Kosaka (1969 IX). *Astron. Astrophys.* 62:143-147.
Keller, H.U., and Meier, R.R. 1976. A cometary hydrogen model for arbitrary observational geometry. *Astron. Astrophys.* 52:272-281.
Keller, H.U., and Meier, R.R. 1980. On the Lα isophotes of Comet West (1976 VI). *Astron. Astrophys.* 81:210-214.
Meier, R.R.; Opal, C.B.; Keller, H.U.; Page, T.L.; and Carruthers, G.R. 1976. Hydrogen production rates from Lyman-α images of Comet Kohoutek (1973 XII). *Astron. Astrophys.* 52:283-290.
Mendis, D.A., and Ip, W.-H. 1976. The neutral atmospheres of comets. *Astrophys. Space Sci.* 39:335-385.
Mies, F.H. 1974. Ultraviolet fluorescent pumping of OH 18-centimeter radiation in comets. *Astrophys. J.* 191:L145-L148.
Mount, G.H.; Rottman, G.J.; and Timothy, J.G. 1980. The solar spectral irradiance 1200-2550 Å at solar maximum. *J. Geophys. Res.* 85:4271-4274.

Opal, C.B., and Carruthers, G.R. 1977a. Carbon and oxygen production rates for Comet Kohoutek (1973 XII). *Astrophys. J.* 211:294-299.

Opal, C.B., and Carruthers, G.R. 1977b. Lyman-alpha observations of Comet West (1975n). *Icarus* 31:503-509.

Opal, C.B.; Carruthers, G.R.; Prinz, D.K.; and Meier, R.R. 1974. Comet Kohoutek: Ultraviolet images and spectrograms. *Science* 185:702-705.

Oppenheimer, M., and Downey, C.J. 1980. The effect of solar-cycle ultraviolet flux variations on cometary gas. *Astrophys. J.* 241:L123-L127.

Schleicher, D.G., and A'Hearn, M.F. 1981. OH fluorescence in comets. Submitted to *Astrophys. J.*

Smith, A.M.; Stecher, T.P.; and Casswell, L. 1980. Production of carbon, sulfur and CS in Comet West. *Astrophys. J.* 242:402-410.

Weaver, H.A.; Feldman, P.D.; Festou, M.C.; and A'Hearn, M.F. 1981a. Water production models for Comet Bradfield (1979 X). *Astrophys. J.* In press.

Weaver, H.A.; Feldman, P.D.; Festou, M.C.; A'Hearn, M.F.; and Keller, H.U. 1981b. Observations of faint comets with IUE. Submitted to *Icarus*.

Whipple, F.L. 1950. A comet model. I. The acceleration of Comet Encke. *Astrophys. J.* 111:375-394.

Whipple, F.L. 1951. A comet model. II. Physical relations for comets and meteors. *Astrophys. J.* 113:464-474.

LABORATORY STUDIES OF PHOTOCHEMICAL AND SPECTROSCOPIC PHENOMENA RELATED TO COMETS

WILLIAM M. JACKSON
Howard University

The relationship between laboratory observations and cometary phenomena is reviewed, with particular attention to how laboratory studies must be done to obtain all data needed for astronomical purposes. It is pointed out how supersonic expansion of the gas evaporating from the nucleus can affect the photochemical lifetime and the energy partitioning among the fragments. The use of the CS radical as a remote sensing probe of conditions in the inner coma is also discussed.

Spectroscopy and astrophysics have been intimately connected over the years Most detailed astrophysical data comes from spectroscopic or visual investigation of the astronomical body, and even in the latter case spectroscopy is generally needed to interpret the observed information. This is particularly true for comets because all we know about them depends on interpretation of observed cometary emissions; detailed spectroscopic information is continually needed to interpret these emissions. An illustration of the synergistic effect between observations and laboratory studies is the identification of H_2O^+ in comets (Benvenuti and Wurm 1974). The emissions ascribed to this species had been observed in other comets but were unassignable because detailed spectroscopic information did not exist. Only because Herzberg and Lew (1974) had observed and analyzed H_2O^+ spectra in the laboratory was it possible to assign the observations, providing important evidence indicating water as the major constituent in comets.

Astronomers generally recognize the connection of comets with spectroscopy but are only beginning to appreciate the connection with photochemistry. Even Wurm (1965), who first postulated that photodissociation was responsible for the radicals, ions, and atoms observed in comets, recanted his original theory. It is generally accepted that species such as H, OH, S, CN, O (^{1}D), O (^{3}P), and CS are produced by photodissociation of a suitable parent molecule. Other species such as the ions may be produced by chemical reactions of the initial ions that result from photoionization of some parent molecules. It is therefore important to understand the details of the photochemical process so that comprehensive models of comets can be developed for comparison with observations.

In this chapter, I will review the spectroscopic and photochemical information as it relates to comets. Since this book contains detailed reviews on infrared observations, I will concentrate on laboratory information related to visible and ultraviolet measurements. In these spectral regimes we are concerned with electronic transitions of fragments and ions produced from some parent molecule. The reasons that certain information is needed to interpret observed cometary phenomena are emphasized.

I. SPECTROSCOPIC MEASUREMENTS

The principal molecular emissions observed in comets are summarized in Table I. Other than CO emission, all of the observed species are neutral or ionized products from some neutral parent compound. A principal piece of information needed to interpret cometary observations is the radiative lifetime of the observed emission, to determine the probability for a particular transition and thus estimate the total amount of material responsible for the observed radiation. This is essential for any comprehensive cometary model.

As Table I shows, detailed information exists for most observed molecular emissions. The references quoted represent the most recent reports of experimental determination of particular lifetimes and contain detailed discussion of the errors involved in these determinations. The table's column for comments and the following explanation illustrate some current difficulties.

First, it is often not appreciated that the highest resolution should be used to measure the lifetime of molecular emissions for greatest accuracy. High resolution ensures that the experimenter is isolating the molecular band of interest and prevents contamination of the observations by emissions from other species. Also, most cometary emissions for heteronuclear molecules arise from the lowest rotational levels of an excited electronic state. These levels are coupled to the higher rotational levels of the lower electronic state by the selection rule $J = 0, \pm 1$. If these upper levels are perturbed or predissociated by another electronic state, profound changes in the lifetime of the state can occur as the rotational energy is changed. It is essential that

TABLE I
Molecular Emissions Observed in Comets

Molecule	Transition	v'	Lifetime (ns)	Comments	References
C_2	$X^1\Sigma_g^+ - D^1\Sigma_u^+$ Mulliken bands	(0–3)	18.1	3–0→0	Curtis et al. 1976
	$a^3\Pi_u \leftarrow d^3\Pi_g$ Swan system	0–6	123–140	low resolution depends on v' and has not been done for $J < 7$	,,
CH	$X^2\Pi \leftarrow A^2\Delta$	0	543		Brzozowski et al. 1976
		1	520		,,
		2	355	predissociated	,,
		2	360	predissociated	,,
CN	$X^2\Sigma \leftarrow B^2\Sigma$	0	64–66	depends on N'	Jackson 1974b
			71.3	perturbed level	,,
	$X^2\Sigma^+ \leftarrow A^2\Pi$	0	8400–6400		Conley et al. 1980
		1–10	3800–4500	depends on v'	Duric et al. 1978
CO	$X^1\Sigma^+ \leftarrow A^2\Pi$	10.9			Provorov et al. 1977
		19.6		distorted by mixing	,,
CS	$X^1\Sigma \leftarrow A^1\Pi$	0	230		Carlson et al. 1979
		1	285		,,
OH	$X^2\Pi \leftarrow A^2\Sigma$	0	740	$J' < 29$	Brzozowski et al. 1978
		0	100	$J' = 100$	,,
		1	780	$J' < 19$	,,
		1	110	$J' = 19$	,,
		2	150	$J' = 0$	,,

Molecule	Transition	v'	Lifetime (ns)	Comment	Reference
NH	$X^3\Sigma^- \leftarrow A^3\Pi$	0	404	$J' < 31$	Smith et al. 1976
		0	96	$J' = 31$	"
		1	413	$J' < 24$	"
		1	41	$J' = 24$	"
CO$^+$	$b^1\Sigma^+ \rightarrow X^3\Sigma^-$		230,000	reflects variation of $Re(v')/Re$	Masanet et al. 1978
	$X^2\Sigma^+ \leftarrow A^2\Pi$	0	3820		Holland and Maier 1972
CH$^+$	$X^1\Sigma^+ \leftarrow A^1\Pi$	0	815	laser-induced fluorescence with an ion trap	Grieman et al. 1981
H$_2$O$^+$	$\widetilde{X}^2B_1 \widetilde{A}^2A_1$		10,500		Mohlmann et al. 1978
OH$^+$	$X^3\Sigma^- \leftarrow A^3\Pi_i$		2500		"
N$_2^+$	$X^2\Sigma_g^+ \leftarrow B^2\Sigma_u^+$		62.4	$v' = 0$, $15 \leq N^1$	Erman and Larsson 1979
			65	$v'+1$, $10 \leq N^1$	
			70	$v' = z$	
			75	$v' = 3$	
C$_3$	$\widetilde{X}^1\Sigma_g^+ \leftarrow A^1\Pi_u$		183	Various vibrational levels	Becker et al. 1979
CO$_2^+$	$X^2\Pi \leftarrow A^2\Pi_u$		108.3		Smith et al. 1975
	$X^2\Pi \, B^2\Sigma^+$		117.4		
NH$_2$	$\widetilde{X}^2B_1 \leftarrow \widetilde{A}^2A_1$		10,000	some small variation with vibrational level	Halpern et al. 1975

TABLE II
Lifetime of CN $A^2\Pi$ State

Vibrational State	Lifetime (μs)	Reference
$A^2\Pi_{1/2}(v'=0, N=4,5)$	8.4 ± 2	Conley et al. 1980
$A^2\Pi_{3/2}(v'=0, N=1,2,3)$	6.2 ± 0.9	
$A^2\Pi (v'=1)$	7.293	Jeunehomme 1965
Average over $v'=1-3$	3.5 ± 0.4	Wentink et al. 1964

lifetime measurements are made for rotational levels actually observed in comets.

The C_2 radical has been an enigma in comets for years; no satisfactory mechanism has been proposed for its formation. The recent observation of the Mulliken bands which originate from the ground singlet state permits an estimate of the relative population of the singlet and triplet levels of C_2. The results (A'Hearn and Feldman 1981) suggest that the column density of the singlet state is 4 to 7 times smaller than that of the triplet state, if their destruction rates are the same. Theoretical calculations (Kirby and Liu 1979) indicate that the upper state of the Mulliken system may be predissociated by a repulsive state. This will make the destruction rate of the singlet state higher than that of the triplet state and would require the ratio of column densities to be reevaluated. The C_2 radical is a homonuclear diatomic molecule; in comets this means that the upper rotational levels of the ground state are populated by radiative pumping. It is therefore important to measure the radiative lifetime of the Mulliken bands as a function of these upper rotational and vibrational levels when this predissociation is searched for.

The CN radical is another molecule for which complete information is unavailable about the radiative lifetime of an excited electronic level. While there have been several measurements of the radiative lifetime of the $A^2\Pi$ state, only one paper (Conley et al. 1980) has appeared on the lifetime of the individual rotational levels of the lowest vibrational state $v'' = 0$. This work, summarized in Table II, indicates that the lifetime may be a function of the spin components of the doublet state.

Finally, the most recent measurements for the lifetime of the various molecular emissions were determined using a high-voltage deflection technique. These measurements may not provide the best method to determine the lifetime of the ions since the measuring technique can sweep the ions out of the viewing zone. This was recently demonstrated (Grieman et al. 1981) in the case of the CH^+ ion, in which laser-induced fluorescence measurements in an ion trap gave a much longer lifetime than the results obtained by the high voltage deflection (HVDFL) technique. Careful reexamination of the experimental technique for ionic species may be necessary in light of the lack of a suitable production mechanism to explain their formation in comets.

In summary, the spectroscopy of most of the identified molecular emissions in comets is reasonably well understood, but there are some specific problems. Spectroscopy will be tied to comets so long as there are unidentified lines in their spectra.

II. PHOTOCHEMISTRY

There are now several sources (Okabe 1978; Leone 1981) of detailed information on the photochemistry of some simple parent molecules postulated to occur in comets. Other reviews (Jackson 1972,1974a,1976) discuss laboratory photochemistry and its relationship with comets. In this section only newer information and experimental techniques will be presented. The type of detailed photochemistry data required to adequately interpret cometary data is emphasized.

Increasing observational evidence from ground, satellite, and rocket observations suggests that photodissociation of parent molecules is the principal mechanism producing most observed radicals. A'Hearn (1981) has shown that the consistent ratios of production rates of observed radicals for a large number of comets with a variety of heliocentric distances preclude explanation by chemical reactions of the observed species in comets. More detailed observational data on comets impel the laboratory scientist to supply more detailed information about the photochemical processes to interpret cometary observations. This is a formidable task because the conditions in comets require understanding all aspects of photochemistry. The Sun's radiation is a broadband light source, so the wavelength dependence of the various primary photochemical processes must be determined. This is illustrated by the recent study of the photolysis of H_2O at 121.6 nm. Previous broadband flash photolysis results measured between 105.0 and 145.0 nm had suggested that the branching ratio between the formation of atomic and molecular hydrogen was 9/1 (Stief et al. 1975). The latest result (T.G. Slanger, personal communication 1981) on the direct photolysis of H_2O at 121.6 nm gives this ratio as 0.82, a difference of a factor of 10.

It has been shown that the wavelength region that must be studied will vary with the molecule under consideration (Jackson 1972). In general, if a molecule can be dissociated at wavelengths longer than 150 nm with an appreciable cross section, then this will be the most important photochemical region because of the large increase in solar output at longer wavelengths. If the molecule can only absorb photons below 150 nm, then most dissociation will occur at Lyman α.

The dynamics of the photodissociation process, i.e., how the excess pho-

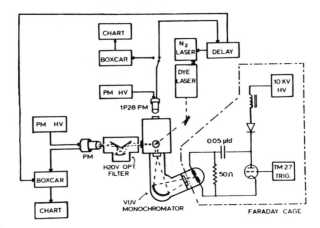

Fig. 1. Tunable vacuum ultraviolet flash lamp apparatus with a tunable dye laser as a detector.

tochemical energy is distributed among the various degrees of freedom of the products, must also be measured as a function of wavelength. This information determines the translational recoil velocity of a given photochemical fragment which in turn determines the shape of the coma. To obtain dynamical information, the photochemical study must be performed at pressures low enough so that collisions do not change the quantum state distribution of the fragment before it is detected. This requires a pulsed light source with good time resolution and a pulsed fluorescence excitation source. Since a pulsed light source with FWHM $<$ 1 microsec emits little light, and the pressure of the parent molecule is low, only a few photochemical products will be formed. To measure quantum state distribution, a sensitive detection method must be available. Such a method has been developed (Miller et al. 1978) which relies on the high detection sensitivity of fluorescence induced by a pulsed tunable dye laser. The apparatus designed for this purpose is shown in Fig. 1. It consists of a special light source with a vacuum ultraviolet (VUV) flash lamp and a monochromator which places the broadband light source at the focal point of a corrected concave holographic grating. With this light source photochemical dynamics can be determined as a function of photolysis wavelength.

The dye laser crosses the light beam which exits from the monochromator along the slit axis. The fluorescence excited as the dye laser is tuned through a transition is collected by a large lens with a filter behind it and detected with a photomultiplier tube. The signal generated is then processed electronically. An example of a spectrum obtained at a particular VUV wavelength is shown in Fig. 2; it shows that CN radicals produced in the X state from the photolysis of C_2N_2 at 158 nm are rotationally excited. Similar spectra of the A state fragment are shown in Fig. 3. These radicals are also

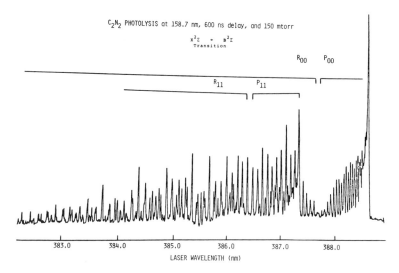

Fig. 2. Laser induced fluorescence spectrum of the CN radical in the X state, taken under collisionless conditions.

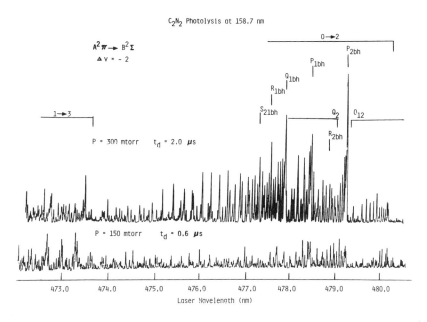

Fig. 3. Laser induced fluorescence spectra of the A state of CN taken at long and short delays. The short delays show no band head, indicating that the radical is rotationally hot when produced.

rotationally hot when produced. Analysis of this kind of data can determine the dynamics of the photodissociation process. The method can also be used to obtain branching ratios for production of electronically excited and ground-state radicals as a function of wavelength of the excited state.

All radicals and ions observed in comets emit resonance fluorescence, so the apparatus described should provide the information needed to analyze cometary observations. But it presently seems that even this apparatus may not supply all the needed information; not all cometary radicals can be explained as a simple photodissociation product of a parent molecule. It has been suggested (Jackson 1976) that the C_2, C_3 and NH radicals are produced by the following scheme: parent + $h\nu$ → (daughter + $h\nu$) → granddaughter. The analysis of some recent cometary data supports this suggestion (Yamamato 1980). In the above scheme, the daughter radical is generally an unstable intermediate, which makes it more difficult to determine the photodissociation dynamics since a suitable source must now be found to generate unstable species in concentrations high enough to be dissociated with the tunable VUV flash lamp. Further, we do not even know in which region of the spectra to attempt to photodissociate most of the likely daughter radicals since their complete absorption spectra is not determined. However, theoretical calculations (R. Saxon, personal communication 1980) are underway to determine energy levels of some of the most likely daughter molecules which could serve as parents for the granddaughter radicals. The new availability of eximer lasers may yield concentrations of daughter radicals high enough to be studied.

Another problem with directly determining the dynamics of photodissociation of the parent molecules responsible for forming cometary radicals is that some possible parents have very weak absorption lines in the regions of interest. If the pressure is kept low enough to determine the reaction dynamics, there are not enough radicals produced to be detected by the dye laser. We are currently designing an apparatus with a light source of higher output per unit bandwidth.

When these problems are overcome, we must still determine the effect of temperature on the dynamics of photodissociation. Seldom are cometary molecules at the same temperature as the ambient temperature of most laboratory measurements. Evaporation from the icy nucleus of the comet keeps the temperature in the 100 to 200 K range, > 100 K below the temperature of most laboratory measurements. As the flow velocity of the gas increases by expansion from subsonic near the nucleus to supersonic at 10 to 100 comet radii (R_c), the temperature is reduced even more by conversion of random and internal motions of the gas into directed motion; see Fig. 4. This latter effect has recently been used in the laboratory to simplify the spectra of complex molecules. Figure 5 is a laboratory illustration of how the spectra can change as the temperature of the molecules is lowered. Ip and Mendis (1976) have shown that the supersonic flow field in comets will be estab-

LABORATORY STUDIES OF PHOTOCHEMISTRY 489

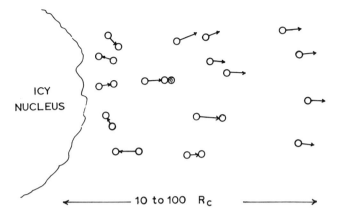

Fig. 4. Schematic diagram of flow in comets. Near the nucleus there are collisions, but within 10 to 100 comet radii (R_c) the flow becomes supersonic.

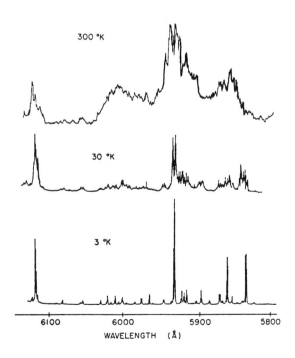

Fig. 5. Laboratory spectra taken from the fluorescence of NO_2 under supersonic flow conditions.

lished within 10 to 100 R_c. Few molecules will have undergone photodissociation within this time, so the recoil energy of photodissociation fragments produced in this region cannot be expected to heat the coma.

There are other sources that might heat the gas in this inner collision region and prevent the parent molecules from being cooled by supersonic expansion. The photochemical fragments produced 10^3 to 10^5 km from the nucleus cannot, however, reach this inner collision region. This region intercepts only a small solid angle compared to the total solid angle available to the recoiling fragments. For example, for a comet with a radius 5 km and a collision region 500 km in diameter, the maximum solid angle subtended by the collision region for a parent molecule with mean scale length 10^3 km is 6%. Only 6% of the molecules formed at this distance of the nucleus can ever intercept the inner coma region. The solid angle subtended by the collision region will also decrease as the radius of the nucleus decreases and as the parent's scale length increases. Therefore the energy of photodissociation cannot heat the parent molecules in the inner coma region.

There are further objections to heating of the gas in the inner collision region by collisions with photochemical fragments. Once the flow is established in the inner region, the parent molecules move with velocity on the order of 1 to 1.5 km s^{-1}. If the recoil velocity is not substantially higher than this, the forward motion of the parent molecules will carry the recoil fragments along with them. Molecules produced with a fragment recoil energy of 1 km s^{-1} will all be scattered forward merely by the motion of the parent molecule; this will reduce even further the number of molecules that can be scattered back into the inner coma region. This also illustrates why photodissociation dynamics is important in modeling the chemistry of the inner coma; without a knowledge of dynamics one cannot determine the effects of photochemical recoil energy on the rest of the molecules in the coma.

Considerable dust is released from the nucleus of the comet as it approaches the Sun. If the dust is silicate or another refractory material, it will be quickly heated by solar radiation to temperatures higher than that of the icy nucleus, resulting in hot dust particles moving with cool gas. These particles could supply heat to the gas and thus prevent supersonic expansion from cooling of parent molecules. The problem with this mechanism is that eventually, within a few hundred km from the surface of the nucleus, the dust will flow along the same streamlines as the gas. The hot particles will lose their random velocity component, so heating the gas by the dust starts the supersonic expansion at a temperature higher than if an additional heat source did not exist. It will not prevent the streamlines from developing nor provide an appreciable random component to the gas molecules. We could expect that the dust heating the gas might increase the terminal velocity but not prevent the streaming that naturally occurs as the gas flows out from the nucleus. This streaming cools the gas and prevents random collisions between gas molecules.

The parent molecules streaming from the icy nucleus will be further cooled due to the long time between collisions. Even without supersonic expansion, a gas will be cooled at the low densities in comets because the frequency of collisions will be unable to support a temperature. At 100 R_c the average time between collisions will be 1 to 100 millisec, comparable to the time for radiation decay between rotational and vibrational levels. The combination of supersonic expansion and radiative cooling makes it likely that parent molecules in comets will have fairly low internal temperature before photodissociation.

What are the possible consequences of internal cooling of the molecules before dissociation? There is good evidence that dynamics and electronic branching ratios may change with internal temperature. Photodissociation cross sections (Zittle and Little 1980) can change by almost two orders of magnitude. The energy may also be a function of internal excitation of the parent molecule. The latter effect is expected to be more prominent in molecules with fundamental bending frequencies of only a few hundred cm^{-1}.

If the molecules are cooled by expansion to the lowest rotational levels, the photochemical lifetimes will have to be redetermined. There are several reasons for this. First, simplification of the absorption spectra will cause shifting of the most important wavelength region responsible for photodissociation. Second, the absorption cross section will be much higher since the transition moment is associated with only one level, not spread over many. This in turn will reduce the photochemical lifetime. On the other hand, the absorption line width will be reduced so the number of photons which can be absorbed will decrease, increasing the photochemical lifetime. Finally, as the absorption spectra is simplified, when the molecule is cooled the photodissociation lifetime becomes much more susceptible to the heliocentric velocity of the comet. It is clear from Fig. 5 that the spectra of a cool molecule is more linelike, similar to a diatomic molecule. The solar radiation is also linelike, and both of these effects can combine so that the photodissociation lifetime exhibits a Swings effect.

Rocket and satellite observations (Jackson et al. 1979; Smith et al. 1980) of comets have identified both CS and S emissions. While not the most prominent emissions, the CS and S emissions are the most spatially concentrated. Both emissions occur very near the nucleus of the comet (Feldman et al. 1980), so the parent molecule of CS and S must have an extremely short lifetime. An ideal candidate is the CS_2 molecule with a photochemical lifetime of 100 s, an order of magnitude shorter than any other postulated parent compound. It can also explain the observation of both cometary species. Using the computed photochemical lifetime of the CS_2 molecule in a Haser model (Jackson et al. 1981) the observed spatial profiles of CS and S have been fitted; CS_2 is almost certainly the parent of both these species.

Given that CS_2 is the most likely parent of CS, its detailed photochemistry is important since CS_2 should provide an ideal probe of the conditions

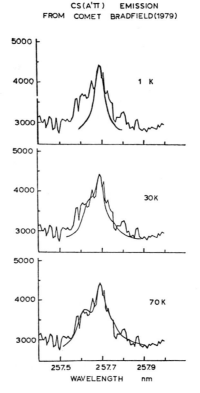

Fig. 6. A fit of the CS high-resolution spectra of Comet Bradfield 1979 X, with best fit at 70 K.

around the nucleus. This radical has a permanent dipole moment, so if no collisions occur then the high-resolution spectra will be determined by a radiative transfer model. If the collisions are important, the high-resolution spectra must be fitted with a model taking this into account; a preliminary effort is illustrated by the results shown in Fig. 6, the high-resolution spectra of CS obtained on Comet Bradfield 1979 X along with computed spectra based on Boltzman temperature 1, 30, and 70 K. It is clear that the 70 K spectra provide a better fit to the observed data than those of lower temperatures. Does this temperature reflect the ambient temperature of the gas? If so, the ambient gas temperature is very low. Or does the temperature represent a balance between radiative decay of the vibrationally and rotationally excited CS radicals known to be produced in the primary photochemical process (Yang et al. 1980; Butler et al. 1980), and radiative and collisional pumping of the radical? This question will only be answered after detailed measurements of the photochemical dynamics, taking into account factors discussed above, are made. We also need better high-resolution spectra of this molecule from a variety of comets. Experiments or observations with the Space Shuttle or the Space Telescope might gain this information. Both these tasks are formidable, but solution of the problem would permit astronomers to remotely probe conditions within the inner coma by measuring high-resolution spectra of just one species.

III. SUMMARY

Most spectroscopic information needed to interpret the molecular emissions in comets is available, but care must be taken to ensure that transition probabilities are appropriate for rotational levels observed in comets. Collection of detailed photochemical information needed to interpret cometary phenomena is in an early stage. No reported study can yet be completely translated to an environment appropriate for comets. The effect of internally cooling the molecule must still be established; this could have profound implications about cometary phenomena observed.

Acknowledgment: The author gratefully acknowledges the support from NASA's Planetary and Space Science program.

REFERENCES

A'Hearn, M.F. 1981. Chemical abundances in comets. In *Comets and the Origin of Life,* ed. C. Ponnamperuma (Dordrecht, Holland: D. Reidel), pp. 53-61.

A'Hearn, M.F., and Feldman, P.D. 1981. Carbon in Comet Bradfield (1979 1). *Astrophys J.* In press.

Becker, K.H.; Tatarczyk, T.; and Radic-Peric, J. 1979. Lifetime measurements of electronically excited $C_3(^1\pi_u)$ radicals in different vibrational states. *Chem. Phys. Letts.* 60:502-506.

Benvenuti, P., and Wurm, K. 1974. Spectroscopic observations of Comet Kohoutek (1973) f). *Astron. Astrophys.* 31:121-122.

Brzozowski, J.; Bunker, P.; Elander, N.; and Erman, P. 1976. Predissociation effects in the A, B, and C states of methylidyne and the interstellar formation rate of methylidyne via inverse predissociation. *Astrophys. J.* 207:414-424.

Brzozowski, J.; Erman, P.; and Lyrra, M. 1978. Precision estimates of the predissociation rates of the OH A $^2\Sigma$ states ($v'\leqslant 2$). *Phys. Scripta.* 17:507-511.

Butler, J.D.; Drozdoski, W.S.; and McDonald, J.R. 1980. 193 nm laser dissociation of CS_2: Prompt emission from CS and internal energy distribution of CS ($X^1\Sigma^+$). *Chem. Phys.* 50:413-421.

Carlson, T.A.; Copley, J.; Duric, N.; Erman, P.; and Larsson, M. 1979. Time resolved studies of collisional transfer and radiative decay of the CS $A^2\Pi$ state. *Chem. Phys.* 42:81-87.

Conley, C.; Halpern, J.B.; Woods, J.; Vaughn, C.; and Jackson, W.M. 1980. Laser excitation of the CN $B^2\Sigma^+ \leftarrow A^2\Pi$ 0-0 and 1-0 bands. *Chem. Phys. Letts.* 73:224-227.

Curtis, L.; Engman, B.; and Erman, P. 1976. High resolution lifetime studies of the $d^3\Pi_g$, $C^1\Pi_g$ and $D^1\Sigma_u^+$ states in C_2 with applications to estimates of the solar carbon abundances. *Phys. Scripta.* 13:270-274.

Duric, N.; Erman, P.; and Larsson, M. 1978. The influence of collisional transfers and perturbations on the measured A and B state lifetimes in CN. *Phys. Scripta.* 18:39-46.

Erman, P., and Larsson, M. 1979. Time resolved studies of the interaction between the A and B states in molecular nitrogen (1+). *Phys. Scripta.* 20:582-586.

Feldman, P.D.; Weaver, H.A.; Festou, M.C.; A'Hearn, M.F.; Jackson, W.M.; Donn, B.D.; Rahe, J.; Smith, A.M.; and Benvenuti, P. 1980. IUE observations of the UV spectrum of Comet Bradfield. *Nature* 286:132-135.

Grieman, F.J.; Mahan, B.H.; O'Keefe, A.; and Winn, J.S. 1981. Laser induced fluorescence of trapped molecular ions: The $CH^+ A^1\Pi \leftarrow X^1\Sigma^+$ system. In *High Resolution Spectroscopy,* Faraday Society discussions, Vol. 71. In press.

Halpern, J.; Hancock, G.; Lenzi, M.; and Welge, K.H. 1975. Laser induced fluorescence from $NH_2(^2A_1)$. State selected radiative lifetimes and collisional de-excitation rates. *J. Chem. Phys.* 63:4808-4816.

Herzberg, G., and Lew, H. 1974. Tentative identification of the H_2O^+ ion in Comet Kohoutek. *Astron. Astrophys.* 31:123-124.

Holland, R.F., and Maier, W.B. 1972. Study of the $A \rightarrow X$ transitions in N_2^+ and CO^+. *J. Chem. Phys.* 56:5229-5246.

Ip, W.H., and Mendis, A. 1976. The structure of cometary ionospheres: 1. H_2O dominated comets. *Icarus* 28:389-400.

Jackson, W.M. 1972. Photochemistry in the atmospheres of comets. *Mol. Photochem.* 4:135-151.

Jackson, W.M. 1974a. Laboratory observations of the photochemistry of parent molecules: A review. *The Study of Comets,* eds. B. Donn, M. Mumma, W. Jackson, M. A'Hearn, and R. Harrington (Washington: NASA SP-393), pp. 670-702.

Jackson, W.M. 1974b. Laser measurements of the radiative lifetime of the $B^2\Sigma^+$ state of CN. *J. Chem. Phys.* 61:4177-4182.

Jackson, W.M. 1976. The photochemical formation of cometary radicals. *J. Photochem.* 5:107.

Jackson, W.M.; Rahe, J.; Donn, B.; Smith, A.M.; Keller, H.U.; Benvenuti, P.; Delsemme, A.H.; and Owen, T. 1979. The ultraviolet spectra of Comet Seargent 1978m. *Astron. Astrophys.* 73:L7-L9.

Jackson, W.M.; Halpern, J.B.; Feldman, P.D.; and Rahe, J. 1981. Analysis of carbon monosulfide in the UV spectrum of Comet Bradfield (1979 l). Submitted to *Astron. Astrophys.*

Jeunehomme, M. 1965. Oscillator strength of the CN red system. *Chem. Phys.* 57:4086-4088.

Kirby, K., and Liu, B. 1979. The valence states of C_2: A configuration interaction study. *J. Chem. Phys.* 70:893-900.

Leone, S.R. 1981. Photofragment dynamics. In *Advances in Chemical Physics: Dynamics of the Excited States,* ed. K. Lawley (New York: Interscience Wiley). In press.

Masanet, J.; Lalo, C.; Durand, G.; and Vermeil, C. 1978. Radiative lifetimes of the $b^1\Sigma^+$ state of NH, ND radicals. *Chem. Phys.* 33:123-130.

Miller, G.E.; Halpern, J.B.; and Jackson, W.M. 1978. Tunable VUV photofragment monochromator. *Applied Optics* 17:2821-2824.

Mohlmann, G.R.; Bhuani, K.K.; DeHeer, F.J.; and Tusrbuchi, S. 1978. Lifetimes of the vibronic A^2A_1 states of H_2O^+ and of the $^3\Pi_i(v'=0)$ state of OH^+. *Chem. Phys.* 31:273-280.

Okabe, H. 1978. *Photochemistry of Small Molecules* (New York: Wiley and Sons).

Provorov, A.C.; Stoicheff, B.P.; and Wallace, S. 1977. Fluorescence studies in CO with tunable VUV laser radiation. *J. Chem. Phys.* 67:5393-5394.

Smith, A.J.; Read, F.H.; and Imhof, R.E. 1975. Measurements of the lifetimes of ionic excited states using the inelastic electron-photon delayed coincidence technique. *J. Phys. B.* 8:2869-2879.

Smith, A.M.; Stecher, T.P.; and Casswell, L. 1980. Production of carbon, sulfur and CS in Comet West. *Astrophys. J.* 242:402-410.

Smith, W.H.; Brzozowski, J.; and Erman, P. 1976. Lifetime studies of the NH molecule: New predissociations, the dissociation energy, and interstellar diatomic recombination. *J. Chem. Phys.* 64:4628-4683.

Stief, L.; Payne, W.; and Klemm, B. 1975. A flash photolysis-resonance fluorescence study of the formation of $O(^1D)$ in the photolysis of water and the reaction of $O(^1D)$ with H_2, Ar, and He. *J. Chem. Phys.* 62:4000-4008.

Wentink, T.W. Jr.; Isaacson, L.; and Morreal, J. 1964. Radiative lifetime of the $^2\Pi$ state of the CN red system. *J. Chem. Phys.* 41:278-279.

Wurm, K. 1965. The head (or coma) of comets: Introductory report. *Nature et Origine des Comêtes,* L'Universite de Liége 37:119-128.

Yamamato, T. 1980. On the photochemical formation of CN, C_2, and C_3. Radiation in cometary comae. *Inst. Space Aeronautical Sci.,* Univ. of Tokyo, ISAS RN 133:1-24.

Yang, S.C.; Freedman, A.; Kawasaki, M.; and Bershohn, R. 1980. Energy distribution of the fragments produced by photodissociation of CS_2 at 193 nm. *J. Chem. Phys.* 72:4058-4062.

Zittle, P.F., and Little, D.D. 1980. Photodissociation of vibrationally excited ozone. *J. Chem. Phys.* 72:5900-5905.

PHOTOCHEMICAL PROCESSES IN THE INNER COMA

W.F. HUEBNER, P.T. GIGUERE, and W.L. SLATTERY
Los Alamos National Laboratory

Very little evidence exists on the physical structure and chemical composition of comet nuclei. Thus, one purpose of modeling the coma is to characterize the chemical and physical properties of the nucleus by fitting model calculations to observations. Models for inner coma chemistry are reviewed. The most advanced models consider time-dependent chemical kinetics of over 100 species with hundreds of chemical and photolytic reactions in an outstreaming, constantly diluting coma gas. Solar ultraviolet radiation and opacity of the coma, fluid dynamic flow (going over to free particle flow in the outer coma), and chemical kinetics must be linked intimately in a realistic model. The physics relevant to the coma's chemistry is summarized. Interaction of solar radiation with the coma and photolytic and chemical processes are described in the context of a chemical reaction network. Formation and destruction of a few observed species are traced through such a network as examples. Parallel reaction paths, i.e., different reactions with the same end product, are very important because: (1) The uncertainty of the final product abundance tends to be smaller than predicted from the uncertainty of the rate coefficients from any one reaction path; (2) different reaction paths dominate in different parts of the coma or at different heliocentric distances of the comet. Also this second reason indicates that it may not be possible to greatly simplify reaction networks. Model compositions assuming the origin of comets in the outer presolar nebula or in a companion fragment of the parent interstellar cloud yield abundances of C_2, C_3, and CN that agree well with observations both in heliocentric distance of the comet and distance into the coma. This is not the case (with present models) if the composition is near chemical equilibrium, as expected if the comets originated near the giant planets. Comparison of model results with observations indicate some severe restrictions on the "interstellar" composition of the nucleus. The ratio of CO^+ to H_2O^+ column density is still not in agreement with the only reduced observation. Some alternatives are discussed. A list of physical and chemical processes that will improve the models is presented.*

I. DEVELOPMENT AND PROGRESS OF CHEMICAL MODELING OF THE COMA

Knowledge about the physical structure and chemical composition of comet nuclei is essential to understanding the formation and early history of our solar system. But the nucleus, in the sense of Whipple's (1950,1951) icy conglomerate model, is too small to be resolved with groundbased telescopes. All knowledge about the nucleus is based on deductions from observations of the coma and, to a lesser degree, of the tails. Until a mission to a comet is carried out, and in order to plan such a mission, it is crucial to maximize interpretation of information from earth-based coma observations. Thus, modeling the coma has two objectives: (1) to provide insight, i.e., to understand the important processes and how they are related, and (2) to obtain numerical results. Coma modeling will be important in interpreting comet mission data.

Physical and chemical processes in the coma are intimately related to each other and the physical structure and chemical composition of the nucleus. The suggestion by Wurm (1943) that comets contain chemical compounds such as H_2O, NH_3, C_2N_2, CH_4, N_2, and CO_2 opened the possibility of chemical reactions. To explain the presence of CO and radicals such as CH, NH, and CN in the coma, Delsemme and Swings (1952) proposed that the nucleus contains solid hydrates of CH_4, CO_2, etc. These hydrates (clathrates) have approximately equal vapor pressures and so would be released into the coma at about the same rate. (The vapor pressure of pure CH_4 ice is several orders of magnitude larger than that for H_2O, so CH_4 would be released at much larger heliocentric distance than H_2O.) Donn and Urey (1956,1957) proposed that exothermic chemical reactions could explain observed brightness outbursts in comets. To obtain explosive energy release, they suggested a modification in the chemical composition: a mixture with stable molecules as well as reactive species such as free radicals and unstable molecules.

The great impetus for theoretical consideration of chemical reactions in the coma did not occur until the number densities in the coma were established as high enough. Biermann and Trefftz (1964) were the first to indicate this in their analysis of the observed emission lines from $O(^1D)$. (For additional evidence for high gas density from theoretical considerations of energy balance, dust production, and analysis of CN rotational levels, see Whipple and Huebner (1976) and references therein.) Qualitative arguments for chemical reactions between neutral radicals in the inner coma were given by Jackson and Donn (1966). Such reactions can now be shown to be important only in large comets.

With the discovery that detected species in interstellar clouds can be explained by chemical reactions involving ions, it became clear that analogous reactions should occur in the comae of comets. Aikin (1974) first applied ion gas-phase reactions to a water dominated comet. He predicted a significant production of H_3O^+, making it the most abundant ion to distances of 5000 km in the coma. He also pointed out the importance of the proton transfer reaction (a special case of positive ion-atom interchange reactions) of H_3O^+ with ammonia

$$H_3O^+ + NH_3 \rightarrow NH_4^+ + H_2O \cdot \qquad (1)$$

Abundant ammonia could shift the dominant ion in the innermost coma from H_3O^+ to NH_4^+. Neither H_3O^+ nor NH_4^+ have been identified in comets; they are important subjects for future spectroscopic or mass-spectroscopic searches. Contemporary with Aikin's work, Biermann and Dierksen (1974) invoked ion chemistry in their discussion of ion formation, predicting formation of HCO^+ and H_3O^+.

However, the papers of Oppenheimer (1975,1976) stimulated large-scale investigations using chemical reaction networks of many species. He assumed a complete steady state for the chemistry in the coma (negligible change in intercepted solar flux by the motion of the comet radially to the Sun while gas flows from the nucleus to the boundary of the collision zone, $\sim 10^4$ km, which corresponds to $\sim 10^4$ s), and chemical reactions so fast that changes of species concentration (dilution of the gas as it streams outward in the coma) have no effect. The first assumption is justifiable as long as $r/v_r > 3 \times 10^4$ s, where r is heliocentric distance of the comet and v_r is its radial velocity component. But the second assumption is not satisfactory for all molecular species in the coma; the concentration of the reactants may reduce faster than local chemical steady state can be established. In this first model the coma was also physically very simple: an isothermal, optically thin gas with constant outstream velocity.

Shimizu (1975), in an investigation similar to Oppenheimer's, assumed that the coma up to 100 km above the nucleus is optically thick to dissociating and ionizing solar ultraviolet radiation. He argued that large amounts of NH_3 and CH_4 would produce large amounts of carbon compounds; but since this is contrary to observations, NH_3 and CH_4 must only be present in small quantities. Recent calculations starting with significant fractions of NH_3 and

CH_4 do not show overproduction of C_2, C_3, CN, CH, and CH^+. Perhaps Shimizu's chemical reaction network, his rate constants, and the physics of the coma were oversimplified in these first attempts. His suggestion that the observed 8 km s^{-1} hydrogen component comes from electron dissociative recombination of H_3O^+

$$e + H_3O^+ \to OH + H + H \qquad (2)$$

is interesting, and should be pursued and tested.

Ip and Mendis (1976a, 1977) considered steady-state solutions for a small reaction network involving species related to H_2O and CO. They approximated coma attenuation of solar ultraviolet radiation with a wavelength independent optical depth, and based the outstream velocity (von Mises 1958) on a polytropic equation of state. Their first model was for an H_2O dominated comet with 10% CO; the second (CO-rich) model assumed ratios of $CO:H_2O$ of 1:1 and also 10:1. Their important innovation was the inclusion of an internal ionization source, an electric current generated in the comet's tail and closing through the inner coma (Ip and Mendis 1975, 1976b; see also rebuttal to large currents by Ershkovich 1978). Yet, their collisional ionization and dissociation time scales for H_2O and CO induced by the electric current are pure guesses (Ip and Mendis 1976a); no measured or theoretical rate coefficients or cross sections were used. Without internal ionization sources, only solar ultraviolet radiation initially causes ionization (charge exchange and similar reactions only change the relative abundance of CO^+ and H_2O^+). In that case their model with the internal ionization source "turned off" gives a result for $CO^+:H_2O^+ \lesssim 1$, which does not represent comets Humason 1962 VIII or Morehouse 1908 III nor recent quantitative observations of Comet Kohoutek 1973 XII (Wyckoff and Wehinger 1976). However, with the assumed current the ratio can be significantly > 1; even a water dominated comet could look like a CO^+-rich comet if the assumed ionization time scales are roughly correct. The model of Ip and Mendis for CO dominated comets with internal ionization source could qualitatively explain the morphology of comets like comets Humason 1962 VIII and Morehouse 1908 III. Another ionization mechanism supplementary to solar ultraviolet, but not quite as internal as the electric current, is the Alfvén ionization (Alfvén 1960; Huebner 1961).

The first attempt at large scale (136 reactions among 30 chemical species) time dependent chemical kinetics modeling of the coma, taking into account dilution of species concentrations as the gas flows radially outward in the coma, was reported by Huebner (1977). A stiff differential equation solver was used in this model's computer program (Gear 1971), and photolysis was calculated in small wavelength bins from 1 Å to the visible range. This model started with a mixture of H_2O and CO_2 and was expanded by Giguere and Huebner (1978) to include effects of coma ultraviolet opacity and to calculate results for four initial compositions containing H_2O, NH_3, CH_4, and CO_2. These compositions correspond to chemical equilibrium or near chemical equilibrium compositions appropriate to comet formation in the Jupiter to Neptune region. At this point the model contained 98 species and allowed 441 photo and chemical reactions. An isothermal (200 K) gas with constant outstream velocity (~ 1 Mach) was assumed. For a nucleus of 1 km radius and the comet at 1 AU heliocentric distance, the model underproduced CO^+, C_2, C_3, HCN, and N_2^+. Production of CN was marginal for one composition (the "chemical equilibrium" composition containing large amounts of CH_4) when compared to observations of comets Bennett 1970 II and Kohoutek 1973 XII. This model was further expanded by Huebner and Giguere (1980) to allow 537 reactions. But the main emphasis was on improved modeling of photolytic reactions that now included 74 branches for 28 species. Of these the photodissociative ionization (PDI) of CO-bearing molecules like

$$CO_2 + h\nu \rightarrow O + CO^+ + e \qquad (3)$$

was found to be an important contributor to the production of CO^+ deep in the coma where observers claim it is produced. It was shown that PDI is also important for the production of CN, C_2 and C_3 through the precursor reactions

$$CH_4 + h\nu \rightarrow CH_3^+ + H + e \qquad (4)$$

and

$$CO_2 + h\nu \rightarrow C^+ + O_2 + e \cdot \qquad (5)$$

Other improvements included direct determination of the gas production at the surface of the nucleus from energy balance and equation of state considerations, and better calculation of the outstream velocity assuming an isentropic expansion of the coma gas (von Mises 1958). A new mixture containing CO in place of CO_2 was also tried. Results of the calculations indicated that predicted CN column densities lie well within the range of observation at 1 AU. But as before, C_2 and C_3 were greatly underproduced and the ratio of $H_2O^+:CO^+$ was ~ 1. There were also indications that production of CN at larger heliocentric distances would fall short of observed values. These problems should be reinvestigated in the light of recent measurements (Masuoka and Samson 1980, Samson et al. 1981) indicating that the PDI cross sections are much larger than those used by Huebner and Giguere.

Feldman (1978) modeled atomic carbon production in the coma of Comet West 1976 VI. He introduced a slow (thermalized) and a fast (non-thermalized) component of photodissociation products to approximate the effect of excess energy of solar photons beyond the dissociation threshold. Shimizu (1976) had pointed out the effect of excess energy on the heating of the coma and hence on the increased outstream speed. The reaction network in Feldman's model was extremely limited.

Mitchell et al. (1981) and Biermann et al. (1982, an extension of the work of Huebner and Giguere; see also Huebner 1981) have independently given results for coma chemistry based on "interstellar" compositions of the nucleus, i.e., mixtures consistent with compositions expected if comets formed in the outer part of the presolar nebula or in a companion fragment of the parent interstellar cloud. The reason for using an "interstellar" composition was that in their models equilibrium or near-equilibrium chemical composition of the nucleus (expected if comets formed near the giant planets) underproduced C_3 by several powers of ten, underproduced C_2, and produced CN only marginally if NH_3 was not abundant in the nucleus; NH_2 and possibly NH were overproduced. Also, production of CN as a function of heliocentric distance predicted by model calculations did not agree with observations. It is possible but not obvious that these results might change with further improvement of the chemistry and physics of the models. For additional arguments in favor of an "interstellar" composition see Biermann et al. (1982).

The model of Biermann et al. (1982) assumes that the relative abundance of H:O is 2 (as is typical in comets) and the relative abundance of C:O:N is solar. This does not completely restrict the molecular composition; many combinations of molecules can be formed from this relative abundance. A composition relatively rich in H_2O (43%) and another with less water (33%) but more CO-bearing molecules (43%) were investigated. The second mixture was intended to model a coma in which H_2O^+:CO^+ is $\ll 1$. Wyckoff and Wehinger (1976) had derived a ratio of ~ 0.03 (for a correction of their original value see Huebner and Giguere 1980) from observations of Comet Kohoutek 1973 XII at 0.5 AU heliocentric distance. The ratio obtained from the H_2O-poor model was only ~ 0.5. A much smaller ratio of H_2O to CO-bearing species would be needed to obtain the low H_2O^+:CO^+ ratio found by Wyckoff and Wehinger, but such a modification cannot be obtained with the assumed constraints on the elemental abundances. In both the water-rich and water-poor calculations, the CN, C_2, C_3, and NH_2 average column densities agree well with data reported for several comets at various heliocentric distances and diaphragm ranges on the coma by A'Hearn (1975, personal communication 1980), A'Hearn et al. (1980), and to a lesser degree Sivaraman et al. (1981). The comparison is shown in Table I. The column densities of NH_2 reported by A'Hearn and coworkers are of particular importance since they greatly restrict the amount of NH_3 in the nucleus and thus limit the production of CN by chemical reactions. To obtain results consistent with observations, all molecules containing NH_2, CN, C_2, and C_3 in the nucleus had to be present only in trace amounts (total $< 2\%$).

The Mitchell et al. chemical reaction network is comprised of 1192 reactions among 128 species. As with Biermann et al. (1982), Huebner and Giguere (1980), and Giguere and Huebner (1978), collisional ionization is not

TABLE I

Some Comparisons of Observed and Model
Calculated Average Column Densities

Species	Comet	Heliocentric Distance (AU)	Distance into Coma (10^4 km)	Log Column Density Observed[a] (cm^{-2})	Model[b] (cm^{-2})
CN	West	0.6	5	12.3	12.7
		1.0	5	11.8	12.3
C_2	West	0.6	5	12.3	12.3
		0.6	1	13.6[c]	12.6
		1.0	5	12.2	11.6
	Kohoutek	1.0	3	11.5	11.6
C_3	West	0.6	5	12.0	12.0
		1.0	5	12.2	11.7
NH_2	Kohoutek	1.0	3	10.8	10.7
CH	Kohoutek	0.6[d]	5	11.5	10.7

[a]From A'Hearn (personal communication, 1980; A'Hearn 1975; and A'Hearn et al. 1980).
[b]"Interstellar."
[c]Sivaraman et al. 1979.
[d]Postperihelion.

calculated. Although Mitchell et al. consider even more reactions than Biermann et al. (1982), they make the additional simplifying assumption of constant temperature and coma expansion velocity 0.6 km s^{-1} at all distances from the nucleus. Effects of coma ultraviolet opacity are not mentioned in their paper. Coma abundances are determined by their one-dimensional time dependent computer code assuming spherical coma symmetry and purely radial expansion.

The goal of Mitchell et al. is similar to that of Huebner and coworkers: characterization of the nucleus by comparison of observed coma abundances with model results. As with Biermann et al. (1982), these ice compositions are carried over into the gas phase at the nuclear surface. Their composition is qualitatively similar to the "interstellar" composition of Biermann et al. (1982). Their choice of composition was also motivated by the poor agreement of near-equilibrium mixtures with coma observations, but Mitchell et al. make an attempt to relate their interstellar composition to present observations and theory of the interstellar medium. As they state, uncertainties in relating this composition to actual interstellar conditions include relative depletions in the solid phase of the more volatile gases such as CO and N_2, and the nature of possible recombination of atoms and radicals over the lifetime of the comet. In general accord with the results of Huebner and coworkers, Mitchell et al. found that their near-equilibrium initial mixture of H_2O, CO_2, CH_4, and NH_3 produces amounts of C_2 and CN in "reasonable agreement" with observations, but seriously underproduces C_3 and overproduces NH_2. The NH_2 comparison is again based on the key observations of A'Hearn et al. (1980). The chemical pathways leading to CN and C_2 in Mitchell et al.'s scheme for near-equilibrium comet composition show some differences from the pathways of Huebner and coworkers, but their overall schemes and results are similar. Mitchell et al.'s interstellar composition produced the species C_2, OH, CH, CN, and C_3 in "satisfactory agreement" with observations, but again overproduced NH_2. While two-body reactions are still significant in their CH and C_2 synthesis, the formation of these radicals is greatly facilitated by complex parents and C_3 production is increased significantly, but it is still about one power of ten too low.

In their first two mixtures overproduction of NH_2 is caused by overabundance of NH_3 in the initial compositions. Computations were made with modified compositions with NH_3 sharply reduced and N_2 increased. Again, the observed low NH_2 abundance is a crucial constraint, and determining the NH_2 photolytic lifetime is very important. Mitchell et al. did not describe their results for ions. Though the models by Mitchell et al. and Biermann et al. (1982) are similar, it is still significant that the two independent groups are reaching the same general conclusions.

Ultraviolet solar flux variations can induce changes in coma chemistry and ultraviolet fluorescence. However, the order of magnitude changes predicted by Oppenheimer and Downey (1980) from the solar cycle appear too large if applied to the entire coma since the combined effect of fluorescence rate and fluorescing daughter production rate is partially offset by the lifetime of the daughter species. In the case of parent molecules whose production does not depend on ultraviolet solar flux, the increase in ultraviolet fluorescence rate is effectively canceled by the reduction in lifetime.

II. SUMMARY OF PHYSICS AND CHEMISTRY

Experience from modeling the coma shows that chemistry and physics must be treated together. Density, temperature, and velocity profiles have a profound effect on the chemistry. Physical models of the inner coma have been presented by Dolginov and Gnedin (1966), Shul'man (1969a,b), Mendis et al. (1972), Ip and Mendis (1974), and Wallis (1974). Interaction of coma gas with solar wind modifies the column density profiles; this has been studied and modeled by Biermann et al. (1967,1974), Brosowski and Schmidt (1967), Brosowski and Wegmann (1973), Wallis (1973), Wallis and Ong (1976), Ip (1980), Mendis (1981) and Schmidt and Wegmann (see their chap-

TABLE II

Expected Reality Compared with Long Term Model Approximations

Reality	Model
Nucleus probably of irregular shape.	Spherical shape.
Heterogeneous nucleus (spotty surface, pockets of volatiles under surface).	Homogeneous nucleus.
Nucleus composed of volatile and nonvolatile frozen gases and dust.	Nucleus composed only of volatiles.
Coma distorted by solar wind and radiation pressure.	Spherically symmetric coma.
Coma attenuation of solar ultraviolet depending on angle of incidence.	One-dimensional opacity effects in most models (though two-dimensional effects have been considered).
Flares, jets, and envelopes in the coma.	Uniform, isotropic outgassing.

ter in this book). A brief summary of most of these papers and others is given by Whipple and Huebner (1976).

Modeling always requires simplifying idealizations; some that seem long-term or nearly permanent are summarized in Table II. Despite these simplifications, coma modeling is complex. Processes considered in various models (in particular see Huebner and Giguere 1980) are summarized and outlined next.

Solar radiation incident on the comet nucleus is partly reflected, partly absorbed; the reflection increases with increasing albedo. Some absorbed radiation is reemitted in the infrared, the rest of the energy changes the frozen gases from solid to vapor (sublimation). At large heliocentric distance most absorbed radiation is reradiated in the infrared, at smaller distance the equilibrium temperature is high enough so that most of the energy goes into sublimation of surface materials; the relationship is nonlinear. This has been described by Squires and Beard (1961).

From energy balance and the equation of state (the Clausius-Clapeyron equation coupled to the ideal gas equation) one can calculate equilibrium surface temperature T_0, gas pressure p_0, gas density n_0, gas production rate per unit surface area Z, and sound speed v_s. The visual albedo A and the infrared emissivity ϵ are assumed parameters in such calculations. The mean latent heat L, which also enters as a parameter, is determined from the assumed constituent frozen gases in the nucleus, as are the mean molecular weight M and the adiabatic exponent γ (ratio of specific heats). Typical values are $A = 0.3$, $\epsilon = 0.7$, $L \cong 8$ kcal mol^{-1}, $\gamma \cong 1.35$, $M \cong 22$, and, for a nucleus with radius $R_0 = 2.5$ km at heliocentric distance 1 AU, $T \cong 150$ K, $Z \cong 3 \times 10^{17}$ molecules cm^{-2} s^{-1}, $n_0 \cong 4.5 \times 10^{13}$ molecules cm^{-3} and $v_s \cong 0.3$ km s^{-1}.

The conservation laws of energy, momentum, and mass determine the fluid dynamics of assumed isotropic outflow. Using the adiabatic exponent as the polytropic exponent, the fluid flow in the isentropic approximation obeys von Mises' (1958) solution of supersonic flow. The asymptotic value of the supersonic outflow velocity, typically 0.7 km s^{-1}, is attained at several tens to a hundred R_0 above the surface. It is important to note that because of mass conservation, the gas density varies as

$$n(R) = n_0 \ [v_0/v(R)] \ (R_0/R)^2 \ . \qquad (6)$$

This deviates markedly from a R^{-2} variation near the nucleus where the velocity $v(R)$ changes rapidly.

From the density-distance relationship, photo cross sections of the coma gas constituents, and their relative abundances the attenuation of solar ultraviolet radiation can be determined as a function of wavelength. Cross-section data measured at room temperature are available for most potentially important mother molecules in the coma. The amount of solar ultraviolet radiation that penetrates the coma to the point of interest is important for determining dissociation and ionization which initiate chemical reactions in the inner coma. At low temperatures most molecules are in their ground state and very few are in excited states. Predissociation and autoionization are most affected by this situation as pointed out by Jackson (see his chapter in this book). But since the absorption cross section of molecules in general increases with wavelength in the ultraviolet region, attenuation of solar ultraviolet tends to make this long wavelength region of the cross sections less effective in the innermost coma where cooling of the coma gas is most likely to occur. At the same location in the coma, photo transitions from the ground state to dissociation and ionization continua are less affected by ultraviolet opacity. Thus, detailed wavelength-dependent ultraviolet opacity tends to compensate for discrepancies of effective cross sections from temperature induced excited states.

Reaction processes important for initiating and maintaining chemical reactions are summarized in Table III. The least important processes are listed near the bottom. There usually are many more chemical reactions than species. Since some reactions proceed fast and others slowly or not at all until reactants build up from other reactions, it is important to use a stiff differential equation solver technique for the coupled differential equations.

Model calculations assume that the processes occur in a thin shell of coma gases as it expands and moves outward. Therefore, chemical reactions take place in a continually diluting gas exposed to a continually increasing

TABLE III

Chemical Reaction Processes with Examples

Photodissociation [$h\nu + H_2O \rightarrow H + OH$]
Photoionization [$h\nu + CO \rightarrow CO^+ + e$]
Photodissociative ionization [$h\nu + CO_2 \rightarrow O + CO^+ + e$]
Electron impact dissociation [$e + N_2 \rightarrow N + N$]
Electron impact ionization [$e + CO \rightarrow CO^+ + 2e$]
Electron impact dissociative ionization [$e + CO_2 \rightarrow O + CO^+ + 2e$]
Positive ion-atom interchange [$CO^+ + H_2O \rightarrow HCO^+ + OH$]
Positive ion charge transfer [$CO^+ + H_2O \rightarrow H_2O^+ + CO$]
Electron dissociative recombination [$C_2H^+ + e \rightarrow C_2 + H$]
3-Body positive ion-neutral association [$C_2H_2^+ + H_2 + M \rightarrow C_2H_4^+ + M$]
Neutral rearrangement [$N + CH \rightarrow CN + H$]
3-Body neutral recombination [$C_2H_2 + H + M \rightarrow C_2H_3 + M$]
Radiative electronic state deexcitation [$O(^1D) \rightarrow O(^3P) + h\nu$]
Electron impact electronic state quenching [$e + O(^1D) \rightarrow e + O(^3P)$]
Radiative recombination [$e + H^+ \rightarrow H + h\nu$]
Radiation stabilized positive ion-neutral association [$C^+ + H \rightarrow CH^+ + h\nu$]
Radiation stabilized neutral recombination [$C + C \rightarrow C_2 + h\nu$]
Neutral-neutral associative ionization [$CH + O \rightarrow HCO^+ + e$]

---

solar ultraviolet radiation flux. The practical details are illustrated in Fig. 1. The time step for the chemical reactions is much smaller than the fluid dynamic time step at which density and attenuated solar flux are recalculated. Fluid dynamic time steps are approximately logarithmic. Only at large distances from the nucleus does the chemical time step approach the fluid dynamic time step. It is important to recognize that in this procedure chemical steady state may not be reached; dilution of the gas may freeze-in some species that might have reacted further.

Next we describe the processes not yet investigated in a self-consistent model. Internal ionization sources have already been mentioned; modeling them self-consistently is very difficult. But there are other processes only partially explored. Temperature and velocity profiles can have a pronounced effect on the chemistry. As the gas expands away from the nucleus it will cool, but when it reaches zones where the coma is becoming optically thin dissociative heating will sharply increase its temperature. Excess energy of solar photons beyond the dissociation threshold has been calculated for many potential cometary molecules (Huebner and Carpenter 1979); this energy leaves dissociation products in internally excited states and also gives them kinetic energy. Some internal energy of excitation will be radiated away. However, kinetic energy has two effects; it increases the outstream velocity

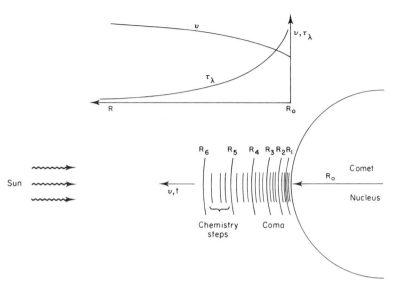

Fig. 1. Schematic representation for combining physics and chemistry. Time steps for chemical kinetics in the coma are small. Time steps for hydrodynamics are larger (R_1, R_2, ... ~ logarithmic). Many species do not reach chemical steady state in a hydrodynamic time step. Opacity and optical depth τ_λ are calculated at each hydrodynamic step as a function of λ in bins of width $\Delta\lambda$. Velocity of outstreaming coma gas is supersonic and increases with distance.

(and therefore reduces the density), and increases the temperature (on the order of 1000 K) which will decrease dissociative electron recombination but increase chemical reactions that require an activation energy or are endothermic. Assuming equipartition, some fraction f_t of the excess energy will go into translation and rotation, depending on the number of degrees of freedom for a molecule. In the collision dominated region this translational energy will be shared by all molecules and cause a temperature rise

$$\Delta T = \frac{2}{3k} \sum_i f_{ti} E_i n'_i / n \qquad (7)$$

where n'_i is the number density of dissociating species i, and E_i is their excess energy in dissociation. In the collision-free outer coma the excess translational energy cannot be shared with other molecules, and differential effects of outstream velocities for dissociated species with different molecular weights will result. This is consistent with the various velocity components observed for atomic hydrogen. Exothermic reactions in the inner coma will tend to increase temperature, while endothermic reactions in the intermediate, hotter coma will reduce temperature. Thus chemical reactions moderate the temperature though their net energies are generally smaller than those of dissociation. Solar photons with energy in excess of ionization potentials will also create hot electrons which reduce dissociative recombination but contribute to dissociation and ionization through impact processes. Photodissociation in the outer coma beyond the collision zone will not affect chemistry but will change the column density profile of the decay products; these products will be in free particle flow and cannot share excess energy.

Since dust grains are abundant in most comae and their temperature is usually higher than that of the coma gas (they cannot radiate efficiently at wavelengths larger than their size), catalytic chemical reactions may be important. Because dust grains have much larger mass than molecules, grains are very inefficient in increasing the kinetic energy of the gas but very efficient in randomizing directions of molecular velocities in the rest frame of the grain. Closely related to dust grains are the conjectured icy grains and clathrates (Huebner and Weigert 1966; Delsemme and Miller 1971; Ip and Mendis 1974; Delsemme and Swings 1952). These grains require modeling an extended source. The possible effect of such a source on the $H_2O^+:CO^+$ ratio was discussed by Huebner and Giguere (1980).

Finally, equilibrium conditions at the coma-nucleus interface depend on the spin orientation of the nucleus as described by Cowan and A'Hearn (1979). Refined models that consider this effect must also include the thermal inertia of the nuclear surface layer which causes a lag in vaporization, and the nongravitational forces on the nucleus.

III. CONCLUSIONS

Chemistry is important in the comae of active and large comets. Many reaction paths must be considered in chemical modeling. One path to a particular product may dominate at one place in the coma or heliocentric distance while a different path dominates at another. Since the processes compete, the chemical reaction networks in present models cannot be simplified significantly. A complex network with chemical production by different, nearly parallel paths also has the advantage that the results will be less sensitive to uncertainties in rate constants. Examples of leading competing reactions producing or destroying C_2, C_3, CN, H_2O^+, and CO^+ (the latter at 1 AU and 3 AU heliocentric distance) based on the interstellar compositions adopted by Biermann et al. (1981) are illustrated in Figs. 2 through 7; note the many closely competing destruction mechanisms for CO^+. Only some of the important reactions are indicated.

The photodissociation cross sections for C_3 and C_2H important for the production of C_2 (see Fig. 2) are not known. Approximate photodissociation rates for C_3 and C_2 have been obtained from their ranges in comet comae. The photodissociation rate of C_2H is a guess based on the H-C bond energy. Similar situations exist for some other radicals; although many processes have been measured or evaluated theoretically for common molecules such as H_2O, CO_2, CH_4, NH_3, H_2CO, etc., this is not always the case for radicals

PHOTOCHEMICAL PROCESSES IN THE INNER COMA 509

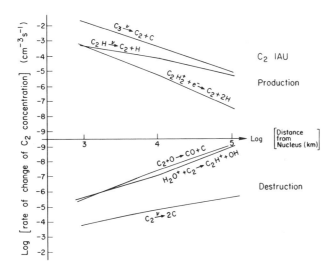

Fig. 2. Logarithm of time rate of production and destruction of C_2 number density versus logarithm of distance from nucleus into coma at 1 AU heliocentric distance. Only representative samples of the most important processes are shown. An "interstellar" composition as adopted by Biermann et al. (1981) was assumed for the nucleus.

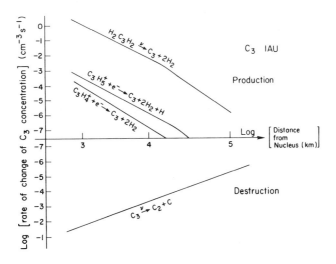

Fig. 3. Same as Fig. 2, except for C_3.

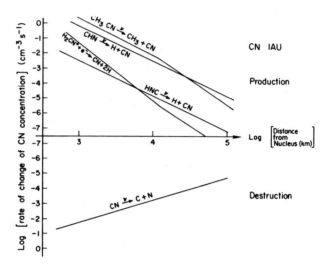

Fig. 4. Same as Fig. 2, except for CN.

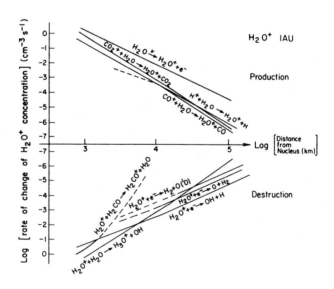

Fig. 5. Same as Fig. 2, except for H_2O^+. Dashes indicate that the time rate of change is uncertain because other processes dominate at various distances in the coma.

PHOTOCHEMICAL PROCESSES IN THE INNER COMA 511

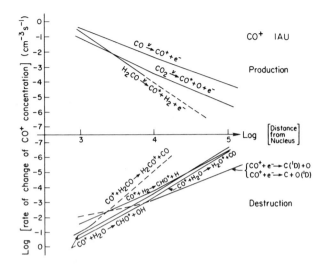

Fig. 6. Same as Figs. 2 and 5, except for CO^+.

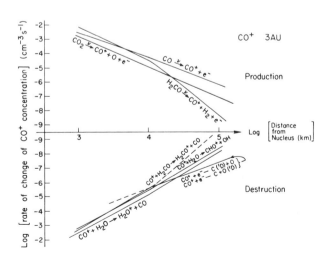

Fig. 7. Same as Fig. 6 except at 3 AU heliocentric distance. Note that competing processes are different at 3 AU than at 1 AU.

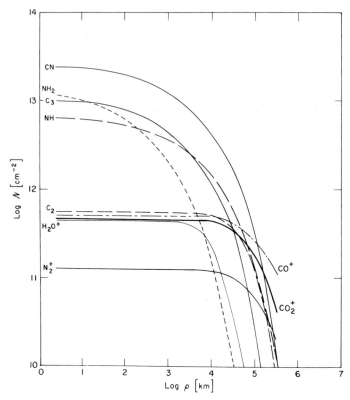

Fig. 8. Logarithm of column density versus logarithm of distance from nucleus, for a few species from a typical model calculation containing trace amounts of interstellar molecules. Note the difference in shape of the C_2 curve compared to other neutral species. Heliocentric distance of the comet is 1 AU, radius of the nucleus is 2.5 km.

derived from them. In general the total rate coefficients for electron dissociative recombination (an important process for coma chemical kinetics) have been measured for many molecular ions, but the dissociation branching ratio is usually not known. Neutral-neutral reactions are generally quite slow, but they can be the dominant process in the innermost coma of large comets with high density and large optical thickness severely restricting photolytic processes. Lists of chemical reactions and rate coefficients with references and comments can be found in Giguere and Huebner (1978), Huebner and Giguere (1980), and Biermann et al. (1981).

H_3O^+ is predicted with absolute certainty in the coma. If a hydrocarbon precursor such as CH_4 is in the nucleus, then HCO^+, C_2H_2, and C_2H_4 are also predicted. If NH_3 is sufficiently abundant ($\gtrsim 10\%$) then NH_4^+ will be in the coma. However, observations of NH_2 and chemical modeling make it doubtful that NH_3 is present by more than a few percent.

Figure 8 shows some column densities from chemical modeling. Among the neutral species, CN has the strongest abundance and shows the largest extent in the coma. It is closely followed by C_3. NH_2 has only a short range and therefore does not contribute effectively to the column density averaged over a large coma; C_2 does. This sequence is consistent with observations of typical bright comets. Note that the column density profile of C_2 is very flat, different from that of other neutral species. A flat profile for C_2 has also been reported by Barker (1981) for Comet Bradfield 1979 X. Observational column densities are absolutely necessary for comparison with models; most other reduction of observed data is model dependent and less useful.

The comet models most consistent with observations indicate that only trace amounts (total amount 2%) of molecules bearing CN, C_2, C_3, and NH_2 can be present in the nucleus. This appears to be consistent with interstellar compositions, but since the amounts are so small a more thorough investigation is needed.

Acknowledgments. We thank W.M. Jackson for valuable comments and suggestions he made during the refereeing of this manuscript. This research was supported in part by funds from the NASA Planetary Atmospheres Program and was performed under the auspices of the U.S. Department of Energy.

REFERENCES

A'Hearn, M.F. 1975. Spectrophotometry of Comet Kohoutek (1973f = 1973 XII). *Astron. J.* 80:861-875.
A'Hearn, M.F.; Hanisch, R.J.; and Thurber, C.H. 1980. Spectrophotometry of Comet West. *Astron. J.* 85:74-80.
Aikin, A.C. 1974. Cometary coma ions. *Astrophys. J.* 193:263-264.
Alfvén, H. 1960. Collision between a nonionized gas and a magnetized plasma. *Rev. Mod. Phys.* 32:710-712.
Barker, E.S. 1981. Spatially resolved observations of the inner coma of Bradfield (1979ℓ). Submitted to *Icarus.*
Biermann, L.; Brosowski, B.; and Schmidt, H.U. 1967. The interaction of the solar wind with a comet. *Solar Phys.* 1:254-284.
Biermann, L., and Dierksen, G. 1974. On the chemical constitution of cometary nuclei. *Origins Life* 5:297-301.
Biermann, L.; Giguere, P.T.; and Huebner, W.F. 1982. A model of a comet coma with interstellar molecules in the nucleus. *Astron. Astrophys.* In press.
Biermann, L.; Lüst, Rh.; and Wegmann, R. 1974. Some new results on the plasmadynamical processes near comets. *Astron. Acta* Suppl. 18:113-118.
Biermann, L. and Trefftz, E. 1964. Über die Mechanismen der Ionisation und der Anregung in Kometenatmosphären. *Zs. f. Astrophys.* 59:1-28.
Brosowski, B. and Schmidt, H.U. 1967. Kometen im Sonnenwind. *Z. Angew. Math. Mech.* 47:T 140-T 143.
Brosowski, B. and Wegmann, R. 1973. Numerische Behandlungen eines Kometenmodells. *Methoden Verfahren Math. Phys.* 8:125-145.
Cowan, J.J. and A'Hearn, M.F. 1979. Vaporization of comet nuclei: Light curves and life times. *Moon and Planets* 21:155-171.
Delsemme, A.H. and Miller, D.C. 1971. Physico-chemical phenomena in comets-III. The continuum of Comet Burnham (1960 II). *Planet. Space Sci.* 19:1229-1257.

Delsemme, A.H., and Swings, P. 1952. Hydrates de gaz dans les noyaux cométaires et les grains interstellaires. *Ann. d'Astrophys.* 15:1-6.

Dolginov, A.Z., and Gnedin, Yu. N. 1966. A theory of the atmosphere of a comet. *Icarus* 5:64-74.

Donn, B., and Urey, C.H. 1956. On the mechanism of comet outbursts and the chemical composition of comets. *Astrophys. J.* 123:339-342.

Donn, B., and Urey, C.H. 1957. Chemical heating processes in astronomical objects. *Mém. Soc. Roy. Sci. Liège,* Ser. 4, 18:124-132.

Ershkovich, A.I. 1978. The comet tail magnetic field: Large or small? *Mon. Not. Roy. Astron. Soc.* 184:755-758.

Feldman, P.D. 1978. A model of carbon production in a cometary coma. *Astron. Astrophys.* 70:547-553.

Gear, C.W. 1971, *Numerical Initial Value Problems in Ordinary Differential Equations,* (Englewood Cliffs, N.J.:Prentice Hall).

Giguere, P.T., and Huebner, W.F. 1978. A model of comet comae. I. Gas-phase chemistry in one dimension. *Astrophys. J.* 223:638-654.

Huebner, W.F. 1961. Relations between plasma physics and astrophysics. Comment about comet tails. *Rev. Mod. Phys.* 33:498.

Huebner, W.F. 1977. Chemistry of the inner coma: A progress report. In *Comets, Asteroids, Meteorites,* ed. A.H. Delsemme (Toledo, Ohio: Univ. of Toledo), pp. 57-59.

Huebner, W.F. 1981. Chemical kinetics in the coma. In *College Park Colloquia on Chemical Evolution, Colloquium V, Comets and the Origin of Life,* ed. C. Ponnamperuma (Dordrecht, Holland: D. Reidel Publ.), pp. 91-103.

Huebner, W.F., and Carpenter, C.W. 1979. Solar photo rate coefficients. *Los Alamos Sci. Lab.* Report LA-8085-MS.

Huebner, W.F., and Giguere, P.T. 1980. A model of comet comae. II. Effects of solar photodissociative ionization. *Astrophys. J.* 238:753-762.

Huebner, W.F., and Weigert, A. 1966. Eiskörner in der Koma von Kometen. *Zs. f. Astrophys.* 64:185-201.

Ip, W-H. 1980. Cometary atmospheres. I. Solar wind modification of the outer ion coma. *Astron. Astrophys.* 92:95-100.

Ip, W-H., and Mendis, D.A. 1974. Neutral atmospheres of comets: A distributed source model. *Astrophys. Space Sci.* 26:153-166.

Ip, W-H., and Mendis, D.A. 1975. The cometary magnetic field and its associated electric currents. *Icarus* 26:457-461.

Ip, W-H., and Mendis, D.A. 1976a. The structure of cometary ionospheres. 1. H_2O dominated comets. *Icarus* 28:389-400.

Ip, W-H., and Mendis, D.A. 1976b. The generation of magnetic fields and electric currents in cometary plasma tails. *Icarus* 29:147-151.

Ip, W-H., and Mendis, D.A. 1977. The structure of cometary ionospheres. 2. CO-rich comets. *Icarus* 30:377-384.

Jackson, W.M., and Donn, B. 1966. Collisional processes in the inner coma. *Mém. Soc. Roy. Sci. Liège,* Ser. 5, 12:133-140.

Masuoka, T., and Samson, J.A.R. 1980. Dissociative and double photoionization of CO_2 from threshold to 90 Å. *J. Chemie Physique* 77:623-630.

Mendis, D.A. 1981. Interaction of comets with the interplanetary medium. In *College Park Colloquia on Chemical Evolution, Colloquium V, Comets and the Origin of Life,* ed. C. Ponnamperuma (Dordrecht, Holland: D. Reidel Publ.), pp. 71-89.

Mendis, D.A.; Holzer, T.E.; and Axford, W.I. 1972. Neutral hydrogen in cometary comas. *Astrophys. Space Sci.* 15:313-325.

von Mises, R. 1958. *Mathematical Theory of Compressible Fluid Flow,* (New York: Academic Press), pp. 73-77.

Mitchell, G.F.; Prasad, S.S.; and Huntress, W.T. 1981. Chemical model calculations of C_2, C_3, CH, CN, OH and NH_2 abundances in cometary comae. *Astrophys. J.* 244:1087-1093.

Oppenheimer, M. 1975. Gas phase chemistry in comets. *Astrophys. J.* 196:251-259.

Oppenheimer, M. 1976. Gas phase chemistry in comets. In *The Study of Comets,* eds. B. Donn, M. Mumma, W. Jackson, M. A'Hearn, and R. Harrington, (Washington, D.C.: NASA SP-393), pp. 753-756.

Oppenheimer, M., and Downey, C.J. 1980. The effect of solar-cycle ultraviolet flux variations on cometary gas. *Astrophys. J.* 241:L123-L127.

Samson, J.A.R.; Masuoka, T.; and Huntress, W.T. 1981. The production rate of C^+ from the photoionization of CO and CO_2. *Geophys. Res. Letters* 8:405-408.

Shimizu, M. 1975. Ion chemistry in the cometary atmosphere. *Astrophys. Space Sci.* 36:353-361.

Shimizu, M. 1976. The structure of cometary atmospheres. I: Temperature distribution. *Astrophys. Space Sci.* 40:149-155.

Shul'man, L.M. 1969a. Hydrodynamics of the circum-nuclear region of a comet. *Astrometry Astrophys.* No. 4, ed. V.P. Konopleva, NASA Tech. Transl. TT F-599, pp. 85-99 (1970).

Shul'man, L.M. 1969b. The physical conditions in the boundary layer of a cometary nucleus. *Astrometry Astrophys.* No. 4, ed. V.P. Konopleva, NASA Tech. Transl. TT F-599, pp. 100-109 (1970).

Sivaraman, K.R.; Babu, G.S.D.; Bappu, M.K.V.; and Parthasarathy, M. 1979. Emission band and continuum photometry of Comet West (1975n)–I. Heliocentric dependence of the flux in the emission bands and the continuum. *Mon. Not. Roy. Astron. Soc.* 189:897-906.

Squires, R.E., and Beard, D.B. 1961. Physical and orbital behavior of comets. *Astrophys. J.* 133:657-667.

Wallis, M.K. 1973. Weakly-shocked flows of the solar wind plasma through atmospheres of comets and planets. *Planet. Space Sci.* 21:1647-1660.

Wallis, M.K. 1974. Hydrodynamics of the H_2O comet. *Mon. Not. Roy. Astron. Soc.* 166:181-189.

Wallis, M.K., and Ong, R.S.B. 1976. Cooling and recombination processes in cometary plasma. In *The Study of Comets,* eds. B. Donn, M. Mumma, W. Jackson, M. A'Hearn, and R. Harrington, (Washington, D.C.: NASA SP-393), pp. 856-875.

Whipple, F. 1950. A comet model. I. The acceleration of Comet Encke. *Astrophys. J.* 111:375-394.

Whipple, F. 1951. A comet model II. Physical relations for comets and meteors. *Astrophys. J.* 113:464-474.

Whipple, F., and Huebner, W.F. 1976. Physical processes in comets. *Ann. Rev. Astron. Astrophys.* 14:143-172.

Wurm, K. 1943. Die Natur der Kometen. *Mitt. Hamburger Sternwarte* 8:Nr. 51.

Wyckoff, S., and Wehinger, P.A. 1976. Molecular ions in comet tails. *Astrophys. J.* 204:604-615.

PART V
Ion Tails and Solar Wind Interactions

OBSERVATIONS AND DYNAMICS OF PLASMA TAILS

JOHN C. BRANDT
NASA-Goddard Space Flight Center

Photographs of comet tails reveal a constantly changing array of structural features such as rays, streamers, knots, kinks, helices, condensations, and disconnected tails. I briefly review the history of observation and theory. The visual emission of the CO^+ molecule has been used in the past to locate the magnetized cometary plasma. Before about 1980 the general approach to the structure in cometary plasma tails had not concentrated on the connection between the various features. Work on structure and evolution now tends to regard this subject as a coordinated sequence of phenomena, with, we hope, a specific physical cause. A tentative morphology of plasma tail evolution has been developed; it can be interpreted as resulting from interaction of a comet with sector boundaries in the solar wind, utilizing magnetic reconnection. Thus a regular forcing function appears to produce the systematic evolution of the plasma tail. The study of plasma tail evolution is greatly hampered by the lack of continuous coverage on comets for a week or more. Some extended coverage was possible in 1908 with Comet Morehouse, because of its high declination, and isolated time sequences have been obtained. Nevertheless, our morphological view of the plasma tail is based on fragmentary data. The appearance of P/Halley in 1985-86 presents an unparalleled opportunity to obtain extensive coordinated photographic sequences which should give basic cometary plasma morphology a sound observational basis.

I. HISTORY

The development of astronomical photography around the turn of this century enabled the plasma tails of comets to be seen in detail for the first time. The photographs of comets Borrelly 1903 IV, Daniel 1907 IV, Morehouse 1908 III, and P/Halley revealed an array of changing structural detail. We now know that the photographs are dominated by emission from the bands of CO^+ around 4200 Å and that the emission serves as an indicator of the magnetized, cometary plasma. Figures 1 through 5 [a] illustrate the detail often found in plasma tails. The structures shown in the first five figures refer primarily to CO^+ emission in the blue.

The terminology of features observed in plasma tails can be confusing and Figs. 1 through 5 are used as a brief introduction. Knots are small regions of enhanced brightness observed in almost all comet tails; condensations are also regions of enhanced brightness; the terms are sometimes used interchangeably, but a condensation is often a larger cloud-like structure. Rays are the thin rectilinear features most easily seen when inclined at a large angle to the tail axis. The rays merge to produce the larger bundles or streamers forming the main plasma tail. Kinks are seen as sharp bends, usually in streamers. Helices are the wavy features that look like corkscrews. A disconnection event (DE) is a major condensation that has become detached from the region of the comet's head. DE's are discussed extensively in this chapter.

Figure 1 shows Comet Kohoutek 1973 XII on 11 January 1974. The distance from the head to the "Swan Cloud" near the end of the tail is ~ 0.1 AU. The plasma tail appears above the structureless dust tail and clearly shows knots, streamers, a kink (where the main tail meets the Swan Cloud), and the Swan Cloud itself, probably a terminal phase of a DE. Figure 2 shows Comet West 1976 VI on 9 March 1976. The plasma tail is the structured feature at a 2-o'clock orientation from the head. The length of the tail on this photograph is $\sim 7°$. Many streamers are shown which themselves contain condensations. Figure 3 shows Comet Kohoutek on 13 January 1974; a so-called helical structure and condensations are clearly visible. Figure 4 shows photographs of Comet West on 1 April 1976, taken (top) at 10:04 UT and (bottom) at 11:14:30 UT. A kink and its tailward motion (measured at 97 km s^{-1}) are shown. Figure 5 clearly shows tail rays as they lengthen and turn toward the tail axis of Comet Kobayashi-Berger-Milon 1975 IX on 31 July 1975. [Several DE's are shown in Figs. 7, 8, 9, and 11.]

Many of the features described exhibit some form of tailward motion. The apparent speeds of knots, condensations, kinks, helices, and DE's can range from tens of km s^{-1} near the head to > 100 km s^{-1} at 10^6 km down the tail. The speeds quoted may refer to bulk motion of material or to waves;

---

[a]Illustrations credited as "JOCR Photograph" are from the Joint Observatory for Cometary Research, operated by NASA-Goddard Space Flight Center and the New Mexico Institute of Mining and Technology.

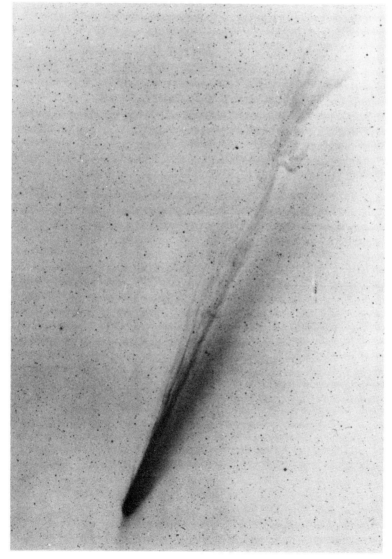

Fig. 1. Comet Kohoutek 1973 XII on 11 January 1974. (JOCR Photograph.)

Fig. 2. Comet West 1976 VI on 9 March 1976. (JOCR Photograph.)

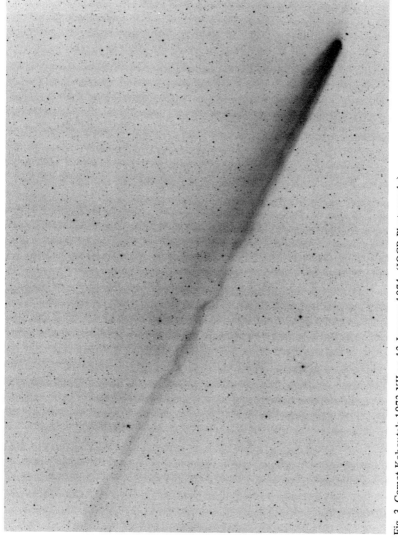

Fig. 3. Comet Kohoutek 1973 XII on 13 January 1974. (JOCR Photograph.)

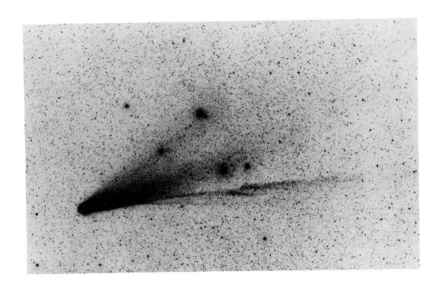

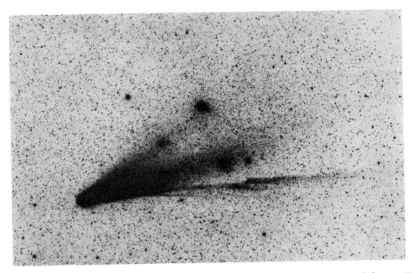

Fig. 4. Comet West 1976 VI on 1 April 1976. Photographs are taken ~1 hr apart. (JOCR Photograph.)

there is no general agreement on the choice. Additional details about plasma tails can be found in the review articles by Brandt (1968) and by Brandt and Mendis (1979).

Many of the features in plasma tails were thoroughly and accurately described by Barnard (1908a,b,1909) and speeds of moving features were measured by Curtis (1910). Descriptions of systematic changes in the appear-

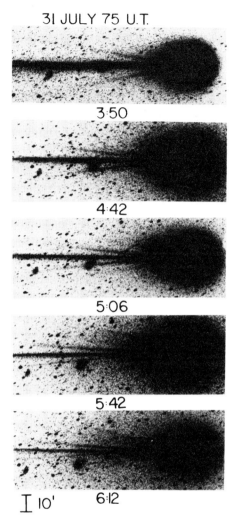

Fig. 5. An example of folding tail rays in Comet Kobayashi-Berger-Milon 1975 IX. (JOCR Photograph.)

ance of the tail including DE's were available in the early literature (Barnard 1920; also see below). These descriptions seem prescient with respect to the current convergence toward a basic morphology. Orientations of plasma tails were extensively investigated by Hoffmeister (1943). The importance of the solar wind (Biermann 1951,1952,1953) and its magnetic field (Alfvén 1957) in understanding cometary plasma was recognized. The existence of the tail rays is regarded as *prima facie* evidence for a cometary magnetic field and tail rays are almost surely captured from the solar wind. It was widely believed that the capture process would produce a tail field greatly enhanced over the solar-wind value (e.g., see Brandt and Mendis 1979), but the current trend considers the tail field of the same order of magnitude as the solar-wind field, as advocated by Ershkovich (1978).

Despite many good observational (e.g., Lüst 1962; Jockers 1979) and theoretical efforts (Marotschnik 1963; Ioffe 1966a,b; Biermann et al. 1967; Ioffe 1968a,b; Biermann and Lüst 1972; Wallis 1973; Ershkovich 1980; chapter by Schmidt and Wegmann in this book), it can be questioned whether we are truly converging toward an understanding of plasma phenomena in comets. A useful first step would be to establish the basic morphology; then observations would focus on the relevant details, and theoretical work could first focus on understanding the broad picture. As this process progresses toward a general understanding, topics such as physical processes in the ion tails (chapter by Ip and Axford in this book), physics of the cometary ionosphere (Mendis and Houpis 1981), helical structures in the tail (Ershkovich 1980; Krishan and Sivaraman 1981), folding and evolution of tail rays (Miller 1981; Ershkovich 1981), the role of turbulence (Buti 1981), and many other details (Jockers 1981) would be integrated into the overall picture.

II. BASIC MORPHOLOGY

In my opinion, there are two features essential to a general understanding of the plasma tail phenomena.

1. *The ray-folding phenomenon.* Tail rays lengthen and turn toward the tail axis, coalescing to form the main tail (Fig. 5). The measured rate in this sequence is $\sim 3^\circ$ hr^{-1} with respect to the tail axis. Only a few observational sequences show this phenomenon clearly (e.g., Wurm and Maffei 1961; Wurm and Mammano 1964; Wurm 1968; Moore 1977; Brandt and Mendis 1979), implying that the process does not always occur. Orlov (1944,1945) gave a separate classification (I_0) to tails displaying this phenomenon and noted its relative rarity. The qualitative explanation of the ray-folding phenomenon has been given by Alfvén (1957) in terms of field-line loading and deceleration in the cometary ionosphere, as shown schematically in Fig. 6.

2. *Disconnection events.* Occasionally, the entire plasma tail separates from the comet and drifts away. The DE is followed by a renewal of the plasma tail; the cyclic nature and predictability of the phenomenon was noted long ago.

> "It seems that this comet shows a recurrence of phases which has not been shown hitherto. Starting from a stellar nucleus and a very narrow tail, a single streamer, there is an expansion of the comet which on the side towards the sun forms parabolic envelopes, and these are brought round to the tail, forming a sort of fan-shape; and we note that the angles between the rays of the fan change in the course of a night; then, as time goes on, the fan-shaped streamers coalesce into the tail. We then get a

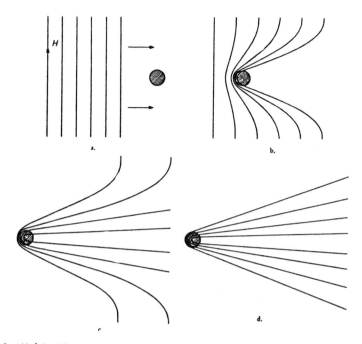

Fig. 6. Alfvén's (1957) explanation of folding tail rays.

streaming tail which breaks up into fluocculent masses and appears to be in a state of great disturbance, and this goes off apparently into space and the comet starts fresh again. That sequence of events happened three or four times in the course of our history of the comet at somewhat irregular intervals."

(Astronomer Royal,[a] 1908)

"One broad generalization was, however, indicated by Mr. Davidson. The formation of the tail seems to be intermittent rather than continuous. There seem to be at intervals convulsions or explosions in the nucleus, producing big bunches or lumps of tail which travel away and leave the comet with a small tail for a time ... The signs of a coming convulsion can be recognized, and were, indeed, recognized by Mr. Davidson in a striking manner at the meeting. In the pictures shown by him he traced the history from a quiescent period on October 1 to his last photograph on the night of October 6, which, he remarked, showed that an outburst was imminent. He had not then seen Prof. Barnard's picture, taken six hours later than his own. When this beautiful photograph was subsequently put on the screen, it showed the outburst thus predicted!"

(*Observatory*, 1908)

---

[a]W.H.M. Christie.

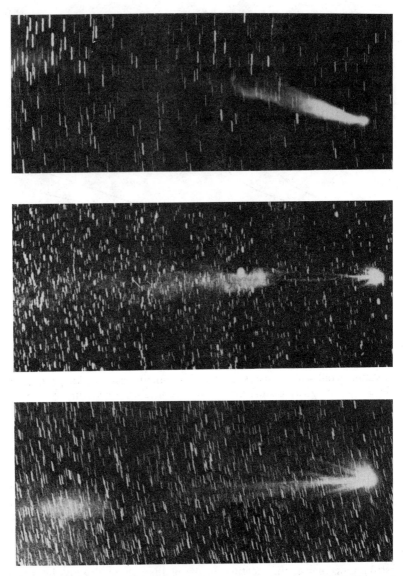

Fig. 7. Development of a disconnection event in Comet Morehouse 1908 III on a time scale of days. (Yerkes Observatory.)

A DE in Comet Morehouse is shown on two different time scales in Figs. 7 and 8. Figure 7 shows Comet Morehouse on 30 September (top), 1 October (middle), and 2 October (bottom) 1908. Figure 8 shows detail on 30 September 1908, with exposures taken at GMT times of 18:50 (top), 20:00 (middle), and 21:45 (bottom). At present, there are 72 known DE (Niedner

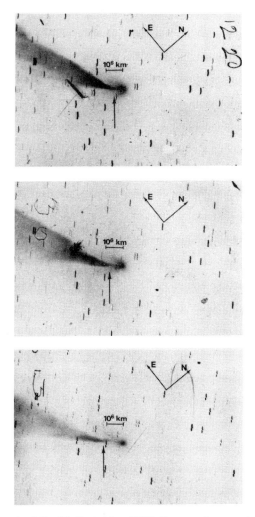

Fig. 8. Development of the 30 September 1908 disconnection event in Comet Morehouse 1908 III on a time scale of hours. (Indiana University.)

1981). Additional DE's are shown in Fig. 9 as follows: upper left—Comet Borrelly, 24 July 1903 (Yerkes Observatory); upper right—P/Halley, 13 May 1910 (Lowell Observatory); lower left—P/Halley, 6 June 1910 (Yerkes Observatory); lower right—Comet Bennett 1970 II, 4 April 1970 (Hamburg Observatory). Note that the use of the term disconnection event is rooted in the observational history of this subject. The DE's have been specifically described as "disconnected" (Barnard 1908a,1909), "broken off" (Barnard 1903), "detached" (Barnard 1893), "thrown off" (Barnard 1908b), "discarded" (Barnard 1920), and "rejected" (Barnard 1920).

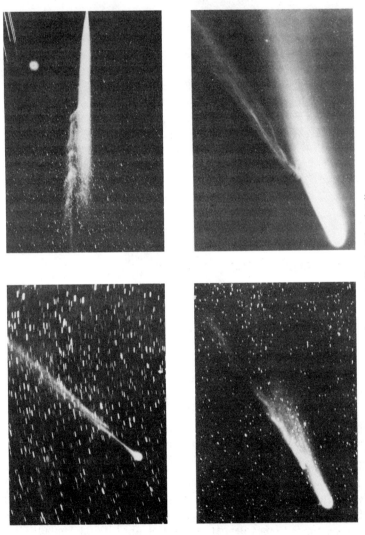

Fig. 9. Examples of disconnection events in comets. See text for details.

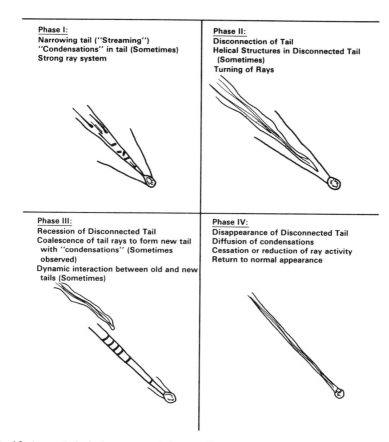

Fig. 10. A morphological sequence of plasma tails.

The available data have been collected (Niedner and Brandt 1980) into a schematic morphological sequence (Fig. 10), illustrated by a sequence of photographs of Comet Kohoutek (Fig. 11). The scale bars in Fig. 11 represent 10^6 km and the bracket in the third panel from the top indicates an area possibly showing arcade loops which may correspond to the cross-tail condensations shown in Phase III of Fig. 10. Every comet is not expected to display every detail of the sequence.

III. QUALITATIVE MODEL

Niedner and Brandt (1978,1979) have noticed the preference for DE's to be associated with sector boundaries; this result seems well established. The cyclic occurrence of magnetic reversals (or sector boundaries) in the solar wind could well provide the physical cause for the cyclic effect observed in comets. This is the basis of the qualitative model shown in Fig. 12; sector boundaries are marked and shaded areas indicate the visual plasma tail.

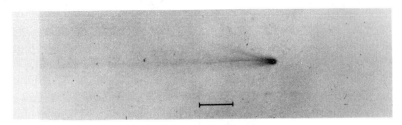

1974 JANUARY 9,2^{h}18^{m}30^s UT (UNIVERSAL TIME)
JOINT OBSERVATORY FOR COMETARY RESEARCH (JOCR)

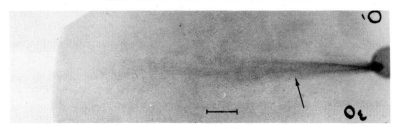

JANUARY 9,8^{h}59^{m}30^s UT
TOKYO ASTRONOMICAL OBSERVATORY

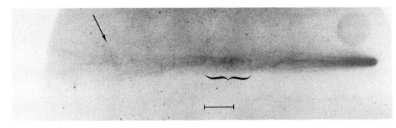

JANUARY 10,9^{h}05^{m}30^s UT
TOKYO ASTRONOMICAL OBSERVATORY

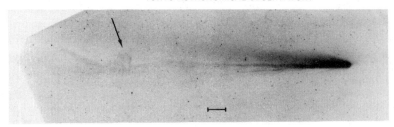

JANUARY 11,2^{h}41^{m}40^s UT
JOCR

Fig. 11. Photographic summary of a disconnection event in Comet Kohoutek 1973 XII illustrating the morphology presented in Fig. 10. The scale bars represent 10^6 km and the bracket in the third panel from the top indicates an area possibly showing arcade loops.

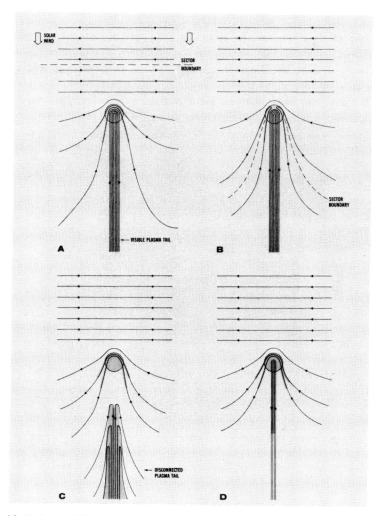

Fig. 12. Basic model for explaining disconnection events in terms of the interaction with sector boundaries in the solar wind.

The basic approach here is borrowed from Alfvén (1957), with the inclusion of polarity reversals in the solar-wind magnetic field and invoking the process of magnetic reconnection. The stages of the model presented in Fig. 12 are as follows.

A. The cometary plasma is in its normal state, but with a sector boundary approaching.
B. The sector boundary encounters the comet and the field of new polarity (opposite to the old field draped around the comet) is pressed into the

ionosphere. Reconnection occurs and the old field begins to be stripped from the comet.
C. The old field is completely stripped away, disconnection occurs, and formation of a tail with the new field begins.
D. The disconnected tail drifts away and the new tail builds up until the field value in the ionosphere reaches a saturation value.

The cycle repeats at the next sector boundary encounter. Time scales for some of the processes follow: the interval between sector-boundary encounter and the DE itself is ~ 18 hr; the time for the tail rays to close and form the new tail is ~ 15 to 25 hr; the entire DE process is expected to occur at each sector-boundary crossing or $\sim$ once per week.

This qualitative model explains or is consistent with:

1. Field capture and its relative rarity;
2. Generation of several kinds of fine structure;
3. Disconnection events.

In addition, the cometary phenomena have a physical forcing function that, at least in principle, produces them. The physical model requires that reconnection operates in the cometary ionosphere; this has been studied by Niedner et al. (1981).

The qualitative model implies an energy release associated with sector boundary passage. Niedner (1980) may have found evidence for this in brightness variations observed in some comets. Clearly, as our understanding of the reconnection physics in comets improves, so should our ability to predict the consequences and the electromagnetic signature of the event.

Similar physics may be involved in other solar system objects; the prime candidate is Venus (see the chapter by Russell et al. in this book). These authors have noted that the magnetotail lobes on Venus change polarity after the passage of a sector boundary, so there is a high probability that the basic cometary scenario, including reconnection, occurs on Venus.

The model can be observationally and theoretically criticized, but the criticism should be in the context of attempting to establish and understand a basic morphology for the interaction of the solar wind with comets.

IV. FUTURE STUDY

The basic need in this subject is more useful observational data, namely closely spaced sequences of wide-angle photographs. Jockers (1979, 1981) has been a pioneer in this area. This need was recognized long ago:

"From these considerations it is evident that every active comet should be photographed as often as possible. Photographs made on the same

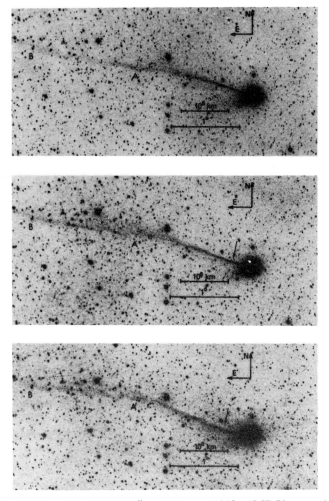

Fig. 13. Comet Bradfield 1979ℓ on 6 February 1980. (JOCR Photograph.)

night will be especially valuable, and these should be made as frequently as possible during the night. It is from such photographs alone that the progressive changes can be followed intelligently. The changes from day to day in a comet are so very great that it is usually not possible to connect certainly any phenomena of one night with that of the night before, as no part of the tail would be likely to live over the interval. But from successive photographs on the same night—especially when the comet is bright enough to permit short exposures—the actual changes may be observed and measured with certainty."

(Barnard, 1905)

A recent example of rapid variations in Comet Bradfield (1979ℓ) is shown in Fig. 13. The UT times of mid-exposure on 6 February 1980, are as follows: 2:32:30 (top); 2:48:00 (middle); and 3:00:00 (bottom). The position angle of the inner tail (arrows) changed by $10°$ in < 0.5 hr (Brandt et al. 1980b). The need for near continuous coverage is obvious. An excellent opportunity to carry this out will occur during the 1985-1986 apparition of P/Halley. Perhaps the best chance would involve use of the imaging system during the cruise phase of the Halley Intercept Mission. The product could be a near continuous record of the large-scale structure of P/Halley over weeks or even months.

Even if there is not an adequate space mission to P/Halley to study large-scale structure and solar-wind interaction, an organization to optimize the science return from the next apparition has been set up, the International Halley Watch (IHW) (Brandt et al. 1980a). Support of the IHW in establishing a network of wide-field cameras is essential to progress in understanding the large-scale structure of the cometary plasma.

REFERENCES

Alfvén, H. 1957. On the theory of comet tails. *Tellus* 9:92-96.
Astronomer Royal 1908. Meeting of Royal Astronomical Society. *Observatory* 31:432.
Barnard, E.E. 1893. Changes in Comet Brooks. *Brit. Astron. Assoc. J.* 4:59-63.
Barnard, E.E. 1903. Photographic observations of Borrelly's comet and explanation of the phenomenon of the tail on July 24, 1903. *Astrophys. J.* 18:210-217.
Barnard, E.E. 1905. On the anomalous tails of comets. *Astrophys. J.* 22:253-254.
Barnard, E.E. 1908a. Comet c 1908 (Morehouse). *Astrophys. J.* 28:292-299.
Barnard, E.E. 1908b. Photographic observations of Comet c 1908. Second paper. *Astrophys. J.* 28:384-388.
Barnard, E.E. 1909. Photographic observations of Comet c 1908 (Morehouse). Third paper. *Astrophys. J.* 29:65-71.
Barnard, E.E. 1920. On Comet 1919b and on the rejection of a comet's tail. *Astrophys. J.* 51:102-108.
Biermann, L. 1951. Kometenschweife und Solare Korpuskularstrahlung. *Zs. f. Ap.* 29:274-286.
Biermann, L. 1952. Über den Schweif des Kometen Halley im Jahre 1910. *Zs. Naturforschung* 7a:127-136.
Biermann, L. 1953. Physical processes in comet tails and their relation to solar activity. *Mém. Soc. Roy. Sci. Liège* (Ser. 4) 13:291-302.
Biermann, L.; Brosowski, B.; and Schmidt, H.U. 1967. The interaction of the solar wind with a comet. *Solar Phys.* 1:254-284.
Biermann, L., and Lüst, Rh. 1972. Some new results on the plasma-dynamical processes near comets. *MPI-PAE/Astro.* 52:1-9.
Brandt, J.C. 1968. The physics of comet tails. *Ann. Rev. Astron. Astrophys.* 6:267-286.
Brandt, J.C.; Friedman, L.D.; Newburn, R.L.; and Yeomans, D.K., eds. 1980a. *The International Halley Watch.* JPL NASA TM 82181.
Brandt, J.C.; Hawley, J.D.; and Niedner, M.B. 1980b. A very rapid turning of the plasma-tail axis of Comet Bradfield 1979ℓ on 1980 February 6. *Astrophys. J.* 241:L51-L54.
Brandt, J.C., and Mendis, D.A. 1979. The interaction of the solar wind with comets. In *Solar System Plasma Physics,* Vol. II, eds. C.F. Kennel, L.J. Lanzerotti, and E.N. Parker (Amsterdam: North Holland Publ. Co.), pp. 253-292.

Buti, B. 1981. Role of high-frequency-turbulence in cometary plasma tails. *Icarus*. In press.
Curtis, H.D. 1910. Photographs of Halley's Comet made at Lick Observatory. *Publ. Astron. Soc. Pacific.* 22:117-130.
Ershkovich, A.I. 1978. The comet tail magnetic field: Large or small? *Mon. Not. Roy. Astron. Soc.* 184:755-758.
Ershkovich, A.I. 1980. Kelvin-Helmholtz instability in Type-1 comet tails and associated phenomena. *Space Sci. Rev.* 25:3-34.
Ershkovich, A.I. 1981. On the folding phenomenon of comet tail rays. *Icarus*. In press.
Hoffmeister, C. 1943. Physikalische Untersuchungen an Kometen. I. Die Beziehungen des Primaren Schweifstrahls zum Radiusvektor. *Zs. f. Astrophys.* 22:265-285.
Ioffe, Z.M. 1966a. Comets in the solar wind. *Sov. Astron.* 10:138-142.
Ioffe, Z.M. 1966b. Comets in the solar wind. II. *Sov. Astron.* 10:517-519.
Ioffe, Z.M. 1968a. Comets in the solar wind. III. *Sov. Astron.* 11:668-671.
Ioffe, Z.M. 1968b. Some magnetohydrodynamic effects in comets. *Sov. Astron.* 11:1044-1047.
Jockers, K.K. 1979. An atlas of the tail of Comet Kohoutek 1973 XII. *Max-Planck-Inst. Aeron.* W-100-79-37.
Jockers, K.K. 1981. Plasma dynamics in the tail of Comet Kohoutek 1973 XII. *Icarus* 47:397-411.
Krishan, V., and Sivaraman, K.R. 1981. Peculiarities in the ionic tail of Comet Ikeya-Seki (1965f). *Icarus*. In press.
Lüst, Rh. 1962. Die Bewegung und Form von Strukturen im Schweif des Kometen Mrkos 1957 d. *Zs. f. Astrophys.* 54:67-97.
Marotschnik, L.S. 1963. Interaction between solar corpuscular streams and cometary atmospheres. II. "Collapsing" envelopes. Radio-frequency emission. *Sov. Astron.* 7:384-390.
Mendis, D.A., and Houpis, H.L.F. 1981. The cometary ionosphere. *Icarus*. In press.
Miller, F.D. 1981. Configurations of evolving plasma tail rays. *Bart J. Bok 75th Anniversary Symp. Vol.* In press.
Moore, E.P. 1977. Cometary ray closing rate. *Bull. Amer. Astron. Soc.* 9:618 (abstract).
Niedner, M.B. 1980. Interplanetary gas. XXV. A solar wind and interplanetary magnetic field interpretation of cometary light outbursts. *Astrophys. J.* 241:820-829.
Niedner, M.B. 1981. Interplanetary gas. XXVII. A catalogue of disconnection events in cometary plasma tails. *Astrophys. J. Suppl.* In press.
Niedner, M.B., and Brandt, J.C. 1978. Interplanetary gas. XXIII. Plasma tail disconnection events in comets: Evidence for magnetic field line reconnection at interplanetary sector boundaries? *Astrophys. J.* 223:655-670.
Niedner, M.B., and Brandt, J.C. 1979. Interplanetary gas. XXIV. Are cometary plasma tail disconnections caused by sector boundary crossings or by encounters with high-speed streams? *Astrophys. J.* 234:723-732.
Niedner, M.B., and Brandt, J.C. 1980. Structures far from the head of Comet Kohoutek. II. A discussion of the Swan Cloud of January 11 and of the general morphology of cometary plasma tails. *Icarus* 42:257-270.
Niedner, M.B.; Ionson, J.A.; and Brandt, J.C. 1981. Interplanetary gas. XXVI. On the reconnection of magnetic fields in cometary ionospheres at interplanetary sector boundary crossings. *Astrophys. J.* 245:1159-1169.
Observatory 1908. From an Oxford note-book. *Observatory* 31:471.
Orlov, S.W. 1944. The nature of comets. In Russian. See *Astr. Jber. 46,* Ref. No. 6515.
Orlov, S.W. 1945. The comet head and the classification of comet forms. In Russian. See *Astr. Jber. 46,* Ref. No. 6517.
Wallis, M.K. 1973. Solar wind interaction with H_2O comets. *Astron. Astrophys.* 29:29-36.
Wurm, K. 1968. Structure and kinematics of cometary Type I tails. *Icarus* 8:287-300.
Wurm, K., and Maffei, P. 1961. Über einige Schweifaufnahmen des Kometen Morehouse 1908 III. *Zs. f. Astrophys.* 52:294-299.
Wurm, K., and Mammano, A. 1964. The axes of the Type I tails of comets. *Icarus* 3:1-7.

PLASMA FLOW AND MAGNETIC FIELDS IN COMETS

H.U. SCHMIDT and R. WEGMANN
Max-Planck-Institut für Physik und Astrophysik

The plasma flow and field amplification resulting from interaction of the solar wind with a comet are studied by model calculations for a set of parameters covering the rates $G = 7 \times 10^{26}$ to 2×10^{30} s^{-1} of total cometary gas production. Two surfaces of discontinuity, the shock front and the contact surface, are discussed. The influence of the embedded magnetic field is investigated first in an axisymmetric model since the transverse field in the solar wind changes its direction frequently. A three-dimensional MHD-calculation then shows how the magnetic field induces an asymmetry in the plasma flow. A sudden change of direction in the interplanetary field produces streamerlike structures in the density distribution of the cometary ions.

Magnetohydrodynamic models are used to calculate the plasma flow and magnetic field near comets. The flow is similar to that of the solar wind past planets, but in comets several peculiarities appear, mainly due to the large atmosphere extending to a distance of $\sim 10^6$ km, and to the prevailing role of ionization processes. At large distances from the nucleus the solar wind is perturbed by contamination with cometary ions and a bow shock is formed. A tangential discontinuity, the contact surface, separates the plasma flow originating from the nucleus from the onstreaming solar wind which is braked by the admixture of slow and heavy cometary ions. In the transition region between the shock front and the stagnation point the magnetic field is piled up and amplified. This has important consequences for the dynamics of the flow, since it enhances the effective stagnation pressure, depletes the plasma, and may form conspicuous density structures and flux ropes.

Furthermore, one can study the relative influence on the overall flow field of specific processes such as photoionization, charge exchange, and collisional ionization, or dissociation, or elastic exchange of momentum with neutral particles. Some of these features have been studied by numerical hydrodynamic and magnetohydrodynamic model calculations for the steady flow in the region outside the ionopause. The effect of a sudden change in the direction of the transverse interplanetary field has been simulated by fitting two stationary solutions.

For earlier reviews in this field see Biermann (1974) or Brandt and Mendis (1979). In this chapter we report mainly on our own numerical calculations. In Sec. I we introduce the basic assumptions common to these models, together with several simplifications; Sec. II discusses the mechanisms that control the stand-off distances of the bow shock and the stagnation point on the contact surface; in Sec. III we present a set of rotationally symmetric hydrodynamic models for different rates of gas supply from the nucleus. The next two sections deal with the specific effects of the interplanetary magnetic field. First we study an axisymmetric model, where only the effects of the field averaged over all directions are taken into account. This averaging is justified in that the interplanetary field changes its direction frequently compared with the travel time of the field through the coma. Second, we present the result of a fully three-dimensional calculation which may be used to explain streamers. In Sec. VI we discuss some open questions.

I. BASIC ASSUMPTIONS

The general flow pattern is depicted in Fig. 1. The cometary nucleus is a source of radially expanding gases, which are gradually ionized. Thus the comet forms an extended source of ions in the supersonic solar wind. Even a contamination with 1% of heavy slow cometary ions brakes the solar wind so strongly that a shock front must evolve at a large distance from the nucleus (Biermann et al. 1967). In the subsonic regime behind the shock the wind is further braked as it is saturated with cometary particles, and finally reaches a stagnation point where the ram pressure of the solar wind balances the ram pressure of the expanding purely cometary plasma streaming from the nucleus. The stagnation point lies on a paraboloid contact surface which separates the streamlines originating on the Sun from those originating on the nucleus. This surface constitutes a tangential discontinuity in the flow and may be called a cometary ionopause. Under stable conditions and for perfect conductivity it cannot be penetrated by solar magentic fields or particles, unless they undergo a charge exchange. Possible violations due to the interchange instability and finite resistivity will be discussed in Secs. II and VI.

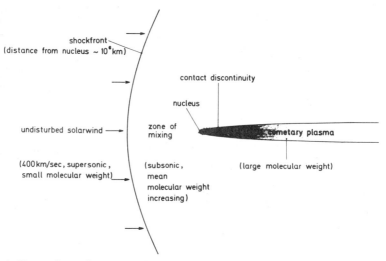

Fig. 1. Flow regimes of cometary plasma.

To describe these processes we proposed a simplified magnetohydrodynamic model (Schmidt and Wegmann 1980). Inherent in this model is the equilibrium which establishes a unique mass velocity **u** and temperature T in the plasma so that we can deal with macroscopic variables like pressure p, mean molecular weight μ, mass density $\rho = N\, m_p \mu$ where m_p is the proton mass, and number density of ions N. We write the inviscid magnetohydrodynamic equations for perfect electrical conductivity and adiabatic changes (see e.g., Landau and Lifshitz 1960, p. 213) as conservation laws for mass, momentum, energy, and magnetic flux, modified by source terms on the right hand side:

$$\frac{\partial}{\partial t}\rho + \mathrm{div}\,\rho\mathbf{u} = A\, m_c \qquad (1)$$

$$\frac{\partial}{\partial t}\rho\mathbf{u} + (\mathbf{u}\cdot\mathrm{grad})\,\rho\mathbf{u} + \rho\mathbf{u}\,\mathrm{div}\,\mathbf{u} + \mathrm{grad}\,p - \frac{1}{4\pi}(\mathrm{curl}\,\mathbf{B})\times\mathbf{B} = A\, m_c \mathbf{w} \qquad (2)$$

$$\frac{\partial}{\partial t}\left(\frac{\rho u^2}{2} + \frac{p}{\gamma-1} + \frac{B^2}{8\pi}\right) + \mathrm{div}\left[\mathbf{u}\left(\frac{\rho u^2}{2} + \frac{\gamma p}{\gamma-1} + \frac{B^2}{4\pi}\right) - \frac{1}{4\pi}(\mathbf{u}\cdot\mathbf{B})\,\mathbf{B}\right]$$
$$= A\, m_c \frac{w^2}{2} \qquad (3)$$

$$\frac{\partial}{\partial t}\mathbf{B} - \mathrm{curl}\,\mathbf{u}\times\mathbf{B} = 0 \qquad (4)$$

where γ is ratio of specific heats and **B** magnetic field. The right hand sides represent the gain of mass, momentum, and energy by the insertion of new ions. These ions are formed by photoionization at a constant rate σ from the originally neutral particles evaporated from the nucleus. We assume that these neutral molecules all have the same mass m_c and expand radially from the comet with a constant velocity w. The source term A for the particle density is proportional to the density of the neutrals N_n and is at a distance R from the nucleus given by

$$A = \sigma N_n = \frac{G \sigma}{4 \pi w R^2} \exp\left(-\frac{\sigma R}{w}\right) \qquad (5)$$

where G is total number of molecules per second released from the comet.

These equations are to be solved with these boundary conditions: far from the nucleus towards the Sun the flow becomes homogeneous with the density ρ_w, velocity u_w and Mach number Ma_w of the unperturbed solar wind; along an assumed rigid contact surface the normal components of velocity and magnetic field vanish.

Ionization mechanisms besides photoionization, e.g. charge exchange and ionization by electron impact, can be included by a suitable modification of the source terms. Other important microscopic processes, e.g. the heating due to surplus energies gained by the particles in ionization and dissociation processes, the exchange of momentum between plasma particles and the neutral background in elastic collisions, the cooling of electrons in inelastic collisions which lead to radiative loss, etc., could be incorporated at least in a similar global way. Unless otherwise noted we shall assume here a constant rate of photoionization and a radiative cooling of the electrons so effective that the electrons do not contribute to ionization or gas pressure.

The most stringent limitation of a magnetohydrodynamic model lies in the assumption that the plasma is in local thermodynamical equilibrium. Collisions occur extremely rarely beyond the innermost part of the coma. With a gas-kinetic cross section $Q = 3 \times 10^{-15}$ cm^2 the exospheric radius R_e which may be defined by the equation

$$\int_{R_e}^{\infty} N_n Q \, dR = 1 \qquad (6)$$

is even for a big comet ($G = 10^{30}$ s^{-1}, $w = 1$ km s^{-1}) only 3×10^4 km. Nevertheless, the common velocity for solar and cometary particles can be established quickly. The velocity components perpendicular to the field are adjusted within a period of gyration and the component parallel to the field adjusted within the time scale for the two-stream instability. The effectiveness of the latter has not yet been investigated for cometary conditions. For a discussion of the observed strong coupling of heavier ions to the speed of the solar wind see Neugebauer (1974), Feldman (1979), and the literature cited therein. The assumption of a common temperature is a simplification necessary to avoid the integration of a complete set of cometary velocity distributions for longitudinal motions as well as gyration with source terms and adiabatic changes. For a discussion of some of the deviations from equilibrium we refer the reader to the chapter by Ip and Axford in this book. In the solar wind instabilities seem to equalize particle thermal velocities but not temperatures of hydrogen and helium; see e.g. Feldman (1979). Only at the higher densities at which Coulomb collisions become effective are deviations from equilibrium in temperature, and in flow velocity, gradually suppressed (Neugebauer and Feldman 1979). Yet even when, in the outer collisionless cometary region, the heavy cometary ions picked up by the solar wind keep their initial high speed relative to the solar wind $u + w$, as thermal speed of gyration, the conservation laws (see Eqs. 1 through 4) are not expected to be grossly violated. For example, in the idealized 2-dimensional flow with purely transverse field of fixed direction and gyration as the only thermal motion, any solution for $\gamma = 2$ describes the flow exactly, whether the partial pressures of the different constituents contributing to p are isothermal or not. Obviously in collisionless three-dimensional flow deviations become possible and even the choice of γ becomes rather arbitrary. Spreiter et al. (1966) have shown that the hydromagnetic flow around a planetary magnetosphere does not depend much on the choice of γ; for flow past nonmagnetic planets see Spreiter et al. (1970). We assume that the electrons do not get their full share of energy in the collisionless bow shock, and that they are cooled very efficiently in the coma by collisions with ions that radiate the excitation energies (see Biermann and Trefftz 1964 for a discussion of these processes). Therefore the electrons are treated as a cool component whose effects on the hydro- and thermo-dynamical behavior are negligible.

II. BOW SHOCK AND STAGNATION

The stationary flow of the originally supersonic and homogeneous solar wind around a source of cometary particles expanding with supersonic velocity from the nucleus contains three free surfaces:

1. The tangential discontinuity or contact surface in the cometary ionopause, which separates streamlines of cometary origin from streamlines of the solar wind and has a subsolar stagnation point on both sides;
2. The hyperboloidal bow shock where the solar wind changes from supersonic to subsonic;
3. Another shock inside the cometary ionosphere where the supersonic expansion off the nucleus changes to subsonic stagnation flow.

An attempt to determine the ionopause and the ionospheric shock simultaneously by a boundary layer approach was made by Houpis and Mendis (1980).

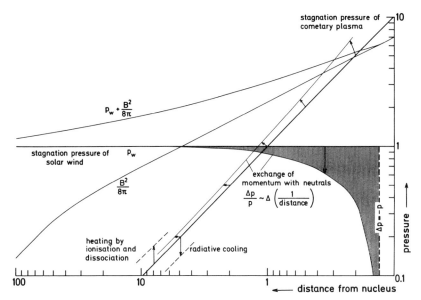

Fig. 2. Pressure balance at stagnation (see Sec. I in the text).

In view of our limited knowledge of the physical processes in the cometary ionosphere we restricted our own calculations to the flow regime outside the ionopause. Thus, we avoid some numerical problems with the large differences in sizes and characteristic speeds in the flow regions. We choose as the inner boundary a fixed contact surface on which the normal components of velocity and magnetic field vanish. This contact surface of the plasma flow should closely resemble the ionopause. We assume it to be a paraboloid. The stand-off distance of the subsolar stagnation point must be derived from the balance of the estimated stagnation pressures inside and outside the ionopause. The different processes affecting this balance are illustrated schematically in Fig. 2. On both sides we can assume that the stagnation streamline crosses the shock at right angles. The stagnation pressure p_W of the solar wind can be calculated for given γ from gas dynamics, when $\rho_W u_W^2$ and the Mach number Ma_W of the unperturbed solar wind are known (Landau and Lifshitz 1959, p. 458). For hypersonic solar wind ($Ma_W \to \infty$) and $\gamma = 2$ the result is

$$p_W = 0.84\, \rho_W u_W^2 \ . \tag{7}$$

The source term (Eq. 2) reduces this stagnation pressure continuously, but not by as much as an order of magnitude. The numerical results give $p_W = 0.6\, \rho_W u_W^2$ for a wide range of parameters (see Table I in Sec. III). For the stagnation pressure p_c of the cometary plasma we assume by analogy to Eq. (7),

$$p_c = 0.84\, \rho_c w^2 = 0.84\, \frac{G m_c}{4\pi R^2 w} \cdot \frac{\sigma R}{w} \cdot w^2 \; . \tag{8}$$

Since $p_c w^2$ varies as R^{-1} the stagnation pressure (Eq. 8) would be correct only at the position of the inner shock itself; i.e. the finite thickness of the layer between the inner shock and ionopause is neglected and Eq. (8) gives only a lower limit. Houpis and Mendis (1980) have estimated the layer thickness for a point source of cometary ions but no estimates exist for the realistic case of an extended source of ions. From equilibrium between the two pressures (Eqs. 7 and 8) we get an estimate of the stand-off distance

$$R_S = \frac{G \sigma m_c}{4\pi \rho_w u_w^2} \; . \tag{9}$$

There are several other effects which can modify this estimate. The stagnation pressure of the cometary plasma can be enlarged due to the surplus energy of dissociation products (see chapter by Huebner et al. in this book) or photoelectrons, or it can be reduced by cooling of the electrons which by collisional excitation gives rise to radiative losses.

In Fig. 2 we arbitrarily assumed a heating which enlarges the pressure p_c by one order of magnitude over $0.84\, \rho_c w^2$. Hence also R_S is enlarged by a factor of 10. The enlarged distance R_S is used as unit in the abscissa of Fig. 2. The same R_S is also drawn as a guideline in Fig. 3 for the stand-off distance of the stagnation point. The crosses represent the actual choice of stand-off distances for the 4 hydrodynamic solutions discussed in the next section.

Furthermore, in the region dominated by collisions, inside the exospheric radius defined by Eq. (6), the neutral particles expanding radially with velocity w will transfer momentum to the ions in the stagnation flow. This will be effected on both sides of the contact discontinuity by elastic collisions with cross sections of order $Q_{ni} = 3 \times 10^{-15}$ cm^2 (Biermann 1974). Here we want to give an order of magnitude estimate of this additional outward volume force $\mathbf{f}_{ni}$ not incorporated in Eq. (2) and not accounted for in the numerical calculations discussed in Secs. III through V. This force can be written (Chapman and Cowling 1952) as a product of particle densities N and N_n, particle mass m_c, bulk velocity vector of the neutrals relative to the ions $\mathbf{w} - \mathbf{u}$, cross-section Q_{ni}, and relative speed of the colliding particles averaged over their thermal distributions v_{rel} (the term $N_n Q_{ni} v_{rel}$ is often called collision frequency):

$$\mathbf{f}_{ni} = N N_n m_c (\mathbf{w} - \mathbf{u}) Q_{ni} v_{rel} \; . \tag{10}$$

On the stagnation line $|\mathbf{w} - \mathbf{u}| = w + u$. For a crude estimate we can assume $\sigma R/w \ll 1$ in Eq. (5), so that $N_n = G/4\pi R^2 w$ and $v_{rel} = w + u + v_{th}$, where the thermal velocity v_{th} is connected with the pressure of the ions by

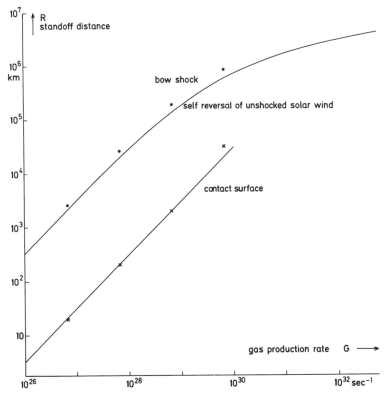

Fig. 3. Standoff distances of bow shock and stagnation point. Dots, calculated bow shocks; crosses, assumed stagnation points.

$p = N\, m_c v^2_{\text{th}}$. In stagnation flow the volume force Eq. (10) can only be balanced by a positive pressure gradient dp/dR (or by an inward Lorentz force), which in turn influences the standoff distance of the stagnation point (Eq. 9). With our estimates this pressure gradient in the stagnation region on both sides of the contact surface amounts to

$$\frac{dp}{dR} = p\, M(M+1)\, Q_{ni}\, \frac{G}{4\pi R^2 w} \qquad (11)$$

where M is the ratio $(w + u)/v_{\text{th}}$. On the solar side numerical models (see Sec. III) exhibit a plateau $M(1 + M) = 0.2$ over a wide range in R so that we can integrate Eq. (11) with $\int_R^\infty \frac{dR}{R^2} = \frac{1}{R}$. On the cometary side we assume $u \ll w = v_{\text{th}}$ or $M(1 + M) = 2$ due to the thermalization by the shock. Assuming somewhat arbitrarily a thin shock layer $\int_{R_{\text{shock}}}^R \frac{dR}{R^2} = \frac{\Delta R}{R^2} = \frac{0.1}{R}$, we get the same amount of change in the pressure on both sides

$$\Delta p = \pm 0.2 \, \frac{G \, Q_{ni}}{4\pi R \, w} \, p \qquad (12)$$

plus for the inside and minus for the outside of the contact surface. In a more rigorous treatment one would also consider that in the elastic collisions the ions release thermal energy to the neutrals. With R_c as the critical distance where with $\Delta p = -p$ the pressure of the solar wind would drop to zero, we get the ratio

$$\frac{R_c}{R_s} = 0.2 \, \frac{Q_{ni} \rho_w u_w^2}{w \, \sigma \, m_c} = 1 \qquad (13)$$

for typical values of the parameters. This ratio is independent of cometary gas production and even of distance from the Sun, as both ρ_w and σ must vary with the inverse square of that distance.

Finally there is the influence of the transverse magnetic field which contributes its own pressure $B^2/8\pi$ along the contact surface. In ideally conducting axisymmetric flow with div $n\mathbf{u} = 0$ a relation

$$B^2 u/n = \text{const} \qquad (14)$$

holds on the axis (Lees 1964). In our model div $n\mathbf{u} = 0$ applies for the density of solar protons $n = N_p$. Equation (14) implies a singularity in B if N_p stays finite and u vanishes. The pileup in solar protons can be relaxed by expansion in both transverse directions, but the pileup in magnetic field is relaxed only by transverse expansion perpendicular to the field. Whereas in the unperturbed solar wind the gas and magnetic pressures are comparable, the gas pressure dominates immediately behind the shock by ~1 order of magnitude. Therefore Lorentz forces can be neglected. Purely gas-dynamic calculations for this region (see Sec. III) show that u/R is approximately constant on the axis, so we get an increase of the magnetic pressure $B^2/8\pi$ proportional to $1/R$ until it competes again with the almost constant gas pressure.

From this point on, the field influences the flow. The curved field lines exert a volume force toward the nucleus included in Eq. (2). An axisymmetric magnetohydrodynamic calculation (see Sec. IV) yields u proportional to $R^{1/2}$ as soon as $B^2/8\pi$ is comparable to p. Therefore the magnetic pressure now increases roughly as $1/R^{1/2}$ and the total pressure at the stagnation point reaches values substantially above the stagnation pressure of the gas dynamic solution (Eq. 7). This increase is due to a systematic transport of momentum along the field lines from the passing solar wind to the axis.

But the Lorentz force in Eq. (2) leads also to an asymmetric expansion in the flow which further relaxes the magnetic tensions. A three-dimensional magnetohydrodynamic calculation (see Sec. V) therefore yields a weaker increase of the magnetic pressure. The total pressure at the stagnation point in this case also reaches values above the gas dynamic stagnation pressure (Eq. 7). This action of the Lorentz force is achieved only because the radius of curvature of the field lines in a comet remains comparable to the stand-off distance from the nucleus over several orders of magnitude from the bow shock to the contact surface. Such action is not possible in a planetary ionosphere like Venus' because the planetary gravity forces the stand-off distances of the bow shock and of the ionopause to be comparable.

Lipatov (1978) calculated the field amplification for the ionosphere of Venus and discussed the field diffusion caused by finite conductivity. In a comet collisions of the electrons with ions and especially with neutral particles will also cause finite conductivity in the inner coma. Therefore one should consider the diffusion of magnetic field across the contact surface. This process may be further enhanced by the interchange instability (Bernstein et al. 1958) and it will somewhat reduce the increase of total pressure. The criterion for this instability is fulfilled because of the curvature of the dominating magnetic field on the outside, and because of the positive pressure gradient dp/dR (Eq. 11) due to the momentum exchange on the inside. The instability may even cast doubt on the existence of a stationary contact surface. It is obvious that a future rigorous theory of contact surface and stagnation flow must include in detail all the aforementioned processes.

We obtain an estimate for a close lower limit of the stand-off distance R_b of the bow shock by considering the Fanno-curve (see Massey 1979, p. 418) for the change of the almost plane-parallel supersonic flow before the shock with increase of mean molecular weight caused by the dominating source term (Eq. 1). This Fanno-curve ends with a vertical tangent in a turning point or self-reversal where the supersonic branch meets a subsonic branch. This point corresponds to an increase in mean molecular weight by a factor $\gamma^2/(\gamma^2 - 1) = 4/3$ over the value in the hypersonic limit (Biermann et al. 1967). Hence the supersonic flow can only carry an added relative mass load of 1/3. This critical mass load is obtained by contamination with $\sim 1\%$ of cometary ions. Assuming that the flow remains parallel before the shock, the added mass is equal to the integrated source term in Eq. (1). Hence we obtain for the distance R_{sr} of the self reversal the equation

$$\frac{\rho_w u_w}{\gamma^2 - 1} = \int_{R_{sr}}^{\infty} \frac{Gm_c}{4\pi R^2} \frac{\sigma}{w} \exp(-\frac{\sigma R}{w}) dR = \frac{Gm_c \sigma^2}{4\pi w^2} \int_{R_{sr} \cdot \sigma/w}^{\infty} \frac{e^{-y}}{y^2} dy . \quad (15)$$

For $R_{sr} \ll w/\sigma$ the integral on the right hand side is about $w/\sigma R_{sr}$, hence the linear relation (Wallis 1973b):

$$R_{sr} = \frac{Gm_c \sigma (\gamma^2 - 1)}{4\pi w \rho_w u_w} . \quad (16)$$

For $G \sim 10^{30}$ s^{-1} a dependence $R_{sr} \sim G^{1/2}$ is more appropriate, in accordance with the findings of Biermann et al. (1967), while for larger values of G the exhaustion of neutrals beyond the ionization scale length w/σ leads to a logarithmic growth of R_{sr} with G. The actual distance R_b of the bow shock must be larger than R_{sr}. This lower limit of R_b is only applicable as long as the flow before the shock is very nearly plane-parallel. Model calculations (Brosowski and Wegmann 1973) confirm the nearly parallel flow in Eq. (1). Wallis (1973a) has shown that only for a fictitious source term (Eq. 1) that falls off more slowly than $R^{-1.1}$ would the divergence of the supersonic flow suffice to cause a gradual transition to subsonic flow without a bow shock.

In Fig. 3 we show how the distance R_{sr} calculated from Eq. (15) depends on G, and compare it with the calculated bow shock distances for four hydrodynamic models described in Sec. III. The distance R_b is very close to R_{sr}. This indicates that in fact the bow shock is caused by contamination with cometary ions rather than by the obstacle of the cometary ionopause, whose dimension R_s is $\geqslant 2$ orders of magnitude $< R_b$.

If Eqs. (1) and (3) are rewritten in dimensionless variables $\rho^* = \rho/\rho_w$, $u^* = u/u_w$, $p^* = p/\rho_w u_w^2$, $R^* = \sigma R/w$, the result is that in regions with $u \gg w$ the influence of the comet on the hydrodynamic flow with $B = 0$ is caused only by the source term in the continuity Eq. (1). Therefore the flow outside the stagnation region depends only on the dimensionless parameter

$$P = \frac{m_c G \sigma^2}{4\pi w^2 \rho_w u_w} \qquad (17)$$

which is the ratio of the fluxes of cometary molecules and solar protons at distance w/σ from the nucleus. At the limit $P \to 0$ this dependence reduces to a linear scaling, so that asymptotically the solution depends only on R^*/P.

III. HYDRODYNAMIC MODELS

A set of purely hydrodynamic, axisymmetric models has been calculated for four values of gas production rate G covering 4 orders of magnitude, with the following values of the other parameters:

comet: $m_c = 23.3$ m_p, $\sigma = 10^{-6}$ s^{-1}, $w = 1$ km s^{-1}

solar wind: $\rho_w = 5\, m_p$ cm^{-3}, $u_w = 350$ km s^{-1}, $Ma_w = 10$ (18)

interplanetary field: $B = 0$.

The chosen value of m_c is the average molecular weight in a mixture of molecules H_2O, CO, CO_2 and N_2 with ratios 100:15:15:10.

TABLE I

Shock Parameters and Subsonic Pressures
for Different Gas Production Rates

Particle Production Rate G (s^{-1})	Bow Shock Distance R_b (km)	Mean Molecular Weight μ	Mach Number Ma	Max. Pressure on Stagnation Line p_{max} (10^{-9} dyn cm^{-2})	Stagnation Pressure P_{stag} (10^{-9} dyn cm^{-2})
7×10^{26}	2.76×10^3	1.23	1.7	5.9	5.5
7×10^{27}	2.6×10^4	1.22	1.7	5.9	5.5
7×10^{28}	1.97×10^5	1.21	1.7	5.9	5.4
7×10^{29}	0.86×10^6	1.17	2.0	6.2	5.7

In Table I we list the calculated bow shock distances R_b (see also Fig. 3), the mean molecular weight μ at the shock, the Mach number Ma before the shock, the maximum pressure on the stagnation line p_{max}, and the actual pressure at the stagnation point p_{stag}.

The first calculations of Biermann et al. (1967) with a one-dimensional approximation yielded rather strong shocks with Mach numbers almost as high as the undisturbed $Ma_w = 10$. Wallis (1973a,b) on the other hand argued in favor of a weak or even nonexistent shock and calculated the subsonic flow on the axis of symmetry assuming incompressibility. The two-dimensional calculations of Brosowski and Wegmann (1973) confirmed the weak shock hypothesis.

In Figs. 4 through 7 we present contour profiles in a meridional plane for number densities of protons and cometary ions, velocity, gas pressure, and temperature. The ion density charts resemble those of cometary gas tails closely, as do the velocity charts. The pressure charts illustrate strong accelerating gradients and also a substantial inclination between these and the gradients of mass density, almost at right angles. This implies a curl in the electromotive forces which can induce azimuthal magnetic flux (Biermann and Schlüter 1951)

$$\frac{\partial \mathbf{B}}{\partial t} - \operatorname{curl} \mathbf{u} \times \mathbf{B} = -\frac{c m_p}{2 e \rho^2} \operatorname{grad} \rho \times \operatorname{grad} p \qquad (19)$$

where c is velocity of light and e electronic charge. But if we insert numbers we see that even for the smallest comet with $G = 7 \times 10^{26}$ s^{-1}, less than $1\gamma = 10^{-5}$ gauss is induced within the convective time scale. Therefore the effect can be neglected.

The temperature chart illustrates shock heating to coronal temperatures for solar particles and tremendous cooling in the coma and tail because of the

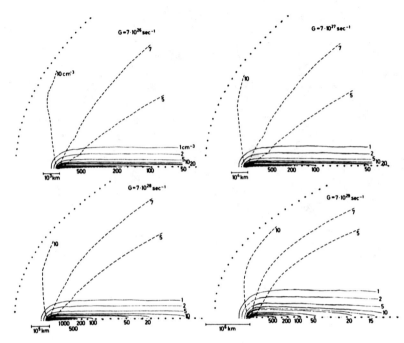

Fig. 4. Hydrodynamic solutions, contour maps in meridional plane: solid lines, proton density; dashed lines, ion density; broken lines, bow shock and contact surface.

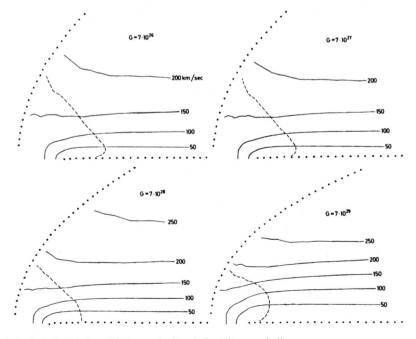

Fig. 5. As in Fig. 4: solid lines, velocity; dashed lines, sonic line.

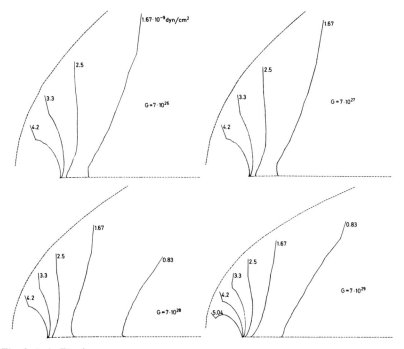

Fig. 6. As in Fig. 4: gas pressure.

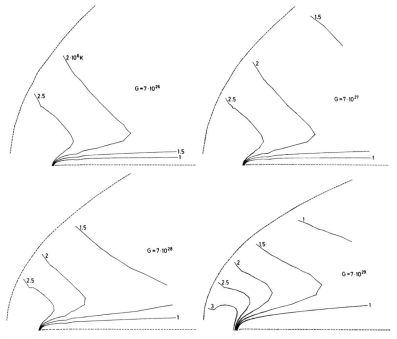

Fig. 7. As in Fig. 4: temperature.

distribution of energy among the bulk of cometary ions. But even with this crude apportioning we get some hot cometary ions behind the shock near the axis. The hot isotherms move close to the nucleus.

IV. SYMMETRIZED MAGNETOHYDRODYNAMIC MODELS

The purely hydrodynamic description becomes insufficient at ~0.1 times the distance of the bow shock from the nucleus, since here the pileup of magnetic flux in the axisymmetric and nearly incompressible subsonic flow produces a magnetic pressure that can compete with stagnation pressure. The Lorentz force of this field has important dynamic consequences. As the field lines are heavily bent around the stagnation region, the relatively weak gradient in field strength toward the nucleus necessarily implies a strong volume force in the same direction. This additional momentum imparted to the flow along the axis originates from the momentum of the solar wind passing around the sides of the comet. The magnetic field itself transports this momentum towards the axis via its Maxwellian stress. To determine this and other dynamic effects of the enhanced magnetic fields near the nucleus we integrate magnetohydrodynamic models. Since travel from the bow shock to the stagnation point takes 10 to 20 hr we must allow for the fact that the interplanetary field changes direction (but not strength) suddenly by ~1 rad in intervals of < 1 hr (Burlaga 1971). Assuming a transverse field which changes direction at random so frequently that the flow cannot adjust to the azimuthal forces, we can still describe the average flow pattern as axisymmetric. We express the instantaneous field B in cylindrical coordinates r, ϕ, z with first order harmonics

$$B_r = \mathrm{B}_r \cos(\phi-\chi)$$
$$B_z = \mathrm{B}_z \cos(\phi-\chi) \qquad (20)$$
$$B_\phi = \mathrm{B}_\phi \sin(\phi-\chi)$$

where χ is the angle between the meridional plane parallel to the instantaneous field and the plane $\phi = 0$. The coefficients B are functions of r and z only. When $u\phi \equiv 0$, the field Eq. (4) assures that no other harmonics can be generated if the field originally satisfies Eq. (20). Now the Lorentz forces and the magnetic terms in the energy equation, all quadratic in B, can be averaged over a uniform stochastic distribution in χ and give nonvanishing contributions. Only the average azimuthal Lorentz force vanishes. For details of this method see Schmidt and Wegmann (1980). Results of calculations with this model are shown in Fig. 8 for the streamlines and the isochrones, identical with the average field lines, and in Fig. 9 for the distribution of gas pressure, velocity, and mass density. The parameters of these solutions are:

comet: $G = 2 \times 10^{30} \mathrm{s}^{-1}, m_c = 15\, m_p, \sigma = 10^{-6} \mathrm{s}^{-1}, w = 1\ \mathrm{km\ s}^{-1}$

solar wind: $\rho_w = 5\, m_p\ \mathrm{cm}^{-3}, u_w = 400\ \mathrm{km\ s}^{-1}, Ma_w = 10$

interplanetary field: $B = 0, 5, 10 \gamma\ (1\gamma = 10^{-5}\ \mathrm{gauss})$. $\qquad (21)$

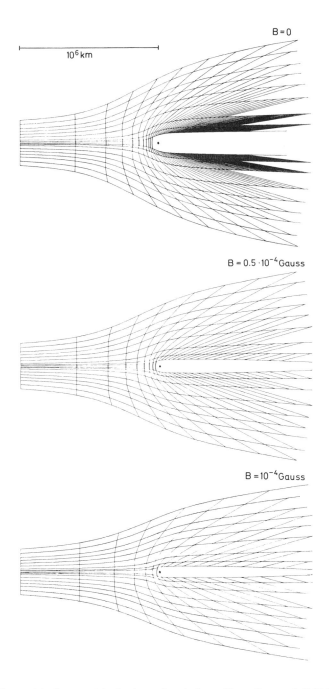

Fig. 8 Symmetrized magnetohydrodynamic solutions. Streamlines and field lines (isochrones $\Delta t = 1$ hr) start at the bow shock on the left. The region near the contact surface is omitted for clarity.

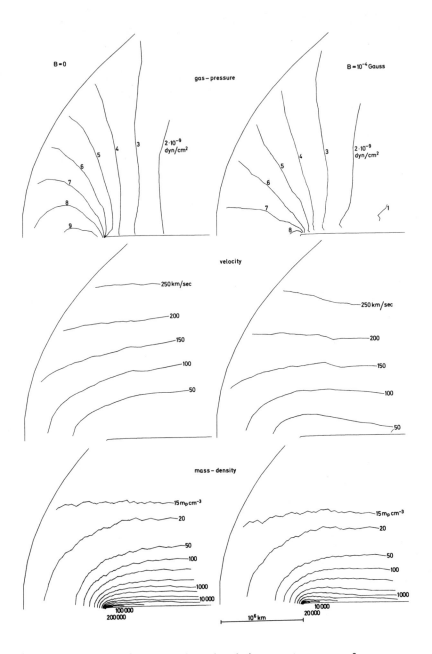

Fig. 9. Symmetrized magnetohydrodynamic solutions: contour maps of gas pressure, velocity, and mass density in a meridional plane: broken lines, bow shock and contact surface.

From the comparison of the isochrones at intervals of 1 hr in Fig. 9 it is apparent that the magnetic field accelerates the flow in the inner region and the tail. The travel time along the innermost streamlines in Fig. 9 is about 4 d for zero field and 17 hr for 10-γ field. Also, as one might expect, the streamlines are pushed towards the axis by the Lorentz forces. The gas pressure and density charts in Fig. 9 show a depletion of matter near the axis, which due to the rotational symmetry of the model is necessary to relax the pileup of magnetic flux. Such an effect of the interplanetary field near obstacles is observed in the case of Venus (Elphic et al. 1980). The field amplification is very effective in this axisymmetric case. The maximum field strength at the stagnation point reaches 67 and 94 γ for interplanetary values of 5 and 10 γ respectively. The corresponding values of total pressure are 3.1 and 5.9 $\times$ 10^{-8} dyn cm^{-2} as compared to a gas dynamic stagnation pressure, (Eq. 7), of 1.1×10^{-8}. These maxima are very steep and have a half width of the same order as the standoff distance of the stagnation point.

V. A THREE-DIMENSIONAL MODEL CALCULATION

We have solved the fully three-dimensional MHD-equations for a homogeneous interplanetary field of 5 γ and the other parameters as in the last section. Results are shown in Figs. 10 through 12. The plots for field strength B (Fig. 12a) in meridional planes parallel ($\theta = 0°$) and perpendicular ($\theta = 90°$) to the direction of the interplanetary field show that in this asymmetric case there is also an amplification of the field near the stagnation point, where the

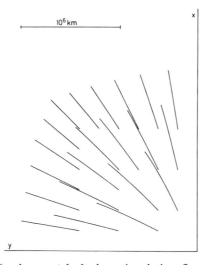

Fig. 10. Three-dimensional magnetohydrodynamic solution. Some streamlines are projected into the plane perpendicular to the comet axis. The interplanetary field is parallel to the x-axis.

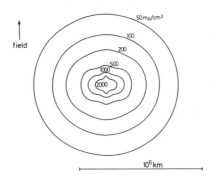

Fig. 11. Three-dimensional magnetohydrodynamic solution. Density is given in a cross section though the tail 8×10^5 km behind the nucleus. Direction of the interplanetary field is indicated.

field reaches a maximum of 43 γ. The total pressure at this point is 2×10^{-8} dyne cm^{-2}, about twice the gas dynamic stagnation pressure (Eq. 7). But this region of strong field and pressure is as small as in the axisymmetric model of Sec. IV. The field strength in the tail depends strongly on the azimuthal angle.

By investigation of several phenomena in cometary tails, different estimates for the field strength have been derived and can be compared with our results in Secs. IV and V. Ershkovich (1978,1981) argued in favor of rather weak fields (not much stronger than the interplanetary field), while Hyder et al. (1974) and Ip and Mendis (1976) suggest fields of 100 to 1500 γ in the tail. From laboratory experiments Podgorny et al. (1980) concluded that the tail field should be 3 to 5 times stronger than the interplanetary field. Lipatov (1978) obtained in his numerical calculations for the ionosphere of Venus a field amplification by a factor of 2 to 10, and also discussed the effect of field diffusion due to finite conductivity.

The contours for B_z/B show how the field lines are bent around the stagnation region; they remain nearly plane curves. The component B_y perpendicular to the plane is always $< 10\%$ of B. Figure 10 shows that the streamlines lean towards the meridional plane perpendicular to the field direction since the Lorentz forces try to relax the magnetic tensions by lateral expansion. The result of this transport towards the plane of symmetry perpendicular to the field is clearly visible in Fig. 11. Only the small nose of the innermost density contour very near the plane of symmetry parallel to the field reminds us that here the Lorentz force cannot contribute to the tailward acceleration. The matter is moved only by the gas pressure gradient which is particularly weak here as can be concluded from the low β and the isotropic distribution of the total pressure $p + B^2/8\pi$ (see Fig. 12). Thus the asymmetry imposed by the field onto the flow field is mainly due to the curvature force $(1/4\pi)$ (B grad)B, which tries to push the matter towards the meridional plane perpendicular to the field.

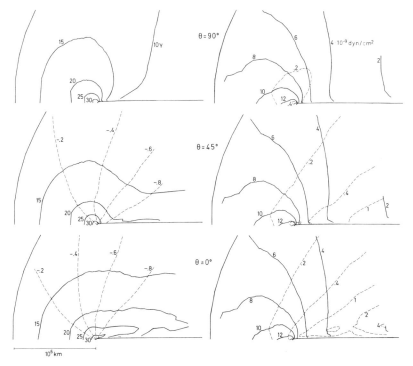

Fig. 12. Three-dimensional magnetohydrodynamic solution, contour maps of field and pressure in three meridional planes which are inclined by an angle θ with respect to the interplanetary field. Left: solid lines, B; dashed lines, B_z/B (z is the coordinate along the comet axis). Right: solid lines, $p + B^2/8\pi$; dashed lines, $B^2/8\pi p$.

In order to simulate the reaction of the cometary tail to a sudden 90° change in direction of the interplanetary field, we fit two three-dimensional stationary solutions of different field directions along an isochronic surface. The isotropy in the total pressure suggests that this fit may be made along an axisymmetric surface, obtained by rotation of an isochrone calculated from the axisymmetric solution of Sec. IV. The resulting time dependent model is then used to integrate column densities along lines of sight perpendicular and parallel to the original magnetic field at different epochs 0 hr, 10 hr, and 12 hr after the new field touches the bow shock. These column densities are represented as isophote charts in Fig. 13. The first projection clearly simulates a streamer as can be judged from comparison with the isophotes of Comet Morehouse 1908 III in Fig. 14 (from Högner and Richter 1969), whereas in the other orientation only a slight bend in the isophotes can be seen at the location of the common isochronic surface. We believe that the resemblance in the first case gives a strong hint that the streamers visible in a cometary tail present a detailed chart of the directional changes in the mag-

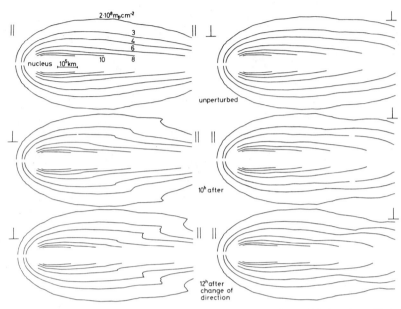

Fig. 13. Three-dimensional magnetohydrodynamic streamer model showing effect o sudden change in the field direction. Isophotes are for line of sight perpendicular (left) and parallel (right) to the original field: ∥, field perpendicular to line of sight; ⊥, field parallel to line of sight.

Fig. 14. Isophotes of Comet Morehouse 1908 III (from Högner and Richter 1969).

netic field which passed the comet within 10 to 20 hr (Schmidt 1974). These streamers will be aligned with the average field lines as given in Fig. 8. Miller (1979, 1981) has compared these field lines with observed tail rays.

VI. DISCUSSION

The model streamer just described moves towards the tail axis and finally drapes the contact surface before it vanishes. The lifetime of the visible structure is equal to the travel time in the laminar flow in the model of Sec. V. There are reasons to believe that the magnetic structure may form a discrete flux rope which remains much longer with the comet. After dispersion of the previous magnetic field the discontinuity will reach the contact surface at the stagnation point. From this time on the curvature of the new magnetic field outside and the positive pressure gradient dp/dR inside (see Eq. 11) will cause an interchange instability (Bernstein et al. 1958). Any stabilizing effect of the cometary gravity is negligible since the mass of the cometary nucleus is small ($< 10^{18}$ g). This instability will be most effective for the narrow bundle of strong magnetic field in the immediate neighborhood of the stagnation point. This bundle will penetrate into the ionosphere with approximately Alfvén speed and a longer living flux rope will be formed. Its magnetic flux will still extend across the ionopause into the tail and the interplanetary medium. Here it is embedded into other younger plasma that slides more freely along the ionopause. The flow of the surrounding plasma is nonaxisymmetric because of the embedded magnetic field. Since the interplanetary field changes its direction frequently (Burlaga 1971), it is likely that the flux rope is inclined with respect to the field in the passing plasma by a nonzero angle α. We can assume that there is shear in the nonaxisymmetric part of the velocity, as in the meridional velocities (see Fig. 5). Then the nonaxisymmetric flow exerts a torque on the flux rope which has different signs on the opposite strands. Therefore the rope is twisted systematically by an amount varying as $\sin 2\alpha$. This twist finally induces an electric current. The average interplanetary field is in the ecliptic plane, so the angle α is on the average the inclination of the flux rope relative to the ecliptic, and the induced current should be correlated with this inclination. Similar twisted flux ropes have been observed in the ionosphere of Venus (Elphic et al. 1980; see also the chapter by Russell et al. in this book and the literature cited therein). Measurements of these flux ropes might test our hypothesis.

Acknowledgments: We wish to thank M. Neugebauer and B. Goldstein for helpful comments.

REFERENCES

Bernstein, I.B.; Frieman, E.A.; Kruskal, M.D.; and Kulsrud, R.M. 1958. An energy principle for hydromagnetic stability problems. *Proc. Roy. Soc. London* A 244:17-40.

Biermann, L. 1974. Interaction of a comet with the solar wind. In *Solar Wind III*, ed. C.T. Russell (Los Angeles: Univ. of California), pp. 396-414.

Biermann, L.; Brosowski, B.; and Schmidt, H.U. 1967. The interaction of the solar wind with a comet. *Solar Phys.* 1:254-284.

Biermann, L., and Schlüter, A. 1951. Cosmic radiation and cosmic magnetic fields II. Origin of cosmic magnetic fields. *Phys. Rev.* 82:863-868.

Biermann, L., and Trefftz, E. 1964. Uber die Mechanismen der Ionisation und der Anregung in Kometenatmosphären. *Zs. f. Astrophys.* 59:1-28.
Brandt, J.C., and Mendis, D.A. 1979. The interaction of the solar wind with comets. In *Solar System Plasma Physics*, eds. C.F. Kennel, L.J. Lanzerotti, and E.N. Parker (Amsterdam: North Holland) Vol. 2, pp. 253-292.
Brosowski, B., and Wegmann, R. 1973. Numerische Behandlung eines Kometenmodells. *Meth. Verf. Math. Phys.* 8:125-145.
Burlaga, L.F. 1971. Nature and origin of directional discontinuities in the solar wind. *J. Geophys. Res.* 76:4360-4365.
Chapman, S., and Cowling, T.G. 1952. *The Mathematical Theory of Non-uniform Gases*, 2nd edition (London: Cambridge Univ. Press).
Elphic, R.C.; Russell, C.T.; Slavin, J.A.; and Brace, L.H. 1980. Observations of the dayside ionopause and ionosphere of Venus. *J. Geophys. Res.* 85:7679-7696.
Ershkovich, A.I. 1978. The comet tail magnetic field: Large or small? *Mon. Not. Roy. Astron. Soc.* 184:755-758.
Ershkovich, A.I. 1981. On the folding phenomenon of comet tail rays. Submitted to *Icarus*.
Feldman, W.C. 1979. Kinetic processes in the solar wind. In *Solar System Plasma Physics*, eds. E.N. Parker, C.F. Kennel, L.J. Lanzerotti (Amsterdam: North Holland), Vol. 1, pp. 321-344.
Högner, W., and Richter, N. 1969. Isophotometrischer Atlas der Kometen (Leipzig: Barth).
Houpis, H.L.F., and Mendis, D.A. 1980. Physicochemical and dynamical processes in cometary ionospheres. 1. The basic flow profile. *Ap. J.* 239:1107-1118.
Hyder, C.L.; Brandt, J.C.; and Roosen, R.G. 1974. Tail structures far from the head of Comet Kohoutek. I. *Icarus* 23:601-610.
Ip, W.-H., and Mendis, D.A. 1976. The generation of magnetic fields and electric currents in cometary plasma tails. *Icarus* 29:147-151.
Landau, L.D., and Lifshitz, E.M. 1959. Fluid Mechanics (Oxford: Pergamon Press).
Landau, L.D., and Lifshitz, E.M. 1960. Electrodynamics of Continuous Media (Oxford: Pergamon Press).
Lees, L. 1964. Interaction between the solar plasma wind and the geomagnetic cavity. *Amer. Inst. Aero. Astronaut. J.* 2:1576-1582.
Lipatov, A.S. 1978. The induced magnetosphere of Venus. *Cosmic Res.* 16:346-349.
Massey, B.S. 1979. Mechanics of Fluids. 4th ed. (New York: Van Nostrand).
Miller, F.D. 1979. Comet Tago-Sato-Kosaka 1969 IX: Tail structure 25 December 1969 to 12 January 1970. *Icarus* 37:443-456.
Miller, F.D. 1981. Configurations of evolving plasma tail rays. Submitted to *Icarus*.
Neugebauer, M.M. 1974. Relation of solar wind fluctuations to differential flow between protons and alphas. In *Solar Wind III*, ed. C.T. Russell (Los Angeles: Univ. of California), pp. 33-34.
Neugebauer, M.M., and Feldman, W.C. 1979. Relation between superheating and superacceleration of helium in the solar wind. *Solar Phys.* 63:201-205.
Podgorny, I.M.; Dubinin, E.M.; and Israelevich, P.L. 1980. Laboratory simulation of the induced magnetospheres of comets and Venus. *Moon and Planets* 23:323-338.
Schmidt, H.U. 1974. The flash-up time scale of plasma structures in cometary heads. *Mitt. Astron. Ges.* 35:248-250.
Schmidt, H.U., and Wegmann, R. 1980. MHD-calculations for cometary plasmas. *Comp. Phys. Comm.* 19:309-326.
Spreiter, J.R.; Summers, A.L.; and Alksne, A.A. 1966. Hydromagnetic flow around the magnetosphere. *Planet. Sp. Sci.* 14:223-253.
Spreiter, J.R.; Summers, A.L.; and Rizzi, A.W. 1970. Solar wind flow past nonmagnetic planets—Venus and Mars. *Planet. Sp. Sci.* 18:1281-1299.
Wallis, M.K. 1973a. Weakly-shocked flows of the solar wind plasma through atmospheres of comets and planets. *Planet. Sp. Sci.* 21:1647-1660.
Wallis, M.K. 1973b. Solar wind interaction with H_2O comets. *Astron. Astrophys.* 29:29-36.

SOLAR WIND INTERACTION WITH COMETS: LESSONS FROM VENUS

C.T. RUSSELL, J.G. LUHMANN, R.C. ELPHIC
University of California at Los Angeles

and

M. NEUGEBAUER
Jet Propulsion Laboratory

While Venus itself is in many respects most unlike a comet, it is very comet-like in its interaction with the solar wind. A bow shock is expected to deflect and heat the solar wind before it reaches Venus, or the comet proper. Charge exchange and photoionization are expected to play an important role in the interaction at both Venus and comets. However, at Venus we expect most of the effects to occur post-bow shock while at comets these processes can be very important well upstream of the bow shock. Both comets and Venus have magnetic tails which are not intrinsic; they occur because of the mass-loading of the field by cold heavy ions. The Venus tail appears to be either striated or very dynamic and thus is quite similar to a cometary tail. Plasma clouds are seen above the Venus ionosphere which may be the Venus analog of cometary tail rays. Strong flows are found in the Venus ionosphere because of strong day-night asymmetries. Such asymmetry should not be present in comets unless there is strong coupling to the solar wind. One surprise from the Venus data is the presence of twisted coils of field called magnetic flux ropes. These often nearly force-free fields are in some senses analogs of solar prominences and are possible energy storage locations. Occasionally, when the dynamic pressure of the solar wind is high, large ionospheric magnetic fields are observed. This phenomenon may be a simple manifestation of the coupling of the solar wind and ionosphere when the dynamic pressure reaches a critical level. Also the cometary interaction should be sensitive to changes in the solar wind dynamic pressure both as the comet changes heliocentric distance, and as the solar wind varies with time. Further, as the rate of outgassing of the comet changes during its perihelion passage, we expect the nature of the interaction to change.

The solar wind interacts with a wide variety of objects on its journey through the solar system. When it intercepts the Moon and presumably the asteroids, it simply is absorbed by the dust and rocks on the surface. However, the interaction with the other objects of the solar system is much more complex, whether this interaction is dominated by the intrinsic magnetic field of the planet, as it is at Mercury, Earth, Jupiter, and Saturn, or by the planetary ionosphere, as it is at Venus and possibly at Mars. A planetary magnetosphere is an efficient, but not perfect, deflector of the solar wind mass and energy flux. Since the solar wind flows much faster than the velocity of pressure waves in the plasma, a shock front forms which deflects the solar wind plasma around the obstacle, slows it down and heats it. Observations at Venus show that an ionosphere can also be an efficient deflector of the solar wind and here too a strong bow shock is formed.

Comets differ from planets in two important respects. They are smaller and they have little self-gravitation. However, the former and the latter are in some senses compensatory since the gases which evaporate from the comet as it is heated during perihelion passage can escape to great distances from the nucleus and can affect the solar wind to even greater distances and inject more mass into the solar wind than can a planet of much greater size.

In this chapter we examine the data that have been obtained on the solar wind interaction with the planet Venus for the purpose of comparison with some of the similar processes that may be occurring in comets. Of course, each comparison and analogy should be examined critically in terms of the known behavior of comets. On the other hand, it is easier to scale the behavior of a comet from that of a planetary object like Venus than from the behavior of a laboratory device in which the range of parameter space available for investigation is restricted and the relevant scale lengths are much different than in the cometary-solar wind interaction.

I. BOW SHOCK

The Venus bow shock was detected on the Venera 4 and 6 and Mariner 5 and 10 missions (Dolginov et al. 1969; Gringauz et al. 1970; Bridge et al. 1967; Ness et al. 1974) and extensively mapped with Venera 9 and 10 (Verigin et al. 1978). The bow shock was somewhat closer to the planet than might be expected if all of the solar wind were deflected by the planet (Russell 1977) and was closer to Venus than the Martian bow shock was to Mars. This latter fact was used to infer an intrinsic field at Mars (Verigin et al. 1978). However, when Pioneer Venus was inserted into orbit in December 1978, the location of the bow shock at the terminator was 35% farther from the planet than it had been in 1975-76 and in fact it had assumed a position very similar to that found at Mars (Slavin et al. 1979). The most likely cause of this variation is a change in the solar extreme ultraviolet (EUV) flux with the solar cycle such as that postulated to explain the variability in the ionopause position (Wolff et al. 1979). A similar solar-cycle variation is suggested for comets by Oppenheimer and Downey (1980), who estimate that the photoionization rates of cometary H_2O and CO may have varied by a factor of ~ 3 between 1976 and 1979. Thus the Venus data provide observational evidence that the predicted variation may in fact occur in comets.

Figure 1 shows the magnetic profile through 6 Venus bow shock traversals. The Venus bow shock is somewhat weaker than the terrestrial bow shock. This weakness is evident in the size of the magnetic field jump at Venus which is $\sim 30\%$ less than the field jump at similar solar zenith angles (SZA) at Earth (Russell et al. 1979). It is also evident in the weakness of the shock overshoot in the Venus data. In the terrestrial shock the magnitude of the magnetic field just downstream of the shock often overshoots and achieves magnitudes of up to 50% higher than occur farther downstream. Such large overshoots are seldom seen in the Venus data (Russell et al. 1981a). The reason for this weakness is not clear. We do not expect much solar wind interaction with the neutral atmosphere upstream of the shock, although there probably is some. What may be significant here is downstream absorption of the flow. Closer to the planet where the neutral density is higher, charge exchange with neutral hydrogen and oxygen can lead to fast neutrals which pass into the atmosphere and become absorbed by the planet rather than deflected. This post-shock momentum removal reduces the amount of deflection needed to be produced by the shock and hence the shock should be weaker. However, the amount of absorption expected is small (Gombosi et al. 1980).

The greater scale lengths in cometary atmospheres are expected to result in mass addition to the flow upstream of the shock which will slow the incoming flow and hence weaken the shock. The effect is the same as at Venus, but is expected to have a different cause. The theoretical point of view as of 1981 is that the position and strength of a cometary bow shock depend mainly on the upstream mass-loading and are nearly independent of the details of the cometary ionospheric obstacle, such as its size and shape (see the chapter by Schmidt and Wegmann in this book). Perhaps the lesson to be learned from the Venus observations is that present theories about bow shock location and strength are known to be inadequate at Venus and are thus suspect in other applications.

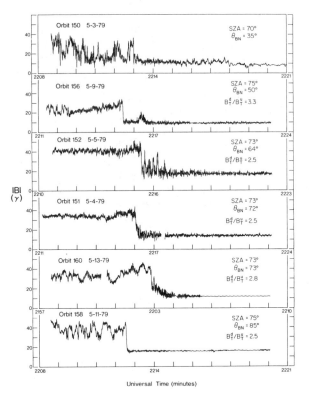

Fig. 1. Magnetic field profiles during six Venus shock crossings for a variety of angles between the interplanetary magnetic field and the shock normal ranging from 35° to 85°. The field jump ratio downstream (d) to upstream (u) is the tangential component (t) measured in the shock plane. The solar zenith angle (SZA) of these shocks ranges from 70° to 75°.

The last bow shock property that should be noted is the dependence of shock structure on the orientation of the interplanetary magnetic field relative to the shock normal. This dependence is seen both at Earth and at Venus (Greenstadt 1976; Russell et al. 1979). Figure 1 illustrates the change in the structure of the magnetic field across the Venus shock for different directions of the interplanetary field. Parallel and quasi parallel shocks in which the field is oriented $\lesssim 45°$ from the shock normal tend to be much more turbulent than shocks with angles between the field and normal $> 45°$. In the region of the bow shock where the magnetic field is nearly along the local shock normal, the shock does little to heat or deflect the solar wind flow, while the shock in regions where the field is nearly perpendicular to the shock normal strongly affects the solar wind (Greenstadt et al. 1977). This effect spoils the cylindrical symmetry of the solar wind interaction. It should only be important for a comet to the extent that the bow shock influences the behavior of the solar wind interaction. For example, the rate of charge exchange between the shocked solar wind and cometary neutral molecules could be different behind the parallel and perpendicular parts of the shock. If the bow shock is weak, the field direction probably will not have an important effect; but if a strong shock forms, as it might on occasion, then this asymmetry could have noticeable effects.

Finally, we note that Venus like the strongly magnetized planets, exhibits a variety of wave phenomena upstream of the bow shock that are presumably associated with energetic ion and electron beams reflected by the bow shock (Hoppe and Russell 1981). In fact, MHD (magnetohydrodynamic) waves are seen upstream of all planetary bow shocks but have not been reported for the weaker interplanetary shocks. If beams of particles are reflected or accelerated by the shock wave in front of a comet, these particles could further modify the properties of the incoming solar wind.

II. THE MAGNETOSHEATH

The post-shock plasma at Venus in many respects resembles the terrestrial magnetosheath. The flow pattern is consistent with deflection around a blunt obstacle, as expected behind a standing shock in supersonic fluid flow (Mihalov et al. 1980). One significant difference from the Earth's magnetosheath, however, is the consistent presence of strong magnetic fields in the inner sheath, to the extent that the pressure just above the ionosphere is almost entirely magnetic (Elphic et al. 1980b). Figure 2 illustrates this effect for a near-terminator pass of Pioneer Venus. In this figure, the magnitude of the magnetic field and the electron density as determined by the Langmuir probe are plotted versus time. The region of enhanced field strength in the inner magnetosheath has been called the Venus mantle; within the mantle the electron energy spectrum slowly changes from that typical of the solar wind to that typical of the cooler ionosphere (Spenner et al. 1980).

This pile-up of magnetic field outside the Venus ionosphere is reminiscent of many theoretical models of the interaction of the solar wind with comets (e.g., Alfvén 1957; Harwit and Hoyle 1962; Ioffe 1968). In these models the convected solar wind field lines become hung up at the stagnation point of the flow around the comet's contact surface (equivalent to the ionopause of Venus, discussed in Sec. III), while the more remote parts of the field lines continue on downstream, still frozen into the solar wind. The end result is a draping of the field around the comet. The Pioneer Venus observations of the direction of the magnetic field in the Venus magnetosheath are entirely consistent with such a model, as demonstrated in Fig. 3.

The properties of the magnetosheath of Venus have been modeled by Spreiter and Stahara (1980) using a gas dynamic code. To first order, the code replicates the observations seen by Pioneer Venus during sample orbits near the terminator. Some important differences are observed however. In

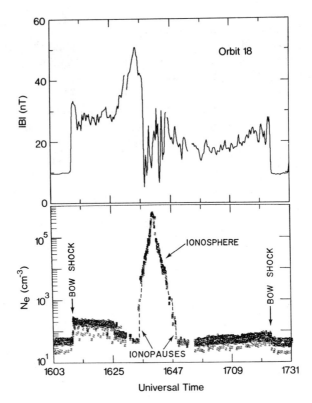

Fig. 2. Magnetic field magnitude and electron density as measured by the Goddard Space Flight Center Langmuir probe on Pioneer Venus orbit 18 as the satellite passed through periapsis. The increase in magnetic field strength and electron density behind the bow shock inbound and outbound are clearly seen as is the buildup in field strength just outside the dayside (inbound) ionopause. The outbound ionopause crossing occurs at night and no such buildup is seen.

particular these authors were unable to reproduce simultaneously the magnitude of the jump in the magnetic field strength and the location of the bow shock. Secondly, the maximum field strength observed just outside the ionopause was significantly ($\sim 30\%$) greater than predicted. These differences are probably the effect of non-gasdynamic processes. MHD effects were not included in this code nor were effects of photoionization and charge exchange. While these are possibly of minor importance at Venus under normal solar wind conditions, these effects are expected to play a major role in cometary interactions. Zhuang et al. (1981) have found, for example, that changes in the shape of the terrestrial magnetopause due to MHD effects in the magnetosheath are inversely proportional to the square of the Alfvén Mach number. Thus in the neighborhood of a comet where the Mach number is expected to be low because of mass-loading of the solar wind by the

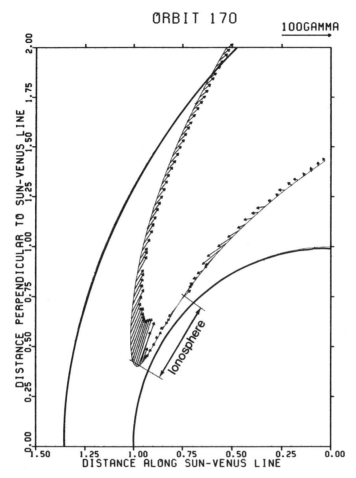

Fig. 3. Magnetic field vectors plotted along the trajectory of the Pioneer Venus orbiter during a near subsolar periapsis passage in solar cylindrical coordinates. The plane of the figure is the plane containing the Venus-Sun line and the Venus-spacecraft line. This figure shows the buildup in field strength and the draping of the field just above the ionopause and the sudden drop in field strength inside the ionosphere.

extended hydrogen corona, full MHD calculations must be performed like those described by Schmidt and Wegmann (1980; see also their chapter).

There is at least one major difference, however, between the MHD model of a cometary magnetosheath and the observed properties of the magnetosheath of Venus. Schmidt and Wegmann (see their chapter) argue that the magnetic pressure at the stagnation point of the flow around a cometary contact surface can greatly exceed the stagnation pressure of the upstream solar wind; the proposed mechanism is the extraction of momentum from the downstream solar wind via magnetic stresses along the connected field lines.

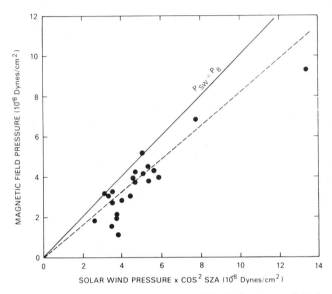

Fig. 4. The magnetic field pressure observed just outside the ionopause plotted versus the solar wind pressure corrected by the cosine squared of the solar zenith angle (Brace et al. 1980).

Such a phenomenon is not observed at Venus. Figure 4 shows the relation between the magnetic pressure just outside the Venus ionopause and the upstream solar wind pressure multiplied by a $\cos^2$ (SZA) factor to account for oblique incidence away from the axis of symmetry. In no case does the magnetic pressure exceed the solar wind pressure. Perhaps the pressure is limited by the slippage of field lines over and under the planet, and the slippage rate may be different at a comet. Also the radius of curvature of the field lines may be important. This subject clearly requires further analysis.

The neutral atmosphere of Venus extends out into the magnetosheath. Ions created by either photoionization or charge exchange in this region probably escape from the planet under the action of either the electric field in the magnetosheath which causes the newly created ions to drift away or the MHD forces from the stress of the magnetic field draped across the planet. In addition to H^+, H_2^+, and He^+, which are major constituents only above the ionopause, O^+ is also produced in substantial quantity ($\sim 10^{24}$ ions s^{-1} above 260 km) in spite of its much smaller scale height, as observed by measurement of the neutral atmospheric composition (Niemann et al. 1980). The large O^+ production above the ionopause is partly caused by the presence of a superthermal component of O. (A. Nagy and T. Cravens, personal communication, 1980).

It is obvious that the early simplified concepts of the mass-loading process at Venus are incorrect. In fact it is often very difficult to define a boundary separating ionospheric plasma that is gravitationally bound from plasma that is moving fast enough to escape from the planet since the Pioneer Venus orbit is not instrumented to measure plasma flows from ~ 10 km s^{-1} to ~ 100 km s^{-1}. The principal indirect observations pertaining to such flow are:

1. The ion mass spectrometer on Pioneer Venus often detects superthermal ions, with energies in the range 30 to 80 eV, in one or more layers a few hundred km thick outside the ionopause (Taylor et al. 1980).

2. On one occasion, the plasma analyzer saw relatively cool ($\sim 3 \times 10^5$ K) O$^+$ ions flowing at nearly the bulk speed of protons in the sheath (Mihalov et al. 1980). If these ions had been created from neutrals in a fast magnetosheath flow and then accelerated up to the sheath flow speed, their thermal velocities would be expected to be roughly the same as the flow speed (~ 240 km s^{-1}), which corresponds to a temperature of 5×10^7 K. On the other hand, if the magnetosheath flow near the ionopause is slow, the ions formed there will be cool. Furthermore, the angular distribution of these O$^+$ ions was narrower than that of the protons, again suggesting the acceleration of a parcel of ionospheric plasma.

3. On $\sim 20\%$ of the ionospheric passes, the electron temperature (Langmuir) probe detects plasma clouds in the magnetosheath with densities and temperatures typical of upper ionospheric plasma (Brace et al. 1981). The observations that (a) these clouds are penetrated by the magnetosheath magnetic field, (b) they contain a secondary electron population with temperatures higher than those observed in the ionosphere, and (c) the ion thermal speed is unlikely to be high enough for the ionosphere to expand and contract rapidly enough to account for the observations suggest that these clouds are probably physically detached from the ionosphere (Brace et al. 1980,1981). Brace et al. (1981) estimate an upper limit to the pickup rate of the order of 10^{27} ions s^{-1} via the scavenging process, which is about two orders of magnitude greater than the loss rate via ionization of the extended atmosphere.

Figure 5 shows the magnetic field components in essentially solar ecliptic coordinates, the magnetic field strength, the plasma wave activity at 100 Hz, and the electron number density through a plasma cloud and into the ionosphere on Pioneer Venus orbit 601 (Russell et al. 1981c). There are two important points to note about the behavior of the magnetic field in the plasma cloud. First the sunward or X-component reverses in the cloud and second the field strength drops slightly in the cloud. The drop in field strength allows us to estimate the energy density of the plasma and by using the number density we obtain an average ion and electron temperature of 4×10^4 K. The magnetic or Maxwell stress on the cloud due to the field reversal is $\sim 4 \times 10^{-8}$ dynes cm^{-2} which could accelerate the plasma at

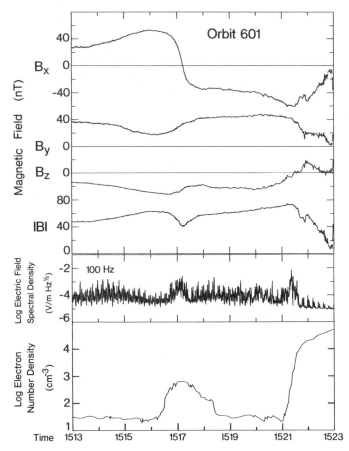

Fig. 5. *Top,* magnetic field components in spacecraft coordinates (essentially solar ecliptic coordinates) and the total field strengths during a passage through a plasma cloud and entry into the ionopause. *Middle,* plasma wave activity in the 100 Hz channel of the electric field detector. *Bottom,* electron density as measured by the Goddard Space Flight Center Langmuir probe (Russell et al. 1981c).

~ 0.7 km s^{-1}. If the cloud started at rest at the subsolar point, it should take ~ 130 s to reach the point of observation and be moving at ~ 90 km s^{-1} at that time. Similar magnetic field structures are seen in other plasma clouds but none as clearly as this one.

The observations of the magnetic structure in this cloud aids our understanding of the properties of the pickup process described above. First, the suprathermal ions at the ionopause could be accelerated by the magnetic field which permeates this region of space. Second, the newly formed ions will be accelerated by the field from rest up to nearly the solar wind speed in not much more than the length of time required to reach the terminator. As observed the temperatures will reflect their temperature at formation in a slow flow and their acceleration to solar wind energies will be directed and not stochastic. Third, this process can lead to substantial ion losses as Brace et al. (1981) suggest; this loss rate, however, depends on geometrical assumptions. If we assume the cloud is 500 km in height and 300 km in width and that there is a corresponding cloud in the south, we obtain a loss rate of 2×10^{25} ions s^{-1}, less than Brace et al.'s upper limit but still quite significant.

It is in this area of mass-loading of the solar wind that Venus and comets differ the most. The cometary atmosphere is so enormous that all but a few percent of the gas escapes beyond the contact surface before undergoing ionization. At Venus most of the ionization takes place below the contact surface (ionopause) and only a small fraction of the neutral atmosphere becomes ionized above the contact surface. A major question at comets is how rapidly newly ionized material is isotropized and thermalized (see the chapter by Ip and Axford in this book); thus far the Pioneer Venus data have not supplied an answer to this question. The plasma wave activity in conjunction with the cloud in Fig. 5 suggests some heating, but not any more than usually seen at the ionopause.

In comparing the mass-loading effects at comets and Venus we should keep in mind the relative sizes and expected production rates. Combi and Delsemme (1980) conclude that when Comet West 1976 VI was at 0.7 AU, its subsolar contact surface was 2×10^4 km from the nucleus, and hence the area of the contact surface would have been ~ 10 times larger than the area of the Venus ionopause. From the observations of Comet West by Feldman and Brune (1976), the total production rate at 0.7 AU was $\sim 4 \times 10^{29}$ ions s^{-1}. This compares with our estimate for Pioneer Venus orbit 601 of 2×10^{25} ions s^{-1} and of Brace et al. of up to 10^{27} ions s^{-1}. The larger cometary production rate should lead to both larger scales and slower velocities for the cometary processes.

A further cometary feature that may be related to mass-loading is the tail ray. The best phenomenological description of the geometry and motion of cometary rays comes from Comet Morehouse 1908 III, which had an extraordinary low dust:gas ratio, allowing plasma forms to be seen unusually clearly. This comet displayed a series of collapsing envelopes whose extremeties evolved into the type of rays observable on many comets (Eddington 1910; Lust 1967). The tailward motions and nearly parabolic shapes of these envelopes clearly suggest that they are a magnetosheath, rather than a solar wind phenomenon. In discussing the plasma cloud shown in Fig. 5, we assumed that it was a thin vertical structure connected to the ionosphere proper supported and accelerated by a U-shaped field configuration. However, it could be ray-like (i.e., tubular) rather than sheet-like; we cannot tell from an *in situ* measurement along a single trajectory. If the Venus plasma clouds are analogous to cometary tail rays, then the rays are not field-aligned structures but rather occur in a plane perpendicular to the plane containing the solar wind and interplanetary magnetic field. In other words if the X-Y-Z coordinate system were the solar ecliptic system with X towards the Sun and Z along the ecliptic pole, and if the interplanetary magnetic field were in the ecliptic plane, i.e., the X-Y plane, then the rays would form in the X-Z plane.

Schmidt and Wegmann (1980; see also their chapter) associate tail rays with the changes in field direction, e.g., tangential discontinuities convected by the comet. However, the plasma cloud of Fig. 5 occurred during a period of moderately steady interplanetary magnetic field. The major preceding interplanetary event was a sudden increase of the solar wind dynamic pressure a few hours previously. This event would have established a lower ionopause and hence increased the mass-loading rate. Perhaps tail-ray formation occurs when a change in the solar wind suddenly increases the mass-loading rate at the site of the tail ray. Certainly more events at Venus and eventually good coordinated solar wind comet studies must be investigated before this possibility is more than just speculation.

III. THE IONOPAUSE

Near the subsolar point (solar zenith angle $\leqslant 45°$), the density of the ionospheric plasma is usually observed to drop abruptly, by several orders of magnitude, within an altitude interval of some 10's of km at a typical altitude of 250-300 km, but sometimes much higher (Brace et al. 1980; Hartle et al. 1980). In the same interval, the horizontal magnetic field is observed to increase with increasing altitude from ~ 0 to 50-100 nT. This boundary is called the ionopause. The magnetic pressure observed in the magnetosheath just outside the ionopause is approximately sufficient to balance the ionospheric pressure observed inside the ionopause (Elphic et al. 1980b).

The altitude of the ionopause decreases as the strength of the magnetic pressure outside the ionopause increases. The magnetic pressure in turn is determined by the solar wind dynamic pressure (Brace et al. 1980). Figure 6 shows magnetic field and electron density altitude profiles for three orbits. When the dynamic pressure of the solar wind is low, the ionopause is at relatively high altitudes and the ionopause is thin. When the dynamic pressure of the solar wind is large the ionosphere is compressed to low altitudes and the ionopause thickens (Elphic et al. 1981). This thickening is presumably a result of collisional coupling of the ions and neutrals discussed below. The relative strength of the solar wind dynamic pressure and the ionospheric thermal pressure is thus very important for the Venus interaction with the solar wind. This same relation underlies essentially all theories of the interaction of the solar wind with comets (e.g., Biermann et al. 1967; Wallis and Ong 1976; chapter by Schmidt and Wegmann).

Near the Venus terminator, the ionospheric plasma normally flows in an antisunward direction at speeds of 2 or 3 km s^{-1}. The day-night ion pressure

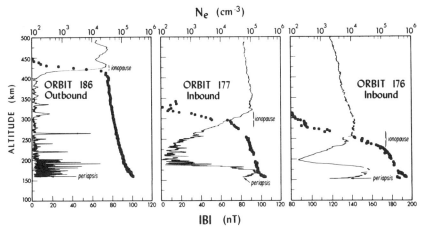

Fig. 6. Altitude profiles of magnetic field and electron number density for three periapsis passes of the Pioneer Venus spacecraft near local noon. Orbit 186 is an example of low solar wind dynamic pressure with a moderately high ionopause and an ionosphere containing flux ropes. Orbit 177 has a higher solar wind dynamic pressure, lower ionopause and moderately strong magnetic field at low altitudes. On orbit 176 the solar wind is stronger still, the ionopause is lower and the low-altitude layer of magnetic field is even stronger than on orbit 177. However, the location of this layer is very similar on both days. Note also the increased thickness of the low-altitude ionopause.

gradient is great enough to account for the observed flow (Knudsen et al. 1981); no momentum flux across the ionopause (arising from either a viscous interaction or from electromagnetic forces) is required. Unless a comet is so active that its coma is optically thick to the photoionizing sunlight, the day-night pressure gradient effects in its ionosphere should be negligible. Yet 17 to 40 km s^{-1} antisunward flows of H_2O^+ ions have been observed within $\sim 10^5$ km of the nuclei of comets Kohoutek 1973 XII (Huppler et al. 1975) and Bennett 1970 II (Delsemme and Combi 1979), although the latter observation might refer to wave motion rather than fluid flow. If the moving plasma is inside the cometary contact surface, the acceleration mechanism must be different from that at Venus. Possible mechanisms include momentum transfer across a contact surface located within the comet's collision zone or tailward deflection at a shock within the ionosphere (Wallis and Dryer 1976; Houpis and Mendis 1980a). An interior shock is neither expected nor reported to exist at Venus.

Unlike momentum, a significant amount of energy is transmitted across the Venus ionopause into the ionosphere. Cravens et al. (1980) have developed a model of the energy balance of the dayside Venus ionosphere based on thermal conduction consistent with a realistic magnetic field which includes fluctuations; they deduce that, during quiet times, an energy flux of

$\sim 4 \times 10^9$ eV cm^{-2} s^{-1} is transferred across the dayside ionopause from the sheath to the ionosphere. Nearly all of this energy goes into heating electrons. Taylor et al. (1979) have shown that a possible mechanism of energy transfer is Landau damping in the ionosphere of whistler waves generated in the ionosheath. From their plasma wave data, Taylor et al. estimate that the energy input to the outer ionosphere varies between 6×10^9 and 1×10^{11} eV cm^{-2} s^{-1}, which corresponds to a range of 1 to 20% of the average solar wind energy flux at the orbit of Venus.

A primary question for a comet is whether or not the pressure of its ionosphere can be sufficiently great to stand off the solar wind at an altitude at which an ionopause can be maintained against destruction by ion-neutral collisions. Biermann et al. (1967) estimated that the subsolar contact surface of a moderately bright comet is located $\sim 10^5$ km sunward of the nucleus. In this model, ion-electron pairs are created in the cometary atmosphere by photoionization and the plasma pressure on the comet side of the contact surface is provided principally by the hot photoelectrons (kT assumed to be 14.5 eV). According to Wallis and Ong (1976), these photoelectrons are rapidly cooled by inelastic collisions with neutral cometary molecules. In the Wallis and Ong model, cometary and solar wind pressures balance at a distance $\sim 10^2$ km sunward of the nucleus, well within the collision zone. This model, however, neglects the outward force on the plasma arising from collisions with the outflowing neutral gas. Observations of the ion profiles of comets Bennett 1970 II and West 1976 VI (Delsemme and Combi 1979; Combi and Delsemme 1980) offer evidence that their contact surfaces may have been located a few times 10^4 km from the nucleus, which is at the outer fringe of the collision zone, between the values calculated by Biermann et al. and by Wallis and Ong. Contact surfaces at altitudes this high can be explained by a substantial source of ionization below the contact surface in addition to photoionization and/or substantial heating or acceleration of the ionospheric plasma. The Pioneer Venus detection of ion and electron heating at the topside ionosphere may thus help us understand some of the discrepancies between comet theory and observations.

IV. THE IONOSPHERE

The dayside ionosphere of Venus, created in the usual manner via illumination of the neutral atmosphere by solar extreme ultraviolet (EUV) radiation, consists mainly of O_2^+ and O^+. The ionosphere has a peak density of $\sim 3 \times 10^5$ cm^{-3} at an altitude of ~ 150 km (Theis et al. 1980) where the composition is primarily O_2^+ (Taylor et al. 1980). The electron and ion temperatures at the density peak are ~ 2000 K (Brace et al. 1980) and 500 K (Knudsen et al. 1980) respectively, giving a peak thermal pressure nkT of 10^{-7} dynes cm^{-2}. The contribution of O^+ becomes dominant at an altitude of ~ 200 km (Nagy et al. 1980; Taylor et al. 1980).

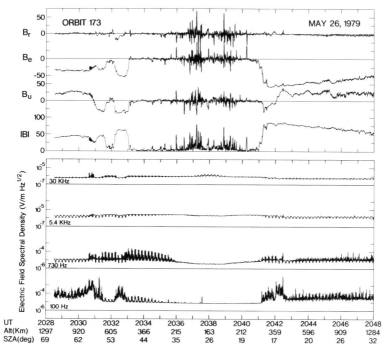

Fig. 7. High-resolution magnetometer data (top panel) and very-low-frequency electric field data (lower panel) through a near subsolar ionospheric passage of Pioneer Venus. The top three traces show the magnetic field in the radial, east, and north directions. The lower four traces show the electric field spectral amplitude in four frequency bands centered on 100 Hz, 730 Hz, 5.4 kHz, and 30 kHz. The periodic signal on these channels is associated with the spin of the spacecraft and the rotation of the electric field antenna through the spacecraft photoelectron cloud and optical shadow. There are strong emissions at 100 Hz near the ionopause but no or few signals associated with the flux ropes, the striated field at high altitudes.

Although the average field in the ionosphere is close to zero when the dynamic pressure of the solar wind is low, moderately strong, filamentary magnetic fields are observed as illustrated in Fig. 7. These structures can be modeled as twisted coils of magnetic field with varying strength and helicity or pitch with distance from the center of the coil. These structures have been termed flux ropes (Russell and Elphic 1979; Elphic et al. 1980a). The structure of these fields sometimes resembles that of a self-balancing or force-free field. At other times the field does not appear to be twisted to the proper degree to be force-free.

The origin of flux ropes is still something of a mystery. One possibility is that they are formed by bundles of magnetosheath magnetic flux which are carried down into the subsolar ionosphere by the magnetic tension exerted ultimately by the solar wind which overcomes the buoyancy of the empty sheath field floating on the ionospheric plasma. Another possible mechanism involves the breakup of regions of large field, which appear in the ionosphere during and following times of high solar wind dynamic pressure. This breakup would occur due to shear in the ionospheric flow, which would also induce twisting of the flux tubes.

Theories attempting to explain the flux rope phenomenon must deal with a number of observed attributes. First, flux ropes are an almost ubiquitous feature of the dayside ionosphere (in the absence of the high fields related to high solar wind pressure). They are observed to have maximum occurrence density at lowest altitudes (~ 160 km), where their peak field strength also appears to maximize. Their scale diameter, like the peak field strength, is a function of altitude, becoming smaller at lower altitudes (Elphic et al. 1980a). This scale size is characteristically a few ion gyroradii. Flux rope orientation also appears to be a function of altitude; structures at low altitudes ($\lesssim 250$ km) appear to have virtually random orientations, while those at higher altitudes tend to be horizontal more frequently. This tendency is especially pronounced near the terminator. Moreover, flux ropes at the terminator, above ~ 250 km, appear to have a mainly sunward or antisunward orientation, as if draped across the dayside.

The currents flowing in flux ropes have a component parallel (or antiparallel) to the field, causing the observed twist or helical structure. This field-aligned current component may be small or large, giving little or great helicity, but it does not appear to be a function of altitude. Defining under what conditions, and in what locations, the field-aligned current is parallel or antiparallel, and where it maximizes, is crucial to distinguishing between various source mechanisms. This is a subject of ongoing research.

It is not obvious that similar fine-scale structure of the magnetic field exists in cometary ionospheres. If flux ropes do form at the cometary ionopause, they would aid in the penetration of the magnetic field deeper into the cometary ionosphere and atmosphere and could lead to greater mass-loading of some of the magnetic flux. If indeed flux rope formation represents erosion at the edges of regions of uniform field, then this mechanism may play a role in the evolution of cometary ion tails. It should be remembered that the twisted fields represent an axial current. If this axial current becomes resistive either through an instability or motion into a resistive medium, it will lead to an electric potential drop. The twist in the field also represents a storage of magnetic energy. The inductance of this twisted field will help drive the axial current even when the current becomes disrupted. Thus, if present, these structures could lead to processes like solar flares or substorms in comets. The energy so released could lead to increased ionization and to heating of the ionospheric plasma. Thus it is conceivable that flux ropes could be in large part responsible for the maintenance of the ionospheric pressure required for a contact surface to exist above the collision zone. Since we do not yet have a full understanding of the role of flux ropes in the physics of the Venus ionosphere, any inferences about possible similar cometary behavior is highly speculative.

The magnetic character of the Venus ionosphere is different than that described above when the solar wind is enhanced. The different nature of the interaction can be attributed to the fact that on some occasions the dynamic pressure of the solar wind incident on the dayside ionosphere exceeds the peak thermal pressure of the ionosphere. The details of this type of interaction are not currently understood, but observations show that a horizontal magnetic field of magnitude about equal to the ionosheath field (50 to 100 γ) can be present at altitudes down to $\sim$ 150 km during and (probably) following the passage of a high dynamic pressure solar stream (Luhmann et al. 1980). If the magnetic field observed at these low altitudes is still connected to the solar wind as the ionosheath field lines are, this low altitude penetration apparently has two important effects:

1. The mass-loading of the solar wind must greatly increase because the entire ionosphere is being created on field lines connected to the solar wind (instead of merely that fraction of the ionosphere above the ionopause);

2. The solar wind field lines are present in a resistive region where collisional interactions between ions and neutrals influence the ion dynamics.

Modeling of the ionosphere on occasions when penetration of the ionosheath field is evident suggests that $\mathbf{J} \times \mathbf{B}$ forces are very important within the ionosphere at these times; the vertical $\mathbf{J} \times \mathbf{B}$ force is principally responsible for balancing the vertical plasma pressure gradient force of the ionosphere (Luhmann et al. 1981a). Some dissipation in the form of Joule heating must occur at these times. Also, the horizontal Lorentz body force on the ionospheric plasma may at these times dominate the pressure gradient force at the terminator, thereby affecting the dynamics of the plasma there. The details of the interaction under conditions of high solar wind dynamic pressure are currently under investigation.

It should be mentioned that the Venus nightside ionosphere is also affected by the solar wind interaction. Even under conditions of low solar wind dynamic pressure, a feature is observed at low ($\sim$ 150 km) altitudes near the antisolar point that includes radially oriented magnetic fields of strength $\sim$ 20 γ (Luhmann et al. 1981b) accompanied at higher altitudes by depletions in the plasma density of the nightside upper ionosphere (Brace et al. 1981). It has been suggested that this feature is the low altitude part of the Venus tail. There are also times when the entire nightside ionosphere is depleted. These occurrences may be related to the change of the interaction that occurs when the solar wind dynamic pressure is high. Perhaps at these times the different magnetic configuration on the nightside consisting of large horizontal magnetic fields affects the normal plasma supply to the nightside. Since we under-

Low Dynamic Pressure

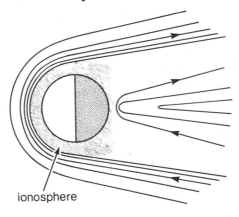

High Dynamic Pressure

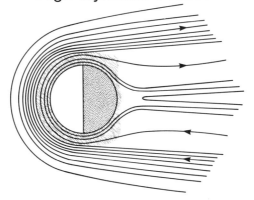

Fig. 8. Schematic representation of the field-line configuration for low solar wind dynamic pressure conditions (top panel), and high solar wind dynamic pressure (bottom panel). During periods of low solar wind pressure the ionosphere (shaded) is essentially field free. During periods of high solar wind dynamic pressure, significant field strengths are observed in the dayside ionosphere extending some distance into the night hemisphere.

stand the physics of the nightside ionosphere poorly, this hypothesis is only speculation.

The observations of the magnetic fields in the dayside and nightside ionosphere and their response to solar wind dynamic pressure are summarized in Fig. 8. When the solar wind dynamic pressure is low the magnetic field does not penetrate the dayside ionosphere, but some may get hung up in the night ionosphere and contribute to the magnetotail. However, when the solar

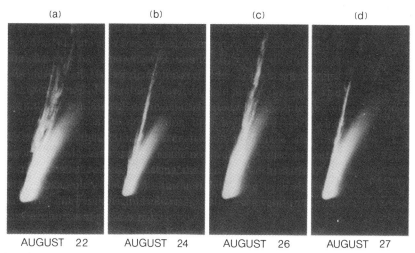

Fig. 9. Comet Mrkos 1957 V photographed with the 122-cm Schmidt telescope. (Courtesy of Hale Observatories.)

wind dynamic pressure is high, the magnetic field permeates the dayside ionosphere and perhaps, because of horizontal gravitationally bound plasma flows, gets dragged to the night side, thereby producing the observed horizontal nighttime ionospheric fields. In short, the interaction is quite different for high and low solar wind dynamic pressure and it appears that ion-neutral coupling in the lower ionosphere has a very important role in this dichotomy.

At comets we would expect a similar dichotomy which also depends on the relative solar wind and ionospheric pressures. But comets might switch back and forth between the two states with greater frequency because of the frequent, large variations in cometary gas emission rates. What might the visible (from Earth) manifestations of a switch between states be? No systematic difference in the magnetosheath flow pattern for the two states has been reported from the Pioneer Venus data; thus any visible effects may be confined to the ionosphere and tail. Because plasma structures within 10^3 to 10^4 km of the nucleus can usually not be resolved, the ion tail may contain the only visible traces of the ionospheric magnetic configuration. Perhaps a skinny rat tail, as in Figs. 9b and 9d, is produced when the external pressure is sufficiently high to drive the interplanetary field into the ionosphere, whereas a wide, highly structured ion tail, as in Figs. 9a and 9c, might result when the ionospheric pressure is sufficiently high to force the interplanetary field to flow around a rather large obstacle.

V. THE MAGNETOTAIL

A distinct magnetic structure can be found behind Venus. Figure 10 shows the three components and total magnetic field and one channel of

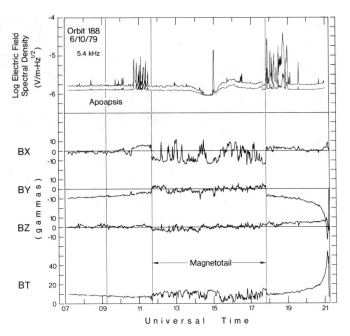

Fig. 10. One-minute averages of the magnetic field observed by the Pioneer Venus orbiter during a passage through the near center of the tail at a distance of 7 to 12 planetary radii behind the planet. The magnetic field components are displayed in solar ecliptic coordinates with the X-direction towards the Sun. The top panel shows the plasma wave spectral amplitude in the 5.4 kHz channel of the electric field experiment.

plasma wave data for a Pioneer Venus orbiter pass almost directly behind Venus. In the center of this interval is a region in which the magnetic field is enhanced and the direction of the magnetic field is principally along the solar direction. Note that the field strength within the tail is on the order of 10 γ. There is no evidence of the $\gtrsim$ 100 γ cometary tail field suggested by Hyder et al. (1974) and by Ip and Mendis (1975). Instead, the Pioneer Venus observations agree with the predictions of Ershkovich (1980) that the strength of a cometary tail field will be, at most, comparable to the strength of the local interplanetary field.

There is a long-standing uncertainty about whether the observed motions of tail features represent mainly fluid flow or the propagation of MHD waves (see review by Brandt and Mendis 1979). If cometary tail fields are as weak as the Venus tail field, the Alfvén speed in a cometary tail will be low, implying that rapidly moving features are probably caused, in large part, by bulk motion of the tail plasma.

Figure 10 shows that, inside the magnetotail-like region, there are many current sheet encounters. Figure 11 shows a high resolution time series and

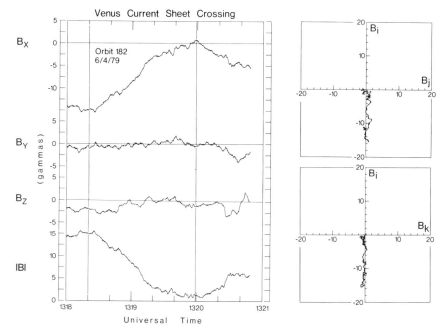

Fig. 11. High-resolution (6 samples per second) measurements in the center of the Venus magnetotail on orbit 182, 6/4/79. The left-hand panel shows the time series in solar ecliptic coordinates. The right-hand panel show two orthogonal hodograms of the evolution of the tip of the magnetic field in the principal axis or minimum variance system. The eigenvalues associated with the analysis were 27.3:0.55:0.17 (nT)^2. The direction of the minimum eigenvector which is associated with the normal to the current sheet was (0.026, -0.958, 0.287) or essentially in the direction of planetary motion. This is the expected direction for a tail caused by a draped interplanetary field lying in the ecliptic plane. When the spacecraft entered the solar wind 8 hr later the magnetic field was very nearly in the ecliptic plane but also exhibited much variability over the next 4 hr. This current layer resembles that often observed in the terrestrial plasma sheet and has the characteristics of a tangential discontinuity.

hodogram of one of these current sheet encounters. The hodogram shows the evolution of the tip of the magnetic field vector during the crossing in the current sheet coordinate system. This current system and others like it resemble the terrestrial current sheet or neutral sheet in the Earth's magnetotail. The multiple current sheets observed are probably due to the flapping of the tail in the solar wind. The area occupied by the magnetotail of Venus at 10 R_V is a cylinder $\sim$ 2 R_V in radius and contains $\sim$ 2 megawebers (Russell et al. 1981b). Some of the field lines probably close in the Venus ionosphere but the normal component across the tail may be sufficient for much of the tail flux to close across the current sheet between the point of observations ($\sim$ 10 R_V back) and the planet.

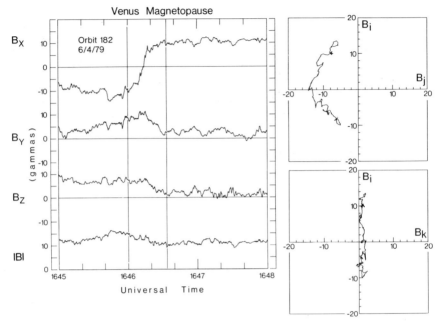

Fig. 12. High-resolution (6 samples per second) magnetic field measurements at the edge of the Venus magnetotail on orbit 182, 6/4/79. The left-hand panel shows the time series in solar ecliptic coordinates. The right-hand panel shows two orthogonal hodograms of the evolution of the tip of the magnetic field in the principal axis system. The eigenvalues were: 63.50:5.20:0.39 (nT)^2. The direction of the eigenvector associated with the minimum eigenvalue was (0.050, -0.511, 0.858). This is approximately the direction expected for the normal to the surface of a cylindrical tail aberrated in the solar wind flow by the planetary motion. The clear rotation of the magnetic field with little change in magnitude is the expected signature of a rotational discontinuity. This signature is often observed at the Earth's magnetopause and is the expected signature of the process called reconnection. The average field magnitude along the minimum variance direction here is small (0.9 nT) and less than the expected uncertainty of 2.5 nT.

The orientation of the field in the tail is as expected for magnetic field lines draped over the planet. When a sector boundary is crossed, the orientation of the magnetotail field reverses. It is not precisely clear how this reversal takes place and whether it is analogous to tail disruptions observed in comets and proposed to take place via reconnection (Niedner and Brandt 1978).

There is some evidence for connection between the tail and magnetosheath fields. Figure 12 shows a time series and a hodogram for a high-latitude magnetopause crossing in the Venus magnetotail. The orientation of this discontinuity is as expected for a tail magnetopause. The current layer consists mainly of a rotation of the field vector and little change in the magnetic field strength, signatures of a rotational discontinuity. A rotational discontinuity across topologically different plasmas will result in merging of the two plasmas. Returning to Fig. 10, we note that outside the discontinuity the Venus tail is bounded by plasma wave activity.

VI. DISCUSSION AND CONCLUSIONS

The nature of the obstacle to the solar wind at Venus and at a comet is quite different. Venus is a large solid object with a thin gravitationally bound atmosphere, whereas a comet, which may be very small, develops a large escaping atmosphere as it passes by the Sun. Nevertheless, the morphology of the interaction is qualitatively similar. A bow shock forms to deflect and heat the flow. There is a contact discontinuity which separates cometary and Venus ions from the solar wind ions. Both Venus and comets have magnetic tails.

Our understanding as of 1981 of the solar wind interaction with Venus is derived from *in situ* observations whereas our cometary knowledge is derived from remote sensing. The former observations aid in the understanding of local processes and the latter of global processes. Thus they are quite complementary. On the other hand, we should not carry the analogy too far for, while the processes may be analogous they are not necessarily scaled in the same manner. For example, the mass-loading at comets is expected to occur over a very large volume surrounding the comet, slowing down the solar wind before it reaches the bow shock and thus weakening the shock. At Venus the same process occurs but apparently principally behind the shock. This process will not slow the upstream solar wind but it does remove momentum flux from the flow which the planetary ionosphere would otherwise have to deflect. Thus the bow shock of Venus need not be as strong as a similar shock at Earth. Another similarity is that both the Venus and cometary interaction seem to be dependent on variations of the EUV flux of the Sun during the solar cycle. However, we cannot be sure if this dependence stems from a similar underlying cause.

The processes of photoionization and charge exchange are very important at Venus and at comets because they lead to mass-loading of the field which slows the convection of the field line where this plasma is created. Since the ends of the field lines in the solar wind are not mass-loaded they continue to move with the solar wind speed stretching out the field in a long tail behind the body. The slowness of the convection near the obstacle leads naturally to the depletion of the hot solar wind plasma since it can flow along field lines. However, calculations for Venus show that charge exchange will eventually deplete the solar wind ions so that the field lines near the obstacle contain only the cold freshly ionized plasma (Gombosi et al. 1981). Thus is formed the magnetic barrier or cushion which puts a cap on the Venus ionosphere. On the other hand, this magnetic cap is moving, albeit slowly, and can scrape away the newly borne ions. This mechanism presumably operates at comets also but on a far grander scale. Our estimate, for one well-defined example at Venus, gives a loss rate of 2×10^{25} ions s^{-1} whereas cometary loss rates can be at least four orders of magnitude greater.

A very intriguing cometary phenomenon is the formation of tail rays. These features appear to elongate a specific direction and slowly drift towards the center of the tail. It is tempting to speculate that the direction of elongation is a flux tube and the drift to the center of the wake is due to the convection of the plasma to the center of the tail. However, Venus observations suggest that perhaps the magnetic field gathers up plasma at bends in the field and that tail rays form at the magnetic poles, that is in the Sun-comet plane perpendicular to the plane containing the solar wind, the comet, and the interplanetary magnetic field.

The behavior of the ionosphere of Venus shows an interesting dichotomy that may have an analog at comets. At normal or quiet times the magnetic field tends to isolate the solar wind plasma from direct interaction with the Venus ionospheric plasma. As we have discussed above, this isolation is accomplished with the formation of a magnetic barrier. At these times the Venus ionosphere is essentially field free. However, when the solar wind dynamic pressure is high, the ionosphere becomes magnetized, presumably because of stronger coupling of the solar wind to the ionosphere. Changes with solar wind dynamic pressure may also be important at comets, although perhaps in different ways. Houpis and Mendis, for example, have emphasized the possibility that occasionally the solar wind could affect the neutral flow in comets.

One aspect of the Venus interaction that has no analog in terrestrial interaction is the flux rope. Its clearest analog is in the behavior of solar magnetic fields. These flux ropes consist of twisted bundles of flux which at times assume an almost self-balancing configuration in which the inward pressure due to the twist in the field contains the magnetic pressure of the rope. These features are quite small in transverse dimension, being on the order of 10 to 30 km across. These ropes could be caused by several mechanisms. An attractive mechanism is the flute instability at the ionopause in which flux tubes break off and sink in the subsolar ionosphere. It is not clear what role these features could play in a cometary interaction but it is noteworthy that the stressed fields in a flux rope represent stored energy which could be released if the axial currents supporting the twist in the field became disrupted.

In closing we note that the present and continuing observations of the solar wind interaction with Venus should provide new insight into the solar wind interaction with comets for years to come. This chapter represents only the first step in that direction. Similarly too the global observations of comets provides a complementary data base from which we can improve our understanding of the solar wind interaction with Venus. However, the two sets of data will become even more closely linked in 1986 when P/Halley passes between Venus and the Sun.

Acknowledgments. We are grateful to the many Pioneer Venus team members who have shared their ideas and data with us. We are especially grateful to L. Brace, T. Cravens, J. Mihalov, A. Nagy, F. Scarf, and J. Slavin. The work at the University of California at Los Angeles was supported by a contract from the National Aeronautics and Space Administration. The work at the Jet Propulsion Laboratory, California Institute of Technology represents one phase of research supported by another NASA contract.

REFERENCES

Alfvén, H. 1957. On the theory of comet tails. *Tellus* 9:92-96.

Biermann, L.; Brosowski, B.; and Schmidt, H.U. 1967. The interaction of the solar wind with a comet. *Solar Phys.* 1:254-284.

Brace, L.H.; Theis, R.F.; and Hoegy, W.R. 1981. Plasma clouds above the ionopause of Venus and their implications. Submitted to *Planet. Space Sci.*

Brace, L.H.; Theis, R.F.; Hoegy, W.R.; Wolfe, J.H.; Mihalov, J.D.; Russell, C.T.; Elphic, R.C.; and Nagy, A.F. 1980. The dynamic behavior of the Venus ionosphere in response to solar wind interactions. *J. Geophys. Res.* 85:7663-7678.

Brandt, J.C., and Mendis, D.A. 1979. The interaction of the solar wind with comets. In *Solar System Plasma Physics,* Vol. II, eds. C.F. Kennel, L.J. Lanzerotti, and E.N. Parker (Amsterdam: North Holland), pp. 253-292.

Bridge, H.S.; Lazarus, A.J.; Synder, C.W.; Smith, E.J.; Davis, L.; Coleman, P.J., Jr.; and Jones, D.E. 1967. Plasma and magnetic fields observed near Venus. *Science* 158:1669-1673.

Combi, M.R., and Delsemme, A.H. 1980. Brightness profiles of CO^+ in the ionosphere of Comet West (1976 VI). *Astrophys. J.* 238:318-387.

Cravens, T.E.; Gombosi, T.I.; Kozyra, J.; Nagy, A.F.; Brace, L.H.; and Knudsen, W.C. 1980. Model calculations of the dayside ionosphere of Venus: Energetics. *J. Geophys. Res.* 85:7778-7786.

Delsemme, A.H., and Combi, M.R. 1979. $O(^1D)$ and H_2O^+ in Comet Bennett 1970 II. *Astrophys. J.* 228:330-337.

Dolginov, Sh. Sh.; Yeroshenko, Y.G.; and Davis, L. 1969. On the nature of the magnetic field near Venus. *Kosmich, Issled.* 7:747-752.

Eddington, A.S. 1910. The envelopes of Comet Morehouse (1908c). *Mon. Not. Roy. Astron. Soc.* 70:442-458.

Elphic, R.C.; Russell, C.T.; Luhmann, J.G.; and Scarf, F.L. 1981. The Venus ionopause current sheet: Scale size and controlling factors. Submitted to *J. Geophys. Res.*

Elphic, R.C.; Russell, C.T.; Slavin, J.A.; and Brace, L.H. 1980a. Observations of the dayside ionopause and ionosphere of Venus. *J. Geophys. Res.* 85:7679-7696.

Elphic, R.C.; Russell, C.T.; Slavin, J.A.; Brace, L.H.; and Nagy, A.F. 1980b. Ionopause of Venus–Pioneer Venus observations. *Geophys. Res. Letters.* 7:561-564.

Ershkovich, A.I. 1980. Kelvin-Helmholtz instability in type 1 comet tails and associated phenomena. *Space Sci. Rev.* 25:3-34.

Feldman, P.D., and Brune, W.H. 1976. Carbon production in Comet West 1975n. *Astrophys. J.* 209:L45-L48.

Gombosi, T.I.; Cravens, T.E.; Nagy, A.F.; Elphic, R.C.; and Russell, C.T. 1980. Solar wind absorption by Venus. *J. Geophys. Res.* 85:7747-7753.

Gombosi, T.I.; Horanyi, M.; Cravens, T.E.; Nagy, A.F.; and Russell, C.T. 1981. The role of charge exchange in the solar wind absorption by Venus. Submitted to *Nature.*

Greenstadt, E.W. 1976. Phenomology of the Earth's bow shock system. A summary description. In *Magnetospheric Particles and Fields,* ed. B.M. McCormac (Dordrecht, Holland: D. Reidel Publ. Co.), p. 13.

Greenstadt, E.W.; Russell, C.T.; Formisano, V.; Hedgecock, P.C.; Scarf, F.L.; Neugebauer, M.; and Holzer, R.E. 1977. Structure of the quasi-parallel, quasi-laminar bow shock. *J. Geophys. Res.* 82:651-666.

Gringauz, K.I.; Bezrukikh, V.V.; Volkov, G.I.; Musatov, L.S.; and Breus, T.K. 1970. Interplanetary plasma disturbances near Venus determined from Venera-4 and Venera-6 data. *Kosmich. Issled.* 8:431-436.

Hartle, R.E.; Taylor, H.A., Jr.; Bauer, S.J.; Brace, L.H.; Russell, C.T.; and Daniell, R.E., Jr. 1980. Dynamical response of the dayside ionosphere of Venus to the solar wind. *J. Geophys. Res.* 85:7739-7746.

Harwit, M., and Hoyle, F. 1962. Plasma dynamics in comets. II. Influence of magnetic fields. *Astrophys. J.* 135:875-882.

Hoppe, M.M., and Russell, C.T. 1981. On the nature of ULF waves upstream of planetary bow shocks. *Advances in Space Research* 1:327-332.

Houpis, H.L.F., and Mendis, D.A. 1980a. Physicochemical and dynamical processes in cometary ionospheres. I. The basic flow profile. *Astrophys. J.* 239:1107-1118.

Houpis, H.L.F., and Mendis, D.A. 1980b. On the development and global oscillation of cometary ionospheres. Preprint, Univ. California, San Diego.

Huppler, D.; Reynolds, R.J.; Roesler, F.L.; Scherb, F.; and Trauger, J. 1975. Observations of Comet Kohoutek (1973f) with a ground-based Fabry-Perot spectrometer. *Astrophys. J.* 202:276-282.

Hyder, C.L.; Brandt, J.C.; and Roosen, R.G. 1974. Tail structure far from the head of Comet Kohoutek I. *Icarus* 23:601-610.

Ioffe, Z.M. 1968. Some magnetohydrodynamic effects in comets. *Soviet Phys.-Astron.* 11:1044-1047.

Ip, W.-H., and Mendis, D.A. 1975. The cometary magnetic field and its associated electric currents. *Icarus* 26:457-461.

Knudsen, W.C.; Spenner, K.; and Miller, K.L. 1981. Anti solar acceleration of ionospheric plasma across the Venus terminator. *Geophys. Res. Letters* 8:241-244.

Knudsen, W.C.; Spenner, K.; Miller, K.L.; and Novak, V. 1980. Transport of ionospheric O^+ ions across the Venus terminator and its implications. *J. Geophys. Res.* 85:7803-7810.

Luhmann, J.G.; Elphic, R.C.; and Brace, L.H. 1981a. Large-scale current systems in the dayside Venus ionosphere. *J. Geophys. Res.* 86:3509-3514.

Luhmann, J.G.; Elphic, R.C.; Russell, C.T.; Mihalov, J.D.; and Wolfe, J.H. 1980. Observations of large scale steady magnetic fields in the dayside Venus ionosphere. *Geophys. Res. Letters* 7:917-920.

Luhmann, J.G.; Elphic, R.C.; Russell, C.T.; Slavin, J.A.; and Mihalov, J.D. 1981b. Observations of large-scale steady magnetic fields in the nightside Venus ionosphere and near wake. *Geophys. Res. Letters* 8:517-520.

Lüst, R. 1967. Bewegung von Strukturen in der Koma und im Schweif des Kometen Morehouse. *Zs. f. Ap.* 65:236-250.

Mihalov, J.D.; Wolfe, J.H.; and Intriligator, D.S. 1980. Pioneer Venus plasma observations of the solar wind Venus interaction. *J. Geophys. Res.* 85:7613-7624.

Nagy, A.F.; Cravens, T.E.; Smith, S.G.; Taylor, H.A., Jr.; and Brinton, H.C. 1980. Model calculations of the dayside ionosphere of Venus: Ionic composition. *J. Geophys. Res.* 85:7795-7801.

Ness, N.F.; Behannon, K.W.; Lepping, R.P.; Whang, Y.C.; and Schatten, K.H. 1974. Magnetic field observations near Venus: Preliminary results from Mariner 10. *Science* 183:1301-1306.

Niedner, M.B., Jr., and Brandt, J.C. 1978. Interplanetary gas, XXIII. Plasma tail disconnection events in comets: Evidence for magnetic field reconnection at interplanetary sector boundaries. *Astrophys. J.* 223:655-670.

Niemann, H.B.; Kasprzak, W.T.; Hedin, A.E.; Hunten, D.M.; and Spencer, N.W. 1980. Mass spectrometric measurements of the neutral gas composition of the thermosphere and exosphere of Venus. *J. Geophys. Res.* 85:7817-7827.

Oppenheimer, M., and Downey, C.J. 1980. The effect of solar-cycle ultraviolet flux variations on cometary gas. *Astrophys. J.* 241:L123-L127.

Russell, C.T. 1977. The Venus bow shock: Detached or attached? *J. Geophys. Res.* 82:625-631.

Russell, C.T., and Elphic, R.C. 1979. Observation of magnetic flux ropes in the Venus ionosphere. *Nature* 279:616-618.

Russell, C.T.; Elphic, R.C.; and Slavin, J.A. 1979. Pioneer magnetometer observations of the Venus bow shock. *Nature* 282:815-816.

Russell, C.T.; Hoppe, M.M.; and Livesey, W.A. 1981a. Overshoots in planetary bow shocks. Submitted to *Nature*.

Russell, C.T.; Luhmann, J.G.; Elphic, R.C.; and Scarf, F.L. 1981b. The distant bow shock and magnetotail of Venus: Magnetic field and plasma wave observations. *Geophys. Res. Letters*. In press.

Russell, C.T.; Luhmann, J.G.; Elphic, R.C.; Scarf, F.L.; and Brace, L.H. 1981c. Magnetic field and plasma wave observations in a plasma cloud at Venus. Submitted to *Geophys. Res. Letters*.

Schmidt, H.U., and Wegmann, R. 1980. MHD calculations for cometary plasmas. *Computer Phys. Commun.* 19:309-326.

Slavin, J.A.; Elphic, R.C.; and Russell, C.T. 1979. A comparison of Pioneer Venus and Venera bow shock observations: Evidence for a solar cycle variation. *Geophys. Res. Letters* 6:905-908.

Spenner, J.; Knudsen, W.C.; Miller, K.L.; Novak, V.; Russell, C.T.; and Elphic, R.C. 1980. Observation of the Venus mantle, the boundary region between solar wind and ionosphere. *J. Geophys. Res.* 85:7655-7662.

Spreiter, J.R., and Stahara, S.S. 1980. Solar wind flow past Venus: Theory and comparisons. *J. Geophys. Res.* 85:7715-7738.

Taylor, H.A., Jr.; Brinton, H.C.; Bauer, S.J.; Hartle, R.E.; Cloutier, P.A.; Daniell, R.E., Jr. 1980. Global observations of the composition and dynamics of the ionosphere of Venus: Implications for the solar wind interaction. *J. Geophys. Res.* 85:7765-7777.

Taylor, W.W.L.; Scarf, F.L.; Russell, C.T.; and Brace, L.H. 1979. Absorption of whistler mode waves in the ionosphere of Venus. *Science* 205:112-114.

Theis, R.F.; Brace, L.H.; and Mayr, H.G. 1980. Empirical models of the electron temperature and density in the Venus ionosphere. *J. Geophys. Res.* 85:7787-7794.

Verigin, M.I.; Gringauz, K.I.; Gombosi, T.; Breus, T.K.; Bezrukikh, V.V.; Remizov, A.P.; and Volkov, G.I. 1978. Plasma near Venus from the Venera 9 and 10 wide angle analyzer data. *J. Geophys. Res.* 83:3721-3728.

Wallis, M.K., and Dryer, M. 1976. Sun and comets as sources in an external flow. *Astrophys. J.* 205:895-899.

Wallis, M.K., and Ong, R.S.B. 1976. Cooling and recombination processes in cometary plasma. In *The Study of Comets*, eds. B. Donn, M. Mumma, W. Jackson, M. A'Hearn, and R. Harrington (Washington: NASA SP-393), pp. 856-876.

Wolff, R.S.; Goldstein, B.E.; and Kumar, S. 1979. A model of the variability of the Venus ionopause altitude. *Geophys. Res. Letters* 6:353-356.

Zhuang, H.-C.; Walker, R.J.; and Russell, C.T. 1981. The influence of the interplanetary magnetic field and thermal pressure on the location of the magnetopause. Submitted to *J. Geophys. Res.*

THEORIES OF PHYSICAL PROCESSES IN THE COMETARY COMAE AND ION TAILS

W.-H. IP and W.I. AXFORD
Max-Planck-Institut für Aeronomie

A number of physical processes are involved in the comet-solar wind interaction. In the outer region of the cometary coma, there is the pickup of the new cometary ions by the solar wind. Space probe observations must explore the questions concerning the manner in which the cometary ions are being assimilated into the inward streaming plasma flow and what their resulting velocity distributions are, even though certain limits can be set by theoretical arguments. The formation of narrow ion rays, as the cometary plasma is being swept downstream, is a most enigmatic phenomenon in cometary physics. MHD calculations have established the relation between these filamentary structures and the time-variation of the interplanetary magnetic fields. However, the importance of other physical effects cannot yet be ruled out. Another basic issue in cometary plasmas concerns the nature of the ionospheric outflow from the central source. Two possible flow patterns (supersonic wind versus subsonic flow) are discussed here. In the case of the subsonic model, there could exist a broad region of lateral plasma streaming between the inward contaminated solar wind and the outward expanding ionospheric flow. The draped interplanetary magnetic fields on the one hand would affect the dynamics of the cometary ions in the vicinity of the comet head, and on the other, it would form a large-scale current system in the ion tail. Reconfiguration of the tail-aligned magnetic fields would cause the production of energetic charged particles of keV energies as well as a large change in the morphology of the ion tails. Another factor which can have strong impact on the dynamical behaviors of ion tails is charge-exchange ionization by the solar wind plasma. Sharp increase in such an ionization effect in the solar wind stream-stream interaction regions may play a role in ion tail disturbances. Though theoretical in nature, this review also covers both comet observations and laboratory simulation experiments connected with the topic.

The problems of the interaction of comet and solar wind cover a lot of ground. Not only is the whole process extremely complicated, but the basic physical nature also changes drastically according to the gas production rates of the comets and their distances from the Sun. For a comet with H_2O molecules as the main outgassing component, its atmospheric activity is extremely low beyond 2 AU and the solar wind interaction can be characterized as lunar type; i.e., the solar wind could have direct access to the nuclear surface and no shock would form since the comet constitutes a negligible obstacle to the solar wind. Such a process may also apply to short-period comets with very low gas production rates. On the other hand, for a comet with large gas production rate, either because of its intrinsic gas production rate or because of its proximity to the Sun, the ionized cometary gas could attain supersonic speed, resulting in strong interaction with the solar wind plasma. Between these two extremes we have the situation that the solar wind inflow is countered by a subsonic ionospheric outflow from the central source. These three possibilities are illustrated in Fig. 1. If we add the volatile-rich (i.e. CO and CO_2) comets to this consideration there may be even more varieties as far as radial variation is concerned. Here we would limit ourselves mostly to the solar wind interaction of a medium-bright comet like P/Halley at a solar distance of 1 AU. After its return in 1986 P/Halley will probably become the best known comet with its atmospheric environment thoroughly explored by space probes. Comparison of theoretical studies of this type of comet with the results from the *in situ* measurements should be most fruitful.

The study of cometary plasmas, of course, is characterized by the fact that cometary ions, CO^+ in particular, can be used as a tracer for the dynamical process. This effect also underlines the close relationship between the ion composition and plasma dynamics. Beginning with its initial expansion to the final pickup process upstream in the solar wind, the chemical effects and the solar wind interaction in the coma are to a large extent coupled. So far not much attention has been paid to this problem, but we expect rapid progress in this area. Theoretical considerations are further handicapped by the fact that most model calculations only apply to steady-state conditions, whereas cometary atmospheres and plasmas display many nonstationary phenomena (e.g., expanding halos, plasma envelopes and ion rays). We are sometimes forced to explain many of the observations with what we have learned from steady-state models in lieu of better treatments. Therefore we must take precaution in accepting the theoretical interpretations too easily. In this re-

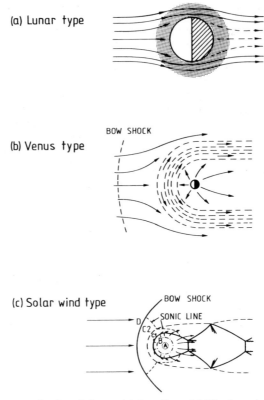

Fig. 1. Three types of solar wind-comet interactions. (a) The lunar-type interaction or weak interaction is pertinent to comets with very small gas production rates like the short-period comets or even large comets but at great solar distances. The solar wind plasma could have direct access to the nucleus. (b) In the Venus-type interaction the cometary atmosphere constitutes an obstacle sufficiently large to produce a bow shock at the front side. The ionosphere could only marginally standoff the contaminated solar wind and these two plasmas are separated by a thick layer of magnetosheath. (c) The solar wind type interaction or strong interaction is appropriate for comets with very large gas production rates such that the ionospheric outflow may attain supersonic speed and form an inner shock before being diverted to lateral flows. The supersonic wind pattern is from Wallis and Dryer (1976).

view, we use a *standard* comet-solar wind interaction model to describe some of the basic physical processes in ion comae and ion tails. In areas in which we have very little understanding, various alternatives are considered. These topics include the assimilation of cometary ions in the outer ion coma and the structure of the inner ionosphere. Besides theoretical models, we are also aided by comet observations and laboratory simulation experiments.

Another source of input concerning cometary plasma environment comes from the observations of Venus by the Pioneer Venus spacecraft. The findings that the interplanetary magnetic field is highly compressed just ahead of the ionopause and that the ionosphere is not field-free (Elphic et al. 1980) have important implications on ion coma structure of comets. Indeed, we cannot rule out the scenario that the flow pattern and magnetic field morphology in the inner ion coma of P/Halley at 1 AU is much like those observed for Venus. (Further information on the Venus-solar wind interaction and its implication for comets can be found in this book in the chapter by Russell et al.)

The content of this chapter is arranged in such a way that it will be largely complementary to the chapters by Brandt (*Observations and Dynamics of Plasma Tails*) and by Schmidt and Wegmann (*Plasma Flow and Magnetic Fields in Comets*). But there will also be overlaps in areas of common interest.

I. THE OUTER COMA

The first step of solar wind interaction with a comet is its penetration into the extensive halo of the neutral hydrogen atoms (Keller 1976) and oxygen atoms, presumably, from the photodissociation of H_2O molecules (Keller 1971; Mendis et al. 1972). These neutral fragments may have a range of expansion speeds due to the energy excess gained in their production process. For example, several velocity components have been determined for the hydrogen atoms 4 to 20 km s^{-1}; see Meier et al. (1976). These high speeds as compared with the 0.5 to 1 km s^{-1} usually assumed for the initial expansion of the neutral gas, probably result from photodissociation processes like

$$H_2O + h\nu \rightarrow H + OH + \epsilon \qquad (1)$$

with $\epsilon \sim 1.8$ eV (Schul'man 1972; Festou 1978). By the same token, energetic neutral C and O atoms, with ejection speed ~ 5 km s^{-1}, may be created by the photodissociation of CO and other ion-molecule reactions (Feldman 1978). So far, in the model of solar wind-comet interaction it is generally assumed that the cometary atmosphere is characterized by molecules of a single expansion speed, a photodissociation time scale as well as ionization time scale. In a more detailed study, this assumption would have to be modified to take into account the different components of the neutral gas in the coma.

A. Solar Wind Mass Loading

That mass loading of the solar wind plasma, as a result of pickup of cometary ions from the neutral coma, would lead to the formation of a bow shock a few 10^5 to 10^6 km ahead of the cometary nucleus is perhaps now common knowledge (Marochnik 1963a,b,1964; Axford 1964; Beard 1966; Biermann et al. 1967). As estimated by Axford (1964), the solar wind may accrete no more than just a few percent (by number) of the heavy ions before

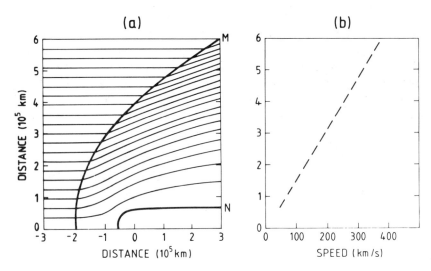

Fig. 2. The axially symmetric flow pattern of cometary plasmas from stationary MHD calculations (Biermann et al. 1974). The original diagram (a) is for a comet with gas production rate of 10^{30} molecules s^{-1} and solar distance at 1 AU (bow shock distance = 1.3×10^6 km). Here we have made the scaling for a comet with a gas production rate a factor of 10 smaller. The approximate flow speed variation along the M-N line perpendicular to the comet-Sun axis is shown in (b). Note the almost constant gradient over the width of the tail ($\sim 6 \times 10^5$ km).

reaching the shock transition. The remaining question then is how strong would such a shock characterized by the mass loading effect be. At this point both analytical and numerical models seem to converge on the opinion that, as far as steady-state interaction is concerned, the shock would be weak with the Mach number $\lesssim 2$ (Wallis 1973a,b; Brosowski and Wegmann 1972; Schmidt 1978; Schmidt and Wegmann 1980). This conclusion applies only to medium-bright to bright comets at a solar distance of ~ 1 AU with gas production rate between 10^{29} and 10^{30} molecules s^{-1}. For other solar distances and gas production rates, the shock structure may be stronger or weaker. In any event, much about the physical process of solar wind interaction can still be learned by referring to the *standard* model. The magnetohydrodynamic (MHD) calculations reported by Schmidt and Wegmann (1980) are particularly useful here because they present a three-dimensional view of the configuration of the outer ion coma. This is exemplified by the flow pattern of the solar wind-cometary plasma illustrated in Fig. 2. One should note that the velocity gradient across the tail with a width of 4×10^5 km is almost constant near the head. There is no strong velocity shear nor discontinuity within a short distance on the order of $\sim 10^4$ km. Except for the jumps at the shock front, other physical parameters also have smooth behaviors. In Fig. 3 are presented the approximated mass density, the thermal pressure, Mach num-

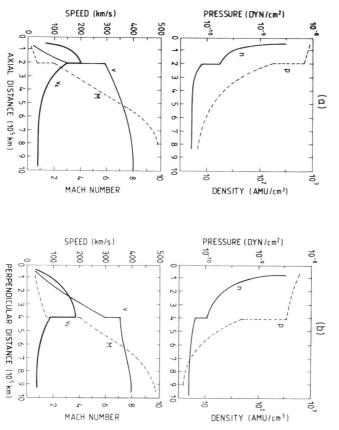

Fig. 3. The variations of the thermal pressure p, total ion mass density n, bulk flow speed V, thermal speed V_s, and Mach number M in the coma of a P/Halley-type comet at 1 AU solar distance. (a) Along the comet-Sun line, the shock transition (with $M \sim 2$) occurs at a distance of 2×10^5 km from the nucleus. After the shock the flow has become subsonic with $M \sim 0.7$. (b) Along the direction perpendicular to the comet-Sun line and passing through the comet head, the shock transition is now at 4×10^5 km and the flow is still supersonic ($M \sim 1.5$) after the shock. Smooth transition to subsonic flow is attained at a distance of $\sim 2.5 \times 10^5$ km. These diagrams are the scaling approximations from the stationary hydrodynamic calculations by Schmidt and Wegmann (1980) as reported in Biermann et al. (1974).

ber, the bulk flow speed, and the thermal speed in the directions along and perpendicular to the Sun-comet axis.

The buildup of the plasma density as the distance to the cometary nucleus decreases is partly produced by the continuous accretion of the cometary ions along the streamline and partly caused by the slowdown of the plasma flow. In the subsonic region, there will be divergence of the inward flow channeling the cometary plasma into the tailward direction. To a certain

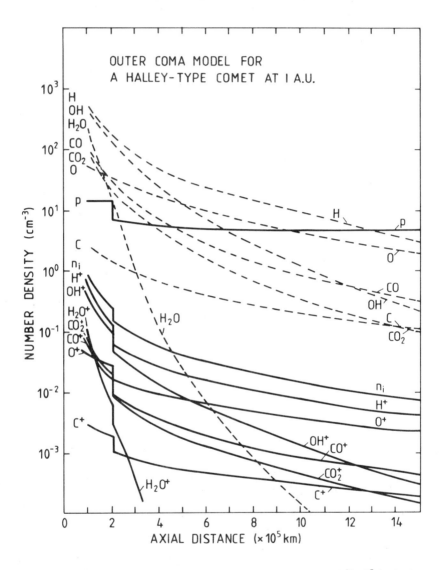

Fig. 4. The main neutral and ion compositions in the outer coma ($r \geqslant 10^5$ km) of a P/Halley-type comet at 1 AU solar distance. The original compositional abundance has the ratio of $[H_2O]:[CO_2]:[CO] = 1:0.15:0.15$. For an H_2O-rich comet as modeled here, the solar wind would be slowed down essentially by the O^+ and OH^+ heavy ions from the photodissociation of the H_2O molecules. (Note the short scale length of the H_2O molecules.) In contrast, direct pickup of the molecular ions may play the dominant role in the case of a CO/CO_2-rich comet. See Ip (1980a) for more details.

extent, the flow speed u and the cross-section A of the stream tube follow the relation of an incompressible fluid, i.e., uA = const. Interesting analytical results have been obtained by Wallis (1973a,b) by adopting this relation. This relation, however, breaks down close to the nucleus.

In principle, some idea about the ion composition in the outer coma may be obtained by substituting the computed velocity profile into the continuity equation

$$\frac{1}{A}\frac{d}{ds}(n_i u A) = q_i - \ell_i \tag{2}$$

for ion species i along the stream tube of the flow. Here q_i and ℓ_i are respectively the source and sink terms. In Fig. 4 we present such a model of the outer coma along the axial direction. The neutral coma which acts as a source for the cometary ions is also depicted. That the outer neutral coma is dominated by photodissociation products like O, OH, and H is directly reflected in the ion abundance before the shock front. Both O^+ and OH^+ should therefore contribute most to the mass loading effect. This is the situation for a comet with gas production rate dominated by H_2O molecules. In the case of a CO/CO_2-rich comet, the situation would be somewhat different because the photodissociation times of CO and CO_2 are comparable to those of photoionization. This would mean the solar wind interaction process is determined in part by the direct pickup of the parent molecules, (CO_2 and CO), as well as by their corresponding daughters. This is one example showing that the chemical nature of comets could affect the overall cometary fluid dynamics.

B. Cometary Plasma Assimilation

During their inward streaming, the solar wind protons and helium ions are subject to charge-exchange loss with the neutral atmosphere. One byproduct would be singly ionized helium ions before their final neutralization. Here it may be worthwhile to point out that there are basically three types of ions in the contaminated solar wind; the heavy ions purely of cometary origin, the protons contaminated by the cometary hydrogen ions, and the He^{++} and He^+ ions from the solar wind. Tracking of these three populations throughout the outer coma would be useful in sorting out the detailed physical mechanisms involved in the momentum and energy transfer between the solar wind and cometary gas. This brings us to the next topic of interest, namely, how are the new cometary ions being assimilated in the inward flow.

In the supersonic region, the slowdown of the plasma flow is accompanied by heating of the ions. This effect is well demonstrated by the increase of the thermal speed V_s depicted in Fig. 3. (A combination of the bulk flow speed reduction and the increase of the thermal speed results in the formation of a weak shock.) Schmidt and Wegmann (1980) obtained these profiles by assuming complete thermalization of the new cometary ions picked up by the solar wind. Questions, however, have been raised concerning the presence of energetic cometary ions (O^+, CO^+, CO_2^+, etc.) with thermal speed typical of the local solar wind speed. As described by Wallis (1973a,b), just after their ionization, the new cometary ions would follow cycloidal trajectories with the gyration centers moving with the solar wind. For example, in the case of the interplanetary magnetic fields perpendicular to the solar wind velocity, besides producing a very peculiar velocity distribution, these new ions also have a peaked 90° pitch angle distribution superimposed on the isotropic Maxwellian distribution of the background plasma.

C. Fast Relaxation Limit

One possible picture for the presence of cometary ions, newly picked up by the solar wind flow (with transverse magnetic field), is illustrated in Fig. 5a. As shown here, the $V_\perp$–distribution of the new ions resulting from adiabatic motion (see Sec. I.D for more details) forms a hump in the ion distribution function, and such a configuration is very unstable, for instance, against transverse cyclotron instabilities (Wallis 1971). According to linear calculation and computer simulation the energy relaxation time scale is just a few times that of the ion gyroperiod (see Okuda and Hasegawa 1969).

There are two steps involved in the collective wave-particle interactions thermalizing the energetic new ions. First, there is pitch angle scattering making isotropic the ion flux with a time scale τ_α, and then there is the energy equipartition process equalizing the thermal energy of the new ions with that of the cold background plasma with a time scale τ_K. These two processes may operate simultaneously or separately depending on the ratio of τ_α/τ_K. In fact, as the inward streaming of the solar wind continues, the energetic ions created at large upstream distances will be further accelerated if $\tau_K > \tau_f$ (the transit time of the solar wind flow through the outer coma). To what extent these suprathermal particle populations would influence the overall budgets of energy and pressure balance is yet to be explored. The lower limit of this flux of energetic particles can be estimated by considering the extreme case that $\tau_\alpha \sim \tau_K \ll \tau_f$.

The rapid thermalization of photoions created in the solar wind as a result of interpenetration of the solar wind plasma with the interstellar neutral gas or planetary exospheres has been further considered by several authors (Wu and Davidson 1972; Hartle and Wu 1973; Hartle et al. 1973). Some of their formulations can be used in the present context. Assuming that the interplanetary magnetic field B_0 is uniform and constant, then in the solar wind frame, the newly ionized ions (from photoionization or charge exchange) would have an initial speed $v_{0\parallel}$ parallel to B_0 and an initial speed $v_{0\perp}$ perpendicular to the magnetic field B_0.

At the beginning of the pickup process, the background ambient plasma (of number density n_s) may be assumed to be protons only and the cometary ions represented by one single species of mass m_{ic}. If our consideration is

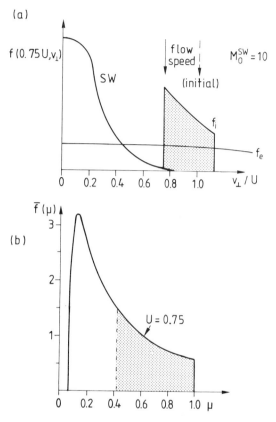

Fig. 5. (a) Adiabatic model of ion distribution. The solar wind electron and ion, and cometary ion distribution functions are shown over gyration velocity $V_\perp$ for a 25% decelerated solar wind with transverse magnetic field. Because of the conservation of the first adiabatic invariant, cometary ions accreted at large upstream distance could have $V_\perp > U$ (solar wind speed). (From Wallis 1971.) (b) Normalized cometary ion distribution function $\bar{f}(\mu)$ in the adiabatic model. The distribution corresponding to $U = 0.75$ of the original solar wind speed is indicated by the shaded area.

further restricted to electrostatic and electromagnetic waves in the intermediate ($\Omega_p \ll |\omega| \ll \Omega_e$) and low-frequency ($|\omega| \ll \Omega_p$) range, the maximum linear growth rate of transverse electromagnetic waves propagating along the ambient magnetic field is given in s^{-1} by (Wu and Davidson 1973)

$$\gamma_K \sim \frac{\omega_{in}}{\sqrt{2}} \left(\frac{v_{o\perp}}{c}\right) \qquad (3)$$

where ω_{in} is the plasma frequency of the energetic new ions ($\gamma_K \to 0$ as $n_{in} \to 0$); and

$$\gamma_K = \frac{1}{4}\left(\frac{\pi}{2}\right)^{1/2} \omega_{in} \left(\frac{v_{0\|}}{v_{es}}\right) \qquad (4)$$

for electrostatic waves (Hartle and Wu 1973); where v_{es} is the speed in s^{-1} of the solar wind electrons. For the sake of illustration we may consider the following example. At an axial distance of 5×10^5 km ahead of the nucleus of a medium-bright comet, the solar wind speed is not yet significantly reduced and the time scale for thermalization can be expressed in s as

$$\tau_K = \gamma_K^{-1} \sim \left(\frac{m_{ic}}{m_p}\right)^{1/2} (n_{in})^{-1/2} \qquad (5)$$

where m_p is the proton mass and n_{in} is in cgs unit. To calculate the number density of the energetic new ions we can equate the production rate to the thermalization rate, i.e.

$$\frac{n_o}{\tau_i} = \frac{n_{in}}{\tau_K}. \qquad (6)$$

Here n_o is the number density of the parent neutral density (e.g., H and O atoms) and τ_i is the ionization rate due to photoionization and charge exchange. Taking $n_o(H) \sim 30$ cm^{-3}, $\tau_i \sim 1.3 \times 10^7$ s we find then $\tau_K \sim 100$ s and $n_{ic}(H^+) \sim 2 \times 10^{-4}$ cm^{-3}. This value may be compared with the number density of $\sim 2 \times 10^{-2}$ cm^{-3} for the thermal population. Similar procedure can be applied to the oxygen ions. With $n_o(0) \sim 10$ cm^{-3}, $\tau_i \sim 5 \times 10^6$ s, we have $\tau_K \sim 200$ s and $n_{ic}(0^+) \sim 4 \times 10^{-4}$ cm^{-3}. The thermal oxygen ions have a number density of 10^{-2} cm^{-3} at 5×10^5 km. Thus the consequence of rapid thermalization via plasma collective effects is to limit the number density of the energetic new ions gyrating at the local solar wind speed to be $\sim 1\%$ that of the thermal ions.

D. Slow Relaxation Limit

There is always the doubt that nonlinear effects would tend to saturate the wave activities and hence the above estimates for the thermalization time scales should be considered as lower limits. Wu and Hartle (1974) have discussed that a finite spread in the initial pickup velocity ($v_{0\|}$ and $v_{0\perp}$) would suppress the instabilities. Another effect as reported by Curtis (1981) is that changing the orientation angle between the interplanetary magnetic field and the solar wind direction from $\sim 90°$ to $< 5°$-$10°$ could lengthen the growth rates of plasma instabilities by several orders of magnitude. The assimilation process thus might be time-variable. The proportion of the hot cometary ions, as a result, might be far more than just a few percent. If pitch-angle scattering is efficient, these new ions will be lost from the front side by streaming along the open field lines. If not, they would contribute significantly to the general pressure balance in the cometary plasma flows. The adiabatic model considered by Wallis and Ong (1975) can be used to approximate the pickup ion velocity distribution in the slow relaxation limit. To begin with, the first adiabatic invariant $\mu = v_\perp^2/B$ is assumed to be conserved and since uB = const. (the magnetic field is assumed to be transverse to the solar wind velocity) we have

$$\mu = \frac{mv_\perp^2}{B} \ \alpha \ v_\perp^2 u \ . \tag{7}$$

Because of the assumption of transverse field, $\mu = u^3$ for the new ions. Expressing the distribution function f as a function of μ, then in the supersonic flow regime the one-dimension fluid equation of the cometary ions of density ρ_i can be written as

$$\frac{d}{ds}(\rho_i u f) = q_i \delta(\mu - u^3) \tag{8}$$

where q_i is the source term. A combination of Eq. (8) with the total continuity equation of the total plasma flow yields

$$\frac{d(\rho_i u f)}{d(\rho u)} = \delta(\mu - u^3) \tag{9}$$

which in turn leads to a distribution function of the form

$$f(u,\mu) = \frac{1}{\rho_i u} \int_1^u \frac{d}{du'}(\rho u') \delta(\mu - u'^3) \, du' \ . \tag{10}$$

For a one-dimensional accretional flow, the plasma density (ρ), flow velocity (u) and the adiabatic index γ are related by the simple expression (Biermann et al. 1967)

$$\rho u - \frac{2\gamma}{\gamma+1} u^{-1} + \frac{\gamma-1}{\gamma+1} u^{-2} = 0 \tag{11}$$

and this equation can be substituted into Eq. (10) to obtain an analytical form of $f(u,\mu)$, i.e.

$$f(u,\mu) = \frac{2}{3} \frac{\gamma}{\gamma+1} \frac{1}{\rho_i u} \frac{\mu^{1/3} - u^\star}{u^{1/2}} \ ; \ u^3 < \mu < 1 \tag{12}$$

where $u^\star = 1 - \gamma^{-1}$. The function of $\bar{f}(\mu) = (\mu^{1/3} - u^\star)/u^{5/3}$ is illustrated in Fig. 5b showing the extreme example of slow relaxation.

The actual relaxation process in the cometary plasma is likely to be a combination of these two extreme situations of cometary ion assimilation. While new ions accreted would produce a hump in the ion velocity distribution, those implanted in the solar wind at large upstream distance would be subject to higher degree of thermalization. The complexity of the plasma effect in a multi-component plasma flow do not yet allow a simple answer to this interesting problem and the whole topic of time-dependent solar wind assimilation of cometary ions is still very much an open issue to be explored by *in situ* spacecraft measurements.

After crossing the shock front at a distance of $\sim 2 \times 10^5$ km, the cometary ions will soon become the dominant ions as the solar wind proton number density remains more or less constant. Also, instead of heating, the plasma flow is now subject to cooling to accommodate the rapid mass accretion in the subsonic region (see Fig. 3). The newly created ions now would have a gyration energy less than the average thermal energy. Therefore, besides the process of pitch angle isotropization, other mechanisms will also be involved in equalizing the ion thermal temperatures (or thermal speeds). As the shocked solar wind is expected to be rather turbulent, cyclotron-resonant scattering or Landau damping should be effective in such an energy transfer process.

E. Charge-Exchange Termination of the Plasma Inflow

The final step in the neutral coma control of the basic structure of the solar wind-comet interaction begins $\sim 10^4$ km from the cometary nucleus where the cometary plasma assimilated upstream will suffer rapid charge exchange with the cometary neutrals (Wallis and Ong 1976). Two associated effects would take place here. First, the hot ions would be substituted by new ions of low temperature via charge exchange and second, the reduction of thermal pressure would allow further compression of the flux tubes, hereby enhancing the magnetic field strength in this ion-neutral coupling zone (see Fig. 1).

II. ION TAIL STRUCTURES

To a certain extent, the channeling of the cometary plasmas from the front side to the tailward region depends on the draping of the interplanetary magnetic fields by the cometary ionosphere. The model cometary magnetic field first proposed by Alfvén (1957) has been essentially confirmed by MHD numerical calculations (Schmidt and Wegmann 1980,1981) and laboratory simulation experiments (Dubinin et al. 1980; Podgorny et al. 1979). The recent detection of a magnetotail structure near Venus may be cited as further supporting evidence for this idea (Eroshenko 1979; Slavin et al. 1980). Alfvén also raised the point that the turning of the ion rays sometimes observed may actually reflect the motion of the magnetic flux tubes. Indeed, as described by Wurm (1963), the structure of the ion tail is often characterized by the appearance of a system of symmetric pairs of ion rays, with diameters ranging between 10^3 and a few 10^4 km, folding towards the main tail axis (see Fig. 6). When the ion rays were first formed with an inclination of $\sim 60°$ relative to the central axis, the angular speed was high with the

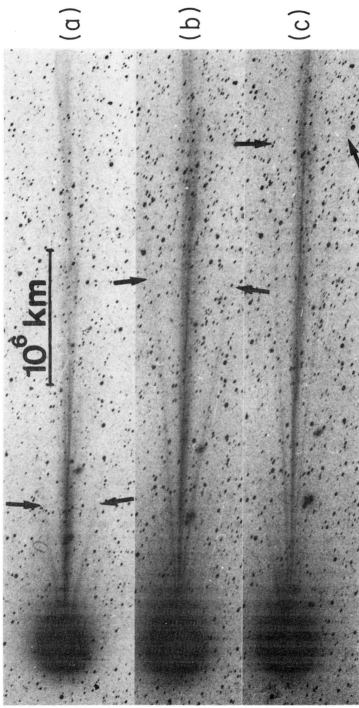

Fig. 6. The morphology of the folding effect of a symmetric pair of ion rays, observed in Comet Kobayashi-Bergen-Milon 1975 IX on Joint Observatory for Cometary Research photographs. (Courtesy of K. Jockers.)

linear speed ~ 50 km s^{-1}. But near the end of the closure, the perpendicular speed is extremely low, no more than a few km s^{-1}. By fitting the linear feature of the ion rays with the folding time scale of ~ 15 to 25 hr, Ness and Donn (1965) derived that the tail-aligned flow speed must vary linearly across a width of 1 to 2×10^6 km. This early result is in agreement with the velocity profiles obtained from stationary MHD calculations (see Fig. 2).

A. Mechanisms of Ion Ray Formation

As far as the origin of the inhomogeneous streamer structures is concerned, Schmidt and Wegmann (1980) have suggested that the ion rays are related to the occurrence of tangential discontinuities in the interplanetary magnetic field with a frequency of ~ 1 hr^{-1}. Miller (1979) has compared the theoretical picture of the flux tube movement with the observations of a plume of several nearly parallel rays by one side of Comet Tago-Sato-Kosaka 1969 IX, and he concluded that the model of the plume arising from a system of nearly coplanar rays fits the theoretical expectation. [Note that the principle of ion-ray formation with the narrow filaments of plasma density enhancement but low field strength separated from regions of opposite properties (i.e., low density but large magnetic field) is somewhat similar to the model proposed by Ness and Donn (1965) in which the ion rays are said to be the plasma sheets sandwiched between two regions of opposite magnetic polarities. The sweeping of interplanetary magnetic fields by the comets, which is intrinsic in this early model, is now considered to be of great significance on the dynamics of ion tails. See Sec. IV.]

The study of Comet Tago-Sato-Kosaka 1969 IX by Miller (1979), incidentally, has also highlighted a great variety of ion-ray structures observed in comets. Besides the formation of one or two pairs of symmetric tail rays as shown in Fig. 6, examples can be found in which the ion tails in fact are made up of a bundle of densely packed narrow streamers (and sometimes ion tails of more or less uniform structures, without tail rays, were observed). Fig. 7a is a photograph of Comet Ikeya 1963 I. Numerous ion rays can be discerned here. An example of asymmetric side-ray formation is also given in Fig. 7b.

The morphology of these two types of ion-ray structures is further delineated in Fig. 8. One essential feature here is that the symmetric ion-ray pairs appear to emanate from the central source region while the parallel side rays apparently have their roots in different points along the central tail axis. It is not yet clear whether this difference is caused simply by viewing geometry or by physical effects. In any event, the comparison of the plume of side-ray formation with the quiet-time symmetric features modeled by Schmidt and Wegmann (1981) is not necessarily appropriate. As discussed by Niedner et al. (1978), Niedner and Brandt (1978), Ip (1980c) and Jockers (1981), the formation of side rays may be related to disturbed conditions in the solar wind (e.g., a sudden change in the flow direction of the solar wind or some

COMETARY PLASMA PROCESSES

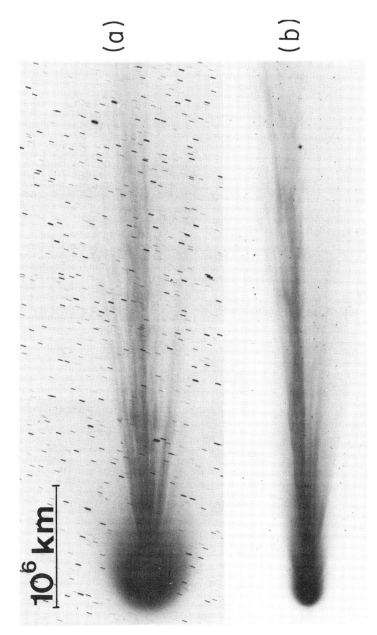

Fig. 7. Further examples of symmetric and asymmetric tail rays. (a) Comet Ikeya 1963 I: Photograph taken at Boyden Observatory on 24 February 1963, 1835 UT by E.H. Geyer. (b) Comet Tago-Sato-Kosaka 1969 IX: Photograph taken at Cerro Tololo Interamerican Observatory on 2 January 1970, 0121 UT by F.K. Miller. Note the anti-solar direction is vertical. (Courtesy of K. Jockers.)

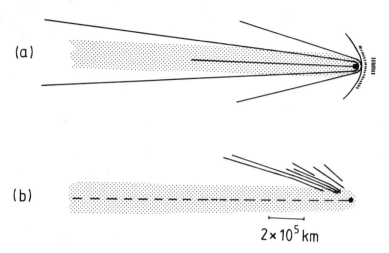

Fig. 8. Summary of the formation processes of ion rays. (a) A possible time sequence is shown for the evolution of a contracting envelope into a symmetric pair of tail rays (after Wurm 1968). The total folding time is on the order of 15 to 20 hr. This could also be the snapshot of a number of envelopes and ion rays in different stages of development (cf. Figs. 6 and 7a). (b) The structure of a plume of parallel side rays rooted to one side of the ion tail (after Miller 1979). Such an asymmetric feature could be the precursor of a tail disturbance event as described in Fig. 20 (cf. Fig. 7b).

compression effect). Caution thus must be exercised in making direct comparison of the observations.

The Schmidt-Wegmann theory of the ion-ray formation relies heavily on the time-variation of the interplanetary magnetic field direction. And we may suppose that, during the disturbed conditions of the solar wind when magnetic field fluctuation is high, the effect proposed by Schmidt and Wegmann could be valid; and some modification of their model would also yield a physical explanation to the side-ray formation near the head.

Other different mechanisms also have been suggested to explain the ion rays. For example, Beard (1981) has advocated that the enhancement of ion density in these narrow structures could be caused by the impact ionization effect of keV electrons produced by a strong shock (Axford 1964; Beard 1966). In this interesting model, it is argued that because of the effect of solar radiation pressure the radial distribution of the cometary molecules (CO in this case) in the sunward direction is substantially reduced (i.e., $n(CO) \propto r^{-a}$ with $a > 2$). And because of the sharp radial gradient in the mass addition rate to the solar wind accretion flow, a strong shock will form (Wallis 1973a,b) generating energetic protons and keV electrons. One important point in this mechanism is that the keV electrons streaming along the magnetic field lines are capable of producing enhanced ionization in a short time scale ($\lesssim 1$ hr) and in a very localized region (the lateral width of the ion rays could be as small as 10^3 km; see Wurm 1963). Furthermore, the rapid lengthening of the ion rays as they bend backward can be explained by the fast motion of the ionizing electrons. On the other hand, the crucial assumption that CO molecules, as a result of solar radiation pressure, should be confined within a distance of $\sim 10^4$ km within the coma is somewhat problematic as even the density distributions of C_2 and CN, which define the optical coma, have larger scale lengths ($\sim 3 \times 10^5$ km).

Adopting the strong-shock hypothesis but seeking a different physical cause, Houpis and Mendis (1981a,b) have proposed that because of collisional coupling between the neutral gas and the cometary ion flow, the neutral comae of CO_2/CO-rich comets would stand off the solar wind at close distances to the nuclei as in the case of Comet Humason 1962 VIII at a heliocentric distance of ~ 2.8 AU. (A similar effect is not found for H_2O-rich comets in their model.) As the relevant length scale is on the order of 10^3 to 10^4 km, these authors then agreed that such confinement of a neutral atmosphere should lead to a strong shock with Mach number $M \sim 10$ instead of ~ 2. The corresponding processes of electron acceleration to keV energies at the shock front and the resulting rapid ionization of the neutral atmosphere are essentially the same as those proposed by Beard (1981). In short then, it is suggested that the appearance of ion rays in Comet Humason 1962 VIII and perhaps other bright comets is related to the formation of strong shock in a CO_2/CO-rich comet. One possible drawback of this novel idea is that, because of the low degree of ionization in the coma, it is generally true that the ionospheric flow is guided by the expansion of the neutral gas instead of the other way around. The assessment by Houpis and Mendis (1981b) is therefore not without fault. Besides, as discussed earlier, narrow ion rays are quite common in the ion tails of comets that are definitely not CO_2/CO-rich.

B. Contracting Envelopes and Tail Rays

Because of the weak emissions from the cometary neutrals and dust, the plasma structures as traced by the CO^+ ions in the comae and ion tails of Comet Morehouse 1908 III and Comet Humason 1962 VIII can be identified more easily as compared with other comets. Comet Morehouse 1908 III, in particular, has provided some tantalizing observational material. A case in point is the so-called receding envelopes studied by Eddington (1910), Lüst (1962), and Wurm (1968). According to these authors, photographic plates of close time-sequence show ion structures in the form of parabolic envelopes forming at a projected distance ~ 1 to 1.5×10^5 km from the optical center (Fig. 9). Immediately after its rapid formation the envelope would always start to shrink towards the center with a speed of ~ 10 km s^{-1}. After 3 to 4 hr the gradually sharpened structure of the plasma envelope would slow down to a much smaller speed. And as it recedes to a distance of 5×10^4 km it would intermingle with the coma emission and become unrecognizable. At one time, three or more parabolic envelopes were observed to be in different

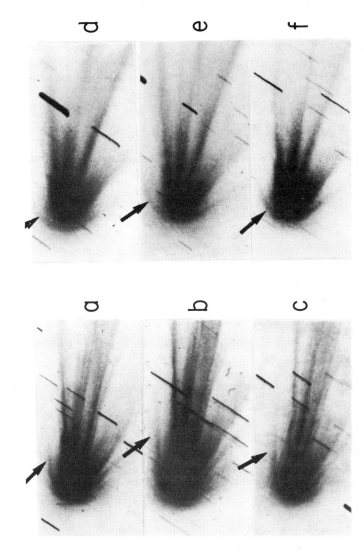

Fig. 9. Photographs of Comet Morehouse 1908 III in time sequence (3-4 Jan. 1908). The development of a contracting envelope is indicated by the arrows. Because of the weak emissions of the neutrals and dust particles in the coma, these fine scale features of CO^+ ions can be traced very close to the head in this case (cf. Figs. 6 and 7). The foreshortening of the ion tail is caused by the relative orbital orientations between the Earth and the comet. (After Royal Greenwich Observatory photographs and prepared by K. Jockers.)

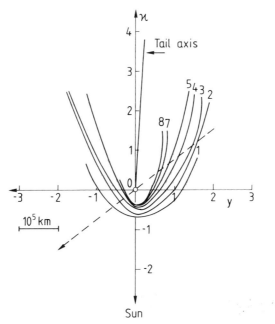

Fig. 10. The development of a very well-defined envelope observed on 21-22 Oct. 1908 in Comet Morehouse 1908 III. The time interval between (1) and (8) is 5.25 hr. The orbital direction of the comet is indicated by the dashed line. (After Lüst 1962.)

stages of collapse. A composite picture of the formative sequence of receding envelopes has been depicted by Lüst (1962). As shown in Fig. 10 there is a clear tendency for the envelopes to lengthen into ray-like symmetric structures at the final stage. A further comparison of the structure of the tail rays and envelopes is illustrated in Fig. 11. In this sketch, the tail rays are all convergent towards a point just ahead of the optical center. This effect certainly reinforces the impression that the contracting envelopes are the precursors of folding ion rays. The frequency of formation of new envelopes ($\sim$ 1/hr), incidentally, is comparable with the frequency of new ion-ray generation during the active period of ion tails.

Schmidt and Wegmann (1981) have matched the near-symmetric ion ray structure in Comet Morehouse 1908 III to their MHD model of streamers resulting from directional changes of the interplanetary magnetic field, while as mentioned previously, Miller (1979) has compared their theoretical results with the asymmetric plume of parallel side rays in Comet Tago-Sato-Kosaka 1969 IX.

From a morphological point of view, the parabolic envelopes and even the ion rays could be the projection of a thin paraboloidal shell of density enhancement (see Eddington 1910; Ip 1980a; Miller 1981). From radio source occultation measurements of Comet Kohoutek 1973 XII and Comet

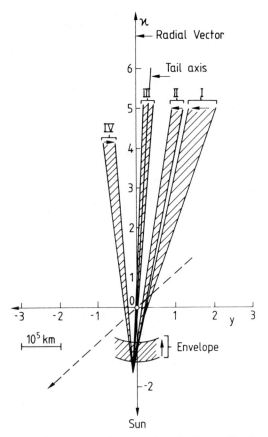

Fig. 11. Movements of 4 ion rays and an envelope in Comet Morehouse 1908 III as summarized from photographs taken on 3-4 October. The arrows indicate the direction of motions of the structures and the displacements are represented by the shaded regions. (After Lüst 1962.)

West 1976 VI, Wright and Nelson (1979) have deduced that the cometary plasma in the tail could be confined in a cresent-like hollow sheath. Such an axially asymmetric (or symmetric) configuration is therefore a viable alternative to the approximately linear structure of the ion rays as favored by Schmidt and Wegmann (1981). On the other hand, these two models are not necessarily contradictory to each other. If at large upstream distances a paraboloidal shell of plasma condensation were formed in relation to the solar wind variation, the structure would be most likely transformed into several bands of high field–low density and low field–high density structures once the amplified magnetic field is strong enough. Along the same line, some of the issues in classification and interpretation of the different small-scale structures of ion tails have been discussed by Miller (1981).

From an observational point of view, the straightness of the ion rays (before their complete closure) appears to indicate that the plasma flows in the filaments are quite stable against Kelvin-Helmholtz instability or other hydromagnetic instabilities. (The cometary ion rays of enhanced plasma density would most likely have a lower velocity than those in the more tenuous regions.) In fact, wide ion rays have been observed to split into several components or into several preexisting narrow streamers (Miller 1979). A similar phenomenon is the apparent transformation of one contracting envelope into a number of folding ion rays (e.g. Comet Morehouse 1908 III; see Wurm 1968). From this point of view, the fine-scale structure of the ion rays must still be resolved.

C. The Central Ion Tail

The final step of ion-ray formation is dictated by its merging with the central ion tail. Briefly speaking, the visible lengths of type I tails are generally of the order of 1 to 2×10^7 km. But ion tails of much greater lengths have been observed (3×10^8 km for the Great Comet in 1843; see Richter 1963; Wurm 1963). The widths rarely exceed 10^6 km, and in most cases the visible part of the ion tail was found to be no more than a few times 10^5 km wide. Therefore, cometary plasma in the main part of the ion tail should be moving with a speed $\lesssim 100$ km s^{-1} according to Fig. 2. This result is in agreement with the measurements of the speeds of ion condensations in different parts of the ion tail of Comet Tago-Sato-Kosaka 1969 IX by Jockers et al. (1972).

A combination of theoretical modeling and observations thus indicate that the cometary ions, swept up by the solar wind in the outer coma, would be the fast-moving tips of the ion rays; whereas the slow-moving plasma originating in the inner coma within a radial distance $\lesssim 10^4$ km would form the central streamer coma (see Fig. 8a). Following this scenario of production of ion rays, the central ion tail can be supposed to be continuously fed by the folding ion rays composed of cold plasma from the central source region (see Fig. 8). It is interesting to note here that, according to Wurm (1963,1968), it is not uncommon to find the absence of long central tails as if the ion source were cutoff before the complete closure of the ion rays. Miller (1979) has also provided examples of such ion tail morphology. The governing mechanism for such depletion of CO$^+$ ions and other cometary ions is still obscure. Perhaps this phenomenon is related to whether or not magnetic connection to the central source region can be maintained; that is, if the draped field lines at times can move across the coma in a time scale relatively shorter than the folding time interval of the ion rays, continuous channeling of ionized gas to the central axis will be hindered. Disconnection of the field lines could have a similar effect. On the other hand, the absence of central tail streamers may be simply due to an enhanced dissociation of the cometary ions there. These various possibilities remain to be investigated.

III. THE INNER ION COMA AND ION TAIL

Within a distance of $\sim 10^3$ km, the ion-neutral collisional time scale is short compared to the transit time of the expanding gas, and the structure and dynamics of the ion coma in this region are significantly influenced by ion-molecule reactions as well as by photochemistry (Aikin 1974; Shimizu 1975,1976; Oppenheimer 1975; Giguere and Huebner 1978; Huebner and Giguere 1980; Ip 1980a; Prasad and Huntress 1980; Mitchell et al. 1981; Huntress et al. 1981). In the following we shall argue that the chemical composition of the inner ionosphere and ion tail could be closely coupled to the flow dynamics. But first we need to deal with the question of the contact surface separating the contaminated solar wind inflow and the outward expanding plasma flow from the central source.

For the sake of orientation we may consider the variation of the solar wind ram pressure for three different solar wind conditions, e.g., low-speed stream, high-speed stream, and the leading edge of a high-speed stream. These dynamic pressures may be counteracted by the magnetic pressure in a magnetosheath, the ram pressure of a supersonic ionospheric outflow, or the thermal pressure of a subsonic flow. The required magnetic field strength B_s, and ion number densities (n_1 for supersonic flow and n_2 for subsonic flow) are listed in Table I. While B_s varies between 60 γ and 170 γ, $n_1 \sim 5 \times 10^4$ to 4×10^5 cm^{-3}, if the supersonic flow speed is ~ 1 km s^{-1} and $n_2 \sim 1$ to 8×10^4 cm^{-3}, if $k(T_e + T_i) \sim 1$ eV [$n_2 \sim 1$ to 8×10^5 cm^{-3}, if $k(T_e + T_i) \sim 0.1$ eV].

Assuming equilibrium between photoionization and recombination the radial variation of the ion number density in the inner coma can be expressed as

$$n_i = \frac{(Q_0/4\pi v_0 \alpha_e \tau_i)^{\frac{1}{2}}}{R} \quad (13)$$

where Q_0 is the total gas production rate, α_e the electronic dissociative recombination coefficient, and τ_i the photoionization time scale. We have then $n_i \sim 10^4$ cm^{-3} for $R = 10^3$ km and $n_i \sim 10^5$ cm^{-3} for $R = 10^2$ km, if $\alpha_e \sim 2 \times 10^{-7}$ cm^{3}s^{-1} and $\tau_i \sim 10^6$ s. The corresponding values for a bright comet with $Q_0 \sim 10^{30}$ molecules s^{-1} are a factor of 3 larger. Therefore it is difficult for a medium-bright comet like P/Halley to stand off the inward streaming plasma flow at a distance $> 10^3$ km upstream of the cometary nucleus, even during periods of low solar wind momentum flux. However, a brighter comet with higher gas production rate might be able to do so.

A. Ionospheric Sunward Expansion

In the following consideration, the term contact surface will be used loosely to mean the position along the Sun-comet axis at which the external plasma flow mixed with the solar wind and the outward ionospheric flow are

TABLE I
Plasma Parameters in Different Solar Wind Regions

Solar Wind and Cometary Plasma Parameters		Low-speed Stream $n_p = 8$ cm^{-3} $v_{sw} = 340$ km s^{-1}	High-speed Stream $n_p = 3$ cm^{-3} $v_{sw} = 750$ km s^{-1}	Leading Edge of a High-speed Stream $n_p = 60$ cm^{-3} $v_{sw} = 350$ km s^{-1}	Remarks
Solar wind ram pressure	$P_S = m_p n_p v_{sw}^2$ (dyn/cm^2)	1.6×10^{-8}	2.9×10^{-8}	1.2×10^{-7}	
Solar wind flux	$F_S = n_p v_{sw}$ (cm^{-2} s^{-1})	2.7×10^8	2.3×10^8	2.1×10^9	
Compressed magnetic field	$B = (8\pi P_S)^{1/2}$ Gauss	6.3×10^{-4}	8.5×10^{-4}	1.7×10^{-3}	
Cometary ion number density in supersonic outflow	$n_1 = \dfrac{P_S}{m_i v_i^2}$ (cm^{-3})	4.7×10^4	8.5×10^4	3.5×10^5	$v_i = 1$ km s^{-1}
		4.7×10^3	8.5×10^3	3.5×10^4	$v_i = 3$ km s^{-1}
Cometary ion number density in subsonic outflow	$n_2 = \dfrac{P_S}{k(T_e + T_i)}$ (cm^{-3})	10^5	1.8×10^5	7.5×10^5	$k(T_e + T_i) = 0.1$ eV
		10^4	1.8×10^4	7.5×10^4	$k(T_e + T_i) = 1$ eV
Ionization rate by impact[a]	CO	1.7×10^{-7} s^{-1}	1.4×10^{-7} s^{-1}	1.3×10^{-6} s^{-1}	photoionization rate[b] 3.1×10^{-7} s^{-1}
Ionization and charge-exchange	H_2O	5.8×10^{-7} s^{-1}	4.8×10^{-7} s^{-1}	4.4×10^{-6} s^{-1}	photoionization rate[b] 3.3×10^{-7} s^{-1}

[a] From Siscoe and Mukherjee (1972).
[b] From Huebner and Giguere (1980).

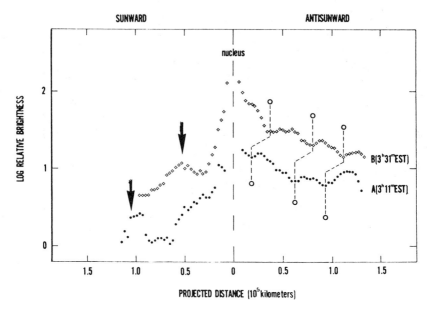

Fig. 12. The brightness profiles of H_2O^+ in Comet Bennett 1970 II plotted as log of the relative brightness versus projected distance. Filled circles, profile from spectrum A; small open circles, profile from spectrum B; large open circles, three major troughs shifted profile. Profile A has been arbitrarily shifted downward for easy comparison. The upstream structures marked by arrows may be related to the contracting envelope as observed in Comet Morehouse 1908 III. (From Delsemme and Combi 1979.)

divided. As discussed below in Sec. III.B this region could be rather broad and diffuse. In the observations of Comet Kohoutek 1973 XII when it was at a solar distance of 0.9 AU, Wyckoff and Wehinger (1976) found that the ion emissions are not visible at a distance of 5×10^3 km on the sunward side whereas the tail features extend to larger distance. This observation might then give us the upper limit of the position of the contact surface defining the sunward dimension of the inner ion coma of Comet Kohoutek 1973 XII at 0.9 AU. Another determination of the spatial distribution of the ion emission comes from the work of Delsemme and Combi (1979) on the H_2O^+ brightness profiles of Comet Bennett 1970 II. As shown in Fig. 12 the sunward side extension of the H_2O^+ emission is limited to within 5×10^4 km. The difference between these two length scales may be partly due to the larger gas production rate of Comet Bennett 1970 II (which is $\sim 10^{30}$ s^{-1} at 1 AU as compared with a value of 10^{29} s^{-1} for Comet Kohoutek 1973 XII) and partly due to the smaller solar distance of Comet Bennett ($r = 0.56$ AU) at the time of observations.

The monochromatic intensity profile of CO^+ (3-0) band emission at 4020 Å, with continuum substracted, when Comet West 1976 VI was at a solar distance of 0.5 AU has been presented by Wyckoff (1981) indicating an abrupt change in the slope of the brightness profile at a distance of $\sim 10^5$ km ahead of the nucleus. This position of CO^+ brightness cutoff may be interpreted as the location of the bow shock inside of which the ion density will rise sharply until reaching the contact surface. Examination of the brightness profiles of the (2-0) band of the ($A^2\pi$–$X^2\Sigma$) system of CO^+ from spectrograms of the same bright comet have enabled Combi and Delsemme (1980) to draw a number of tentative conclusions about the structure of one single ionosphere over a range of solar distances $r \sim 0.44$ to 0.84 AU: first, the radius of the contact surface varies between 10^4 and 3×10^4 km with $r_c \alpha r^{-1}$; second, the estimated production rate of CO^+ with the ionosphere assumed to have the configuration of a hemisphere in the sunward direction and a cylinder in the antisunward direction is $Q_i \sim 10^{27}$ s^{-1} inside the contact surface, and $Q_i \alpha r^{-4.6 \pm 1.0}$. The assumption that the radius of the hemispherical contact surface should be the same as the lateral radius of the cylindrical ion tail allows them to approximate the positions of the contact surfaces even though they are not obvious in the brightness profiles in the sunward direction. (This is perhaps a reasonable method in order to avoid the problem with continuum contamination, though it is highly model-dependent.)

The above are more or less all the information concerning the structures of the cometary ionospheres that groundbased observations so far have been able to supply. Not much can be said about the details of the expansion of the ionospheric flow. In the following we attempt to review several possible ion coma models in the light of recent theoretical work and the Pioneer Venus results.

As mentioned at the beginning of this chapter, the strongest possible flow pattern of the ionospheric plasmas occurs if the cometary ions acquire supersonic speed. This radial streaming would then be diverted into lateral flow by the formation of an inner shock in analogy to the process of solar wind-interstellar wind interaction (Wallis and Dryer 1976). This problem has been treated analytically by Houpis and Mendis (1980) considering hypersonic expansion [Mach number $M \gg \sqrt{2/(\gamma-1)}$ where γ is the ratio of the specific heats] of ionized gas from a central point source. For a comet similar to P/Halley (at 1 AU) they found that the position of the stagnation point dividing the contaminated solar wind inflow and the ionospheric outflow is $\gtrsim 4 \times 10^3$ km. The basic structure of such an inner shock model is summarized in Fig. 13. As a result of the hypersonic approximation, the shock layer between the contact surface and the inner shock is relatively small (~ 0.23 of the nuclear distance to the stagnation point).

This treatment is illuminating in many ways since we expect such a process to occur when the gas production rate is very strong. In the specific case of P/Halley at 1 AU, the situation is a little different since the ram pressure of the ionospheric flow is not sufficient to balance the momentum

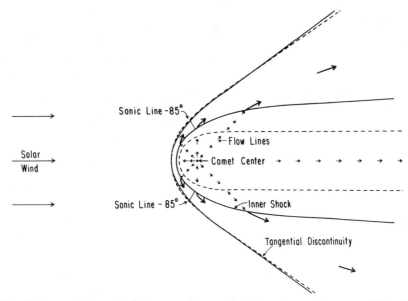

Fig. 13. A comet inner shock structure from analytical treatment. The outward flow is assumed to be hypersonic and the plasma density variation is proportional to $1/r^2$. (From Houpis and Mendis 1980.)

flux of a low-speed stream even at $R \sim 10^3$ km. The assumption of a central point source is also somewhat problematic because continuous mass addition is one characteristic of the ionospheric outflow. Neglect of this term could lead to a large difference in the flow model. Another, more direct effect is that, within a few thousand km of the nucleus, ion-molecule chemistry is very important in determining the ion composition as well as the energetics. Besides charge rearrangement, exothermic ion-molecule reactions like

$$CO^+ + CO_2 \rightarrow CO + CO_2^+ + 0.22 \text{ eV} \qquad (14)$$

could yield an addition of kinetic energy on the order of $0.1 - 1$ eV per reaction. This would mean that the plasma outflow, besides mass addition, is also subject to heating. The tendency is for the flow to approach transonic or even subsonic expansion. As in the case of the outer coma, the effect of the neutral gas is to reduce the Mach number of the outward plasma flow.

We must then consider the possibility that the outflow is essentially subsonic and that the solar wind dynamic pressure is counteracted by thermal pressures of the ions and electrons. Depending on the ion and electron temperatures, this so-called standoff distance can be estimated to be $\sim 10^2$ to 10^3 km during low-speed interplanetary conditions. As suggested by the Pioneer Venus observations of the time-variation of the ionopause position, this standoff distance perhaps would also fluctuate widely in response to changes in the solar wind dynamic pressure.

B. Contact Surface or Buffer Zone

Wallis and Dryer (1976) have discussed the structure of the stagnation point flow pattern between two counter-streaming gas flows with the addition of a source in between. They argued that to accommodate the extra source, the contact surface should be broadened into a thick layer in which the plasma flow is mainly in the lateral direction. Combining this picture with the Venus-type interaction model illustrated in Fig. 1, we can see that a magnetosheath of piled-up magnetic field ($B_c \gtrsim 50\,\gamma$) may play the role of the contact surface layer. On the one hand, it acts as a cushion braking both the inward and outward ion flows, and on the other, channels the cometary ions created there into the tail region. Though the bow shock position of Venus is relatively insensitive to solar wind conditions, the Pioneer Venus observations discovered that the ionopause altitude responds sharply to solar wind momentum flux changes until reaching a certain lower limit (Brace et al. 1980). Briefly speaking, the external solar wind dynamic pressure is transmitted to the ionospheric boundary via the magnetic field pressure, and the adjustment of the pressure balance at the ionopause allows the interface to oscillate with large amplitude. The Venus ionopause altitude may increase to an extremely large value (> 1000 km) in the case of low solar wind momentum flux, but the exponential dependence of the ionospheric thermal pressure plus the effect of neutral-ion collisional coupling limits the ionopause altitude to > 290 km. For cometary ionospheres, the size of the neutral-ion collisional coupling region can be approximated to $r_c \sim Q_0 \sigma / 4\pi V_0$ (see Biermann 1974). For $Q_0 \sim 10^{29}$ molecules s^{-1}, $\sigma \sim 2 \times 10^{-15}$ cm^2 and $V_0 \sim 1$ km s^{-1}, we have $r_c \sim 2 \times 10^3$ km. In principle, penetration of the contaminated solar wind plasma much farther in is impossible since most of it would suffer rapid charge-exchange loss due to ion-molecule interaction with the H_2O molecules (Wallis and Ong 1976). While the distance of 2×10^3 km may define the position of an outer contact surface, an inner one (r_c') is determined by the balance between the curvature force of the magnetic fields and the ion-neutral momentum transfer rate. This can be approximated as

$$\frac{B^2}{4\pi r_c'} = k\, n_i n_n m_i V_n \qquad (15)$$

where k is the rate coefficient of ion-H_2O reactions and n_n is the neutral number density. The ion number density n_i can be calculated by assuming photoionization equilibrium with the electron dissociative recombination coefficient α_e taken to be 2×10^{-7} cm^3 s^{-1} and the ionization time scale τ_i taken to be 10^6 s.

If the neutral speed is 1 km s^{-1}, we have $r_c' \sim 10^3$ km for $B \sim 50\,\gamma$. (In this approximation $r_c' \propto B^{-1}$; consequently the boundary defining the outward ionospheric flow may recede to a lower altitude in the case of solar wind compression.) The dimension of the buffer zone separating the inward plasma flow and outward expanding ionospheric flow therefore may be comparable to the size of the cometary ion coma (See Fig. 1b). The flow speed transverse to the magnetic field should be about the same as the expansion speed of the neutral gas, if not less. But the field-aligned flow speed could be substantially greater as a result of the acceleration effect of the photoelectrons and magnetic pressure gradient (see Sec. III.D).

Besides variation in the solar wind conditions, the structure of the inner ion coma is also subject to the effect of time-variation in the neutral gas production rate. For instance, the expanding halos observed in the coma of P/Halley in its last return after perihelion may be caused by spin-modulation of the gas production rate Q_0 (Bobronikoff 1931; Wurm 1963; Wurm and Mammano 1972; Ip 1981). Suppose in one rotation of the cometary nucleus—the spin period of P/Halley is 10.3 hr according to Whipple (1980)—Q_0 changes by a factor of 5; the size of the ionosphere should also vary accordingly. In the extreme situation, the resulting dynamical oscillation of the ionosphere may allow the ionospheric outflow to alternate between supersonic and subsonic.

C. Ionospheric Instabilities

As far as the dynamical response of the cometary ionosphere to the solar wind condition is concerned, one possible consequence is the occurrence of plasma instabilities at the (inner) contact surface triggered by large changes in solar wind dynamic conditions. In connection with the tail disruption events reported by Niedner and Brandt (1978), Ip and Mendis (1978) have proposed that during compression by high-speed stream, infiltration of the solar wind plasma inflow could be induced by the so-called flute instability at the contact surface. In an idealized situation, the dense plasma (the cometary ionosphere) can be pictured as being confined by a magnetosheath. The curvature force of the piled-up magnetic field then provides the effective gravity required for a gravitational instability. For disturbances of small wavelengths, the finite Larmor radius effect stabilizes the system; however, in the long-wavelength regime stabilization can be produced only by short-circuiting the charge separation electric field at the boundary surface by a rapid movement of the electrons parallel to the field lines. In the absence of such an effect, the growth time of this flute instability in the long-wavelength limit is given by (Ichimaru 1973)

$$\tau_f = \left[\frac{1}{N}\frac{dN}{dx}\frac{2k(T_e + T_i)}{m_i R_f}\right]^{-1/2} \sim \left[\frac{Lm_i R_f}{2k(T_e + T_i)}\right]^{-1/2} \quad (16)$$

Here $\frac{dN}{dx}$ is the gradient of the plasma density across the surface and R_f is the radius of curvature of the field lines, m_i is the ion mass, $k(T_e + T_i)$ is the thermal pressure of an ion-electron pair, and L is the scale length for the density variation, i.e. $L \sim N/(dN/dx)$.

Taking $L \sim R_f \sim 10^3$ km, $m_i \sim 20\, m_p$ (proton mass) and $k(T_e + T_i) \sim 1$ eV, we obtain $\tau_f \sim 300$ s. This time scale is comparable to, if not shorter than, the time scales of other physical processes (e.g., the travel time of a sound wave or Alfvén wave is typically $\tau_A \sim 200$, if $n_i \sim 10^3$ cm^{-3} and $B \sim 50\,\gamma$). Therefore the contact surface may be considered as being marginally stable to flute instability even during low-speed solar wind conditions. At the impact of a high-speed stream, the scale lengths of L and R_f will be reduced, thus leading to a shorter growth rate τ_f. The above treatment is necessarily very crude due to the large number of uncertainties.

The formation of twisted flux ropes inside the ionosphere of Venus (Russell and Elphic 1979; Elphic et al. 1980; chapter by Russell et al.) has been interpreted as the result of Kelvin-Helmholtz instability at the ionopause intermixing the magnetized solar wind plasma with the ionosphere plasma (Wolff et al. 1980). The sporadic occurrence of the flute instability (if indeed possible) would have a similar effect, allowing small flux tubes to penetrate inside the contact surface. Therefore close to the nucleus ($R \lesssim 10^3$ km), there might exist magnetic inhomogeneities of high field strength ($B \sim 50\,\gamma$). The resulting $\mathbf{J} \times \mathbf{B}$ force and Joule heating may play an important role in the dynamics and energetics of the inner ion coma.

From this point of view, the ion composition and density distribution of the inner coma are also likely to be modified by the flow pattern. This problem is particularly serious on the sunward side (even without the introduction of magnetic fields), as strong divergence of the stream tube is introduced into the flow pattern. On the tailward side the treatment is relatively simpler with the ion flow approaching cylindrical geometry. A consistent description including the proper chemistry (i.e., photochemistry and ion-molecule reactions) and fluid dynamics (i.e., chemical and photoelectron heating, ion-neutral momentum coupling) should make a very interesting study.

D. Ionospheric Tailward Expansion

There are two recent observations which have direct bearing on the morphology and dynamical expansion of the inner ion tail. From correlation of the time variation of the H_2O^+ brightness distribution along the projected tail distance of Comet Bennett 1970 II, Delsemme and Combi (1979) detected patterns moving downstream with a speed of ~ 17 km s^{-1} at an average distance of 0.8×10^5 km (see Fig. 12). Interpreting this as due to mass motion, they derived an acceleration of ~ 130 cm s^{-2}. The actual acceleration may be substantially less since waves traveling with the Alfvén speed is another possibility. A more direct method may be to determine the Doppler-shift velocity of cometary ions as attempted by Huppler et al. (1975). Using a Fabry-Perot spectrometer fitted to the McMath solar telescope, these workers have observed the Hα, H_2O^+ and [OI] emission lines from Comet Kohoutek 1973 XII during its perihelion passage when the

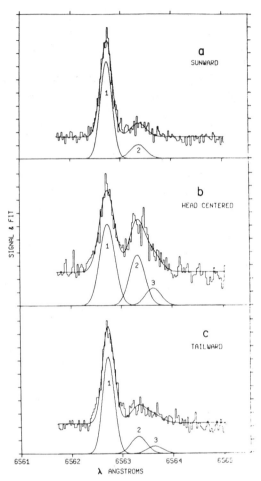

Fig. 14. Observations of Comet Kohoutek 1973 XII. With a Fabry-Perot spectrometer, spectral scans near Hα were taken on 26 Jan. 1974 with a circular 6.6 field of view. Scan (a) is the sum of scans taken at 0254 UT and 0428 UT and centered 5.5 (2.1 × 10^5) sunward of the comet head. The cometary Hα line (component 2) is redshifted from the geocoronal Hα line (component 1) at 6562.74 Å due to the comet-Earth relative velocity. Scan (b), taken at 0350 UT centered on the comet head, shows the H_2O^+ feature (component 3). Scan (c) is the sum of scans taken at 0307 UT and 0416 UT with the field centered 5.5 tailward of the comet head. The third component is again evident. The identification of this third component with an H_2O^+ emission feature indicates that the H_2O^+ ions were moving in a tailward direction with a velocity of 20 to 40 km s^{-1} with respect to the comet nucleus. (From Huppler et al. 1975.)

Sun-comet distances varied between 1 and 0.4 AU. Comparison of the observations in the sunward, head-centered, and tailward regions of the cometary coma indicated that there is a third component in addition to the cometary Hα and the geocoronal Hα line emissions (see Fig. 14) in the head and tailward side. If this feature is interpreted to be from H_2O^+ (7-0) lines redshifted by ~ 0.3 Å, the H_2O^+ ions would have a tailward velocity in the range of 20 to 40 km s^{-1} at a distance of 2×10^5 km. These observations would require an average acceleration of ~ 100 cm s^{-2} which is in agreement with the deduction by Delsemme and Combi (1979). However, note that a substantially smaller tailward flow speed ($\lesssim 4$ km s^{-1}) is derived for Comet West 1976 VI by Combi and Delsemme (1980).

There are several effects which may account for this tailward acceleration. The presence of small-scale turbulence in the ion tail of Comet Kohoutek 1973 XII—as indicated by radio wave fluctuations in a radio source occultation observation (Ananthakrishnan et al. 1975)—have led Lee and Wu (1979) to believe that the hydromagnetic Kelvin-Helmholtz instability, which is responsible for the observed turbulence, could excite Alfvén waves accelerating the cometary ions. The resulting plasma acceleration is estimated to be 30 to 300 cm s^{-2}. Another is the (thermal or magnetic) pressure gradient built up between the comet head and the tailward side. Taking the magnetic pressure gradient effect as an example (see Ioffe 1968), we can write the acceleration as $a = -1/\rho \nabla [B^2/8\pi]$. Together with the condition of magnetic flux conservation $B \cdot \ell^2 = $ const., where ℓ is the distance from the comet head, we have $a \sim B^2/2\pi\rho\ell \sim 60$ cm s^{-2}, if $B \sim 10$ γ, $\rho \sim 3.4 \times 10^{-23}$ g cm^{-3} (for 100 H_2O^+ ions cm^{-3} and $\ell \sim 10^5$ km). As this acceleration mechanism depends sensitively on the three-dimensional magnetic field configuration, the actual value may be substantially smaller than the value from this order-of-magnitude approximation. Another acceleration effect concerns the $\mathbf{J} \times \mathbf{B}$ Lorentz force acting on the inner tail ions as a result of the magnetic curvature force. In effect, it is this part of the electrodynamic coupling of the cometary ions to the solar wind which partially determines the cross-tail current system to be discussed in the next section.

IV. THE MAGNETOTAIL/CURRENT SYSTEM

In addition to the structure of contracting envelopes, Wurm (1963,1968) and Wurm and Rahe (1969) have also made the interesting observation that the inner coma with a radius of 1 to 2×10^3 km is the site of the strongest CO^+ emission with the ions concentrated in several sunward-pointing jets or streamers (see Fig. 15). They have also argued that the ion rays in the tail could actually connect to these ionospheric streamers. The more probable explanation appears to be that the plasma envelopes and rays are determined by the solar wind inflow through the cometary coma whereas the jets—if they are indeed made up of ions and not neutrals or dust particles—emanating from the central source are related to the flow pattern of the expanding ionospheric outflow. A projection effect could then produce an overlapping of these two classes of structures. In any event, study of the inner coma of Comet Morehouse 1908 III and other comets has led Wurm to believe that,

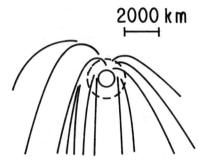

Fig. 15. Diagram of streamer structures in the vicinity of the nucleus of Comet Morehouse 1908 III on 14 Oct. 1908, after Wurm and Mammano (1967). According to the authors, these filamentary structures contain the strongest ion emissions. From the rapid time scale of the brightness fluctuation ($\tau \sim 10^{3.5}$ s) the corresponding ionization time scale should also be relatively short.

besides photoionization and solar wind charge-exchange, there must be in addition a different mechanism (the so-called internal ionization source) operating in the inner coma to produce rapid ionization.

One possibility for such an ionization effect is the aurora-type activities resulting from precipitation of energetic particles (Ip and Mendis 1976a,b; Mendis 1978; Ip 1979). As in the case of the planetary magnetosphere, this process requires discharge of the current system in the ion tail. One possible consequence of such an ionization effect would be to increase the ion density in the ion coma so that enough ionospheric pressure could be built up to stand off the solar wind. In other words, the solar wind-comet interaction in a global scale might have a direct effect on the very structure and dynamics of the inner coma.

A. The Cross-tail Current System

While the dayside current system is induced by the momentum transfer between the solar wind plasma and the cometary ions picked up upstream, the cross-tail current system downstream is generated by the continuous acceleration of the cometary plasma along the ion tail. This process can be best demonstrated by the simulation experiments performed by Podgorny et al. (1979).

Using the principle of limited simulation, the solar wind interaction process was simulated by ejecting an artificial solar wind towards a wax ball. The plasma speed was $\sim 10^7$ cm s^{-1}, electron temperature $T_e \sim 15$ eV, ion temperature $T_i \sim 5$ eV, ion density $n_i \sim 10^{13}$ cm^{-3}, the frozen-in magnetic field (pointing perpendicular to the velocity vector) $B \sim 10$ gauss and the flow duration was ~ 20 μs. The rapidly evaporated wax products act as an obstacle to the plasma flow. Associated with this plasma interaction, a magnetic tail structure is formed.

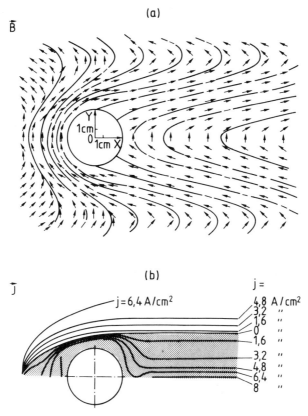

Fig. 16. (a) The field configuration in the magnetic tail of the artificial comet found in the model experiment. The arrows show the direction of the magnetic field vectors, the solid lines show the field lines. (b) The corresponding distribution of the current density. These magnetic field and current distributions may be characteristic of the cometary ionosphere close to the nucleus. (From Podgorny et al. 1979.)

The topology of the resulting magnetic fields is given in Fig. 16. The pileup of the magnetic field at the front side is clearly shown. The curvature of the magnetic field behaves in the manner that at the boundary (in the case of a comet) the solar wind will be decelerated whereas on the central axis, the ion tail will be accelerated. It confirms the theoretical idea that similar to the magnetotail of the Earth, a crosstail current system with a current sheet separating two tail regions of opposite magnetizations should be present (Ness 1965; Ness and Donn 1966; Ip and Mendis 1976b; Mendis 1978; Ip 1979). The deduced distribution of the current density in the tail of the artificial comet is also shown. For a first order approximation, the idealized crosstail current may be assumed to assume a Θ-shape configuration as depicted in Fig. 17. If the tail-aligned field is B_t, the current density j in the current sheet is given by the Maxwellian relation

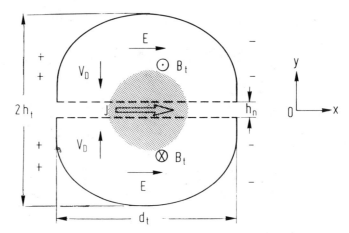

Fig. 17. Idealized cometary crosstail current system. The two regions of opposite magnetizations are separated by a current sheet as in the case of the Earth's magnetotail. The resulting Lorentz **J** X **B** force will act in such a way that the ion tail plasmas are accelerated tailward.

$$\nabla \times B_t = \frac{4\pi}{c} j \qquad (17)$$

under steady state conditions. The sheet current density integrated through the thickness of the current sheet (L_n) then will be

$$j_s = \frac{cB_t}{4\pi} \qquad (18)$$

and the total crosstail current flowing through the central current sheet may be estimated to be

$$I_t = \frac{c}{4\pi} B_t L_t \qquad (19)$$

where L_t is the length of the ion tail. The crucial question here is what is the strength of the tail-aligned magnetic field. As argued by Ershkovich (1980), the consideration of pressure balance across the tail boundary dictates that the average value of B_t is at most comparable to the interplanetary magnetic field, i.e., 5-10 γ at a solar distance of 1 AU. Taking L_t to be 10^7 km, and $B_t \sim 10 \gamma$, we have $I_t \sim 2 \times 10^8$ A.

The distribution of such a crosstail current need not be uniform and is dependent on the dynamical evolution of the magnetic field and thermal pressure structures along the tail. In other words, a self-consistent description of the tail current system and the corresponding acceleration effect requires the simultaneous consideration of the variations of the plasma pressure and magnetic field pressure across the tail rather than a simple introduction of a current sheet into the ion tail.

In any event, to explore the **J** × **B** acceleration of cometary ions we adopt the simplest possible model (Ip 1980b). First, the ion tail system is assumed to be two-dimensional with z pointing in the axial direction and the magnetic field **B** can be described by having a constant x-component

$$B_x = \text{const.} \tag{20}$$

and a z-component

$$B_z = B_z(\infty)\tanh(x/L_n) \tag{21}$$

where L_n is the scale width of the current sheet. From the above equations we can see that B_z satisfies the conditions that $B_z \to 0$ as $x \to 0$ and that $B_z \to B_z(\infty)$ as $x \to \infty$ (see Tandberg-Hanssen 1974). If the cometary plasma is further assumed to be isothermal and if there is pressure equilibrium in the x-direction, we have

$$\frac{dP}{dx} = -\frac{B_z}{4\pi kT}\left[\frac{dB_z}{dx} - \frac{dB_x}{dz}\right] \tag{22}$$

and

$$\rho = \frac{m_i B_z(\infty)^2}{8\pi kT}\left[1 - \tanh^2\left(\frac{x}{L_n}\right)\right] \tag{23}$$

where k is the Boltzmann constant, T the plasma temperature, and m_i the ion mass (note that for $B_z(\infty) = 10\,\gamma$, $kT = 1$ eV, the ion number density $n_i \sim 300$ cm^{-3} at $x \to 0$).

Hence the Lorentz force **J** × **B** can be written as

$$|\mathbf{J} \times \mathbf{B}| = \frac{1}{4\pi} B_x \frac{d}{dx} B_z = \frac{B_x B_z(\infty)}{4\pi L_n}\left[1 - \tanh^2\left(\frac{x}{L_n}\right)\right] \tag{24}$$

and the acceleration due to the curvature force would be

$$a = 2\left(\frac{B_x}{B_z(\infty)}\right)\left(\frac{kT}{m_i}\right)\left(\frac{1}{L_n}\right). \tag{25}$$

The Lorentz acceleration of the cometary ions is therefore dependent on the ratio of the magnetic field component normal to, and parallel to the current sheet, as well as the thermal pressure kT and the width of the crosstail current sheet. As far as the inner ion tail is concerned, L_n may be taken to be $\sim 10^3$ km and $a \gtrsim 10$ to 20 cm s^{-2}, if $B_x/B_z(\infty) > 10^{-3}$ and $kT \sim 1$ eV. This simple calculation indicates that the curvature force of the magnetic field associated with the crosstail current may be very efficient in accelerating the cometary ions.

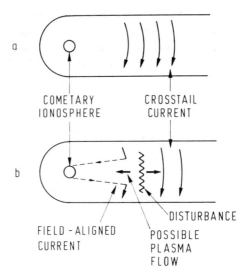

Fig. 18. Schematic drawing of the cometary "substorm" circuit. (a) Flowing of the crosstail current across the current sheet is steady. (b) Partial interruption of the crosstail current produces a tail-aligned current discharging through the cometary ionosphere. It has been suggested that dissipation of the magnetic energy stored in the ion tail in this way (Ip and Mendis 1976b; Ip 1979) may provide the ionization source for the internal ionization mechanism advocated by Wurm (1963).

B. Current Disruption and Reconnection Effect

When the tail current system was first proposed (Ip and Mendis 1976b) it was recognized that in the case of current disruption (or reconnection) there could be field-aligned current discharging through the cometary head (see Fig. 18). Such transient effects may be able to accelerate the charged particles to high energy as observed in the Earth's magnetotail (Krimigis and Sarris 1979). The maximum energy that might be derived from this process can be written as $\Phi = V_A B_t L_0$ where V_A is the Alfvén speed in the ion tail, B_t the tail field, and L_0 the width of the ion tail. For $B_t \sim 10\,\gamma$, ion number density $n_i \sim 100\,\mathrm{cm}^{-3}$ and $L_0 \sim 10^5$ km, we obtain $\Phi \sim 5$ keV. The flux of such energetic particles might be very low, however, and their contribution to the ionization effect in the ionosphere and the ion tail must be considered more carefully.

At this point, we should note that circumstantial evidence for a fast ionization effect in the inner coma comes from Wurm's (1963) report that the time scale for the fluctuation of ion emission within a region of few thousand km around the cometary nucleus could be as short as $10^{3.5}$s. But a rapid time variation of this scale may be accounted for by the fluctuation of the gas production rate, if the size of the inner coma is no larger than a few thousand km. In any event, the occurrence of reconnection in the ion tail

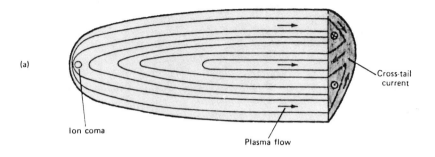

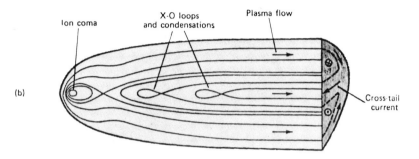

Fig. 19. Possible effect of reconfiguration of magnetic fields in the ion tail. (a) The quiet time structure is shown with the crosstail current without inhomogeneities in the crosstail current sheet. (b) Breakup of the current sheet which on one hand would lead to a field-aligned current discharging into the ionosphere, and on the other, to the formation of X- and O-type loops would perhaps result in the enhancement of plasma density in the vicinity of the O-type loops.

(and elsewhere) is an intriguing possibility, and the associated effect of particle acceleration and enhanced ionization should not be ruled out. If there is disruption of the crosstail current, would visible structures be formed in the ion tail? Morrison and Mendis (1978) have argued that the small-scale structures sometimes observed close to the tail axis of comets may be the consequence of the tearing mode instability. In Fig. 19 is presented a possible scenario for the magnetic field reconfiguration. It can be seen that the $\mathbf{J} \times \mathbf{B}$ force in the O-type loop would tend to focus the cometary plasma in this region whereas the cometary ions in the vicinity of the X-point would be dispersed. Thus condensation of ionized material may grow in the ion tail in this manner.

C. Ion Tail Disturbances

Other mechanisms have also been proposed to explain the origin of ion tail structures. The Kelvin-Helmholtz (K-H) instability is the best studied example (Ioffe 1970; Ershkovich 1980). The propagation of helical structures along the ion tails has been connected to the kink mode K-H instability. A much studied case is the ion tail disturbance in Comet Kohoutek 1973 XII on 19 and 20 January 1974. As shown in Fig. 20, the ion tail displays the typical structures of a kink and helical waves. Close examination indicates that the helical structure has already started well within 10^6 km of the head. According to the results of MHD calculations (cf. Fig. 2) no steep velocity shear required to excite the K-H instability is established at this distance. Also this particular event has been found to be well correlated with the interaction of the comet within the compression region of a high-speed solar wind (Niedner et al. 1978; Jockers 1981). However, it should be made clear that even though solar wind disturbances (and not velocity discontinuity) probably play the decisive role in initiating the ion tail perturbation, the analysis of the K-H instability and its nonlinear stabilization by Ershkovich and Chernikov (1973) and Ershkovich (1980) is nevertheless a valid treatment of the subsequent evolution of the ion tail.

Detailed analysis of the above Comet Kohoutek event can be found in Niedner et al. (1978), Niedner and Brandt (1978) and Jockers (1981). There seems to be agreement that the occurrence of the kink (or joint) is connected with a sudden change in the solar wind direction; as pointed out by Jockers, the wavy structure along this part (segment B in Fig. 20) is consistent with the effect of sideway action of the cometary plasma by the solar wind (see also Ip 1980c). However, opinions differ as to what mechanism is exactly responsible for the asymmetric side rays, especially those hanging out like a curtain from segment B. Niedner and Brandt (1978) have made the interesting suggestion that such a phenomenon could be associated with the crossing of an interplanetary sector boundary by the comet; reconnection of the new magnetic field lines with those piled up from the old sector would cause cometary plasmas to be gradually eroded from the comet head (see also Niedner and Brandt 1979,1980; Niedner et al. 1981; chapter by Brandt). The side rays therefore represent the reconnected magnetic flux tubes with pointing-direction determined by the local solar wind flow. If the segment B side rays are indeed the signature of ion tail reconnection at the front side, Jockers (1981) then questions why all of them appear to be connected to the ion tail. In his opinion, this feature is by no means consistent with the reconnection process. Instead, two alternatives have been suggested. First, the side rays could reflect the density enhancement resulting from propagation of compression waves along the ion tail; and second, the side rays could be caused by leakage of matter from the tail condensations; however Jockers finds no definite proof for either of these effects.

Note that while the side rays forming close to the head (see Fig. 7b) do not have the problem of an ion source, the "tail" side rays which appear at large distances far down the tail do. Working along the line of the hypothesis of matter leakage, we may suggest the occurrence of K-H sausage mode instability—which could break the cylindrical structure of the ion tail into individual clouds (Ershkovich and Chernikov 1973)—as another possibility. If

COMETARY PLASMA PROCESSES 627

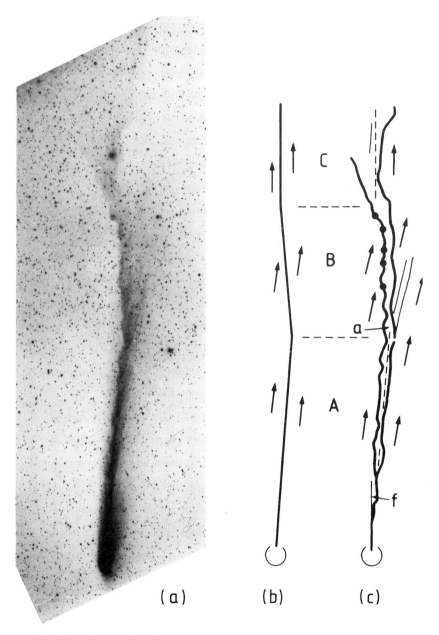

Fig. 20. (a) An ion tail disturbance of Comet Kohoutek 1973 XII photographed on 20 Jan. 1974 UT 0156; (b) and (c) delineation of the gross structure and relation with the solar wind flow. This picture shows the formation of a kink (segment B) and large-scale turbulence along the ion tail. One additional feature is the parallel side rays emanating from one side of the ion tail (most conspicuous from the kink) like a curtain. Photographs taken a few hours earlier (i.e. 19 Jan. 1974 UT 1827) also display straight side rays connected to the head. (From Jockers 1981.)

this process takes place, reconnection of the tail-aligned magnetic fields in this segment could also occur. As a matter of fact, the magnetic field topology may be rather complicated even in the nonlinear evolution of the K-H kink mode instability (Bateman 1978); confinement and leakage of cometary ions in such plasma configuration may be worthy of attention.

In the front-side reconnection model of Niedner and Brandt the final step of eliminating the last old magnetic flux tubes from the cometary ion coma is represented by the disconnection of the ion tail, i.e., the entire ion tail would be uprooted from the comet head. Even though the evidence for a tail disconnection is not overwhelming in the case of the 19-20 January event of Comet Kohoutek 1973 XII, a number of clear examples in other comets have been listed in Niedner and Brandt (1978). Observational as well as theoretical studies have been carried out by Niedner and co-workers to support the reconnection scheme (see the chapter by Brandt in this book). On the other hand, a somewhat different interpretation stressing the possible role of the dynamical effect of the compression region of high-speed streams have been made (Ip and Mendis 1976a,b; Ip 1980c). And interaction with interplanetary shock waves have been considered by several authors (Jockers and Lüst 1973; Burlaga et al. 1973; Wurm and Mammano 1972). Besides dynamical processes, ionization effect may be important as well. As indicated in Table I, solar wind charge-exchange effect may play a crucial role here as the corresponding ionization rate of the cometary ions increases sharply by a factor of ~ 10 in the compression region. The ion condensation, secondary tails or detached tails sometimes observed thus could be simply the tailward convection of the cometary plasma clouds produced in the solar wind region with high charge-exchange ionization rate—without the necessity of invoking magnetic field reconnection. Perhaps consideration should be given to more than a single physical ingredient if the complicated processes of large-scale ion tail disturbances are to be fully understood. In any event the idea of Niedner and Brandt in a certain sense is essentially correct since large-scale reconfiguration of the magnetotail and the crosstail current system should occur, one way or another, during a sector crossing.

V. CONCLUSION

In this chapter we have considered several processes that may be important in the comet-solar wind interaction. First, in the outer ion coma, the initial stage is characterized by the pickup of the new cometary ions from the extensive neutral coma and their subsequent thermalization. The details of the physical mechanisms are not known. Whether or not strong plasma wave activity is associated with the momentum transfer and assimilation of the cometary ions into the solar wind flow will hopefully be answered by space probes. The shocked plasma flow is expected to be turbulent and the front-side coma will certainly act as a source of Alfvén waves propagating tailward along the draped magnetic fields. Whether these waves help accelerate the cometary ions as in the case of the solar wind is another interesting issue to be explored by magnetometer measurements. Without this first-hand knowledge of the plasma environment in the comet head, we are limited to discussion of various qualitative descriptions. Interpretations of the formation of ion rays are one case in point. Since we know neither the ion density, ion and electron temperatures, and magnetic field nor the approximate configuration (e.g., coplanar linear structure or conical envelopes), not many constraints can be put on the models suggested so far. Indeed, even the strength of the shock is still a matter for debate. The opinions vary from the idea that the shock may be as strong as the Earth's bow shock (Beard 1981) to that there may be no shock at all (Wallis 1971). A similar degree of uncertainty exists in our understanding of the structure of the inner ion coma. The generally held idea is that the ionospheric plasma will be coupled to the supersonic expansion of the neutral gas until it reaches an inner shock counteracting the contaminated solar wind inflow (Wallis and Dryer 1976; Houpis and Mendis 1980). We have stressed here that the ionospheric expansion may be subsonic. Furthermore, the introduction of a magnetic field into the ionosphere could have a significant effect in decelerating the expanding flow via the $\mathbf{J} \times \mathbf{B}$ force. In this scenario, the ionospheric flow of P/Halley at 1 AU solar distance may be approximated by an incompressible subsonic flow within a sunward distance of $\sim 10^3$ km (determined by the balance between the magnetic curvature force and ion-neutral friction). And then there is a wide region, perhaps on the order of 10^3 km, in which the ion flow is basically lateral along the field direction as in the case of the polar wind. Such broadening of the hydrodynamic contact surface into a layer of source flow has been suggested by Wallis and Dryer (1976), and this configuration may be present even for the supersonic case; that is, the distance between the inner shock and the stagnation point may be much larger than that estimated by Houpis and Mendis (1980). Needless to say, the flow dynamics of the inner coma is one problem that is still in the early stage of development and some of the issues could be clarified by more quantitative treatments.

As the photochemistry and fast ion-molecule reactions within 10^3 km of the nucleus could be closely coupled to the overall ionospheric dynamics and vice versa, some modifications to the present chemical models of the inner ion coma may be necessary.

The observations by Wyckoff and Wehinger (1976), Delsemme and Combi (1979), and Huppler et al. (1975) have been useful in the diagnosis of cometary ionospheric structures, but more observations (and analyses) of this kind are clearly needed to confirm reported results and to explore new phenomena. For instance, the emission features observed in the head and in the tail by Huppler et al. (1975) do not vary the assigned red-shift values. If these are indeed due to the H_2O^+ ions, does that mean these ions have been already accelerated to a velocity of 20 to 40 km s^{-1} in the vicinity of the head and then that a constant speed is maintained as the cometary plasma moves tailward? Next we have the tantalizing observations by Wyckoff and Wehinger (1976) that in the ion tail shortly behind the nucleus ($\ell \lesssim 10^4$ km) of Comet Kohoutek 1973 XII the abundance of the CO^+ ions is much higher than that of H_2O^+ ions whereas the CH^+ and N_2^+ ions have abundances comparable to the H_2O^+ abundance. Photochemical calculations using spherically symmetric coma models have difficulty in explaining this observational result. If this not caused by an instrumental effect, it could have very interesting implications for the physico-chemistry in the cometary ionosphere. This is especially true in view of the fact that Miller (1980) has measured the surface brightness ratio of H_2O^+ emission relative to the CO^+ emission to be high in the tail of Comet Kohoutek 1973 XII. One extreme solution would be to allow the gas evaporated from the darkside to be enriched in molecules more volatile than H_2O. This may be possible if the temperature difference between the dayside and darkside surfaces is large so that the volatile material not trapped in the clathrate hydrates (i.e., $CO \cdot 6H_2O$ and $CH_4 \cdot 6H_2O$) would be freely evaporated from the darkside where the H_2O production rate is low. This volatile (CO, CH_4, N_2, etc.)- enriched gas would contribute to the abundances of the CO^+, N_2^+ and CH^+ ions in the tail immediate behind the nucleus. Even if this is not the right interpretation of the observations by Wyckoff and Wehinger (1976), similar effects, though to a lesser degree, must take place because anisotropic cometary outgassing is a well-known effect.

How the cometary ion tails would respond to solar wind disturbances continues to be a most interesting problem in cometary plasma physics. As in the study of the Earth's magnetosphere, determination of the solar wind plasma parameters that are the most important factors in controlling the tail activity is not necessarily straightforward. The issue may be even more complicated as the cometary activity may be partially influenced by the time-variation of the outgassing rate. For example, sporadic outbursts may directly contribute to tail disturbances due to the sharp increase in the ion production rate. Hopefully many of the outstanding questions will be answered by the *in situ* observations of P/Halley in 1986. We would then be in the position to construct quantitative and realistic models of comet-solar wind interactions rather than pondering various qualitative ideas.

Acknowledgments. We thank A.I. Ershkovich, H.U. Keller, K. Jockers, D.A. Mendis, F.D. Miller, H.U. Schmidt and M.K. Wallis for interesting discussions and, especially, K. Jockers for all the kind help in providing the pertinent photographic material of cometary ion tails and insight in their interpretation. Comments from D.B. Beard, M. Neugebauer and M.B. Niedner have been useful in the preparation of this review.

REFERENCES

Aikin, A.C. 1974. Cometary coma ions. *Astrophys. J.* 193:263-264.
Alfvén, H. 1957. On the theory of comet tails. *Tellus* 9:92-96.
Ananthakrishnan, S.; Bhandari, S.M.; and Rao, A.P. 1975. Occultation of radio source PKS 2025-15 by Comet Kohoutek (1973). *Astrophys. Space Sci.* 37:275-282.
Axford, W.I. 1964. The interaction of solar wind with comets. *Planet. Space Sci.* 12:719-720.
Bateman, G. 1978. *MHD Instabilities.* (Cambridge, Massachusetts:The MIT press), Ch. 9.
Beard, D.B. 1966. The theory of type I comet tails. *Planet. Space Sci.* 14:303-311.
Beard, D.B. 1981. Cometary tails. *Astrophys. J.* 245:743-752.
Biermann, L. 1974. Interaction of a comet with the solar wind. In *Solar Wind Three*, ed. C.T. Russell (Los Angeles: IGPP, UCLA Publication), pp. 396-414.
Biermann, L.; Brosowski, B.; and Schmidt, H.U. 1967. The interaction of the solar wind with a comet. *Solar Phys.* 1:254-283.
Biermann, L.; Lüst, Rh.; and Wegmann, R. 1974. Some new results on the plasma-dynamical processes near comets. *Astronautica Acta* 18 (Suppl.):113-118.
Bobrovnikoff, N.T. 1931. Halley's comet in its apparition of 1909-1911. *Publ. Lick Observatory* 37, Part II:403-405.
Brace, L.H.; Theis, R.F.; Hoegy, W.R.; Wolfe, J.H.; Mikhalov, J.D.; Russell, C.T.; Elphic, R.C.; and Nagy, A.F. 1980. The dynamic behavior of the Venus ionosphere in response to solar wind interaction. *J. Geophys. Res.* 85:7663-7678.
Brosowski, B., and Wegmann, R. 1972. Numerische Behandlung eines Kometenmodells. *Max-Planck-Inst. für Physik u. Astrophysik, Report, MPI/PAE-Astro 46.*
Burlaga, L.F.; Rahe, J.; Donn, B.; and Neugebauer, M. 1973. Solar wind interaction with Comet Bennett (1969i). *Solar Phys.* 30:211-222.
Combi, M.R., and Delsemme, A.H. 1980. Brightness profiles of CO^+ in the ionosphere of Comet West (1976 VI). *Astrophys. J.* 238:381-387.
Curtis, S.A. 1981. Solar wind pickup of ionized Venus exosphere atoms. *J. Geophys. Res.* 86:4715-4720.
Delsemme, A.H., and Combi, M.R. 1979. O(^{1}D) and H_2O^+ in Comet Bennett 1970. II. *Astrophys. J.* 228:330-337.
Dubinin, E.M.; Israelevich, P.L.; Podgorny, I.M.; and Shkolnikova, S.I. 1980. Magnetic tail and electrodynamic forces in the Comet Halley tail (Analysis of laboratory and observation data). *Space Res. Inst.*, Moscow, USSR.
Eddington, A.S. 1910. The envelope of Comet Morehouse (1908 c), *Mon. Not. Roy. Astron. Soc.* 70:442-458.
Elphic, R.C.; Russell, C.T.; Slavin, J.A.; and Brace, L.H. 1980. Observations of the dayside ionopause and ionosphere of Venus. *J. Geophys. Res.* 85:7679-7696.
Eroshenko, E.G. 1979. Unipolar induction effects in the magnetic tail of Venus. *Cosmic Res.* 17:77.
Ershkovich, A.I. 1980. Kelvin-Helmholtz instability in type-1 comet tails and associated phenomena. *Space Sci. Rev.* 25:3-34.
Ershkovich, A.I., and Chernikov, A.A. 1973. Non-linear waves in type-1 comet tails. *Planet. Space Sci.* 21:663-670.
Feldman, P.D. 1978. A model of carbon production in a cometary coma. *Astron. Astrophys.* 70:547-553.
Festou, M. 1978. L'hydrogène atomique et le radical oxhydril dans les comètes. Doctoral Thesis, Paris, France.
Giguere, P.T., and Huebner, W.F. 1978. A model of comet comae. I. Gas-phase chemistry in one dimension. *Astrophys. J.* 223:638-654.
Hartle, R.E.; Ogilvie, K.W.; and Wu, C.S. 1973. Neutral and ion-exosphere in the solar wind with applications to Mercury. *Planet. Space Sci.* 21:2181-2191.
Hartle, R.E., and Wu, C.S. 1973. Effects of electrostatic instabilities on planetary and interstellar ions in the solar wind. *J. Geophys. Res.* 78:5802-5807.
Houpis, H.L.F., and Mendis, D.A. 1980. Physicochemical and dynamical processes in cometary ionospheres. I. The basic flow profile. *Astrophys. J.* 239:1107-1118.
Houpis, H.L.F., and Mendis, D.A. 1981a. On the development and global oscillations of cometary ionospheres. *Astrophys. J.* 243:1088-1102.

Houpis, H.L.F., and Mendis, D.A. 1981b. The nature of the solar wind interaction with CO_2/CO-dominated comets. *Moon and Planets* 25:95-104.

Huebner, W.F., and Giguere, P.T. 1980. A model of comet comae. II. Effects of solar photodissociative ionization. *Astrophys. J.* 238:753-762.

Huntress, W.T. Jr.; Prasad, S.S.; and Mitchell, G.F. 1981. Chemical models of cometary comae. *Icarus* (special Comet issue). In press.

Huppler, D.; Reynolds, R.J.; Roesler, F.L.; Scherb, F.; and Trauger, J. 1975. Observations of Comet Kohoutek (1973 f) with a ground-based Fabry-Perot spectrometer. *Astrophys. J.* 202:276-282.

Ichimaru, S. 1973. *Basic Principles of Plasma Physics—A Statistical Approach*. (Reading, Massachusetts: Benjamin), p. 173.

Ioffe, Z.M. 1968. Some magnetohydrodynamic effects in comets. *Sov. Phys.-Astron.* 11:1044-1047.

Ioffe, Z.M. 1970. A mechanism for generating Alfvén waves in comets. *Sov. Phys.-Astron.* 13:1042-1043.

Ip, W.-H. 1979. Currents in the cometary atmosphere. *Planet. Space Sci.* 27:121-125.

Ip, W.-H. 1980a. Cometary atmospheres, I. Solar wind modification of the outer ion coma. *Astron. Astrophys.* 92:95-100.

Ip, W.-H. 1980b. On the acceleration of cometary plasma. *Astron. Astrophys.* 81:260-262.

Ip, W.-H. 1980c. On the dynamical response of a cometary ion tail to a solar-wind event. *Astrophys. J.* 238:388-393.

Ip, W.-H. 1981. Expanding haloes in cometary comae. *Nature* 289:269-271.

Ip, W.-H., and Mendis, D.A. 1976a. The cometary magnetic field and its associated electric currents. *Icarus* 26:457-461.

Ip, W.-H., and Mendis, D.A. 1976b. The generation of magnetic fields and electric currents in cometary plasma tails. *Icarus* 29:147-151

Jockers, K. 1981. Plasma dynamics in the tail of Comet Kohoutek 1973 XII. *Icarus* (special Comet issue). In press.

Jockers, K., and Lüst, Rh. 1973. Tail pecularities in Comet Bennett caused by solar wind disturbances. *Astron. Astrophys.* 26:113-121.

Jockers, K.; Lüst, Rh.; and Nowak, Th. 1972. The kinematical behaviour of Comet Tago-Sato-Kosaka 1969 IX. *Astron. Astrophys.* 21:199-207.

Keller, H.U. 1971. Wasserstoff als Dissoziationsprodukt in Kometen. *Mitt. der Astron. Gesell.* 30:143-148.

Keller, H.U. 1976. The interpretations of ultraviolet observations of comets. *Space Sci. Rev.* 18:641-684.

Krimigis, S.M., and Sarris, E.T. 1979. Energetic particle bursts in the Earth's magnetotail. In *Dynamics of the Magnetosphere,* ed. S.I. Akasofu (Dordrecht, Holland: D. Reidel), pp. 599-630.

Lee, L.C., and Wu, C.S. 1979. On small-scale turbulence in cometary tails. *Astrophys. J.* 228:935-938.

Lüst, Rh. 1962. Bewegung von Strukturen in der Koma und im Schweif des Kometen Morehouse. *Z. Astrophys.* 65:236-250.

Marochnik, L.S. 1963a. Interaction of solar corpuscular streams with cometary atmospheres, I. Shock waves in comets. *Soviet Astron.-AJ* 6:828-832.

Marochnik, L.S. 1963b. Interaction between solar corpuscular streams and cometary atmospheres, II. "Collapsing" envelopes and radio-frequency emission. *Soviet Astron.-AJ* 7:384-390.

Marochnik, L.S. 1964. Magnetohydrodynamic phenomena in comets and their convection with geoactive currents. *Soviet Physics USPEKHI* 7:80-100.

Meier, R.R.; Opal, C.B.; Keller, H.U.; Page, T.L.; and Carruthers, G.R. 1976. Hydrogen production rates from Lyman-α images of Comet Kohoutek (1973 XII). *Astron. Astrophys.* 52:283-290.

Mendis, D.A. 1978. On the hydromagnetic model of comets. *Moon and Planets* 18:361-369.

Mendis, D.A.; Holzer, T.; and Axford, W.I. 1972. Neutral hydrogen in cometary comas. *Astrophys. Space Sci.* 15:313-325.

Miller, F.D. 1979. Comet Tago-Sato-Kosaka 1969 IX: Tail structure 25 December 1969 to 12 January 1970. *Icarus* 37:443-456.
Miller, F.D. 1980. H_2O^+ in the tails of 13 comets. *Astron. J.* 85:468-473.
Miller, F.D. 1981. Plasma envelopes, rays, and plumes of comets. In *Bart J. Bok 75th Anniversary Colloquium*, ed. R.E. White (Tucson: Pachart). In press.
Mitchell, G.F.; Prasad, S.S.; and Huntress, W.T. 1981. Chemical model calculations of C_2, C_3, CH, CN, OH, and NH_2 abundances in cometary comae. *Astrophys. J.* 244:1087-1093.
Morrison, P.J.; and Mendis, D.A. 1978. On the fine structure of cometary plasma tails. *Astrophys. J.* 226:350-354.
Ness, N.F. 1965. The Earth magnetic tail. *J. Geophys. Res.* 70:2989-3005.
Ness, N.F., and Donn, B.D. 1965. Concerning a new theory of type-1 comet tails. *Mém. Soc. Roy. Liège*, Sér. 5, 12:141-144.
Niedner, M.B. Jr., and Brandt, J.C. 1978. Interplanetary gas. XXIII. Plasma tail disconnection events in comets: Evidence for magnetic field line reconnection at interplanetary sector boundaries. *Astrophys. J.* 223:655-670.
Niedner, M.B. Jr., and Brandt, J.C. 1979. Interplanetary gas. XXIV. Are cometary plasma tail disconnections caused by sector boundary crossings or by encounters with high-speed streams? *Astrophys. J.* 234:723-732.
Niedner, M.B.; Ionson, J.A.; and Brandt, J.C. 1981. Interplanetary gas. XXVI. On the reconnection of magnetic fields in cometary ionospheres at interplanetary sector boundary crossings. *Astrophys. J.* 245. 1159-1169.
Niedner, M.B. Jr.; Rothe, E.D.; and Brandt, J.C. 1978. Interplanetary gas. XXII. Interaction of Comet Kohoutek's ion tail with the compression region of a solar-wind corotating stream. *Astrophys. J.* 221:1014-1025.
Okuda, H., and Hasegawa, A. 1969. Computer experiments on plasma instabilities due to anisotropic velocity distributions. *Phys. Fluids.* 12:676-686.
Oppenheimer, M. 1975. Gas phase chemistry in comets. *Astrophys. J.* 196:251-259.
Podgorny, I.M.; Dubinin, E.M.; Potanin, Yu.N.; and Shkolnikova, S.I. 1979. Simulation of cometary magnetic tails. *Astrophys. Space Sci.* 61:369-374.
Prasad, S.S., and Huntress, W.T. Jr. 1980. A model for gas phase chemistry in interstellar clouds: I. The basic model, library of chemical reactions, and chemistry among C, N, and O compounds. *Astrophys. J. Suppl.* 43:1-35.
Russell, C.T., and Elphic, R.C. 1979. Observations of flux tubes in the Venus ionosphere. *Nature* 279:616-618.
Schmidt, H.U. 1978. Action of interplanetary magnetic fields on comets. *ESA Workshop on Cometary Missions*, (SOL(78)14), ESOC, Darmstadt.
Schmidt, H.U., and Wegmann, R. 1980. MHD-calculations for cometary plasmas. *Computer Phys. Commun.* 19:309-326.
Schu'lman, L.M. 1972. *Dinamika Kometnikh Atmosfer-Neitralnic Gaz*. Chap 3, (Kiev, U.S.S.R.), Ch. 3.
Shimizu, M. 1975. Ion chemistry in the cometary atmosphere. *Astrophys. Space Sci.* 36:353-361.
Shimizu, M. 1976. The structure of cometary atmospheres. II. Ion distribution. *Astrophys. Space Sci.* 40:243-251.
Slavin, J.A.; Elphic, R.C.; Russell, C.T.; Scarf, F.L.; Wolfe, J.H.; Mihalov, J.D.; Intriligator, D.S.; Brace, L.H.; Taylor, H.A.; and Daniell, R.E. Jr. 1980. The solar wind interaction with Venus: Pioneer Venus observations of bow shock location and structure. *J. Geophys. Res.* 85:7625-7641.
Tandberg-Hanssen, E. 1974. *Solar Prominences*, (Dordrecht, Holland: D. Reidel).
Wallis, M.K. 1971. Shock-free deceleration of the solar wind? *Nature* 233:23-25.
Wallis, M.K. 1973a. Solar wind interaction with H_2O comets. *Astron. Astrophys.* 29:29-36.
Wallis, M.K. 1973b. Weakly-shocked flows of the solar wind plasma through atmospheres of comets and planets. *Planet. Space Sci.* 21:1647-1660.
Wallis, M.K., and Dryer, M. 1976. Sun and comets as sources in an external flow. *Astrophys. J.* 205:895-899.
Wallis, M.K., and Ong, R.S.B. 1975. Strongly-cooled ionizing plasma flows with application to Venus. *Planet. Space Sci* 23:713-721.

Wallis, M.K., and Ong, R.S.B. 1976. Cooling and recombination processes in cometary plasmas. In *The Study of Comets,* eds. B. Donn, M. Mumma, W. Jackson, M. A'Hearn, and R. Harrington (Washington: NASA SP-393), pp. 856-876.

Wolff, R.S.; Goldstein, B.E.; and Yeates, C.M. 1980. The onset and development of Kelvin-Helmholtz instability at the Venus ionosphere. *J. Geophys. Res.* 85:7697-7707.

Wright, C.S., and Nelson, G.J. 1979. Comet plasma densities deduced from refraction of occulted radio sources. *Icarus* 38:123-135.

Wu, C.S., and Davidson, R.C. 1972. Electromagnetic instabilities produced by neutral particle ionization in interplanetary space. *J. Geophys. Res.* 77:5399-5406.

Wu, C.S., and Hartle, R.E. 1974. Further remarks on plasma instabilities produced by ions born in the solar wind. *J. Geophys. Res.* 79:283-285.

Wurm, K., and Mammano, A. 1972. Contributions to the kinematics of type I tails of Middlehurst, and G.P. Kuiper (Chicago: Univ. Chicago Press), pp. 573-617.

Wurm, K. 1968. Structure and kinematics of cometary type 1 tails. *Icarus* 8:287-300.

Wurm, K., and Mammano, A. 1967. Dissoziation und Ionisation in Kometen. *Icarus* 6:281-291.

Wurm, K., and Mammano, A. 1972. Contributions to the kinematics of Type I tails of comets. *Astrophys. Space Sci.* 18:273-286.

Wurm, K., and Rahe, J. 1969. Type I tail structures of comets within the inner coma region. *Icarus* 11:408-412.

Wyckoff, S. 1981. Ground-based cometary spectroscopy. *Astron. Preprint Series,* Physics Dept., Arizona State Univ.

Wyckoff, S., and Wehinger, P.A. 1976. Molecular ions in comet tails. *Astrophys. J.* 204:604-615.

PART VI
Origin, Evolution, and Interrelations

DYNAMICAL HISTORY OF THE OORT CLOUD

Paul R. Weissman
Jet Propulsion Laboratory

The hypothesis of a cloud of comets surrounding the solar system and extending out to interstellar distances as suggested by Oort has been successful in explaining the observed distribution of orbits of the long-period comets. Oort demonstrated that the motion of comets in the cloud is controlled by perturbations from random passing stars. Numerical studies employing Monte Carlo techniques have further expanded our understanding of the dynamics of comets in the Oort Cloud. The population of the cloud has been depleted over the history of the solar system. If comets formed in orbits near the outer planets and were subsequently ejected to the cloud, then only about 15% of the original population still remains. However, if comets formed farther from the Sun in satellite fragments of the primordial solar nebula, then the depletion is less severe; $\sim 70\%$ of the initial population survives. Estimates of the current cloud population range between 1.2 and 2.0×10^{12} comets, roughly an order of magnitude greater than Oort's original estimate. Loss mechanisms from the cloud are: diffusion of cometary perihelia into the planetary region where planetary perturbations will eject the comets from the solar system; diffusion of cometary aphelia to distances beyond the Sun's sphere of influence ($\sim 2 \times 10^5$ AU); and direct ejection due to close encounters with passing stars. The numerical simulations are used to find the distributions of perihelia, aphelia, and energy for comets in the present Oort Cloud, and to comment on the population of interstellar comets.

I. THE OORT HYPOTHESIS

The modern dynamical theory for long-period comets originates with Oort's (1950) classic paper. He identified the source of long-period comets as a vast cloud of several times 10^{11} comets surrounding the solar system and extending roughly halfway to the nearest stars. The motion of comets in the cloud is dominated by perturbations from random passing stars which have effectively randomized the distributions of orbital elements. These same stellar perturbations cause the perihelia of the orbits to diffuse into the planetary region where the comets are observed.

Oort calculated the mean thermal velocity in the cloud to be 136 m s^{-1} at 10^5 AU from the Sun, equal to the escape velocity at that distance. He showed that the perturbations were great enough to randomize the inclinations of the comets remaining in the cloud. He also demonstrated that the perturbations were able to provide a continuous flux of new comets from the cloud into the planetary region ($q < 15$ AU, where q is perihelion distance) and that the perihelion distribution of the new comets would be uniform with q. Oort estimated that 5 to 9% of the cloud population had been ejected due to close encounters with stars passing directly through the cloud. An additional 13% had been lost by diffusion of perihelia back into the planetary region where perturbations by the major planets would then control the motion, ejecting the comets on hyperbolic trajectories or capturing them to short-period orbits.

Oort's accomplishment in linking the observed distribution of orbits to his theory of a huge cloud surrounding the solar system is more remarkable considering the very small and incomplete sample of accurate orbits then available. Table 1 of his 1950 paper gives the distribution of original inverse semimajor axes $1/a_0$ for a mere 19 observed long-period comets. Far more complete data are now available; see, e.g., Table I in the Appendix by Marsden and Roemer in this book. Fig. 1 shows the $1/a_0$ distribution for 190 long-period comets as given by Marsden et al. (1978). The original semimajor axis a_0 is the value of the semimajor axis of a comet's orbit prior to entering the planetary region (< 40 AU) and being perturbed by the planets, and refers to the barycenter of the solar system rather than the Sun. Plotting the number of comets versus $1/a_0$ is equivalent to plotting them as a function of orbital energy where the energy $E = -\mu/2a_0$ ($\mu = GM$, the gravitational mass of the solar system). The large number of orbits between 0 and 0.1×10^{-3} AU^{-1} in Fig. 1 are the dynamically new comets from the Oort Cloud. According to Oort's theory these comets are making their first pass through the planetary region. On subsequent passes planetary perturbations scatter the comets in $1/a$, either bringing them to shorter period orbits or ejecting them from the solar system on hyperbolic orbits.

A number of the comets in Fig. 1 have weakly hyperbolic orbits ($1/a_0 < 0$). These are generally thought to result from errors in calculation of the orbital elements, compounded by nongravitational forces on the comets

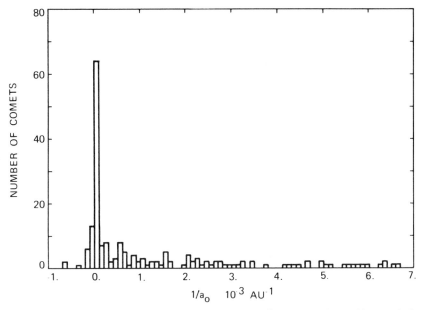

Fig. 1. Distribution of original inverse semimajor axes $1/a_0$ for the observed long-period comets. The original orbit is that derived for the comet before it enters the planetary region and becomes subject to planetary perturbations.

tending to make the orbits appear more hyperbolic than they are. True extra-solar comets would be expected to approach the solar system with hyperbolic excess velocities V_∞ comparable to the Sun's velocity relative to the neighboring stars. A V_∞ of 20 km s^{-1} would give a $1/a_0$ value of -0.45 AU^{-1}. The most negative observed value in Fig. 1 is -7.27×10^{-4} corresponding to a hyperbolic velocity of only 0.8 km s^{-1}. The problem of hyperbolic comets is discussed below.

The remaining comets in Fig. 1 are those diffused to smaller semimajor axes as a result of planetary perturbations, mainly by Jupiter. Oort showed that this was the expected distribution of $1/a_0$, but that the number of observed comets making returns was less than expected from the theory. He attributed this to the ability of new comets to produce anomalously bright comae and thus increase their probability of discovery over that for returns when they will have dimmed. Subsequent work by Kendall (1961), Whipple (1962), and Weissman (1979) on modeling the dynamics of long-period comets from the Oort Cloud and the resulting $1/a_0$ distribution has strengthened the evidence for Oort's hypothesis.

Additional evidence on the Oort Cloud and the evolution of long-period comets is shown in Fig. 2 which is a scatter diagram in $1/a_0$ and q for observed long-period comets. The dynamically new comets from the cloud are seen as a horizontal band across the bottom of the figure at near-zero

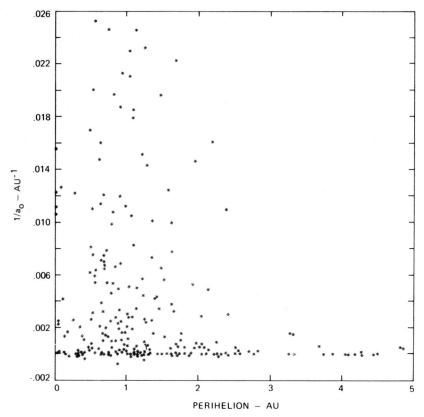

Fig. 2. Scatter diagram for the observed long-period comets as a function of original inverse semimajor axis and perihelion distance. The Oort Cloud is shown by the horizontal band of comets across the figure at near-zero $1/a_0$.

$1/a_0$. Planetary perturbations cause the comets to random-walk in $1/a$, diffusing up and down in the reference frame of Fig. 2. Note that almost all observed comets with perihelia > 2.8 AU are Oort Cloud members ($1/a_0 < 100 \times 10^{-6}$ AU^{-1}). The visibility of these comets must result from sublimation of surface materials more volatile than water ice, which begins to sublimate only within 3 AU. On later returns comets with $q > 3$ AU are not observed because their surfaces are depleted of these highly volatile materials. This tends to confirm Oort's idea that new comets are able to produce more extensive comae.

Also seen in Fig. 2 is a lack of long-period comets with $q < 0.4$ AU and $1/a_0 > 0.0025$ AU^{-1}, which is even more striking given that the four comets with near-zero perihelion and $1/a_0 \sim 0.010$ to 0.016 AU^{-1} belong to the Kreutz family of sungrazing comets and thus are probably fragments of a single parent comet. Weissman (1979) has shown that this lack of older, small-q comets cannot be explained dynamically and must result from formation of nonvolatile crusts on the cometary nuclei which cut off sublimation and render them unobservable.

Oort's work was anticipated in part by Öpik (1932) who studied the effect of stellar perturbations on very long-period orbits. Öpik, however, drastically overestimated the Sun's sphere of influence, and failed to recognize the connection between the observed distribution of cometary orbits and that which the theory would predict. Russell (1935) developed Öpik's work further and speculated that comets might be preserved in orbits with perihelia beyond Jupiter's orbit, but he questioned this hypothesis because of "the enormous number of comets at large perihelion distance" required.

II. RECENT WORK

After Oort and Öpik, stellar perturbations on cometary orbits have been studied by Sekanina (1968a, 1968b), Faintich (1971), Yabushita (1972), Rickman (1976), and Weissman (1980). These authors have used different approaches, both analytical and statistical, to examine the nature of the perturbations and some first order effects on the dynamics of comets in the Oort Cloud. The typical method is to treat the stellar perturbation as an instantaneous velocity impulse on the orbit while the comet is near aphelion. This is a good approximation because the typical stellar encounter velocity is ~ 20 km s^{-1} and a comet in the Oort Cloud has velocity ~ 100 m s^{-1}. The magnitude of the perturbation is given by

$$\Delta V = 2\, GM_*/DV_* \qquad (1)$$

where GM_* is gravitational mass of the star, V_* is encounter velocity, and D is minimum comet-star distance. For values of $M_* = M_\odot$ and $V_* = 20$ km s^{-1}, $\Delta V = 43$ cm s^{-1} at an encounter distance of 1 pc (2×10^5 AU). The velocity impulse is directed along the vector from the comet to the star at closest approach.

The frequency of encounters between the Oort Cloud and passing stars can be estimated by

$$N = \pi R^2 \rho_* V_\odot T \qquad (2)$$

where N is number of encounters, R is radius of the cross-section considered, ρ_* is local density of stars in the solar neighborhood, $V_\odot$ is the Sun's velocity relative to the local stars, and T is time. Inserting numbers one finds $N = 5.1 \times 10^{-6}\, R^2 T$ with T in yr and R in pc. Thus, ~ 5 stars pass within 1 pc of the Sun (or any comet in the Oort Cloud) every 10^6 yr. If this number is constant over the history of the solar system then 2.3×10^4 stellar encounters at ≤ 1 pc have occurred.

Using Eqs. (1) and (2) as their starting point the authors listed above further analyzed the stellar perturbations problem. One key calculation is that of total velocity perturbation over the history of the solar system, with estimates ranging between 110 and 150 m s^{-1}. Equating these estimates to escape velocity as a function of solar distance, the limit on the radius of the present Oort Cloud is $\sim 10^5$ AU; comets at greater distance have been perturbed by passing stars to escape.

Theories on the origin of the Oort Cloud are generally divided between formation of comets among the outer planets and ejection to current heliocentric distances, and formation in satellite fragments of the collapsing proto-solar nebula. Oort originally suggested that comets formed among the asteroids and were ejected by Jupiter perturbations. Kuiper (1951) pointed out that icy bodies would only condense at distances > 3 AU from the Sun, and thus the comets were probably planetesimals ejected from the Jupiter-Saturn zone. More recently Safronov (1972) and Fernandez (1978) have shown that Jupiter and Saturn perturbations tend to eject material on hyperbolic orbits and not to distant ellipses, whereas Uranus and Neptune typically perturb material into distant but captured orbits. The problem of how comets were transported to their present heliocentric distances is eliminated by the assumption that they formed in their current orbits, in satellite fragments of the primordial solar nebula, as described by Cameron (1973). Capture of comets from interstellar space is generally dismissed as too inefficient a mechanism to produce the Oort Cloud.

Previous studies of the dynamics and evolution of the Oort Cloud have been limited. Weissman (1977), Schreur (1979), and Fernandez (1980), have begun to examine the problem employing Monte Carlo simulation techniques. Weissman showed that the perihelion distribution of new comets entering the planetary region from the Oort Cloud is uniform with respect to perihelion distance and that the comets are limited to a very narrow range in $1/a$. Schreur studied changes in orbital elements for a hypothetical cloud of 10^5 comets under the influence of close encounters with 100 stars, and demonstrated that perturbations tend to randomize the orbits within the Oort Cloud and leave little trace of initial element distributions. Fernandez showed that comets ejected to the cloud from the Uranus-Neptune zone random walk in perihelion distance until they reenter the planetary region and are influenced by planetary perturbations.

In the following sections a more sophisticated version of the Monte Carlo model used in Weissman (1977) is developed and used to study the dynamics of comets in the Oort Cloud. The number of comets in the cloud as a function of time, the percentage fraction that evolves to each of several possible end-states, and the distributions of orbital elements are of interest. The model is also used to estimate the present and original population of the cloud and to examine implications for different theories of cometary origin. The problem of interstellar comets is considered with respect to the flux of interstellar comets produced by the solar system and the possibility of observing these or other interstellar comets.

III. MODEL DESCRIPTION

Weissman (1977) modeled the total stellar perturbations on comets in the Oort Cloud as a single, massive velocity perturbation occurring at aphelion of some initial orbit representative of the comet's origin. Though as a first-order approximation this is acceptable, it fails to account for the random walk that each comet's orbital elements undergo, and the possibility that one step of the random walk brings the comet to an end-state which removes it from the cloud. In this chapter the stellar perturbations are modeled as a velocity impulse at aphelion of each orbit of the comet.

Over the history of the solar system an Oort Cloud comet should make $\sim 10^3$ revolutions about the Sun. Since the typical period of an Oort Cloud comet is 4×10^6 yr and since ~ 20 to 40 stars will pass within 1 pc of the cloud in that time, a statistical approach is preferable to an attempt to integrate the perturbations for each random passing star.

Weissman (1980) has shown that the rms velocity perturbation on the comet is given in m s^{-1} by

$$\Delta V_{rms} = 1.7 \times 10^{-3} \, T^{\frac{1}{2}} \qquad (3)$$

where T is the period of the orbit in yr. For 4.5×10^9 yr this leads to a total rms perturbation of 113 m s^{-1}, in good agreement with Faintich (1971) who found $\Delta V_{rms} = 120$ m s^{-1}. Faintich concluded that the cloud beyond 10^4 AU from the Sun had been thermalized over the history of the solar system.

Using Eq. (3) the typical perturbation for one orbit would be 3.4 m s^{-1}, enough to raise the perihelion distance of an initial orbit with $q = 1$ AU to > 24 AU, assuming an aphelion distance of 5×10^4 AU. In the Monte Carlo model the velocity perturbation occurs at aphelion of each orbit and is assumed randomly oriented in space. The value ΔV_{rms} is determined from the period of the orbit and is assumed to be the rms value of a Maxwellian velocity distribution from which the actual velocity perturbation is randomly selected.

Three dynamic end-states are possible for comets in the cloud. The first is ejection from the cloud on a hyperbolic orbit due to a close stellar encounter. Weissman (1980) showed that $\sim 9\%$ of the cloud has been ejected over 4.5×10^9 yr by single encounters with stars passing through it. The second loss mechanism is perturbation of the comet's aphelion distance to values outside the Sun's sphere of influence. Chebotarev (1965, 1966) showed that perturbations by the galactic nucleus limited the aphelia of orbits still bound to the Sun to 2×10^5 AU. This end-state will be referred to as "stellar" in discussion below.

The last end-state is diffusion of the comet's perihelion into the planetary region. Weissman (1979) showed that comets passing within the orbit of Jupiter will be removed from the system, with a mean lifetime of 6×10^5 yr measured from the time of the first perihelion passage. The principal individual loss mechanisms are hyperbolic ejection due to planetary perturbations (64%), random disruption (27%), and formation of nonvolatile crusts on the nuclei (7%). Minor end-states include capture to short-period orbit, diffusion of aphelia to distances greater than 2×10^5 AU (stellar loss, the second end-state listed above), and perturbation of perihelia to within the solar Roche limit. For the current model all of these possible end-states are lumped together under the heading of "planetary" loss.

However, a small fraction of the comets entering the planetary region will be returned to orbits with aphelia in the Oort Cloud, 5.8% per perihelion passage for comets subject to Jupiter perturbations and 34.7% for Saturn-perturbed comets. Typical planetary perturbations of Uranus and Neptune are so small that virtually all comets with perihelia outside the Jupiter and Saturn zones will return to the cloud.

In the present Monte Carlo simulation the Jupiter and Saturn zones were defined as 1.5 times each planet's semimajor axis. For comets passing within 7.8 AU of the Sun, 5.8% were randomly selected and allowed to return to the cloud and continue their orbital evolution. For Saturn the respective numbers were 14.3 AU and 34.7%. The 1.5 factor is based on studies of planetary perturbations by Everhart (1968) and represents the distance at which the perturbations drop off sharply.

No physical loss mechanisms are expected to operate while the comets are in the Oort Cloud. Collisions between comets in the cloud are insignificant. Even stars passing through the cloud will typically not collide with more than a few comets. The comets are assumed to be stored in the cloud far from any process that can remove or destory them, other than orbital dynamics. Figure 3 is a flow diagram for the Monte Carlo simulation. Each comet is started in an orbit with perihelion distance representative of its origin and aphelion in the Oort Cloud. The random stellar perturbation is applied at aphelion and new orbit elements calculated. The new elements are tested to see if they fall into one of the possible end-states. If so, the simulation is halted and the time and loss mechanism are noted. If not, the comet moves through one full orbit, the time increases by the orbit period, and the cycle repeats. If the time exceeds the age of the solar system, 4.5×10^9 yr, the simulation is halted and the comet is counted as member of the present Oort Cloud. By following large numbers of hypothetical comets through the model, the statistics of various end-states and the current cloud can be determined. Input to the program includes: initial perihelia and aphelia, total rms velocity perturbation over the history of the solar system, radius of the Sun's sphere of influence, number of comets to be run (typical cases use 10^4 hypothetical comets), and a starting seed for the random number generator.

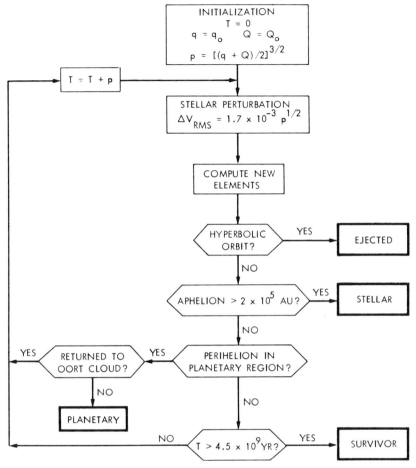

Fig. 3. Flow diagram for the numerical simulation model. Calculations are shown by rectangles, decision points by hexagons, and end-states by heavy black boxes.

IV. MONTE CARLO RUNS

With three independent variables in the Monte Carlo model, i.e., perihelion, aphelion, and total velocity perturbation, many combinations are possible, and so it is necessary to limit the cases to those of particular interest and those helpful to understanding the dynamics of the Oort Cloud.

To study the effect of initial perihelion distance, six cases were run with perihelia of 20, 100, 200, 10^3, 2×10^3, and 10^4 AU, initial aphelion distance of 4×10^4 AU, and total rms velocity perturbation of 120 m s^{-1}. The perihelia were chosen as representative of cometary formation among the outer planets or in satellite fragments of the primordial solar nebula. The aphelion distance was chosen so that the comets enter the planetary region with mean inverse semimajor axes approximately the same as noted by Marsden et al. (1978) for observed new comets. The total rms velocity perturbation was based on work by Faintich (1971), Schreur (1979), and Weissman (1980). To investigate the dependence on aphelion distance, cases were run for aphelia between 2×10^4 and 1.2×10^5 AU at 10^4 AU intervals, plus additional cases at 2.5×10^4 and 1.5×10^5 AU to improve the data resolution. Perihelia of 20 and 200 AU and total velocity perturbation of 120 m s^{-1} were used. Finally, to study the effect of the magnitude of the total perturbation, cases were run for perihelia of 20 AU, aphelia of 4×10^4 AU, and total rms velocity perturbations from 20 to 300 m s^{-1} at 20 m s^{-1} intervals.

V. PERIHELION CASES

Results for the six perihelion cases are shown in Fig. 4 and summarized in Table I which gives the fraction of comets in each end-state and the fraction surviving after 4.5×10^9 yr. Also listed are the mean inverse semimajor axes for comets entering the planetary region during the last 5×10^8 yr of the simulation. It is evident that a substantial fraction of the

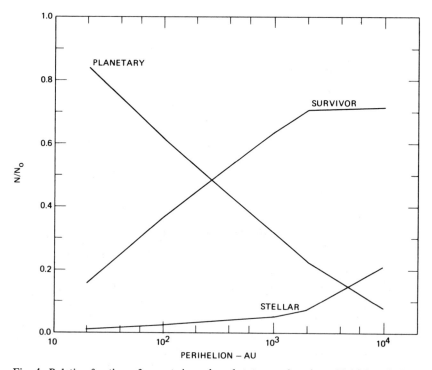

Fig. 4. Relative fraction of comets in each end-state as a function of initial perihelion distance. There was no significant loss to hyperbolic ejection. Initial aphelion value is 4×10^4 AU and $\Delta V_{rms} = 120$ m s^{-1}.

TABLE I

Oort Cloud End-States for Perihelion Cases

Initial Perihelion (AU)	Ejected	Stellar	Planetary	Survivor	Mean $1/a$ of New Comets (10^{-6} AU^{-1})
20	0.0	0.009	0.834	0.157	42
100	0.0	0.024	0.609	0.367	42
200	0.0	0.033	0.520	0.447	42
1000	0.0	0.051	0.313	0.636	42
2000	0.0	0.073	0.222	0.705	41
10^4	0.0	0.212	0.075	0.713	38

Oort Cloud has been lost over the history of the solar system. In the minimum perihelion case less than a sixth of the original cloud is still bound to the solar system. For comets started at small perihelia, diffusion into the planetary region dominates as the loss mechanism, whereas for cases of larger perihelia the dominant end-state is loss to large aphelion orbits. Surprisingly, hyperbolic ejection does not figure prominently as a loss mechanism (a few ejections are typically seen in each run), partly because stellar perturbations greatly affect the angular momentum of the orbit but not the energy. Weissman (1980) showed that the typical velocity perturbation results in a change in $1/a$ of only 0.05%. As the comets random-walk in $1/a$, this is enough to raise the aphelion distances to $> 2 \times 10^5$ AU ($1/a < 10^{-5}$ AU^{-1}) but not enough to immediately eject them ($1/a < 0$). The mean values of $1/a_0$ for new comets entering the planetary region correspond closely with the value of 46.3×10^{-6} AU^{-1} found by Marsden et al. (1978)[1] for observed orbits of new long-period comets after correcting for the effects of nongravitational perturbations.

Figure 5 shows the number of comets remaining in the cloud versus time, for four of the six perihelion cases. For comets with initial perihelia near the planetary region, the orbits rapidly evolve to Jupiter- or Saturn-crossing orbits and are lost to planetary perturbations, random disruption, or other processes. Only a small fraction of the comets started in small perihelia orbits survive long enough in the cloud to evolve to large semimajor axes and loss to interstellar space.

The large number of comets in the small initial perihelion cases returning to the planetary region during the early history of the solar system may provide a source for the primordial bombardment of planetary surfaces which ended ~ 3.5 to 4.0×10^9 yr ago. For initial $q = 20$ AU, the flux of comets

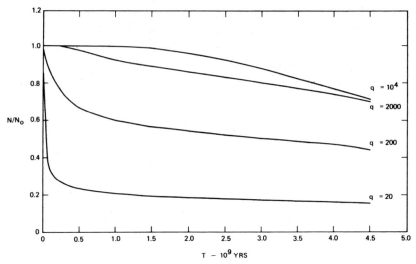

Fig. 5. Fraction of the initial cloud population surviving versus time for four initial perihelion distances. Rapid depletion of the cloud due to diffusion back into the planetary region for small initial q is evident. For large q the loss is more gradual, occurs later in time, and results more from diffusion of orbits to large aphelion distances.

through the planetary region during the first 10^8 yr of the integration is a factor of 10^3 times the current flux predicted by the model. After 5×10^8 yr the flux drops to ~ 11 times the current value and after 10^9 yr, to < 5 times the current flux of new long-period comets. These figures' possible statistical error is $\sim 8\%$. As the initial perihelion distance in the simulation increases, the magnitude of the primordial bombardment decreases. For $q = 100$ AU the flux in the first 10^8 yr is ~ 190 times the current flux, for $q = 200$ AU that factor drops to ~ 80, and for $q = 10^3$ AU the flux is never > 10 times the current predicted flux.

It is also possible to use numerical simulation to estimate the current population of the Oort Cloud. Everhart (1967) has shown that, after accounting for observational selection effects, $\sim 8 \times 10^3$ long-period comets have passed within 4 AU of the Sun between 1840 and 1967. Examination of the orbits of the long-period comets shows that about 25 to 30% are new in the Oort-Schmidt (1951) sense ($1/a_0 < 100 \times 10^{-6}$ AU^{-1}). This then gives a rate of 4.7 new comets per AU of perihelion distance per year. This rate can be related to the rate of hypothetical comets entering the planetary region during the final 5×10^8 yr of the Monte Carlo simulation, and to the population of the modeled cloud at that time. The results are shown in Table II. The numbers in Table II are roughly an order of magnitude greater than Oort's original estimate of 1.9×10^{11} comets in the cloud. However Oort's estimate was based on an observed flux of 97 new comets per century (uncor-

TABLE II

Current and Original Population of the Oort Cloud

Initial Perihelion (AU)	Current Population $\times 10^{12}$	Original Population $\times 10^{12}$
20	1.2	7.6
100	1.5	4.1
200	1.4	3.2
1000	1.2	2.0
2000	1.4	2.0
10^4	1.7	2.4

rected for observational selection effects) within 1.5 AU of the Sun, or 0.65 comets per AU per year. If Everhart's (1967) value for the flux is used instead, Oort's estimate would increase to 1.4×10^{12} comets, in excellent agreement with the figures in Table II. The statistical error of the figures in Table II is 10% for the 20 AU case and decreases to 6% for the 10^4 AU case.

Using the figures in Table I for the fraction of the cloud which survives after 4.5×10^9 yr, and the current population estimates in Table II, it is possible to estimate the original population of the cloud as shown in Table II. If it is assumed that the average comet has a radius of one kilometer and density of one g cm^{-3}, then its mass is 4.2×10^{15} g, the total mass of the current Oort Cloud is ~ 1 M$_\oplus$, and the original cloud mass was between 2 and 6 M$_\oplus$.

An additional result from the numerical simulation is the distributions of orbital elements in the present Oort Cloud. Figure 6 shows the distributions of perihelion and aphelion distances and orbital energy in the cloud for the case of initial $q = 20$ AU. For the perihelion distribution a significant fraction of the comets have evolved to large perihelion distances as a result of the random stellar perturbations, but the majority remains in orbits with relatively small perihelia compared to the dimensions or the cloud. The distribution for aphelion distances is considerably narrower than the perihelion distribution, though some orbits do evolve to very large aphelia. The lack of orbits with aphelion distance $Q < 4.5 \times 10^4$ AU is partly a result of limitations in the numerical model, which allows aphelia to evolve outwards but not inwards. However, because most perturbations occur when the comets are near aphelion, there is in fact little opportunity for the orbits to evolve inward. This slow outward diffusion of the cloud was also noted by Schreur (1979). Figure 6 also shows the distribution of orbital energy for the

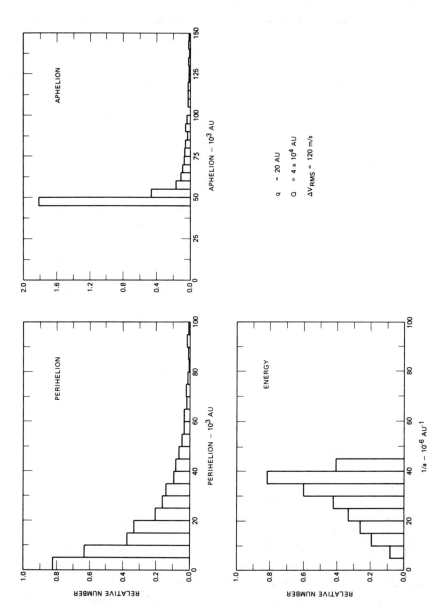

Fig. 6. Orbital element distributions for the surviving population in the Oort Cloud, for the $q = 20$ AU case. The comets are well dispersed in perihelion and aphelion distance (initially 4×10^4 AU) and inverse semimajor axis (initially 50×10^{-6} AU^{-1}).

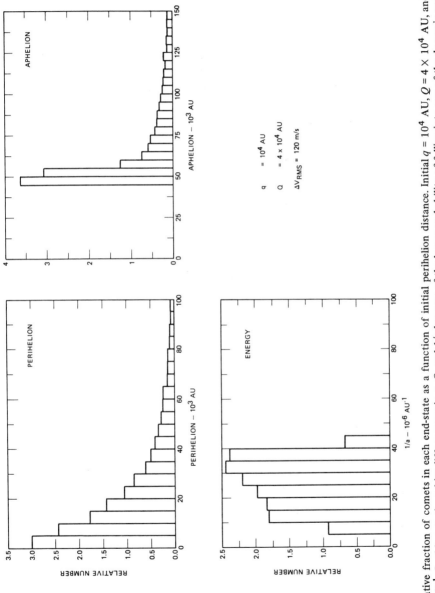

Fig. 7. Relative fraction of comets in each end-state as a function of initial perihelion distance. Initial $q = 10^4$ AU, $Q = 4 \times 10^4$ AU, and $1/a = 40 \times 10^{-6}$ AU^{-1}. In this case the orbits diffuse more in q, Q, and $1/a$ because of the lower probability of falling into one of the dynamic end-states.

comets in the cloud, plotted as a function of inverse semimajor axis where $1/a = -2E\mu^{-1}$. Again, the numerical model does not generally allow comets to evolve to larger values of $1/a$ (more negative energies) but few comets are expected to do so.

Figure 7 shows the same distributions of orbital elements, but for initial $q = 10^4$ AU. The statistics of the distributions are somewhat better than for Fig. 6 because they are based on ~ 4 times the number of survivors in numerical simulation. In the case of the perihelion distribution the comets have again diffused to larger values of q, but in this case also to smaller values of q than the initial 10^4 AU. Diffusion to larger aphelia is also seen, and is more pronounced than for the initially small q case because these comets are more likely to evolve dynamically without falling into the planetary end-state. The same is true for the energy distribution in which more comets have evolved to smaller $1/a$. There is some indication that, given sufficient time and a lack of end-states at both near-zero and large values of $1/a$, the orbits in the cloud would evolve to a state of equipartitioned energy. Note that although in this case the orbits were all started with $1/a = 40 \times 10^{-6}$ AU^{-1}, a number of orbits have evolved to larger values of $1/a$ due to instances of the perihelia diffusing inward while retaining reasonably constant aphelion distance.

VI. APHELION CASES

Results for cases in which the initial aphelion distance of the comets was varied with all other parameters held constant are shown in Fig. 8. The same data is shown in the upper row as a function of initial orbital energy, and in the lower row as a function of aphelion distance. As opposed to Fig. 4 with all end-states plotted together in Fig. 8, each is shown in successive columns. The loss states for the cloud are dominated by planetary loss at small initial values of the aphelion and stellar loss at large initial aphelia. Also, the initial perihelion distance is important for cases in which planetary loss is significant, but is relatively unimportant where stellar loss dominates. This occurs because for stellar loss, perturbation of the aphelion distance beyond some limiting value, the two perihelia used are not significantly different. Stellar loss is slightly lower for the $q = 20$ AU cases because comets are likely to be removed more rapidly by planetary loss than for the $q = 200$ AU cases, thus making them unavailable to stellar loss.

However, for planetary loss the initial perihelion distance has a significant effect, as shown in Sec. IV, because the small q comets have less distance to diffuse in velocity space before falling into the loss cone around near-zero transverse velocity, as shown by Weissman (1980).

One notable result is that any comets whose initial aphelion distance is $> 1.2 \times 10^5$ AU are lost from the solar system after 4.5×10^9 yr. As shown in Sec. IV, comets from initially smaller aphelia have diffused out to these and

OORT CLOUD DYNAMICS

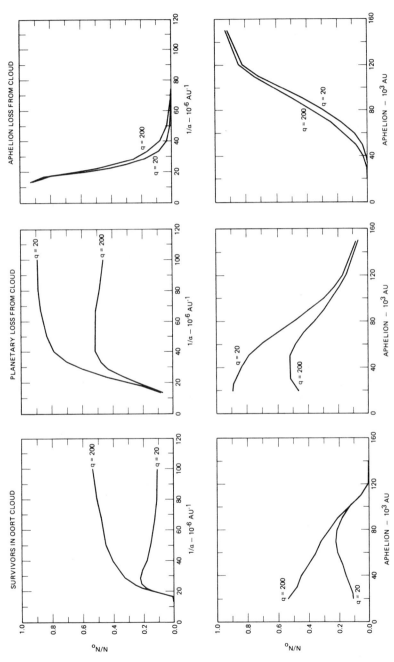

Fig. 8. End-states for comets evolving in the Oort Cloud as a function of initial aphelion distance, shown for initial perhelia of 20 and 200 AU. Each set of data is shown first as a function of initial $1/a$, and below as a function of initial aphelion distance.

greater distances. Thus the stellar perturbations cause a continual outward diffusion of orbits whereby the outer shells of the cloud are eventually stripped away. At smaller initial aphelion values stellar loss becomes insignificant because the orbits are too tightly bound to the solar system. Also, it has been shown by several authors (see Sec. II) that at these distances the stellar perturbation drops off because the impulse the passing star exerts on the comet is the same as the impulse it exerts on the Sun, leaving little or no net difference. Thus the number of survivors at small initial aphelion distance may be greater than that shown in Fig. 8.

These small-aphelion comets represent a largely unsampled part of the cloud population. They do not experience stellar perturbations significant enough to dump their perihelia into the inner solar system. For those which might have originated in the Uranus-Neptune zone, it is possible that subsequent passages through this zone would perturb the orbits to higher aphelia, thus making these comets available to stellar perturbations. Planetary perturbations by Uranus or Neptune are able to do this even without very close encounters. However, if comets originated beyond the planetary region altogether then there is nothing known which can perturb their aphelion distances farther from the Sun, or bring their perihelia into the planetary zone.

VII. VELOCITY CASES

Results for the cases in which the total rms velocity perturbation was varied are shown in Fig. 9. Even for very low values of the total perturbation, a substantial fraction of the cloud has been lost, mostly by diffusion into the planetary region. This dominance by planetary loss is partly because all cases were started at an initially small perihelion distance of $q = 20$ AU. As one moves to larger values of the total perturbation the other loss states also become important. Stellar loss turns on at $\sim \Delta V_{rms} = 100$ m s^{-1} and rises rapidly with the magnitude of the perturbation. At ~ 200 m s^{-1} hyperbolic ejections begin to become significant though only reaching 2% of the total cloud population at 300 m s^{-1}. The number of survivors in the cloud goes to zero for $\Delta V_{rms} > 250$ m s^{-1}. This is of course a statistical result based on a sample size for the Monte Carlo simulation of 10^4 hypothetical comets. However, assuming the earlier estimate of the current cloud mass as of ~ 1 $M_\oplus$ to be correct, it would be hard to consider as physically real an initial cloud mass several times 10^4 $M_\oplus$. The estimates of ΔV_{rms} ranging from ~ 110 to 140 m s^{-1} provide a more physically reasonable number of cloud survivors.

The relatively small role played by hyperbolic ejections throughout the simulation cases is somewhat puzzling, particularly in light of estimates by Oort and Weissman that up to 9% of the cloud populations has been ejected due to close encounters with stars passing through the cloud. In part the

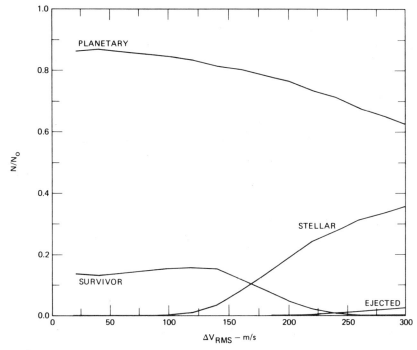

Fig. 9. End-states for comets evolving in the Oort Cloud as a function of the magnitude of the total rms velocity perturbation. Initial perihelion and aphelion values are 20 AU and 4×10^4 AU, respectively.

problem can be attributed to the statistical tendency of the Monte Carlo simulation to emphasize average perturbation over more extreme cases. However, also, it appears that the Maxwellian velocity distribution for the velocity impulses tends to underestimate the larger perturbations and that the true distribution has a significantly longer tail. Clearly, additional study is needed in this area.

VIII. DISCUSSION

Although other theories have been proposed for the source of long-period comets, the weight of evidence favors Oort's hypothesis of a cloud of some 10^{12} comets surrounding the solar system and extending out to $\sim 10^5$ AU. Previous studies have shown that the $1/a_0$ distribution for long-period comets is consistent with comets having their source in the Oort Cloud and evolving dynamically under the influence of planetary perturbations. This chapter has demonstrated that the dynamical evolution of the cloud is consistent with information from observations of the long-period comets, and that estimates of the population necessary to produce the observed flux of dynamically new comets lead to physically reasonable values for current and initial cloud mass.

An unfortunate aspect of the problem is that the processes that have randomized the Oort Cloud have left little evidence of the cloud's origin. There is no clear indication of whether the comets formed among the outer planets of the solar system or in satellite fragments of the solar nebula. The problem is compounded because comets entering the planetary region are not necessarily representative of the entire cloud. As shown by Weissman (1977), a wide range of initial perihelia and velocity perturbation conditions all lead to the same uniform perihelion distribution in the planetary region; this chapter confirms this result. Also, due to errors in deriving $1/a_0$ for observed long-period comets and to the small available sample, little information about the actual distribution of $1/a_0$ can be found except for the mean value derived by Marsden et al. (1978).

The cloud is here shown to be a continually evolving object whose population steadily diminishes as comets are lost to various dynamic end-states. Comets are ejected to or they diffuse to interstellar space as a result of stellar perturbations, or enter the planetary region where they are most likely ejected by planetary perturbations. This implies a substantial population of escaped comets filling interstellar space in the solar neighborhood. Weissman (1979) showed that comets are ejected by Jupiter and Saturn with a mean hyperbolic excess velocity of 0.6 km s^{-1}. Based on this work, comets ejected from the cloud by stellar perturbations would have a somewhat lower mean excess velocity. Using estimates for the original number of comets given in Table II, the ejected comets may total as much as 5 $M_\oplus$ of material.

An additional source of interstellar material would be planetesimals ejected during formation of the planets. Fernandez (1978) estimated the ejected mass to be 25 $M_\oplus$, assuming that the Oort Cloud was formed at the same time with a mass of 1.5 $M_\oplus$. The higher figure from this chapter of 6 $M_\oplus$ for the initial cloud gives 100 $M_\oplus$ ejected by the major planets. Safronov (1972) gives somewhat higher estimates, as much as 10^3 $M_\oplus$.

Presumably other stars with planetary systems would also eject large numbers of planetesimals, as well as producing their own Oort Clouds which would further contribute to the population of extra-solar comets. Encounters between interstellar comets and stars would rapidly randomize the cometary velocities. The comets would be expected to approach other stars with typical velocities comparable to the mean stellar velocity in the solar neighborhood, ~ 20 km s^{-1}. As noted in Sec. I, such comets passing through the planetary region should be readily recognizable by their extreme hyperbolic orbits. As noted also, no such comets have ever been observed. Sekanina (1976) used this fact to place an upper limit of 6×10^{-4} $M_\oplus$ pc^{-3} on the space density of comets in the solar neighborhood. This is approximately 200 $M_\oplus$ pc^{-3}, consistent with the estimates of the total ejected material by Fernandez, but not with some of Safronov's higher figures. However, since it is likely that not all stars produce planetary systems and Oort Clouds, the ejected material would be spread over a wider number of stars, possibly bringing even Safronov's numbers within the limit set by Sekanina's work.

Further study of stellar perturbations will contribute to an improved understanding of the dynamics of the Oort Cloud and the possible implications for various theories of solar system formation. Improvement in the statistics for observed long-period comets, along with ability to detect more comets, particularly at large perihelion distances, will also serve to bound the problem. If comets are indeed the best obtainable source of original solar nebula material, then it is essential that as complete an understanding of their dynamical history as possible be developed so as to interpret the cosmochemical record in cometary nuclei.

Acknowledgments. It is a pleasure to thank R. Carlson for motivating the initiation of this study and for many helpful discussions in the course of the work. I also wish to thank R. Nelson, W. Smythe, R. Carlson, and R. Greenberg for their reviews of an earlier draft of this chapter. This work was supported by the Planetary Geophysics and Geochemistry Program and was carried out at the Jet Propulsion Laboratory under contract with the National Aeronautics and Space Administration.

REFERENCES

Cameron, A.G.W. 1973. Accumulation processes in the primitive solar nebula. *Icarus* 18:407-450.

Chebotarev, G.A. 1965. On the dynamical limits of the solar system. *Soviet Astron. J.* 8:787-792.

Chebotarev, G.A. 1966. Cometary motion in the outer solar system. *Soviet Astron. J.* 10:341-344.

Everhart, E. 1967. Intrinsic distributions of cometary perihelia and magnitudes. *Astron. J.* 72:1002-1011.

Everhart, E. 1968. Change in total energy of comets passing through the solar system. *Astron. J.* 73:1039-1052.

Faintich, M.D. 1971. Interstellar Gravitational Perturbations of Cometary Orbits. Ph.D. Dissertation, Univ. of Illinois.

Fernandez, J.A. 1978. Mass removed by the outer planets in the early solar system. *Icarus* 34:173-181.

Fernandez, J.A. 1980. Evolution of comet orbits under the perturbing influence of the giant planets and nearby stars. *Icarus* 42:406-421.

Kendall, D.G. 1961. Some problems in the theory of comets, I and II, *Proc. Fourth Berkeley Symposium on Mathematical Statistics and Probability,* Univ. of California Press 3:99-148.

Kuiper, G.P. 1951. Origin of the solar system. In *Astrophysics,* ed. J.A. Hynek, (New York:, McGraw Hill Book Co.), pp. 357-424.

Marsden, B.G.; Sekanina, Z.; and Everhart, E. 1978. New osculating orbits for 110 comets and analysis of original orbits for 200 comets. *Astron. J.* 83:64-71.

Oort, J.H. 1950. The structure of the cometary cloud surrounding the solar system and a hypothesis concerning its origin. *Bull. Astron. Neth.* 11:91-110.

Oort, J.H., and Schmidt, M. 1951. Differences between new and old comets. *Bull. Astron. Neth.* 12:259-269.

Opik, E.J. 1932. Note on stellar perturbations of nearly parabolic orbits. *Proc. Amer. Acad. Arts and Sciences* 67:169-183.

Richman, H. 1976. Stellar perturbations of orbits of long-period comets and their significance for cometary caputre. *Bull. Astron. Czech.* 27:92-105.

Russell, H.N. 1935. *The Solar System and Its Origin.* (New York: MacMillan Co.).

Safronov, V.S. 1972. *Evolution of the Protoplanetary Cloud and Formation of the Earth and the Planets.* NASA TT F-677.

Schreur, B. 1979. On the Production of Long-Period Comets by Stellar Perturbations of the Oort Cloud. Ph.D. Dissertation, Florida State Univ.

Sekanina, Z. 1968a. On the perturbation of comets by near-by stars: Sphere of action of the solar system. *Bull. Astron. Czech.* 19:223-229.

Sekanina, Z. 1968b. On the perturbation of comets by near-by stars: Encounters of comets with fast moving stars. *Bull. Astron. Czech.* 19:291-301.

Sekanina, Z. 1976. A probability of encounter with interstellar comets and the likelihood of their existence. *Icarus* 27:123-134.

Weissman, P.R. 1977. Initial energy and perihelion distributions of Oort Cloud comets. In *Comets, Asteroids, Meteorites,* ed. A.H. Delsemme, (Toledo, Ohio: Univ. of Toledo), pp. 87-91.

Weissman, P.R. 1979. Physical and dynamical evolution of long-period comets. In *Dynamics of the Solar System,* ed. R.L. Duncombe, (Dordrecht, Holland: D. Reidel Publ. Co.), pp. 277-282.

Weissman, P.R. 1980. Stellar perturbations of the cometary cloud. *Nature* 288:242-243.

Whipple, F.L. 1962. On the distribution of semimajor axes among comet orbits. *Astron. J.* 67:1-9.

Yabushita, S. 1972. Stellar perturbations of orbits of long-period comets. *Astron. Astrophys.* 16:395-403.

EVOLUTION OF LONG- AND SHORT-PERIOD ORBITS

Edgar Everhart
University of Denver

The orbits of short-period comets belong to the class of chaotic orbits. This type, always of inclination less than $35°$, includes near-parabolic orbits with perihelia near Jupiter's orbit or beyond, near-circular orbits at Jupiter or Saturn's distance, some unstable Trojan and horseshoe orbits, temporary satellite captures by Jupiter and Saturn, and the orbits of visible short-period comets. There are frequent changes from one of these forms to another. The long-period orbits, meaning both long-period and intermediate-period orbits of any inclination with small enough perihelia for the comets to be visible, do not interact with the chaotic orbits. We trace the separate evolution of these two types of orbits from the Oort Cloud via stellar and planetary perturbations.

This chapter is intended as a framework for understanding the evolution of comet orbits. The most widely accepted place or origin for comets is the Oort Cloud. This is located in a vast region surrounding the Sun and extending out to 50,000 AU, 1/5 the distance to alpha Centauri, at present the nearest star to the Sun. Most comets orbiting within this region have perihelia greater than 30 AU (Neptune's distance) and thus do not enter the planetary region; these objects would suffer no dissipation over billions of years. According to Oort (1950), passing stars perturb the perihelion distances of these comets-to-be causing them to enter the planetary region with a small perihelion distance as a new comet on its first apparition.

Oort's description now has wide support. In their study of the original orbits of some 200 comets Marsden et al. (1978) find a concentration of reciprocal semimajor axes $1/a \sim 25 \times 10^{-6}$ AU^{-1} as would be expected from new comets entering in such a way from Oort's cloud. Early work on stellar perturbations on comets by Öpik (1932), and Oort (1950), and more recent studies of increasing detail by Yabushita (1972), Everhart (1978), Weissman (this book), and Fernandez (1981) have given a fairly complete picture of the action of stars on comets. Thus, stellar perturbations acting on comets in the Oort Cloud can account for the influx of new comets. Comets at enormous distances within the cloud move more slowly than passing stars and so a single comet rounding its aphelion turn can be perturbed by several passing stars. Apparently gravitational impulses from passing stars do not ordinarily greatly change the heliocentric total energy of the comets but they do change their perihelion distance. The comets in the cloud whose perihelia are just beyond 30 AU are most likely to be perturbed to small perihelia. It is quite possible for a single passing star to change the perihelion of a comet from 30 or 40 AU to 1 or 2 AU, whereupon it becomes a new long-period comet.

Oort's cloud is seen conceptually in Fig. 1. All comets in this region have slightly negative total energies and are weakly bound to the Sun. It is possible for a passing star to cause a comet to gain energy and be lost to interstellar space, but this rarely happens. Most comets leave the Oort Cloud by reduction of their perihelion distance by a stellar perturbation. In Fig. 1 the dashed arrows are transitions caused by stellar perturbations and the solid arrows are transitions by planetary perturbations. The statements about the evolution paths given here are largely based on my experience in numerically integrating hundreds of thousands of hypothetical comet orbits interacting with the solar systm. The detailed numerical experiments are described in Everhart (1969, 1972, 1973a,b, 1978). Figure 1 is a condensation of these experiments.

There are three channels for leaving the Oort Cloud, of which the most complicated is that on the left entitled "Chaotic Orbits." These comets start as new or near-parabolic with small inclinations and perihelia of between 5 and 30 AU. Such comets are ordinarily invisible by reason of their large perihelion distances and suffer no solar dissipation for the same reason. About half of these receive a positive energy perturbation by a major planet and leave the solar system as interstellar comets after their first apparition, never to return. The other half of these new comets lose energy; their aphelia decrease and their periods become shorter. Gravitational energy perturbations by the planets change the total energy of the comets, but (except for the short-period orbits) do not change their perihelia. The comets which happen to lose energy enter complicated orbits. These unstable orbits exhibit a wild variety of temporary forms as seen in my numerical integrations (Everhart 1973a,b). These include unstable horseshoe and Trojan orbits associated with Jupiter or Saturn (some Jupiter Trojans are stable), new orbit

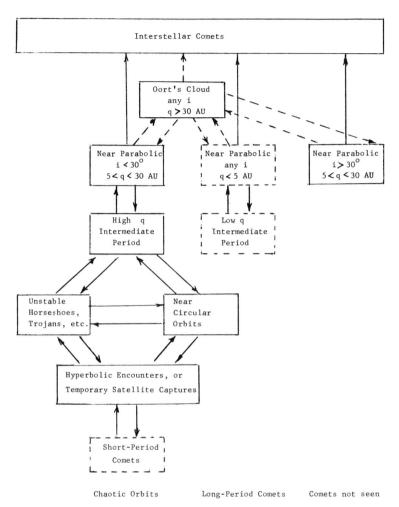

Fig. 1. The evolution of comets. Solid arrows indicate transitions caused by planetary perturbations and dashed arrows refer to the effects of stellar perturbations. A dashed box around an orbit form indicates that dissipation of comets by the Sun's heat is important for that orbit.

forms called generalized horseshoe and Trojan orbits, near-circular or elliptical orbits of large size (such as the orbit of P/Schwassmann-Wachmann 1 and that of Chiron), temporary satellite-capture orbits, and others, as well as the orbits of the visible short-period comets. All but this last form are at perihelion distances of ~ 5 AU or more, protected from the Sun's heat and for the most part invisible.

A comet in a near-circular orbit at Jupiter's distance can become entangled in a temporary satellite-capture orbit around Jupiter, and then emerge in an orbit like that of the visible short-period comets. The overall pattern is found in Everhart (1972); the temporary satellite-capture portions are described in Everhart (1973a; see especially Sec. VIII and Fig. 10) and Carusi and Valsecchi (1981). This latter paper shows that of 22 short-period comets studied, seven had undergone temporary satellite capture within the past 100 yr. One of these particular captures had been found earlier by Kazimirchak-Polonskaya and Shaporev (1976), who have worked extensively on such problems. In fact, many short-period comets are discovered immediately after entanglements with Jupiter.

The region of the short-period comets is shown dashed on Fig. 1 to indicate a leaky box because of solar dissipation. Short-period comets can dissipate entirely by vaporization or lose all their volatiles and become dead comets like Hidalgo, or can be perturbed in a random-walk process in the direction whence they came. Most that do not dissipate completely will eventually be lost to interstellar space. A very few will be perturbed to large aphelia, then a passing star may increase their perihelia so that they reenter the Oort Cloud.

The region of near-parabolic comets from the Oort Cloud of any inclination and low perihelia (center box, Fig. 1) is shown dashed because these are the visible long-period comets, subject to dissipation at each perihelion passage by the Sun. Very few return to the Oort Cloud. On the first passage planetary perturbations send half out to interstellar space and bring half to orbits of intermediate period. The evolution is slow; the comets dissipate entirely or are ultimately thrown out on hyperbolic orbits, so they do not have time to work their way to orbits of very short periods.

It is an old idea, perhaps dating back to Laplace and certainly treated by Tisserand (1889) and Callandreau (1892), that short-period comets are created when long-period comets pass close to Jupiter and drastically lose energy. Indeed Tisserand and Callandreau illustrate the possibility of this event. However, Newton (1893) and Everhart (1969) show that such events are too rare to account for the short-period comets. One quarter of all short-period comets would be in retrograde orbits under this mechanism, a result quite contrary to observations. There is no likely evolutionary channel connecting the visible long-period comets to the visible short-period comets.

The third channel for leaving Oort's cloud leads to the near-parabolic comets of large perihelia, 5-30 AU, and large inclination. These comets evolve showly and are never visible. Sooner or later almost all will be thrown to interstellar space on hyperbolic orbits by planetary perturbations; a very few will reenter the Oort Cloud.

Figure 1 and the foregoing discussion may suggest that the problems of comet evolution are solved. I should mention some of the unsolved areas.

1. It is my opinion, adopted in Fig. 1, that comets originate in the Oort Cloud. This place of origin is not conceded by all cosmologists or celestial

mechanicians. Some half dozen competing ideas are listed in my paper (Everhart 1973b).

2. There is some question whether the near-parabolic flux of new comets that feeds the chaotic orbit column in Fig. 1 is sufficient to account for the number of short-period comets observed. A paper by Joss (1973) answers "no" to this question and another by Delsemme (1973) answers "yes." My own work (Everhart 1977) enlarges the numbers of near-parabolic comets by showing how Neptune, Uranus, and Saturn all capture new comets and feed them to Jupiter's sphere of action, but still does not answer the quantitative question.

3. There seem to be too few observed long-period comets of perihelia greater than 2 AU in definitely elliptical orbits, in comparison to what a Monte Carlo model of planetary perturbations would predict. Practically all the observed large-q, long-period comets are new; too few are old with definitely elliptical orbits. This could be explained by assuming that comets, on their first appearance, are quite bright and likely to be discovered, but on their second appearance thousands of years later, they are much fainter and less likely to be discovered. My discussion of this (Everhart 1978) offers no satisfactory solution because the amount of fading required appears unreasonable. On the other hand Weissman (1978) apparently solves this problem with a rather complicated model of the dissipation effects.

Finally we should note that all channels to interstellar space are oneway. The idea that if comets exit to interstellar space, then steady-state equilibrium requires that just as many must reenter on the same channel, is answered in that steady-state has not been established. The 4.5 billion years since the solar system formed are not 1% of the time required to reach steady-state. There are too few comets to populate all the position-momentum boxes in interstellar space; we are not likely to see the lost comets thrown into space by the planetary systems of other stars and they are not likely to receive our lost comets. This conclusion regarding a non-steady-state distribution of comets in interstellar space is based primarily on the fact that not one comet has yet been observed whose orbit when it first approached the solar system was hyperbolic, although many comets leave our system on hyperbolic orbits (see Marsden et al. 1978). A truly hyperbolic comet may be discovered, but this will not change the conclusion that the gain-loss of comets from and to interstellar space is far from equilibrium.

Acknowledgment. I gratefully acknowledge support for my work by the National Science Foundation.

REFERENCES

Callandreau, O. 1892. Étude sur la théorie des comètes périodiques. *Ann. Obs. Paris Mém.* 20:B.1.-B.64.
Carusi, A., and Valsecchi, G.B. 1981. Temporary satellite captures by Jupiter. *Astron. Astrophys.* In press.
Delsemme, A.H. 1973. Origin of short-period comets. *Astron. Astrophys.* 29:377-381.
Everhart, E. 1969. Close encounters of comets and planets. *Astron. J.* 74:735-750.
Everhart, E. 1972. The origin of short-period comets. *Astrophys. Lett.* 10:131-135.
Everhart, E. 1973a. Horseshoe and Trojan orbits associated with Jupiter and Saturn. *Astron. J.* 78:316-328.
Everhart, E. 1973b. Examination of several ideas of comet origins. *Astron. J.* 78:329-337.
Everhart, E. 1977. The evolution of comet orbits as perturbed by Uranus and Neptune. In *Comets, Asteroids, Meteorites*, ed. A.H. Delsemme, (Toledo, Ohio: Univ. of Toledo), pp. 99-104.
Everhart, E. 1978. The shortage of comets in elliptical orbits. In *Dynamics of the Solar System*, ed. R.L. Duncombe, (Dordrecht, Holland: D. Reidel Publ. Co.), pp. 273-276.
Fernandez, J.A. 1980. Evolution of comet orbits under the perturbing influence of the giant planets and nearby stars. *Icarus* 42:406-421.
Joss. P.C. 1973. On the origin of short-period comets. *Astron. Astrophys.* 25:271-273.
Kazimirchak-Polonskaya, E.I., and Shaporev, S.D. 1976. Orbital evolution of P/Lexell (1770I) and some regularities of cometary orbits' transformation in Jupiter's sphere of action. *Sov. Astron. J.* 53:1306-1314.
Marsden, B.G.; Sekanina, Z.; and Everhart, E. 1978. New osculating orbits for 110 comets and analysis of original orbits for 200 comets. *Astron. J.* 83:64-71.
Newton, H.A. 1893. On the capture of comets by planets, especially their capture by Jupiter. *Mem. Nat. Acad. Sciences* (Washington) Vol. VI, pp. 8-23.
Oort, J.H. 1950. The structure of the cometary cloud surrounding the solar system and a hypothesis concerning its origin. *B.A.N.* 11:91-110.
Öpik, E.J. 1932. Note on stellar perturbations of nearly parabolic orbits. *Proc. Amer. Acad. Arts Sci.* 67:171-183.
Tisserand, M.F. 1889. Sur la théorie de la capture des comètes périodiques. *Bull. Astronomique* 6:242-292.
Weissman, P.R. 1978. Physical and dynamic evolution of long-period comets. In *Dynamics of the Solar System*, ed. R.L. Duncombe, (Dordrecht, Holland: D. Reidel Publ. Co.), pp. 277-282.
Yabushita, S. 1972. Stellar perturbations of orbits of long-period comets. *Astron. Astrophys.* 16:395-403.

DO COMETS EVOLVE INTO ASTEROIDS?
EVIDENCE FROM PHYSICAL STUDIES

JOHAN DEGEWIJ
Jet Propulsion Laboratory

and

EDWARD F. TEDESCO
The University of Arizona

Physical observations of low-activity periodic comets and asteroids in distant and peculiar orbits are reviewed. In particular we discuss magnitude studies and reflection spectra at visible and infrared wavelengths. While it may never be possible to prove unambiguously that a cometary nucleus has been observed, the fainter a given comet is, the more likely it is that some light reflected from a solid surface will be observed. Studies of magnitudes obtained by Roemer suggest that at certain periods the nuclei of P/Arend-Rigaux and P/Neujmin 1 may be photometrically resolved. We show that this may also be the case for P/Schwassmann-Wachmann 1, but that it is less likely for P/Tempel 2. At times of minimum activity, the optical spectra of P/Arend-Rigaux and P/Schwassmann-Wachmann 1 resemble those of RD-type asteroids like 944 Hidalgo. In order to progress with physical studies of cometary nuclei, an increased effort should be made to obtain magnitudes and other physical observations of comets during stages of minimal activity.

Numerous papers have been written concerning the possible evolution of comets into asteroids (e.g., Kresák 1973,1979; Marsden 1970,1971). In the past the emphasis has been on studies involving brightness, morphology, and dynamical properties. Following a review of these studies we will discuss photometric and spectrophotometric observations of distant comets of low activity. Further we will address the assumption that, at large heliocentric distances, a substantial fraction of a comet's observed brightness may be due to light reflected from the solid surface of the nucleus. Finally we will compare the observed physical properties of comets with the same data for asteroids and evaluate the evidence concerning their possible interrelation.

I. THE DYNAMICAL EVIDENCE

In this section we summarize major results from dynamical studies which bear on the question of whether the precursors for some asteroids and satellites were comets. We will restrict ourselves to the short-period comets ($P < 200$ yr) as possible precursors. The dynamical history of short-period comets is discussed by Everhart in this book. The reviews summarized below will guide the reader to more detailed treatments.

Marsden (1970,1971,1972) reviewed the important parameters for interpreting dynamical results, namely the size of the nongravitational effects and the closest-approach distance to Jupiter. Most periodic comets deviate from their predicted position (computed assuming gravitational forces alone are responsible for their orbital motion) due to the action of a jet force caused by sublimation of surface volatiles. The magnitude of the nongravitational effect is thus related to the level of gas and dust production in relation to the total mass of the nucleus as well as to the position of the comet's rotational angular momentum vector and its precession, if any (Sekanina 1981a). Marsden noted that some comets have no detectable nongravitational effect, which he interpreted as indicating low outgassing activity or a large nuclear mass. Marsden also emphasized the importance of the closest-approach distance to Jupiter. All but two short-period comets have approached Jupiter to within 0.9 AU during the past 200 yr (Marsden 1971). The numbered asteroids, on the other hand, do not come within 1.1 AU of Jupiter and their orbits are stable for at least the next 10^4 yr. 944 Hidalgo and 2060 Chiron are the only known exceptions. Hidalgo passed within 0.4 AU of Jupiter in 1673 while Chiron is in a chaotic orbit such that its past and future orbital evolution cannot be accurately determined (Everhart 1979). Marsden noticed that the two comets which avoid close approaches to Jupiter, P/Arend-Rigaux and P/Neujmin 1, also displayed the lowest activity among the comets in his sample.

A qualitative statistical approach for distinguishing between the orbits of comets and asteroids was used by Kresák (1980). Kresák used the minimum approach distance to Jupiter, the aphelion distance Q, and the Tisserand

DO COMETS EVOLVE INTO ASTEROIDS?

invariant T. T, while strictly speaking not an invariant during the encounter with Jupiter (cf., Everhart 1976), is nevertheless approximately constant before and after such encounters and thus serves as a powerful discriminant for the type of orbit (cf., Everhart 1976 and discussion following it). For objects with inclinations of zero deg, curves of constant T_0 (T for $i = 0$) and perihelion distance q may be plotted in a graph of semimajor axis a versus eccentricity e (Fig. 1). This figure, taken from Kresák (1979) shows a and e for all then known short-period comets and asteroids. Kresák emphasized the importance of the $T_0 = 3$ curve (running from upper left to lower right) in distinguishing between cometary and asteroidal type orbits. Objects on this curve have, at the sphere of influence of Jupiter, an asymptotic encounter velocity, relative to Jupiter, equal to zero. He argued that approaches to Jupiter are possible only if $T_0 \leqslant 3$, and even then, only if stable librations do not prevent it.

The lettered areas in the figure show Kresák's classification of different orbital and physical configurations and demonstrates the importance of the $T_0 = 3$ borderline dividing nearly all active from extinct objects. We caution against an overinterpretation of this figure. It shows only two of Kresák's parameters (T_0 and Q), omitting the minimum distance of approach to Jupiter.

Of interest are Kresák's discussions on which comets are potential candidates for evolving into extinct objects in asteroid-like orbits. Dynamical studies of the comets of the quasi-Hilda type like P/Oterma, P/Gehrels 3 and P/Smirnova-Chernykh, show that a temporary capture into this type of orbit is relatively frequent. However, a transition into a stable Hilda-type orbit would require nongravitational forces during a long period without encounters with Jupiter. Kresák argued further that comets P/Arend-Rigaux and P/Neujmin 1 are in stable orbits avoiding approaches with Jupiter, but that their low values of T will prevent them from evolving into typical asteroid orbits. He concluded that the best candidate for such an evolution is P/Encke, and argued that after complete extinction, this comet may become indistinguishable from an Apollo asteroid.

Kresák offers little concerning the opposite question, can we recognize possible ex-comets among the asteroids. There is of course one outstanding case: 944 Hidalgo (Marsden 1970,1971); but the known asteroids in the main belt are in very stable orbits, most probably dating back to the origin of the solar system. The Earth-approaching asteroids (Apollo, Amor, and Aten objects) do not satisfy Kresák's assumptions, because the gravitational influence of the inner planets makes use of the Tisserand criterion with respect to Jupiter invalid. The dynamical arguments for a certain fraction of the Earth-approaching asteroids being extinct cometary nuclei are given elsewhere; the most recent reviews are those by Shoemaker et al. (1979) and Wasson and Wetherill (1979). The evidence from physical studies is given in Sec. II.E.

Recent studies on the possibility of capture by Jupiter were made by

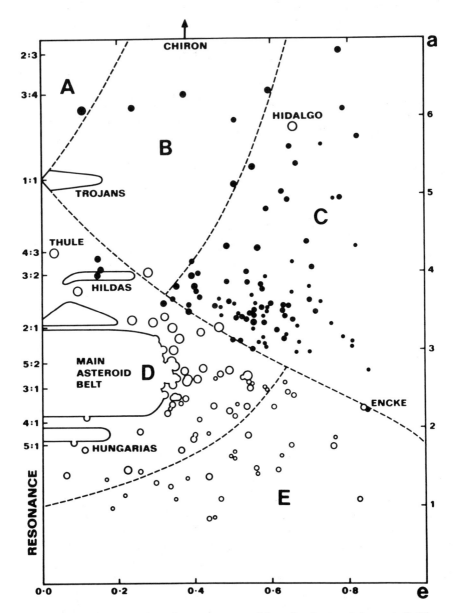

Fig. 1. A plot of semimajor axis a versus eccentricity e for short-period comets (solid circles) and asteroids (open circles). A is the transjovian region, B Jupiter's domain of weak cometary activity, C Jupiter's domain of strong cometary activity, D the minor planet region, and E the Apollo region. The sizes of the circles are related to the absolute magnitudes. With the assumption of a geometric albedo of 0.10, diameters range between ~ 1 km (smallest circles) and >30 km (largest circles). (From Kresák 1979.)

DO COMETS EVOLVE INTO ASTEROIDS? 669

Everhart (1973a,b), Kazimirchak-Polonskaya and Shaporev (1976), and Carusi and Valsecchi (1979,1981). Carusi and Valsecchi (1981) found that of 22 short-period comets, 7 had been short-lived satellites of Jupiter for part of the past 200 yr. For example, they found that P/Gehrels 3 was captured three times. The third time it stayed in an elliptical orbit around Jupiter for 7 yr. Everhart (1979), however, excludes the possibility of a permanent capture without the action of nongravitational forces. Several physical models to provide such nongravitational forces are discussed by Pollack et al. (1979), with emphasis on the gas drag experienced during passage through a primordial circumplanetary nebula.

II. PHYSICAL STUDIES

In this section we discuss physical observations of low-activity comets. The emphasis will be on short-period comets because they are the most likely source objects for some asteroids (see Sec. I). We will focus our discussion on physical observations of faint ($V \gtrsim 16$ mag) and distant ($r \gtrsim 2$ AU) short-period comets. These objects are more likely to be relatively quiescent, so that the assumption that a substantial fraction of the light is reflected from a solid surface has the highest probability of being correct. We will concentrate on studies in continuum spectral windows, i.e., those spectral regions unaffected, or only slightly affected, by gaseous emissions.

First we summarize the magnitude estimates and morphology obtained from direct photography. Next we summarize narrow and broadband spectrophotometry at visual and infrared wavelengths for comets displaying relatively little activity. We then discuss rotational properties deduced from variations in morphology and compare the results with those obtained for the smaller asteroids and finally we examine the efficacy of using thermal radiometry to determine albedos of cometary nuclei.

A. Magnitudes and Morphology

The most extensive account of the behavior of brightness and morphology of faint and distant comets is given in numerous papers by Roemer. Reports can be found in the *Publications of the Astronomical Society of the Pacific* between 1955 and 1971. A series of astrometric papers in the *Astronomical Journal* (Jeffers et al. 1954; Jeffers and Roemer 1955; Roemer 1956,1965; Roemer and Lloyd 1966; Roemer et al. 1966) also include reports on magnitudes and morphology. For 1972 and early 1973, monthly reports are given in *Mercury*. More observations are given in the *Quarterly Journal of the Royal Astronomical Society*. The last reported year is 1975.

Estimates of the diameters of periodic comets given by Roemer (1966), using the faintest observed magnitudes, together with a wide range of estimated albedos (p_V = 0.02–0.70), range between ∼ 0.5 and 20 km. The only exception is P/Schwassmann-Wachmann 1 that (if its albedo is 0.02) has a diameter of 75 km, and thus appears to be the largest comet known. Additional evidence for P/Schwassmann-Wachmann being a large object comes from the fact that, in spite of its frequent outbursts, its orbit can be fit without the introduction of nongravitational terms (Herget 1968).

It is remarkable that the morphology of several comets reported by Roemer, changes appreciably on relatively short time scales. Without a significant change in a comet's distance from the Sun, its appearance can change from "condensed" to "stellar" or "diffuse" in weeks and sometimes even in days. It is not possible to distinguish between gas and dust comae because often blue sensitive 103a-O plates were used without filters.

A table of magnitudes for short-period comets observed at large distances was compiled by Kresák (1973). He used Roemer's more recent reports, IAU circulars, and other sources to compile a well-documented table with all parameters related to observations of 54 short-period comets at extreme distances. For 35 comets he used magnitudes obtained by Roemer. We can expect that many comets in this table were still active at large distances, and that therefore in most cases Kresák's magnitude estimates are lower limits. It will be important to update this table.

Kresák used only one observation at the maximum heliocentric distance at which the comet was observed to estimate its absolute magnitude. Sekanina (1976a), in his review of magnitude studies of selected comets, emphasized that this is a weak basis for inferring sizes and models for cometary nuclei.

While all comets display some activity, i.e., comae, gas and/or dust tails, outbursts, etc., some comets show very little activity. Sekanina (1976a) plotted the photographic nuclear magnitudes obtained by Roemer, Barnard, and van Biesbroeck, for P/Arend-Rigaux and P/Neujmin 1, at unit distance to Earth and Sun. The measurements follow an asteroidal phase relation within the expected errors (see Figs. 2 and 3). Sekanina concluded that these comets have optically thin comae and very probably do show their solid nuclear surfaces (in Sekanina's words: "... the nuclei are photometrically resolved"). He also suggested that P/Encke may eventually become as inactive as P/Arend-Rigaux and P/Neujmin 1. In Sec. I we summarized the dynamical arguments which led Marsden to conclude that both comets were nearly devolatilized.

However, Sekanina reports on a frustrating lesson learned from the observations by Roemer of the "new" (i.e., $1/a_{\text{orig}} < 0.0001$ AU^{-1}) Comet Kohoutek 1973 XII. Three observations before perihelion, with r ranging between 2.1 and 4.4 AU, refer to a nearly stellar condensation with a constant (within 0.3 mag) absolute magnitude, while a year later after perihelion passage, the absolute magnitude was suddenly 3 mag fainter. Sekanina explains this by assuming the presence of a circumnuclear cloud of debris that disappeared after perihelion passage; however, it is also possible that the

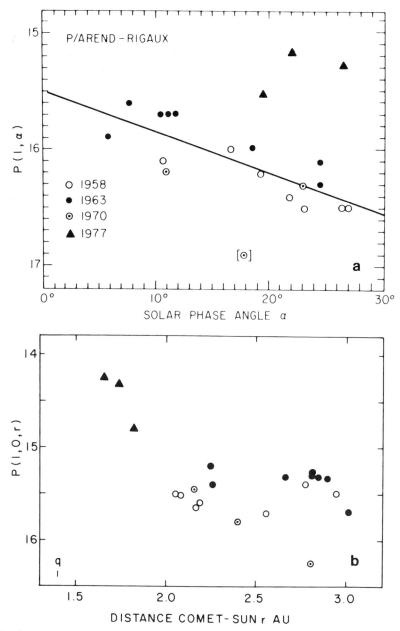

Fig. 2. (a) Reduced photographic nuclear magnitudes, $P(1,\alpha)$ versus phase angle for P/Arend-Rigaux. The magnitudes in 1958, 1963, and 1970 are from Roemer and coworkers (for references see text). The magnitudes in 1977 were measured by Degewij and Gradie (see text). The straight line is a least-squares fit by Sekanina, excluding the bracketed and 1977 observations. (Adapted from a figure by Sekanina, 1976a.) (b) P magnitude reduced to unit distance from Earth and Sun and zero degree phase angle (using a phase coefficient of 0.039 mag deg^{-1}) versus heliocentric distance.

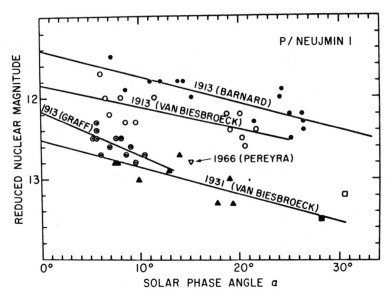

Fig. 3. Reduced magnitudes versus phase angle, for P/Neujmin 1. Visual magnitude estimates obtained in 1913 are combined with photographic magnitudes obtained in 1931 and 1966. The straight lines are least-squares fits by Sekanina. Systematic differences in the magnitudes are attributable to differences in the effective wavelengths and the telescopes used. (Adapted from a figure by Sekanina 1976a.)

surface could have been covered with large areas of high-albedo material, which the comet lost after perihelion passage, exposing a darker underlying surface.

It is possible to estimate the near-surface albedo by using plausible physical models for the vaporization mechanism (Delsemme and Rud 1973; Sekanina 1976a). However, the real magnitude of the nucleus remains an essential input parameter. Delsemme and Rud assumed that water vaporization controls the production rate of gas, and they computed Bond albedos for comets Tago-Sato-Kosaka 1969 IX and Bennett 1970 II of 0.63 and 0.66, respectively. They used Roemer's nuclear magnitudes obtained at r = 2 to 2.5 AU for Comet Tago-Sato-Kosaka, and r = 3 to 4.5 AU for Comet Bennett. Sekanina (1976a) argued that Delsemme and Rud may have over-estimated the brightness of the cometary nucleus. The high albedo is incompatible with the large quantity of dust observed in Comet Bennett. Sekanina estimated that the actual magnitude of the nuclei of both comets was probably 2 to 3 mag fainter than the observed "nuclear" magnitudes.

A completely independent estimate of this magnitude bias can be obtained from the study of size distributions of impact craters on the Earth and Galilean satellites (Shoemaker and Wolfe 1982). These authors attempted to identify the source of the impacting bodies and their relative numbers. They used Roemer's faintest nuclear magnitudes and derived cumulative frequency distributions of absolute nuclear mag for 42 short-period comets and 23 long-period comets. The mag distribution for the short-period comets is exponential to absolute photographic mag 16, and for the long-period comets to mag 13. With the assumption of a blue albedo of 0.03 the corresponding nuclear diameters are 6.4 km and 25 km. Shoemaker and Wolfe concluded that the slope of the short-period comet nuclear diameter distribution is consistent with the inferred diameter distribution of the impacting bodies that formed the bright ray craters on Ganymede.

At Jupiter, Shoemaker and Wolfe estimate that 10 to 30% of the present cratering production rate is due to long-period comets, 50% to extinct comets, and the remaining part due to active short-period comets. They conclude that relatively few short-period comets impact the Earth: roughly one third of the present crater production is due to long-period comets and the other two thirds to Earth-crossing asteroids. They give several arguments in favor of the low albedo they adopted and conclude that contamination by an unresolved coma makes long-period comets on the average appear too bright by ~ 2.3 mag. This is in agreement with the conclusion reached by Sekanina (1976a).

It is clear that magnitude studies are the first essential step in the process of acquiring direct physical information on cometary nuclei. If the magnitude behavior during several apparitions is well documented using a high-precision, repeatable technique (e.g., photoelectric photometry or two-dimensional detectors using suitably chosen filters), then a magnitude measurement may serve as an indicator of the state of activity.

The discussion below is based on a combination of Roemer's photographic nuclear magnitudes with more recent photoelectric measurements. We assume that it is possible to combine the two data sets when the images are stellar in appearance; it is clear that the two data sets are incompatible as soon as a bright coma is apparent. The relation $B = P + 0.1$ (Gehrels 1970) was used to transform the photographic magnitudes to the Johnson *UBV* system and we assume $B - V = 0.64$ to obtain V magnitudes. Roemer (1976) used three different telescopes to obtain her extensive data set; her published observations were obtained with: (a) the 102-cm f/6.8 reflector of the U.S. Naval Observatory in Flagstaff between 27 April 1957 and 1 November 1965; (b) 154-cm f/13.5 Catalina reflector of the Lunar and Planetary Laboratory between 15 October 1966 and 25 December 1976; and (c) the 229-cm f/9 reflector of the Steward Observatory at Kitt Peak between 7 October 1969 and 27 December 1976.

Magnitude measurements of P/Arend-Rigaux during those times when its images appear stellar (Fig. 2) allow us to look for possible systematic differences among the magnitudes obtained with the three telescopes. It appears that the data agree reasonably well. There may be a systematic difference between the 102-cm data obtained in 1958 and 1963 but it is barely significant. During the 1977 apparition a weak coma was apparent. Photoelectric measurements were obtained by Degewij with the 154-cm Catalina telescope with an aperture of 15 arcsec on 20.35 October 1977 (V = 16.03 ± 0.11, $B-V$ = 0.76 ± 0.09, $U-B$ = 0.15 ± 0.08, α = 19°.1) and 17.23 November 1977 (V = 15.28 ± 0.03, $B-V$ = 0.83 ± 0.04, $U-B$ = 0.19 ± 0.09, α = 26°.5). Gradie observed the comet on 3.28 November 1977 with the 229-cm Steward reflector (V = 15.45 ± 0.02, $B-V$ = 0.74 ± 0.04, $U-B$ = 0.26 ± 0.01, α = 22°.2). These photoelectric data are also plotted in Fig. 2 but cannot really be compared with Roemer's data because of the apparent coma. However, the fact that the measurements of the comet obtained at small heliocentric distances when it displayed a coma (Fig. 2b) are brighter than those obtained at larger heliocentric distances when no coma was observed, is consistent with the idea that this comet is less active, and perhaps even inactive, at large heliocentric distances.

P/Tempel 2 is more active than P/Arend-Rigaux. Its reduced magnitudes vary by $\sim$ 1 mag. Roemer observed it during several apparitions between 1961 and 1972. A campaign organized by Newburn at the Jet Propulsion Laboratory, provided a series of photoelectric measurements obtained in late 1978 and early 1979; for a summary see Barker et al. (1982). All reduced magnitudes are plotted in Fig. 4, where we have labeled those data points for which a coma was observed. A puzzling aspect of this figure is that the coma was present during the faintest observations by Roemer in 1962, yet no coma was reported during the brighter observations in later years. Note also the absence of a clear faint plateau consisting of data points without reported coma. It is therefore unlikely that any of these observations represents an observation of the bare nucleus of P/Tempel 2. Barker et al. could not find it on 26 March 1979 UT (r = 3.46 AU, α = 15°.3); their limit of V = 22 corresponds to $V(1,0)$ = 16.5. An outburst (without apparent coma) at 3.2 AU is clearly visible in Fig. 4.

P/Schwassmann-Wachmann 1 has been observed extensively by Roemer between 1957 and 1975. The early results are discussed by Roemer (1958,1962). Her published magnitudes, reduced to unit distance from Earth and Sun, are plotted in Fig. 5a. There is no clear correlation between reduced magnitude and activity. There is evidence for a faint plateau, marked with a dashed line having the shape of the mean opposition effect observed for asteroids (Gehrels and Tedesco 1979). This plateau is $\sim$ 1 to 2 mag brighter than the limit for the telescopes Roemer has used.

There may be a slight correlation between coma activity and heliocentric distance. There exist about equal numbers of observations of P/Schwassmann-Wachmann 1 at distances on either side of 6.4 AU (Fig. 5b). 27 of 44 observations, with $\bar{r} \sim$ 5.9 AU, displayed significant activity whereas only 16 of 40 observations, with $\bar{r} \sim$ 6.8 AU, did so. Hence the comet appears to be relatively less active at larger heliocentric distances. It is, however, possible that this is a selection effect in that a given level of activity is more difficult

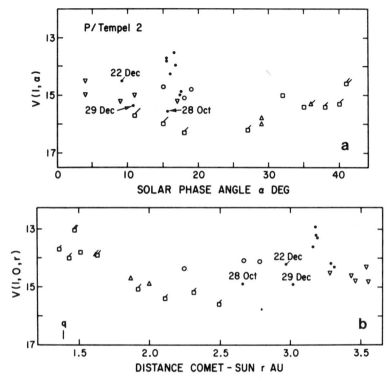

Fig. 4. Brightness behavior of P/Tempel 2 constructed from photographic observations by Roemer in 1961 (▽), 1962 (□), 1967 (△), 1972 (○), and from photoelectric observations (●) by several teams (Barker et.al. 1982) between 28 October 1978 and 26 March 1979. The appearance of a coma is marked with ⌁ for a weak coma; ⌁ for a brighter coma. (a) V magnitude reduced to unit distance from Earth and Sun (assuming an inverse-square law for the variation of brightness with distance) versus solar phase angle. (b) V magnitude reduced to unit distance from Earth and Sun and zero degree phase angle (using a phase coefficient of 0.039 mag deg^{-1}), versus heliocentric distance. The uncertainties in the photoelectric measurements are on the order of 0.1 mag (Barker et al. 1981), and ∼ 0.2 to 0.3 mag for the photographic measurements (Roemer, personal communication).

to recognize at larger distances. Cochran et al. (1980) and Larson (1980) detected CO$^+$ emission lines in the spectrum of P/Schwassmann-Wachmann 1 both during outbursts and near quiescence. The presence and strength of these lines may be a sensitive parameter for establishing the absence of outgassing activity (cf. Tholen et al. 1981).

B. Visual and Infrared Spectrophotometry

There are very few reports of accurate spectrophotometric measurements of comets using photoelectric techniques (see A'Hearn's chapter in this book;

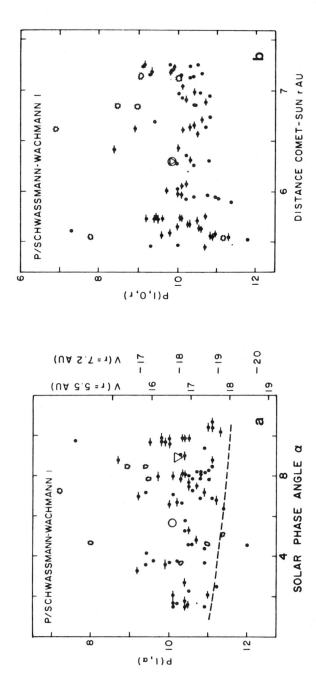

Fig. 5. (a) Reduced photographic nuclear magnitudes, at unit distance from Earth and Sun, versus solar phase angle for P/Schwassmann-Wachmann 1. The magnitudes are from observations made by Roemer and coworkers between 1957 and 1975 (see text for references). Also computed are apparent V magnitudes at opposition. Symbols represent various degrees of activity; ◆: bright and stellar, or very bright coma; ✪: obvious coma; ●: coma very faint or undetected. The lower envelope of the magnitude distribution is marked with a dashed line, which is the mean opposition effect observed for asteroids (Gehrels and Tedesco 1979). The photometric uncertainty is on the order of 0.2 to 0.3 mag (Roemer, personal communication). The symbols ○ (Tholen et al. 1981) and ▽ (Hartmann et al. 1982) refer to photoelectric measurements (see Figs. 7 and 13). (b) Reduced photographic nuclear magnitudes at zero phase angle, versus heliocentric distance for P/Schwassmann-Wachmann 1. The reduced magnitudes are corrected to zero phase angle using a phase coefficient of 0.039 mag deg^{-1}.

Hanner 1980). In this section we discuss the characteristics of comet and asteroid continuum spectra at visual and infrared wavelengths.

Gebel (1970) obtained neutral colors between 0.35 and 0.63 μm for the comae of comets Ikeya-Seki 1968 I and Honda 1968 VI. T.V. Johnson et al. (1971) observed an increase of 10 to 15% in reflectivity between 0.4 and 1.1μm for the coma of Comet Bennett 1970 II. The tails of the comets Arend-Roland 1957 III and Mrkos 1957 V display very red spectra (Liller 1960), the reflectivity at 0.60 μm being 2 to 3 times larger than that at 0.42 μm. It is evident that the comae of these bright comets shield the nucleus, and therefore only a negligible amount of light reflected from the nucleus will contribute to the continuum spectrum. For the remainder of this section we will restrict ourselves to the discussion of observations of low-activity comets, because, if a bare nucleus can ever be observed it will be for objects with negligible activity or for those which come very near the Earth.

At least two types of investigations can be made to assess the amount of contamination to the nuclear reflection spectrum due to the dust coma. The first consists in obtaining colors or spectrophotometry of the seeing disk condensation over a long period of time. This makes it possible to observe changes in the spectrum and brightness correlated with changing reduced magnitude, which in some cases is an indicator of the state of activity.

The second method consists in obtaining spatially resolved spectrophotometry of low-activity comets. Narrowband filters in combination with a panoramic detector may be used (Fig. 6) to distinguish the spectrum of the seeing disk condensation from that of an extended dust-and-gas coma. The spectrum of the coma can be extrapolated and subtracted from that of the seeing disk condensation.

The obvious limitation here is that the seeing disk condensation will in general not be equivalent to the bare nucleus. The best candidates for these types of studies have been discussed in Sec. II.A. P/Arend-Rigaux, P/Tempel 2, and P/Schwassmann-Wachmann 1 and, in addition, P/Neujmin 1 may at some time allow their solid surfaces to be seen through an optically thin coma. In the following we will discuss the appearance of the optical spectra observed for the first three of these objects.

P/Arend-Rigaux was observed by several teams in late 1977, when it was thought it might return as a stellar object without any activity (Sekanina 1976a; Weissman 1978). However, it showed a coma. Allen (1977) and Degewij and Tifft (see Fig. 6) obtained spectra showing the presence of emission bands attributable to CN(0,0), C_3, $C_2(1,0)$, and $C_2(0,0)$. Chapman obtained spectral reflectance measurements of the nuclear regions on 5, 6 and 7 November 1977 UT using a filter spectrophotometer on the 224-cm telescope on Mauna Kea. The average spectrum is given in Fig. 7, and was obtained and reduced with the same techniques and procedures described by Chapman and Gaffey (1979). The standard star used was 10 Tau. Chapman noticed by eye a prominent central condensation but also a prominent coma.

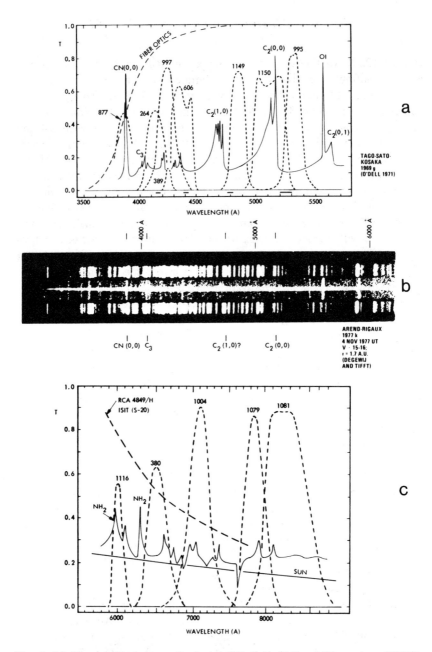

Fig. 6. (a) Transmission curves of selected Kitt Peak National Observatory (KPNO) filters are plotted together with the relative distribution (arbitrary scale) of Comet 1969 IX (O'Dell 1971) and the ultraviolet cutoff of the fiber optics in the KPNO RCA 4849/H ISIT video camera tube. The horizontal bars just above the wavelength scale are windows free from molecular emission lines (Arpigny 1972). (From Degewij 1981.) (b) Spectrum of faint ($V \sim 16$) P/Arend-Rigaux obtained on 4 Nov 1977 UT with the Cassegrain spectrograph behind the 228-cm telescope of the Steward Observatory. (From Degewij 1980.) (c) As for (a) but for the red part of the spectrum.

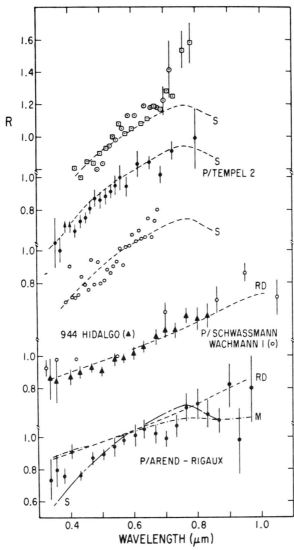

Fig. 7. Reflectance spectra of low-activity comets compared with typical asteroidal spectra (from Fig. 8). The spectra of P/Tempel 2 are from Spinrad et al. (1979): [⊡] 28 Oct 1978 UT (∼0.5 mag); [⊙] 29 Dec 1978 UT (∼0.5 mag); from P.E. Johnson et al. (1981): [●] 22.2 Dec 1978 UT (∼1 mag); and from Barker et al. (1982): [○] 30 Nov 1978 UT. Following each date is our estimate of how much brighter the comet was compared with its faintest observed (nuclear?) brightness at that distance (see Figs. 2, 4, and 5). The 8-color spectrum of P/Schwassmann-Wachmann 1, obtained on 8 Dec 1980 UT (∼1 mag), is from Tholen et al. (1981). The spectrum of P/Arend-Rigaux was obtained on 5, 6, and 7 Nov 1977 UT (∼1 mag) by C.R. Chapman (personal comm.), and the spectrum of 944 Hidalgo is from Chapman and Gaffey (1979). For 944 Hidalgo we omitted the data beyond 0.85 μm due to excessive noise.

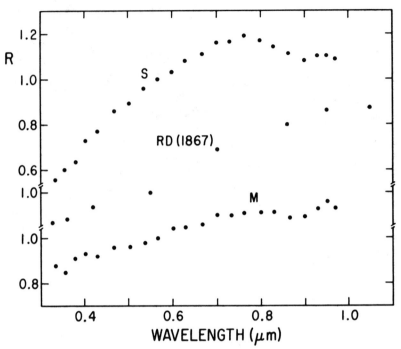

Fig. 8. Reflectance spectra of an average S-type and M-type asteroid. The spectrum of 1867 Deiphobus is representative of a typical RD-type asteroid (Tedesco and Gradie 1981). (From Fig. 1 of Chapman and Gaffey 1979.)

Gradie's photoelectric measurement on 3 November UT (▲ at 22°.2 phase angle in Fig. 2) is ~ 1 mag brighter than the measurements by Roemer. Chapman's spectrum shows a high reflectance value at 0.4 μm, probably caused by emission of CN(0,0) perhaps mixed with C_3. For whatever it is worth, however, this spectrum exhibits a rather linear trend throughout the observed wavelength region, indicating a slightly reddish color. The data are too noisy to distinguish unambiguously between an average M-, C-, or RD-type spectrum (see Figs. 7, 8, and 10). UV data appear to exclude S. A definition of the asteroid types is given by Zellner (1979); there is an error, however, in his table II. The limit on BEND for M types should be ≤ 0.06.

P/Tempel 2 was observed during its 1978-1979 apparition by at least four groups using various types of spectrophotometric techniques (Barker et al. 1982). The photometric behavior has been discussed in Sec. II.A. The spectra are given in Fig. 7 and show a relatively good match with an average S-type asteroid spectrum, except for the data beyond 0.7 μm obtained on 28 October 1978 UT by Spinrad et al. (1979). A study of the reduced magnitudes (Fig. 4), however, shows that the data are expected to be affected by a significant coma. The reduced magnitudes of the comet during the acquisition

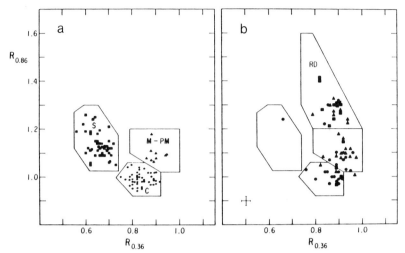

Fig. 9. 0.36 μm reflectivity versus 0.86 μm reflectivity for selected groups of asteroids. (Adapted from a figure in Tedesco and Gradie, 1981.) The reflectivities are relative to the Sun and normalized to 1 at 0.55 μm. Typical uncertainties, expressed as ±1 sigma error bars, are given in the lower left corner of 9b. Main-belt asteroids having unambiguous taxonomic classifications of C (●), S (■), or M (▲) as defined by Zellner (1979) and given in Bowell et al. (1979) are presented in Fig. 9a. This figure serves to define the C, S, and M domains shown in 9b and in Fig. 10. Results for outer-belt asteroids, i.e., Cybele (●), Hilda (▲), and Trojan (■) group members are displayed in 9b which defines the RD field. There is only one known RD type in the main-belt (Tedesco et al. 1981b).

of the spectra, are ∼ 1 mag brighter than the observations by Roemer, during the time when the coma was still visible. Therefore, we conclude that the spectra were affected by a significant coma, which means that Spinrad et al. did not observe the bare nucleus.

The last comet to be discussed is P/Schwassmann-Wachmann 1 (see below). Observations at both visual and infrared wavelengths are available for this comet during near quiescence and outburst. First, however, we will discuss observations of reflectances for bodies in and beyond the outer zones of the main asteroid belt.

Degewij et al. (1980) noted that the *UBV* colors of these objects are similar and cluster in the lower part of the C-field. Broadband (red and infrared) observations of several Trojan asteroids confirmed the discovery by McCord and Chapman (1975) of very high red reflectances for these objects, called RD-type (Degewij and van Houten 1979). Tedesco and Gradie (1981) showed that the S, C, M, and RD asteroidal surface types fall into relatively tight groups in a two-reflectance plot (Fig. 9a,b).

Hartmann et al. (1982) drew from the experience obtained with *JHK* infrared photometry in the detection of icy surfaces (T.V. Johnson and

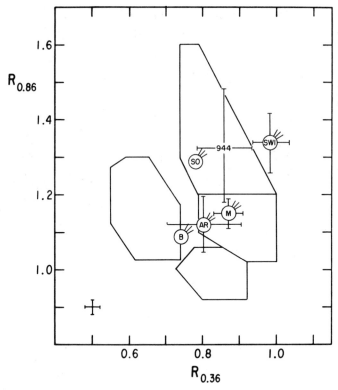

Fig. 10. 0.36 μm reflectivity versus 0.86 μm reflectivity for objects in chaotic orbits. 944 Hidalgo: (944) from Chapman and Gaffey (1979), P/Arend-Rigaux: (AR) from Chapman (personal communication), and P/Stephan-Oterma: (SO), P/Schwassmann-Wachmann 1: (SW), Comet Bowell 1980b: (B), and Comet Meier 1978 XXI: (M) from Tholen et al. (1981).

McCord 1971; Morrison et al. 1976) and rocky surfaces (T.V. Johnson et al. 1975; Chapman and Morrison 1976; Veeder et al. 1978). Because the *JHK* colors are located near major water ice absorptions (see the chapter by Fink and Sill), icy surfaces are located in the lower left of the diagram (Fig. 11). Rocky surfaces have neutral and reddish colors and are located in the upper right part of the diagram. There is a transition area in the middle of the diagram, where the Uranian satellites and S9 Phoebe are located (Cruikshank 1981; Degewij et al. 1980). Comets displaying various degrees of activity and asteroids are in the upper right part of the diagram, and clearly avoid the middle and lower left parts. A'Hearn et al. (1981) addressed this finding with the appropriate title: "Where is the ice in comets?". Hanner (1981), however, calculated that scattering from pure water ice grains with sizes on the order of several μm, will not show the water ice bands. She noted, in addition, that evaporation will cause the ice to disappear at larger distances from the comet

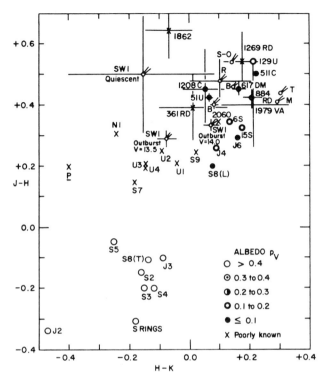

Fig. 11. *JHK* colors of comets, satellites, and asteroids from Hartmann et al. (1982). They point out that comets with varying degrees of coma activity lie within the field defined by objects known to have dark stony surfaces. Comets avoid the lower left part of the diagram occupied by bright icy surfaces. Asteroids are designated by number and spectral type, and satellites are designated by planet initials. The comets plotted are: Tuttle, Meier, Stephan-Oterma, Bowell, Reinmuth 1, and Schwassmann-Wachmann 1; their symbols carry no albedo connotation. The assumed solar colors used were $J-H = 0.29$ and $H-K = 0.06$ (Degewij et al. 1980).

thus masking the 2 μm ice absorption feature by scattering from dust grains.

Preliminary studies of *VJK* colors (Fig. 12) by Hartmann et al. (1982) indicate that the rocky area in the *JHK* diagram may split up into surface types C, S, and RD, with increasing $V-J$ color.

P/Schwassmann-Wachmann 1 is a recent target of active studies. It has been known (Mayall 1941; Jeffers 1946; Walker 1959) to display a solar continuum during outbursts. Cochran et al. (1980) discovered emission lines attributable to CO^+, and confirmed by Larson (1980). Tholen et al. (1981) obtained an 8-color spectrum during near quiescence (~ 1 mag above the faint plateau in Fig. 5a). This visual spectrum is slightly redder than that of 944 Hidalgo and an average RD-type asteroid like 1867 Deiphobus (Figs. 7 and 10). Hartmann et al. (1982) obtained nearly simultaneous *VJHK* pho-

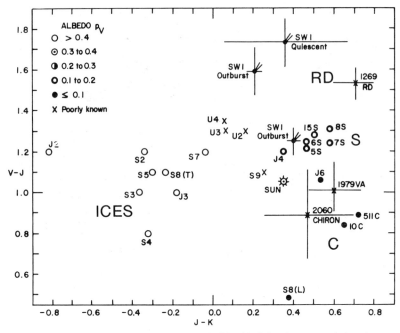

Fig. 12. *VJK* colors for some of the objects in Fig. 11. It has been noted that the stony field, in the upper right of Fig. 11, seems to split up into objects having surfaces of varying *V–J* redness. (From Hartmann et al. 1982.)

tometry. The V magnitude was also ~ 1 mag above the faint plateau (Fig. 5a). The visual and infrared data sets are combined in Fig. 13. Hartmann et al. also measured the RD-type asteroid 1269 Rollandia, and they note (Fig. 12) that both P/Schwassmann-Wachmann 1 (S W 1) and Rollandia are in the same, presumably RD-type, field in the *VJK* diagram. During an outburst it was found that the *JHK* colors of S W 1 on 2 consecutive days were bluer than those of other active comets; possibly a weak ice band may be present. It appears from these preliminary data that S W 1 may have an RD-type spectrum during near quiescence.

Cochran et al. (1981) obtained spatially resolved spectrophotometry of S W 1 during the outburst and near quiescence for a limited wavelength domain between 0.35 and 0.6 μm. By assuming Mie scattering they could model the spectrum obtained during quiescence with a distribution of relatively large (0.9 μm) particles having a small (0.01 μm) variance. As the outburst progressed, more of the smaller (0.20 μm) particles also having a small (0.02 μm) variance were required. They did not find evidence for the particles growing smaller as they moved outward from the nucleus.

The pertinent question, do distant comets ever become quiescent, was addressed by Sekanina (1981c) who noted that the appearance of the coma

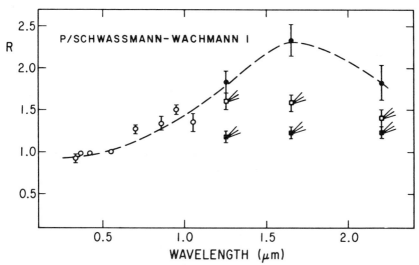

Fig. 13. Visual and infrared reflection spectra of P/Schwassmann-Wachmann 1 normalized at 0.55 μm, during outbursts on 5 and 7 Feb 1981 UT, □ and ■, respectively, and during near quiescence [$B(1,\alpha) = 10.4$, $\alpha = 8°.9$ on 8 Dec 1980 UT], ●. (From Hartmann et al. 1982.) Also plotted are measurements at visual wavelengths reported by Tholen et al. (1981) obtained on 30 March 1981 UT when $B(1,\alpha) = 10.2$, and $\alpha = 5°.7$ (○), i.e., when the reduced magnitudes were similar to those observed on 8 Dec 1980 (see Fig. 5).

of the "new" Comet Bowell (1980b), during the time it moved from 7.3 AU to 3.4 AU, continued to be the same. Sekanina speculated that "the particles observed in the coma of this new comet might be left-over pristine material that has never been in contact with the nucleus surface or a product of the comet's erratic activity associated with its chemical instability at temperatures below ~ 40 K that could be stimulated by cosmic ray and/or ultraviolet light irradiation of the surface layer during the comet's long stay in the Oort Cloud." Theoretical modeling is certainly needed to estimate the time it takes the particles to dissipate so that the comet behaves spectrophotometrically as an asteroid.

If comets have cores of nonvolatile material then, after they become devolatilized, what can we expect them to look like? The answer to this question depends on where the cometary cores condensed out of the solar nebula (Everhart 1973b) and on whether significant amounts of short-lived radionuclides (e.g., ^{26}Al) were incorporated into them early enough to cause at least some of them to melt. If all comets formed beyond, say, the orbit of Saturn, and if their interiors never melted, then it is quite possible that cometary cores do not exist. If, however, some comets, or at least their nonvolatile cores, either formed closer to the Sun or were subjected to radiogenic heating then such cores may exist (Wilkening 1979). It therefore seems interesting to compare the physical properties of asteroids, especially small and distant ones, with those of bare cometary nuclei. Since both classes of objects are faint, the most feasible types of observations to make are broadband spectrophotometry, *VJHK* photometry, and, where possible, thermal radiometry.

Figure 9 adopted from Tedesco and Gradie (1981), presents what are in effect two-color plots of various groups of objects. Figure 9a shows the locations in $R_{0.36}$, $R_{0.86}$ parameter space of the major main-belt asteroid types: the C-type, centered near 0.85, 1.0; the S-type, centered near 0.65, 1.15; and the M- and PM-types, centered near 0.92, 1.08. Figure 9b clearly shows the trend with increasing semimajor axis of the taxonomic types of outer-belt asteroids from C to PM to RD.

Note that the fraction of C-types decreases monotonically among the Cybele, Hilda, and Trojan groups, where only 56%, 34%, and 22%, respectively, lie in the C domain. The RD-type distribution is the mirror image of this; RD-types constitute 9% of the Cybele population, 36% of the Hilda, and 72% of the Trojan population. Tedesco and Gradie (1981) concluded that these trends in taxonomic types implied that each of these groups of asteroids formed over a restricted range of heliocentric distances.

Figure 10 shows that $R_{0.36}$ and $R_{0.86}$ for P/Arend-Rigaux (derived from Chapman's spectrum presented in Fig. 7) and comets Bowell 1980b, Meier 1978 XXI, P/Stephan-Oterma, and P/Schwassmann-Wachmann 1, (from Tholen et al. 1981) are similar to that of 944 Hidalgo and the outer-belt asteroids in general. Since all of these comets are known to have had comae at the time of the photometric observations, it is probable that the spectra are of light reflected from a dust coma consisting of irregularly shaped particles of ice and/or rock. In principle it may be possible, given spectra over a sufficiently wide range of phase angles and wavelengths, to model the size and shape distributions and refractive indices of the particles comprising these comae. In practice, however, this may prove impossible, if only for the reason that particle distribution of the dust coma is likely to vary with time.

Even if comets do not have monolithic cores it appears that most of their surfaces are covered by dust (Sekanina 1981b). Since asteroid surfaces are also dust covered, it is possible that the nonvolatile component of cometary dust is compositionally similar to that found on the surfaces of outer-belt asteroids. It is curious that the reflection spectra of cometary dust comae so closely resembles that of certain asteroids (Fig. 7; Tholen et al. 1981). If in the future we are certain that spectrophotometry of a bare nucleus has been obtained, comparison with asteroid spectra will be essential. Until then, little physical information regarding the compositions of cometary nuclei can be extracted from spectrophotometry of cometary comae, beyond studying in an empirical fashion the variation of such spectra from comet to comet and over a period of time for a given comet.

C. Rotation Periods and Shapes

Most authors agree that the asteroids represent a collisionally evolved population; for a review see Davis et al. (1979). Comets, on the other hand, spend essentially all of their lives in regions of space where collisions with particles significantly larger than dust grains are probably exceedingly rare. Nevertheless, cometary spin rates are expected to be altered by the selective evaporation of volatiles from their afternoon hemispheres when they are near the Sun (see Whipple's chapter in this book). It is therefore of interest to see how the spin rates of these two classes of objects compare.

Until very recently spin rates had been reported for only a handful of comets. Due primarily to Whipple and Sekanina that situation has changed dramatically in the past two years (chapter by Whipple). The median cometary rotation period M is found by Whipple to be 15.0 hr. Whipple compares this value to that for 41 asteroids, mostly Earth-approaching objects, with diameters ≤ 40 km for which he finds $M = 6.8$ hr. It has been noted, however, that the lightcurve sample for small asteroids, from which their rotation periods are determined, is biased by inclusion of data from photographic lightcurves and Earth-approaching asteroids (Tedesco and Zappalá 1980). We have therefore compared the cometary results with those for the 13 photoelectrically observed main-belt asteroids with diameters < 60 km. The median rotation period for this class is 10.5 hr. Clearly more lightcurves of small asteroids, from both the Earth-approaching and main-belt populations, are needed.

Shapes of asteroids can be inferred, at least crudely, from knowledge of their rotational lightcurve amplitudes while the shapes of cometary nuclei can be estimated from variations in the comet's nongravitational motions (Sekanina 1981b). This latter work is just beginning, however, and so no comparison between asteroid and comet body shapes is yet available. Likewise we feel that it is premature to conclude that cometary spin rates are significantly lower than those of asteroids of similar size.

D. Estimating Albedos from Thermal Radiometry

In general, radiometric observations of comets have been limited to studies of the 10 μm and 18 μm silicate emission features in bright comets as functions of activity and heliocentric distance (Chapter by Ney in this book). Recent work by Campins et al. (1981), and Campins and Hanner (see their chapter) has attempted to determine the albedos of the dust grains but the possible presence of icy grains mixed in with the silicate grains (Hanner 1980) complicates the interpretation.

If the range in albedo of the dust particles in periodic comets Encke, Stephan-Oterma, and Chernykh ($0.02 \leq A \leq 0.10$) determined by Campins et al. is accurate, then it appears that this dust is as dark as that found on most asteroid surfaces. Since reflection spectra of several periodic and nonperiodic comets have recently been demonstrated to bear a strong resemblance to those of dark asteroids located in the outer part of the asteroid belt (Tholen et al. 1981), it is tempting to conclude that the dust in cometary comae is similar to that found on the surfaces of these asteroids. Modeling based on observations covering a large range of wavelengths and phase angles will be necessary, however, before such a conclusion should be drawn.

Thermal radiometry of cometary nuclei for purposes of deducing their albedos in an analogous fashion to that used so successfully on asteroids faces several unique problems. The standard model described by Morrison and Lebofsky (1979) has been calibrated via observations of bodies with surface regoliths whose thermal properties may differ considerably from those of cometary nuclei. Since the diameters of cometary nuclei are believed to be on the order of 1 km, they must be relatively close ($\lesssim 2$ AU) to both the Earth and Sun to be detected at all (with the possible exception of P/Schwassmann-Wachmann 1). However, at these heliocentric distances all comets can be expected to be rather active and hence thermal emission from dust grains will interfere with the interpretation of the radiometry. It therefore appears doubtful that radiometrically determined diameters of cometary nuclei will be obtainable.

E. Extinct Comet Candidates

Asteroid 944 Hidalgo is perhaps the most widely accepted candidate for being an extinct cometary nucleus (Marsden 1971; Kresák 1973; Shoemaker and Wolfe 1982). This conclusion is based on Hidalgo's short-period comet-like orbit and not on its physical properties. Figure 7 shows that the 0.33 to 0.9 μm reflection spectrum of Hidalgo closely resembles that of the RD class of asteroids, the dominant type among the Trojan asteroids. Assuming that the average albedo of 0.03 found by Tedesco and Gradie (1981) for the RD types corresponds to Hidalgo's albedo, which has not been directly measured, then Hidalgo's diameter is 60 km. While this is rather large for a cometary nucleus it is not unreasonably so since the periodic comet Schwassman-Wachmann 1, if also covered with low albedo material, may have a diameter on the order of 75 km (Roemer 1966).

Furthermore, Hidalgo shows a strong variation in *UBV* colors over its surface. The amplitude is 6 times larger than that found for most asteroids (Degewij et al. 1979). Physical studies at visual and infrared wavelengths during its perihelion apparition in 1990 ($V \sim 14.5$) are essential to understanding the nature of this object. It is possible that the color spot is related to an early phase of strong outgassing activity. On the other hand, the fact that Hidalgo's reflection spectrum is typical of that found for asteroids in the outer belt and especially among the Trojan asteroids, could mean that it may be an escaped Trojan even if the details of possible escape mechanisms are presently unknown or require *ad hoc* assumptions; we know of only one such object. We are therefore forced to conclude that either Hidalgo is an escaped Trojan asteroid, or an asteroid originating elsewhere in the solar system, or the extinct nuclei of a comet covered with RD-type material.

2060 Chiron has been described by Kowal et al. (1979) as being a large inactive cometary nucleus, a minor planet, or a member of a distinct group of objects to which Saturn's small satellite Phoebe also belongs. Hartmann et al. (1981) concluded that the infrared colors are inconsistent with those of outer planetary satellites whose surfaces are covered by relatively clean ices. They speculate that the surface material is dark, carbonaceous-like silicate dust, perhaps containing some admixture of icy grains. If this is true, then Chiron has a diameter of 310 to 400 km (assuming p_V = 0.05-0.03) making it the eighth largest asteroid.

Earth-approaching asteroids (Atens, Apollos, and deep Mars crossers) are in orbits of short dynamical lifetimes (i.e., $\lesssim 3 \times 10^7$ yr) and hence require a source from which they can be replenished on a similar time scale. Since known mechanisms for relocating main-belt asteroids onto Earth-approaching orbits are not efficient enough to account for the numbers of such objects inferred to exist, it has been proposed that the majority of these objects are defunct, short-period comets (Öpik 1963; Shoemaker et al. 1979; Wetherill 1979). Based on physical observations obtained through mid-1981 we would be hard pressed to present a well-documented case for any Earth-approaching asteroid's having physical properties, as deduced from measured albedos, *UBV* colors, 8-color, or 24-color spectrophotometry, significantly different from those of main-belt asteroids. Tedesco et al. (1981c) using a sample of 32 (of the 80 known) Earth-approaching asteroids with measured *UBV* colors (Fig. 14) noted that the lack of *a priori* knowledge regarding the physical properties of extinct cometary cores places severe constraints on any interpretations. Nevertheless, they concluded that extinct cometary cores are either (1) indistinguishable from asteroids, or (2) mostly small ($D \sim 1$ km), dark ($p_V \lesssim 0.06$) objects that remain to be discovered, or (3) the fraction of near-earth objects which are extinct cometary cores is much smaller than presently believed.

McFadden (1981), based on 24-color spectrophotometry of 15 Earth-approaching asteroids, concluded that these asteroids are composed of basically the same kinds of minerals found in the surface material of main-belt asteroids. Sekanina (1976b), and more recently Drummond (1981), noted associations between certain meteor streams and several Earth-approaching asteroids. According to Drummond the best associations exist for 2101 Adonis and 2201 1947 XC, neither of which have had physical observations made on them. Even if a real association exists between an asteroid and a meteor stream, this should not be considered to be proof that the asteroid is of cometary origin; it is possible that the associated meteor stream originated from the most recent collision suffered by that asteroid. In this case, however, we might expect that the physical characteristics of these asteroidal

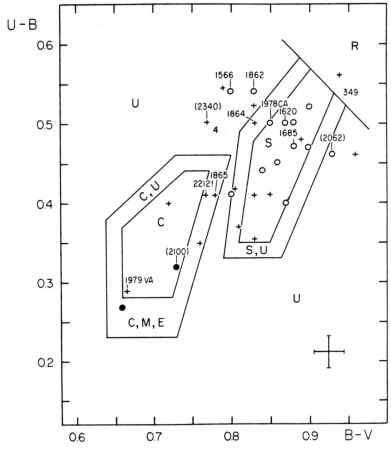

Fig. 14. Two-color plot for 32 Earth-approaching asteroids from Tedesco et al. (1981c): Atens (numbers within parenthesis), Apollos (numbered) and deep Mars-crossers. Open circles are used for objects with albedos >0.065, solid circles for objects with albedos ≤0.065, and (+) symbols where the albedo is unknown. A (!) following a number indicates an association with a radar meteor stream as reported by Drummond (1981). The boxes defining the locations of the various asteroid taxonomic types in UBV space are as given in Bowell et al. (1978).

meteors might differ from those of cometary meteors (see the chapter by Wetherill and ReVelle in this book). It is curious that among those asteroids proposed by various authors as candidates for extinct cores, only 2101 Adonis appears common to all lists (e.g., Anders 1971; Marsden 1971; Sekanina 1971; Kresák 1979).

If extinct cores exist (now beginning to be doubted [Whipple 1981]) and if, as seems likely, their sizes are small and albedos low, we may yet find them among the Earth-approaching asteroids.

III. CONCLUSIONS AND RECOMMENDATIONS FOR FUTURE WORK

Based on available data we are forced to conclude that no extinct cometary nucleus, either among the asteroids or elsewhere, has yet been recognized as such. It is equally unlikely that an uncontaminated bare cometary nucleus has ever been observed with any compositionally sensitive technique. Eventually reflection spectra of the solid surface of a comet will be obtained. We present below several types of observations which will help establish the relevance of such interpretations once they are made.

Physical observations of distant asteroids and low-activity comets are sparse. As a first step in beginning to understand the quiescent cometary nucleus it is essential to monitor the magnitude and morphology of low-activity comets, in order to compile a high-precision photometric data base. A phase curve (reduced magnitudes versus solar phase angle) is essential for estimating cometary activity from a single magnitude measurement, although a low-dispersion spectrum is undoubtedly more useful in detecting the presence of a gas coma. As discussed in Sec. II.A, much work has already been done in this area but, as demonstrated in Figs. 2, 3, and 4, such phase curves would benefit from the higher precision of photoelectric techniques.

Spectrophotometry at visual wavelengths has, in most cases, been limited to observations shortward of 0.7 μm. It is essential to go to longer wavelengths in order to obtain data which can be interpreted in a mineralogical context. At red wavelengths, measurements of faint comets are limited to the use of wide-band filters, such as those employed in the 8-color asteroid system described by Tedesco et al. (1981a). The separation attainable with this system is graphically illustrated in Fig. 10. Spectrophotometric observations obtained over a large range of phase angles may, in addition to providing a phase function, help establish whether a bare nucleus has been observed by enabling measurements of the degree of spectral reddening to be compared with those expected for solid surfaces (Gradie and Veverka 1982). In this context it would also be useful to know how the spectra of dust comae change with solar phase angle.

The water ice bands, marginally indicated with *JHK* filters, have not been observed in bright comet comae (A'Hearn et al. 1981); this result, however, is not unexpected (see chapter by Campins and Hanner in this book). To date there exists only one low precision *VJHK* measurement and one 8-color spectrum of P/Schwassmann-Wachmann 1 obtained when the comet was nearly quiescent. More observations are needed at these wavelengths, preferably with a simultaneous *V* measurement obtained along with the *JHK* measurements; we could then judge the state of activity, and make comparisons with asteroids over a longer wavelength range.

Another possibility which promises to provide a direct view of a cometary nucleus is to take advantage of those times when comets pass close to the Earth. On these occasions it may be possible to resolve the nucleus since the coma is so spread out that its surface brightness is very low.

We conclude with a reference to the chapter by Kamoun and coworkers in this book who discuss radar observations of nearby comets. It may be possible to obtain direct determination of the diameter of a cometary nucleus if the rotation period and orientation of the rotation axis are known. The search to physically characterize quiescent cometary nuclei has just begun; at this point we can only hope that it is not quixotic.

Acknowledgments. We appreciate the use of unpublished observations of P/Arend-Rigaux made available by C.R. Chapman, who in turn thanks the Institute for Astronomy of the University of Hawaii for allocation of time on the 224-cm telescope, and members of McCord's group for assistance. J.C. Gradie provided us with his measurement of P/Arend-Rigaux. We thank E. Everhart, C.R. Chapman, L.A. Lebofsky, G.J. Veeder, E. Roemer, and Z. Sekanina for critical comments. We have benefited from discussions with Z. Sekanina, who provided us with his unpublished calculations on P/Arend-Rigaux and P/Neujmin 1. R. Wolfe kindly made available her compilation of magnitude measurements of P/Schwassmann-Wachmann 1 obtained by E. Roemer, while D. Yeomans provided ephemerides for the early observations of P/Schwassmann-Wachmann 1. This work was supported by several NASA grants to the University of Arizona, and by a NASA contract to the Jet Propulsion Laboratory, California Institute of Technology.

REFERENCES

Anders, E. 1971. Interrelations of meteorites, asteroids and comets. In *Physical Studies of Minor Planets,* ed. T. Gehrels (Washington: NASA SP-267), pp. 429-446.
A'Hearn, M.F.; Dwek, E.; and Tokunaga, A.T. 1981. Where is the ice in comets? Submitted to *Astrophys. J.*
Allen, D.A. 1977. *IAU Circ.* No. 3142.
Arpigny, C. 1972. Comet spectra. In *Comets: Scientific Data and Missions,* eds. G.P. Kuiper and E. Roemer (Tucson: Lunar and Planetary Laboratory, Univ. of Arizona), pp. 84-111.
Barker, E.; Tedesco, E.F.; Zellner, B.; and Rybski, P. 1982. A synthesis of photometric observations of P/Tempel 2 during its 1978/79 apparition. In preparation.
Bowell, E.; Chapman, C.R.; Gradie, J.C.; Morrison, D.; and Zellner, B. 1978. Taxonomy of asteroids. *Icarus* 35:313-335.
Bowell, E.; Gehrels, T.; and Zellner, B. 1979. Magnitudes, colors, types and adopted diameters of the asteroids. In *Asteroids,* ed. T. Gehrels (Tucson: Univ. of Arizona Press), pp. 1108-1129.
Campins, H.; Gradie, J.; Lebofsky, M.; and Rieke, G. 1981. Infrared observations of faint comets. In *Modern Observational Techniques for Comets.* (Pasadena: Jet Propulsion Lab.), pp. 83-92.
Carusi, A., and Valsecchi, G.B. 1979. Numerical simulations of close encounters between Jupiter and minor bodies. In *Asteroids,* ed. T. Gehrels (Tucson: Univ. of Arizona Press), pp. 391-416.
Carusi, A., and Valsecchi, G.B. 1981. Temporary satellite captures by Jupiter, *Astron. Astrophys.* In press.
Chapman, C.R., and Gaffey, M.J. 1979. Spectral reflectances of the asteroids. In *Asteroids,* ed. T. Gehrels (Tucson: Univ. of Arizona Press), pp. 1064-1089.

Chapman, C.R., and Morrison, D. 1976. *JHK* photometry of 433 Eros and other asteroids. *Icarus* 28:91-94.
Cochran, A.L.; Barker, E.S.; and Cochran, W.D. 1980. Spectrophotometric observations of P/Schwassmann-Wachmann 1 during outburst. *Astron. J.* 85:474-477.
Cochran, A.L.; Cochran, W.D.; and Barker, E.S. 1981. Spectrophotometry of Comet Schwassmann-Wachmann 1. II. Its color and CO^+ emission. Submitted to the *Astrophys. J.*
Cruikshank, D.P. 1981. The satellites of Uranus. In *Uranus and the Outer Planets*, ed. G.E. Hunt (Cambridge: Cambridge Univ. Press). In press.
Davis, D.R.; Chapman, C.R.; Greenberg, R.; Weidenschilling, S.J.; and Harris, A.W. 1979. Collisional evolution of asteroids: Populations, rotations, and velocities. In *Asteroids*, ed. T. Gehrels (Tucson: Univ. of Arizona Press), pp. 528-557.
Degewij, J. 1980. Spectroscopy of faint asteroids, satellites, and comets. *Astron. J.* 85:1403-1412.
Degewij, J. 1981. Spectrophotometry of faint comets: The asteroid approach. In *Modern Observational Techniques for Comets*. (Jet Propulsion Lab.), pp. 63-69.
Degewij, J.; Cruikshank, D.P.; and Hartmann, W.K. 1980. Near-infrared colorimetry of J6 Himalia and S9 Phoebe: A summary of 0.3 to 2.2 μm reflectances. *Icarus* 44:541-547.
Degewij, J.; Tedesco, E.F.; and Zellner, B. 1979. Albedo and color contrasts on asteroid surfaces. *Icarus* 40:364-374.
Degewij, J., and van Houten, C.J. 1979. Distant asteroids and outer Jovian satellites. In *Asteroids*, ed. T. Gehrels (Tucson: Univ. of Arizona Press), pp. 417-435.
Delsemme, A.H., and Rud, D.A. 1973. Albedos and cross sections for the nuclei of comets 1969 IX, 1970 II, and 1971 II. *Astron. Astrophys.* 28:1-6.
Drummond, J.D. 1981. Theoretical meteor radiants of Apollo, Amor, and Aten asteroids. *Icarus*. In press.
Everhart, E. 1973a. Horseshoe and Trojan orbits associated with Jupiter and Saturn. *Astron. J.* 78:316-328.
Everhart, E. 1973b. Examination of several ideas of comet origins. *Astron. J.* 78:329-337.
Everhart, E. 1976. The evolution of comet orbits: A review. In *The Study of Comets*, eds. B. Donn, M. Mumma, W. Jackson, M. A'Hearn, and R. Harrington (Washington: NASA SP-393), pp. 445-461.
Everhart, E. 1979. Chaotic orbits in the solar system. In *Asteroids*, ed. T. Gehrels (Tucson: Univ. of Arizona Press), pp. 283-288.
Gebel, W.L. 1970. Spectrophotometry of comets 1967n, 1968b, and 1968c. *Astrophys. J.* 161:765-777.
Gehrels, T. 1970. Photometry of asteroids. In *Surfaces and Interiors of Planets and Satellites*, ed. A. Dollfus (London: Academic Press), pp. 317-374.
Gehrels, T., and Tedesco, E.F. 1979. Minor planets and related objects. XXVII. Asteroid magnitudes and phase relations. *Astron. J.* 84:1079-1087.
Gradie, J., and Veverka, J. 1982. Wavelength dependence of phase coefficients. Submitted to *Icarus*.
Hanner, M.S. 1980. Physical characteristics of cometary dust from optical studies. In *Solid Particles in the Solar System*, eds. I. Halliday and B.A. McIntosh (Dordrecht, Holland: D. Reidel), pp. 223-236.
Hanner, M.S. 1981. On the detectability of icy grains in the comae of comets. Submitted to *Icarus* (special Comet issue).
Hartmann, W.K.; Cruikshank, D.P.; and Degewij, J. 1982. Remote comets and related asteroids: *VJHK* photometry and composition of surface materials. Submitted to *Icarus*.
Hartmann, W.K.; Cruikshank, D.P.; Degewij, J.; and Capps, R.W. 1981. Surface materials on unusual planetary object Chiron. Submitted to *Icarus* (special Comet issue).
Herget, P. 1968. Revised orbit of Comet Schwassmann-Wachmann 1. *Astron. J.* 73:729-730.
Jeffers, H.M. 1946. Comet notes. *Publ. Astron. Soc. Pacific* 58:61-62.

Jeffers, H.M., and Roemer, E. 1955. Observations of comets and asteroids. *Astron. J.* 60:440-442.
Jeffers, H.M.; Vasilevskis, S.; and Roemer, E. 1954. Observations of comets and asteroids. *Astron. J.* 59:305-307.
Johnson, P.E.; Smith, D.W.; and Shorthill, R.W. 1981. An outburst of Comet Tempel 2 observed spectrophotometrically. *Nature* 289:155-156.
Johnson, T.V.; Lebofsky, L.A.; and McCord, T.B. 1971. Comet Bennett 1969i: Narrowband filter photometry 0.3 to 11 microns. *Publ. Astron. Soc. Pacific* 83:93-98.
Johnson, T.V.; Matson, D.L.; Veeder, G.J.; and Loer, S.J. 1975. Asteroids: Infrared photometry at 1.25, 1.65, and 2.2 microns. *Astrophys. J.* 197:527-531.
Johnson, T.V., and McCord, T.B. 1971. Spectral geometric albedo of the Galilean satellites, 0.3 to 2.5 microns. *Astrophys. J.* 169:589-594.
Kazimirchak-Polonskaya, E.I., and Shaporev, S.D. 1976. Orbital evolution of P/Lexell (1770I) and some regularities of cometary orbit's transformation in Jupiter's sphere of action. *Astron. J. (U.S.S.R.)* 53:1306-1314.
Kowal, C.T.; Liller, W.; and Marsden, B.G. 1979. The discovery and orbit of (2060) Chiron. In *Dynamics of the Solar System*, ed. R.L. Duncombe (Dordrecht, Holland: D. Reidel), pp. 245-250.
Kresák, L. 1973. Short-period comets at large heliocentric distances. *Bull. Astron. Inst. Czech.* 24:264-283.
Kresák, L. 1979. Dynamical interrelations among comets and asteroids. In *Asteroids*, ed. T. Gehrels (Tucson: Univ. of Arizona Press), pp. 289-309.
Kresák, L. 1980. Dynamics, interrelations and evolution of the systems of asteroids and comets. *Moon and Planets* 22:83-98.
Larson, S.M. 1980. CO˙ in comet Schwassmann-Wachmann 1 near minumum brightness. *Astrophys. J.* 238:L47-L48.
Liller, W. 1960. The nature of the grains in the tails of comets 1956h and 1957d. *Astrophys. J.* 132:867-882.
Marsden, B.G. 1970. On the relationship between comets and minor planets. *Astron. J.* 75:206-217
Marsden, B.G. 1971. Evolution of comets into asteroids? In *Physical Studies of Minor Planets*, ed. T. Gehrels (Washington: NASA SP-267), pp. 413-421.
Marsden, B.G. 1972. Nongravitational effects on comets: The current status. In *The Motion, Evolution of Orbits, and Origin of Comets*, eds. G.A. Chebotarev, E.I. Kazimirchak-Polonskaya, and B.G. Marsden (Dordrecht, Holland: D. Reidel), pp. 135-143.
Mayall, N.U. 1941. Note on the spectrum of Comet Schwassmann-Wachmann 1 (1925b). *Publ. Astron. Soc. Pacific* 53:340-341.
McCord, T., and Chapman, C.R. 1975. Asteroids: Spectral reflectance and color characteristics. II. *Astrophys. J.* 197:781-790.
McFadden, L.A. 1981. Near-earth asteroids: 1981 perspectives based on reflectance spectroscopy. *Bull. Amer. Astron. Soc.* In press.
Morrison, D.; Cruikshank, D.P.; Pilcher, C.B.; and Rieke, G.H. 1976. Surface compositions of the satellites of Saturn from infrared photometry. *Astrophys. J.* 207:L213-L216.
Morrison, D., and Lebofsky, L. 1979. Radiometry of asteroids. In *Asteroids*, ed. T. Gehrels (Tucson: Univ. of Arizona Press), pp. 184-205.
O'Dell, C.R. 1971. Spectrophotometry of comet 1969g (Tago-Sato-Kosaka). *Astrophys. J.* 164:511-519.
Öpik, E.J. 1963. The stray bodies in the solar system. Part 1. Survival of cometary nuclei and the asteroids. *Adv. Astron. Astrophys.* 2:219-262.
Pollack, J.B.; Burns, J.A.; and Tauber, M.E. 1979. Gas drag in primordial circumplanetary envelopes: A mechanism for satellite capture. *Icarus* 37:587-611.
Roemer, E. 1956. Observations of comets. *Astron. J.* 61:391-394.
Roemer, E. 1958. An outburst of Comet Schwassmann-Wachmann 1. *Publ. Astron. Soc. Pacific* 70:272-278.
Roemer, E. 1962. Activity in comets at large heliocentric distance. *Publ. Astron. Soc. Pacific* 74:351-365.
Roemer, E. 1965. Observations of comets and minor planets. *Astron. J.* 70:397-402.
Roemer, E. 1966. The dimensions of cometary nuclei. *Mém. Soc. Roy. Sci. Liège* 12, Ser. 5:23-28.

Roemer, E. 1976. Luminosity and astrometry of comets: A review. In *The Study of Comets*, eds. B. Donn, M. Mumma, W. Jackson, M. A'Hearn, and R. Harrington (Washington: NASA SP-393), pp. 380-409.
Roemer, E., and Lloyd, R.E. 1966. Observations of comets, minor planets, and satellites. *Astron. J.* 71:443-457.
Roemer, E.; Thomas, M.; and Lloyd, R.E. 1966. Observations of comets, minor planets, and Jupiter VIII. *Astron. J.* 71:591-601.
Sekanina, Z. 1971. A core-mantle model for cometary nuclei and asteroids of possible cometary origin. In *Physical Studies of Minor Planets*, ed. T. Gehrels (Washington: NASA SP-267), pp. 423-428.
Sekanina, Z. 1976a. A continuing controversy: Has the cometary nucleus been resolved? In *The Study of Comets*, eds. B. Donn, M. Mumma, W. Jackson, M. A'Hearn, and R. Harrington (Washington: NASA SP-393), pp. 537-585.
Sekanina, Z. 1976b. Statistical model of meteor streams. IV. A study of radio streams from the synoptic year. *Icarus* 27:265-321.
Sekanina Z. 1981a. Rotation and precession of cometary nuclei. *Ann. Rev. Earth Planet. Sci.* 9:113-145.
Sekanina, Z. 1981b. Large-scale nucleus surface topography and outgassing pattern analysis of periodic Comet Swift-Tuttle. Submitted to the *Astron. J.*
Sekanina, Z. 1981c. Comet Bowell 1980b: An active-looking dormant object? Submitted to the *Astron. J.*
Shoemaker, E.M.; Williams, J.G.; Helin, E.F.; and Wolfe, R.F. 1979. Earth-crossing asteroids: Orbital classes, collision rates with Earth, and origin. In *Asteroids*, ed. T. Gehrels (Tucson: Univ. of Arizona Press), pp. 253-282.
Shoemaker, E.M., and Wolfe, R.F. 1982. Cratering time scales for the Galilean satellites. In *The Satellites of Jupiter*, ed. D. Morrison (Tucson: Univ. of Arizona Press). In press.
Spinrad, H.; Stauffer, J.; and Newburn, R.L. Jr. 1979. Optical spectrophotometry of Comet Tempel 2 far from the sun. *Publ. Astron. Soc. Pacific* 91:707-711.
Tedesco, E.F., and Gradie, J.C. 1981. Taxonomy of outer-belt asteroids and the distribution of taxonomic types from 1.8 to 5.2 AU. Submitted to *Icarus*.
Tedesco, E.F.; Tholen, D.J.; and Zellner, B. 1981a. The eight-color asteroid system (ECAS) survey: The standard stars. To be submitted to *Astron. J.*
Tedesco, E.F.; Tholen, D.J.; and Zellner, B. 1981b. Eight-color asteroid system (ECAS) spectrophotometry for 400 asteroids. To be submitted to *Astron. J.*
Tedesco, E.F.; Tholen, D.J.; Zellner, B.; Williams, J.G.; and Veeder, G.J. Jr. 1981. Physical observations of near-earth asteroids. *Bull. Amer. Astron. Soc.* In press.
Tedesco, E.F., and Zappalá, V. 1980. Asteroid rotational properties: Correlations and selection effects. *Icarus* 40:33-50.
Tholen, D.J.; Tedesco, E.F.; and Larson, S.M. 1981. Broad-band spectrophotometry of cometary comae. *Bull. Amer. Astron. Soc.* In press.
Veeder, G.J.; Matson, D.L.; and Smith, J.C. 1978. Visual and infrared photometry of asteroids. *Astron. J.* 83:651-663.
Walker, M.F. 1959. Observations of Comet Schwassmann-Wachmann 1 during an outburst. *Publ. Astron. Soc. Pacific* 71:169-170.
Wasson, J.T., and Wetherill, G.W. 1979. Dynamical, chemical and isotopic evidence regarding the formation locations of asteroids and meteorites. In *Asteroids*, ed. T. Gehrels (Tucson: Univ. of Arizona Press), pp. 926-974.
Weissman, P.R. 1978. Observations of a near-extinct cometary nucleus. *Bull. Amer. Astron. Soc.* 10:588 (abstract).
Wetherill, G.W. 1979. Steady state populations of Apollo-Amor objects. *Icarus* 37:96-112.
Whipple, F.L. 1950. A comet model. I. The acceleration of Comet Encke. *Astrophys. J.* 111:375-394.
Whipple, F.L. 1981. The nature of comets. In *Comets and the Origin of Life* ed. C. Ponnamperuma (Greenbelt: Univ. of Maryland Press), pp. 1-20.
Wilkening, L.L. 1979. The asteroids: Accretion, differentiation, fragmentation, and irradiation. In *Asteroids*, ed. T. Gehrels (Tucson: Univ. of Arizona Press), pp. 61-74.
Zellner, B. 1979. Asteroid taxonomy and the distribution of the compositional types. In *Asteroids*, ed. T. Gehrels (Tucson: Univ. of Arizona Press), pp. 783-806.

COMETS AND ORIGIN OF LIFE

C. PONNAMPERUMA and E. OCHIAI
University of Maryland

The observation of carbon compounds in comets is suggestive of chemical evolution. However, it appears unlikely that prebiological processes would have gone much further than the stage of monomer formation. The proposition that comets may be a vehicle for panspermia is not warranted by currently available data.

I. CHEMICAL EVOLUTION AND THE ORIGIN OF LIFE

According to the Oparin-Haldane hypothesis (Oparin 1924; Haldane 1929) of chemical evolution, life is considered to be an inevitable result of the evolutionary process in the universe. The synthesis, accumulation, and organization of molecules of biological importance are postulated as a preamble to the emergence of life on Earth (Bernal 1959; Ponnamperuma 1977). Laboratory experiments have clearly demonstrated that the constituents of the nucleic acids and proteins can be synthesized by abiotic processes. Evidence for such syntheses has been obtained by the analysis of carbonaceous chondrites (Shimoyama et al. 1979; Kotra 1981). The question which is uppermost in this discussion is whether cometary bodies will have on them any evidence in support of the concepts of chemical evolution (Ponnamperuma 1981).

In considering the role of comets in the origin of life, the following specific questions immediately come to mind: (i) Are organic compounds present in comets? (ii) Are organic molecules synthesized in comets? (iii) Does polymerization occur in comets? (iv) Can life arise in comets? and (v) Can comets be a vehicle for panspermia?

II. ORGANIC COMPOUNDS IN COMETS

According to the best imformation currently available, a comet may be considered to be dirty snow ball or icy conglomerate (Whipple 1950, 1951, 1981). The ice consists largely of frozen H_2O, NH_3, and CO_2. In addition, silicates are considered to be an important component of the comet nucleus. When a comet comes into the inner solar system, the temperature of the comet is raised and volatile compounds and dust are blown off mainly by the solar wind.

The spectroscopic studies of this coma and tail have detected a number of smaller inorganic and organic molecules and ions. These include C, C_2, C_3, CH, CN, CO, CS, HCN, CH_2CN, H, NH, NH_2O, OH, H_2O, S, C^+, CO^+, CO_2^+, CH^+, H_2O^+, OH^+, N_2^+, CN^+, Na, K, Ca, V, Mn, Fe, Co, Ni, Cu, Ca^+ and silicates (Delsemme 1981; see also his chapter in this book). Silicates have been identified by infrared reflection spectra. These are observed only when a comet comes into the inner solar system.

Are these molecules present in the comet when it is at low temperatures and in the outer solar system where it is presumed to have originated? Are there parent compounds from which these smaller molecules are derived when the comet comes near the Sun? Both possibilities appear to be feasible. Compounds similar to those listed above are found in the interstellar medium. The interstellar molecules observed include H_2, OH, CH^+, CO, CN, NO, C_2, SO, NS, SiO, SiS, HCO^+, N_2H^+, H_2O, HCN, C_2H, HCO, HNO, OCS, H_2S, SO_2, NH_2, HCHO, C_2H_2, C_3N, HNCO, H_2CS, HNCS, CH_4, C_4H, C_3HN, $H_2C_2O_2$, NH_2CN, CH_2NH, HCOOH, CH_3CN, NH_2CHO, CH_3OH, CH_3SH, $CH_2-C{\equiv}CH$, CH_3CHO, CH_3NH_2, $H_2C=CHCN$, $HC{\equiv}C-C{\equiv}C-CN$, $HCOOCH_3$, C_2H_5OH, C_2H_5CN, CH_3OCH_3, $HC{\equiv}C-C{\equiv}C-C{\equiv}C-CN$ and $HC{\equiv}C-C{\equiv}C-C{\equiv}C-CN$ (Oro et al. 1981). If the comets were formed as a result of accretion of the presolar nebular gas, it is not surprising to find similar compounds in both the comets and the interstellar medium.

It is equally possible, however, that most of these organic molecules could have arisen from more stable organic molecules trapped in the ice by the action of solar ultraviolet and particle radiation. For example, H_2O would produce H_2O^+, OH, O and H; CO_2: CO_2^+, CO^+, O, etc.; HCHO: CH, CO, CO^+; CH_3CN: CN; C_2H_2: C_2; CS_2: CS (Delsemme 1981); or $CH_4 + NH_3 + h\nu \to$ HCN, CN; $CO_2 + CH_4 + h\nu \to C_3$, C_2H (Huebner 1981; see also his chapter in this book).

There is ample evidence to establish the fact that organic molecules of smaller molecular weights and simpler compositions are present in comets.

III. PREBIOTIC ORGANIC SYNTHESIS IN COMETS

The next question of importance to the origin of life is whether or not organic synthesis similar to the terrestrial prebiotic organic synthesis is taking place or took place in comets. The presence of such compounds as HCN, HCHO, CH_3CHO and H_2O in the cometary gas indicates that the more complex compounds which can be derived from these simpler compounds may also be present in the comet nucleus. These compounds include amino acids, carbohydrates, purines and pyrimidines. They have never been observed directly in comets.

There are, however, a few indications strongly suggestive of their presence. First, it should be pointed out that the cometary dust is similar in composition to the carbonaceous chondrites, especially those of Type CI. A comet as a whole contains even more carbon than the CI chondrites and therefore the origins of comets may be similar to that of CI chondrites. As carbonaceous chondrites contain amino acids, carboxylic acids (Nagy 1975) and bases (Stoks and Schwartz 1981), comets are also likely to contain amino acids, bases and other biologically relevant organic compounds.

Further evidence in support of this concept comes from simulation experiments. Irradiation of icy mixtures of H_2O, CH_4 (or HCHO) and NH_3 by high-energy electrons (Berger 1961; Oro 1963) and carbon ion beam ($^{14}C^+$) (Lemmon 1979) produced significant amounts of nonvolatile organic compounds including amino acids, carboxylic acids, purines and pyrimidines. A photolysis of a frozen mixture of $CO-CH_4-NH_3-H_2O$ (Greenberg 1981; see also his chapter in this book) produced HCHO, HCOOH, $HCONH_2$ and C_3O_2 among others. Similar results were also reported by Khare and Sagan (1973).

The cometary organic compounds have been estimated to represent as high as 5% of the nucleus mass (Kajmakov and Matvejev 1979). When were these compounds synthesized? One possibility is that they were produced in the solar nebula before accretion (Greenberg 1981; his chapter). The second possibility is that they were produced after the formation of a comet by the radiation of ^{26}Al and other presolar isotopes which were present at the earlier stages of solar system evolution (Irvine et al. 1980). Cosmic rays could be another source of energy which caused such organic synthesis in comets (Donn 1976).

The presence of ^{26}Al in the early solar system was inferred because some meteorites contain anomalous levels of ^{26}Mg. According to Irvine et al. (1980) and Wallis (1980), the decay of ^{26}Al and other isotopes could have created a liquid environment within a cometary nucleus for at least the first 10^7 yr. This radiation is, however, like a double-edged sword. It could synthesize complex organic molecules but at the same time destroy them. Thus, the net effect of this source is rather questionable. If the radiation-damaging effect of ^{26}Al and others were too strong, the organic compounds inherited from the presolar nebula might have also been destroyed during the process of accretion.

IV. POLYMER FORMATION IN COMETS

Once the monomers necessary for biological macromolecules are synthesized, the next step in chemical evolution is the formation of polymeric compounds, polynucleotides and polypeptides. Such polymerization requires the interaction of the monomers. It seems highly unlikely in view of the prevailing harsh conditions in comets that the formation of the polymers took place or is taking place in comets. The temperature is low. Liquid water may not exist to help in the interaction of these molecules. Radiation by cosmic rays would destroy any polymer formed. Therefore, chemical evolution probably has not proceeded beyond the monomer productions on comets.

There is, however, a possibility that monomeric compounds were not necessarily the first to be formed but rather were derived from complex polymeric compounds. Such complex organic compounds have been reported to form in a variety of simulation experiments involving CH_4/NH_3, $CH_4/NH_3/H_2O$ or HCN (e.g., Sagan and Khare 1979; Ferris et al. 1981).

V. SIGNIFICANCE OF COMETS FOR THE ORIGINS OF LIFE ON EARTH

It can be concluded from the argument advanced so far that it is highly unlikely that life formed on comets. It is, however, possible that comets could have made some or significant contribution to the formation of life on Earth.

First of all, let us compare the elemental composition of chondrites, comets and Earth (Table I). It is to be noted that the elemental composition of comets is very close to the cosmic one; this implies that comets are very primitive objects. The Earth is very much depleted of the two biologically essential elements, carbon and nitrogen, as compared to the cosmic abundance.

It has been estimated from the lunar cratering rates (Wetherill 1975; Oro et al. 1980; Pollack and Yung 1980; Chang 1979) that the Earth could have accreted 10^{22} to 10^{23} g of cometary matter. Cometary collisions with the Earth could have happened much more frequently at an earlier stage of the Earth's formation than today. Taking the higher figure of 10^{23} g and making a reasonable assumption that comets contain carbon up to $\sim 10\%$ of their mass, Oro et al. estimated that as much as 10^{22} g of carbon were added to the surface of the primitive Earth by cometary collisions (Oro et al. 1980; Oro and Lazcano-Araujo 1981). This figure (10^{22} g) is indeed comparable to the estimated amount of carbon buried in the Earth's sedimentary shell, i.e., 1.9×10^{22} g (Hunt 1972, 1977) and 1.2×10^{22} g (Schidlowski 1978).

Most of these cometary carbon compounds would probably be decomposed upon collision. If only 1% of the cometary organic compounds survived, the comets could still have contributed about 10^{20} g of carbon

TABLE I
Elemental Compositions of the Universe, Comets, Chondrites, and Earth[a]

Element	Cosmic Abundance[b]	Cometary Abundance[b]	% Cosmic	Chondrites[c]	% Cosmic	Earth[c]	% Cosmic
H	2.66×10^{10}	2.6×10^7	0.1	—	—	1.4×10^5	5×10^{-4}
C	1.17×10^7	4.9×10^6	42	2.00×10^3	0.002	7.0×10^3	0.07
N	2.31×10^6	1.15×10^6	50	9.0×10^1	5×10^{-3}	2.0×10^1	10^{-3}
O	1.84×10^7	1.84×10^7	100	3.7×10^6	20	3.5×10^6	20
S	5.0×10^5	5.0×10^5	100	1.1×10^5	20	1×10^5	20
Mg	1.06×10^6	1.06×10^6	100	9.4×10^5	90	8.9×10^5	90
Si	1.00×10^6	1.00×10^6	100	1.00×10^6	100	1.00×10^6	100
Fe	9.0×10^5	9.0×10^5	100	6.9×10^5	77	1.35×10^6	150

[a] Relative to Si = 10^6 atoms.
[b] Delsemme (1981).
[c] Brownlow (1979).

compounds. This figure is two orders of magnitude greater than the total mass of the present biota. From estimations of this kind, Oro et al. (1980) and Oro and Lazcano-Araujo (1981) concluded that comets could have contributed significantly towards the processes of chemical evolution necessary for the emergence of life on the Earth.

VI. CAN COMETS BE A VEHICLE FOR PANSPERMIA?

Hoyle and Wickramasinghe (1978,1979,1980,1981) have recently suggested that comets may have been responsible for introducing life on the Earth. The idea is (i) that life is a cosmic phenomenon, and (ii) that the comet is a likely vehicle for panspermia.

The idea of panspermia is not new, and is indeed a possibility as far as it means that life is a cosmic phenomenon. The main objection to the Hoyle-Wickramasinghe contention is that life did not originate on the Earth but rather arrived here from somewhere in the universe. Let us consider their basic assumptions. Their conviction is based on a statistical improbability of the spontaneous formation of even a few enzymes, let alone a number of them. To quote their argument: "the probability of finding N such enzymes by random assembly is $1:(20)^{10 N}$. It is easily seen that we obtain a number of trials exceeding the number of all the atoms in all the stars in the whole universe even before we come to the $N = 100$" (Hoyle and Wickramasinghe 1981). They go on to say, one has to assume that life is a cosmic phenomenon. Their understanding of "cosmic phenomenon" appears to be equivalent to nonterrestrial. There are a few basic fallacies in this argument. If their quoted argument is correct, then life should not have emerged anywhere in the universe; it is simply statistically impossible. Their basic assumption and statistical argument must be wrong.

One of their implicit assumptions appears to be that atoms are used only once, i.e., they are not utilized again once they form a compound. Atoms and molecules are, in fact, in a dynamic state. Atoms are continually assembled to form compounds, and compounds are dissociated into atoms. Likewise amino acids condense together to form proteins, and proteins would be decomposed into amino acids. The same atom can thus be utilized over and over again to form different compounds. If this was not the case, the universe would have been filled with possible nonfunctional molecules.

Secondly, the specific protein formation may not be or rather should not be entirely statistical. On the one hand, each individual amino acid has a different reactivity. If the formation of protein is aided by an autocatalytic type of mechanism, for example through hypercycles (Eigen and Schuster 1977,1978), the statistical dilemma would be overcome. Hoyle and Wickramasinghe's proof that the universe is full of cellulosic material which is meant to indicate the presence of bacteria, is tenuous to say the least. First of all, whether cellulose alone is the only molecule that matches the observed infrared spectra is highly questionable. It is likely that any number of similar and dissimilar compounds would do the job. Besides, cellulose is not the main component of the bacterial cell wall. The possibility that microorganisms originated, propagated and preserved their vitality, even as spores, in the hostile environment of space is hardly reasonable. Material has to be concentrated highly to develop such an organized system as life, and organisms have to be fairly concentrated in order to be able to propagate. These conditions do not seem to prevail in the interstellar space and comets. It is well known that organisms would be damaged by ultraviolet, cosmic ray, and other radiations.

The Earth, on the other hand, seems to provide a comfortable environment for life. It is not improbable that other planets exist somewhere in the universe which have conditions similar to Earth. Therefore, panspermia although a possibility, does not necessarily exclude the terrestrial origin of life.

Our studies of comets thus lead us to the conclusion that organic molecules related to chemical evolution may be common in the solar system, but the conditions necessary for the origin of life occur only under restricted conditions requiring an environment less hostile than that provided by comets.

REFERENCES

Berger, R. 1961. The proton irradiation of methane, ammonia and water at 77°K. *Proc. Nat. Acad. Sci.* 47:1434-1436.
Bernal, J.D. 1959. The problem of stages in biopoesis. In *Origin of Life on the Earth*, ed. A.I. Oparin (London: Pergamon), pp. 38-53.
Brownlow, A.H. 1979. *Geochemistry*, (New York: Prentice-Hall).
Chang, S. 1979. Comets: Cosmic connections with carbonaceous meteorites, interstellar molecules and the origin of life. In *Space Missions to Comets*, eds. M. Neugebauer, D.K. Yeomans, J.C. Brandt, and R.W. Hobbs, (Washington: NASA SP-2089), p. 59.
Delsemme, A.H. 1981. Are comets connected to the origin of life? In *Comets and the Origin of Life*, ed. C. Ponnamperuma (Dordrecht, Holland: D. Reidel), pp. 141-159.
Donn, B.D. 1976. The nucleus: Panel discussion. In *The Study of Comets*, eds. B. Donn, M. Mumma, W. Jackson, M. A'Hearn, and R. Harrington (Washington: NASA SP-393), pp. 611-621.
Eigen, M., and Schuster, P. 1977. The hypercycle. *Naturwis.* 64:541-565.
Eigen, M., and Schuster, P. 1978. The hypercycle. *Naturwis.* 65:7-41, 341-369.
Ferris, J.P.; Edelson, E.H.; Auyeung, J.M.; and Joshi, P.C. 1981. Structural studies on HCN oligomers. *J. Mol. Evol.* 17:69-77.
Greenberg, J.M. 1981. Chemical evolution of interstellar dust and the composition of comets. In *Comets and the Origin of Life*, ed. C. Ponnamperuma (Dordrecht, Holland: D. Reidel), pp. 111-127.
Haldane, J.B.S. 1929. The origin of life. *Rationalist Annual.* 148:3-10.
Hoyle, F., and Wickramasinghe, N.C. 1978. *Lifecloud*. (London: J.M. Dent & Sons).
Hoyle, F., and Wickramasinghe, N.C. 1979. On the nature of interstellar grains. *Astrophys. Space Sci.* 66:77-90.
Hoyle, F., and Wickramasinghe, N.C. 1980. Organic material and the 2.5-4 μm spectra of galactic sources. *Astrophys. Space Sci.* 72:183-188.
Hoyle, F., and Wickramasinghe, N.C. 1981. Comets—a vehicle for panspermia. In *Comets and the Origin of Life*, ed. C. Ponnamperuma (Dordrecht, Holland: D. Reidel), pp. 227-239.

Huebner, W.F. 1981. Chemical kinetics in the coma. In *Comets and the Origin of Life,* ed. C. Ponnamperuma (Dordrecht, Holland: D. Reidel), pp. 91-103.

Hunt, J.M. 1972. Distribution of carbon in crust of earth. *Am. Assoc. Pet. Geol. Bull.* 56:2273-2277.

Hunt, J.M. 1977. Distribution of carbon as hydrocarbons and asphaltic compounds in sedimentary rocks. *Am. Assoc. Pet. Geol. Bull.* 61:100-104.

Irvine, W.M.; Leschine, S.B.; and Schloerb, F.P. 1980. Comets and the origin of life. *Origin of Life* 11:27-32.

Kajmakov, E.A., and Matvejev, I.N. 1979. The role of organic compounds in comets. *A.F. Ioffe Phys. Tech. Inst. Center* 628 Leningrad; cited in Oro and Lazcano-Araujo (1981).

Khare, N., and Sagan, C. 1973. Experimental interstellar organic chemistry. In *Molecules in Galactic Environment,* ed. M. Gordon, and L. Snyder (New York: Wiley), pp. 399-408.

Kotra R. 1981. Organic Analysis of Antartic Carbonaceous Chondrites. Ph.D. Dissertation. University of Maryland.

Lemmon, R.M. 1979. Personal communication, cited in Oro and Lazcano-Araujo (1981).

Nagy, B. 1975. *Carbonaceous Chondrites.* (Amsterdam: Elsevier).

Oparin, A.I. 1924. *Proischogdenie Zhizni.* (Moscovsky: Robotchii).

Oro, J. 1963. Synthesis of organic compounds by high-energy electrons. *Nature* 197:971-974.

Oro, J.; Holzer, G.; and Lazcano-Araujo, A. 1980. The contribution of cometary volatiles to the primitive earth. In *COSPAR: Life Sciences and Space Research XVIII,* ed. R.M. Holmquist (New York: Pergamon Press), pp. 67-82.

Oro, J., and Lazcano-Araujo, A. 1981. Cometary material and the origins of life on earth. In *Comets and the Origin of Life,* ed. C. Ponnamperuma (Dordrecht, Holland: D. Reidel), pp. 191-225.

Pollack, J.B., and Yung, Y.L. 1980. Origin and evolution of planetary atmospheres. *Ann. Rev. Earth Planet. Sci.* 8:425-487.

Ponnamperuma, C. 1977. Cosmochemistry and the origin of life. In *Origins of Life, the Earth and the Universe,* ed. Mitsubishi-Kasei Inst. of Life Sciences (Tokyo: Heibonsha), pp. 35-70.

Ponnamperuma, C. Ed. 1981. *Comets and the Origin of Life, Proceedings of the Fifth College Park Colloquium on Chemical Evolution,* (Dordrecht, Holland: D. Reidel).

Sagan, C., and Khare, N. 1979. Tholins: Organic chemistry of interstellar grains and gas. *Nature* 277:102-107.

Schidlowski, M. 1978. Evolution of earth's atmsophere: Current state and exploratory concepts. In *Origin of Life,* ed. H. Noda (Tokyo: Japan Soc. Sci. Press), p. 3.

Shimoyama, A.; Ponnamperuma, C.; and Yanai, K. 1979. Amino acids in Yamato carbonaceous chondrite from Antarctica. *Nature* 282:394-396.

Stoks, P.G., and Schwartz, A.W. 1981. Nitrogen-heterocyclic compounds in meteorites: Significance and mechanisms of formation. *Geochim. Cosmochim. Acta* 45:563-569.

Wallis, M.K. 1980. Radiogenic melting of primordial comet interiors. *Nature* 284:431-433.

Wetherill, G.W. 1975. Late heavy bombardment of the moon and terrestrial planets. *Proc. Lunar Sci. Conf.* 6:1539-1561.

Whipple, F.L. 1950. A comet model I, the acceleration of Comet Encke. *Astrophys. J.* 111:375-394.

Whipple, F.L. 1951. A comet model II, physical relations for comets and meteors. *Astrophys. J.* 113:464-474.

Whipple, F.L. 1981. The nature of comets. In *Comets and the Origin of Life,* ed. C. Ponnamperuma (Dordrecht, Holland: D. Reidel), pp. 1-20.

PART VII
Appendix

BASIC INFORMATION AND REFERENCES

BRIAN G. MARSDEN
Harvard-Smithsonian Center for Astrophysics

and

ELIZABETH ROEMER
University of Arizona

I. GENERAL: MORPHOLOGY

A. Nucleus

Most of the mass of a typical comet is contained in a compact body, the nucleus, with diameter in the range 0.1 to 10 km. The composition includes ices and clathrates, through which are dispersed grains of meteoroidal (rocky) material (Whipple and Huebner 1976). It is possible that some nuclei include a rocky core that could survive dispersal of the volatile material and be observable as an inert asteroidal body. Although no direct determination of the mass has been made on the basis of gravitational perturbations of another body by a comet, other evidence suggests masses in the range 10^{13} g to perhaps as high as 10^{19} g (chapter by Donn and Rahe).

Optical resolution of the nucleus is exceedingly difficult, either at great distance when activity is minimal, or very close to the Earth, when the surface brightness of the inner coma may be low. All observations of the nucleus of active[a] comets are to be understood as referring to a "nuclear condensation" that includes a contribution from a volume around the nucleus that is substantially larger than that of the nucleus alone. Noting that an object of diameter 725 km at a distance of 1 AU from the observer subtends an angle of 1 arcsec, it should be realized that a comet nucleus is almost invariably far short of optical resolution.

B. Coma

As a comet approaches the Sun, ices and clathrates are vaporized, leading to an outflow from the nucleus of neutral gas in which small dust grains may be entrained. This diffuse material forms the coma, the apparent dimensions of which depend on the distance of the comet from the Sun and the species that contribute to the wavelength band in which observations are made. Development of the gaseous coma typically begins at a heliocentric distance of ~ 3 AU, the CN radical generally being the first to be observed. Most comets, however, are detectably nonstellar in appearance, if observed with sufficient resolution, even at heliocentric distances as large as 8 AU.

Neutral radicals, dust, ice coated grains, atoms, and molecular ions are all present in the coma of an active comet at heliocentric distances ≤ 1 AU. The inner visible coma, which overlaps the dust coma, includes the molecular or chemical coma, with typical radius of the order of 10^4 km, and the coma composed of radicals, which may extend to distances from the nucleus of a few times 10^5 km (Whipple and Huebner 1976). The coma composed of atomic hydrogen, observable in ultraviolet light of Lyman α, may extend to 10^7 km.

The coma photographed in blue light (3000 to 4900 Å), to which emissions from the CN and OH radicals are the strongest contributors, tends to be nearly circular in outline. Comae photographed in yellow light (4900 to 6200 Å), to which dust scattering and C_2 Swan band emissions are strong contributors, appear characteristically asymmetric, with a definite boundary on the sunward side and greater extension in the antisolar direction lagging the prolonged orbital radius vector. Asymmetric comae of distinctive shape have been characteristic of a number of comets that never approached closer to the Sun than some 3 to 4 AU (Roemer 1962).

The nucleus and coma together are referred to as the head of a comet.

C. Tails

Long tails are perhaps the most striking feature of the most active comets, particularly those of small perihelion distance.

Type I tails, composed of molecular ions (e.g., CO^+, CO_2^+) are readily photographed in blue light and are distinguished by intricate and rapidly

---

[a]Active comets are defined here as those comets in which substantial amounts of gases and dust are being released from the nucleus, whether or not a coma is optically resolvable.

changing ray structure. Such tails may begin to develop as active comets approach to within ~ 1.5 AU of the Sun. The rapidly changing structure is caused by interaction between the cometary plasma and the solar wind (see Brandt and Mendis 1979 and the chapter by Brandt). The axes of type I tails are typically nearly straight and lag the prolonged orbital radius vector by only a few degrees. Tail lengths may approach 10^7 or even 10^8 km.

Type II tails, composed mainly of dust, are yellow in color and more readily visible than the blue ion tails. Some comets display both Type I and Type II tails simultaneously. Dust tails often show a pronounced curvature, even as seen in projection on the sky. The particles of which these tails are composed move independently in space under the combined forces of gravity and solar radiation pressure. Tail axes extend generally away from the Sun, but are distinctly curved, lagging the prolonged orbital radius vector by increasing angles as the heliocentric distance of each portion of the tail increases. Lengths of Type II tails of active comets near the Sun may approach 10^6 or 10^7 km. Dust or ice grain tails have been observed in comets at heliocentric distances of 4 to 5 AU and even beyond (Roemer 1962; Sekanina 1975).

Anomalous tails, sometimes seen in projection on the sunward side of the comet nucleus, are composed of submillimeter- and millimeter-size grains closely confined to the orbital plane, located on the outside of the orbit well behind the Sun-comet direction (Sekanina 1976a). Such tails are detected only during short intervals around the time when the Earth passes through the plane of the comet orbit. The aspect of the tail changes rapidly from night to night, the interval of good visibility usually lasting no more than a week or two. Such tails are typically spike-like near the exact time of orbit-plane passage, as for example, observed in Comet Arend-Roland 1957 III, on 25 April 1957.

II. SPACE DISTRIBUTION

When comets are close to the Sun and active, they are seen constantly to lose material into the coma and tail. Their lifetimes in the inner solar system are thus clearly limited to a much shorter interval than the age of the solar system as a whole. The observationally established source from which the comet population is replenished is the Oort Cloud, a reservoir of comets far from the Sun, in which comets can be maintained in pristine condition indefinitely.

The evidence for the existence of the Oort Cloud is an observed clustering of the reciprocal of the semimajor axes of the orbits upon which comets in long-period orbits approach the region of the principal planets $1/a_{\rm orig}$ ~ 4×10^{-5} AU^{-1} (Oort 1950; Marsden and Sekanina 1973; Marsden et al. 1978). This clustering in $1/a$ corresponds to a clustering of orbital periods in the range of 10^6 to 10^7 yr. Planetary perturbations during even a single passage of a comet through the region of the principal planets would tend either to eject the comet from the solar system, or to move the value of $1/a$ toward larger values by around $+5 \times 10^{-4}$ AU^{-1} (rms), depending somewhat on the perihelion distance. Comets with $1/a_{\rm orig} < 1 \times 10^{-4}$ AU^{-1} are generally defined as dynamically "new"[a]. See Weissman's chapter in this book for a review of this subject. Of observed long-period comets, about half are new in the Oort sense, and half have made previous passages near enough to the Sun for their orbits to have been affected by perturbations due to the principal planets. It should be understood that the Oort Cloud does *not* constitute a region of enhanced space density of comets at the edge of the solar system; the space density of comets is almost certainly greatest in the region near the Sun where they are observable directly.

Repeated passages of comets deflected inward from the Oort Cloud into the region of the principal planets not only leads, through gravitational perturbations, to additional losses of comets from the solar system, but also to a slow diffusion of some survivors into progressively smaller orbits. An end product is the group of short-period comets, defined as including those with periods < 200 yr. Subgroups include the short-period comets of the Halley type, with periods between 20 and 200 yr, and the Jupiter-family comets, with periods < 20 yr. The motions of the latter, in particular, are strongly affected by perturbations during repeated passages near Jupiter.

The most recent catalog of comet orbits (Marsden 1979) includes orbital elements for 1027 cometary apparitions observed up to the end of 1978. These elements refer to 658 individual comets, of which 113 have orbits of short period. Some 72 of these periodic comets have been observed at two or more apparitions, while 41 have been observed only at the discovery apparition. Of the 545 comets of long period, the orbits of 285 were listed as parabolic, observational data being insufficient to support determination of general orbits. Of the remainder, osculating orbits for 162 comets were elliptical and 98 were slightly hyperbolic.

Patterns in the distributions of various of the osculating orbital elements of both long-period and short-period comets are well known (Porter 1963; Marsden 1974). Some patterns are clearly properties of the particular populations of comets, but others are strongly influenced by observational selection effects, particularly of discoveries. For a review of this topic see the chapter by Kresák.

III. NEWLY-DISCOVERED COMETS

A. Discovery

Comets are discovered photographically by professional astronomers and visually by amateurs in approximately equal numbers. The professional discoveries are almost always made when the comets are near opposition; such objects generally remain faint and outside the Earth's orbit. The amateur discoveries are commonly made when the comets are at elongations of sub-

---

[a] See Table I in Sec. III. H.

stantially less than 90° from the Sun and geometrically located inside the orbit of the Earth; for a few weeks these comets exhibit such dramatic increases in brightness that when they pass near opposition they may be too faint to be photographed with the largest instruments. Techniques for conducting visual searches for comets have been described by Bortle (1981a), among others.

B. Confirmation

Whether a suspected discovery is made photographically or visually, there are several precautions an observer should take to establish the reality of the object. Comparison with sky charts determines whether a diffuse object is a galaxy, star cluster, or planetary nebula. This is a straightforward procedure if a supposed photographic discovery is being compared with a deeper photographic chart of the sky, but the visual observer using a hand-drawn star atlas must realize that bright nebulae are often missing from such atlases or that he may have made a marginal observation of an asterism—a group of close, faint stars individually below the nominal magnitude limit of the atlas.

Care must be taken that the supposed comet is not simply some artifact, such as a photographic defect or the visual or photographic ghost image of a bright star or planet just outside the field of the telescope. Photographic plates should always be taken in duplicate, preferably with the telescope moved slightly between exposures. Visual ghosts can generally be identified as such by jostling the telescope, changing the eyepiece, or shifting to a different telescope.

It is most important to verify that the object is moving in a regular manner with respect to the stars and at an appropriate speed. Multiple photographic exposures a few minutes apart should serve this purpose. The visual observer should be able to detect motion in about half an hour by switching to a high-power eyepiece and repeatedly relating the object's position to a recognizable pattern of faint stars. Comets found visually generally have a daily motion of a degree or so; even faint photographic discoveries rarely move at $< 5'$ per day. Conditions may render it impossible for the observer to confirm a discovery until 24 or even 48 hr later. There exists the chance that another observer will independently discover and announce the comet in the meantime; this is rather rare, and it is much more probable that the early announcement of an inadequately confirmed discovery will lead to the embarrassment of the observer.

C. Recording

If the observer is sure a comet has been found, its position must be determined as accurately as possible. In the case of a visual discovery this should be done by comparison with a suitably calibrated star chart, not from simply reading the setting circles of the telescope. A photographic discoverer should follow the normal procedures of photographic astrometry (see Sec. III. E below), although for a rapid initial announcement the measuring process can be simplified by using a ruler or grid rather than a measuring engine. The comet's position should be determined at two or more specified times. For a photographic discovery the time is normally that of the middle of the exposure, and the measurement refers to the middle of any trailed images. If the comet appears only as a fuzzy trail on a single exposure, the positions of both ends should be determined; the sense of motion will be ambiguous, although Miller (1965) has noted that in such a case the end of the trail tends to be less dense than the beginning.

The observer should also estimate the comet's brightness, normally by giving the total magnitude m_1 (see Sec. V). A general description of the comet's appearance is also in order, in particular the presence or absence of central condensation or tail.

D. Announcement

The information collected in Sec. III.C should be communicated by telegram (or TWX/telex or cablegram) to the International Astronomical Union's Central Telegram Bureau. The number for TWX/telex users is (U.S.A.) 710-320-6842 (answerback ASTROGRAM CAM). Telegrams and cablegrams should include in the address "TWX 7103206842 ASTROGRAM CAM" and optionally "CENTRAL BUREAU FOR ASTRON CAMBRIDGEMASS". The telegram can be supplemented (but not supplanted) by making a telephone call to (U.S.A.) (617) 864-5758 and leaving a 30-second recorded message.

To guard against transmission errors and to ensure that relevant information is not accidentally omitted it is recommended that the information be sent according to the official IAU code. This is described in detail in IAU *Inf. Bull.* No. 46, pp. 12-16 and briefly in the following example:

BRADFIELD	Name of discoverer.
COMET	Type of object.
BRADFIELD	Name of observer.
19501	Equinox to which the comet's position is referred; final 1 signifies that the position is approximate.[a]
01217	Date of observation (1980 Dec. 17).
751//	Time of observation (.751) in units of 10^{-5} day (UT); / signifies a suppressed digit.
16200	Right ascension ($16^h 20^m 0$).
13616	Declination ($-36°16'$); for positive declination the initial digit should be 2.
01068	Magnitude 06. The first zero in this group signifies a redundant digit. The 1 signifies that the magnitude is total

---

[a]In the case of accurate position, the final digit of this group is a 2 and the three groups giving the right ascension, declination and magnitude would be extended to four: 16195 73413 61542 62137 (signifying $16^h 19^m 57\overset{s}{.}34$, $-36°15'42\overset{''}{.}6$, nuclear magnitude $m_2 = 13$, comet diffuse with condensation, no tail).

BASIC INFORMATION AND REFERENCES

	(m_1). The 8 indicates that the comet is diffuse with central condensation and has a tail $< 1°$; add 1 if the tail is $> 1°$; subtract 1 if no tail at all is observed; subtract 3 if no condensation is observed.
2035/	Daily motion in right ascension ($+03\overset{m}{.}5$).
20018	Daily motion in declination ($+00°18'$); the daily motion can be omitted and two separate observations transmitted instead if desired.
67070	The last five digits of the sum of the previous groups, with / counted as zero.
30884	The last five figures of the sum of the groups giving right ascension, declination and magnitude.
BRADFIELD	Communicator; if not known to the Central Bureau, the discoverer should also supply his full name, address and telephone number.

It is also useful for the discoverer to provide more details of his observation by mail, using the address:

> Central Bureau for Astronomical Telegrams
> Smithsonian Astrophysical Observatory
> 60 Garden Street
> Cambridge MA 02138
> U.S.A.

After a report has been verified, the Central Bureau will assign a year-and-letter designation to a comet discovery and relay the information to the astronomical community by IAU telegram and postcard IAU *Circulars*.

E. Astrometry

The process of photographic astrometry involves the accurate measurement of the images of the comet and reference stars on a photographic plate. The measurement should be made with a conventional screw-type measuring engine or optical device capable of recording relative positions of images on the plate with accuracy $\sim 1\ \mu m$. For the resulting position of the comet in the sky one should aim at an accuracy of $1''$ to $2''$; the accuracy will be diminished if the focal length of the telescope is very small and the exposure such that the images (especially of the comet) are large and irregular.

If (α_i, δ_i) represent the right ascension and declination of a reference star, and (A, D) an adopted position for the center of the plate, then the corresponding standard coordinates (ξ_i, η_i) of the reference star are given by

$$\xi_i = H_i^{-1} \cos \delta_i \sin (\alpha_i - A)$$
$$\eta_i = H_i^{-1} [\sin \delta_i \cos D - \cos \delta_i \sin D \cos (\alpha_i - A)] \qquad (1)$$

where
$$H_i = \sin \delta_i \sin D + \cos \delta_i \cos D \cos (\alpha_i - A). \tag{2}$$

The measured coordinates (x_i, y_i) on the plate are related to the standard coordinates by equations of the form

$$\xi_i - x_i L^{-1} = ax_i + by_i + c$$
$$\eta_i - y_i L^{-1} = a'x_i + b'y_i + c' \tag{3}$$

where L is the adopted focal length of the telescope. Determination of the plate constants a, b, c, a', b', c' requires the simultaneous solution of Eqs. (3) using several reference stars. Although a minimum of three stars will suffice, a least-squares solution using 4 to 8 stars is strongly recommended. To a first approximation it will be found that $a = b'$ and $a' = -b$. The standard coordinates (ξ_0, η_0) for the comet then follow from its measured coordinates (x_0, y_0) by further application of Eqs. (3), and the resulting right ascension and declination (α_0, δ_0) are obtained from

$$\tan (\alpha_0 - A) = \Delta_0^{-1} \xi_0$$
$$\tan \delta_0 = \Gamma_0^{-1} (\sin D + \eta_0 \cos D) \tag{4}$$

where, except very near a pole, $|\alpha_0 - A| \ll 90°$ and

$$\Delta_0 = \cos D - \eta_0 \sin D$$
$$\Gamma_0 = (\xi_0^2 + \Delta_0^2)^{1/2} > 0. \tag{5}$$

Unless the observation is made with a long-focus reflector, suitable reference stars can be identified in, for example, the *SAO Catalogue*. The equinox 1950.0 positions given there need be adjusted only for proper motion, and the resulting coordinates for the comet will also be referred to equinox 1950.0. The reduction of exposures with long-focus reflectors requires either the intermediary of field plates with a wide-field instrument or use of the *Astrographic Catalogue*; for further information on the *Astrographic Catalogue* see Van Biesbroeck (1963).

The "dependence" method for reducing astrometric observations is popular with many astronomers. However, it contains fewer checks, and modern improvements in computing machinery tend to negate its simplicity, which is most significant when used in the form given by Comrie (1929). For further information on both procedures, see for example Smart (1931).

The resulting 1950.0 coordinates for the comet, quoted to $0\overset{s}{.}01$ (except at very high declinations) for α_0 and to $0\overset{''}{.}1$ for δ_0, together with the time of mid-exposure quoted to 0.00001 day, should be communicated to the

Central Bureau for Astronomical Telegrams for publication in the IAU *Circulars* and *Minor Planet Circulars/Minor Planets and Comets (MPC's)* and inclusion in the machine-readable file of observations. Rapidly-reduced, early positions of new discoveries should be telegraphed as described in Sec. III. D.

F. Ephemerides

An initial determination of the orbit of a comet is possible in principle when three accurate measurements of the comet's position suitably separated in time are available. Since some 80% of all new comets discovered (and > 95% of comets discovered visually) have approximately parabolic orbits, the initial determination usually assumes that the orbit is exactly parabolic. Methods of orbit determination are described in detail in Herget (1948), Väisälä and Oterma (1951) and Dubyago (1961).

The computation of an ephemeris position at any time requires first the calculation of the comet's heliocentric position vector (x, y, z). This is given by

$$x = P_x r \cos v + Q_x r \sin v$$
$$y = P_y r \cos v + Q_y r \sin v \tag{6}$$
$$z = P_z r \cos v + Q_z r \sin v$$

where the P's and Q's (the corresponding R's are not required) are conveniently obtained from the product of rotation matrices:

$$\begin{pmatrix} P_x & P_y & P_z \\ Q_x & Q_y & Q_z \\ R_x & R_y & R_z \end{pmatrix} = \begin{pmatrix} \cos \omega & \sin \omega & 0 \\ -\sin \omega & \cos \omega & 0 \\ 0 & 0 & 1 \end{pmatrix} \begin{pmatrix} 1 & 0 & 0 \\ 0 & \cos i & \sin i \\ 0 & -\sin i & \cos i \end{pmatrix} \begin{pmatrix} \cos \Omega & \sin \Omega & 0 \\ -\sin \Omega & \cos \Omega & 0 \\ 0 & 0 & 1 \end{pmatrix} \begin{pmatrix} 1 & 0 & 0 \\ 0 & \cos \epsilon & \sin \epsilon \\ 0 & -\sin \epsilon & \cos \epsilon \end{pmatrix}$$

the argument of perihelion ω, longitude of the ascending node Ω and inclination i (generally for equinox 1950.0) being provided by the orbit computer and the obliquity of the ecliptic ϵ taking on the 1950.0 value of $23°.445789$.

In the case of a parabolic orbit, the remaining quantities required in Eqs. (6), involving radius vector r and true anomaly v, follow from

$$r \cos v = q (1 - \tau^2)$$
$$r \sin v = 2q\tau \tag{8}$$

where q is the comet's perihelion distance, and $\tau = \tan 1/2\, v$ is related to the time $t - T$ since perihelion passage by the cubic equation

$$k (t - T) = (2q^3)^{1/2} (\tau + 1/3\, \tau^3). \tag{9}$$

The crux of the process lies in the solution of this equation for τ at various times t. If distances are measured in AU and time in days, $k = 0.01720209895$.

The comet's geocentric position vector can now be derived from the heliocentric vector by

$$\xi = x + X$$
$$\eta = y + Y \tag{10}$$
$$\zeta = z + Z$$

where (X, Y, Z), the Sun's geocentric position vector (again for equinox 1950.0), can be taken from the 7-figure tabulation at 4-day intervals from 1800 to 2000 by Herget (1953) or calculated (to 5 digit accuracy) from the expressions by Doggett et al. (1977).

Finally, the heliocentric and geocentric distances r and Δ, and right ascension and declination α and δ follow from

$$r = (x^2 + y^2 + z^2)^{1/2}$$
$$\Delta = (\xi^2 + \eta^2 + \zeta^2)^{1/2} \tag{11}$$
$$\tan \alpha = \eta/\xi$$
$$\tan \delta = \zeta/\Gamma$$

where $\sin \alpha$ has the sign of η, $\cos \alpha$ that of ξ, and

$$\Gamma = (\xi^2 + \eta^2)^{1/2} > 0. \tag{12}$$

The above procedure describes the production of a so-called geometric ephemeris from specified parabolic orbital elements T, q, ω, Ω, i. Precise comparison with astrometric observations is not directly possible because of the effects of:

(a) the irregular rotation of the Earth;
(b) the aberration due to the finite velocity of light;
(c) the parallactic displacement of the observer from the center of the Earth.

The effect of (a) is the smallest of the three and can be allowed for by noting that the determination of (X, Y, Z) is in Ephemeris Time (ET) (usually for 0 hr on a date when the Julian Date (JD) leaves, upon division by 10, a remainder of 0.5); interpolation in an ephemeris should therefore be for a Universal Time (UT) of observation adjusted by

$$ET - UT = +0.00059 + 0.0000119 \, (U - 1980.0) \tag{13}$$

where this expression is reasonably valid for year U from 1960 onward.

Effect (b) can amount to 0".5 or more but was ignored in most published ephemerides until late 1979. After a resolution of IAU Commission 20 at that time, compensation is now made for the effect. It is accomplished by noting that the time t utilized in Eq. (9) should be antedated by Δ/c, c being the velocity of light. Thus, one should consider instead of t, the time in terms of the units specified

$$t' = t - 0.005776 \Delta. \tag{14}$$

The determination of t' and the final solution through Eqs. (11) requires an initial approximate solution for Δ using time t in Eq. (9). However, note that computation of the Sun's geocentric vector required in Eqs. (10) must be for the observed time t, not for the time t'. An ephemeris corrected in this manner is termed an *astrometric* ephemeris. (Strictly speaking, it is an *astrographic* ephemeris, for the small difference, $\lesssim 0".3$, due to the so-called elliptic aberration terms in the catalogued 1950.0 star positions has been ignored; the difference will become meaningless with the introduction of the Fifth Fundamental Catalogue (FK5) and subsequent catalogues that eliminate these terms.) The IAU resolution also permits the publication of *apparent* ephemerides, in which the positions have been adjusted by stellar aberration, nutation and precession from the true values referred to a standard mean equinox to the apparent values referred to the true equinox of date.

Allowance for the effect (c) of parallax is not practicable in published ephemerides, but it can conveniently be taken into account in specialized ephemerides by replacing the vector (X, Y, Z) in Eqs. (10) by

$$\begin{aligned} X' &= X - \rho' \cos \phi' \cos \theta \\ Y' &= Y - \rho' \cos \phi' \sin \theta \\ Z' &= Z - \rho' \sin \phi' \end{aligned} \tag{15}$$

where

$$\theta = 241°.40 + 0.9856091 (t - t_0) + 360 t_1 + \lambda \tag{16}$$

is the local sidereal time (again adjusted to correspond to equinox 1950.0) at time t at geographic longitude (east of Greenwich) λ, geocentric latitude ϕ' and geocentric distance ρ'; λ and the quantities $\rho' \cos \phi'$ and $\rho' \sin \phi'$ are conveniently tabulated (in astronomical units under the headings Δ_{xy} and ΔZ) for major observatories in the *American Ephemeris and Nautical Almanac* for 1980 and earlier and on *MPC* Nos. 4766-4768; t_0 is the epoch 1968 May 24.0 UT = JD 2440000.5, and t_1 is the UT fraction of a day.

As further astrometric observations are made of a comet, successive least-squares differential corrections improve the determination of its orbit. When the observations cover weeks or months, deviations from parabolic motion may be noted, and in the resulting computation of an ephemeris Eqs. (8) and (9) will be replaced by a more complex procedure, described by, for example, Benima et al. (1969). At the same time, it will be necessary to allow for planetary perturbations. That such allowance has been made is indicated by specifying a particular Epoch at which the comet's actual position and velocity vectors are identical with the corresponding values derived from standard equations for unperturbed motion in a conic.

G. Catalogues

The standard catalogue of cometary orbits is that by Marsden (1979). The comets are listed by Roman-numeral designations, which indicate the order of perihelion passage of the comets in a particular year; new designations are published annually in the *MPC's*. In addition to the general table and lists of references and names of comets, there are supplementary tables showing the orbits in order of the values of specific elements. For past comets and the collection of several orbits determined for each comet, refer to an earlier detailed catalogue by Galle (1894).

Vsekhsvyatskij (1958) gives a less definitive catalogue of orbits but includes data on absolute magnitudes and actual descriptions of cometary apparitions. Similar and more detailed data for older comets have been provided by Holetschek (1896). A much older but still useful compilation of descriptive data is that by Pingré (1783,1784). Ho Peng Yoke (1962) gives a useful account of the oriental records of ancient comets; Hasegawa (1980) lists basic data on naked-eye comets. Orbital and descriptive data on recent comets have been included in the annual reports in the *Quarterly Journal of the Royal Astronomical Society*.

H. Long-term Motion

The orbital elements so far discussed relate to conics at specified osculation epochs with the Sun at a focus. The long-term motion of a comet in a nearly-parabolic orbit is better described by a conic with the center of mass of the solar system at a focus. The conversion from heliocentric to long-term barycentric motion is accomplished by integrating the heliocentric motion to heliocentric distance 30 to 40 AU and then making the necessary adjustment to the barycenter. Separate determinations must of course be made for past and future motions, but unless a comet recedes far enough from the Sun that stellar perturbations become significant, the conics thus defined will represent the comet's motion until it makes another passage through the inner solar system. The most significant element in these barycentric orbits is $1/a$, the reciprocal of the semimajor axis; the distribution of data according to this quantity relates to the concept of the Oort Cloud and whether orbits are or are not bound to the solar system.

Everhart and Raghavan (1970) have tabulated for known comets the quantities $u_b = (1/a)_{osc} - (1/a)_{orig}$ and $u_a = (1/a)_{fut} - (1/a)_{osc}$ whereby heliocentric values at the time of perihelion passage (and for an osculation epoch consisting of the nearest occasion when division of the Julian Date by

40 leaves remainder 0.5) can be converted into the barycentric original and future values. A comprehensive and critical list of absolute values of $(1/a)_{\text{orig}}$ and $(1/a)_{\text{fut}}$ has been given by Marsden et al. (1978). This table is modified, rearranged and updated below (see Table I). The 220 entries are here arranged according to increasing values of orig = $(1/a)_{\text{orig}}$, which, like fut = $(1/a)_{\text{fut}}$, is tabulated in units of 10^{-6} AU^{-1}. Comets are specified by both Roman-numeral and letter designation, and their names are also given. The column q gives the perihelion distance (in AU). Cl. indicates the quality class for the orbit, as defined by Marsden et al. (1978); IA orbits are the most reliable, IIB the least. An asterisk indicates that the orbit solution can be improved by solving also for nongravitational parameters, in which case the values of $1/a$ will be increased; it is probable that all the negative values of $(1/a)_{\text{orig}}$ would be eliminated if allowance could properly be made for nongravitational effects. The new entries in the table are from calculations by E. Everhart applied to new osculating orbits calculated by B.G. Marsden, S. Nakano, and R.J. Buckley.

IV. RETURNING COMETS

A. Recovery

A small but significant fraction of new comets are found to have elliptical orbits of moderate to low eccentricity, so these objects will return to perihelion after intervals of a few years. In a given year the numbers of new comets discovered and returning comets recovered are roughly comparable. Recoveries are generally made photographically with long-focus reflectors or Schmidt telescopes when the comets are in the range 17 to 20 mag.

B. Confirmation

Since a comet at recovery is usually inconspicuous and can be some distance from its predicted position, some precautions are in order. The object's motion during the hour or so of the exposure is not a sufficient test, for most returning periodic comets have orbits of low inclination so that both position and motion can resemble those of many faint, unidentified minor planets. A more significant test of the recovery exposure lies in the agreement of the comet's position with the line of variation. The time of perihelion passage is by far the most uncertain orbital element for a comet at a predicted return; an ephemeris will frequently list with α and δ corrections $\Delta\alpha$ and $\Delta\delta$ corresponding to a change of + 1 d in the nominal perihelion time. If the recovery is correct, the comet should be close to the arc passing through the points (α, δ) and $(\alpha + \Delta\alpha, \delta + \Delta\delta)$.

The comet's motion is a necessary test; observation the next night will generally remove any doubt about the recovery. Although faint comets observed under good conditions will almost always be distinguishable from minor planets by the presence of some diffuseness, under marginal or adverse conditions the two classes of objects can look identical, and in rare instances

TABLE I
"Origins" and "Future" Orbits of Long-Period Comets

Comet		q	orig.	fut.	Cl.	Name	Comet		q	orig.	fut.	Cl.	Name	Comet		q	orig.	fut.	Cl.	Name
1976 I	-5q	0.864	- 734	- 1461	IIB	Sato	1962 III	-2c	0.031	+ 25	+ 352	IB	Seki-Lines	1973 IX	-3o	3.842	+ 71	- 109	IB	Gibson
1955 V	-5g	0.885	- 727	- 432	IIA*	Honda	1907 I	-7a	2.052	+ 25	+ 290	IB	Giacobini	1915 II	-5a	1.005	+ 75	+ 960	IA*	Mellish
1959 III	-9d	1.251	- 446	- 1956	IIB	Bester-H.	1903 II	-2d	2.774	- 26	+ 488	IA	Giacobini	1959 II	-8e	1.528	+ 76	+ 244	IA	Burnham-Slaughter
1895 IV	-5c	0.192	- 172	+ 457	IB	Perrine	1914 III	-4c	3.747	+ 27	- 1	IA	Neujmin	1941 VIII	-1d	0.875	+ 78	+ 190	IB	van Gent
1971 V	-1a	1.233	- 142	- 488	IIA*	Toba	1902 IV	-2b	0.401	+ 27	+ 865	IB	Perrine	1966 V	-6b	2.385	+ 78	+ 591	IIA	Kilston
1960 II	-9k	0.504	- 135	- 587	IIA*	Burnham	1948 II	-8a	1.500	+ 28	+ 53	IB	Mrkos	1954 V	-5b	4.496	+ 82	- 21	IB	Abell
1953 III	-2f	0.778	- 125	- 283	IB	Mrkos	1905 IV	-6b	3.340	+ 28	+ 519	IB	Kopff	1890 II	-0a	1.908	+ 89	+ 126	IB	Brooks
1940 IV	-0e	1.062	- 124	- 1123	IB	Okabayasi-Honda	1978 XXI	-8f	1.137	- 29	- 1023	IA	Meier	1980b		3.364	+ 120	-15938	IIA	Bowell
1899 I	-9a	0.327	- 109	- 1253	IB	Swift	1914 V	-3f	1.104	+ 29	+ 63	IA	Delavan	1937 V	-7f	0.863	+ 124	+ 281	IIA	Finsler
1957 III	-6h	0.316	- 98	- 604	IA*	Arend-Roland	1977 IX	-8a	5.606	- 33	- 102	IA	West	1914 II	-4a	1.199	+ 126	+ 1561	IIA	Kritzinger
1968 VI	-8c	1.160	- 82	+ 260	IIA	Honda	1903 IV	-3c	0.330	+ 33	+ 1000	IB	Borrelly	1910 I	-0a	0.129	+ 135	+ 674	IIA	Great comet
1904 IV	-4d	1.882	+ 75	+ 170	IIA	Giacobini	1979 VII	-9d	0.413	+ 33	+ 126	IIA	Bradfield	1976 XII	-7c	5.715	+ 142	+ 97	IB	Lovas
1911 IV	-1g	0.303	+ 74	+ 175	IIA	Beljawsky	1948 V	-8d	2.107	+ 33	+ 31	IA	Pajdusakova-Mrkos	1882 II	-2a	0.061	+ 144	+ 779	IIA	Wells
1898 VIII	-8j	2.285	- 71	+ 620	IB	Chase	1947 VIII	-0k	3.261	+ 34	+ 225	IA	Wirtanen	1908 III	-8c	0.945	+ 174	+ 393	IIA	Morehouse
1932 VII	-2f	1.647	+ 56	- 327	IB	Newman	1925 VI	-5a	4.181	+ 35	+ 128	IA	Shajn-Comas Sola	1847 II		2.116	+ 180	- 453	IIA	Coila
1975 XI	-5p	0.219	+ 56	+ 1218	IIA	Bradfield	1975 VIII	-4c	3.011	+ 36	+ 517	IIA	Lovas	1847 VI	-9a	0.329	+ 212	+ 98	IIB	Mitchell
1942 VIII	-2b	4.313	- 34	+ 282	IA	Oterma	1954 XII	-4d	0.746	+ 36	+ 59	IIA	Kresak-Peltier	1904 I		2.708	+ 227	+ 515	IA	Brooks
1892 VI	-2d	0.976	+ 27	+ 539	IB	Brooks	1892 IX	-6k	5.857	+ 37	+ 189	IA	Lovas	1947 VI	-7h	2.828	+ 234	+ 233	IA	Wirtanen
1849 II	-5d	1.159	- 25	+ 1021	IIA	Goujon	1951 I	-0b	2.572	+ 37	+ 270	IA	Minkowski	1977 XIV	-7m	0.991	+ 234	+ 1314	IA	Kohler
1886 I	-5d	0.642	- 18	- 272	IB*	Fabry	1956 I	-4k	4.077	+ 39	+ 228	IA	Haro-Chavira	1958 III	-8a	1.323	+ 256	+ 1798	IIA	Burnham
1946 I	-6a	1.724	+ 13	+ 373	IIA	Timmers	1925 I	-5c	1.109	+ 40	+ 533	IA	Orkisz	1950 I	-9a	2.553	+ 263	+ 537	IA	Johnson
1947 I	-6k	2.408	+ 1	+ 26	IA	Bester	1959 X	-9a	4.267	+ 40	- 1	IA	Humason	1970 XV	-0g	1.113	+ 283	+ 215	IA	Abe
1941 IV	-0c	0.368	- 2	+ 1570	IIA	Cunningham	1979 VI	-9e	4.687	+ 42	+ 145	IIA	Torres	1977 X	-7q	3.603	+ 314	+ 549	IA	Tsuchinshan
1952 VI	-2b	1.202	+ 2	+ 410	IIA	Peltier	1955 IV	-4h	3.870	+ 42	+ 252	IA	Baade	1973 II	-2j	2.147	+ 320	+ 28	IA	Kojima
1897 I	-6f	1.063	+ 5	- 799	IB	Perrine	1946 VI	-6h	1.136	+ 44	+ 16	IA	Jones	1886 II	-5e	0.479	+ 332	+ 118	IB	Barnard
1974 XII	-4g	6.019	+ 11	+ 569	IA	van den Bergh	1976 XIII	-6g	1.569	+ 45	+ 570	IIA	Harlan	1910 IV	-0b	1.948	+ 474	+ 288	IB	Metcalf
1853 VI		0.307	+ 12	+ 636	IB	Klinkerfues	1932 VI	-2g	2.314	+ 45	+ 240	IA	Geddes	1972 XII	-21	4.861	+ 476	+ 356	IA	Araya
1863 IV	-2a	1.313	+ 14	+ 770	IA	Baeker	1912 II	-2a	0.716	+ 45	+ 363	IB	Gale	1949 IV	-8h	2.517	+ 498	+ 581	IA	Wirtanen
1942 VII	-2a	1.445	+ 16	+ 821	IB	Whipple-B.-K.	1886 IX	-6f	0.663	+ 46	+ 48	IA	Barnard-Hartwig	1969 IX	-9g	0.473	+ 507	+ 431	IB	Tago-Sato-Kosaka
1917 III	-6b	1.686	+ 17	+ 763	IA	Wolf	1889 I	-8e	1.815	+ 48	+ 573	IA	Barnard	1863 I		0.795	+ 521	+ 771	IA	Bruhns
1957 VI	-6c	4.447	+ 17	+ 148	IA	Wirtanen	1954 VIII	-4f	0.677	+ 49	+ 217	IIA	Vozarova	1930 IV	-0b	2.079	+ 524	+ 44	IB	Beyer
1921 II	-1a	1.008	+ 18	+ 351	IA	Reid	1967 II	-6h	0.419	+ 49	+ 889	IIA	Rudnicki	1948 IV	-8g	0.208	+ 525	+ 1937	IIB	Honda-Bernasconi
1944 IV	-0e	2.226	+ 19	+ 522	IA	Van Gent	1972 VIII	-8a	2.511	+ 49	+ 362	IA	Heck-Sause	1973 X	-3k	4.812	+ 531	+ 508	IA	Sandage
1936 I	-5d	4.043	+ 19	+ 281	IA	Van Biesbroeck	1975 II	-2i	6.881	+ 51	+ 48	IB	Schuster	1880 II	-0b	1.814	+ 534	+ 103	IB	Schaeberle
1973 XII	-3f	0.142	+ 20	+ 545	IA	Kohoutek	1900 I	-c	1.332	+ 57	+ 722	IB	Giacobini	1970 II	-9b	1.719	+ 555	+ 896	IA	Kohoutek
1919 V	-9c	1.115	- 20	- 42	IIA	Metcalf	1937 IV	-7b	1.734	+ 62	+ 1400	IB	Whipple	1959 IV	-9e	1.150	+ 593	+ 476	IIA	Alcock
1922 II	-2d	2.259	+ 21	+ 523	IA	Baade	1898 VII	-8c	1.702	+ 68	+ 709	IA	Coddington-Pauly	1900 II	-0b	1.015	+ 610	+ 565	IIB	Borrelly-Brooks
1975 VII	-5i	1.217	- 23	+ 841	IA	Bradfield	1972 IX	-2h	4.276	+ 69	+ 603	IA	Sandage	1927 IV	-7d	3.684	+ 623	+ 1087	IA	Stearns
1925 VII	-5i	1.566	+ 24	+ 322	IA	Van Biesbroeck	1959 IX	-9j	1.253	+ 69	+ 2575	IB	Mrkos	1905 VI	-6a	1.296	+ 630	+ 363	IIB	Brooks
1948 I	-7k	0.748	- 24	- 366	IA	Bester	1954 X	-3g	0.970	+ 70	- 965	IB	Abell	1966 II	-6c	2.019	+ 643	+ 1358	IIA	Barbon

TABLE I (continued)

Comet		q	orig.	fut.	Cl.	Name	Comet		q	orig.	fut.	Cl.	Name	Comet		q	orig.	fut.	Cl.	Name
1974 III	-4b	0.503	+ 690	+ 705	IB*	Bradfield	1907 IV	-7d	0.512	+ 2650	+ 3322	IA	Daniel	1963 I	-3a	0.632	+11389	+10492	IA	Ikeya
1949 IV	-9c	2.058	+ 735	+ 957	IB	Bappu-Bok-Newkirk	1964 IX	-4h	1.259	+ 2721	+ 3641	IB	Everhart	1886 V	-6a	0.270	+12213	+11870	IIA	Brooks
1975 IX	-5h	0.426	+ 821	+ 130	IB	Kobayashi-B.-M.	1826 IV		0.853	+ 2749	+ 2890	IIA	Pons	1882 II	-2b	0.008	+12265	+12791	IIB	Great comet
1844 III		0.855	+ 824	+1322	IIA	Mauvais	1889 IV	-9e	1.040	+ 2899	+ 2849	IIA	Davidson	1961 VIII	-1f	0.681	+13656	+12651	IB	Seki
1968 I	-7n	1.697	+ 842	+ 704	IA	Ikeya-Seki	1953 III	-3a	1.022	+ 2983	+ 3215	IB	Mrkos-Honda	1936 V	-7g	1.953	+14623	+15721	IIB	Hubble
1892 II	-2b	1.971	+ 846	+ 513	IB	Denning	1945 I	-4b	2.411	+ 3058	+ 2385	IA	Vaisala	1930 III	-0c	0.482	+16959	+16217	IIA	Wilk
1980u	-0c	1.658	+ 851	+1315	IB	Panther	1874 III	-4c	0.676	+ 3206	+ 1814	IIA	Coggia	1975 X	-5k	0.838	+17378	+17420	IIB	Suzuki-Saigusa-M.
1890 VI	-9b	1.260	+ 916	+1283	IIB	Denning	1953 I	-2e	1.665	+ 3277	+ 3323	IB	Harrington	1940 IV	-0d	1.082	+18093	+19095	IIA	Whipple-P.
1889 II	-3b	2.256	+ 933	+1268	IB	Barnard	1822 IV		1.145	+ 3481	+ 4133	IB	Pons	1898 I	-8b	1.095	+18475	+17673	IB	Perrine
1893 II		0.675	+ 987	+1791	IIA	Rordame-Quenisset	1925 III	-5b	1.633	+ 3778	+ 4521	IB	Reid	1979 IX	-9i	1.432	+18580	+19287	IIA	Meier
1980t		0.260	+1034	+1943	IIA	Bradfield	1980q	-7e	1.520	+ 3947	+ 4017	IIA	Meier	1964 VIII	-4f	0.822	+18662	+19224	IIB	Ikeya
1864 III		0.931	+1056	+1305	IIA	Donati-Toussaint	1887 IV	-5f	1.394	+ 4276	+ 2975	IIB	Barnard	1861 I		0.921	+18770	+17775	IIB	Thatcher
1903 I	-3a	0.411	+1063	+1030	IIB	Giacobini	1955 IV		1.427	+ 4353	+ 4254	IIA	Bakharev-M.-K.	1861 II		0.822	+19676	+18242	IB	Great comet
1924 I	-4a	1.756	+1076	+ 703	IIA	Reid	1825 IV		1.241	+ 4488	+ 4946	IIA	Pons	1955 III	-5e	0.534	+20013	+20872	IA	Mrkos
1892 I	-2a	1.027	+1187	+1756	IB	Swift	1962 VIII	-1e	2.133	+ 4935	+ 5403	IA	Humason	1931 III	-1b	1.047	+21020	+20128	IB	Nagata
1963 III	-3b	1.537	+1281	+1725	IIA	Alcock	1920 III	-0c	1.148	+ 5023	+ 6896	IIB	Skjellerup	1932 V	-2k	1.037	+22945	+22774	IIB	Peltier-Whipple
1948 XI	-8l	0.135	+1294	+ 518	IIA	Eclipse comet	1811 I	-1b	1.035	+ 5103	+ 4833	IIA	Great comet	1932 I	-2b	1.254	+23175	+22729	IIB	Houghton-Ensor
1873 V	-3d	0.385	+1320	+ 803	IIB	Henry	1881 III		0.735	+ 5461	+ 6314	IIA	Great comet	1840 IV		1.480	+23181	+22965	IIB	Bremiker
1952 I	-1i	0.740	+1348	+ 569	IIB	Wilson-Harrington	1888 V	-8f	1.528	+ 5626	+ 5786	IB	Barnard	1979 X	-9l	0.545	+23184	+22660	IIB	Bradfield
1863 II		1.068	+1489	+ 816	IIB	Klinkerfues	1840 II		1.220	+ 5713	+ 6136	IIB	Galle	1932 X	-2n	1.131	+24562	+24661	IIB	Dodwell-Forbes
1969 I	-8j	3.316	+1502	+1601	IIA	Thomas	1926 I	-6b	1.345	+ 5808	+ 6326	IIB	Blathwayt							
1973 VII	-3e	1.382	+1541	+2090	IA	Kohoutek	1947 IV		0.560	+ 5924	+ 4564	IIA	Rondaninm-Bester							
1863 IV		0.706	+1546	+2544	IIB	Tempel	1911 IV	-1c	0.489	+ 6280	+ 6506	IA	Brooks							
1976 VI	-5n	0.197	+1569	+ 29	IA	West	1911 IV	-1b	0.684	+ 6337	+ 6136	IIB	Kiess							
1971 I	-2e	3.277	+1582	+1311	IIB	Gehrels	1858 VI		0.578	+ 6370	+ 6913	IB	Donati							
1847 III		1.766	+1604	+1860	IIA	Mauvais	1871 IV	-1e	0.691	+ 6579	+ 6952	IIB	Tempel							
1927 IX	-7k	0.176	+1674	+ 907	IIA	Skjellerup-M.	1888 I	-8a	0.699	+ 6648	+ 6290	IIB	Sawerthal							
1975 XII	-5j	1.604	+1964	+1545	IIB	Mori-Sato-F.	1970 II	-9i	0.538	+ 7334	+ 7142	IIA	Bennett							
1957 V	-7d	0.355	+2001	+3082	IB	Mrkos	1943 II		1.354	+ 7352	+ 6146	IB*	Whipple-Fedtke-T.							
1941 IV	-1c	0.790	+2029	+1130	IIB	de Kock-P.	1939 I	-9a	0.716	+ 7813	+ 6716	IIB	Kozik-Peltier							
1877 III	-7c	1.009	+2074	+2422	IIB	Swift-Borrelly-B.	1964 VI	-4c	0.500	+ 8131	+ 9517	IB	Tomita-Gerber-H.							
1863 III	-3e	0.629	+2083	+1395	IIB	Respighi	1936 II	-6a	1.100	+ 8294	+ 7923	IIA	Peltier							
1850 I		1.081	+2155	+1353	IIA	Petersen	1887 III	-7b	1.630	+10004	+10618	IIA	Brooks							
1961 VI	-1d	0.040	+2209	+3674	IIA	Wilson-Hubbard	1976 II	-6a	0.848	+10036	+ 7243	IIA	Bradfield							
1914 IV	-4e	0.713	+2239	+2782	IIA	Campbell	1961 II	-0n	1.062	+10480	+ 9640	IIA	Candy							
1972 III	-2f	0.927	+2297	+2252	IIA	Bradfield	1963 V	-3e	0.005	+10697	+11722	IIA	Pereyra							
1847 I	-1f	0.043	+2438	+2658	IIB	Hind	1935 I	-5a	0.811	+10774	+11387	IIA	Johnson							
1911 VI		0.788	+2491	+2761	IB	Quenisset	1936 III	-6b	0.518	+11007	+11329	IIB	Kaho-Kozik-Lis							
1844 III		0.251	+2589	+4075	IIB	Great comet	1975 VIII	-5f	0.008	+11128	+11613	IB	Ikeya-Seki							
1948 X	-8m	1.273	+2633	+2302	IIA	Bester	1894 II	-4b	0.983	+11206	+10459	IIB	Gale							

a recovery cannot be confirmed until accurate astrometric measurements are made on several nights.

C. Recording

The procedure under Sec. III.C should be followed, except that visual observations are inapplicable, and unless a comet is particularly far from its predicted position or is unusually bright, there is little value in the derivation of an approximate position. If a long-focus reflector is used the magnitude will be closer to an m_2 than to an m_1 (see Sec. V).

D. Announcement

Sec. III.D is applicable, except that there may be no need for the urgency of a telegram. Recovered periodic comets share in the system of cometary year-and-letter designations.

E. Astrometry

Sec. III.E applies.

F. Ephemerides

The process of ephemeris computation is as described in Sec. III.F except that Eqs. (8) are replaced by

$$r \cos v = a(\cos E - e)$$
$$r \sin v = a(1 - e^2)^{1/2} \sin E. \tag{17}$$

Here a and e are the semimajor axis and the orbital eccentricity, and from them the perihelion distance $q = a(1 - e)$ can be derived. The eccentric anomaly E is derived from the time t by successive approximations from the transcendental equation that replaces Eq. (9):

$$k(t - T) = a^{3/2} (E - e \sin E). \tag{18}$$

Note that if E, as used in the first term on the right-hand side, is in degrees, then k and e must be multiplied by $180/\pi = 57.29577951...$, the number of degrees in a radian. The quantity $180\,k/\pi a^{3/2}$ may be denoted by $n°$, the comet's mean motion in deg/day. The remarks in Sec. III. F about antedating by the light time are also applicable.

After recovery of a comet, the departure of observed α and δ from the prediction can be compared with the variations $\Delta\alpha$, $\Delta\delta$ to yield an approximate correction ΔT to the nominal perihelion date. (In practice the Central Telegram Bureau will make an actual orbit solution for a correction to the single element T.) If ΔT is in days, then the remainder of the ephemeris can be adjusted by applying the quantities $\Delta T \cdot \Delta\alpha$ to α and $\Delta T \cdot \Delta\delta$ to δ.

G. Catalogues

Periodic comets are conventionally described as those with revolution periods < 200 yr, and, unlike comets in nearly parabolic orbits, those bearing the name of the same discoverers are distinguished by Arabic numerals. Periodic comets also receive Roman-numeral designations at each observed perihelion passage and the catalogue by Marsden (1979) contains orbital elements for 482 perihelion passages of 113 individual objects. The references in Sec. III. G to Galle (1894), Vsekhsvyatskij (1958) and Holetschek (1896) also apply; Hasegawa (1968) has provided a partial catalogue of returns of periodic comets, including those returns when no observations were made.

H. Long-term Motion

Long-term calculations of the motions of several short-period comets have been made, notably over the interval 1660-2060 by members of the staff of the Institute for Theoretical Astronomy in Leningrad (e.g., Kazimirchak-Polonskaya 1967). The computations are complicated, and to some extent even negated, by the fact that cometary motions are affected by nongravitational forces. Application of these effects is at best accomplished by empirical procedures (e.g. Marsden et al. 1973), and the most reliable results are those extrapolated no more than one revolution following the last available observation. Up-to-date predictions for the returns of comets are regularly given in the annual *Handbook of the British Astronomical Association,* the *MPC's* and the *Nakano Notes* (*Nakano wa Kangaeru noda*, distributed privately by S. Nakano, Sumoto, Japan).

Table II is a low-precision listing of the orbital elements at predicted returns of comets during the years 1982-1999. The 199 entries, arranged in order of T, have been compiled from computations made by R.J. Buckley, E.I. Kazimirchak-Polonskaya, B.G. Marsden, S. Nakano, G. Sitarski, and D.K. Yeomans. The code letter and figure at the far right represents an attempt to estimate the accuracy of each entry, especially the value for T. A1 orbits are of the highest quality, with T accurate to ≤ 0.1 d, with full planetary perturbations applied to good starting orbits and allowance for nongravitational effects if possible or appropriate. Orbits of classes A2, A3 and A4 are progressively less reliable, in part because they involve extrapolation further into the future. Entries of classes B2, B3 and B4 (there are no class B1 orbits) refer to comets that have yet to be observed to return; in some cases the errors in T could amount to several days, although corrections to the later predictions will be possible after a first return has been observed. Finally, the entries of classes C3 and C4 (C1 and C2 are absent) refer to comets that have been observed at > 1 perihelion passage but have failed to appear at recent predicted returns.

The tabulation is intended only to assist observers in preliminary planning. Potential users are strongly advised to examine the usual sources for more current predictions and observations of comets.

TABLE II
Predicted Returns of Short-Period Comets, 1982–1999

Periodic Comet	T			Peri.	Node	Incl.	q	e	P	Code
Grigg-Skjellerup	1982	May	15.0	359.33	212.63	21.14	0.9892	0.6657	5.09	A1
Perrine-Mrkos	1982	May	16.8	166.50	239.91	17.78	1.3032	0.6369	6.80	C3
Vaisala 1	1982	July	30.6	47.93	134.51	11.61	1.7999	0.6334	10.9	A2
d'Arrest	1982	Sept.	14.3	176.97	138.86	19.43	1.2911	0.6248	6.38	A1
Tempel-Swift	1982	Oct.	22.0	163.70	240.03	13.45	1.6049	0.5354	6.42	C3
Churyumov-Gerasimenko	1982	Nov.	12.1	11.32	50.36	7.11	1.3061	0.6292	6.61	A3
Gunn	1982	Nov.	26.9	196.98	67.88	10.38	2.4591	0.3164	6.82	A1
Neujmin 3	1982	Dec.	6.1	145.16	149.69	3.94	2.0595	0.5813	10.9	A2
Pons-Winnecke	1983	Apr.	7.5	172.32	92.75	22.31	1.2540	0.6347	6.36	A1
Arend	1983	May	22.4	46.93	355.63	19.93	1.8569	0.5364	8.02	A2
du Toit-Neujmin-Delporte	1983	June	1.4	115.21	188.41	2.85	1.7082	0.5028	6.37	A2
Tempel 2	1983	June	1.5	190.92	119.16	12.44	1.3814	0.5449	5.29	A1
Oterma	1983	June	18.2	55.94	331.19	1.94	5.4709	0.2430	19.4	A1
Tempel 1	1983	July	9.8	179.04	68.33	10.56	1.4911	0.5209	5.49	A2
Kopff	1983	Aug.	10.3	162.81	120.31	4.72	1.5763	0.5445	6.4	A2
Harrington-Abell	1983	Dec.	2.0	138.57	336.72	10.16	1.7854	0.5388	7.62	A2
Johnson	1983	Dec.	3.3	208.16	116.72	13.68	2.3021	0.3675	6.94	A3
Taylor	1984	Jan.	7.7	355.61	108.19	20.52	1.9613	0.4639	7.00	A4
Crommelin	1984	Feb.	20.2	195.85	250.19	29.10	0.7345	0.9192	27.4	A1
Smirnova-Chernykh	1984	Feb.	21.5	90.87	77.04	6.64	3.5574	0.1462	8.51	B2
Tritton	1984	Mar.	2.7	147.64	300.04	7.03	1.4446	0.5790	6.36	B4
Encke	1984	Mar.	27.7	186.00	334.18	11.93	0.3410	0.8463	3.31	A1
Clark	1984	May	28.9	209.00	59.08	9.51	1.5512	0.5023	5.50	A3
Wolf	1984	May	31.8	162.18	203.51	27.51	2.4153	0.4068	8.21	A1
Faye	1984	July	9.9	203.83	198.98	9.09	1.5935	0.5783	7.34	A1
Tuttle-Giacobini-Kresak	1984	July	28.4	49.51	153.26	9.94	1.1234	0.6430	5.58	A2
Wild 2	1984	Aug.	20.4	40.02	136.07	3.27	1.4940	0.5561	6.18	B3
Wolf-Harrington	1984	Sept.	22.7	186.87	254.21	18.45	1.6158	0.5377	6.54	A1
Neujmin 1	1984	Oct.	8.2	346.84	346.35	14.17	1.5531	0.7756	18.2	A1
Arend-Rigaux	1984	Dec.	1.4	328.91	121.55	17.84	1.4462	0.5987	6.84	A1
Schaumasse	1984	Dec.	7.3	57.38	80.43	11.84	1.2127	0.7033	8.26	A2
Haneda-Campos	1984	Dec.	26.2	305.30	67.27	4.93	1.2212	0.6409	6.27	B3
Tsuchinshan 1	1985	Jan.	2.5	22.77	96.18	10.49	1.5078	0.5745	6.67	A3
Schwassmann-Wachmann 3	1985	Jan.	11.2	198.77	69.27	11.41	0.9365	0.6939	5.35	A4
Honda-Mrkos-Pajdusakova	1985	May	23.9	325.68	88.71	4.22	0.5423	0.8217	5.30	A1
Schuster	1985	June	2.6	355.62	49.99	20.15	1.5345	0.5903	7.25	B3
Russell 1	1985	July	5.3	0.40	230.13	22.66	1.6116	0.5173	6.10	B3
Gehrels 3	1985	July	7.6	226.56	243.52	1.06	3.3954	0.1744	8.34	B3
Kowal 2	1985	July	10.6	189.46	247.08	15.83	1.5025	0.5674	6.47	B3
Tsuchinshan 2	1985	July	21.3	203.16	287.61	6.70	1.7941	0.5024	6.85	A3
Daniel	1985	Aug.	3.8	10.84	68.45	20.14	1.6512	0.5518	7.07	A3
Giacobini-Zinner	1985	Sept.	5.7	172.49	194.71	31.88	1.0282	0.7076	6.59	A1
Giclas	1985	Oct.	3.5	276.42	111.87	7.28	1.8370	0.4951	6.94	B3
Boethin	1986	Jan.	23.2	11.62	25.86	5.76	1.1141	0.7780	11.2	B3
Ashbrook-Jackson	1986	Jan.	24.3	348.81	1.95	12.51	2.3070	0.3960	7.46	A3
Halley	1986	Feb.	9.6	111.85	58.15	162.23	0.5871	0.9673	76.0	A2
Holmes	1986	Mar.	14.1	23.34	327.34	19.19	2.1685	0.4118	7.08	A2
Wirtanen	1986	Mar.	19.8	356.08	81.65	11.67	1.0844	0.6521	5.50	A2
Kojima	1986	Apr.	5.1	348.50	154.09	0.88	2.4147	0.3905	7.89	A3
Shajn-Schaldach	1986	May	27.4	216.50	166.24	6.09	2.3307	0.3896	7.46	A2
Whipple	1986	June	25.0	202.05	181.80	9.94	3.0775	0.2606	8.49	A2
Wild 1	1986	Oct.	1.2	167.86	358.05	19.90	1.9770	0.6471	13.3	A2
Forbes	1987	Jan.	2.5	262.74	22.89	4.67	1.4744	0.5658	6.26	A3
Neujmin 2	1987	Apr.	2.4	214.94	307.15	5.37	1.2714	0.5865	5.39	C3
Jackson-Neujmin	1987	May	24.5	196.60	163.12	14.06	1.4374	0.6526	8.42	A3
Grigg-Skjellerup	1987	June	20.1	359.31	212.63	21.11	0.9933	0.6646	5.10	A4
Russell 2	1987	July	4.6	245.65	44.38	12.53	2.1534	0.4173	7.10	B3
Encke	1987	July	17.4	186.26	334.03	11.93	0.3317	0.8499	3.29	A1
Klemola	1987	July	22.6	154.54	175.78	10.96	1.7727	0.6405	10.9	A1
West-Kohoutek-Ikemura	1987	July	27.5	359.84	83.53	30.58	1.5706	0.5444	6.40	A3
Denning-Fujikawa	1987	Aug.	5.1	338.26	35.86	9.43	0.7625	0.8218	8.85	A3
Gehrels 1	1987	Aug.	13.6	28.47	12.91	9.67	2.9882	0.5100	15.1	B3
Comas Sola	1987	Aug.	18.8	45.52	60.38	12.95	1.8303	0.5698	8.78	A2
Schwassmann-Wachmann 2	1987	Aug.	30.6	357.89	125.66	3.78	2.0714	0.3984	6.39	A2
Wild 3	1987	Aug.	31.7	179.56	71.98	15.46	2.2926	0.3672	6.90	B3
Brooks 2	1987	Oct.	16.8	198.14	176.25	5.55	1.8448	0.4907	6.89	A3
Reinmuth 2	1987	Oct.	25.7	45.46	296.01	6.98	1.9360	0.4565	6.72	A1
Kohoutek	1987	Oct.	29.8	175.79	268.97	5.92	1.7762	0.4976	6.65	A3
Harrington	1987	Oct.	32.0	233.05	118.94	8.66	1.5962	0.5571	6.84	A2
de Vico-Swift	1987	Dec.	7.1	1.92	358.54	6.09	2.1821	0.4252	7.40	A3
Bus	1987	Dec.	11.6	23.97	181.52	2.57	2.1940	0.3713	6.52	B3
Borrelly	1987	Dec.	18.3	353.32	74.75	30.32	1.3567	0.6242	6.86	A2
Reinmuth 1	1988	May	10.0	13.02	119.15	8.14	1.8693	0.5030	7.29	A1
Finlay	1988	June	5.9	322.21	41.74	3.65	1.0944	0.6995	6.95	A3
Tempel 2	1988	Sept.	16.7	191.04	119.12	12.43	1.3834	0.5444	5.29	A1
Longmore	1988	Oct.	12.2	195.70	15.00	24.39	2.4090	0.3414	7.00	A2
du Toit 1	1988	Dec.	25.9	257.05	21.84	18.69	1.2735	0.7879	14.7	A2
Tempel 1	1989	Jan.	4.4	178.98	68.33	10.54	1.4967	0.5197	5.50	A2
d'Arrest	1989	Feb.	4.0	177.07	138.80	19.43	1.2921	0.6246	6.39	A2
Perrine-Mrkos	1989	Mar.	1.4	166.56	239.39	17.82	1.2977	0.6378	6.78	C4
Tempel-Swift	1989	Apr.	12.1	163.42	240.31	13.17	1.5884	0.5391	6.40	C4
Churyumov-Gerasimenko	1989	June	18.8	11.37	50.36	7.11	1.2996	0.6303	6.59	A3
Pons-Winnecke	1989	Aug.	19.9	172.32	92.75	22.27	1.2610	0.6335	6.38	A3
Gunn	1989	Sept.	24.9	196.94	67.87	10.37	2.4716	0.3144	6.84	A3
Brorsen-Metcalf	1989	Sept.	28.9	129.73	310.84	19.33	0.4781	0.9720	70.6	A4

BASIC INFORMATION AND REFERENCES

TABLE II (continued)

Periodic Comet	T		Peri.	Node	Incl.	q	e	P	Code
Lovas	1989	Oct. 9.2	73.62	341.72	12.20	1.6796	0.6141	9.08	B3
du Toit-Neujmin-Delporte	1989	Oct. 18.6	115.35	188.31	2.85	1.7154	0.5017	6.39	A3
Schwassmann-Wachmann 1	1989	Oct. 26.2	49.86	312.12	9.37	5.7718	0.0447	14.9	A1
Gehrels 2	1989	Nov. 3.7	183.55	215.52	6.67	2.3483	0.4098	7.94	B3
Clark	1989	Nov. 28.3	208.93	59.07	9.50	1.5558	0.5013	5.51	A3
Kopff	1990	Jan. 20.3	162.82	120.30	4.72	1.5851	0.5430	6.46	A3
Tuttle-Giacobini-Kresak	1990	Feb. 8.0	61.57	140.88	9.23	1.0680	0.6557	5.46	A3
Sanguin	1990	Apr. 3.3	162.84	181.81	18.72	1.8137	0.6633	12.5	B3
Schwassmann-Wachmann 3	1990	May 18.8	198.78	69.26	11.41	0.9363	0.6940	5.35	A4
Tritton	1990	July 7.2	147.75	299.95	7.05	1.4357	0.5807	6.34	B4
Honda-Mrkos-Pajdusakova	1990	Sept. 12.7	325.76	88.64	4.23	0.5412	0.8219	5.30	A3
Encke	1990	Oct. 28.5	186.25	334.04	11.94	0.3309	0.8502	3.28	A1
Johnson	1990	Nov. 19.0	208.28	116.67	13.66	2.3126	0.3661	6.97	A4
Kearns-Kwee	1990	Nov. 23.2	131.84	315.03	9.01	2.2154	0.4867	8.97	A3
Wild 2	1990	Dec. 17.3	41.54	135.60	3.25	1.5779	0.5410	6.37	B3
Taylor	1990	Dec. 31.2	355.59	108.18	20.55	1.9504	0.4657	6.97	A4
Russell 1	1991	Jan. 4.0	333.74	225.80	17.72	2.1734	0.4387	7.62	B3
Swift-Gehrels	1991	Feb. 22.1	84.83	313.73	9.25	1.3547	0.6917	9.21	A3
Wolf-Harrington	1991	Apr. 4.9	186.98	254.17	18.47	1.6078	0.5391	6.51	A3
Haneda-Campos	1991	Apr. 8.6	305.40	67.21	4.93	1.2249	0.6402	6.28	B3
Van Biesbroeck	1991	Apr. 25.0	134.15	148.44	6.62	2.4010	0.5527	12.4	A3
Arend	1991	May 25.2	47.07	355.50	19.93	1.8500	0.5370	7.99	A3
Harrington-Abell	1991	May 31.9	139.18	336.60	10.25	1.7607	0.5421	7.54	A4
Tsuchinshan 1	1991	Aug. 30.8	22.76	96.19	10.50	1.4975	0.5764	6.65	A3
Wirtanen	1991	Sept. 21.1	356.19	81.60	11.67	1.0830	0.6523	5.50	A2
Arend-Rigaux	1991	Oct. 3.0	329.06	121.45	17.89	1.4378	0.6001	6.82	A3
Faye	1991	Nov. 15.6	203.96	198.88	9.09	1.5934	0.5782	7.34	A3
Kowal 2	1992	Jan. 1.2	189.62	247.01	15.84	1.4995	0.5678	6.46	B4
Chernykh	1992	Jan. 28.0	263.20	129.75	5.08	2.3563	0.5938	14.0	B2
Giacobini-Zinner	1992	Apr. 14.6	172.52	194.68	31.83	1.0340	0.7066	6.61	A3
Tsuchinshan 2	1992	May 20.3	203.15	287.59	6.72	1.7822	0.5044	6.82	A4
Kowal 1	1992	June 1.9	180.49	28.07	4.36	4.6736	0.2344	15.1	B3
Grigg-Skjellerup	1992	July 24.9	359.27	212.64	21.10	0.9946	0.6644	5.10	A4
Smirnova-Chernykh	1992	Aug. 2.9	88.83	76.84	6.63	3.5727	0.1470	8.57	B4
Neujmin 2	1992	Aug. 20.3	214.90	307.17	5.38	1.2655	0.5879	5.38	C4
Wolf	1992	Aug. 28.4	162.30	203.44	27.48	2.4275	0.4056	8.25	A3
Daniel	1992	Aug. 31.6	11.01	68.37	20.13	1.6497	0.5519	7.06	A4
Schuster	1992	Sept. 6.6	355.72	49.92	20.13	1.5392	0.5896	7.26	B3
Giclas	1992	Sept. 18.2	276.54	111.80	7.28	1.8461	0.4938	6.96	B3
Gale	1992	Dec. 17.0	215.44	59.26	10.73	1.2136	0.7581	11.2	C3
Schaumasse	1993	Mar. 5.0	57.46	80.38	11.84	1.2023	0.7048	8.22	A4
Forbes	1993	Mar. 15.8	310.64	333.59	7.16	1.4495	0.5676	6.14	A4
Holmes	1993	Apr. 10.6	23.23	327.34	19.16	2.1767	0.4104	7.09	A4
Vaisala 1	1993	Apr. 29.2	47.36	134.40	11.60	1.7830	0.6347	10.8	A4
Slaughter-Burnham	1993	June 19.9	44.08	345.75	8.15	2.5426	0.5036	11.6	A3
Ashbrook-Jackson	1993	July 13.9	348.68	1.97	12.49	2.3160	0.3950	7.49	A4
Gehrels 3	1993	Nov. 12.1	227.19	243.38	1.06	3.3834	0.1754	8.31	B4
Neujmin 3	1993	Nov. 13.3	146.98	149.77	3.99	2.0015	0.5860	10.6	A4
Shajn-Schaldach	1993	Nov. 16.0	216.52	166.21	6.08	2.3444	0.3877	7.49	A3
West-Kohoutek-Ikemura	1993	Dec. 25.8	359.97	83.48	30.54	1.5768	0.5433	6.42	A3
Schwassmann-Wachmann 2	1994	Jan. 24.1	358.15	125.62	3.76	2.0703	0.3988	6.39	A3
Encke	1994	Feb. 9.4	186.28	334.02	11.94	0.3309	0.8502	3.28	A1
Kojima	1994	Feb. 18.6	348.50	154.15	0.89	2.4000	0.3924	7.85	A3
Tempel 2	1994	Mar. 16.8	194.86	117.58	11.98	1.4835	0.5224	5.48	A1
Bus	1994	June 13.8	23.83	181.53	2.57	2.1841	0.3732	6.51	B4
Tuttle	1994	June 27.0	206.71	269.85	54.69	0.9980	0.8241	13.5	A4
Kohoutek	1994	June 29.0	175.93	268.91	5.91	1.7854	0.4960	6.67	A3
Reinmuth 2	1994	June 29.8	45.92	295.43	6.98	1.8930	0.4641	6.64	A3
Tempel 1	1994	July 3.3	178.87	68.32	10.55	1.4942	0.5202	5.50	A3
Wild 3	1994	July 20.5	179.23	71.94	15.46	2.3002	0.3658	6.91	B4
Harrington	1994	Aug. 23.7	233.45	118.59	8.66	1.5722	0.5613	6.78	A3
Brooks 2	1994	Sept. 1.5	198.00	176.20	5.50	1.8430	0.4910	6.89	A4
Russell 2	1994	Oct. 28.8	249.48	41.75	12.04	2.2792	0.3991	7.39	B4
Borrelly	1994	Oct. 31.8	353.20	74.70	30.30	1.3650	0.6230	6.89	A3
Whipple	1994	Dec. 22.3	201.88	181.79	9.93	3.0938	0.2587	8.53	A3
de Vico-Swift	1995	Apr. 4.8	1.86	358.34	6.08	2.1450	0.4308	7.31	A4
Finlay	1995	May 4.6	323.47	41.42	3.67	1.0358	0.7103	6.76	A4
Clark	1995	May 31.2	208.81	59.07	9.50	1.5525	0.5021	5.51	A4
d'Arrest	1995	July 27.2	178.04	138.30	19.53	1.3460	0.6140	6.51	A4
Tuttle-Giacobini-Kresak	1995	July 61.67	140.82	9.23	1.0652	0.6564	5.46	A4
Tempel-Swift	1995	Aug. 30.5	163.31	240.30	13.18	1.5837	0.5400	6.39	C4
Reinmuth 1	1995	Sept. 1.4	13.25	119.08	8.13	1.8736	0.5025	7.31	A3
Schwassmann-Wachmann 3	1995	Sept. 22.3	198.76	69.28	11.42	0.9330	0.6948	5.34	A4
Jackson-Neujmin	1995	Oct. 6.6	200.35	160.02	13.48	1.3810	0.6614	8.24	A4
Longmore	1995	Oct. 9.4	195.79	14.96	24.40	2.3991	0.3430	6.98	A3
Perrine-Mrkos	1995	Dec. 7.2	166.56	239.91	17.84	1.2928	0.6386	6.77	C4
Honda-Mrkos-Pajdusakova	1995	Dec. 30.1	324.98	89.49	4.11	0.5328	0.8241	5.27	A4
Pons-Winnecke	1996	Jan. 2.7	172.30	92.74	22.30	1.2559	0.6344	6.37	A3
Churyumov-Gerasimenko	1996	Jan. 18.4	11.34	50.36	7.11	1.3000	0.6302	6.59	A3
du Toit-Neujmin-Delporte	1996	Mar. 5.9	115.26	188.25	2.85	1.7197	0.5008	6.39	A3
Denning-Fujikawa	1996	June 4.7	337.41	35.76	9.18	0.7882	0.8183	9.03	A4
Comas Sola	1996	June 11.0	45.79	60.17	12.82	1.8460	0.5680	8.83	A3
Kopff	1996	July 2.2	162.75	120.29	4.72	1.5795	0.5441	6.45	A3
Gunn	1996	July 24.2	196.79	67.86	10.38	2.4619	0.3163	6.83	A3
Tritton	1996	Nov. 2.9	147.63	299.97	7.04	1.4365	0.5805	6.34	B4

TABLE II (continued)

Periodic Comet	T	Peri.	Node	Incl.	q	e	P	Code
Wirtanen	1997 Mar. 14.5	356.37	81.52	11.71	1.0635	0.6568	5.46	A3
Boethin	1997 May 1.1	20.02	16.07	4.99	1.1581	0.7747	11.7	B4
Wild 2	1997 May 7.0	41.67	135.56	3.25	1.5824	0.5402	6.39	B4
Encke	1997 May 23.5	186.28	334.01	11.92	0.3314	0.8500	3.28	A1
Gehrels 2	1997 Aug. 7.6	192.93	209.81	6.25	1.9980	0.4640	7.20	B4
Haneda-Campos	1997 Aug. 14.4	306.94	66.02	4.94	1.2673	0.6319	6.39	B4
Grigg-Skjellerup	1997 Sept. 3.8	359.34	212.61	21.09	0.9968	0.6639	5.11	A4
Wolf-Harrington	1997 Sept. 29.4	187.16	254.04	18.51	1.5818	0.5440	6.46	A3
Johnson	1997 Oct. 31.9	207.97	116.68	13.67	2.3084	0.3672	6.97	A4
Taylor	1997 Dec. 15.8	355.41	108.17	20.55	1.9479	0.4661	6.97	A4
Neujmin 2	1998 Jan. 7.3	214.94	307.11	5.37	1.2701	0.5868	5.39	C4
Tempel-Tuttle	1998 Feb. 27.4	172.52	234.56	162.48	0.9758	0.9056	33.2	A3
Tsuchinshan 1	1998 Apr. 19.4	22.74	96.15	10.50	1.4958	0.5766	6.64	A4
Klemola	1998 May 1.9	154.54	174.85	11.10	1.7545	0.6413	10.8	A3
Kowal 2	1998 May 12.9	194.26	242.74	14.40	1.3458	0.6008	6.19	B4
Arend-Rigaux	1998 July 13.2	330.55	121.05	18.30	1.3683	0.6116	6.61	A3
Russell 1	1998 Aug. 25.3	333.90	225.72	17.71	2.1809	0.4379	7.64	B4
Lovas	1998 Oct. 11.7	74.52	339.29	12.23	1.6920	0.6130	9.14	B3
Giacobini-Zinner	1998 Nov. 23.8	172.55	194.70	31.86	1.0337	0.7066	6.61	A3
Harrington-Abell	1998 Dec. 7.1	139.29	336.58	10.27	1.7491	0.5436	7.50	A4
Tsuchinshan 2	1999 Mar. 8.6	203.24	287.45	6.71	1.7704	0.5064	6.79	A4
Faye	1999 May 6.5	204.99	198.63	9.05	1.6568	0.5683	7.52	A4
Forbes	1999 May 8.5	310.83	333.48	7.16	1.4493	0.5675	6.13	A4
Arend	1999 Aug. 3.0	49.00	354.68	19.17	1.9168	0.5303	8.24	A4
Giclas	1999 Sept. 1.9	276.47	111.76	7.28	1.8448	0.4942	6.97	B4
Tempel 2	1999 Sept. 8.4	194.99	117.54	11.98	1.4817	0.5228	5.47	A2
Kearns-Kwee	1999 Sept. 17.5	127.53	312.32	9.34	2.3374	0.4767	9.44	A3
Schuster	1999 Dec. 16.8	355.84	49.91	20.13	1.5496	0.5880	7.29	B4
Wild 1	1999 Dec. 27.4	167.99	357.83	19.93	1.9609	0.6497	13.2	A3

V. BRIGHTNESS OF COMETS

The activity of a comet, and therefore its brightness, is highly dependent on incident solar radiation, hence on distance from the Sun. The brightness is also dependent, though generally less strongly, on distance of the comet from the observer. The geometry of illumination of the comet or of viewing by the observer may also be a significant factor in some instances. However, it is difficult to define a readily observable parameter that specifies the brightness of a comet in a way directly and unambiguously interpretable in terms of the variety of physical processes that contribute to the comet's luminosity. Two limiting cases have been defined, the "total" magnitude m_1 and the "nuclear" magnitude m_2; all observed broad wavelength band magnitudes of comets lie somewhere between these limits.

A. Total Magnitude

The total magnitude m_1 in principle is a measure of the integrated brightness of nucleus, coma, and tail. An approximation to total magnitude is generally determined visually and thus refers to the wavelength range to which the eye is sensitive. Observations of this general kind have been made for many years and it is because of the long time baseline that their value persists. Either the naked eye, or the smallest optical instrument with which the comet is easily visible, is generally preferred. Comparisons are made with stars of known brightness by one of several methods that have now become rather well defined (see, e.g., Morris 1980; Bortle 1981b). Even so, observational results are systematically affected by a variety of instrumental and other factors (see Meisel and Morris 1976 and their chapter in this book).

Corrections for at least some of the systematic effects are commonly attempted. Collections of data, with or without correction, may be represented by formulae of the form

$$m_1 = H_1 + 2.5\,k \log \Delta + 2.5\,n \log r \tag{19}$$

where m_1 is the apparent total magnitude, H_1 is the absolute total magnitude (corresponding to $r = \Delta = 1$ AU), Δ is the geocentric distance, r is the heliocentric distance, k is a measure of the dependence of magnitude upon geocentric distance, and n is a measure of the sensitivity of the magnitude to heliocentric distance. Generally k is taken equal to 2, corresponding to a dependence of the brightness upon Δ^{-2}, but smaller values have been suggested (Öpik 1963; also see the chapter by Meisel and Morris). The values found for H_1 and for n are not independent, and both are likely to reflect the particular range in r over which observations included in the solution were made. For long-period comets, values of n are often found to be near 4, corresponding to a dependence of the brightness on r^{-4}. Thus when observations do not extend over a sufficient range in r to make a solution for n meaningful, or to make comparisons of H_1 possible on a consistent basis, it is frequent practice to assume $n = 4$. The resulting value of H_1 is then usually designated H_{10}. Values for H_{10} for many comets on a consistent basis have been given by Vsekhsvyatskij (1958,1963,1964,1967,1972) and Vsekhsvyatskij and Il'ichishina (1971). Additional data have been given more recently by Meisel and Morris (1976 and their chapter).

Since the gaseous emission component generally shows a stronger dependence on heliocentric distance than does the continuum, the dust-scattered component of the comet's luminosity, the value found for n is likely to be smaller for dusty comets than for those relatively dust-free. It is also likely to be smaller for comets observed at relatively large heliocentric distance, before full development of the emission spectrum, than for comets observed only relatively close to the Sun.

Some comets show a definite asymmetry in their brightness behavior at corresponding distances from the Sun before and after perihelion passage (Whipple 1978). Comets brighter *before* perihelion passage than under comparable circumstances afterward include P/Encke and Comet Kohoutek 1973 XII. P/Halley is an example of a comet that is brighter *after* perihelion passage than under comparable circumstances before perihelion.

B. Nuclear Magnitude

In principle, the nuclear magnitude m_2 refers only to light scattered from the surface of the solid nucleus. Thus the dependence on geocentric and heliocentric distance and viewing geometry should follow that expected for an inert solid body

$$m_2 = H_2 + 5 \log \Delta + 5 \log r + f(\alpha^\circ) \tag{20}$$

where the last term represents a dependence on phase angle $\alpha°$ appropriate to the nature of the surface of the nucleus.

It has been shown (Sekanina 1976b) that conformity to such a law may be a necessary condition but it is not a sufficient test that photometric resolution of the true nucleus has been achieved.

In practice, "nuclear" magnitudes observed photographically even with large telescopes show a sensitivity to the focal ratio f, being brighter for shorter f-values. Observed nuclear magnitudes are also sensitive to seeing and to density of photographic images; poor seeing and overexposed images lead to greater estimated brightness. It is hard to determine how much fainter true nuclear magnitudes would be than their observational approximation. But the difference between the best observational approximations to "total" and "nuclear" magnitudes for a well-developed active comet may approach 5 to 6 mag. That is, the light scattered from the solid nucleus amounts to $\leqslant 1\%$ of the total from a bright comet.

C. Outbursts

Small fluctuations in brightness of active comets are frequently reported, but because of uncertainties in the photometric systems of different observers it is hard to be certain of the reality of variations of less than a magnitude or two unless a comet is being consistently monitored with the same instrumental system. Nevertheless there can be no doubt that some comets have undergone brightness variations of 2 to 5 mag or more over very short time spans—variations not related to changing heliocentric and geocentric distances. P/Schwassmann-Wachmann 1 has undergone many outbursts amounting to as much as 5 to 8 mag since its discovery at the time of an outburst in 1927 (Whipple 1980). Kresák (1974b) has studied two outbursts of P/Tuttle-Giacobini-Kresák in 1973 in which this small comet brightened by as much as 9 mag. Even very energetic outbursts do not necessarily imply serious disruption of the comet, nor is there a strong correlation between photometric outbursts and splitting. Reviews on this subject have been given by Hughes (1975) and by Sekanina in this book.

D. Secular Fading of Periodic Comets

Although there is agreement that periodic comets must gradually fade because of continuing loss of material at each perihelion passage, the rate of fading, and thus the lifetimes of short-period comets, has been the subject of considerable controversy. Vsekhsvyatskij (1958,1972) has found a mean fading rate of 0.3 to 0.4 mag per revolution for short-period comets, which would correspond to 5 to 6 mag of fading per century for Jupiter-family members. On the other hand, Kresák (1974a) finds that instrumental effects have produced a mean scale shift of 3 mag since the end of the 19th century, and that extreme systematic differences as large as 7 to 8 mag affect data for some comets. In Kresák's view the intrinsic fading amounts to no more than 1 mag per century. Svoren (1979) also finds systematic instrumental effects on maximum apparent brightnesses, but concludes that a fading rate of 0.22 mag per revolution exists in the weighted average for all short-period comets with orbital periods < 28 yr. Some puzzling correlations of fading rates with orbital period and with perihelion distance seem to leave room for question that observational effects have in fact been fully allowed for.

E. Photometric Behavior of "New" versus "Old" Comets

In their studies of the physical characteristics of "new" as compared with more dynamically evolved comets, Oort and Schmidt (1951) concluded that there were patterns distinguishing the two populations in several ways. "New" comets seemed to be more susceptible to disintegration than older ones, "new" comets appeared to have much greater strength in their continuous spectrum as compared with the emission spectrum at heliocentric distances $\gtrsim 1$ AU from the Sun, and "new" comets were characterized by a slower rate of brightening with decreasing heliocentric distance than were dynamically older ones. That is, "new" comets are already more responsive to solar radiation at relatively large distance from the Sun, and further brightening occurs more slowly as the heliocentric distance decreases.

Not all these conclusions have been sustained by more recent investigations of spectral and photometric properties of larger populations of both "new" and "old" comets. Donn (1977) finds that there is a range of continuum-to-emission intensity ratios for comets in all age categories as measured by $(1/a)_{orig}$ and no systematic difference between "new" and "old" comets. Whipple (1978) finds that for all age classes of comets with period > 25 yr, the rate of change of brightness with heliocentric distance is statistically the same *after* perihelion passage. But "new" comets appear to approach perihelion with a smaller rate of brightening than do older comets, a finding consistent with that of Oort and Schmidt. A study by Kresák (1981) suggests that there is a systematic difference in the mean absolute brightness of comets, in the sense that they fade with dynamical age, as might be expected if the one population evolved into the other. But there seems to be no other obvious distinctions in the observable physical properties of comets of different dynamical ages.

It must be noted that different investigators have used different values of $(1/a)_{orig}$ for the cutoff between "new" and "old" comets. An appropriate cutoff value is perhaps $(1/a)_{orig} \simeq 100 \times 10^{-6}$ AU^{-1}. Comets with much larger $(1/a)_{orig}$ are certainly "old"; those with smaller $(1/a)_{orig}$ will generally be "new". Of the bright comets in recent years Comet Kohoutek 1973 XII is properly classified as "new", but comets Bennett 1970 II, West 1976 VI, Seargent 1978 XV (not included in Table I) and Bradfield 1979 X are all to be classified as dynamically evolved.

VI. SPECTRA OF COMETS

The typical pattern of development of the emission spectrum of an active comet upon its approach to the Sun, recognized long ago on the basis of objective-prism and low-resolution slit spectroscopy, has been described by Swings and Haser (1956). First to develop of the principal features in the visible is CN (0-0), which is generally first observed as the comet's distance from the Sun decreases from 3.0 to 2.5 AU. Emissions of C_3 and of NH_2 begin to appear at ~ 2 AU. The Swan bands of C_2 develop as the distance decreases to ~ 1.8 AU, and the D lines of Na may be visible as the heliocentric distance decreases to ~ 0.7 AU. In some comets the sodium lines may reach great strength at small distance from the Sun.

The characteristic emission bands of type I, or ion tails, are those of CO^+, N_2^+, CO_2^+, CH^+, and OH^+ in the blue and of H_2O^+ in the visible and red. These features generally become detectable as soon as the ion tail is formed, typically at distances of 1.5 AU or less from the Sun. Type I tails and weak CO^+ emissions have, however, been detected in a few unusual objects at much greater distances from the Sun.

The excitation mechanism for most features was recognized quite early as resonance-fluorescence with sunlight (see Swings 1956). Strengths of individual rotational lines are affected by Doppler shifting of the Fraunhofer lines of the solar spectrum as seen by the comet according to the radial component of the orbital velocity (Swings effect) or attributable to differential motions in the coma (Greenstein effect).

At large heliocentric distance the typical comet shows only a solar-type continuum. The indication thus is that the head in such comets consists predominantly of solid grains or of polyatomic molecules surrounding the nucleus. In some comets the continuous spectrum may continue to be strong even to quite small heliocentric distance.

A new era of high-dispersion and high spatial resolution cometary spectroscopy was introduced with the appearance of the two bright comets of 1957, Comets Arend-Roland 1957 III and Mrkos 1957 V (Swings 1965). As with the more recent Comet Kohoutek 1973 XII, the discovery of Comet Arend-Roland occurred several months in advance of the optimum time for physical observations, so there was time to plan and to prepare. The opportunities presented by the unexpected appearance later in the same year of Comet Mrkos, a bright comet with distinctly different spectral characteristics, gained from both the planning for and the experience with Comet Arend-Roland.

The daylight sungrazer of 1965, Comet Ikeya-Seki 1965 VIII, provided the first opportunity to confirm the suspected detection by visual spectroscopy of Fe lines in the sungrazer 1882 II by Copeland and Lohse (1882; see also Swings 1956). Identifications were extended to lines of a number of other metals by high-dispersion spectroscopy close to the time of perihelion passage (Preston 1967; Slaughter 1969).

In more recent years development of sensitive detectors, together with the use of spacecraft in astronomical exploration, has permitted extension of the sampled wavelength regions of the cometary spectrum into the vacuum ultraviolet and into the infrared and radio regions. Discovery of the extended atomic hydrogen coma that apparently envelops all active comets, information about the constitution of solid grains, and direct observation of a few of the long-hypothesized parent molecules have been among the benefits of the extended wavelength baseline of cometary spectroscopy and spectrophotometry. Overviews of recent developments are given in the chapters by A'Hearn, Feldman, Ney, and Wyckoff in this book.

REFERENCES

Benima, B.; Cherniack, J.R.; Marsden, B.G.; and Porter, J.G. 1969. The Gauss method for solving Kepler's equation in nearly parabolic orbits. *Publ. Astron. Soc. Pacific* 81:121-129.

Bortle, J.E. 1981a. Comets and how to hunt them. *Sky Teles.* 61:123-125.

Bortle, J.E. 1981b. How to observe comets. *Sky Teles.* 61:210-214.

Brandt, J.C., and Mendis, D.A. 1979. The interaction of the solar wind with comets. In *Solar System Plasma Physics,* Vol. II, eds. C.F. Kennel, L.J. Lanzerotti, and E.N. Parker (Amsterdam: North Holland Publ.), pp. 255-292.

Comrie, L.J. 1929. Note on the reduction of photographic plates. *J. Brit. Astron. Assoc.* 39:203-209.

Copeland, R., and Lohse, J.G. 1882. Spectroscopic observations of comets III and IV, 1881, comet I 1882, and the great comet of 1882. *Copernicus* 2:225-244.

Doggett, L.E.; Kaplan, G.H.; and Seidelmann, P.K. 1977. *Almanac for Computers 1978.* (Washington, D.C.: Nautical Almanac Office, U.S. Naval Observatory).

Donn, B.D. 1977. A comparison of the composition of new and evolved comets. In *Comets, Asteroids, Meteorites,* ed. A.H. Delsemme (Toledo, Ohio: Univ. Toledo), pp. 15-23.

Dubyago, A.D. 1961. *The Determination of Orbits.* (New York: Macmillan)

Everhart, E., and Raghavan, N.A. 1970. Changes in total energy for 392 long-period comets, 1800-1970. *Astron. J.* 75:258-272.

Galle, J.G. 1894. *Verzeichnis der Elemente der bisher berechneten Cometenbahnen.* (Leipzig: Engelmann)

Hasegawa, I. 1968. Catalogue of periodic comets (1967). *Mem. College of Science, Kyoto Univ.,* Ser. Phys., Astrophys, Geophys., Chem. 32:37-83 (Repr. Astron. Kyoto Univ. No. 32).

Hasegawa, I. 1980. Catalogue of ancient and naked-eye comets. *Vistas Astron.* 24:59-102.

Herget, P. 1948. *The Computation of Orbits* (Cincinnati: private publ.)

Herget, P. 1953. Solar coordinates 1800-2000. *Astron. Papers Amer. Ephemeris* 14:1-735.

Holetschek, J. 1896-1917. *Untersuchungen über die Grösse und Helligkeit der Kometen und Ihrer Schweife* 1896 I. *Die Kometen bis zum Jahre 1760.* 63:1-258; 1905 II. *Die Kometen von 1762 bis 1799.* 77:1-107; 1913 III. *Die Kometen von 1801 bis 1835 und auszugsweise auch noch die Helleren bis 1884.* 88:1-116; 1916 IV. *Die Helleren Periodischen Kometen.* 93:1-105; 1917 V. *Die Minder Hellen Periodischen Kometen.* 94:1-114 (Wien: Denkschr. Akad. Wiss.).

Ho Peng Yoke 1962. Ancient and mediaeval observations of comets and novae in Chinese sources. *Vistas Astron.* 5:127-225.

Hughes, D.A. 1975. Cometary outbursts, a brief survey. *Quart. J. Roy Astron. Soc.* 16:410-427.

Kazimirchak-Polonskaya, E.I. 1967. Evolution of the orbit of Comet Wolf 1 during 400 years 1660-2060. *Trudy Inst. Teor. Astron.* (Leningrad) 12:63-85. In Russian.
Kresák, L. 1974a. The aging and the brightness decrease of comets. *Bull. Astr. Inst. Czech.* 25:87-112.
Kresák, L. 1974b. The outbursts of Periodic Comet Tuttle-Giacobini-Kresák. *Bull. Astr. Inst. Czech.* 24:293-304.
Kresák, L. 1981. The lifetimes and disappearance of periodic comets. *Bull. Astr. Inst. Czech.* 32. In press.
Marsden, B.G. 1974. Comets. *Ann. Rev. Astron. Astrophys.* 12:1-21.
Marsden, B.G. 1979. *Catalogue of Cometary Orbits*, 3rd ed. (Cambridge, Mass.: IAU Central Bureau Astron. Telegrams).
Marsden, B.G. and Sekanina, Z. 1973. On the distribution of "original" orbits of comets of large perihelion distance. *Astron. J.* 78:1118-1124.
Marsden, B.G.; Sekanina, Z.; and Everhart, E. 1978. New osculating orbits for 110 comets and analysis of original orbits for 200 comets. *Astron. J.* 83:64-71.
Marsden, B.G.; Sekanina, Z.; and Yeomans, D.K. 1973. Comets and nongravitational forces. V. *Astron. J.* 78:211-225.
Meisel, D.D. and Morris, C.S. 1976. Comet brightness parameters: Definition, determination, and correlations. In *The Study of Comets*, eds. B. Donn, M. Mumma, W. Jackson, M. A'Hearn, and R. Harrington (Washington: NASA SP-393) pp. 410-444.
Miller, W.C. 1965. Photographic discrimination between beginning and end of trail produced by a moving object. *Publ. Astron. Soc. Pacific* 77:391-392.
Morris, C.S. 1980. A review of visual comet observing techniques. I. *Internat. Comet Quart.* 2:69-73.
Oort, J.H. 1950. The structure of the cloud of comets surrounding the solar system and a hypothesis concerning its origin. *Bull. Astron. Inst. Netherlands* 11:91-110 (No. 408).
Oort, J.H., and Schmidt, M. 1951. Differences between new and old comets. *Bull. Astron. Inst. Netherlands* 11:259-270 (No. 419).
Öpik, E.J. 1963. Photometry, dimensions, and ablation rate of comets. *Irish Astron. J.* 6:93-112.
Pingré, A.G. 1783, 1784. Cométographie ou Traité historique et théorique des Comètes. 2 vols. (Paris: Imprimerie royale)
Porter, J.G. 1963. The statistics of comet orbits. In *The Moon, Meteorites and Comets*, eds. B.M. Middlehurst and G.P. Kuiper (Chicago: Univ. Chicago Press), pp. 550-572.
Preston, G. 1967. The spectrum of Comet Ikeya-Seki (1965f). *Astrophys. J.* 147:718-742.
Roemer, E. 1962. Activity in comets at large heliocentric distance. *Publ. Astron. Soc. Pacific* 74:351-365.
Sekanina, Z. 1975. A study of the icy tails of distant comets. *Icarus* 25:218-238.
Sekanina, Z. 1976a. Anomalous tails of comets. I. A review of past edge-on appearances. *Center for Astrophysics Preprint* No. 445.
Sekanina, Z. 1976b. A continuing controversy: Has the cometary nucleus been resolved? In *The Study of Comets*, eds. B.Donn, M. Mumma, W. Jackson, M. A'Hearn, and R. Harrington (Washington: NASA SP-393), pp. 537-585.
Slaughter, C.D. 1969. The emission spectrum of Comet Ikeya-Seki 1965f at perihelion passage. *Astron. J.* 74:929-943.
Smart, W.M. 1931. *Textbook on Spherical Astronomy* (Cambridge: Cambridge Univ. Press). (6th edition, revised by R.M. Green, 1977; Cambridge Univ. Press.)
Svorén, J. 1979. Secular variations in the absolute brightness of short period comets. *Contr. Astron. Obs. Skalnaté Pleso* VIII:105-140.
Swings, P. 1956. The spectra of comets. *Vistas Astron.* 2:958-981.
Swings, P. 1965. Cometary spectra. *Quart. J. Roy. Astron. Soc.* 6:28-69.
Swings, P., and Haser, L. 1956. *Atlas of Representative Cometary Spectra*. Institut d'Astrophysique, Liège ARDC, Technical Final Report, Contract No. AF 61 (514)-628.
Väisälä, Y., and Oterma, L. 1951. Formulae and directions for computing the orbits of minor planets and comets. *Astron.-Opt. Inst. Turku Informo* No. 7.

Van Biesbroeck, G. 1963. Star catalogues and charts. In *Basic Astronomical Data* (Stars and Stellar Systems, Vol. 3), ed. K. Aa. Strand (Chicago: Univ. Chicago Press). pp. 471-480.

Vsekhsvyatskij, S.K. 1958. *Fizicheskie Kharakteristiki Komet.* (Moscow: Gos. Izd. Fiz-Mat. Lit.) English translation 1964. *Physical Characteristics of Comets.* NASA TT-F 80 Israel Program for Scientific Translation (Jerusalem: S. Monson).

Vsekhsvyatskij, S.K. 1963. Absolute magnitudes of 1954-60 comets. *Soviet Astron.-AJ* 6:849-854.

Vsekhsvyatskij, S.K. 1964. Physical characteristics of 1961-1963 comets. *Soviet Astron.-AJ* 8:429-431.

Vsekhsvyatskij, S.K. 1967. Physical characteristics of comets observed during 1961-1965. *Soviet Astron.-AJ* 10:1034-1041.

Vsekhsvyatskij, S.K. 1972. Cometary observations and variations in cometary brightness. In *The Motion, Evolution of Orbits, and Origin of Comets.* eds. G.A. Chebotarev, E.I. Kazimirchak-Polonskaya, and B.G. Marsden. (Dordrecht: D. Reidel Publ. Co.) pp. 9-15.

Vsekhsvyatskij, S.K. and Il'ichishina, N.I. 1971. Absolute magnitudes of comets, 1965-1969. *Soviet Astron.-AJ* 15:310-313.

Whipple, F.L. 1978. Cometary brightness variation and nucleus structure. *Moon Planets* 18:343-359.

Whipple, F.L. 1980. Rotation and outbursts of Comet P/Schwassmann-Wachmann 1. *Astron. J.* 85:305-313.

Whipple, F.L., and Huebner, W.F. 1976. Physical processes in comets. *Ann. Rev. Astron. Astrophys.* 14:143-172.

Glossary, Acknowledgments, and Index

GLOSSARY[a]

a   semimajor axis of an orbit.

Å   Ångstrom = 10^{-10} m.

absolute magnitude   the apparent magnitude that a comet would have if observed at 1 AU both from the Sun and from the Earth, and at zero phase, defined by

$$V = V(1,0) + 5 \log r \Delta + F(\alpha)$$

where r is distance from the Sun and Δ from the Earth, and $F(\alpha)$ is the phase function. Absolute magnitude may refer either to the total light or only to the light scattered from the nucleus; calculated values are not independent of the exponents that characterize the variation rate of observed apparent magnitudes with heliocentric and geocentric distances.

aeon (AE)   1 Gyr = 10^9 yr.

Alfvén Mach number   the ratio of local flow velocity to local Alfvén speed.

---

[a]We have used some definitions from *Glossary of Astronomy of Astrophysics* by J. Hopkins (by permission of the University of Chicago Press, ©1976 by The University of Chicago) and from *Astrophysical Quantities* by C.W. Allen (London: Athlone Press, 1973).

GLOSSARY

Alfvén speed	the speed at which Alfvén waves are propagated along the magnetic field: $V_A = B/(4\pi\rho)^{1/2}$ where B is magnetic field and ρ is plasma density.
Alfvén wave	a transverse wave in a magnetohydrodynamic fluid, in which the driving force is the tension introduced by the magnetic field along the lines of force.
Amor asteroids	asteroids having perihelion distance $1.017 \text{ AU} < q < 1.3$ AU.
amu	atomic mass unit.
anomalous tail	also called antitail and type III tail; a tail with apparent direction in the sky considerably divergent from the projected antisolar direction, in particular a tail that appears to point toward the Sun. Such a tail is essentially a thin sheet of solid particles confined to the orbital plane of the comet, outside the orbit and well behind the Sun-comet direction.
Apollo asteroids	asteroids having semimajor axis $a > 1.0$ AU, and perihelion distance $q < 1.017$ AU.
arcsec	second of arc.
asteroid	a small body or minor planet in heliocentric orbit which has never shown any cometary activity, but appears stellar at groundbased resolution.
asteroid belt	a region of space lying between Mars and Jupiter, where the great majority of asteroids is found.
Aten asteroids	asteroids having semimajor axis $a < 1.0$ AU.
AU	astronomical unit = 1.496×10^{11} m $\cong$ the semimajor axis of the Earth's orbit around the Sun.
B	latitude of perihelion.
blackbody	an idealized body which absorbs radiation of all wavelengths incident on it. (Because it is a perfect absorber, it is also a perfect emitter.) The radiation emitted by a blackbody is a function of temperature only.
blackbody radiation	sometimes called thermal radiation. Radiation whose spectral intensity distribution is that of a blackbody in accordance with Planck's law.
bolide	see meteor.

GLOSSARY

bow shock	a surface or sheet of discontinuity (i.e., abrupt changes in physical conditions) set up in a supersonic flow at which the fluid undergoes a finite decrease in velocity accompanied by increase in pressure, density, temperature, and entropy, as occurs, e.g., in the supersonic solar wind flow about the Earth's magnetosphere.
breccia	rock composed of broken rock fragments cemented together by finer-grained material.
Brownlee particles	interplanetary dust particles collected by aircraft in the Earth's stratosphere.
c	speed of light in a vacuum = 3×10^8 m s^{-1}.
CA	cosmic abundance.
carbonaceous chondrite	a special high-carbon type of chondrite, thus classified:

CI=C1=type I	contains 3.5% C and no chondrules
CM=C2=type II	contains 2.5% C
CV, CO=C3V, C3O= type III	contains 0.5% C

chondrite	a stony meteorite characterized by the presence of chondrules, small spherical grains predominantly composed of iron and magnesium silicates.
clathrate hydrate	a type of solid molecular compound in which one component, such as CH_4, is trapped by van der Waals forces in cavities in the lattice structure of another compound, such as H_2O ice. Such a structure has been suggested for the icy material in comets to explain the nearly simultaneous appearance in the spectra of emission bands of such radicals as CN, CO, and OH as comets approach the Sun.
cm	centimeter.
cm-A	centimeter-Amagat.
color index	the difference in magnitudes between any two spectral regions. Color index is always defined as the short-wavelength magnitude minus the long-wavelength magnitude. In the *UBV* system, the color index for an AO star is defined as $B-V = U-B = 0$; it is negative for hotter stars and positive for cooler ones.

coma	the temporary atmosphere of a cometary nucleus, composed of gas and small solid particles, which may have diameter 10^4 to 10^6 km or more (a Lyman α coma may reach 10^{14} km in diameter). The gas component is generally spherical, but the dust is confined to parabolic envelopes which merge with the type II tail.
comet	an object of small mass ($\sim 10^{10}$ to 10^{16} kg), in heliocentric orbit, characterized by icy volatiles that produce cometary phenomena (e.g. a coma and tail) upon sufficiently close approach to the Sun to cause vaporization of ices. Orbits of comets are often highly elliptic or even nearly parabolic, and many are highly inclined with respect to the orbital plane of the Earth.
comets, family of	an ensemble of comets with nodal distances similar to the orbit radius of a major planet, e.g. Jupiter's family.
comets, group of	an ensemble of comets with closely similar orbits.
comets, nomenclature	when a newly discovered comet is confirmed, the IAU Central Bureau for Astronomical Telegrams assigns an interim designation consisting of the year of discovery followed by a lowercase letter in order of discovery for that year. Frequently the discoverer's name precedes the designation. If a reliable orbit is later established, the comet is given a permanent designation consisting of the year of perihelion passage followed by a Roman numeral in order of perihelion passage. Comets with orbital periods $>$ 200 yr are called "long period"; those with periods $<$ 200 yr are called "short-period" and indicated by P/ preceding the name of the comet.
contact surface	a surface or sheet which separates two fluids. In the context of this book, it refers to the surface dividing the inward streaming flow of solar wind and the outwardly expanding ionospheric plasma of the coma.
contracting envelopes	an inhomogeneous structure of cometary ions sometimes observed to appear in front of the comet head. Soon after formation, these condensations recede towards the nucleus.
Δ	geocentric distance, conventionally given in AU.
Debye	a unit of measure for electric dipole moments.

Debye length	a characteristic distance in a plasma beyond which the electric field of a charged particle is shielded by particles with charges of the opposite sign.
disconnection event (DE)	a major condensation of cometary ions that has become detached from the region of the comet's head (see chapter by Brandt).
DSS	deep sea spherules, particles composed of homogenized melt droplets of $\geqslant 1$ mm meteoroids (see chapter by Fraundorf et al.).
e	eccentricity of an elliptical orbit, the amount by which the orbit deviates from circularity: $e = c/a$, where c is the distance from the center to a focus and a is the semimajor axis.
elongation	the planet-Earth-Sun angle. Eastern elongations appear east of the Sun in the evening; western elongations, west of the Sun in the morning. An elongation of 0° is called conjunction; 180° is called opposition; and 90° is called quadrature.
ephemeris	(pl., ephemerides) a list of computed positions occupied by a celestial body over successive intervals of time.
Ephemeris	or EMP: *Ephemerides of Minor Planets,* published yearly by the Russian Academy of Sciences, Institute of Theoretical Astronomy, Leningrad, U.S.S.R.
EUV	extreme ultraviolet.
eV	electron volt = 1.60×10^{-12} ergs.
fall	a meteorite recovered immediately after falling to the earth, in some cases also applied to meteorites recovered months or years after observations of fall phenomena.
Fifth Fundamental Catalogue (FK5)	the fifth fundamental catalog (in preparation) in a series which began with the FC by A. Auwers. The positions and proper motions of the fundamental stars given in the series have successively served as the reference coordinate system in astronomy.
find	a meteorite which cannot be associated with an observed fall, also applied to weathered meteorites recovered months or years after an observed fall.
fireball	see meteor.

flux ropes	small-scale magnetic structures discovered in the ionosphere of Venus. The magnetic field in these structures appears to have the helical configuration of a bundle of twisted ropes.
Fo_{65}	forsterite concentration 65%. See olivine.
folding ion rays	narrow linear structures of cometary ions in a comet's tail, usually appearing in symmetric pairs emanating from the head. The gradual lengthening of the ion rays is accompanied by folding towards the central tail axis.
Fourier analysis	analysis of a periodic function into its simple harmonic components.
Fourier theorem	the theorem that any finite periodic motion can be analyzed into components, each of which is a simple harmonic motion of definite and determinable amplitude and phase.
FTS	Fourier transform spectroscopy.
FWHM	full width at half maximum.
g	gram.
G	gauss = 1 oersted = 79.58 amp-turn/m.
GeV	giga electron volt = 10^9 eV.
G-factor	fluorescence efficiency, usually expressed as the probability of scattering of a solar photon per unit time per molecule (see chapter by Feldman).
Giotto	spacecraft planned by European Space Agency to flyby P/Halley in 1986.
GMT	Greenwich mean time. See also UT.
gravitational constant	constant of proportionality in the attraction between two unit masses a unit distance apart: $G = 6.668 \times 10^{-8}$ dyne cm^2 g^{-2}.
gaussian gravitational constant	$k = 0.01720209895$ when mass is in terms of the solar mass, distance in AU, and time in mean solar days.
Greenstein effect	anomalous strengths of rotation lines of molecular spectra of the cometary coma, caused by Doppler shifting of the solar spectrum associated with internal motions in the coma (see chapters by Wyckoff and by Feldman).
Gyr	gigayear = aeon = 10^9 yr.

H_o	magnitude at $r = \Delta = 1$ AU (see absolute magnitude).
head	the coma and nucleus of a comet.
i	inclination, the angle between an orbital plane and a reference plane, usually the ecliptic plane in the case of motion around the Sun.
icy grain halo	hypothetical halo of icy particles surrounding the cometary nucleus. Grains of icy particles can be dragged away from a subliming surface by expanding gas, in laboratory simulations; similar phenomena might occur in comets.
IAU	International Astronomical Union.
IAUC	International Astronomical Union Circulars (see Appendix by Marsden and Roemer).
IDP	interplanetary dust particles (see chapter by Fraundorf et al.).
IHW	International Halley Watch, the international program to coordinate observations of P/Halley.
IR	infrared.
IUE	International Ultraviolet Explorer, joint NASA-ESA spacecraft launched into geocentric orbit 26 Jan. 1978.
JD	Julian Day number.
Jy	Jansky = 10^{-26} W m^{-2} Hz^{-1}.
k	Boltzmann constant = 1.38×10^{-16} erg deg^{-1} = 8.62×10^{-5} eV deg^{-1}.
K	degrees Kelvin.
kcal	kilocalorie = 4.185×10^{10} erg.
Kepler's laws	1. Each planetary orbit is an ellipse with the Sun at one focus. 2. (law of areas) Equal areas are swept out by the heliocentric radius vector in equal times. 3. (harmonic law) The square of the period if proportional to the cube of the distance. Newton's generalized formula for the third law is $P^2 = 4\pi^2 a^3 / [G(m_1 + m_2)]$.
keV	kilo electron volt = 10^3 eV.
km	kilometer = 10^3 m.
kpc	kiloparsec = 10^3 pc.

Kreutz sungrazers	a group of comets with closely similar orbital elements and perihelion distance barely greater (in most cases) than the solar radius.
kW	kilowatt = 10^{10} erg s^{-1}.
L	longitude of perihelion (see $\tilde{\omega}$).
L_4, etc.	one of the Lagrangian points preceding Jupiter (see Trojans).
$L_\odot$	solar luminosity = 3.826×10^{33} erg s^{-1}.
Lagrangian point	see Trojans.
Lambert's law	also called cosine law. It states that the intensity of the light emanating in a given direction from a perfectly diffusing surface is proportional to the cosine of the angle of emission measured between the normal to the surface and the emitted light ray.
lightcurve	magnitude values plotted as a function of time. Note that this plot does not necessarily have to show variability.
Lorentz forces	the force affecting a charged particle due to the motion of the particle in electric and magnetic fields. The Lorentz force is $F_L = q(V \times B)$ with q the charge on the moving object; V the velocity of the charged particle; and B the magnetic induction vector.
$M_\odot$	solar mass = 1.989×10^{33} g.
$M_\oplus$	mass of Earth = 5.976×10^{27} g.
μm	1 micrometer = 1 micron = 10^{-6} m.
Mach number	the ratio of the speed of a flow to the local speed of sound.
mag	astronomical magnitude proportional to $-2.5 \log_{10} I$, where I is intensity.
magnetosheath	a layer or sheath of magnetized plasma surrounding a planetary body, produced by interaction of the planetary atmosphere with the solar wind.
me	mean square error.
megaweber	10^6 weber, unit of magnetic flux. (See chapter by Russell et al.)

GLOSSARY

meteor	a "shooting star"—the streak of light in the sky produced by the transit of a meteoroid through the Earth's atmosphere; also the glowing meteoroid itself. The term fireball is sometimes used for a meteor approaching the brightness of Venus; the term bolide for one approaching the brightness of the full Moon.
meteorite	extraterrestrial material which survives passage through the Earth's atmosphere and reaches the Earth's surface as a recoverable object or objects.
meteoroid	a small particle orbiting the Sun in the vicinity of Earth.
MeV	million electron volts = 10^6 eV.
mG	milligauss = 10^{-3} gauss.
MHD	magnetohydrodynamics, the study of the interaction between a magnetic field and an electrically conducting field.
Mie theory	a theory of light scattering by small spherical particles.
MPC	*Minor Planet Circulars* (see Appendix by Marsden and Roemer).
msec	millisecond = 10^{-3} s.
"new" comet	a comet making its first passage through a perihelion point close enough to the Sun so that its motion is subject to significant disturbances caused by the gravitational attractions of the principal planets. See Appendix Sec. III.H, Table I for a listing of comets in order of $1/a_{\text{orig}}$.
nm	nanometer = 10^{-9} m.
nongravitational effects	systematic effects remaining in the motion of comets after all gravitational perturbations by attractions from the principal planets have been allowed for. Such effects are generally associated with a jet effect on the comet's motion caused by asymmetric loss of volatiles and dust into the coma as the rotating nucleus is warmed by sunlight.
nt	nanotesla, 10^{-9} Tesla (see chapter by Russell et al.).
nuclear magnitude	m_2: ideally, a measure of the brightness of a comet resulting only from sunlight scattered from the surface of the solid nucleus. In practice, published nuclear magnitudes may include an unknown component of light from an unresolved coma.

ω	argument of perihelion, angular distance (measured in the plane of the comet's orbit and in the direction of motion) from the ascending node to the perihelion point.
$\tilde{\omega}$	longitude of perihelion = $\Omega + \omega$.
Ω	longitude of ascending node. The angular distance measured in the ecliptic from the vernal equinox to the ascending node of the orbit.
OAO	Orbiting Astronomical Observatory, NASA spacecraft launched 7 Dec. 1968.
obliquity	the angle between a plane's axis of rotation and the pole of its orbit.
OGO	Orbiting Geophysical Observatory, a series of NASA spacecraft.
olivines	common rock-forming silicate minerals in meteorites and other rocks. The ratio of metal oxides (MgO and FeO) to SiO_2 is 2:1 in olivine. Olivines vary continuously in composition from Mg_2SiO_4, *forsterite*, to Fe_2SiO_4, *fayalite*. Iron and magnesium ions freely substitute for one another. A given composition is usually expressed in terms of mole % forsterite; e.g. Fo_{65} means 65% Mg_2SiO_4.
old comet	a periodic comet whose orbit has evolved dynamically due to planetary perturbations, and thus has spent considerable time in the inner solar system. These generally have $1/a_{orig}$ 10^{-4} AU^{-1}. Note that "old" does not denote chronological age.
Oort Cloud	the hypothetical ensemble of comets with aphelion distances in the range $\sim$ 20,000 to 100,000 AU or somewhat greater, and therefore subject to stellar perturbations which occasionally deflect a comet into an orbit with perihelion close enough to the Sun to permit observation. The concept of such an ensemble of comets is based on the observed concentration of values of $1/a_{orig}$ within a narrow range $\sim 40 \times 10^{-6}$ AU^{-1}.
opposition	see elongation.
osculating orbit	the path that a comet would follow if it were subject from a certain moment onward only to the inverse-square attraction of the Sun or other central body. The "moment" is called the epoch of osculation.

GLOSSARY

P	orbital period.
P/	see comets, nomenclature.
parent body	planet- or comet-like solar-system bodies in which meteorites were formed or stored.
parent molecules	precursors of cometary atoms and radicals like H, OH, CN, C_2, and C_3; e.g., H_2O could be the parent molecule of H and OH, HCN that of CN, C_2H_2 that of C_2.
pc	parsec = the distance at which 1 AU subtends 1 arcsec = 206,265 AU = 3.26 lightyear = 3.086×10^{16} m.
perihelion	the point in the cometary orbit closest to the Sun.
perihelion distance	distance between the Sun and a comet at the comet's closest approach to the Sun.
phase angle	α, the angle subtended at the comet or asteroid by the directions to Sun and observer.
Planck's blackbody formula	a formula that determines the distribution of intensity of radiation that prevails under conditions of thermal equilibrium at a temperature T: $B_\nu = (2h\nu^3/c^2) [\exp(h\nu/kT) - 1]^{-1}$ where h is Planck's constant and ν is frequency.
planetesimal	a hypothetical early body of intermediate (perhaps meter to 100 km) size, usually eventually accreted to a large body.
position angle	p.a.; the angle measured from north to east through 360° in the plane of the sky, of one object with respect to another, or of a cometary tail with respect to the nucleus.
Poynting-Robertson effect	a loss of orbital angular momentum by orbiting particles associated with the absorption and reemission of solar radiation, equal to the time for particles to move into the Sun $t = 7.0 \times 10^6 \ r \ p \ a \ q$ yr (radius r in cm, density p in g cm^{-3}, a and q in AU) where a and q are the semimajor axis and perihelion distance of the initial particle orbit.
ppb	parts per billion.
precession	a slow, periodic conical motion of the rotation axis of a nonspherical spinning body subjected to external torques.
p_ν	geometric albedo (q.v.); p_V, geometric albedo with the V filter of the UBV system.

pyroxenes	a group of common rock-forming silicates which have ratios of metal oxides (MgO, FeO or CaO) to SiO_2 of 1:1. These are called the metasilicates. Pure members of this group are $MgSiO_3$, *enstatite,* and $FeSiO_3$, *ferrosilite.* Pure $CaSiO_3$ does not crystallize with the pyroxene structure. Ca does substitute for up to 50% of the Mg and Fe in the pyroxene structure. A truncated triangular composition diagram illustrates these relationships.

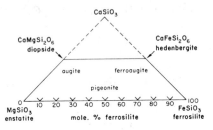

q	perihelion distance (q.v.).
Q	aphelion distance, the most distant point in the cometary orbit from the Sun.
r	heliocentric distance, usually measured in AU.
R	distance from nucleus, or nucleus radius, or as defined by authors.
ρ	density, in g cm^{-3} or kg m^{-3}.
recombination	the process by which a positive ion and an electron join to form a neutral molecule or other neutral particles.
reconnection	merging of two plasma flows of opposite magnetic polarities, with the possibility of conversion of the magnetic field energy into kinetic energy of the plasma.
refractory	describing a material having a high vaporization point or the property of resisting heat.
regolith	the layer of fragmental incoherent rocky debris that nearly everywhere forms the surface terrain, produced by meteoritic impact on the surface of a planet, satellite, or asteroid.
resonance	the selective response of any oscillating system to an external stimulus of the same frequency as the natural frequency of the system, or a simple fraction or multiple thereof.

GLOSSARY

Reynolds number	a dimensionless number ($R = Lv/\nu$, where L is a typical dimension of the system, v is a measure of the velocities that prevail, and ν is the kinematic viscosity) that governs the conditions for the occurrence of turbulence in fluids.
rms	root mean square, the square root of the mean square value of a set of numbers.
Roche limit	the minimum distance at which a satellite influenced by its own gravitation and that of a central mass, about which it describes a Keplerian orbit, can be in equilibrium. For a satellite of negligible mass, zero tensile strength, and the same mean density as its primary, in a circular orbit around its primary, this critical distance at which the satellite will break up is 2.44 times the radius of the primary.
Schmidt telescope	a type of catadioptric reflecting telescope in which the coma produced by a spherical concave primary mirror is compensated for by a thin correcting lens placed at the opening of the telescope tube. It is usually intended for wide field imaging.
solar flare	a violent eruption from the Sun's chromosphere which sends a burst of energetic particles into the interplanetary medium.
solar wind	a radial outflow of energetic charged particles from the solar corona, carrying mass and angular momentum away from the Sun. Mean number density of solar wind 5 per cm^3; mean velocity at Earth 400 km s^{-1}; mean magnetic field 5×10^{-5} gauss; mean electron temperature 20,000 K; mean ion temperature 10,000 K. The Sun ejects $\sim 10^{-13}$ $M_\odot$ yr^{-1} via the solar wind.
ST	Space Telescope, an optical telescope to be launched into geocentric orbit by NASA in 1985.
stagnation point	a point in a field of flow about a body where the fluid particles have zero velocity with respect to the body.
stagnation pressure	pressure at the stagnation point. Also, in compressible flow, the pressure exhibited by a moving gas or liquid brought to zero velocity by an isentropic process.
stand-off distance	the radial distance from the cometary nucleus at which the inward streaming flow of the contaminated solar wind is stopped by the ionospheric flow.

Stoke's law	the law stating that the force which retards a sphere moving through a viscous fluid is directly proportional to the velocity of the sphere, radius of the sphere, and viscosity of the fluid.
STP	standard temperature and pressure: 273 K and 1 atm.
striae	straight structures in a cometary dust tail more nearly parallel to the Sun-comet axis than dust streamers.
Swings effect	anomalous strengths of rotational lines of molecular spectra of comets, caused by Doppler shifting of the solar spectrum associated with orbital motion of the comet (see chapter by Wyckoff).
synchrone	in the Bessel-Bredikhin mechanical theory of comet dust tails, the locus of particles that leave the nucleus at nearly the same time but have a full range of initial velocities and are subject to a range of repulsive forces. Radiation pressure is presumed to be the repulsive force, which consequently depends upon particle size.
syndyne	(syndyname) in the Bessel-Bredikhin mechanical theory of comet dust tails, the locus of particles that leave the nucleus over a time interval but have the same initial velocity and are subject to the same repulsive force. This force is presumed to be solar radiation pressure, so the locus is of particles of equal size.
synodic period	the mean interval of time between two consecutive conjunctions of a planet or the Moon with the Sun, or of a satellite with its primary, as seen from the Earth.
T	time of perihelion passage. The orbital element conventionally used as a basis for calculating the position of a comet in its orbit.
total magnitude	a measure of the integrated brightness of nucleus, coma, and tail, if any, of an active comet: symbol m_1.
TRIAD	Tucson Revised Index of Asteroid Data (see *Asteroids,* ed. T. Gehrels, Univ. Arizona Press, Tucson, 1979, pp. 1011-1154).

GLOSSARY

Trojans	Trojan asteroids occur in two Lagrangian points (those preceding and following Jupiter in its orbit, equidistant from Sun and Jupiter), defined as points in the orbital plane of two massive bodies in circular orbits around a common center of gravity where a third body of negligible mass can remain in equilibrium.
type I	ion tail, a comet tail composed of plasma. This designation is also used for a type of carbonaceous chondrite and for a type of meteor (see chapter by Wetherill and ReVelle).
type II	dust tail, a comet tail composed of small solid particles of dust or ice.
type III	see anomalous tail.
UBV system	a system of stellar magnitudes devised by Johnson and Morgan at the Yerkes Observatory which consists of measuring an object's apparent magnitude through three color filters: the ultraviolet U at 3600 Å; the blue B at 4200 Å; and the "visual" V in the green-yellow spectral region at 5400 Å. It is defined so that, for AO stars, B-V = U-B = 0; it is negative for hotter stars and positive for cooler stars. Filters at other wavelengths are also used and indicated with letters R, I, H, J, K, L, M, etc.
UT	universal time. The measure of time defined as the Greenwich hour angle of the fictitious mean Sun plus 12 hr. Because the motion of the fictitious mean Sun is referred to the vernal equinox, UT is formally defined by a mathematical relation involving sidereal time.
UV	ultraviolet. The spectral region $1850 < \lambda < 4000$ Å.
van der Waals forces	the relatively weak attractive forces operative between neutral atoms and molecules.
VLF	very low frequency.
VUV	vacuum ultraviolet. The spectral region $\lambda < 1850$.
wave number	cm^{-1}, reciprocal of the wavelength (conventionally used instead of wavelength in relation to infrared radiation) wave number $v = 1/\lambda = f/c$ where λ is the wavelength of radiation, f is the frequency, and c the velocity of light.
zodiacal light	a faint glow that extends away from the Sun in the ecliptic plane, caused by scattering of sunlight by interplanetary particles.

ACKNOWLEDGMENTS

The following people helped to make this book possible, in organizing, writing, refereeing or otherwise. The members of the Organizing Committee and Chairpersons are indicated with an asterisk (*).

V. Adams, Lunar and Planetary Laboratory, University of Arizona, Tucson, AZ.
M.F. A'Hearn, Astronomy Program, University of Maryland, College Park, MD.
S. Aiello, Cattedra di Fisica dello Spazio, Università de Firenze, Firenze, Italy.
M. Allen, Division of Geological and Planetary Sciences, California Institute of Technology, Pasadena, CA.
C. Andersson, Onsala Space Observatory, Onsala, Sweden.
J. Arduini, Service d'Aéronomie CNRS, Verrières-le-Buisson, France.
B. Armando, Istituto di Fisica, Università di Lecce, Lecce, Italy.
C. Arpigny, Institut d'Astrophysique, Université de Liège, Cointe Ougrée, Belgium.
W.I. Axford, Max-Planck-Institut für Aeronomie, Heidelberg, West Germany.
H. Balsiger, Physikalisches Institut, Bern Universität, Bern, Switzerland.
E.S. Barker, McDonald Observatory, University of Texas, Austin, TX.
D.B. Beard, Department of Physics and Astronomy, University of Kansas, Lawrence, KS.
R. Becker, Division of Geological and Planetary Sciences, California Institute of Technology, Pasadena, CA.
D.F. Bender, Jet Propulsion Laboratory, Pasadena, CA.
J.T. Bergstralh, Jet Propulsion Laboratory, Pasadena, CA.
J.L. Bertaux, Service d'Aéronomie CNRS, Verrières-le-Buisson, France.
N. Bhandari, Physical Research Laboratory, Ahmedabad, India.
G. Bjoraker, Department of Planetary Sciences, University of Arizona, Tucson, AZ.
J.H. Black, Center for Astrophysics, Cambridge, MA.
J.E. Blamont, Service d'Aéronomie CNRS, Verrières-le-Buisson, France.
D. Bockelée-Morvan, Départment de Radioastronomie, Observatoire de Paris, Meudon, France.
A. Bonetti, Instituto di Fisica, Università Superiore, Firenze, Italy.
D. Bosley, Alltype, Tucson, AZ.
A. Brahic, Section d'Astrophysique, Observatoire de Paris, Meudon, France.
*J.C. Brandt, NASA Goddard Space Flight Center, Greenbelt, MD.
R.A. Brown, Lunar and Planetary Laboratory, University of Arizona, Tucson, AZ.
W.L. Brown, Bell Laboratories, Murray Hill, NJ.

*D.E. Brownlee, Department of Astronomy, University of Washington, Seattle, WA.
W.E. Brunk, NASA Headquarters, Washington, DC.
J.A. Burns, Theoretical and Applied Mechanics, Cornell University, Ithaca, NY.
B. Buti, Physical Research Laboratory, Ahmedabad, India.
H. Campins, Lunar and Planetary Laboratory, University of Arizona, Tucson, AZ.
C.W. Carlson, Space Sciences Laboratory, University of California, Berkeley, CA.
G.R. Carruthers, US Naval Research Laboratory, Washington, DC.
A. Carusi, Laboratorio di Astrofisica Spaziale CNR, Frascati, Italy.
A. Cazenave, CNES-CRGS, Toulouse, France.
C.R. Chapman, Planetary Sciences Institute, Tucson, AZ.
G.P. Chernova, Institute of Astrophysics, Dushanbe, USSR.
E. Ciaccio, Lunar and Planetary Laboratory, University of Arizona, Tucson, AZ.
E. Cirlin, Rockwell International, Thousand Oaks, CA.
B.C. Clark, Planetary Science Laboratory, Martin-Marietta Aerospace, Denver, CO.
D.R. Clay, Jet Propulsion Laboratory, Pasadena, CA.
A.L. Cochran, Astronomy Department, University of Texas, Austin, TX.
W.D. Cochran, Astronomy Department, University of Texas, Austin, TX.
N. Conarro, Lunar and Planetary Laboratory, University of Arizona, Tucson, AZ.
G. Consolmagno, Earth and Planetary Sciences, Massachusetts Institute of Technology, Cambridge, MA.
A. Coradini, Laboratorio di Astrofisica Spaziale CNR, Frascati, Italy.
C. Cosmovici, Laboratorio di Fisica Cosmica, Università di Lecce, Lecce, Italy.
H. Courten, Tucson, AZ.
J.F. Crifo, Service d'Aéronomie CNRS, Verrières-le-Buisson, France.
D.R. Criswell, California Space Institute, University of California at San Diego, La Jolla, CA.
J. Crovisier, Section d'Astrophysique, Observatoire de Paris, Meudon, France.
D. Cruikshank, Institute for Astronomy, University of Hawaii, Honolulu, HI.
A. Cucchiaro, Institut d'Astrophysique, Université de Liège, Cointe-Ougrée, Belgium.
C. Cunningham, Lunar and Planetary Laboratory, University of Arizona, Tucson, AZ.
C.C. Curtis, Department of Physics, University of Arizona, Tucson, AZ.
D.R. Davis, Planetary Sciences Institute, Tucson, AZ.
J. Degewij, Jet Propulsion Laboratory, Pasadena, CA.
*A.H. Delsemme, Department of Physics and Astronomy, University of Toledo, Toledo, OH.
K. Denomy, Lunar and Planetary Laboratory, University of Arizona, Tucson, AZ.
W.A. Deutschmann, Department of Physics and Astronomy, Rose-Hulman Institute, Terre Haute, IN.
N.T. Divine, Jet Propulsion Laboratory, Pasadena, CA.
O.V. Dobrovolsky, Institute of Astrophysics, Dushanbe, USSR.
*B. Donn, NASA Goddard Space Flight Center, Greenbelt, MD.
J.D. Drummond, Physical Science Laboratory, New Mexico State University, Las Cruces, NM.
L. Dunkelmann, Lunar and Planetary Laboratory, University of Arizona, Tucson, AZ.
A.K. Duvall, Steward Observatory, University of Arizona, Tucson, AZ.
E. Dwek, Astronomy Program, University of Maryland, College Park, MD.
P. Eberhardt, Physikalishes Institut, Bern Universität, Bern, Switzerland.
L. Ekelund, Onsala Space Observatory, Onsala, Sweden.
C. Elachi, Jet Propulsion Laboratory, Pasadena, CA.
R.C. Elphic, Institute of Geophysics and Planetary Physics, University of California, Los Angeles, CA.
A. Ershkovich, NASA Goddard Space Flight Center, Greenbelt, MD.
G. Evans, Space Sciences Laboratory, University of Kent, Canterbury, United Kingdom.
E. Everhart, Physics Department, University of Denver, Denver, CO.
C.Y. Fan, Department of Physics, University of Arizona, Tucson, AZ.
J. Farnow, Los Alamos Scientific Laboratory, Los Alamos, NM.
J.A. Farrell, Los Alamos Scientific Laboratory, Los Alamos, NM.
W.G. Fastie, Physics Department, Johns Hopkins University, Baltimore, MD.
*H. Fechtig, Max-Planck-Institut für Kernphysik, Heidelberg, West Germany.
P.D. Feldman, Department of Physics, Johns Hopkins University, Baltimore, MD.

ACKNOWLEDGMENTS

J. Fernandez, Department of Physics, Johns Hopkins University, Baltimore, MD.
M.C. Festou, Department of Atmospheric and Oceanic Sciences, University of Michigan, Ann Arbor, MI.
U. Fink, Lunar and Planetary Laboratory, University of Arizona, Tucson, AZ.
G. Foti, Istituto di Struttura della Materia, Università di Catania, Catania, Italy.
P. Fraundorf, McDonnell Center for Space Science, Washington University, St. Louis, MO.
J.J. Freeman, Corp. Research Department, Monsanto, Co., St. Louis, MO.
C. Froeschlé, Observatoire de Nice, Nice CEDEX, France.
D. Fugate, Steward Observatory, University of Arizona, Tucson, AZ.
*T. Gehrels, Lunar and Planetary Laboratory, University of Arizona, Tucson, AZ.
E. Gerard, Départment de Radioastronomie, Observatoire de Paris, Meudon, France.
D.M. Gibson, Department of Physics, New Mexico Institute of Mining and Technology, Socorro, NM.
H.L. Giclas, Lowell Observatory, Flagstaff, AZ.
R.H. Giese, Bereich Extraterrestrische Physik, Ruhr-Universität Bochum, Bochum, West Germany.
P.T. Giguere, Los Alamos Scientific Laboratory, Los Alamos, NM.
B.E. Goldstein, Jet Propulsion Laboratory, Pasadena, CA.
R.M. Goldstein, Jet Propulsion Laboratory, Pasadena, CA.
J.C. Gradie, Laboratory for Planetary Studies, Cornell University, Ithaca, NY.
M. Grady, Cambridge University, Cambridge, England.
J.M. Greenberg, Laboratory for Astrophysics, University of Leiden, Leiden, The Netherlands.
R. Greenberg, Planetary Sciences Institute, Tucson, AZ.
E. Grün, Max-Planck-Institut für Kernphysik, Heidelberg, West Germany.
B.A.S. Gustafson, Space Astronomy Laboratory, Albany, NY.
M.S. Hanner, Jet Propulsion Laboratory, Pasadena, CA.
B.W. Hapke, Department of Geology and Planetary Sciences, University of Pittsburgh, Pittsburgh, PA.
L.M. Harris, Alltype, Tucson, AZ.
W.K. Hartmann, Planetary Sciences Institute, Tucson, AZ.
H.F. Haupt, Astronomisches Institut der Universität Graz, Graz, Austria.
R. Hellmich, Max-Planck-Institut für Aeronomie, Katlenberg-Lindau, West Germany.
F. Herbert, Lunar and Planetary Laboratory, University of Arizona, Tucson, AZ.
R.W. Hobbs, NASA Goddard Space Flight Center, Greenbelt, MD.
J.M. Hollis, NASA Goddard Space Flight Center, Greenbelt, MD.
H.L.F. Houpis, Physics Department, University of California at San Diego, La Jolla, CA.
R.M. Housely, Rockwell International, Thousand Oaks, CA.
K.C. Hsieh, Physics and Atmospheric Sciences, University of Arizona, Tucson, AZ.
W.F. Huebner, Los Alamos Scientific Laboratory, Los Alamos, NM.
D.W. Hughes, Department of Physics, The University, Sheffield, England.
D.M. Hunten, Department of Planetary Sciences, University of Arizona, Tucson, AZ.
W.T. Huntress Jr., Jet Propulsion Laboratory, Pasadena, CA.
S. Ibadov, Institute of Astrophysics, Dushanbe, USSR.
W.-H. Ip, Max-Planck-Institut für Aeronomie, Heidelberg, West Germany.
W.M. Irvine, Department of Physics and Astronomy, University of Massachusetts, Amherst, MA.
T. Ishii, Tokyo Astronomical Observatory, Tokyo, Japan.
S. Isobe, Tokyo Astronomical Observatory, Tokyo, Japan.
W.M. Jackson, Chemistry Department, Howard University, Washington, DC.
C.A.J. Jamar, Institut d'Astrophysique, Université de Liège, Cointe-Ougrée, Belgium.
D. Jewitt, Division of Planetary Sciences, California Institute of Technology, Pasadena, CA.
K. Jockers, Max-Planck-Institut für Aeronomie, Katlenberg-Lindau, West Germany.
J.R. Johnson, Lunar and Planetary Laboratory, University of Arizona, Tucson, AZ.
R.E. Johnson, Engineering Physics, University of Virginia, Charlottesville, VA.
P.G. Kamoun, Department of Earth and Planetary Sciences, Massachusetts Institute of Technology, Cambridge, MA.
I. Kazès, Départment de Radioastronomie, Observatoire de Paris, Meudon, France.
H.U. Keller, Max-Planck-Institut für Aeronomie, Katlenberg-Lindau, West Germany.

J.F. Kerridge, Institute of Geophysics and Planetary Physics, University of California, Los Angeles, CA.
B.N. Khare, Center for Radio Physics and Space Research, Cornell University, Ithaca, NY.
H.H. Kieffer, US Geological Survey, Flagstaff, AZ.
E.A. King, Geology Department, University of Houston, Houston, TX.
N.N. Kiselev, Institute of Astrophysics, Dushanbe, USSR.
J. Kissel, Max-Planck-Institut für Kernphysik, Heidelberg, West Germany.
J. Klinger, Laboratoire de Glaciologie, Géophysique de l'Environment, Grenoble CEDEX, France.
A.J. Kliore, Jet Propulsion Laboratory, Pasadena, CA.
B. Kneissel, Bereich Extraterrestrische Physik, Ruhr-Universität Bochum, Bochum, West Germany.
A. Korth, Max-Planck-Institut für Aeronomie, Katlenberg-Lindau, West Germany.
*L. Kresák, Astronomical Institute of the Slovak Academy of Sciences, Bratislava, Czechoslovakia.
V. Krishan, Indian Institute of Astrophysics, Bangalore, India.
*D. Lal, Physical Research Laboratory, Ahmedabad, India.
L. Lane, Lunar and Planetary Laboratory, University of Arizona, Tucson, AZ.
L.J. Lanzerotti, Bell Laboratories, Murray Hill, NJ.
S. Larson, Lunar and Planetary Laboratory, University of Arizona, Tucson, AZ.
M. Leake, Lunar and Planetary Laboratory, University of Arizona, Tucson, AZ.
L. Lebofsky, Lunar and Planetary Laboratory, University of Arizona, Tucson, AZ.
M. Lebofsky, Lunar and Planetary Laboratory, University of Arizona, Tucson, AZ.
A.C. Levasseur-Rogourd, Service d'Aéronomie CNRS, Verrières-le-Buisson, France.
*B.I. Levin, Astronomical Council of the Academy of Sciences of USSR, Moscow, USSR.
E. Levy, Lunar and Planetary Laboratory, University of Arizona, Tucson, AZ.
C.F. Lillie, Department of Physics and Astronomy, University of Colorado, Boulder, CO.
J.G. Luhmann, Institute of Geophysics and Planetary Physics, University of California, Los Angeles, CA.
D.J. Malaise, Institut d'Astrophysique, Université de Liège, Cointe-Ougrée, Belgium.
P. Maloney, Lunar and Planetary Laboratory, University of Arizona, Tucson, AZ.
A.A. Manara, Osservatorio Astronomico di Milano, Milano, Italy.
S.P. Maran, NASA Goddard Space Flight Center, Greenbelt, MD.
J.N. Marcus, Comet News Service, Cincinnati, OH.
*B.G. Marsden, Harvard-Smithsonian Center for Astrophysics, Cambridge, MA.
M.A. Matthews, Lunar and Planetary Laboratory, University of Arizona, Tucson, AZ.
M.S. Matthews, Lunar and Planetary Laboratory, University of Arizona, Tucson, AZ.
J.C. McConnell, Kitt Peak National Observatory, Tucson, AZ.
J.A.M. McDonnell, Space Science Laboratory, University of Kent, Canterbury, United Kingdom.
J.F. McKenzie, Danish Space Research Institute, Lyngby, Denmark.
I. McKinnon, NASA Johnson Space Center, Houston, TX.
G. McLaughlin, Lunar and Planetary Laboratory, University of Arizona, Tucson, AZ.
C.E.K. Mees, Observatory, University of Rochester, Rochester, NY.
R.R. Meier, US Naval Research Laboratory, Washington, DC.
D.D. Meisel, Department of Physics and Astronomy, State University College, Geneseo, NY.
D.A. Mendis, Department of Applied Physics and Information Sciences, University of California at San Diego, La Jolla, CA.
J.J. Michalsky, Atmospheric Sciences, Batelle University, Richland, WA.
F.D. Miller, Department of Astronomy, University of Michigan, Ann Arbor, MI.
R.L. Millis, Lowell Observatory, Flagstaff, AZ.
N.Y. Misconi, Space Astronomy Laboratory, University of Florida, Gainesville, FL.
G. Mitchell, Department of Astronomy, St. Mary's University, Halifax, Canada.
E.P. Moore, Joint Observatory for Cometary Research, New Mexico Institute of Mining and Technology, Socorro, NM.
M.H. Moore, Astronomy Program, University of Maryland, College Park, MD.
C.S. Morris, Prospect Hill Observatory, Harvard, MA.

ACKNOWLEDGMENTS

B. Nagy, Department of Geosciences, University of Arizona, Tucson, AZ.
L. Nagy, Department of Geosciences, University of Arizona, Tucson, AZ.
J.S. Neff, Department of Physics and Astronomy, University of Iowa, Iowa City, IA.
R.M. Nelson, Jet Propulsion Laboratory, Pasadena, CA.
*M. Neugebauer, Jet Propulsion Laboratory, Pasadena, CA.
R.L. Newburn Jr., Jet Propulsion Laboratory, Pasadena, CA.
H. Newsom, Lunar and Planetary Laboratory, University of Arizona, Tucson, AZ.
E.P. Ney, Astronomy Department, University of Minnesota, Minneapolis, MN.
M.B. Niedner Jr., NASA Goddard Space Flight Center, Greenbelt, MD.
A.O. Nier, School of Physics and Astronomy, University of Minnesota, Minneapolis, MN.
K. Nishioka, Tokyo Astronomical Observatory, Tokyo, Japan.
E. Ochiai, Laboratory of Chemical Evolution, University of Maryland, College Park, MD.
J.A. Olson, Alltype, Tucson, AZ.
C.B. Opal, McDonald Observatory, University of Texas, Austin, TX.
M. Oppenheimer, Center for Astrophysics, Cambridge, MA.
N. Pailer, Max-Planck-Institut für Kernphysik, Heidelberg, West Germany.
D. Partlow, Department of Geology and Planetary Sciences, University of Pittsburgh, Pittsburgh, PA.
R. Patel, McConnell Center for Space Science, Washington University, Seattle, WA.
G.H. Pettengill, Department of Earth and Planetary Sciences, Massachusetts Institute of Technology, Cambridge, MA.
C.M. Pieters, Department of Geological Sciences, Brown University, Providence, RI.
V. Pirronello, Osservatorio Astrofisico di Catania, Catania, Italy.
C. Ponnamperuma, Laboratory of Chemical Evolution, University of Maryland, College Park, MD.
S.S. Prasad, Jet Propulsion Laboratory, Pasadena, CA.
D.K. Prinz, US Naval Research Laboratory, Washington, DC.
J. Rahe, NASA Goddard Space Flight Center, Greenbelt, MD.
U. Rainer, Max-Planck-Institut für Aeronomie, Heidelberg, West Germany.
A.S.P. Rao, Department of Geology, Osmania University, Hyderabad, India.
R. Reinhard, European Space Technology Center, Noordwijk, The Netherlands.
H.J. Reitsema, Lunar and Planetary Laboratory, University of Arizona, Tucson, AZ.
L. Remy-Battiau, Institut d'Astrophysique, Université de Liège, Cointe-Sclessin, Belgium.
D.O. ReVelle, Department of Physics, Northern Arizona University, Flagstaff, AZ.
N. Richter, Post Ulla am Bahnof, East Germany.
H. Rickman, Astronomiska Observatory, Uppsala, Sweden.
G.H. Rieke, Lunar and Planetary Laboratory, University of Arizona, Tucson, AZ.
S.E. Robinson, Department of Physics and Astronomy, University of Massachusetts, Amherst, MA.
*E. Roemer, Lunar and Planetary Laboratory, University of Arizona, Tucson, AZ.
H. Rosenbauer, Max-Planck-Institut für Aeronomie, Katlenberg-Lindau, West Germany.
C.T. Russell, Institute of Geophysics and Planetary Physics, University of California, Los Angeles, CA.
*R.Z. Sagdeev, Institute for Cosmic Studies, Moscow, USSR.
K. Saito, Tokyo Astronomical Observatory, Tokyo, Japan.
D.G. Schleicher, Astronomy Program, University of Maryland, College Park, MD.
F.P. Schloerb, Astronomy Department, University of Massachusetts, Amherst, MA.
*H.U. Schmidt, Max-Planck-Institut für Physik and Astrophysik, Garching, West Germany.
N. Schneider, Lunar and Planetary Laboratory, University of Arizona, Tucson, AZ.
H.J. Scholl, Astronomisches Rechen-Institut, Heidelberg, West Germany.
G.H. Schwehm, Bereich Extraterrestrische Physik, Ruhr-Universität Bochum, Bochum, West Germany.
R. Schwenn, Max-Planck-Institut für Aeronomie, Katlenberg-Lindau, West Germany.
Z. Sekanina, Jet Propulsion Laboratory, Pasadena, CA.
I.I. Shapiro, Department of Earth and Planetary Sciences, Massachusetts Institute of Technology, Cambridge, MA.
R. Sharp, Lockheed Palo Alto Laboratory, Palo Alto, CA.
E.G. Shelley, Lockheed Palo Alto Laboratory, Palo Alto, CA.

*M. Shimizu, Institute of Space and Aeronautical Sciences, University of Tokyo, Tokyo, Japan.
G.T. Sill, Lunar and Planetary Laboratory, University of Arizona, Tucson, AZ.
A.N. Simonenko, Astronomy Council, USSR Academy of Science, Moscow, USSR.
S.F. Singer, Department of Environmental Sciences, University of Virginia, Charlottesville, VA.
K.R. Sivaraman, Indian Institute of Astrophysics, Bangalore, India.
W.L. Slattery, Los Alamos Scientific Laboratory, Los Alamos, NM.
B.A. Smith, Lunar and Planetary Laboratory, University of Arizona, Tucson, AZ.
P.L. Smith, Center for Astrophysics, Cambridge, MA.
T.S. Smith, Lunar and Planetary Laboratory, University of Arizona, Tucson, AZ.
R. Smoluchowski, Departments of Astronomy and Physics, University of Texas, Austin, TX.
W. Smythe, Jet Propulsion Laboratory, Pasadena, CA.
L.E. Snyder, Astronomy Department, University of Illinois, Urbana, IL.
H. Spinrad, Department of Astronomy, University of California, Berkeley, CA.
J.E. Stanley, Department of Environmental Sciences, University of Virginia, Charlottesville, VA.
J.R. Stauffer, Department of Astronomy, University of California, Berkeley, CA.
N. Stephens, Lunar and Planetary Laboratory, University of Arizona, Tucson, AZ.
A.I. Stewart, Laboratory for Atmospheric and Space Physics, University of Colorado, Boulder, CO.
G. Strazzulla, Osservatorio Astrofisico di Catania, Catania, Italy.
D.F. Strobel, US Naval Research Laboratory, Washington, DC.
M. Swift, Department of Astronomy, St. Mary's University, Halifax, Canada.
P.Z. Takacs, Physics Department, Johns Hopkins University, Baltimore, MD.
J.B. Tatum, Department of Physics, University of Victoria Observatory, Victoria, Canada.
E.F. Tedesco, Lunar and Planetary Laboratory, University of Arizona, Tucson, AZ.
D.J. Tholen, Lunar and Planetary Laboratory, University of Arizona, Tucson, AZ.
D.T. Thompson, Planetary Research Center, Lowell Observatory, Flagstaff, AZ.
A.T. Tokunaga, Institute for Astronomy, University of Hawaii, Honolulu, HI.
P. Turek, Lunar and Planetary Laboratory, University of Arizona, Tucson, AZ.
E. Ungstrup, Danish Space Research Institute, Lyngby, Denmark.
N.G. Utterback, TRW Defense and Space Systems Group, Redondo Beach, CA.
G.B. Valsecchi, Laboratorio di Astrofisica Spaziale CNR, Frascati, Italy.
T.C. Van Flandern, US Naval Observatory, Washington, DC.
V. Vanýsek, Astronomical Institute, Charles University, Prague, Czechoslovakia.
*J. Veverka, Astronomy Department, Cornell University, Ithaca, NY.
F. Vilas, Lunar and Planetary Laboratory, University of Arizona, Tucson, AZ.
J. Wagner, Department of Geological and Planetary Sciences, University of Pittsburgh, Pittsburgh, PA.
R.M. Walker, McDonnell Center for Space Sciences, Washington University, St. Louis, MO.
*M.K. Wallis, Department of Applied Mathematics and Astronomy, University College, Cardiff, England.
J.W. Warner, NASA Marshall Space Flight Center, Huntsville, AL.
H.A. Weaver, Johns Hopkins University, Baltimore, MD.
R. Wegmann, Max-Planck-Institut für Physik und Astrophysik, Garching, West Germany.
S.J. Weidenschilling, Planetary Sciences Institute, Tucson, AZ.
K.W. Weiler, National Science Foundation, Washington, DC.
P.R. Weissman, Jet Propulsion Laboratory, Pasadena, CA.
E. Wells, Department of Geological and Planetary Sciences, University of Pittsburgh, Pittsburgh, PA.
R.A. West, Laboratory of Atmospheric and Space Physics, University of Colorado, Boulder, CO.
*G.W. Wetherill, Carnegie Institution of Washington, Washington, DC.
*F.L. Whipple, Smithsonian Astrophysical Observatory, Cambridge, MA.
I. Williams, Applied Mathematics Department, Queen Mary College, London, England.

ACKNOWLEDGMENTS

W. Wisniewski, Lunar and Planetary Laboratory, University of Arizona, Tucson, AZ.
J.A. Wood, Harvard College Observatory, Center for Astrophysics, Cambridge, MA.
I. Wright, University of Cambridge, Cambridge, England.
*S. Wyckoff, Department of Physics, Arizona State University, Tempe, AZ.
D.K. Yeomans, Jet Propulsion Laboratory, Pasadena, CA.
S. Yngvesson, Electrical Engineering Department, University of Massachusetts, Amherst, MA.
D.T. Young, Physikalisches Institut, Bern Universität, Bern, Switzerland.
B.H. Zellner, Lunar and Planetary Laboratory, University of Arizona, Tucson, AZ.
R.H. Zerull, Bereich Extraterrestrische Physik, Ruhr-Universität Bochum, Bochum, West Germany.

The editor acknowledges the support of NASA Grant NASW-3413 and NSF Grant AST-8010829. The following authors wish to acknowledge specific funds involved in supporting the preparation of their chapters.

A'Hearn, M.F.; NASA Grant NSG-7322
Brownlee, D.E.; NASA Grant NSG-9052
Campins, H.; JPL Contract BP-724887 and NASA Grant NSG-7114
Degewij, J.; JPL NASA Grant NAS7-100
Delsemme, A.H.; NSF Grant AST-80-18919 and NASA Grant NSG-7381
Elphic, R.C.; UCLA NASA Contract NAS2-9491 and JPL NASA Contract NAS7-100
Feldman, P.D.; NASA Grants NGR 21-001-001 and NSG-5393
Fink, U.; NASA Grants GSFC 71-7 (13 and 15) and NSG-7070
Fraundorf, P.; NASA Grant NGL-26-008-067
Hanner, M.S.; JPL Contract BP-724887 and NASA Grant NSG-7114
Jackson, W.M.; NASA Grant NSG-5071
Kamoun, P.G.; NSF Grant PHY78-07760 and NASA Grant NGR 22-002-672
Luhmann, J.G.; UCLA NASA Contract NAS2-9491 and JPL NASA Contract NAS7-100
Neugebauer, M.; UCLA NASA Contract NAS2-9491 and JPL NASA Contract NAS7-100
Ney, E.P.; NASA Grant NSG-2014
Pettengill, G.H.; NSF Grant PHY78-07760 and NASA Grant NGR 22-002-672
Russell, C.T.; UCLA NASA Contract NAS2-9491 and JPL NASA Contract NAS7-100
Sekanina, Z.; NASA Contract NAS7-100
Shapiro, I.I.; NSF Grant PHY78-07760 and NASA Grant NGR 22-002-672
Tedesco, E.F.; JPL NASA Grant NAS7-100
Walker, R.M.; NASA Grant NGL 26-008-067
Weissman, P.R.; NASA Contract NAS7-100
Whipple, F.L.; NASA Grant NSG-7082

INDEX

Ablation, 396
Albedo. *See under* Dust, *and under* Nucleus
Alfvén wave, 619
Andromedids. *See under* Meteor
Antitail. *See under* Tail
Arecibo Observatory, 292, 294 (tab)
Asteroid
  Amor objects, 216, 217, 248, 249, 298, 300, 667
  Apollo objects, 216, 217, 248, 249, 298, 300, 667, 689, 690 (fig)
  Aten objects, 667, 689, 690 (fig)
  C-type, 686
  dust, 380
  families, 72
  Hildas, 668 (fig)
  orbits, 667
  RD-type, 686
  rotation rates, 244, 245, 249
  S-type, 457
  Trojans, 76, 668 (fig)
Asteroids
  132 Althea, 216
  CA 1978, 248
  313 Chaldaea, 313
  321 Florentina, 248
  944 Hidalgo, 662, 666, 668 (fig), 679 (fig), 681 (fig), 688
  434 Hungaria, 668 (fig)
  1566 Icarus, 248
  1204 Renzia, 313
  1269 Rollandia, 684 (fig)
  1978 SB, 217
  279 Thule, 668 (fig)
Astrometry, 713-715
Autoionization. *See under* Ionization mechanisms
Becklin-Neugebauer object (BN), 147, 150 (fig), 152
Bow shock, 542, 547-553, 591, 613
Brightness, fluctuation, 63-66, 246, 426, 427. *See also* Fading
Brownlee particles, 11, 104, 119, 121-124, 336, 338, 347, 372, 378, 383 *ff*
C, 26 (tab), 114 (tab), 463 (tab), 472 (tab), 700 (tab)
C^+, 114 (tab)
C_2, 27 (tab), 114 (tab), 116, 421 (tab), 422 (tab), 425, 440-443, 454 (tab), 455 (tab), 463 (tab), 482 (tab), 484, 488, 500-502, 508, 509 (fig), 512 (fig)
C_3, 27 (tab), 114 (tab), 116, 421 (tab), 422 (tab), 425, 448, 449, 454 (tab), 455 (tab), 483 (tab), 488, 500-502, 508, 509 (fig), 512 (fig)

Ca, 26 (tab), 114 (tab)
Ca^+, 114 (tab)
CH, 27 (tab), 114 (tab), 446, 447, 454 (tab), 455 (tab), 482 (tab), 502 (tab)
CH_4, 90 (tab), 124, 143 (tab), 165, 167 (tab), 168 (tab), 173 (fig), 176 (fig), 186, 188 (fig), 500
$CH_4 \cdot 6H_2O$, 167 (tab)
CH^+, 27 (tab), 483 (tab)
CH_3CN, 27 (tab), 114 (tab), 116
CN, 27 (tab), 100-102, 114-116, 421 (tab), 422 (tab), 425, 443-446, 454 (tab), 455 (tab), 482 (tab), 484, 487 (fig), 500-502, 508, 510 (fig), 512 (fig)
CN^+, 28 (tab), 114 (tab)
CO, 27 (tab), 90 (tab), 101, 104, 114 (tab), 117, 124, 133, 143 (tab), 167 (tab), 168 (tab), 188, 190 (fig), 463 (tab), 472 (tab), 482 (tab), 499
CO^+, 28 (tab), 114 (tab), 421 (tab), 422 (tab), 452, 463 (tab), 483 (tab), 500, 508, 511 (fig), 512 (fig), 613
CO_2, 90 (tab), 101, 116, 143 (tab), 164 *ff*
CO_2^+, 28 (tab), 114 (tab), 463 (tab), 483 (tab), 512 (fig)
Cr, 104, 122 (tab), 123 (tab)
CS, 27 (tab), 114 (tab), 463 (tab), 472 (tab), 482 (tab), 491
CS_2, 463, 491-493
Cu, 27 (tab), 104, 114 (tab)
Carbonaceous chondrite, *See under* Meteorite
Catalina Observatory, 674
Charge-coupled device (CCD), 426, 435
Charge exchange, 541
Chinese chronicles, 59
2060 Chiron, 216, 217, 661, 666, 668 (fig), 689
Chondrite. *See under* Meteorite
Clathrate, 21, 96, 192, 193, 630
Clausius-Clapeyron equation, 504
Coma
  brightness, 420
  chemical reactions, 505-508
  composition, 501, 502 (tab)
  evolution, 473, 474
  general, 20-40, 97-103
  inner, 490, 610-619
  photochemistry, 496 *ff*
  photodissociation, 500
  photometry, 422, 423
  physical processes, 591-600
Comet
  discovery, 710-719
  Jupiter family, 13 (fig), 74-76
  new, 660

[761]

Comet, *cont.*
number of, 57, 649
observation history, 58
orbits, 56 *ff*, 205-209, 710
Comets
Abell 1954 X, 242 (tab)
P/Arend-Rigaux, 5, 418 (tab), 456, 666, 667, 670, 671 (fig), 673, 677-679, 681 (fig)
Arend-Roland 1957 III, 17, 106, 108, 109 (tab), 111-114, 124, 677, 709
P/d'Arrest, 229, 243 (tab), 416, 418 (tab), 421 (tab), 422, 427, 428 (tab)
P/Ashbrook-Jackson, 418 (tab), 421 (tab), 428 (tab)
Baade 1922 II, 235
1955 VI, 241 (tab)
Bakharev-MacFarlane-Krienke 1955 IV, 242 (tab)
Bappu-Bok-Newkirk 1949 IV, 242 (tab)
Bennett 1970 II, 7 (fig), 16, 22, 31, 33 (fig), 46, 91 (tab), 109, 112-114, 117, 118 (fig), 229, 234, 237, 238 (tab), 242 (tab), 324, 334 (tab), 335 (fig), 347, 350, 404, 443, 444, 447, 450, 456, 461, 500, 529, 573, 574, 612, 617, 677
Bester 1947 I, 240 (tab)
1948 I, 240 (tab)
P/Biela, 254, 258 (tab), 262 (tab), 264, 283
P/Boethin, 417 (tab)
P/Borrelly, 231, 520, 529
P/Borsen, 231
Bowell 1980b, 17, 343 (fig), 344 (fig), 457, 681 (fig), 683 (fig), 685
Bradfield 1974 III, 18, 46, 120, 326, 331 (fig), 335 (fig), 336, 349, 350, 417 (tab), 440-442, 444, 450-452
1975 V, 417 (tab)
1975 XI, 417 (tab)
1976 IV, 417 (tab)
1978 VII, 418 (tab)
1978 XVIII, 418 (tab)
1979 VII, 418 (tab), 421 (tab), 536
1979 X, 44, 419 (tab), 422 (tab), 462-464, 466 (fig), 469 (tab), 473 (fig), 475 (tab), 492 (fig), 513, 535
1980t, 328-331, 335 (fig), 336, 419 (tab)
Brooks, 1893 IV, 229
P/Brooks 2, 14, 258 (tab), 262 (tab), 263, 265
Burnham 1958 III, 242 (tab)
1960 II, 41, 235, 236 (tab), 241 (tab)
Campbell 1914 IV, 259 (tab), 277 (tab)
Candy 1961 II, 243 (tab)
Cernis-Petraukis 1980k, 419 (tab)
P/Chernykh, 418 (tab), 421 (tab)
P/Churyumov-Gerasimenko, 294 (tab), 295
Coggia 1874 III, 242 (tab)
P/Comas-Solá, 231, 418 (tab), 428 (tab)
Cunningham 1941 I, 240 (tab)
Daniel 1907 IV, 228, 242 (tab), 520
Davidson 1889 IV, 258 (tab)
DeKock-Paraskevopoulos 1941 IV, 242 (tab)
Donati 1858 VI, 228, 232 (fig), 242 (tab)
P/Encke, 3 *ff*, 91, 93 (fig), 94, 165, 210, 218, 231, 243 (tab), 281, 292, 293 (tab), 295, 300, 312, 334-336, 343-345, 349, 367, 416, 417 (tab), 419 (tab), 421 (tab), 422 (tab), 424, 428 (tab), 451 (fig), 457, 462, 468 (fig), 469 (tab), 475 (tab), 667, 668 (fig), 670
P/Faye, 243 (tab), 418 (tab), 428 (tab)
P/Finlay, 231
Finsler 1937 V, 242 (tab)
P/Forbes, 417 (tab), 428 (tab)
Geddes 1932 VI, 240 (tab)
P/Gehrels 3, 667, 669
P/Giacobini, 258 (tab)
Great Comet 1843, 609
Great September Comet 1882 II, 14, 213, 254, 258 (tab)
Great Southern Comet 1887 I, 14
P/Grigg-Skjellerup, 294 (tab), 295, 418 (tab), 421 (tab), 428 (tab)
P/Halley, 59, 100 (fig), 239, 243 (tab), 261, 294 (tab), 295, 520, 529, 536, 591, 610, 614, 616, 629
P/Haneda-Campos, 418 (tab), 421 (tab)
Harrington 1953 I, 242 (tab)
Honda 1955 V, 11, 241 (tab), 248, 259 (tab), 262 (tab)
1968 VI, 241 (tab), 677
Honda-Bernasconi 1948 IV, 242 (tab)
P/Honda-Mrkos-Pajdusakova, 417 (tab), 428 (tab)
Humason 1962 VIII, 15, 41, 97, 125, 499, 605
Ikeya 1963 I, 443, 603
Ikeya-Seki 1965 VIII, 14, 18, 213, 254, 259 (tab), 324, 677
1968 I, 104, 421 (tab), 426
Jones 1946 VI, 240 (tab)
P/Klemola, 417 (tab)
Kobayashi-Berger-Milon 1975 IX, 17, 18, 24 (fig), 28, 326-328, 331 (fig), 334-337, 350, 417 (tab), 443, 520, 525 (fig), 601 (fig)
Kohler 1977 XIV, 418 (tab), 421 (tab)
Kohoutek 1973 XII, 18, 22, 34 (fig), 41, 43, 46, 96, 97, 116, 117, 259 (tab), 292, 324, 325 (fig), 331-335, 337

INDEX

Comets, Kohoutek, *cont.*
(fig), 346, 349, 350, 352, 353 (tab), 379, 380, 404, 421 (tab), 427, 436, 437, 439 (fig), 442-444, 446, 449, 450, 452, 462, 499, 500, 502 (tab), 521 (fig), 523 (fig), 532 (fig), 573, 607, 612, 617, 626, 627 (fig), 630, 670
Kopff 1905 IV, 231, 259 (tab)
Liais 1860 I, 258 (tab)
Machhloz 1978 XIII, 418 (tab)
Meier 1978 XXI, 343 (fig), 344 (fig), 418 (tab), 421 (tab), 681 (fig), 683 (fig)
  1979 IX, 419 (tab), 421 (tab)
  1980q, 419 (tab), 475 (tab)
Mellish 1915 II, 259 (tab), 262 (tab), 263, 280 (fig)
Minkowski 1951 I, 240 (tab)
Morehouse 1908 III, 15, 41, 42, 97, 124, 242 (tab), 499, 520, 528 (fig), 529 (fig), 557, 558 (fig), 571, 605-608, 619
Mori-Sato-Fujikawa 1975 XII, 417 (tab)
Mrkos 1957 V, 677
P/Neujmin 1, 5, 666, 667, 670, 672 (fig)
P/Olbers, 243 (tab)
Orkicz 1925 I, 240 (tab)
P/Oterma, 667
Pajdusakova-Mrkos 1948 V, 240 (tab)
Pajdusakova-Rotbart-Weber 1946 II, 243 (tab)
Panther 1980u, 419 (tab)
Peltier 1952 VI, 240 (tab)
P/Pons-Brooks, 243 (tab)
P/Pons-Winnecke, 243 (tab)
Reid 1921 II, 240 (tab)
Reinmuth 1, 683 (fig)
P/Reinmuth 2, 243 (tab)
Sato 1976 I, 417 (tab)
Sawerthal 1888 I, 258 (tab)
P/Schaumasse, 243 (tab)
P/Schwassmann-Wachmann 1, 14, 65, 99-101, 124, 206, 228, 231, 243 (tab), 246, 436, 452, 661, 670, 674, 676 (fig), 679 (fig), 681 (fig), 683 (fig), 685 (fig)
P/Schwassmann-Wachmann 2, 94, 95 (fig), 419 (tab), 428 (tab)
P/Schwassmann-Wachmann 3, 231, 418 (tab), 428 (tab)
Seargent 1978 XV, 462, 467 (fig), 469 (tab), 475 (tab)
Seki-Lines 1962 III, 109 (tab), 112 (fig), 113
P/Shajn-Schaldach, 243 (tab)
P/Smirnova-Chernykh, 667
Southern Comet 1947 XII, 259 (tab), 280 (fig)

P/Stephan-Oterma, 31, 32 (fig), 343-345, 419 (tab), 422 (tab), 462, 475 (tab), 681 (fig), 683 (fig)
Suzuki-Saigusa-Mori 1975 X, 62, 417 (tab)
Swift 1899 I, 258 (tab), 262 (tab), 265, 277 (tab), 280 (fig)
P/Swift-Tuttle, 228, 231, 235, 243 (tab)
Tago-Sato-Kosaka 1969 IX, 91 (tab), 242 (tab), 259 (tab), 277 (tab), 278, 441, 443, 444, 446, 461, 602, 609
P/Taylor, 259 (tab), 280 (fig)
P/Tempel 2, 5, 231, 243 (tab), 422, 456, 674, 675 (fig), 679 (fig), 680
Timmers 1946 I, 239, 240 (tab)
P/Tuttle, 343-345, 419 (tab), 422 (tab), 428 (tab), 462, 474, 475 (tab), 683 (fig)
P/Tuttle-Giacobini-Kresák, 15, 206
Van Gent 1941 VIII, 242 (tab)
West 1976 VI, 9, 11-14, 22, 31, 38, 39 (fig), 42-46, 117-120, 199, 253, 254 (fig), 259 (tab), 261 (fig), 263, 265, 273-275, 277-280, 283, 326, 328, 329 (fig), 331 (fig), 334 (tab), 335 (fig), 345, 376, 417 (tab), 421 (tab), 443, 444, 446, 449, 456, 462-465, 500, 502 (tab), 522 (fig), 524 (fig), 574, 608, 612, 619
P/West-Kohoutek-Ikemura, 417 (tab)
Whipple-Bernasconi-Kulin 1942 IV, 240 (tab)
Whipple-Fedtke-Tevzadze 1943 I, 242 (tab), 259 (tab), 277 (tab), 278
Wild 1968 III, 259 (tab)
P/Wild 2, 418 (tab), 421 (tab)
Wirtanen 1957 VI, 209, 259 (tab), 262, 263, 280 (fig)
Condensation, 520
Contact surface, 539, 541, 542, 546, 574, 610, 615, 616
Copernicus, 476
Cosmic ray, 132, 388, 401, 402, 698
Coudé spectrograph, 422, 438
Deep-sea-meteor ablation spherule, 386, 387, 401
Digicon array detector, 422, 436
Disconnection event (DE), 520, 526-531, 628
Draconids. *See under* Meteor
Dust
albedo, 110, 687
asteroid. *See under* Asteroid
circumstellar, 400
composition, 18, 103, 352-355, 387
general, 122 (tab), 370 *ff*, 490
infrared spectroscopy, 403
interplanetary, 383 *ff*
interstellar, 131 *ff*, 372, 400

Dust, *cont.*
  lifetime, 385, 386
  models, 154, 357 *ff*
  morphology, 393-396
  noble gases, 388-393
  silicate, 324, 336, 393
  swarm, 374
  tail, 17, 333, 358, 368, 707
  temperature, 364, 402
  thermal properties, 341 *ff*
Dust-to-gas ratio, 105-114
Echelle spectrograph, 438, 449
Electron impact, 541
Ephemeris, 715-718
Fe, 26 (tab), 104, 114 (tab), 122 (tab), 123 (tab)
Fabry-Perot spectrometer, 437
Fading, 728
Fireball, 297 *ff*
Fluorescence, 35, 438, 449 (tab), 503
Flux rope, 539, 559, 575-577, 584, 617
Fourier transform spectrometer, 437
Frost, 171
Gas, 122 (tab)
Gas production rate, 35-38, 255, 333, 358, 425, 469-473, 500
Geminids. *See under* Meteor
Giacobinids. *See under* Meteor
Giotto, 368
Goldstone Tracking Station, 292, 294 (tab)
Grating spectrometer, 435
Greenstein effect, 100-102, 434, 470
H, 25 (tab), 114 (tab), 115, 117, 121-123, 452, 463 (tab), 472 (tab), 700 (tab)
H_2, 472 (tab)
HCH, 114 (tab)
HCN, 27 (tab), 90 (tab), 116
H_2CO, 90 (tab), 101, 116
^{4}He, 390
H_2O, 90 (tab), 103, 114 (tab), 143 (tab), 157, 165, 167 (tab), 168 (tab), 173 (fig), 176-180, 191 (fig), 194 (fig), 196 (fig), 424, 485, 499, 501
H_2O, evidence for, 462
H_2O, production rate, 474, 475 (tab)
H_2O, sublimation, 91-96
H_2O^+, 28 (tab), 115, 452, 453, 480, 483 (tab), 508, 510 (fig), 512 (fig), 617
H_3O^+, 498, 512
HS^-, 168 (tab)
H_2S, 165, 167 (tab), 168 (tab), 173 (fig), 176 (fig), 181, 184 (fig), 185 (fig)
HEOS-2, 373, 375
Halley Intercept Mission, 536
Halo, 101
Halo method (of determining rotation periods of comet nuclei), 231-237
Haser model, 425, 426, 448, 450, 453, 472, 473

Haystack Observatory, 292
Helios (dust experiment), 376-379
Helix, 520, 625
Hydrodynamic models, 548-552
Hyperbolic ejection, 647, 654, 662
Ice
  general, 167, 175-180
  irradiation, 215
  vaporization models, 10, 89-97, 333, 424, 497-503
Icy grains, 342
Image dissector scanner (IDS), 422, 423, 426, 436
Institute for Theoretical Astronomy, 723
International Astronomical Union's Central Telegram Bureau, 712
International Halley Watch (IHW), 295, 530
International Ultraviolet Explorer (IUE), 447, 462, 474
Interplanetary dust. *See under* Dust
Interstellar
  cloud, 222
  comet, 222, 642, 656, 661 (fig)
  dust. *See under* Dust
  molecules, 133-136, 498, 501, 697
  properties of medium, 135 (tab)
Ionization mechanisms, autoionization, photoionization, 44, 505, 506 (tab), 541, 604, 610
Jet, 8, 230, 313, 367, 619
Jupiter, 76, 80, 180, 644, 660, 661
Jupiter's satellites, 165, 197
K, 26 (tab), 114 (tab)
Kelvin-Helmholtz instability, 619, 625
Kinks, 520, 625
Kirkwood gaps, 312
Kitt Peak National Observatory (KPNO), 23 (fig), 678 (fig)
Knots, 520
Kreutz group. *See* Sun grazers
Lambert absorption coefficient, 174
Leonids. *See under* Meteor
Lick Observatory, 29 (fig), 30 (fig)
Lightcurve, 253, 416
Lorentz force, 545, 546, 556, 623
Luminosity, 215
Lyman α, 22, 102, 333, 476, 485
Lyman H I, 461
Mg, 122 (tab)
Mn, 26 (tab), 104, 114 (tab)
Magnetic field, 538 *ff*
Magnetite, 344, 398
Magnetohydrodynamic models, 552-555, 592-595
Magnitude, 22, 61 (fig), 65, 726-728
Magnitude, photographic, 60 (fig), 65
Magnitude, visual, 60 (fig), 415
Mars, 165, 197

INDEX

Mass loss, 331, 334 (tab)
McDonald Observatory, 457
Meteor, 104, 297 ff, 372, 384
Meteor stream, 72, 301, 312-316, 689
Meteorite
  carbonaceous chondrite, 123 (tab), 308, 310, 312, 316, 317, 336, 355, 372, 387, 388, 397, 404-406, 698
  CI chondrites, 104, 121
  CM chondrites, 104
  CO chondrites, 104
  CV chondrites, 104
  density, 104
  exposure age, 217
  irons, 218
  mass accreted by Earth, 298, 299
  ordinary chondrites, 217
Meteorites
  Allende, 88, 389 (fig)
  Innisfree, 298, 303, 308
  Lost City, 298, 299, 302, 303, 307-309
  Murchison, 389 (fig), 403
  Murray, 308
  Orgueil, 308, 389 (fig)
  Pribram, 298, 299, 303
Meteoroid. *See also* Brownlee particles
  Earth-crossing, 301
  Jupiter-crossing orbit, 301
  origin, 300
  retrograde orbit, 301
Meteors
  Andromedids, 283
  Draconids, 213
  Geminids, 301, 372
  Giacobinids, 372
  Leonids, 301
  Orionids, 301, 313
  Perseids, 301
  Taurids, 10, 218, 299, 301, 311-313
Microcrater, 372, 373, 385, 386
Micrometeorite, 384
Mie theory, 174, 175, 342, 344
Molecular cloud, 132, 152, 155 (tab), 156 (tab)
Molecular emission, 481, 482 (tab)
Mt. Lemmon Observatory, 12 (fig)
N, 120-123, 700 (tab)
N_2, 90 (tab)
N_2^+, 28 (tab), 114 (tab), 483 (tab), 512 (fig)
Na, 26 (tab), 114 (tab), 421 (tab)
NH, 27 (tab), 114 (tab), 422 (tab), 448, 454 (tab), 455 (tab), 483 (tab), 488, 512 (fig)
NH_4^+, 168 (tab)
NH_2, 27 (tab), 114 (tab), 449, 454 (tab), 455 (tab), 483 (tab), 501, 502 (tab), 512 (fig)
NH_3, 90 (tab), 143 (tab), 164 ff, 500
$NH_3 \cdot H_2O$, 167 (tab), 184

NH_4HS, 165, 167 (tab), 173 (fig), 176 (fig), 184, 185 (fig)
Ni, 26 (tab), 104, 114 (tab), 122 (tab), 123 (tab)
NO, 472 (tab)
Narrowband filter, 416-422
NASA Ames Research Center, 385
Nongravitational force, 230, 231, 255
Nucleus
  albedo, 91 (tab)
  composition, 15, 16, 85 ff
  mass, 6, 271, 707
  photometry, 414, 422, 423, 426, 427
  rotation rate, 4-6, 227 ff, 284
  secondary, 257
  shape, 231
  size, 5, 91 (tab), 206-209, 673, 707
  split, 251 ff
  structure, 87-89
  temperature, 364
  tidal effects, 283, 284
  Whipple model, 4, 8, 9, 86, 87, 125, 164, 165, 209-214, 217, 230, 252, 497, 697, 707
O, 25 (tab), 114 (tab), 117, 119-123, 450, 451, 463 (tab), 472 (tab), 700 (tab)
OH, 25 (tab), 27 (tab), 114 (tab), 115, 422 (tab), 447, 448, 454 (tab), 463 (tab), 472 (tab), 482 (tab)
OH^+, 28 (tab), 483 (tab)
Olivine, 397
Oort Cloud, 57, 67, 96, 124, 131, 210, 214, 219-222, 244, 637 ff, 659, 709, 710
Oort-Schmidt classification, 57, 87, 267, 416, 424, 638, 648
Oparin-Haldane hypothesis, 696
Orbiting Astronomical Observatory-2 (OAO-2), 461
Orbiting Geophysical Observatory-5 (OGO-5), 461
Orbit
  asteroid. *See under* Asteroid
  comet. *See under* Comet
  evolution, 659 ff
  horseshoe, 659 ff
  hyperbolic, 206 (tab), 219, 638, 656, 659 ff
  Jupiter family, 219
  long period, 56 ff, 637 ff, 659 ff
  Neptune-crossing, 220
  short period, 56 ff, 659 ff, 666
  Trojan, 659 ff
Origin of life, 696 ff
Orionids. *See under* Meteors
Outburst, 14, 88, 89, 158, 728
Palomar Mountain Observatory, 67
Panspermia, 701
Particles, extraterrestrial. *See* Brownlee particles

766 INDEX

Perihelion distance, 63-66, 246
Perseids. *See under* Meteors
Perturbation, stellar, 641, 643-647, 649, 657, 660
Photochemistry, 485-493. *See also under* Coma
Photodissociation, 506 (tab). *See also under* Coma
Photoelectric detector, 435
Photoionization. *See under* Ionization mechanisms
Photometry
 coma. *See under* Coma
 general, 344, 413 *ff*
 models, 423-426
 nucleus. *See under* Nucleus
Pioneer 10/11, 371
Pioneer Venus, 563, 565, 580, 590, 591, 614
Plasma, 538 *ff*
Plasma tail. *See* ion *under* Tail
Poynting-Robertson effect, 338, 379
Prairie Network, 10, 210, 215, 218, 298, 299, 301
Predissociation, 505
Primitive Earth, 699
Probstein model, 98, 105-114, 362, 363
Pyrrhotite ($Fe_{1-x}S$), 396
Radar observation, 288 *ff*
Radiation pressure, 379
Reticon array detector, 436
S, 26 (tab), 114 (tab), 120-123, 463 (tab), 472 (tab), 491, 700 (tab)
Si, 122 (tab), 123 (tab)
SO_2, 165, 167 (tab), 168 (tab), 191 (fig), 192
Saturn
 rings, 165, 177, 193, 197
 satellites, 165, 198
Sector boundary, 533
Silicate
 dust. *See under* Dust
 emission feature, 324, 346
 enstatite, 397
Silicates, 28 (tab), 103, 114 (tab), 158, 326

Solar
 abundances, 121
 activity, 427
 nebula, 221, 501, 642
 wind, 427, 525, 533, 538 *ff*, 561 *ff*, 591-600, 611 (tab)
Spacelab, 477
Space Shuttle, 477
Space Telescope (ST), 477
Spectrophotometry
 continuum, 455
 general, 433 *ff*
Spectroscopy
 general, 434, 481-485
 infrared, 164 *ff*, 323 *ff*
 ultraviolet, 461 *ff*
Stagnation point, pressure, 539, 542-544
Star formation, 133
Steward Observatory, 674, 678 (fig)
Stoke's law, 385
Stream meteors. *See under* Meteor
Sun grazers (Kreutz group), 68, 72, 207 (tab), 213
Swan bands, 440, 457
Swings effect, 100, 434, 445 (fig), 447, 448
Ti, 104
Tail
 antitail, 326, 709
 dust. *See under* Dust
 ion, 40, 519 *ff*, 588 *ff*
 rays, 520, 525-527, 571, 584, 600-609, 619, 626
 streamers, 520, 558, 602
 type-II. *See* tail *under* Dust
Taurids. *See under* Meteors
Troilite (FeS), 396
Ultraviolet radiation, 140, 505
Ultraviolet spectroscopy. *See under* Spectroscopy
United States Naval Observatory, 239
V, 104, 114 (tab)
Vaporization models. *See under* Ice
Venus, 534, 547, 556, 559, 561 *ff*, 591, 592, 617
Zodiacal dust cloud, 336, 338, 371, 374